Mathematical Applications

for the Management, Life,
and Social Sciences

Hybrid Edition

EDITION 10

Mathematical Applications

for the Management, Life, and Social Sciences

Hybrid Edition

Ronald J. Harshbarger

University of South Carolina, Beaufort

James J. Reynolds

Clarion University of Pennsylvania

BROOKS/COLE
CENGAGE Learning

Australia • Brazil • Japan • Korea • Mexico • Singapore • Spain • United Kingdom • United States

BROOKS/COLE
CENGAGE Learning·

Mathematical Applications for the Management, Life, and Social Sciences,
Tenth Edition
HYBRID EDITION
Ronald J. Harshbarger
James J. Reynolds

Publisher: Richard Stratton

Senior Development Editor: Laura Wheel

Senior Editorial Assistant: Haeree Chang

Media Editor: Andrew Coppola

Senior Marketing Manager: Barb Bartoszek

Marketing Coordinator: Michael Ledesma

Marketing Communications Manager:
    Mary Anne Payumo

Content Project Manager: Cathy Brooks

Art Director: Jill Ort Haskell

Manufacturing Planner: Doug Bertke

Rights Acquisitions Specialist:
    Shalice Shah-Caldwell

Production Service: Lachina Publishing Services

Text Designer: Rokusek Design, Inc.

Cover Designer: Rokusek Design, Inc.

Cover Image: www.shutterstock.com

Compositor: Lachina Publishing Services

For product information and technology assistance, contact us at
Cengage Learning Customer & Sales Support, 1-800-354-9706.
For permission to use material from this text or product,
submit all requests online at **www.cengage.com/permissions.**
Further permissions questions can be emailed to
permissionrequest@cengage.com.

Library of Congress Control Number: 2011932871

Hybrid Student Edition:
ISBN-13: 978-1-133-36483-2
ISBN-10: 1-133-36483-7

Hybrid Instructor's Edition:
ISBN-13: 978-1-133-36494-8
ISBN-10: 1-133-36494-2

Brooks/Cole
20 Channel Center Street
Boston, MA 02210
USA

Cengage Learning is a leading provider of customized learning solutions with office locations around the globe, including Singapore, the United Kingdom, Australia, Mexico, Brazil and Japan. Locate your local office at **international.cengage.com/region.**

Cengage Learning products are represented in Canada by Nelson Education, Ltd.

For your course and learning solutions, visit **www.cengage.com.**

Purchase any of our products at your local college store or at our preferred online store **www.cengagebrain.com.**

Instructors: Please visit **login.cengage.com** and log in to access instructor-specific resources.

Printed in the United States of America
1 2 3 4 5 6 7 15 14 13 12 11

Contents

Preface

Many traditional lecture-based courses are evolving into courses in which all homework and tests are delivered online. In addition, with the rapid growth of distance learning courses, there is an even greater need for course materials that blend traditional print resources with rich media-based tools. A hybrid text is designed to address the needs of these courses through the integration of both print and online components. For this hybrid edition, the end-of-section exercises have been removed from the text and are available exclusively online in Enhanced WebAssign, an easy-to-use online homework system. The result is a smaller, more manageable text for students who will do all of their homework online.

About the Book

To paraphrase English mathematician, philosopher, and educator Alfred North Whitehead, the purpose of education is not to fill a vessel but to kindle a fire. In particular, Whitehead encouraged students to be creative and imaginative in their learning and to continually form ideas into new and more exciting combinations. This desirable goal is not always an easy one to realize in mathematics with students whose primary interests are in areas other than mathematics. The purpose of this text, then, is to present mathematical skills and concepts, and to apply them to ideas that are important to students in the management, life, and social sciences. We hope that this look at the relevance of mathematical ideas to a broad range of fields will help inspire the imaginative thinking and excitement for learning that Whitehead spoke of. The applications included allow students to view mathematics in a practical setting relevant to their intended careers. Almost every chapter of this book includes a section or two devoted to the applications of mathematical topics, and every section contains a number of application examples and problems. Although intended for students who have completed two years of high school algebra or its equivalent, this text begins with a brief review of algebra which, if covered, will aid in preparing students for the work ahead.

Pedagogical Features

In this new edition, we have incorporated many suggestions that reflect the needs and wishes of our users. Important pedagogical features that have characterized previous editions have been retained. They are as follows.

Intuitive Viewpoint.   The book is written from an intuitive viewpoint, with emphasis on concepts and problem solving rather than on mathematical theory. Yet each topic is carefully developed and explained, and examples illustrate the techniques involved.

Flexibility.   At different colleges and universities, the coverage and sequencing of topics may vary according to the purpose of the course and the nature of the student audience. To accommodate alternate approaches, the text has a great deal of flexibility in the order in which topics may be presented and the degree to which they may be emphasized.

Applications.   We have found that integrating applied topics into the discussions and exercises helps provide motivation within the sections and demonstrates the relevance of each topic. Numerous real-life application examples and exercises represent the applicability of the mathematics, and each application problem is identified, so the instructor or student can select applications that are of special interest. In addition, we have found that offering separate lessons on applied topics such as cost, revenue, and profit functions brings the preceding mathematical discussions into clear, concise focus and provides a thread of continuity as mathematical sophistication increases. There are ten such sections in the book, and entire chapters devoted to linear programming and financial applications.

Chapter Warm-ups.   With the exception of Chapter 0, a Warm-up appears at the beginning of each chapter and invites students to test themselves on the skills needed for that chapter. The Warm-ups present several prerequisite problem types that are keyed to the appropriate sections in the upcoming chapter where those skills are needed. Students who have difficulty with any particular skill are directed to specific sections of the text for review. Instructors may also find the Warm-ups useful in creating a course syllabus that includes an appropriate scope and sequence of topics.

Application Previews.   Each section begins with an Application Preview that establishes the context and direction for the concepts that will be presented. Each of these Previews contains an example that motivates the mathematics in the section and is then revisited in a completely worked Application Preview example appearing later in the section.

Comprehensive Exercise Sets.   The overall variety and grading of drill and application exercises in Enhanced WebAssign offer problems for different skill levels, and there are enough challenging problems to stimulate students in thoughtful investigations. Many exercise sets contain critical-thinking and thought-provoking multistep problems that extend students' knowledge and skills.

Extended Applications and Group Projects.   Starting with Chapter 1, each chapter ends with at least two case studies, which further illustrate how mathematics can be used in business and personal decision making. In addition, many applications are cumulative in that solutions require students to combine the mathematical concepts and techniques they learned in some of the preceding chapters.

Graphical, Numerical, and Symbolic Methods.   A large number of real data modeling applications are included in the examples and exercises and are denoted by the header ***Modeling.*** Many sections include problems with functions that are modeled from real data, and some problems ask students to model functions from the data given. These problems are solved by using one or more graphical, numerical, or symbolic methods.

 Graphing Calculators and Excel.   Many examples, applications, Technology Notes, Calculator Notes, and Spreadsheet Notes, denoted by the icon, are scattered throughout the text. Many of these notes reference detailed step-by-step instructions in the new Appendix C (Graphing Calculator Guide) and Appendix D (Excel Guide) and in the Online Excel Guide. Discussions of the use of technology are placed in subsections and examples in many sections, so that they can be emphasized or de-emphasized at the option of the instructor.

The discussions of graphing calculator technology highlight its most common features and uses, such as graphing, window setting, Trace, Zoom, Solver, tables, finding points of intersection, numerical derivatives, numerical integration, matrices, solving inequalities, and modeling (curve fitting). While technology never replaces the mathematics, it does supplement and extend the mathematics by providing opportunities for generalization and alternative ways of understanding, doing, and checking. Some exercises that are better worked with the use of technology—including graphing calculators and Excel—are highlighted with the technology icon. Of course, many additional exercises can benefit from the use of technology, at the option of the instructor. Technology can be used to graph functions and to discuss the generalizations, applications, and implications of problems being studied.

Excel is useful in solving problems involving linear equations; systems of equations; quadratic equations; matrices; linear programming; output comparisons of $f(x)$, $f'(x)$, and

$f''(x)$; and maxima and minima of functions subject to constraints. Excel is also a useful problem-solving tool when studying the Mathematics of Finance in Chapter 6.

Checkpoints.   The Checkpoints ask questions and pose problems within each section's discussion, allowing students to check their understanding of the skills and concepts under discussion before they proceed. Solutions to these Checkpoints appear before the section exercises.

Objective Lists.   Every section begins with a brief list of objectives that outline the goals of that section for the student.

Procedure/Example and Property/Example Tables.   Appearing throughout the text, these tables aid student understanding by giving step-by-step descriptions of important procedures and properties with illustrative examples worked out beside them.

Boxed Information.   All important information is boxed for easy reference, and key terms are highlighted in boldface.

Key Terms and Formulas.   At the end of each chapter, just before the Chapter Review Exercises, there is a section-by-section listing of that chapter's key terms and formulas. This provides a well-organized core from which a student can build a review, both to consult while working the Review Exercises and to identify quickly any section needing additional study.

Review Exercises and Chapter Tests.   At the end of each chapter, a set of Review Exercises offers students extra practice on topics in that chapter. These Reviews cover each chapter's topics primarily in their section order, but without section references, so that students get a true review but can readily find a section for further review if difficulties occur. A Chapter Test follows each set of Review Exercises. All Chapter Tests provide a mixture of problems that do not directly mirror the order of topics found within the chapter. This organization of the Chapter Test ensures that students have a firm grasp of all material in the chapter.

Changes in the Tenth Edition

In the Tenth Edition, we continue to offer a text characterized by complete and accurate pedagogy, mathematical precision, excellent exercise sets, numerous and varied applications, and student-friendly exposition. There are many changes in the mathematics, prose, and art. The more significant ones are as follows.

- Two new Appendices have been added.
  - Appendix C Graphing Calculator Guide, containing detailed step-by-step instructions and examples for operating TI-83 and TI-84 Plus calculators.
  - Appendix D Excel Guide, containing detailed step-by-step instructions and examples for Excel 2003, and for Excel 2007 and 2010.
- References to specific calculator and Excel steps in Appendix C and/or Appendix D are given each time a new technology process is introduced.
- The exposition and example discussions have been streamlined to eliminate repetitions and redundancies.
- Section 2.5 "Modeling; Fitting Curves to Data with Graphing Utilities (optional)" has been extensively rewritten to improve logic and flow.
- Section 3.3 "Gauss-Jordan Elimination: Solving Systems of Equations" has been reorganized to clarify the decisions at each step and how the method extends beyond 3×3 matrices.
- Most of the real-data examples and exercises that are time related have been updated or replaced with current applications.
- Drill Exercises thoughout the text have been revised and reorganized to improve their grading and variety.

■ Additional multistep applications have been added to the exercises.
■ Many images and illustrations that relate to the mathematics topics have been added.

Resources for the Student

Student Solutions Manual (978-1-133-10852-8)

This manual provides complete worked-out solutions to all odd-numbered exercises in the text, giving you a chance to check your answers and ensure you took the correct steps to arrive at an answer.

CourseMate
www.cengagebrain.com

CourseMate brings course concepts to life with interactive learning, study, and exam preparation tools that support the printed textbook. CourseMate for Harshbarger/Reynolds' *Mathematical Applications for the Management, Life, and Social Sciences* includes: an interactive eBook, quizzes, flashcards, Excel guide, graphing calculator guide, videos, and more.

Enhanced WebAssign
www.webassign.net

Enhanced WebAssign is an online homework delivery system. Enhanced WebAssign for Harshbarger/Reynolds' *Mathematical Applications for the Management, Life, and Social Sciences* includes: an interactive eBook, Personal Study Plans, a Show My Work feature, an Answer Evaluator, quizzes, videos, and more.

CengageBrain.com

To access additional course materials and companion resources, please visit www.cengagebrain.com. At the CengageBrain.com home page, search for the ISBN of your title (from the back cover of your book) using the search box at the top of the page. This will take you to the product page where free companion resources can be found.

Resources for the Instructor

Complete Solutions Manual (978-1-133-36435-1)

The Complete Solutions Manual provides worked-out solutions of all exercises in the text. In addition, it contains the solutions of the special features in the text, such as *Extended Applications and Group Projects*.

PowerLecture (978-1-133-10920-4)

This comprehensive CD contains all art from the text in both jpeg and PowerPoint formats, key equations and tables from the text, complete pre-built PowerPoint lectures, Solution Builder, and Diploma Computerized Testing.

Solution Builder

www.cengage.com/solutionbuilder
This online instructor database offers complete worked-out solutions of all exercises in the text. Solution Builder allows you to create customized, secure solutions printouts (in PDF format) matched exactly to the problems you assign in class.

Diploma Computerized Testing

Create, deliver, and customize tests in print and online formats with Diploma, an easy-to-use assessment software containing hundreds of algorithmic questions derived from the text exercises. Diploma is available on the PowerLecture CD.

Enhanced WebAssign
www.webassign.net

Enhanced WebAssign's homework delivery system lets you deliver, collect, grade, and record assignments via the web. Online exercises are assignable in free response, multiple choice, and multi-part formats. New enhancements to the system include: Cengage You-Book interactive eBook, Personal Study Plans, a Show My Work feature, an Answer Evaluator that accepts more mathematically equivalent answers, quizzes, videos, and more.

Cengage YouBook

YouBook is an interactive and customizable eBook! Containing all the content from Harshbarger/Reynolds' *Mathematical Applications for the Management, Life, and Social Sciences*, YouBook features a text edit tool that allows instructors to modify the textbook narrative as needed. With YouBook, you can quickly reorder entire sections and chapters or hide any content you don't teach to create an eBook that perfectly matches your syllabus. You can further customize the text by publishing web links. Additional media assets include: animated figures, video clips, highlighting, notes, and more! YouBook is available in Enhanced WebAssign.

CourseMate
www.cengagebrain.com

A simple way to complement your text and course content with study and practice materials, CourseMate brings course concepts to life with interactive learning, study, and exam preparation tools that support the printed textbook. CourseMate for Harshbarger/Reynolds' *Mathematical Applications for the Management, Life, and Social Sciences* includes: an interactive eBook, quizzes, flashcards, Excel guide, graphing calculator guide, videos, and Engagement Tracker, a first-of-its-kind tool that monitors student engagement. Watch student comprehension soar as your class works with both the printed text and text-specific website.

Acknowledgments

We would like to thank the many people who helped us at various stages of revising this text. The encouragement, criticism, contributions, and suggestions that were offered were invaluable to us. Once again we have been fortunate to have Helen Medley's assistance with accuracy checking of the entire text and answer section during manuscript preparation and on the page proofs. As always, we continue to be impressed by and most appreciative of her work ethic, attention to detail, accuracy, and skill.

For their reviews of draft manuscript and the many helpful comments that were offered, we would like to thank

Priscilla Chaffe-Stengel	*California State University–Fresno*
Hugh Cornell	*University of North Florida*
Jae Ki Lee	*Baruch College*
Joseph McCollum	*Siena College*
Jeffrey Mayer	*Syracuse University*
John O'Neill	*Siena College*
Mari Peddycoart	*Lone Star College–Kingwood*
David Price	*Tarrant County College*
Larry Small	*Pierce College*

Ronald J. Harshbarger
James J. Reynolds

EDITION

10

Mathematical Applications

for the Management, Life, and Social Sciences

Hybrid Edition

Chapter 0

Algebraic Concepts

Khoroshunova Olga/Shutterstock.com

This chapter provides a brief review of the algebraic concepts that will be used throughout the text. You may be familiar with these topics, but it may be helpful to spend some time reviewing them. In addition, each chapter after this one opens with a warm-up page that identifies prerequisite skills needed for that chapter. If algebraic skills are required, the warm-up cites their coverage in this chapter. Thus you will find that this chapter is a useful reference as you study later chapters.

The topics and applications studied in this chapter include the following.

SECTIONS		APPLICATIONS
0.1	**Sets** **Set operations** **Venn diagrams**	Dow Jones Average, jobs growth
0.2	**The Real Numbers** **Inequalities and intervals** **Absolute value**	Income taxes, health statistics
0.3	**Integral Exponents**	Personal income, endangered species
0.4	**Radicals and Rational Exponents** **Roots and fractional exponents** **Operations with radicals**	Richter scale, half-life
0.5	**Operations with Algebraic Expressions**	Revenue, profit
0.6	**Factoring** **Common factors** **Factoring trinomials**	Simple interest, revenue
0.7	**Algebraic Fractions** **Operations** **Complex fractions**	Average cost, advertising and sales

0.1

Sets

A **set** is a well-defined collection of objects. We may talk about a set of books, a set of dishes, a set of students, or a set of individuals with a certain blood type. There are two ways to tell what a given set contains. One way is by listing the **elements** (or **members**) of the set (usually between braces). We may say that a set A contains 1, 2, 3, and 4 by writing $A = \{1, 2, 3, 4\}$. To say that 4 is a member of set A, we write $4 \in A$. Similarly, we write $5 \notin A$ to denote that 5 is not a member of set A.

If all the members of the set can be listed, the set is said to be a **finite set.** $A = \{1, 2, 3, 4\}$ and $B = \{x, y, z\}$ are examples of finite sets. When we do not wish to list all the elements of a finite set, we can use three dots to indicate the unlisted members of the set. For example, the set of even integers from 8 to 8952, inclusive, could be written as

$$\{8, 10, 12, 14, \ldots, 8952\}$$

For an **infinite set,** we cannot list all the elements, so we use the three dots. For example, $N = \{1, 2, 3, 4, \ldots\}$ is an infinite set. This set N is called the set of **natural numbers.**

Another way to specify the elements of a given set is by description. For example, we may write $D = \{x: x \text{ is a Ford automobile}\}$ to describe the set of all Ford automobiles. Furthermore, $F = \{y: y \text{ is an odd natural number}\}$ is read "F is the set of all y such that y is an odd natural number."

EXAMPLE **1**   **Describing Sets**

Write the following sets in two ways.
(a) The set A of natural numbers less than 6
(b) The set B of natural numbers greater than 10
(c) The set C containing only 3

Solution

(a) $A = \{1, 2, 3, 4, 5\}$ or $A = \{x: x \text{ is a natural number less than 6}\}$
(b) $B = \{11, 12, 13, 14, \ldots\}$ or $B = \{x: x \text{ is a natural number greater than 10}\}$
(c) $C = \{3\}$ or $C = \{x: x = 3\}$ ∎

Note that set C of Example 1 contains one member, 3; set A contains five members; and set B contains an infinite number of members. It is possible for a set to contain no members. Such a set is called the **empty set** or the **null set,** and it is denoted by $\varnothing$ or by $\{\ \}$. The set of living veterans of the War of 1812 is empty because there are no living veterans of that war. Thus

$$\{x: x \text{ is a living veteran of the War of 1812}\} = \varnothing$$

Special relations that may exist between two sets are defined as follows.

Relations between Sets

Definition	Example
1. Sets X and Y are **equal** if they contain the same elements.	1. If $X = \{1, 2, 3, 4\}$ and $Y = \{4, 3, 2, 1\}$, then $X = Y$.
2. A is called a **subset** of B, which is written $A \subseteq B$ if every element of A is an element of B. The empty set is a subset of every set. Each set A is a subset of itself.	2. If $A = \{1, 2, c, f\}$ and $B = \{1, 2, 3, a, b, c, f\}$, then $A \subseteq B$. Also, $\varnothing \subseteq A$, $\varnothing \subseteq B$, $A \subseteq A$, and $B \subseteq B$.
3. If C and D have no elements in common, they are called **disjoint.**	3. If $C = \{1, 2, a, b\}$ and $D = \{3, e, 5, c\}$, then C and D are disjoint.

In the discussion of particular sets, the assumption is always made that the sets under discussion are all subsets of some larger set, called the **universal set** *U*. The choice of the universal set depends on the problem under consideration. For example, in discussing the set of all students and the set of all female students, we may use the set of all humans as the universal set.

We may use **Venn diagrams** to illustrate the relationships among sets. We use a rectangle to represent the universal set, and we use closed figures inside the rectangle to represent the sets under consideration. Figures 0.1–0.3 show such Venn diagrams.

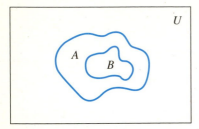

Figure 0.1
B is a subset of *A*; *B* ⊆ *A*.

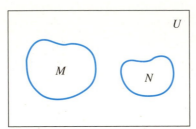

Figure 0.2
M and *N* are disjoint.

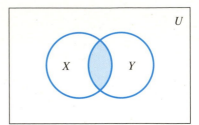

Figure 0.3
X and *Y* are not disjoint.

Set Operations   The shaded portion of Figure 0.3 indicates where the two sets overlap. The set containing the members that are common to two sets is said to be the **intersection** of the two sets.

Set Intersection

The intersection of *A* and *B*, written *A* ∩ *B*, is defined by

$$A \cap B = \{x : x \in A \text{ and } x \in B\}$$

EXAMPLE **2**   **Set Intersection**

(a) If *A* = {2, 3, 4, 5} and *B* = {3, 5, 7, 9, 11}, find *A* ∩ *B*.
(b) Which of *A*, *B*, and *A* ∩ *B* is a subset of *A*?

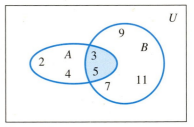

Figure 0.4

Solution

(a) *A* ∩ *B* = {3, 5} because 3 and 5 are the common elements of *A* and *B*. Figure 0.4 shows the sets and their intersection.
(b) *A* ∩ *B* and *A* are subsets of *A*. ■

CHECKPOINT   Let *A* = {2, 3, 5, 7, 11}, *B* = {2, 4, 6, 8, 10}, and *C* = {6, 10, 14, 18, 22}. Use these sets to answer the following.

1. (a) Of which sets is 6 an element?
    (b) Of which sets is {6} an element?
2. Which of the following are true?
    (a) 2 ∈ *A*
    (b) 2 ∈ *B*
    (c) 2 ∈ *C*
    (d) 5 ∉ *A*
    (e) 5 ∉ *B*
3. Which pair of *A*, *B*, and *C* is disjoint?

4. Which of $\varnothing$, A, B, and C are subsets of
   (a) the set P of all prime numbers?
   (b) the set M of all multiples of 2?
5. Which of A, B, and C is equal to $D = \{x: x = 4n + 2 \text{ for natural numbers } 1 \le n \le 5\}$?

The **union** of two sets is the set that contains all members of the two sets.

Set Union

The union of A and B, written $A \cup B$, is defined by

$$A \cup B = \{x: x \in A \text{ or } x \in B \text{ (or both)}\}^*$$

We can illustrate the intersection and union of two sets by the use of Venn diagrams. Figures 0.5 and 0.6 show Venn diagrams with universal set U represented by the rectangles and with sets A and B represented by the circles. The shaded region in Figure 0.5 represents $A \cap B$, the intersection of A and B, and the shaded region in Figure 0.6—which consists of all parts of both circles—represents $A \cup B$.

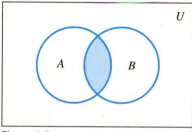

Figure 0.5

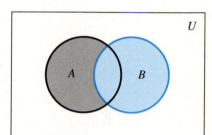

Figure 0.6

EXAMPLE 3   **Set Union**

If $X = \{a, b, c, f\}$ and $Y = \{e, f, a, b\}$, find $X \cup Y$.

Solution
$X \cup Y = \{a, b, c, e, f\}$

EXAMPLE 4   **Set Intersection and Union**

Let $A = \{x: x \text{ is a natural number less than 6}\}$ and $B = \{1, 3, 5, 7, 9, 11\}$.
(a) Find $A \cap B$.
(b) Find $A \cup B$.

Solution
(a) $A \cap B = \{1, 3, 5\}$
(b) $A \cup B = \{1, 2, 3, 4, 5, 7, 9, 11\}$

All elements of the universal set that are not contained in a set A form a set called the **complement** of A.

Set Complement

The complement of A, written A', is defined by

$$A' = \{x: x \in U \text{ and } x \notin A\}$$

*In mathematics, the word *or* means "one or the other or both."

We can use a Venn diagram to illustrate the complement of a set. The shaded region of Figure 0.7 represents A', and the *un*shaded region of Figure 0.5 represents $(A \cap B)'$.

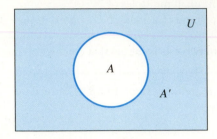

Figure 0.7

EXAMPLE 5 Operations with Sets

If $U = \{x \in N : x < 10\}$, $A = \{1, 3, 6\}$, and $B = \{1, 6, 8, 9\}$, find the following.
(a) A'
(b) B'
(c) $(A \cap B)'$
(d) $A' \cup B'$

Solution
(a) $U = \{1, 2, 3, 4, 5, 6, 7, 8, 9\}$ so $A' = \{2, 4, 5, 7, 8, 9\}$
(b) $B' = \{2, 3, 4, 5, 7\}$
(c) $A \cap B = \{1, 6\}$ so $(A \cap B)' = \{2, 3, 4, 5, 7, 8, 9\}$
(d) $A' \cup B' = \{2, 4, 5, 7, 8, 9\} \cup \{2, 3, 4, 5, 7\} = \{2, 3, 4, 5, 7, 8, 9\}$ ■

CHECKPOINT Given the sets $U = \{1, 2, 3, 4, 5, 6, 7, 8, 9, 10\}$, $A = \{1, 3, 5, 7, 9\}$, $B = \{2, 3, 5, 7\}$, and $C = \{4, 5, 6, 7, 8, 9, 10\}$, find the following.
6. $A \cup B$
7. $B \cap C$
8. A'

EXAMPLE 6 Stocks

Many local newspapers list "stocks of local interest." Suppose that on a certain day, a prospective investor categorized 23 stocks according to whether

- their closing price on the previous day was less than $50/share (set C)
- their price-to-earnings ratio was less than 20 (set P)
- their dividend per share was at least $1.50 (set D).

Of these 23 stocks,

16 belonged to set P	10 belonged to both C and P
12 belonged to set C	7 belonged to both D and P
8 belonged to set D	2 belonged to all three sets
3 belonged to both C and D	

Draw a Venn diagram that represents this information. Use the diagram to answer the following.
(a) How many stocks had closing prices of less than $50 per share or price-to-earnings ratios of less than 20?
(b) How many stocks had none of the characteristics of set C, P, or D?
(c) How many stocks had only dividends per share of at least $1.50?

Solution

The Venn diagram for three sets has eight separate regions (see Figure 0.8(a)). To assign numbers from our data, we must begin with some information that refers to a single region, namely that two stocks belonged to all three sets (see Figure 0.8(b)). Because the region common to all three sets is also common to any pair, we can next use the information about stocks that belonged to two of the sets (see Figure 0.8(c)). Finally, we can complete the Venn diagram (see Figure 0.8(d)).

(a) We need to add the numbers in the separate regions that lie within $C \cup P$. That is, 18 stocks closed under $50 per share or had price-to-earnings ratios of less than 20.

(b) There are 5 stocks outside the three sets C, D, and P.

(c) Those stocks that had only dividends of at least $1.50 per share are inside D but outside both C and P. There are no such stocks. ∎

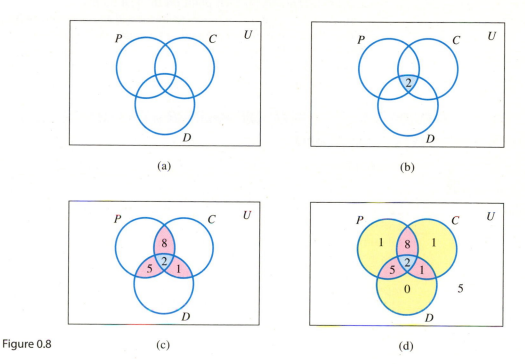

Figure 0.8   (a)   (b)   (c)   (d)

CHECKPOINT SOLUTIONS

1. (a) Sets B and C have 6 as an element.
   (b) None of A, B, or C has $\{6\}$ as an element; $\{6\}$ is itself a set, and the elements of A, B, and C are not sets.

2. (a) True
   (b) True
   (c) False; $2 \notin C$
   (d) False; $5 \in A$
   (e) True

3. A and C are disjoint.

4. (a) $\varnothing \subseteq P$ and $A \subseteq P$
   (b) $\varnothing \subseteq M, B \subseteq M, C \subseteq M$

5. $C = D$

6. $A \cup B = \{1, 2, 3, 5, 7, 9\}$

7. $B \cap C = \{5, 7\}$

8. $A' = \{2, 4, 6, 8, 10\}$

| EXERCISES | 0.1

Access end-of-section exercises online at www.webassign.net

0.2

The Real Numbers

In this text we use the set of **real numbers** as the universal set. We can represent the real numbers along a line called the **real number line.** This number line is a picture, or graph, of the real numbers. Each point on the real number line corresponds to exactly one real number, and each real number can be located at exactly one point on the real number line. Thus, two real numbers are said to be equal whenever they are represented by the same point on the real number line. The equation $a = b$ (a equals b) means that the symbols a and b represent the same real number. Thus, $3 + 4 = 7$ means that $3 + 4$ and 7 represent the same number. Table 0.1 lists special subsets of the real numbers.

| TABLE 0.1 |

SUBSETS OF THE SET OF REAL NUMBERS

	Description	Example (some elements shown)
Natural numbers	{1, 2, 3, . . .} The counting numbers.	
Integers	{. . . , −2, −1, 0, 1, 2, . . .} The natural numbers, 0, and the negatives of the natural numbers.	
Rational numbers	All numbers that can be written as the ratio of two integers, a/b, with $b \neq 0$. These numbers have decimal representations that either terminate or repeat.	
Irrational numbers	Those real numbers that *cannot* be written as the ratio of two integers. Irrational numbers have decimal representations that neither terminate nor repeat.	
Real numbers	The set containing all rational and irrational numbers (the entire number line).	

The following properties of the real numbers are fundamental to the study of algebra.

Properties of the Real Numbers

Let a, b, and c denote real numbers.

1. (Commutative Property) Addition and multiplication are commutative.

$$a + b = b + a \qquad ab = ba$$

2. (Associative Property) Addition and multiplication are associative.

$$(a + b) + c = a + (b + c) \qquad (ab)c = a(bc)$$

3. (Additive Identity) The additive identity is 0.
$$a + 0 = 0 + a = a$$

4. (Multiplicative Identity) The multiplicative identity is 1.
$$a \cdot 1 = 1 \cdot a = a$$

5. (Additive Inverse) Each element a has an additive inverse, denoted by $-a$.
$$a + (-a) = -a + a = 0$$

Note that there is a difference between a negative number and the negative of a number.

6. (Multiplicative Inverse) Each nonzero element a has a multiplicative inverse, denoted by a^{-1}.
$$a \cdot a^{-1} = a^{-1} \cdot a = 1$$

Note that $a^{-1} = 1/a$.

7. (Distributive Law) Multiplication is distributive over addition.
$$a(b + c) = ab + ac$$

Note that Property 5 provides the means to subtract by defining $a - b = a + (-b)$ and Property 6 provides a means to divide by defining $a \div b = a \cdot (1/b)$. The number 0 has no multiplicative inverse, so division by 0 is undefined.

Inequalities and Intervals

We say that a is less than b (written $a < b$) if the point representing a is to the left of the point representing b on the real number line. For example, $4 < 7$ because 4 is to the left of 7 on the real number line. We may also say that 7 is greater than 4 (written $7 > 4$). We may indicate that the number x is less than or equal to another number y by writing $x \le y$. We may also indicate that p is greater than or equal to 4 by writing $p \ge 4$.

EXAMPLE   1   Inequalities

Use $<$ or $>$ notation to write the following.
(a) 6 is greater than 5.   (b) 10 is less than 15.
(c) 3 is to the left of 8 on the real number line.   (d) x is at most 12.

Solution
(a) $6 > 5$ (b) $10 < 15$ (c) $3 < 8$
(d) "x is at most 12" means it must be less than or equal to 12. Thus, $x \le 12$.   ■

The subset of the real numbers consisting of all real numbers x that lie between a and b, excluding a and b, can be denoted by the *double inequality* $a < x < b$ or by the **open interval** (a, b). It is called an open interval because neither of the endpoints is included in the interval. The **closed interval** $[a, b]$ represents the set of all real numbers x satisfying $a \le x \le b$. Intervals containing one endpoint, such as $(a, b]$ and $[a, b)$, are called **half-open intervals**.

We can use $[a, +\infty)$ to represent the inequality $x \ge a$ and $(-\infty, a)$ to represent $x < a$. In each of these cases, the symbols $+\infty$ and $-\infty$ are not real numbers but represent the fact that x increases without bound ($+\infty$) or decreases without bound ($-\infty$). Table 0.2 summarizes three types of intervals.

■ | TABLE 0.2 |

INTERVALS

Type of Interval	Inequality Notation	Interval Notation	Graph
Open interval	$x > a$	(a, ∞)	
	$x < b$	$(-\infty, b)$	
	$a < x < b$	(a, b)	
Half-open interval	$x \geq a$	$[a, \infty)$	
	$x \leq b$	$(-\infty, b]$	
	$a \leq x < b$	$[a, b)$	
	$a < x \leq b$	$(a, b]$	
Closed interval	$a \leq x \leq b$	$[a, b]$	

CHECKPOINT

1. Evaluate the following, if possible. For any that are meaningless, so state.
   (a) $\dfrac{4}{0}$ (b) $\dfrac{0}{4}$
   (c) $\dfrac{4}{4}$ (d) $\dfrac{4-4}{4-4}$

2. For parts (a)–(d), write the inequality corresponding to the given interval and sketch its graph on a real number line.
   (a) $(1, 3)$ (b) $(0, 3]$
   (c) $[-1, \infty)$ (d) $(-\infty, 2)$

3. Express the following inequalities in interval notation and name the type of interval.
   (a) $3 \leq x \leq 6$ (b) $-6 \leq x < 4$

Absolute Value   Sometimes we are interested in the *distance* a number is from the origin (0) of the real number line, without regard to direction. The distance a number a is from 0 on the number line is the **absolute value** of a, denoted $|a|$. The absolute value of any nonzero number is positive, and the absolute value of 0 is 0.

EXAMPLE **2**   **Absolute Value**

Evaluate the following.
(a) $|-4|$ (b) $|+2|$
(c) $|0|$ (d) $|-5 - |-3||$

Solution
(a) $|-4| = +4 = 4$ (b) $|+2| = +2 = 2$
(c) $|0| = 0$ (d) $|-5 - |-3|| = |-5 - 3| = |-8| = 8$ ■

Note that if a is a nonnegative number, then $|a| = a$, *but if a is negative, then $|a|$ is the positive number* $(-a)$. Thus

Absolute Value

$$|a| = \begin{cases} a & if\ a \geq 0 \\ -a & if\ a < 0 \end{cases}$$

In performing computations with real numbers, it is important to remember the rules for computations.

Operations with Real (Signed) Numbers

Procedure	Example
1. (a) To add two real numbers with the same sign, add their absolute values and affix their common sign.	1. (a) $(+5) + (+6) = +11$ $$\left(-\frac{1}{6}\right) + \left(-\frac{2}{6}\right) = -\frac{3}{6} = -\frac{1}{2}$$
(b) To add two real numbers with unlike signs, find the difference of their absolute values and affix the sign of the number with the larger absolute value.	(b) $(-4) + (+3) = -1$ $(+5) + (-3) = +2$ $$\left(-\frac{11}{7}\right) + (+1) = -\frac{4}{7}$$
2. To subtract one real number from another, change the sign of the number being subtracted and proceed as in addition.	2. $(-9) - (-8) = (-9) + (+8) = -1$ $16 - (8) = 16 + (-8) = +8$
3. (a) The product of two real numbers with like signs is positive.	3. (a) $(-3)(-4) = +12$ $$\left(+\frac{3}{4}\right)(+4) = +3$$
(b) The product of two real numbers with unlike signs is negative.	(b) $5(-3) = -15$ $(-3)(+4) = -12$
4. (a) The quotient of two real numbers with like signs is positive.	4. (a) $(-14) \div (-2) = +7$ $+36/4 = +9$
(b) The quotient of two real numbers with unlike signs is negative.	(b) $(-28)/4 = -7$ $45 \div (-5) = -9$

When two or more operations with real numbers are indicated in an evaluation, it is important that everyone agree on the order in which the operations are performed so that a unique result is guaranteed. The following **order of operations** is universally accepted.

Order of Operations

1. Perform operations within parentheses.
2. Find indicated powers ($2^3 = 2 \cdot 2 \cdot 2 = 8$).
3. Perform multiplications and divisions from left to right.
4. Perform additions and subtractions from left to right.

EXAMPLE 3   Order of Operations

Evaluate the following.

(a) $-4 + 3$ (b) $-4^2 + 3$ (c) $(-4 + 3)^2 + 3$ (d) $6 \div 2(2 + 1)$

Solution

(a) -1

(b) Note that with -4^2 the power 2 is applied only to 4, not to -4 (which would be written $(-4)^2$). Thus $-4^2 + 3 = -(4^2) + 3 = -16 + 3 = -13$

(c) $(-1)^2 + 3 = 1 + 3 = 4$

(d) $6 \div 2(3) = (6 \div 2)(3) = 3 \cdot 3 = 9$

CHECKPOINT

True or false:

4. $-(-5)^2 = 25$

5. $|4 - 6| = |4| - |6|$

6. $9 - 2(2)(-10) = 7(2)(-10) = -140$

We will assume that you have a scientific or graphing calculator. Discussions of some of the capabilities of graphing calculators and graphing utilities will be found throughout the text.

Most scientific and graphing calculators use standard algebraic order when evaluating arithmetic expressions. Working outward from inner parentheses, calculations are performed from left to right. Powers and roots are evaluated first, followed by multiplications and divisions, and then additions and subtractions.

CHECKPOINT SOLUTIONS

1. (a) Meaningless. A denominator of zero means division by zero, which is undefined.
   (b) $\frac{0}{4} = 0$. A numerator of zero (when the denominator is not zero) means the fraction has value 0.
   (c) $\frac{4}{4} = 1$
   (d) Meaningless. The denominator is zero.

2. (a) $1 < x < 3$
   (b) $0 < x \le 3$
   (c) $-1 \le x < \infty$ or $x \ge -1$
   (d) $-\infty < x < 2$ or $x < 2$

3. (a) $[3, 6]$; closed interval
   (b) $[-6, 4)$; half-open interval

4. False. $-(-5)^2 = (-1)(-5)^2 = (-1)(25) = -25$. Exponentiation has priority and applies only to -5.

5. False. $|4 - 6| = |-2| = 2$ and $|4| - |6| = 4 - 6 = -2$.

6. False. Without parentheses, multiplication has priority over subtraction.
   $9 - 2(2)(-10) = 9 - 4(-10) = 9 + 40 = 49$.

| EXERCISES | 0.2

Access end-of-section exercises online at www.webassign.net ENHANCED WebAssign

0.3

Integral Exponents

If $1000 is placed in a 5-year savings certificate that pays an interest rate of 10% per year, compounded annually, then the amount returned after 5 years is given by

$$1000(1.1)^5$$

The 5 in this expression is an *exponent*. Exponents provide an easier way to denote certain multiplications. For example,

$$(1.1)^5 = (1.1)(1.1)(1.1)(1.1)(1.1)$$

An understanding of the properties of exponents is fundamental to the algebra needed to study functions and solve equations. Furthermore, the definition of exponential and logarithmic functions and many of the techniques in calculus also require an understanding of the properties of exponents.

For any real number a,

$$a^2 = a \cdot a, \quad a^3 = a \cdot a \cdot a, \quad \text{and} \quad a^n = a \cdot a \cdot a \cdot \ldots \cdot a \quad (n \text{ factors})$$

for any positive integer n. The positive integer n is called the **exponent,** the number a is called the **base,** and a^n is read "a to the nth power."

Note that $4a^n$ means $4(a^n)$, which is different from $(4a)^n$. The 4 is the coefficient of a^n in $4a^n$. Note also that $-x^n$ is not equivalent to $(-x)^n$ when n is even. For example, $-3^4 = -81$, but $(-3)^4 = 81$.

Some of the rules of exponents follow.

Positive Integer Exponents

For any real numbers a and b and positive integers m and n,

1. $a^m \cdot a^n = a^{m+n}$

2. For $a \neq 0$, $\dfrac{a^m}{a^n} = \begin{cases} a^{m-n} & \text{if } m > n \\ 1 & \text{if } m = n \\ 1/a^{n-m} & \text{if } m < n \end{cases}$

3. $(ab)^m = a^m b^m$

4. $\left(\dfrac{a}{b}\right)^m = \dfrac{a^m}{b^m}$ $(b \neq 0)$

5. $(a^m)^n = a^{mn}$

EXAMPLE 1   **Positive Integer Exponents**

Use rules of positive integer exponents to rewrite the following. Assume all denominators are nonzero.

(a) $\dfrac{5^6}{5^4}$

(b) $\dfrac{x^2}{x^5}$

(c) $\left(\dfrac{x}{y}\right)^4$

(d) $(3x^2y^3)^4$

(e) $3^3 \cdot 3^2$

Solution

(a) $\dfrac{5^6}{5^4} = 5^{6-4} = 5^2$

(b) $\dfrac{x^2}{x^5} = \dfrac{1}{x^{5-2}} = \dfrac{1}{x^3}$

(c) $\left(\dfrac{x}{y}\right)^4 = \dfrac{x^4}{y^4}$

(d) $(3x^2y^3)^4 = 3^4(x^2)^4(y^3)^4 = 81x^8y^{12}$

(e) $3^3 \cdot 3^2 = 3^{3+2} = 3^5$

For certain calculus operations, use of negative exponents is necessary in order to write problems in the proper form. We can extend the rules for positive integer exponents to all integers by defining a^0 and a^{-n}. Clearly $a^m \cdot a^0$ should equal $a^{m+0} = a^m$, and it will if $a^0 = 1$.

Zero Exponent

For any nonzero real number a, we define $a^0 = 1$. We leave 0^0 undefined.

In Section 0.2, we defined a^{-1} as $1/a$ for $a \neq 0$, so we define a^{-n} as $(a^{-1})^n$.

Negative Exponents

$$a^{-n} = (a^{-1})^n = \left(\dfrac{1}{a}\right)^n = \dfrac{1}{a^n} \qquad (a \neq 0)$$

$$\left(\dfrac{a}{b}\right)^{-n} = \left[\left(\dfrac{a}{b}\right)^{-1}\right]^n = \left(\dfrac{b}{a}\right)^n \qquad (a \neq 0, \quad b \neq 0)$$

EXAMPLE 2   **Negative and Zero Exponents**

Write the following without exponents.

(a) $6 \cdot 3^0$

(b) 6^{-2}

(c) $\left(\dfrac{1}{3}\right)^{-1}$

(d) $-\left(\dfrac{2}{3}\right)^{-4}$

(e) $(-4)^{-2}$

Solution

(a) $6 \cdot 3^0 = 6 \cdot 1 = 6$ (b) $6^{-2} = \dfrac{1}{6^2} = \dfrac{1}{36}$

(c) $\left(\dfrac{1}{3}\right)^{-1} = \dfrac{3}{1} = 3$ (d) $-\left(\dfrac{2}{3}\right)^{-4} = -\left(\dfrac{3}{2}\right)^4 = \dfrac{-81}{16}$

(e) $(-4)^{-2} = \dfrac{1}{(-4)^2} = \dfrac{1}{16}$

As we'll see in the chapter on the mathematics of finance (Chapter 6), negative exponents arise in financial calculations when we have a future goal for an investment and want to know how much to invest now. For example, if money can be invested at 9%, compounded annually, then the amount we must invest now (which is called the present value) in order to have $10,000 in the account after 7 years is given by $10,000(1.09)^{-7}$. Calculations such as this are often done directly with a calculator.

Using the definitions of zero and negative exponents enables us to extend the rules of exponents to all integers and to express them more simply.

Rules of Exponents

For real numbers a and b and *integers* m and n,

1. $a^m \cdot a^n = a^{m+n}$ 2. $a^m/a^n = a^{m-n}$ $(a \neq 0)$
3. $(ab)^m = a^m b^m$ 4. $(a^m)^n = a^{mn}$
5. $(a/b)^m = a^m/b^m$ $(b \neq 0)$ 6. $a^0 = 1$ $(a \neq 0)$
7. $a^{-n} = 1/a^n$ $(a \neq 0)$ 8. $(a/b)^{-n} = (b/a)^n$ $(a, b \neq 0)$

Throughout the remainder of the text, we will assume all expressions are defined.

EXAMPLE 3   Operations with Exponents

Use the rules of exponents and the definitions of a^0 and a^{-n} to simplify the following with positive exponents.

(a) $2(x^2)^{-2}$ (b) $x^{-2} \cdot x^{-5}$

(c) $\dfrac{x^{-8}}{x^{-4}}$ (d) $\left(\dfrac{2x^3}{3x^{-5}}\right)^{-2}$

Solution

(a) $2(x^2)^{-2} = 2x^{-4} = 2\left(\dfrac{1}{x^4}\right) = \dfrac{2}{x^4}$

(b) $x^{-2} \cdot x^{-5} = x^{-2-5} = x^{-7} = \dfrac{1}{x^7}$

(c) $\dfrac{x^{-8}}{x^{-4}} = x^{-8-(-4)} = x^{-4} = \dfrac{1}{x^4}$

(d) $\left(\dfrac{2x^3}{3x^{-5}}\right)^{-2} = \left(\dfrac{2x^8}{3}\right)^{-2} = \left(\dfrac{3}{2x^8}\right)^2 = \dfrac{9}{4x^{16}}$

CHECKPOINT

1. Complete the following.
   (a) $x^3 \cdot x^8 = x^?$ (b) $x \cdot x^4 \cdot x^{-3} = x^?$
   (c) $\dfrac{1}{x^4} = x^?$ (d) $x^{24} \div x^{-3} = x^?$
   (e) $(x^4)^2 = x^?$ (f) $(2x^4y)^3 = ?$
2. True or false:
   (a) $3x^{-2} = \dfrac{1}{9x^2}$ (b) $-x^{-4} = \dfrac{-1}{x^4}$ (c) $x^{-3} = -x^3$
3. Evaluate the following, if possible. For any that are meaningless, so state. Assume $x > 0$.
   (a) 0^4 (b) 0^0 (c) x^0 (d) 0^x (e) 0^{-4} (f) -5^{-2}

EXAMPLE 4 **Rewriting a Quotient**

Write $(x^2y)/(9wz^3)$ with all factors in the numerator.

Solution

$$\frac{x^2y}{9wz^3} = x^2y\left(\frac{1}{9wz^3}\right) = x^2y\left(\frac{1}{9}\right)\left(\frac{1}{w}\right)\left(\frac{1}{z^3}\right) = x^2y \cdot 9^{-1}w^{-1}z^{-3}$$
$$= 9^{-1}x^2yw^{-1}z^{-3}$$

EXAMPLE 5 **Rewriting with Positive Exponents**

Simplify the following so all exponents are positive.

(a) $(2^3x^{-4}y^5)^{-2}$

(b) $\dfrac{2x^4(x^2y)^0}{(4x^{-2}y)^2}$

Solution

(a) $(2^3x^{-4}y^5)^{-2} = 2^{-6}x^8y^{-10} = \dfrac{1}{2^6} \cdot x^8 \cdot \dfrac{1}{y^{10}} = \dfrac{x^8}{64y^{10}}$

(b) $\dfrac{2x^4(x^2y)^0}{(4x^{-2}y)^2} = \dfrac{2x^4 \cdot 1}{4^2x^{-4}y^2} = \dfrac{2}{4^2} \cdot \dfrac{x^4}{x^{-4}} \cdot \dfrac{1}{y^2} = \dfrac{2}{16} \cdot \dfrac{x^8}{1} \cdot \dfrac{1}{y^2} = \dfrac{x^8}{8y^2}$

CHECKPOINT SOLUTIONS

1. (a) $x^3 \cdot x^8 = x^{3+8} = x^{11}$

   (b) $x \cdot x^4 \cdot x^{-3} = x^{1+4+(-3)} = x^2$

   (c) $\dfrac{1}{x^4} = x^{-4}$

   (d) $x^{24} \div x^{-3} = x^{24-(-3)} = x^{27}$

   (e) $(x^4)^2 = x^{(4)(2)} = x^8$

   (f) $(2x^4y)^3 = 2^3(x^4)^3y^3 = 8x^{12}y^3$

2. (a) False. $3x^{-2} = 3\left(\dfrac{1}{x^2}\right) = \dfrac{3}{x^2}$

   (b) True. $-x^{-4} = (-1)\left(\dfrac{1}{x^4}\right) = \dfrac{-1}{x^4}$

   (c) False. $x^{-3} = \dfrac{1}{x^3}$

3. (a) $0^4 = 0$

   (b) Meaningless. 0^0 is undefined.

   (c) $x^0 = 1$ since $x \neq 0$

   (d) $0^x = 0$ because $x > 0$

   (e) Meaningless. 0^{-4} would be $\dfrac{1}{0^4}$, which is undefined.

   (f) $-5^{-2} = (-1)\left(\dfrac{1}{5^2}\right) = \dfrac{-1}{25}$

| EXERCISES | 0.3

Access end-of-section exercises online at **www.webassign.net**

0.4

Radicals and Rational Exponents

Roots   A process closely linked to that of raising numbers to powers is that of extracting roots. From geometry we know that if an edge of a cube has a length of x units, its volume is x^3 cubic units. Reversing this process, we determine that if the volume of a cube is V cubic units, the length of an edge is the cube root of V, which is denoted

$$\sqrt[3]{V} \text{ units}$$

When we seek the **cube root** of a number such as 8 (written $\sqrt[3]{8}$), we are looking for a real number whose cube equals 8. Because $2^3 = 8$, we know that $\sqrt[3]{8} = 2$. Similarly, $\sqrt[3]{-27} = -3$ because $(-3)^3 = -27$. The expression $\sqrt[n]{a}$ is called a **radical,** where $\sqrt{}$ is the **radical sign,** n the **index,** and a the **radicand.** When no index is indicated, the index is assumed to be 2 and the expression is called a **square root;** thus $\sqrt{4}$ is the square root of 4 and represents the positive number whose square is 4.

Only one real number satisfies $\sqrt[n]{a}$ for a real number a and an odd number n; we call that number the **principal nth root** or, more simply, the **nth root.**

For an even index n, there are two possible cases:

1. If a is negative, there is no real number equal to $\sqrt[n]{a}$. For example, there are no real numbers that equal $\sqrt{-4}$ or $\sqrt[4]{-16}$ because there is no real number b such that $b^2 = -4$ or $b^4 = -16$. In this case, we say $\sqrt[n]{a}$ is not a real number.

2. If a is positive, there are two real numbers whose nth power equals a. For example, $3^2 = 9$ and $(-3)^2 = 9$. In order to have a unique nth root, we define the (principal) nth root, $\sqrt[n]{a}$, as the *positive* number b that satisfies $b^n = a$.

We summarize this discussion as follows.

nth Root of a	The **(principal) nth root** of a real number is defined as

$$\sqrt[n]{a} = b \quad \text{only if} \quad a = b^n$$

subject to the following conditions:

	$a = 0$	$a > 0$	$a < 0$
n even	$\sqrt[n]{a} = 0$	$\sqrt[n]{a} > 0$	$\sqrt[n]{a}$ not real
n odd	$\sqrt[n]{a} = 0$	$\sqrt[n]{a} > 0$	$\sqrt[n]{a} < 0$

When we are asked for the root of a number, we give the principal root.

EXAMPLE 1 Roots

Find the roots, if they are real numbers.

(a) $\sqrt[6]{64}$ (b) $-\sqrt{16}$ (c) $\sqrt[3]{-8}$ (d) $\sqrt{-16}$

Solution

(a) $\sqrt[6]{64} = 2$ because $2^6 = 64$ (b) $-\sqrt{16} = -(\sqrt{16}) = -4$ (c) $\sqrt[3]{-8} = -2$

(d) $\sqrt{-16}$ is not a real number because an even root of a negative number is not real.   ■

Fractional Exponents   In order to perform evaluations on a calculator or to perform calculus operations, it is sometimes necessary to rewrite radicals in exponential form with fractional exponents.

We have stated that for $a \geq 0$ and $b \geq 0$,

$$\sqrt{a} = b \quad \text{only if} \quad a = b^2$$

This means that $(\sqrt{a})^2 = b^2 = a$, or $(\sqrt{a})^2 = a$. In order to extend the properties of exponents to rational exponents, it is necessary to define

$$a^{1/2} = \sqrt{a} \quad \text{so that} \quad (a^{1/2})^2 = a$$

Exponent $1/n$	For a positive integer n, we define

$$a^{1/n} = \sqrt[n]{a} \quad \text{if} \quad \sqrt[n]{a} \text{ exists}$$

Thus $(a^{1/n})^n = a^{(1/n) \cdot n} = a$.

Because we wish the properties established for integer exponents to extend to rational exponents, we make the following definitions.

Rational Exponents

For positive integer n and any integer m (with $a \neq 0$ when $m \leq 0$ and with m/n in lowest terms):

1. $a^{m/n} = (a^{1/n})^m = (\sqrt[n]{a})^m$

2. $a^{m/n} = (a^m)^{1/n} = \sqrt[n]{a^m}$ if a is nonnegative when n is even.

Throughout the remaining discussion, we assume all expressions are real.

EXAMPLE 2 **Radical Form**

Write the following in radical form and simplify.
(a) $16^{3/4}$ (b) $y^{-3/2}$ (c) $(6m)^{2/3}$

Solution

(a) $16^{3/4} = \sqrt[4]{16^3} = (\sqrt[4]{16})^3 = (2)^3 = 8$

(b) $y^{-3/2} = \dfrac{1}{y^{3/2}} = \dfrac{1}{\sqrt{y^3}}$

(c) $(6m)^{2/3} = \sqrt[3]{(6m)^2} = \sqrt[3]{36m^2}$ ■

EXAMPLE 3 **Fractional Exponents**

Write the following without radical signs.

(a) $\sqrt{x^3}$ (b) $\dfrac{1}{\sqrt[3]{b^2}}$ (c) $\sqrt[3]{(ab)^3}$

Solution

(a) $\sqrt{x^3} = x^{3/2}$ (b) $\dfrac{1}{\sqrt[3]{b^2}} = \dfrac{1}{b^{2/3}} = b^{-2/3}$ (c) $\sqrt[3]{(ab)^3} = (ab)^{3/3} = ab$ ■

Our definition of $a^{m/n}$ guarantees that the rules for exponents will apply to fractional exponents. Thus we can perform operations with fractional exponents as we did with integer exponents.

EXAMPLE 4 **Operations with Fractional Exponents**

Simplify the following expressions.
(a) $a^{1/2} \cdot a^{1/6}$ (b) $a^{3/4}/a^{1/3}$ (c) $(a^3 b)^{2/3}$ (d) $(a^{3/2})^{1/2}$ (e) $a^{-1/2} \cdot a^{-3/2}$

Solution

(a) $a^{1/2} \cdot a^{1/6} = a^{1/2 + 1/6} = a^{3/6 + 1/6} = a^{4/6} = a^{2/3}$
(b) $a^{3/4}/a^{1/3} = a^{3/4 - 1/3} = a^{9/12 - 4/12} = a^{5/12}$
(c) $(a^3 b)^{2/3} = (a^3)^{2/3} b^{2/3} = a^2 b^{2/3}$
(d) $(a^{3/2})^{1/2} = a^{(3/2)(1/2)} = a^{3/4}$
(e) $a^{-1/2} \cdot a^{-3/2} = a^{-1/2 - 3/2} = a^{-2} = 1/a^2$ ■

CHECKPOINT

1. Which of the following are *not* real numbers?
   (a) $\sqrt[3]{-64}$ (b) $\sqrt{-64}$ (c) $\sqrt{0}$ (d) $\sqrt[4]{1}$ (e) $\sqrt[5]{-1}$ (f) $\sqrt[8]{-1}$

2. (a) Write as radicals: $x^{1/3}, x^{2/5}, x^{-3/2}$

   (b) Write with fractional exponents: $\sqrt[4]{x^3} = x^?$, $\dfrac{1}{\sqrt{x}} = \dfrac{1}{x^?} = x^?$

3. Evaluate the following.

    (a) $8^{2/3}$ (b) $(-8)^{2/3}$ (c) $8^{-2/3}$ (d) $-8^{-2/3}$ (e) $\sqrt[15]{7^1}$

4. Complete the following.

    (a) $x \cdot x^{1/3} \cdot x^3 = x^?$ (b) $x^2 \div x^{1/2} = x^?$ (c) $(x^{-2/3})^{-3} = x^?$

    (d) $x^{-3/2} \cdot x^{1/2} = x^?$ (e) $x^{-3/2} \cdot x = x^?$ (f) $\left(\dfrac{x^4}{y^2}\right)^{3/2} = ?$

5. True or false:

    (a) $\dfrac{8x^{2/3}}{x^{-1/3}} = 4x$ (b) $(16x^8 y)^{3/4} = 12x^6 y^{3/4}$

    (c) $\left(\dfrac{x^2}{y^3}\right)^{-1/3} = \left(\dfrac{y^3}{x^2}\right)^{1/3} = \dfrac{y}{x^{2/3}}$

Operations with Radicals

We can perform operations with radicals by first rewriting in exponential form, performing the operations with exponents, and then converting the answer back to radical form. Another option is to apply directly the following rules for operations with radicals.

Rules for Radicals	Example
Given that $\sqrt[n]{a}$ and $\sqrt[n]{b}$ are real,*	
1. $\sqrt[n]{a^n} = (\sqrt[n]{a})^n = a$	1. $\sqrt[5]{6^5} = (\sqrt[5]{6})^5 = 6$
2. $\sqrt[n]{a} \cdot \sqrt[n]{b} = \sqrt[n]{ab}$	2. $\sqrt[3]{2}\sqrt[3]{4} = \sqrt[3]{8} = \sqrt[3]{2^3} = 2$
3. $\dfrac{\sqrt[n]{a}}{\sqrt[n]{b}} = \sqrt[n]{\dfrac{a}{b}}$ $(b \neq 0)$	3. $\dfrac{\sqrt{18}}{\sqrt{2}} = \sqrt{\dfrac{18}{2}} = \sqrt{9} = 3$

*Note that this means $a \geq 0$ and $b \geq 0$ if n is even.

Let us consider Rule 1 for radicals more carefully. Note that if n is even and $a < 0$, then $\sqrt[n]{a}$ is not real, and Rule 1 does not apply. For example, $\sqrt{-2}$ is not a real number, and

$$\sqrt{(-2)^2} \neq -2 \quad \text{because} \quad \sqrt{(-2)^2} = \sqrt{4} = 2 = -(-2)$$

We can generalize this observation as follows: If $a < 0$, then $\sqrt{a^2} = -a > 0$, so

$$\sqrt{a^2} = \begin{cases} a & \text{if } a \geq 0 \\ -a & \text{if } a < 0 \end{cases}$$

This means

$$\sqrt{a^2} = |a|$$

EXAMPLE 5 Rules for Radicals

Simplify:

    (a) $\sqrt[3]{8^3}$ (b) $\sqrt[5]{x^5}$ (c) $\sqrt{x^2}$ (d) $\left[\sqrt[7]{(3x^2 + 4)^3}\right]^7$

Solution

(a) $\sqrt[3]{8^3} = 8$ by Rule 1 for radicals

(b) $\sqrt[5]{x^5} = x$ (c) $\sqrt{x^2} = |x|$ (d) $\left[\sqrt[7]{(3x^2 + 4)^3}\right]^7 = (3x^2 + 4)^3$ ■

Up to now, to *simplify* a radical has meant to find the indicated root. More generally, a radical expression $\sqrt[n]{x}$ is considered simplified if x has no nth powers as factors. Rule 2 for radicals $(\sqrt[n]{a} \cdot \sqrt[n]{b} = \sqrt[n]{ab})$ provides a procedure for simplifying radicals.

EXAMPLE 6   **Simplifying Radicals**

Simplify the following radicals; assume the expressions are real numbers.
(a) $\sqrt{48x^5y^6}$ $(y \geq 0)$ (b) $\sqrt[3]{72a^3b^4}$

Solution

(a) To simplify $\sqrt{48x^5y^6}$, we first factor $48x^5y^6$ into perfect-square factors and other factors. Then we apply Rule 2.

$$\sqrt{48x^5y^6} = \sqrt{16 \cdot 3 \cdot x^4xy^6} = \sqrt{16}\sqrt{x^4}\sqrt{y^6}\sqrt{3x} = 4x^2y^3\sqrt{3x}$$

(b) In this case, we factor $72a^3b^4$ into factors that are perfect cubes and other factors.

$$\sqrt[3]{72a^3b^4} = \sqrt[3]{8 \cdot 9a^3b^3b} = \sqrt[3]{8} \cdot \sqrt[3]{a^3} \cdot \sqrt[3]{b^3} \cdot \sqrt[3]{9b} = 2ab\sqrt[3]{9b}$$ ■

Rule 2 for radicals also provides a procedure for multiplying two roots with the same index.

EXAMPLE 7   **Multiplying Radicals**

Multiply the following and simplify the answers, assuming nonnegative variables.
(a) $\sqrt[3]{2xy} \cdot \sqrt[3]{4x^2y}$ (b) $\sqrt{8xy^3z}\sqrt{4x^2y^3z^2}$

Solution

(a) $\sqrt[3]{2xy} \cdot \sqrt[3]{4x^2y} = \sqrt[3]{2xy \cdot 4x^2y} = \sqrt[3]{8x^3y^2} = \sqrt[3]{8} \cdot \sqrt[3]{x^3} \cdot \sqrt[3]{y^2} = 2x\sqrt[3]{y^2}$
(b) $\sqrt{8xy^3z}\sqrt{4x^2y^3z^2} = \sqrt{32x^3y^6z^3} = \sqrt{16x^2y^6z^2}\sqrt{2xz} = 4xy^3z\sqrt{2xz}$ ■

Rule 3 for radicals $(\sqrt[n]{a}/\sqrt[n]{b} = \sqrt[n]{a/b})$ indicates how to find the quotient of two roots with the same index.

EXAMPLE 8   **Dividing Radicals**

Find the quotients and simplify the answers, assuming nonnegative variables.
(a) $\dfrac{\sqrt[3]{32}}{\sqrt[3]{4}}$ (b) $\dfrac{\sqrt{16a^3x}}{\sqrt{2ax}}$

Solution

(a) $\dfrac{\sqrt[3]{32}}{\sqrt[3]{4}} = \sqrt[3]{\dfrac{32}{4}} = \sqrt[3]{8} = 2$

(b) $\dfrac{\sqrt{16a^3x}}{\sqrt{2ax}} = \sqrt{\dfrac{16a^3x}{2ax}} = \sqrt{8a^2} = 2a\sqrt{2}$ ■

Rationalizing   Occasionally, we wish to express a fraction containing radicals in an equivalent form that contains no radicals in the denominator. This is accomplished by multiplying the numerator *and* the denominator by the expression that will remove the radical from the denominator. This process is called **rationalizing the denominator.**

EXAMPLE 9   **Rationalizing Denominators**

Express the following with no radicals in the denominator. (Rationalize each denominator.)
(a) $\dfrac{15}{\sqrt{x}}$ (b) $\dfrac{2x}{\sqrt{18xy}}$ $(x, y > 0)$ (c) $\dfrac{3x}{\sqrt[3]{2x^2}}$ $(x \neq 0)$

Solution

(a) We wish to create a perfect square under the radical in the denominator.

$$\frac{15}{\sqrt{x}} \cdot \frac{\sqrt{x}}{\sqrt{x}} = \frac{15\sqrt{x}}{x}$$

(b) $\dfrac{2x}{\sqrt{18xy}} \cdot \dfrac{\sqrt{2xy}}{\sqrt{2xy}} = \dfrac{2x\sqrt{2xy}}{\sqrt{36x^2y^2}} = \dfrac{2x\sqrt{2xy}}{6xy} = \dfrac{\sqrt{2xy}}{3y}$

(c) We wish to create a perfect cube under the radical in the denominator.

$$\frac{3x}{\sqrt[3]{2x^2}} \cdot \frac{\sqrt[3]{4x}}{\sqrt[3]{4x}} = \frac{3x\sqrt[3]{4x}}{\sqrt[3]{8x^3}} = \frac{3x\sqrt[3]{4x}}{2x} = \frac{3\sqrt[3]{4x}}{2}$$ ∎

CHECKPOINT

6. Simplify:

  (a) $\sqrt[7]{x^7}$ (b) $[\sqrt[5]{(x^2+1)^2}]^5$ (c) $\sqrt{12xy^2} \cdot \sqrt{3x^2y}$

7. Rationalize the denominator of $\dfrac{x}{\sqrt{5x}}$ if $x \neq 0$.

It is also sometimes useful, especially in calculus, to *rationalize the numerator* of a fraction. For example, we can rationalize the numerator of

$$\frac{\sqrt[3]{4x^2}}{3x}$$

by multiplying the numerator and denominator by $\sqrt[3]{2x}$, which creates a perfect cube under the radical:

$$\frac{\sqrt[3]{4x^2}}{3x} \cdot \frac{\sqrt[3]{2x}}{\sqrt[3]{2x}} = \frac{\sqrt[3]{8x^3}}{3x\sqrt[3]{2x}} = \frac{2x}{3x\sqrt[3]{2x}} = \frac{2}{3\sqrt[3]{2x}}$$

CHECKPOINT SOLUTIONS

1. Only *even* roots of negatives are not real numbers. Thus part (b) $\sqrt{-64}$ and part (f) $\sqrt[8]{-1}$ are not real numbers.

2. (a) $x^{1/3} = \sqrt[3]{x}$, $x^{2/5} = \sqrt[5]{x^2}$, $x^{-3/2} = \dfrac{1}{x^{3/2}} = \dfrac{1}{\sqrt{x^3}}$

  (b) $\sqrt[4]{x^3} = x^{3/4}$, $\dfrac{1}{\sqrt{x}} = \dfrac{1}{x^{1/2}} = x^{-1/2}$

3. (a) $8^{2/3} = (\sqrt[3]{8})^2 = 2^2 = 4$ (b) $(-8)^{2/3} = (\sqrt[3]{-8})^2 = (-2)^2 = 4$

  (c) $8^{-2/3} = \dfrac{1}{8^{2/3}} = \dfrac{1}{4}$ (d) $-8^{-2/3} = -\left(\dfrac{1}{8^{2/3}}\right) = -\dfrac{1}{4}$

  (e) $\sqrt[15]{71} = (71)^{1/15} \approx 1.32867$

4. (a) $x \cdot x^{1/3} \cdot x^3 = x^{1+1/3+3} = x^{13/3}$ (b) $x^2 \div x^{1/2} = x^{2-1/2} = x^{3/2}$

  (c) $(x^{-2/3})^{-3} = x^{(-2/3)(-3)} = x^2$ (d) $x^{-3/2} \cdot x^{1/2} = x^{-3/2+1/2} = x^{-1}$

  (e) $x^{-3/2} \cdot x = x^{-3/2+1} = x^{-1/2}$ (f) $\left(\dfrac{x^4}{y^2}\right)^{3/2} = \dfrac{(x^4)^{3/2}}{(y^2)^{3/2}} = \dfrac{x^6}{y^3}$

5. (a) False. $\dfrac{8x^{2/3}}{x^{-1/3}} = 8x^{2/3} \cdot x^{1/3} = 8x^{2/3+1/3} = 8x$

  (b) False. $(16x^8y)^{3/4} = 16^{3/4}(x^8)^{3/4}y^{3/4} = (\sqrt[4]{16})^3 x^6 y^{3/4} = 8x^6 y^{3/4}$

  (c) True.

6. (a) $\sqrt[7]{x^7} = x$

  (b) $[\sqrt[5]{(x^2+1)^2}]^5 = (x^2+1)^2$

  (c) $\sqrt{12xy^2} \cdot \sqrt{3x^2y} = \sqrt{36x^3y^3} = \sqrt{36x^2y^2 \cdot xy} = \sqrt{36x^2y^2}\sqrt{xy} = 6xy\sqrt{xy}$

7. $\dfrac{x}{\sqrt{5x}} = \dfrac{x}{\sqrt{5x}} \cdot \dfrac{\sqrt{5x}}{\sqrt{5x}} = \dfrac{x\sqrt{5x}}{5x} = \dfrac{\sqrt{5x}}{5}$

| EXERCISES | 0.4

0.5

Operations with Algebraic Expressions

In algebra we are usually dealing with combinations of real numbers (such as 3, 6/7, and $-\sqrt{2}$) and letters (such as x, a, and m). Unless otherwise specified, the letters are symbols used to represent real numbers and are sometimes called **variables.** An expression obtained by performing additions, subtractions, multiplications, divisions, or extractions of roots with one or more real numbers or variables is called an **algebraic expression.** Unless otherwise specified, the variables represent all real numbers for which the algebraic expression is a real number. Examples of algebraic expressions include

$$3x + 2y, \quad \frac{x^3y + y}{x - 1}, \quad \text{and} \quad \sqrt{x - 3}$$

Note that the variable x cannot be negative in $\sqrt{x - 3}$ and that $(x^3y + y)/(x - 1)$ is not a real number when $x = 1$, because division by 0 is undefined.

Any product of a real number (called the **coefficient**) and one or more variables to powers is called a **term.** The sum of a finite number of terms with nonnegative integer powers on the variables is called a **polynomial.** If a polynomial contains only one variable x, then it is called a polynomial in x.

Polynomial in x

The general form of a **polynomial in x** is

$$a_n x^n + a_{n-1} x^{n-1} + \cdots + a_1 x + a_0$$

where each coefficient a_i is a real number for $i = 0, 1, 2, \ldots, n$. If $a_n \neq 0$, the **degree** of the polynomial is n, and a_n is called the **leading coefficient.** The term a_0 is called the **constant term.**

Thus $4x^3 - 2x - 3$ is a third-degree polynomial in x with leading coefficient 4 and constant term -3. If two or more variables are in a term, the degree of the term is the sum of the exponents of the variables. The degree of a nonzero constant term is zero. Thus the degree of $4x^2y$ is $2 + 1 = 3$, the degree of $6xy$ is $1 + 1 = 2$, and the degree of 3 is 0. The **degree of a polynomial** containing one or more variables is the degree of the term in the polynomial having the highest degree. Therefore, $2xy - 4x + 6$ is a second-degree polynomial.

A polynomial containing two terms is called a **binomial,** and a polynomial containing three terms is called a **trinomial.** A single-term polynomial is a **monomial.**

Operations with Algebraic Expressions

Because monomials and polynomials represent real numbers, the properties of real numbers can be used to add, subtract, multiply, divide, and simplify polynomials. For example, we can use the Distributive Law to add $3x$ and $2x$.

$$3x + 2x = (3 + 2)x = 5x$$

Similarly, $9xy - 3xy = (9 - 3)xy = 6xy$.

Terms with exactly the same variable factors are called **like terms.** We can add or subtract like terms by adding or subtracting the coefficients of the variables. Subtraction of polynomials uses the Distributive Law to remove the parentheses.

EXAMPLE 1   **Combining Polynomials**

Compute (a) $(4xy + 3x) + (5xy - 2x)$ and (b) $(3x^2 + 4xy + 5y^2 + 1) - (6x^2 - 2xy + 4)$.

Solution

(a) $(4xy + 3x) + (5xy - 2x) = 4xy + 3x + 5xy - 2x = 9xy + x$

(b) Removing the parentheses yields

$$3x^2 + 4xy + 5y^2 + 1 - 6x^2 + 2xy - 4 = -3x^2 + 6xy + 5y^2 - 3$$ ■

Using the rules of exponents and the Commutative and Associative Laws for multiplication, we can multiply and divide monomials, as the following example shows.

EXAMPLE **2** **Products and Quotients**

Perform the indicated operations.

(a) $(8xy^3)(2x^3y)(-3xy^2)$　　　　　(b) $-15x^2y^3 \div (3xy^5)$

Solution

(a) $8 \cdot 2 \cdot (-3) \cdot x \cdot x^3 \cdot x \cdot y^3 \cdot y \cdot y^2 = -48x^5y^6$

(b) $\dfrac{-15x^2y^3}{3xy^5} = -\dfrac{15}{3} \cdot \dfrac{x^2}{x} \cdot \dfrac{y^3}{y^5} = -5 \cdot x \cdot \dfrac{1}{y^2} = -\dfrac{5x}{y^2}$ ■

Symbols of grouping are used in algebra in the same way as they are used in the arithmetic of real numbers. We have removed parentheses in the process of adding and subtracting polynomials. Other symbols of grouping, such as brackets, [ ], are treated the same as parentheses.

When there are two or more symbols of grouping involved, we begin with the innermost and work outward.

EXAMPLE **3** **Symbols of Grouping**

Simplify $3x^2 - [2x - (3x^2 - 2x)]$.

Solution

$$\begin{aligned} 3x^2 - [2x - (3x^2 - 2x)] &= 3x^2 - [2x - 3x^2 + 2x] \\ &= 3x^2 - [4x - 3x^2] \\ &= 3x^2 - 4x + 3x^2 = 6x^2 - 4x \end{aligned}$$ ■

By the use of the Distributive Law, we can multiply a binomial by a monomial. For example,

$$x(2x + 3) = x \cdot 2x + x \cdot 3 = 2x^2 + 3x$$

We can extend the Distributive Law to multiply polynomials with more than two terms. For example,

$$5(x + y + 2) = 5x + 5y + 10$$

EXAMPLE **4** **Distributive Law**

Find the following products.

(a) $-4ab(3a^2b + 4ab^2 - 1)$　　　　(b) $(4a + 5b + c)ac$

Solution

(a) $-4ab(3a^2b + 4ab^2 - 1) = -12a^3b^2 - 16a^2b^3 + 4ab$

(b) $(4a + 5b + c)ac = 4a \cdot ac + 5b \cdot ac + c \cdot ac = 4a^2c + 5abc + ac^2$ ■

The Distributive Law can be used to show us how to multiply two polynomials. Consider the indicated multiplication $(a + b)(c + d)$. If we first treat the sum $(a + b)$ as a single quantity, then two successive applications of the Distributive Law gives

$$(a + b)(c + d) = (a + b) \cdot c + (a + b) \cdot d = ac + bc + ad + bd$$

Thus we see that the product can be found by multiplying $(a + b)$ by c, multiplying $(a + b)$ by d, and then adding the products. This is frequently set up as follows.

Product of Two Polynomials

Procedure	Example
To multiply two polynomials:	Multiply $(3x + 4xy + 3y)$ by $(x - 2y)$.
1. Write one of the polynomials above the other.	1. $3x + 4xy + 3y$ $x - 2y$
2. Multiply each term of the top polynomial by each term of the bottom one, and write the similar terms of the product under one another.	2. $3x^2 + 4x^2y + 3xy$ $-6xy - 8xy^2 - 6y^2$
3. Add like terms to simplify the product.	3. $3x^2 + 4x^2y - 3xy - 8xy^2 - 6y^2$

EXAMPLE 5   The Product of Two Polynomials

Multiply $(4x^2 + 3xy + 4x)$ by $(2x - 3y)$.

Solution

$$4x^2 + 3xy + 4x$$
$$2x - 3y$$
$$\overline{8x^3 + 6x^2y + 8x^2}$$
$$-12x^2y - 9xy^2 - 12xy$$
$$\overline{8x^3 - 6x^2y + 8x^2 - 9xy^2 - 12xy}$$

Because the multiplications we must perform often involve binomials, the following special products are worth remembering.

Special Products

A. $(x + a)(x + b) = x^2 + (a + b)x + ab$
B. $(ax + b)(cx + d) = acx^2 + (ad + bc)x + bd$

It is easier to remember these two special products if we note their structure. We can obtain these products by finding the products of the First terms, Outside terms, Inside terms, and Last terms, and then adding the results. This is called the FOIL method of multiplying two binomials.

EXAMPLE 6   Products of Binomials

Multiply the following.
(a) $(x - 4)(x + 3)$
(b) $(3x + 2)(2x + 5)$

Solution

$\qquad\qquad\qquad$ First $\quad$ Outside $\quad$ Inside $\quad$ Last
(a) $(x - 4)(x + 3) = (x^2) + (3x) + (-4x) + (-12) = x^2 - x - 12$
(b) $(3x + 2)(2x + 5) = (6x^2) + (15x) + (4x) + (10) = 6x^2 + 19x + 10$

Additional special products are as follows:

Additional Special Products

C. $(x + a)^2 = x^2 + 2ax + a^2$ $\qquad$ binomial squared
D. $(x - a)^2 = x^2 - 2ax + a^2$ $\qquad$ binomial squared
E. $(x + a)(x - a) = x^2 - a^2$ $\qquad$ difference of two squares
F. $(x + a)^3 = x^3 + 3ax^2 + 3a^2x + a^3$ $\qquad$ binomial cubed
G. $(x - a)^3 = x^3 - 3ax^2 + 3a^2x - a^3$ $\qquad$ binomial cubed

EXAMPLE 7   Special Products

Multiply the following.

(a) $(x + 5)^2$ (b) $(3x - 4y)^2$
(c) $(x - 2)(x + 2)$ (d) $(x^2 - y^3)^2$
(e) $(x + 4)^3$

Solution

(a) $(x + 5)^2 = x^2 + 2(5)x + 25 = x^2 + 10x + 25$
(b) $(3x - 4y)^2 = (3x)^2 - 2(3x)(4y) + (4y)^2 = 9x^2 - 24xy + 16y^2$
(c) $(x - 2)(x + 2) = x^2 - 4$
(d) $(x^2 - y^3)^2 = (x^2)^2 - 2(x^2)(y^3) + (y^3)^2 = x^4 - 2x^2y^3 + y^6$
(e) $(x + 4)^3 = x^3 + 3(4)(x^2) + 3(4^2)(x) + 4^3 = x^3 + 12x^2 + 48x + 64$ ■

CHECKPOINT

1. Remove parentheses and combine like terms: $9x - 5x(x + 2) + 4x^2$
2. Find the following products.
    (a) $(2x + 1)(4x^2 - 2x + 1)$ (b) $(x + 3)^2$
    (c) $(3x + 2)(x - 5)$ (d) $(1 - 4x)(1 + 4x)$

All algebraic expressions can represent real numbers, so the techniques used to perform operations on polynomials and to simplify polynomials also apply to other algebraic expressions.

EXAMPLE 8   Operations with Algebraic Expressions

Perform the indicated operations.

(a) $3\sqrt{3} + 4x\sqrt{y} - 5\sqrt{3} - 11x\sqrt{y} - (\sqrt{3} - x\sqrt{y})$
(b) $x^{3/2}(x^{1/2} - x^{-1/2})$
(c) $(x^{1/2} - x^{1/3})^2$
(d) $(\sqrt{x} + 2)(\sqrt{x} - 2)$

Solution

(a) We remove parentheses and then combine the terms containing $\sqrt{3}$ and the terms containing $x\sqrt{y}$.

$$(3 - 5 - 1)\sqrt{3} + (4 - 11 + 1)x\sqrt{y} = -3\sqrt{3} - 6x\sqrt{y}$$

(b) $x^{3/2}(x^{1/2} - x^{-1/2}) = x^{3/2} \cdot x^{1/2} - x^{3/2} \cdot x^{-1/2} = x^2 - x$
(c) $(x^{1/2} - x^{1/3})^2 = (x^{1/2})^2 - 2x^{1/2}x^{1/3} + (x^{1/3})^2 = x - 2x^{5/6} + x^{2/3}$
(d) $(\sqrt{x} + 2)(\sqrt{x} - 2) = (\sqrt{x})^2 - (2)^2 = x - 4$ ■

In later chapters we will need to write problems in a simplified form so that we can perform certain operations on them. We can often use division of one polynomial by another to obtain the simplification, as shown in the following procedure.

Division of Polynomials

Procedure	Example
To divide one polynomial by another:	Divide $4x^3 + 4x^2 + 5$ by $2x^2 + 1$.
1. Write both polynomials in descending powers of a variable. Include missing terms with coefficient 0 in the dividend.	1. $2x^2 + 1\overline{)4x^3 + 4x^2 + 0x + 5}$

Division of Polynomials (continued)

Procedure	Example

2. (a) Divide the highest-power term of the divisor into the highest-power term of the dividend, and write this partial quotient above the dividend. Multiply the partial quotient times the divisor, write the product under the dividend, and subtract, getting a new dividend.

(b) Repeat until the degree of the new dividend is less than the degree of the divisor. Any remainder is written over the divisor and added to the quotient.

2. (a)
$$
\begin{array}{r}
2x \\
2x^2 + 1 \overline{\smash{)}4x^3 + 4x^2 + 0x + 5}\\
\underline{4x^3 + 2x}\\
4x^2 - 2x + 5
\end{array}
$$

(b)
$$
\begin{array}{r}
2x \ + 2 \\
2x^2 + 1 \overline{\smash{)}4x^3 + 4x^2 + 0x + 5}\\
\underline{4x^3 + 2x}\\
4x^2 - 2x + 5\\
\underline{4x^2 + 2}\\
- 2x + 3
\end{array}
$$

Degree $(-2x + 3) <$ degree $(2x^2 + 1)$

Quotient: $2x + 2 + \dfrac{-2x + 3}{2x^2 + 1}$

EXAMPLE 9   Division of Polynomials

Divide $(4x^3 - 13x - 22)$ by $(x - 3)$, $x \neq 3$.

Solution

$$
\begin{array}{r}
4x^2 + 12x + 23 \\
x - 3 \overline{\smash{)}4x^3 + 0x^2 - 13x - 22}\\
\underline{4x^3 - 12x^2 }\\
12x^2 - 13x - 22\\
\underline{12x^2 - 36x }\\
23x - 22\\
\underline{23x - 69}\\
47
\end{array}
$$

$0x^2$ is inserted so that each power of x is present.

The quotient is $4x^2 + 12x + 23$, with remainder 47, or

$$4x^2 + 12x + 23 + \frac{47}{x - 3}$$ ■

CHECKPOINT

3. Use long division to find $(x^3 + 2x + 7) \div (x - 4)$.

CHECKPOINT SOLUTIONS

1. $9x - 5x(x + 2) + 4x^2 = 9x - 5x^2 - 10x + 4x^2$
   $$= -x^2 - x$$
   Note: without parentheses around $9x - 5x$, multiplication has priority over subtraction.

2. (a)
$$
\begin{array}{r}
4x^2 - 2x + 1 \\
2x + 1 \overline{\smash{)}8x^3 - 4x^2 + 2x }\\
\underline{8x^3 }\\
4x^2 - 2x + 1\\
\underline{+ 1}\\
\end{array}
$$

   (b) $(x + 3)^2 = x^2 + 2(3x) + 3^2 = x^2 + 6x + 9$
   (c) $(3x + 2)(x - 5) = 3x^2 - 15x + 2x - 10 = 3x^2 - 13x - 10$
   (d) $(1 - 4x)(1 + 4x) = 1 - 16x^2$ Note that this is different from $16x^2 - 1$.

3.
$$
\begin{array}{r}
x^2 + 4x + 18 \\
x - 4 \overline{\smash{)}x^3 + 0x^2 + 2x + \ 7}\\
\underline{x^3 - 4x^2 }\\
4x^2 + 2x + \ 7\\
\underline{4x^2 - 16x }\\
18x + \ 7\\
\underline{18x - 72}\\
79
\end{array}
$$

   The answer is $x^2 + 4x + 18 + \dfrac{79}{x - 4}$.

| **EXERCISES** | **0.5**

Access end-of-section exercises online at **www.webassign.net**

0.6

Factoring

Common Factors We can factor monomial factors out of a polynomial by using the Distributive Law in reverse; $ab + ac = a(b + c)$ is an example showing that a is a monomial factor of the polynomial $ab + ac$. But it is also a statement of the Distributive Law (with the sides of the equation interchanged). The monomial factor of a polynomial must be a factor of each term of the polynomial, so it is frequently called a **common monomial factor.**

EXAMPLE 1 Monomial Factor

Factor $-3x^2t - 3x + 9xt^2$.

Solution

1. We can factor out $3x$ and obtain
$$-3x^2t - 3x + 9xt^2 = 3x(-xt - 1 + 3t^2)$$

2. Or we can factor out $-3x$ (factoring out the negative will make the first term of the polynomial positive) and obtain
$$-3x^2t - 3x + 9xt^2 = -3x(xt + 1 - 3t^2)$$ ■

If a factor is common to each term of a polynomial, we can use this procedure to factor it out, even if it is not a monomial. For example, we can factor $(a + b)$ out of the polynomial $2x(a + b) - 3y(a + b)$ and get $(a + b)(2x - 3y)$. The following example demonstrates the **factoring by grouping** technique.

EXAMPLE 2 Factoring by Grouping

Factor $5x - 5y + bx - by$.

Solution

We can factor this polynomial by the use of grouping. The grouping is done so that common factors (frequently binomial factors) can be removed. We see that we can factor 5 from the first two terms and b from the last two, which gives
$$5(x - y) + b(x - y)$$

This gives us two terms with the common factor $x - y$, so we get
$$(x - y)(5 + b)$$ ■

Factoring Trinomials   We can use the formula for multiplying two binomials to factor certain trinomials. The formula

$$(x + a)(x + b) = x^2 + (a + b)x + ab$$

can be used to factor trinomials such as $x^2 - 7x + 6$.

EXAMPLE   3   **Factoring a Trinomial**

Factor $x^2 - 7x + 6$.

Solution

If this trinomial can be factored into an expression of the form

$$(x + a)(x + b)$$

then we need to find a and b such that

$$x^2 - 7x + 6 = x^2 + (a + b)x + ab$$

That is, we need to find a and b such that $a + b = -7$ and $ab = 6$. The two numbers whose sum is -7 and whose product is 6 are -1 and -6. Thus

$$x^2 - 7x + 6 = (x - 1)(x - 6)$$ ■

A similar method can be used to factor trinomials such as $9x^2 - 31x + 12$. Finding the proper factors for this type of trinomial may involve a fair amount of trial and error, because we must find factors a, b, c, and d such that

$$(ax + b)(cx + d) = acx^2 + (ad + bc)x + bd$$

Another technique of factoring is used to factor trinomials such as those we have been discussing. It is useful in factoring more complicated trinomials, such as $9x^2 - 31x + 12$. This procedure for factoring second-degree trinomials follows.

Factoring a Trinomial

Procedure	Example
To factor a trinomial into the product of its binomial factors:	Factor $9x^2 - 31x + 12$.
1. Form the product of the second-degree term and the constant term.	1. $9x^2 \cdot 12 = 108x^2$
2. Determine whether there are any factors of the product of Step 1 that will sum to the middle term of the trinomial. (If the answer is no, the trinomial will not factor into two binomials.)	2. The factors $-27x$ and $-4x$ give a sum of $-31x$.
3. Use the sum of these two factors to replace the middle term of the trinomial.	3. $9x^2 - 31x + 12 = 9x^2 - 27x - 4x + 12$
4. Factor this four-term expression by grouping.	4. $9x^2 - 31x + 12 = (9x^2 - 27x) + (-4x + 12)$ $= 9x(x - 3) - 4(x - 3)$ $= (x - 3)(9x - 4)$

In the example just completed, note that writing the middle term ($-31x$) as $-4x - 27x$ rather than as $-27x - 4x$ (as we did) will also result in the correct factorization. (Try it.)

EXAMPLE 4 **Factoring a Trinomial**

Factor $9x^2 - 9x - 10$.

Solution

The product of the second-degree term and the constant is $-90x^2$. Factors of $-90x^2$ that sum to $-9x$ are $-15x$ and $6x$. Thus

$$9x^2 - 9x - 10 = 9x^2 - 15x + 6x - 10$$
$$= (9x^2 - 15x) + (6x - 10)$$
$$= 3x(3x - 5) + 2(3x - 5) = (3x - 5)(3x + 2)$$

We can check this factorization by multiplying.

$$(3x - 5)(3x + 2) = 9x^2 + 6x - 15x - 10$$
$$= 9x^2 - 9x - 10$$

Some special products that make factoring easier are as follows.

Special Factorizations

The perfect-square trinomials:

$$x^2 + 2ax + a^2 = (x + a)^2$$
$$x^2 - 2ax + a^2 = (x - a)^2$$

The difference of two squares:

$$x^2 - a^2 = (x + a)(x - a)$$

EXAMPLE 5 **Special Factorizations**

(a) Factor $25x^2 - 36y^2$.
(b) Factor $4x^2 + 12x + 9$.

Solution

(a) The binomial $25x^2 - 36y^2$ is the difference of two squares, so we get

$$25x^2 - 36y^2 = (5x - 6y)(5x + 6y)$$

These two factors are called binomial **conjugates** because they differ in only one sign.

(b) Although we can use the technique we have learned to factor trinomials, the factors come quickly if we recognize that this trinomial is a perfect square. It has two square terms, and the remaining term $(12x)$ is twice the product of the square roots of the squares $(12x = 2 \cdot 2x \cdot 3)$. Thus

$$4x^2 + 12x + 9 = (2x + 3)^2$$

Most of the polynomials we have factored have been second-degree polynomials, or **quadratic polynomials.** Some polynomials that are not quadratic are in a form that can be factored in the same manner as quadratics. For example, the polynomial $x^4 + 4x^2 + 4$ can be written as $a^2 + 4a + 4$, where $a = x^2$.

EXAMPLE 6 **Polynomials in Quadratic Form**

Factor (a) $x^4 + 4x^2 + 4$ and (b) $x^4 - 16$.

Solution

(a) The trinomial is in the form of a perfect square, so letting $a = x^2$ will give us

$$x^4 + 4x^2 + 4 = a^2 + 4a + 4 = (a + 2)^2$$

Thus

$$x^4 + 4x^2 + 4 = (x^2 + 2)^2$$

(b) The binomial $x^4 - 16$ can be treated as the difference of two squares, $(x^2)^2 - 4^2$, so

$$x^4 - 16 = (x^2 - 4)(x^2 + 4)$$

But $x^2 - 4$ can be factored into $(x - 2)(x + 2)$, so

$$x^4 - 16 = (x - 2)(x + 2)(x^2 + 4)$$

CHECKPOINT

1. Factor the following.
   (a) $8x^3 - 12x$ (b) $3x(x^2 + 5) - 5(x^2 + 5)$ (c) $x^2 - 10x - 24$
   (d) $x^2 - 5x + 6$ (e) $4x^2 - 20x + 25$ (f) $100 - 49x^2$
2. Consider $10x^2 - 17x - 20$ and observe that $(10x^2)(-20) = -200x^2$.
   (a) Find two expressions whose product is $-200x^2$ and whose sum is $-17x$.
   (b) Replace $-17x$ in $10x^2 - 17x - 20$ with the two expressions in (a).
   (c) Factor (b) by grouping.
3. True or false:
   (a) $4x^2 + 9 = (2x + 3)^2$ (b) $x^2 + x - 12 = (x - 4)(x + 3)$
   (c) $5x^5 - 20x^3 = 5x^3(x^2 - 4) = 5x^3(x + 2)(x - 2)$

A polynomial is said to be factored completely if all possible factorizations have been completed. For example, $(2x - 4)(x + 3)$ is not factored completely because a 2 can still be factored out of $2x - 4$. The following guidelines are used to factor polynomials completely.

Guidelines for Factoring Completely

Look for: Monomials first.
Then for: Difference of two squares.
Then for: Trinomial squares.
Then for: Other methods of factoring trinomials.

EXAMPLE 7 **Factoring Completely**

Factor completely (a) $12x^2 - 36x + 27$ and (b) $16x^2 - 64y^2$.

Solution
(a) $12x - 36x + 27 = 3(4x^2 - 12x + 9)$ Monomial
                 $= 3(2x - 3)^2$ Perfect Square
(b) $16x^2 - 64y^2 = 16(x^2 - 4y^2)$
                 $= 16(x + 2y)(x - 2y)$

Factoring the difference of two squares immediately would give $(4x + 8y)(4x - 8y)$, which is not factored completely (because we could still factor 4 from $4x + 8y$ and 4 from $4x - 8y$).

CHECKPOINT SOLUTIONS

1. (a) $8x^3 - 12x = 4x(2x^2 - 3)$
   (b) $3x(x^2 + 5) - 5(x^2 + 5) = (x^2 + 5)(3x - 5)$
   (c) $x^2 - 10x - 24 = (x - 12)(x + 2)$
   (d) $x^2 - 5x + 6 = (x - 3)(x - 2)$
   (e) $4x^2 - 20x + 25 = (2x - 5)^2$
   (f) $100 - 49x^2 = (10 + 7x)(10 - 7x)$
2. (a) $(-25x)(+8x) = -200x^2$ and $-25x + 8x = -17x$
   (b) $10x^2 - 17x - 20 = 10x^2 - 25x + 8x - 20$
   (c) $= (10x^2 - 25x) + (8x - 20)$
                   $= 5x(2x - 5) + 4(2x - 5)$
                   $= (2x - 5)(5x + 4)$
3. (a) False. $4x^2 + 9$ cannot be factored. In fact, sums of squares cannot be factored.
   (b) False. $x^2 + x - 12 = (x + 4)(x - 3)$
   (c) True.

Access end-of-section exercises online at www.webassign.net

0.7

Algebraic Fractions

Evaluating certain limits and graphing rational functions require an understanding of algebraic fractions. The fraction 6/8 can be reduced to 3/4 by dividing both the numerator and the denominator by 2. In the same manner, we can reduce the algebraic fraction

$$\frac{(x + 2)(x + 1)}{(x + 1)(x + 3)} \quad \text{to} \quad \frac{x + 2}{x + 3}$$

by dividing both the numerator and the denominator by $x + 1$, if $x \neq -1$.

Simplifying Fractions

We *simplify* algebraic fractions by factoring the numerator and denominator and then dividing both the numerator and the denominator by any common factors.*

EXAMPLE   1   Simplifying a Fraction

Simplify $\dfrac{3x^2 - 14x + 8}{x^2 - 16}$ if $x^2 \neq 16$.

Solution

$$\frac{3x^2 - 14x + 8}{x^2 - 16} = \frac{(3x - 2)(x - 4)}{(x - 4)(x + 4)}$$

$$= \frac{(3x - 2)\overset{1}{\cancel{(x - 4)}}}{\underset{1}{\cancel{(x - 4)}}(x + 4)} = \frac{3x - 2}{x + 4}$$

Products of Fractions   We can multiply fractions by writing the product as the product of the numerators divided by the product of the denominators. For example,

$$\frac{4}{5} \cdot \frac{10}{12} \cdot \frac{2}{5} = \frac{80}{300}$$

which reduces to 4/15.

We can also find the product by reducing the fractions before we indicate the multiplication in the numerator and denominator. For example, in

$$\frac{4}{5} \cdot \frac{10}{12} \cdot \frac{2}{5}$$

we can divide the first numerator and the second denominator by 4 and divide the second numerator and the first denominator by 5, which yields

$$\frac{\overset{1}{\cancel{4}}}{\underset{1}{\cancel{5}}} \cdot \frac{\overset{2}{\cancel{10}}}{\underset{3}{\cancel{12}}} \cdot \frac{2}{5} = \frac{1}{1} \cdot \frac{2}{3} \cdot \frac{2}{5} = \frac{4}{15}$$

---

*We assume that all fractions are defined.

Product of Fractions

> We *multiply* algebraic fractions by writing the product of the numerators divided by the product of the denominators, and then reduce to lowest terms. We may also reduce prior to finding the product.

EXAMPLE 2   Multiplying Fractions

Multiply:

(a) $\dfrac{4x^2}{5y} \cdot \dfrac{10x}{y^2} \cdot \dfrac{y}{8x^2}$

(b) $\dfrac{-4x + 8}{3x + 6} \cdot \dfrac{2x + 4}{4x + 12}$

Solution

(a) $\dfrac{4x^2}{5y} \cdot \dfrac{10x}{y} \cdot \dfrac{y}{8x^2} = \dfrac{\overset{1}{\cancel{4x^2}}}{\underset{1\cdot 1}{\cancel{5y}}} \cdot \dfrac{\overset{2}{\cancel{10x}}}{y^2} \cdot \dfrac{\overset{1}{\cancel{y}}}{\underset{2}{\cancel{8x^2}}} = \dfrac{1}{1} \cdot \dfrac{2x}{y^2} \cdot \dfrac{1}{2} = \dfrac{x}{y^2}$

(b) $\dfrac{-4x + 8}{3x + 6} \cdot \dfrac{2x + 4}{4x + 12} = \dfrac{-4(x - 2)}{3(x + 2)} \cdot \dfrac{2(x + 2)}{4(x + 3)}$

$\qquad = \dfrac{\overset{-1}{\cancel{-4}}(x - 2)}{3\cancel{(x + 2)}} \cdot \dfrac{2\overset{1}{\cancel{(x + 2)}}}{\underset{1}{\cancel{4}}(x + 3)} = \dfrac{-2(x - 2)}{3(x + 3)}$ ■

Quotients of Fractions

In arithmetic we learned to divide one fraction by another by inverting the divisor and multiplying. The same rule applies to division of algebraic fractions.

EXAMPLE 3   Dividing Fractions

(a) Divide $\dfrac{a^2b}{c}$ by $\dfrac{ab}{c^2}$.

(b) Find $\dfrac{6x^2 - 6}{x^2 + 3x + 2} \div \dfrac{x - 1}{x^2 + 4x + 4}$.

Solution

(a) $\dfrac{a^2b}{c} \div \dfrac{ab}{c^2} = \dfrac{a^2b}{c} \cdot \dfrac{c^2}{ab} = \dfrac{\overset{a \cdot 1}{\cancel{a^2b}}}{\underset{1}{\cancel{c}}} \cdot \dfrac{\overset{c}{\cancel{c^2}}}{\underset{1 \cdot 1}{\cancel{ab}}} = \dfrac{ac}{1} = ac$

(b) $\dfrac{6x^2 - 6}{x^2 + 3x + 2} \div \dfrac{x - 1}{x^2 + 4x + 4} = \dfrac{6x^2 - 6}{x^2 + 3x + 2} \cdot \dfrac{x^2 + 4x + 4}{x - 1}$

$\qquad = \dfrac{6(x - 1)(x + 1)}{(x + 2)(x + 1)} \cdot \dfrac{(x + 2)(x + 2)}{x - 1}$

$\qquad = 6(x + 2)$ ■

CHECKPOINT

1. Reduce: $\dfrac{2x^2 - 4x}{2x}$

2. Multiply: $\dfrac{x^2}{x^2 - 9} \cdot \dfrac{x + 3}{3x}$

3. Divide: $\dfrac{5x^2(x - 1)}{2(x + 1)} \div \dfrac{10x^2}{(x + 1)(x - 1)}$

Adding and Subtracting Fractions

If two fractions are to be added, it is convenient that both be expressed with the same denominator. If the denominators are not the same, we can write the equivalents of each of the fractions with a common denominator. We usually use the least common denominator (LCD) when we write the equivalent fractions. The **least common denominator** is the lowest-degree variable expression into which all denominators will divide. If the denominators are polynomials, then the LCD is the lowest-degree polynomial into which all denominators will divide. We can find the least common denominator as follows.

Finding the Least Common Denominator

Procedure	Example
To find the least common denominator of a set of fractions:	Find the LCD of $\dfrac{1}{x^2 - x}, \dfrac{1}{x^2 - 1}, \dfrac{1}{x^2}$.
1. Completely factor each denominator.	1. The factored denominators are $x(x - 1)$, $(x + 1)(x - 1)$, and $x \cdot x$.
2. Identify the different factors that appear.	2. The different factors are x, $x - 1$, and $x + 1$.
3. The LCD is the product of these different factors, with each factor used the maximum number of times it occurs in any one denominator.	3. x occurs a maximum of twice in one denominator, $x - 1$ occurs once, and $x + 1$ occurs once. Thus the LCD is $x \cdot x(x - 1)(x + 1) = x^2(x - 1)(x + 1)$.

The procedure for combining (adding or subtracting) two or more fractions follows.

Adding or Subtracting Fractions

Procedure	Example
To combine fractions:	Combine $\dfrac{y - 3}{y - 5} + \dfrac{y - 23}{y^2 - y - 20}$.
1. Find the LCD of the fractions.	1. $y^2 - y - 20 = (y - 5)(y + 4)$, so the LCD is $(y - 5)(y + 4)$.
2. Write the equivalent of each fraction with the LCD as its denominator.	2. The sum is $\dfrac{(y - 3)(y + 4)}{(y - 5)(y + 4)} + \dfrac{y - 23}{(y - 5)(y + y)}$.
3. Add or subtract, as indicated, by combining like terms in the numerator over the LCD.	3. $= \dfrac{y^2 + y - 12 + y - 23}{(y - 5)(y + 4)}$ $= \dfrac{y^2 + 2y - 35}{(y - 5)(y + 4)}$
4. Reduce the fraction, if possible.	4. $= \dfrac{(y + 7)(y - 5)}{(y - 5)(y + 4)} = \dfrac{y + 7}{y + 4}$, if $y \neq 5$.

EXAMPLE 4 Adding Fractions

Add $\dfrac{3x}{a^2} + \dfrac{4}{ax}$.

Solution

1. The LCD is a^2x.

2. $\dfrac{3x}{a^2} + \dfrac{4}{ax} = \dfrac{3x}{a^2} \cdot \dfrac{x}{x} + \dfrac{4}{ax} \cdot \dfrac{a}{a} = \dfrac{3x^2}{a^2x} + \dfrac{4a}{a^2x}$

3. $\dfrac{3x^2}{a^2x} + \dfrac{4a}{a^2x} = \dfrac{3x^2 + 4a}{a^2x}$

4. The sum is in lowest terms.

EXAMPLE 5 Combining Fractions

Combine $\dfrac{y - 3}{(y - 5)^2} - \dfrac{y - 2}{y^2 - 4y - 5}$.

Solution

$y^2 - 4y - 5 = (y - 5)(y + 1)$, so the LCD is $(y - 5)^2(y + 1)$. Writing the equivalent fractions and then combining them, we get

$$\frac{y - 3}{(y - 5)^2} - \frac{y - 2}{(y - 5)(y + 1)} = \frac{(y - 3)(y + 1)}{(y - 5)^2(y + 1)} - \frac{(y - 2)(y - 5)}{(y - 5)(y + 1)(y - 5)}$$

$$= \frac{(y^2 - 2y - 3) - (y^2 - 7y + 10)}{(y - 5)^2(y + 1)}$$

$$= \frac{y^2 - 2y - 3 - y^2 + 7y - 10}{(y - 5)^2(y + 1)}$$

$$= \frac{5y - 13}{(y - 5)^2(y + 1)}$$ ■

Complex Fractions

A fractional expression that contains one or more fractions in its numerator or denominator is called a **complex fraction.** An example of a complex fraction is

$$\frac{\dfrac{1}{3} + \dfrac{4}{x}}{3 - \dfrac{1}{xy}}$$

We can simplify fractions of this type using the property $\dfrac{a}{b} = \dfrac{ac}{bc}$, with c equal to the LCD of *all* the fractions contained in the numerator and denominator of the complex fraction.

For example, all fractions contained in the preceding complex fraction have LCD $3xy$. We simplify this complex fraction by multiplying the numerator and denominator as follows:

$$\frac{\dfrac{1}{3} + \dfrac{4}{x}}{3 - \dfrac{1}{xy}} = \frac{3xy\left(\dfrac{1}{3} + \dfrac{4}{x}\right)}{3xy\left(3 - \dfrac{1}{xy}\right)} = \frac{3xy\left(\dfrac{1}{3}\right) + 3xy\left(\dfrac{4}{x}\right)}{3xy(3) - 3xy\left(\dfrac{1}{xy}\right)} = \frac{xy + 12y}{9xy - 3}$$

EXAMPLE **6** **Complex Fractions**

Simplify $\dfrac{x^{-3} + x^2y^{-3}}{(xy)^{-2}}$ so that only positive exponents remain.

Solution

$$\frac{x^{-3} + x^2y^{-3}}{(xy)^{-2}} = \frac{\dfrac{1}{x^3} + \dfrac{x^2}{y^3}}{\dfrac{1}{(xy)^2}}, \quad \text{LCD} = x^3y^3$$

$$= \frac{x^3y^3\left(\dfrac{1}{x^3} + \dfrac{x^2}{y^3}\right)}{x^3y^3\left(\dfrac{1}{x^2y^2}\right)} = \frac{y^3 + x^5}{xy}$$ ■

CHECKPOINT

4. Add or subtract:

(a) $\dfrac{5x - 1}{2x - 5} - \dfrac{x + 9}{2x - 5}$

(b) $\dfrac{x + 1}{x} + \dfrac{x}{x - 1}$

5. Simplify $\dfrac{\dfrac{y}{x} - 1}{\dfrac{y}{x} - \dfrac{x}{y}}$.

Rationalizing Denominators

We can simplify algebraic fractions whose denominators contain sums and differences that involve square roots by rationalizing the denominators. Using the fact that $(x + y)(x - y) = x^2 - y^2$, we multiply the numerator and denominator of an algebraic fraction of this type by the conjugate of the denominator to simplify the fraction.

EXAMPLE **7** | **Rationalizing Denominators**

Rationalize the denominators.

(a) $\dfrac{1}{\sqrt{x} - 2}$ (b) $\dfrac{3 + \sqrt{x}}{\sqrt{x} + \sqrt{5}}$

Solution

Multiplying $\sqrt{x} - 2$ by $\sqrt{x} + 2$, its conjugate, gives the difference of two squares and removes the radical from the denominator in (a). We also use the conjugate in (b).

(a) $\dfrac{1}{\sqrt{x} - 2} \cdot \dfrac{\sqrt{x} + 2}{\sqrt{x} + 2} = \dfrac{\sqrt{x} + 2}{(\sqrt{x})^2 - (2)^2} = \dfrac{\sqrt{x} + 2}{x - 4}$

(b) $\dfrac{3 + \sqrt{x}}{\sqrt{x} + \sqrt{5}} \cdot \dfrac{\sqrt{x} - \sqrt{5}}{\sqrt{x} - \sqrt{5}} = \dfrac{3\sqrt{x} - 3\sqrt{5} + x - \sqrt{5x}}{x - 5}$ ∎

CHECKPOINT | 6. Rationalize the denominator: $\dfrac{\sqrt{x}}{\sqrt{x} - 3}$.

CHECKPOINT SOLUTIONS

1. $\dfrac{2x^2 - 4x}{2x} = \dfrac{2x(x - 2)}{2x} = x - 2$

2. $\dfrac{x^2}{x^2 - 9} \cdot \dfrac{x + 3}{3x} = \dfrac{x^2 \cdot (x + 3)}{(x + 3)(x - 3) \cdot 3x} = \dfrac{x}{3(x - 3)} = \dfrac{x}{3x - 9}$

3. $\dfrac{5x^2(x - 1)}{2(x + 1)} \div \dfrac{10x^2}{(x + 1)(x - 1)} = \dfrac{5x^2(x - 1)}{2(x + 1)} \cdot \dfrac{(x + 1)(x - 1)}{10x^2}$

   $= \dfrac{(x - 1)^2}{4}$

4. (a) $\dfrac{5x - 1}{2x - 5} - \dfrac{x + 9}{2x - 5} = \dfrac{(5x - 1) - (x + 9)}{2x - 5} = \dfrac{5x - 1 - x - 9}{2x - 5}$

   $= \dfrac{4x - 10}{2x - 5} = \dfrac{2(2x - 5)}{2x - 5} = 2$

   (b) $\dfrac{x + 1}{x} + \dfrac{x}{x - 1}$ has LCD $= x(x - 1)$. Thus

   $\dfrac{x + 1}{x} + \dfrac{x}{x - 1} = \dfrac{x + 1}{x} \cdot \dfrac{(x - 1)}{(x - 1)} + \dfrac{x}{x - 1} \cdot \dfrac{x}{x}$

   $= \dfrac{x^2 - 1}{x(x - 1)} + \dfrac{x^2}{x(x - 1)} = \dfrac{x^2 - 1 + x^2}{x(x - 1)} = \dfrac{2x^2 - 1}{x(x - 1)}$

5. $\dfrac{\dfrac{y}{x} - 1}{\dfrac{y}{x} - \dfrac{x}{y}} = \dfrac{\left(\dfrac{y}{x} - 1\right)}{\left(\dfrac{y}{x} - \dfrac{x}{y}\right)} \cdot \dfrac{xy}{xy} = \dfrac{\dfrac{y}{x} \cdot xy - 1 \cdot xy}{\dfrac{y}{x} \cdot xy - \dfrac{x}{y} \cdot xy}$

   $= \dfrac{y^2 - xy}{y^2 - x^2} = \dfrac{y(y - x)}{(y + x)(y - x)} = \dfrac{y}{y + x}$

6. $\dfrac{\sqrt{x}}{\sqrt{x} - 3} = \dfrac{\sqrt{x}}{\sqrt{x} - 3} \cdot \dfrac{\sqrt{x} + 3}{\sqrt{x} + 3} = \dfrac{x + 3\sqrt{x}}{x - 9}$

| EXERCISES | 0.7

Access end-of-section exercises online at **www.webassign.net**

KEY TERMS AND FORMULAS

Section	Key Terms	Formulas
0.1	Sets and set membership	
	Natural numbers	$N = [1, 2, 3, 4 \ldots]$
	Empty set	$\varnothing$
	Set equality	
	Subset	$A \subseteq B$
	Universal set	U
	Venn diagrams	
	Set intersection	$A \cap B$
	Disjoint sets	$A \cap B = \varnothing$
	Set union	$A \cup B$
	Set complement	A'
0.2	Real numbers	
	Subsets and properties	
	Real number line	
	Inequalities	
	Intervals and interval notation	
	Closed interval	$a \le x \le b$ or $[a, b]$
	Open interval	$a < x < b$ or (a, b)
	Absolute value	
	Order of operations	
0.3	Exponent and base	a^n has base a, exponent n
	Zero exponent	$a^0 = 1, a \neq 0$
	Negative exponent	$a^{-n} = \dfrac{1}{a^n}$
	Rules of exponents	
0.4	Radical	$\sqrt[n]{a}$
	Radicand, index	radicand $= a$; index $= n$
	Principal nth root	$\sqrt[n]{a} = b$ only if $b^n = a$ and $a \ge 0$ and $b \ge 0$ when n is even
	Fractional exponents	$a^{1/n} = \sqrt[n]{a}$
		$a^{m/n} = \sqrt[n]{a^m} = (\sqrt[n]{a})^m$
	Properties of radicals	
	Rationalizing the denominator	

Section	Key Terms	Formulas
0.5	Algebraic expression Variable Constant term Coefficient; leading coefficient Term Polynomial Degree Monomial Binomial Trinomial Like terms Distributive Law Binomial products Division of polynomials	 $a_n x^n + \cdots + a_1 x + a_0$ $a(b + c) = ab + ac$
0.6	Factor Common factor Factoring by grouping Special factorizations Difference of squares Perfect squares Conjugates Quadratic polynomials Factoring completely	 $a^2 - b^2 = (a + b)(a - b)$ $a^2 + 2ab + b^2 = (a + b)^2$ $a^2 - 2ab + b^2 = (a - b)^2$ $a + b;\ a - b$ $ax^2 + bx + c$
0.7	Algebraic fractions Numerator Denominator Reduce Product of fractions Quotient of fractions Common denominator Least common denominator (LCD) Addition and subtraction of fractions Complex fraction Rationalize the denominator	

REVIEW EXERCISES

1. Is $A \subseteq B$ if $A = \{1, 2, 5, 7\}$ and $B = \{x: x \text{ is a positive integer}, x \leq 8\}$?

2. Is it true that $3 \in \{x: x > 3\}$?

3. Are $A = \{1, 2, 3, 4\}$ and $B = \{x: x \leq 1\}$ disjoint?

In Problems 4–7, use sets $U = \{1, 2, 3, 4, 5, 6, 7, 8, 9, 10\}$, $A = \{1, 2, 3, 9\}$, and $B = \{1, 3, 5, 6, 7, 8, 10\}$ to find the elements of the sets described.

4. $A \cup B'$

5. $A' \cap B$

6. $(A' \cap B)'$

7. Does $(A' \cup B')' = A \cap B$?

8. State the property of the real numbers that is illustrated in each case.

   (a) $6 + \dfrac{1}{3} = \dfrac{1}{3} + 6$ (b) $2(3 \cdot 4) = (2 \cdot 3)4$

   (c) $\dfrac{1}{3}(6 + 9) = 2 + 3$

9. Indicate whether each given expression is one or more of the following: rational, irrational, integer, natural, or meaningless.

   (a) π (b) $0/6$ (c) $6/0$

10. Insert the proper sign ($<$, $=$, or $>$) to replace each $\square$.

   (a) $\pi \,\square\, 3.14$ (b) $-100 \,\square\, 0.1$ (c) $-3 \,\square\, -12$

For Problems 11–18, evaluate each expression. Use a calculator when necessary.

11. $|5 - 11|$

12. $44 \div 2 \cdot 11 - 10^2$

13. $(-3)^2 - (-1)^3$

14. $\dfrac{(3)(2)(15) - (5)(8)}{(4)(10)}$

15. $2 - [3 - (2 - |-3|)] + 11$

16. $-4^2 - (-4)^2 + 3$

17. $\dfrac{4 + 3^2}{4}$

18. $\dfrac{(-2.91)^5}{\sqrt{3.29^5}}$

19. Write each inequality in interval notation, name the type of interval, and graph it on a real number line.
   (a) $0 \le x \le 5$
   (b) $x \ge -3$ and $x < 7$
   (c) $(-4, \infty) \cap (-\infty, 0)$

20. Write an inequality that represents each of the following.
   (a) $(-1, 16)$
   (b) $[-12, 8\}$
   (c)
   $\quad -3 \quad -2 \quad -1 \quad 0 \quad 1$

21. Evaluate the following without a calculator.
   (a) $\left(\dfrac{3}{8}\right)^0$
   (b) $2^3 \cdot 2^{-5}$
   (c) $\dfrac{4^9}{4^3}$
   (d) $\left(\dfrac{1}{7}\right)^3 \left(\dfrac{1}{7}\right)^{-4}$

22. Use the rules of exponents to simplify each of the following with positive exponents. Assume all variables are nonzero.
   (a) $x^5 \cdot x^{-7}$
   (b) x^8/x^{-2}
   (c) $(x^3)^3$
   (d) $(y^4)^{-2}$
   (e) $(-y^{-3})^{-2}$

For Problems 23–28, rewrite each expression so that only positive exponents remain. Assume all variables are nonzero.

23. $\dfrac{-(2xy^2)^{-2}}{(3x^{-2}y^{-3})^2}$

24. $\left(\dfrac{2}{3}x^2y^{-4}\right)^{-2}$

25. $\left(\dfrac{x^{-2}}{2y^{-1}}\right)^2$

26. $\dfrac{(-x^4y^{-2}z^2)^0}{-(x^4y^{-2}z^2)^{-2}}$

27. $\left(\dfrac{x^{-3}y^4z^{-2}}{3x^{-2}y^{-3}z^{-3}}\right)^{-1}$

28. $\left(\dfrac{x}{2y}\right)\left(\dfrac{y}{x^2}\right)^{-2}$

29. Find the following roots.
   (a) $-\sqrt[3]{-64}$
   (b) $\sqrt{4/49}$
   (c) $\sqrt[7]{1.9487171}$

30. Write each of the following with an exponent and with the variable in the numerator.
   (a) $\sqrt{x}$
   (b) $\sqrt[3]{x^2}$
   (c) $1/\sqrt[4]{x}$

31. Write each of the following in radical form.
   (a) $x^{2/3}$
   (b) $x^{-1/2}$
   (c) $-x^{3/2}$

32. Rationalize each of the following denominators and simplify.
   (a) $\dfrac{5xy}{\sqrt{2x}}$
   (b) $\dfrac{y}{x\sqrt[3]{xy^2}}$

In Problems 33–38, use the properties of exponents to simplify so that only positive exponents remain. Assume all variables are positive.

33. $x^{1/2} \cdot x^{1/3}$

34. $y^{-3/4}/y^{-7/4}$

35. $x^4 \cdot x^{1/4}$

36. $1/(x^{-4/3} \cdot x^{-7/3})$

37. $(x^{4/5})^{1/2}$

38. $(x^{1/2}y^2)^4$

In Problems 39–44, simplify each expression. Assume all variables are positive.

39. $\sqrt{12x^3y^5}$

40. $\sqrt{1250x^6y^9}$

41. $\sqrt[3]{24x^4y^4} \cdot \sqrt[3]{45x^4y^{10}}$

42. $\sqrt{16a^2b^3} \cdot \sqrt{8a^3b^5}$

43. $\dfrac{\sqrt{52x^3y^6}}{\sqrt{13xy^4}}$

44. $\dfrac{\sqrt{32x^4y^3}}{\sqrt{6xy^{10}}}$

In Problems 45–62, perform the indicated operations and simplify.

45. $(3x + 5) - (4x + 7)$

46. $x(1 - x) + x[x - (2 + x)]$

47. $(3x^3 - 4xy - 3) + (5xy + x^3 + 4y - 1)$

48. $(4xy^3)(6x^4y^2)$

49. $(3x - 4)(x - 1)$

50. $(3x - 1)(x + 2)$

51. $(4x + 1)(x - 2)$

52. $(3x - 7)(2x + 1)$

53. $(2x - 3)^2$

54. $(4x + 3)(4x - 3)$

55. $(2x^2 + 1)(x^2 + x - 3)$

56. $(2x - 1)^3$

57. $(x - y)(x^2 + xy + y^2)$

58. $\dfrac{4x^2y - 3x^3y^3 - 6x^4y^2}{2x^2y^2}$

59. $(3x^4 + 2x^3 - x + 4) \div (x^2 + 1)$

60. $(x^4 - 4x^3 + 5x^2 + x) \div (x - 3)$

61. $x^{4/3}(x^{2/3} - x^{-1/3})$

62. $(\sqrt{x} + \sqrt{a - x})(\sqrt{x} - \sqrt{a - x})$

In Problems 63–73, factor each expression completely.

63. $2x^4 - x^3$

64. $4(x^2 + 1)^2 - 2(x^2 + 1)^3$

65. $4x^2 - 4x + 1$

66. $16 - 9x^2$

67. $2x^4 - 8x^2$

68. $x^2 - 4x - 21$

69. $3x^2 - x - 2$

70. $x^2 - 5x + 6$

71. $x^2 - 10x - 24$

72. $12x^2 - 23x - 24$

73. $16x^4 - 72x^2 + 81$

74. Factor as indicated: $x^{-2/3} + x^{-4/3} = x^{-4/3}(?)$

75. Reduce each of the following to lowest terms.
   (a) $\dfrac{2x}{2x + 4}$
   (b) $\dfrac{4x^2y^3 - 6x^3y^4}{2x^2y^2 - 3xy^3}$

In Problems 76–82, perform the indicated operations and simplify.

76. $\dfrac{x^2 - 4x}{x^2 + 4} \cdot \dfrac{x^4 - 16}{x^4 - 16x^2}$

77. $\dfrac{x^2 + 6x + 9}{x^2 - 7x + 12} \div \dfrac{x^2 + 4x + 3}{x^2 - 3x - 4}$

78. $\dfrac{x^4 - 2x^3}{3x^2 - x - 2} \div \dfrac{x^3 - 4x}{9x^2 - 4}$

79. $1 + \dfrac{3}{2x} - \dfrac{1}{6x^2}$

80. $\dfrac{1}{x - 2} - \dfrac{x - 2}{4}$

81. $\dfrac{x + 2}{x^2 - x} - \dfrac{x^2 + 4}{x^2 - 2x + 1} + 1$

82. $\dfrac{x - 1}{x^2 - x - 2} - \dfrac{x}{x^2 - 2x - 3} + \dfrac{1}{x - 2}$

In Problems 83 and 84, simplify each complex fraction.

83. $\dfrac{x - 1 - \dfrac{x - 1}{x}}{\dfrac{1}{x - 1} + 1}$

84. $\dfrac{x^{-2} - x^{-1}}{x^{-2} + x^{-1}}$

85. Rationalize the denominator of $\dfrac{3x - 3}{\sqrt{x} - 1}$ and simplify.

86. Rationalize the numerator of $\dfrac{\sqrt{x} - \sqrt{x - 4}}{2}$ and simplify.

APPLICATIONS

87. **Job effectiveness factors** In an attempt to determine some off-the-job factors that might be indicators of on-the-job effectiveness, a company made a study of 200 of its employees. It was interested in whether the employees had been recognized for superior work by their supervisors within the past year, whether they were involved in community activities, and whether they followed a regular exercise plan. The company found the following.

> 30 answered "yes" to all three
> 50 were recognized and they exercised
> 52 were recognized and were involved in the community
> 77 were recognized
> 37 were involved in the community but did not exercise
> 95 were recognized or were involved in the community
> 95 answered "no" to all three

(a) Draw a Venn diagram that represents this information.
(b) How many exercised only?
(c) How many exercised or were involved in the community?

88. **Health insurance coverage** The percent of the U.S. population covered by an employment-based health insurance plan can be approximated by the expression

$$-0.75x + 63.8$$

where x is the number of years past 2000 (*Source:* U.S. Census Bureau). Use this expression to estimate the percent covered by such a plan in the year 2015.

89. **Poiseuille's law** The expression for the speed of blood through an artery of radius r at a distance x from the artery wall is given by $r^2 - (r - x)^2$. Evaluate this expression when $r = 5$ and $x = 2$.

90. **Future value** If an individual makes monthly deposits of $100 in an account that earns 9% compounded monthly, then the future value S of the account after n months is given by the formula

$$S = \$100\left[\frac{(1.0075)^n - 1}{0.0075}\right]$$

(a) Find the future value after 36 months (3 years).
(b) Find the future value after 20 years.

91. **Health care** According to the American Hospital Association, the trends since 1975 indicate that the number of hospital beds B (in thousands) is related to the number of hospitals H by the formula

$$B = 176.896(1.00029)^H$$

(a) In 2004 the number of hospitals had fallen to 5759. How many beds does the equation estimate were available?
(b) If the number of hospitals reaches 5000, what does the equation predict for the number of beds?

92. **Severe weather ice makers** Thunderstorms severe enough to produce hail develop when an upper-level low (a pool of cold air high in the atmosphere) moves through a region where there is warm, moist air at the surface. These storms create an updraft that draws the moist air into subfreezing air above 10,000 feet. The strength of the updraft, as measured by its speed s (in mph), affects the diameter of the hail h (in inches) according to

$$h = 0.000595s^{1.922} \quad \text{or equivalently} \quad s = 47.7h^{0.519}$$

(*Source:* National Weather Service).
(a) What size hail is produced by an updraft of 50 mph?
(b) When a storm produces softball-sized hail (about 4.5 inches in diameter), how fast is the updraft?

93. **Loan payment** Suppose you borrow $10,000 for n months to buy a car at an interest rate of 7.8% compounded monthly. The size of each month's payment R is given by the formula

$$R = \$10,000\left[\frac{0.0065}{1 - (1.0065)^{-n}}\right]$$

(a) Rewrite the expression on the right-hand side of this formula as a fraction with only positive exponents.
(b) Find the monthly payment for a 48-month loan. Use both the original formula and your result from (a). (Both formulas should give the same payment.)

94. **Environment** Suppose that in a study of water birds, the relationship between the number of acres of wetlands A and the number of species of birds S found in the wetlands area was given by

$$S = kA^{1/3}$$

where k is a constant.
(a) Express this formula using radical notation.
(b) If the area is expanded by a factor of 2.25 from 20,000 acres to 45,000 acres, find the expected increase in the number of species (as a multiple of the number of species on the 20,000 acres).

95. **Profit** Suppose that the total cost of producing and selling x units of a product is $300 + 4x$ and the total revenue from the sale of x units is $30x - 0.001x^2$. Find the difference of these expressions (total revenue minus total cost) to find the profit from the production and sale of x units of the product.

96. *Business loss* Suppose that a commercial building costs $1,450,000. After it is placed into service, it loses 0.25% of its original value in each of the x months it is in service. Write an expression for the value of the building after x months.

97. *Revenue* The revenue for a boat tour is $600 - 13x - 0.5x^2$, where x is the number of passengers above the minimum of 50. This expression will factor into two binomials, with one representing the total number of passengers. Factor this expression.

98. *Pollution: cost-benefit* Suppose that the cost C (in dollars) of removing $p\%$ of the pollution from the waste water of a manufacturing process is given by

$$C = \frac{1,200,000}{100 - p} - 12,000$$

(a) Express the right-hand side of this formula as a single fraction.
(b) Find the cost if $p = 0$. Write a sentence that explains the meaning of what you found.
(c) Find the cost of removing 98% of the pollution.
(d) What happens to this formula when $p = 100$? Explain why you think this happens.

99. *Average cost* The average cost of producing a product is given by the expression

$$1200 + 56x + \frac{8000}{x}$$

Write this expression with all terms over a common denominator.

| **0** | **CHAPTER TEST** |

1. Let $U = \{1, 2, 3, 4, 5, 6, 7, 8, 9\}$, $A = \{x: x \text{ is even and } x \geq 5\}$, and $B = \{1, 2, 5, 7, 8, 9\}$. Complete the following.
   (a) Find $A \cup B'$.
   (b) Find a two-element set that is disjoint from B.
   (c) Find a nonempty subset of A that is not equal to A.

2. Evaluate $(4 - 2^3)^2 - 3^4 \cdot 0^{15} + 12 \div 3 + 1$.

3. Use definitions and properties of exponents to complete the following.
   (a) $x^4 \cdot x^4 = x^?$
   (b) $x^0 = ?$, if $x \neq 0$
   (c) $\sqrt{x} = x^?$
   (d) $(x^{-5})^2 = x^?$
   (e) $a^{27} \div a^{-3} = a^?$
   (f) $x^{1/2} \cdot x^{1/3} = x^?$
   (g) $\dfrac{1}{\sqrt[3]{x^2}} = \dfrac{1}{x^?}$
   (h) $\dfrac{1}{x^3} = x^?$

4. Write each of the following as radicals.
   (a) $x^{1/5}$
   (b) $x^{-3/4}$

5. Simplify each of the following so that only positive exponents remain.
   (a) x^{-5}
   (b) $\left(\dfrac{x^{-8}y^2}{x^{-1}}\right)^{-3}$

6. Simplify the following radical expressions, and rationalize any denominators.
   (a) $\dfrac{x}{\sqrt{5x}}$
   (b) $\sqrt{24a^2b}\sqrt{a^3b^4}$
   (c) $\dfrac{1 - \sqrt{x}}{1 + \sqrt{x}}$

7. Given the expression $2x^3 - 7x^5 - 5x - 8$, complete the following.
   (a) Find the degree.
   (b) Find the constant term.
   (c) Find the coefficient of x.

8. Express $(-2, \infty) \cap (-\infty, 3]$ using interval notation, and graph it.

9. Completely factor each of the following.
   (a) $8x^3 - 2x^2$
   (b) $x^2 - 10x + 24$
   (c) $6x^2 - 13x + 6$
   (d) $2x^3 - 32x^5$

10. Identify the quadratic polynomial from among (a)–(c), and evaluate it using $x = -3$.
    (a) $2x^2 - 3x^3 + 7$
    (b) $x^2 + 3/x + 11$
    (c) $4 - x - x^2$

11. Use long division to find $(2x^3 + x^2 - 7) \div (x^2 - 1)$.

12. Perform the indicated operations and simplify.
    (a) $4y - 5(9 - 3y)$
    (b) $-3t^2(2t^4 - 3t^7)$
    (c) $(4x - 1)(x^2 - 5x + 2)$
    (d) $(6x - 1)(2 - 3x)$
    (e) $(2m - 7)^2$
    (f) $\dfrac{x^6}{x^2 - 9} \cdot \dfrac{x - 3}{3x^2}$
    (g) $\dfrac{x^4}{3^2} \div \dfrac{9x^3}{x^6}$
    (h) $\dfrac{4}{x - 8} - \dfrac{x - 2}{x - 8}$
    (i) $\dfrac{x - 1}{x^2 - 2x - 3} - \dfrac{3}{x^2 - 3x}$

13. Simplify $\dfrac{\dfrac{1}{x} - \dfrac{1}{y}}{\dfrac{1}{x} + y}$.

14. In a nutrition survey of 320 students, the following information was obtained.

    145 ate breakfast
    270 ate lunch
    280 ate dinner
    125 ate breakfast and lunch
    110 ate breakfast and dinner
    230 ate lunch and dinner
    90 ate all three meals

    (a) Make a Venn diagram that represents this information.
    (b) How many students in the survey ate only breakfast?
    (c) How many students skipped breakfast?

15. If $1000 is invested for x years at 8% compounded quarterly, the future value of the investment is given by

$$S = 1000\left(1 + \frac{0.08}{4}\right)^{4x}$$

What will be the future value of this investment in 20 years?

Campaign Management

A politician is trying to win election to the city council, and as his campaign manager, you need to decide how to promote the candidate. There are three ways you can do so: You can send glossy, full-color pamphlets to registered voters of the city; you can run a commercial during the television news on a local cable network; and/or you can buy a full-page ad in the newspaper.

Two hundred fifty thousand voters live in the city, and 36% of them read the newspaper. Fifty thousand voters watch the local cable network news, and 30% of them also read the newspaper.

You also know that the television commercial would cost $40,000, the newspaper ad $27,000, and the pamphlets mailed to voters 90 cents each, including printing and bulk-rate postage.

Suppose that the success of the candidate depends on your campaign reaching at least 125,000 voters and that, because your budget is limited, you must achieve this goal at a minimum cost. What would be your plan and the cost of that plan?

If you need help devising a method of solution for this problem, try answering the following questions first.

1. How many voters in the city read the newspaper but do not watch the local cable television news?
2. How many voters read the newspaper or watch the local cable television news, or both?
3. Complete the following chart by indicating the number of voters reached by each promotional option, the total cost, and the cost per voter reached.

	Number of Voters Reached	Total Cost	Cost per Voter Reached
Pamphlet			
Television			
Newspaper			

4. Now explain your plan and the cost of that plan.

Joy Brown/Shutterstock.com

CHAPTER 1

Linear Equations and Functions

A wide variety of problems from business, the social sciences, and the life sciences may be solved using equations. Managers and economists use equations and their graphs to study costs, sales, national consumption, or supply and demand. Social scientists may plot demographic data or try to develop equations that predict population growth, voting behavior, or learning and retention rates. Life scientists use equations to model the flow of blood or the conduction of nerve impulses and to test theories or develop new models by using experimental data.

Numerous applications of mathematics are given throughout the text, but most chapters contain special sections emphasizing business and economics applications. In particular, this chapter introduces two important applications that will be expanded and used throughout the text as increased mathematical skills permit: supply and demand as functions of price (market analysis); and total cost, total revenue, and profit as functions of the quantity produced and sold (theory of the firm).

The topics and applications discussed in this chapter include the following.

SECTIONS	APPLICATIONS
1.1 **Solutions of Linear Equations and Inequalities in One Variable**	Future value of an investment, voting, normal height for a given age, profit
1.2 **Functions** **Function notation** **Operations with functions**	Income taxes, stock market, mortgage payments
1.3 **Linear Functions** **Graphs** **Slopes** **Equations**	Depreciation, U.S. banks, pricing
1.4 **Graphs and Graphing Utilities** **Graphical solutions of linear equations**	Water purity, women in the work force
1.5 **Solutions of Systems of Linear Equations** **Systems of linear equations in three variables**	Investment mix, medicine concentrations, college enrollment
1.6 **Applications of Functions in Business and Economics**	Total cost, total revenue, profit, break-even analysis, demand and supply, market equilibrium

41

Prerequisite Problem Type	For Section	Answer	Section for Review
Evaluate: (a) $2(-1)^3 - 3(-1)^2 + 1$ (b) $3(-3) - 1$ (c) $14(10) - 0.02(10^2)$ (d) $\dfrac{3-1}{4-(-2)}$ (e) $\dfrac{-1-3}{2-(-2)}$ (f) $\dfrac{3(-8)}{4}$ (g) $2\left(\dfrac{23}{9}\right) - 5\left(\dfrac{2}{9}\right)$	1.1 1.2 1.3 1.5 1.6	(a) -4 (b) -10 (c) 138 (d) $\dfrac{1}{3}$ (e) -1 (f) -6 (g) 4	0.2 Signed numbers
Graph $x \le 3$.	1.1		0.2 Inequalities
(a) $\dfrac{1}{x}$ is *undefined* for which real number(s)? (b) $\sqrt{x-4}$ is a real number for which values of x?	1.1, 1.2	(a) Undefined for $x = 0$ (b) $x \ge 4$	0.2 Real numbers 0.4 Radicals
Identify the coefficient of x and the constant term for: (a) $-9x + 2$ (b) $\dfrac{x}{2}$ (c) $x - 300$	1.1, 1.3 1.5, 1.6	*Coeff.* *Const.* (a) -9 2 (b) $\dfrac{1}{2}$ 0 (c) 1 -300	0.5 Algebraic expressions
Simplify: (a) $4(-c)^2 - 3(-c) + 1$ (b) $[3(x + h) - 1] - [3x - 1]$ (c) $12\left(\dfrac{3x}{4} + 3\right)$ (d) $9x - (300 + 2x)$ (e) $-\dfrac{1}{5}[x - (-1)]$ (f) $2(2y + 3) + 3y$	1.1, 1.2 1.3, 1.5 1.6	(a) $4c^2 + 3c + 1$ (b) $3h$ (c) $9x + 36$ (d) $7x - 300$ (e) $-\dfrac{1}{5}x - \dfrac{1}{5}$ (f) $7y + 6$	0.5 Algebraic expressions
Find the LCD of $\dfrac{3x}{2x + 10}$ and $\dfrac{1}{x + 5}$.	1.1	$2x + 10$	0.7 Algebraic fractions

1.1

Solutions of Linear Equations and Inequalities in One Variable

| APPLICATION PREVIEW |

Using data from 1980 and projected to 2050, the number of Hispanics in the U.S. civilian noninstitutional population is given by

$$y = 0.876x + 6.084$$

millions, where x is the number of years after 1980 (*Source:* U.S. Census Bureau). To find the year when this population will equal 36.74 million, we solve $36.74 = 0.876x + 6.084$. (See Example 2.) We will discuss solutions of linear equations and inequalities in this section.

Equations   An **equation** is a statement that two quantities or algebraic expressions are equal. An equation such as $3x - 2 = 7$ is known as an equation in one variable. The x in this case is called a **variable** because its value determines whether the equation is true. For example, $3x - 2 = 7$ is true only for $x = 3$. Finding the value(s) of the variable(s) that make the equation true—that is, finding the **solutions**—is called **solving the equation.** The set of solutions of an equation is called a **solution set** of the equation. The variable in an equation is sometimes called the **unknown.**

Equations that are true for all values of the variables are called **identities.** The equation $2(x - 1) = 2x - 2$ is an example of an identity. Equations that are true only for certain values of the variables are called **conditional equations** or simply **equations.**

Two equations are said to be **equivalent** if they have exactly the same solution set. We can often solve a complicated linear equation by finding an equivalent equation whose solution is easily found. We use the following properties of equality to reduce an equation to a simple equivalent equation.

Properties of Equality

Properties	Examples
Substitution Property	
The equation formed by substituting one expression for an equal expression is equivalent to the original equation.	$3(x - 3) - \frac{1}{2}(4x - 18) = 4$ is equivalent to $3x - 9 - 2x + 9 = 4$ and to $x = 4$. We say the solution set is $\{4\}$, or the solution is 4.
Addition Property	
The equation formed by adding the same quantity to both sides of an equation is equivalent to the original equation.	$x - 4 = 6$ is equivalent to $x - 4 + 4 = 6 + 4$, or to $x = 10$. $x + 5 = 12$ is equivalent to $x + 5 + (-5) = 12 + (-5)$, or to $x = 7$.
Multiplication Property	
The equation formed by multiplying both sides of an equation by the same nonzero quantity is equivalent to the original equation.	$\frac{1}{3}x = 6$ is equivalent to $3\left(\frac{1}{3}x\right) = 3(6)$, or to $x = 18$. $5x = 20$ is equivalent to $(5x)/5 = 20/5$, or to $x = 4$. (Dividing both sides by 5 is equivalent to multiplying both sides by $\frac{1}{5}$.)

Solving Linear Equations   If an equation contains one variable and if the variable occurs to the first degree, the equation is called a **linear equation in one variable.** The following procedure is based on three properties of equality and with it we can solve any linear equation in one variable.

Solving a Linear Equation

Procedure	Example
To solve a linear equation in one variable:	Solve $\dfrac{3x}{4} + 3 = \dfrac{x-1}{3}$.
1. If the equation contains fractions, multiply both sides by the least common denominator (LCD) of the fractions.	1. LCD is 12. $$12\left(\frac{3x}{4} + 3\right) = 12\left(\frac{x-1}{3}\right)$$
2. Remove any parentheses in the equation.	2. $9x + 36 = 4x - 4$
3. Perform any additions or subtractions to get all terms containing the variable on one side and all other terms on the other side.	3. $9x + 36 - 4x = 4x - 4 - 4x$ $5x + 36 = -4$ $5x + 36 - 36 = -4 - 36$ $5x = -40$
4. Divide both sides of the equation by the coefficient of the variable.	4. $\dfrac{5x}{5} = \dfrac{-40}{5}$ gives $x = -8$
5. Check the solution by substitution in the original equation.	5. $\dfrac{3(-8)}{4} + 3 \overset{?}{=} \dfrac{-8-1}{3}$ gives $-3 = -3$ ✔

EXAMPLE 1 Solving Linear Equations

(a) Solve for z: $\dfrac{2z}{3} = -6$ (b) Solve for x: $\dfrac{3x+1}{2} = \dfrac{x}{3} - 3$

Solution

(a) Multiply both sides by 3.

$$3\left(\frac{2z}{3}\right) = 3(-6) \quad \text{gives} \quad 2z = -18$$

Divide both sides by 2.

$$\frac{2z}{2} = \frac{-18}{2} \quad \text{gives} \quad z = -9$$

Check: $\dfrac{2(-9)}{3} \overset{?}{=} -6$ gives $-6 = -6$ ✔

(b) $\dfrac{3x+1}{2} = \dfrac{x}{3} - 3$

$$6\left(\frac{3x+1}{2}\right) = 6\left(\frac{x}{3} - 3\right) \qquad \text{Multiply both sides by the LCD, 6.}$$

$$3(3x+1) = 6\left(\frac{x}{3} - 3\right) \qquad \text{Simplify the fraction on the left side.}$$

$$9x + 3 = 2x - 18 \qquad \text{Distribute to remove parentheses.}$$

$$7x = -21 \qquad \text{Add } (-2x) + (-3) \text{ to both sides.}$$

$$x = -3 \qquad \text{Divide both sides by 7.}$$

Check: $\dfrac{3(-3)+1}{2} \overset{?}{=} \dfrac{-3}{3} - 3$ gives $-4 = -4$ ✔

When using an equation to describe (model) a set of data, it is sometimes useful to let a variable represent the number of years past a given year. This is called **aligning the data**. Consider the following example.

EXAMPLE **2**   **U.S. Hispanic Population | APPLICATION PREVIEW |**

Using data from 1980 and projected to 2050, the number of Hispanics in the U.S. civilian noninstitutional population is given by $y = 0.876x + 6.084$ millions, where x is the number of years after 1980 (*Source:* U.S. Census Bureau). According to this equaton, in what year will the Hispanic population equal 36.74 million?

Solution

To answer this question, we solve

$$36.74 = 0.876x + 6.084$$
$$30.656 = 0.876x$$
$$34.995 \approx x$$

Checking reveals that $36.74 \approx 0.876(34.995) + 6.084.$ ✔

Thus the number of Hispanics in the United States is estimated to be approximately 36.74 million in 2015. ■

Fractional Equations   A **fractional equation** is an equation that contains a variable in a denominator. It is solved by first multiplying both sides of the equation by the least common denominator (LCD) of the fractions in the equation. Some fractional equations lead to linear equations. Note that the solution to any fractional equation *must* be checked in the original equation, because multiplying both sides of a fractional equation by a variable expression may result in an equation that is not equivalent to the original equation. If a solution to the fraction-free linear equation makes a denominator of the original equation equal to zero, that value cannot be a solution to the original equation. Some fractional equations have no solutions.

EXAMPLE **3**   **Solving Fractional Equations**

Solve for x: (a) $\dfrac{3x}{2x + 10} = 1 + \dfrac{1}{x + 5}$ (b) $\dfrac{2x - 1}{x - 3} = 4 + \dfrac{5}{x - 3}$

Solution

(a) First multiply each term on both sides by the LCD, $2x + 10$. Then simplify and solve.

$$(2x + 10)\left(\frac{3x}{2x + 10}\right) = (2x + 10)(1) + (2x + 10)\left(\frac{1}{x + 5}\right)$$

$$3x = (2x + 10) + 2 \text{ gives } x = 12$$

Check: $\dfrac{3(12)}{2(12) + 10} \stackrel{?}{=} 1 + \dfrac{1}{12 + 5}$ gives $\dfrac{36}{34} = \dfrac{18}{17}$ ✔

(b) First multiply each term on both sides by the LCD, $x - 3$. Then simplify.

$$(x - 3)\left(\frac{2x - 1}{x - 3}\right) = (x - 3)(4) + (x - 3)\left(\frac{5}{x - 3}\right)$$

$$2x - 1 = (4x - 12) + 5 \text{ or } 2x - 1 = 4x - 7$$

Add $(-4x) + 1$ to both sides.

$$-2x = -6 \text{ gives } x = 3$$

The value $x = 3$ gives undefined expressions because the denominators equal 0 when $x = 3$. Hence the equation has no solution. ■

1. Solve the following for x, and check.

(a) $4(x - 3) = 10x - 12$ (b) $\dfrac{5(x - 3)}{6} - x = 1 - \dfrac{x}{9}$

(c) $\dfrac{x}{3x - 6} = 2 - \dfrac{2x}{x - 2}$

Linear Equations with Two Variables

The steps used to solve linear equations in one variable can also be used to solve linear equations in more than one variable for one of the variables in terms of the other. Solving an equation such as the one in the following example is important when using a graphing utility.

EXAMPLE 4   Solving an Equation for One of Two Variables

Solve $4x + 3y = 12$ for y.

Solution
No fractions or parentheses are present, so we subtract $4x$ from both sides to get only the term that contains y on one side.

$$3y = -4x + 12$$

Dividing both sides by 3 gives the solution.

$$y = -\frac{4}{3}x + 4$$

Check: $4x + 3\left(\dfrac{-4}{3}x + 4\right) \overset{?}{=} 12$

$4x + (-4x + 12) = 12$ ✔

2. Solve for y: $y - 4 = -4(x + 2)$

EXAMPLE 5   Profit

Suppose that the relationship between a firm's profit P and the number x of items sold can be described by the equation

$$5x - 4P = 1200$$

(a) How many units must be produced and sold for the firm to make a profit of $150?
(b) Solve this equation for P in terms of x.
(c) Find the profit when 240 units are sold.

Solution
(a) $5x - 4(150) = 1200$
$5x - 600 = 1200$
$5x = 1800$
$x = 360$ units

Check: $5(360) - 4(150) = 1800 - 600 = 1200$ ✔

(b) $5x - 4P = 1200$
$5x - 1200 = 4P$
$P = \dfrac{5x}{4} - \dfrac{1200}{4} = \dfrac{5}{4}x - 300$

(c) $P = \dfrac{5}{4}x - 300$

$P = \dfrac{5}{4}(240) - 300 = 0$

Because $P = 0$ when $x = 240$, we know that profit is $0 when 240 units are sold, and we say that the firm **breaks even** when 240 units are sold. ■

Stated Problems   With an applied problem, it is frequently necessary to convert the problem from its stated form into one or more equations from which the problem's solution can be found. The following guidelines may be useful in solving stated problems.

Guidelines for Solving Stated Problems

1. Begin by reading the problem carefully to determine what you are to find. Use variables to represent the quantities to be found.
2. Reread the problem and use your variables to translate given information into algebraic expressions. Often, drawing a figure is helpful.
3. Use the algebraic expressions and the problem statement to formulate an equation (or equations).
4. Solve the equation(s).
5. Check the solution in the problem, not just in your equation or equations. The answer should satisfy the conditions.

EXAMPLE 6   Investment Mix

Jill Ball has $90,000 to invest. She has chosen one relatively safe investment fund that has an annual yield of 10% and another, riskier one that has a 15% annual yield. How much should she invest in each fund if she would like to earn $10,000 in one year from her investments?

Solution

We want to find the amount of each investment, so we begin as follows:

Let $x =$ the amount invested at 10%, then
$90,000 - x =$ the amount invested at 15% (becuse the two investments total $90,000)

If P is the amount of an investment and r is the annual rate of yield (expressed as a decimal), then the annual earnings $I = Pr$. Using this relationship, we can summarize the information about these two investments in a table.

	P	r	I
10% investment	x	0.10	$0.10x$
15% investment	$90,000 - x$	0.15	$0.15(90,000 - x)$
Total investment	90,000		10,000

The column under I shows that the sum of the earnings is

$$0.10x + 0.15(90,000 - x) = 10,000$$

We solve this as follows.

$$0.10x + 13,500 - 0.15x = 10,000$$
$$-0.05x = -3500 \quad \text{or} \quad x = 70,000$$

Thus the amount invested at 10% is $70,000, and the amount invested at 15% is $90,000 - 70,000 = 20,000$. To check, we return to the problem and note that 10% of $70,000 plus 15% of $20,000 gives a yield of $7000 + $3000 = $10,000. ✔ ■

Linear Inequalities   An **inequality** is a statement that one quantity is greater than (or less than) another quantity. The inequality $3x - 2 > 2x + 1$ is a first-degree (linear) inequality that states that the left member is greater than the right member. Certain values of the variable will satisfy the inequality. These values form the solution set of the inequality. For example, $x = 4$ is in the solution set of $3x - 2 > 2x + 1$ because $3 \cdot 4 - 2 > 2 \cdot 4 + 1$. On the other hand, $x = 2$ is not in the solution set because $3 \cdot 2 - 2 \not> 2 \cdot 2 + 1$. *Solving* an inequality means finding its solution set, and two inequalities are *equivalent* if they have the same solution set. As with equations, we find the solutions to inequalities by finding equivalent inequalities from which the solutions can be easily seen. We use the following properties to reduce an inequality to a simple equivalent inequality.

Properties of Inequalities

Properties	Examples
Substitution Property	
The inequality formed by substituting one expression for an equal expression is equivalent to the original inequality.	$5x - 4x < 6$ $x < 6$ The solution set is $\{x: x < 6\}$.
Addition Property	
The inequality formed by adding the same quantity to both sides of an inequality is equivalent to original inequality.	$2x - 4 > x + 6$ $2x - 4 + 4 > x + 6 + 4$ $2x > x + 10$ $2x + (-x) > x + 10 + (-x)$ $x > 10$
Multiplication Property I	
The inequality formed by multiplying both sides of an inequality by the same *positive* quantity is equivalent to the original inequality.	$\dfrac{1}{2}x > 8 \qquad\qquad 3x < 6$ $\dfrac{1}{2}x(2) > 8(2) \qquad 3x\left(\dfrac{1}{3}\right) < 6\left(\dfrac{1}{3}\right)$ $x > 16 \qquad\qquad x < 2$
Multiplication Property II	
The inequality formed by multiplying both sides of an inequality by the same *negative* number and reversing the direction of the inequality symbol is equivalent to the original inequality.	$-x < 6 \qquad\qquad -3x > -27$ $-x(-1) > 6(-1) \qquad -3x\left(-\dfrac{1}{3}\right) < -27\left(-\dfrac{1}{3}\right)$ $x > -6 \qquad\qquad x < 9$

For some inequalities, several operations are required to find their solution sets. In this case, the order in which the operations are performed is the same as that used in solving linear equations.

EXAMPLE 7   Solving an Inequality

Solve the inequality $2(x - 4) < \dfrac{x - 3}{3}$.

Solution

$$2(x - 4) < \frac{x - 3}{3}$$

$$6(x - 4) < x - 3 \qquad \text{\color{red}Clear fractions.}$$

$$6x - 24 < x - 3 \qquad \text{\color{red}Remove parentheses.}$$

$$5x < 21 \qquad \text{\color{red}Perform additions and subtractions.}$$

$$x < \frac{21}{5} \qquad \text{\color{red}Multiply by } \frac{1}{5}.$$

Check: If we want to check that this solution is a reasonable one, we can substitute the integer values around 21/5 into the original inequality. Note that $x = 4$ satisfies the inequality because

$$2[(4) - 4] < \frac{(4) - 3}{3}$$

but that $x = 5$ does not because

$$2[(5) - 4] \not< \frac{(5) - 3}{3}$$

Thus $x < 21/5$ is a reasonable solution. ✔   ■

We may also solve inequalities of the form $a \le b$. This means "a is less than b or a equals b."

EXAMPLE **8**   **Solving an Inequality**

Solve the inequality $3x - 2 \le 7$ and graph the solution.

Solution

This inequality states that $3x - 2 = 7$ or $3x - 2 < 7$. By solving in the usual manner, we get $3x \le 9$, or $x \le 3$. Then $x = 3$ is the solution to $3x - 2 = 7$ and $x < 3$ is the solution to $3x - 2 < 7$, so the solution set for $3x - 2 \le 7$ is $\{x: x \le 3\}$.

The graph of the solution set includes the point $x = 3$ and all points $x < 3$ (see Figure 1.1).

Figure 1.1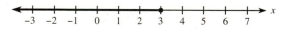

■

CHECKPOINT   Solve the following inequalities for y.

3. $3y - 7 \le 5 - y$ 4. $2y + 6 > 4y + 5$ 5. $4 - 3y \ge 4y + 5$

EXAMPLE **9**   **Normal Height for a Given Age**

For boys between 4 and 16 years of age, height and age are linearly related. That relation can be expressed as

$$H = 2.31A + 31.26$$

where H is height in inches and A is age in years. To account for natural variation among individuals, normal is considered to be any measure falling within ±5% of the height obtained from the equation.* Write as an inequality the range of normal height for a boy who is 9 years old.

---

*Adapted from data from the National Center for Health Statistics

Solution

For a 9-year-old boy, the height from the formula is $H = 2.31(9) + 31.26 = 52.05$ inches. To be considered of normal height, H would have to be within $\pm5\%$ of 52.05 inches. That is, the boy's height H is considered normal if $H \geq 52.05 - (0.05)(52.05)$ and $H \leq 52.05 + (0.05)(52.05)$. We can express this range of normal heights by the compound inequality

$$52.05 - (0.05)(52.05) \leq H \leq 52.05 + (0.05)(52.05)$$

or

$$49.45 \leq H \leq 54.65$$

CHECKPOINT SOLUTIONS

1. (a) $\quad 4(x - 3) = 10x - 12$
$\qquad 4x - 12 = 10x - 12$
$\qquad\quad -6x = 0$
$\qquad\qquad x = 0$
Check: $4(0 - 3) \overset{?}{=} 10 \cdot 0 - 12$
$\qquad\qquad -12 = -12$ ✔

(b) $\dfrac{5(x - 3)}{6} - x = 1 - \dfrac{x}{9}$

LCD of 6 and 9 is 18.

$18\left(\dfrac{5x - 15}{6}\right) - 18(x) = 18(1) - 18\left(\dfrac{x}{9}\right)$
$3(5x - 15) - 18x = 18 - 2x$
$15x - 45 - 18x = 18 - 2x$
$\qquad -45 - 3x = 18 - 2x$
$\qquad\qquad -63 = x$

Check: $\dfrac{5(-63 - 3)}{6} - (-63) \overset{?}{=} 1 - \dfrac{(-63)}{9}$
$5(-11) + 63 \overset{?}{=} 1 + 7$
$\qquad\qquad 8 = 8$ ✔

(c) $\dfrac{x}{3x - 6} = 2 - \dfrac{2x}{x - 2}$

LCD is $3x - 6$, or $3(x - 2)$.

$(3x - 6)\left(\dfrac{x}{3x - 6}\right) = (3x - 6)(2) - (3x - 6)\left(\dfrac{2x}{x - 2}\right)$

$x = 6x - 12 - 3(2x)$ or $x = 6x - 12 - 6x$
$x = -12$

Check: $\dfrac{-12}{3(-12) - 6} \overset{?}{=} 2 - \dfrac{2(-12)}{-12 - 2}$

$\dfrac{-12}{-42} \overset{?}{=} 2 - \dfrac{12}{7}$

$\dfrac{2}{7} = \dfrac{2}{7}$ ✔

2. $y - 4 = -4(x + 2)$
$y - 4 = -4x - 8$
$\quad y = -4x - 4$

3. $3y - 7 \leq 5 - y$
$\quad 4y - 7 \leq 5$
$\qquad 4y \leq 12$
$\qquad\quad y \leq 3$

4. $2y + 6 > 4y + 5$
$\qquad 6 > 2y + 5$
$\qquad 1 > 2y$
$\qquad y < \dfrac{1}{2}$

5. $4 - 3y \geq 4y + 5$
$\qquad 4 \geq 7y + 5$
$\qquad -1 \geq 7y$
$\qquad y \leq -\dfrac{1}{7}$

| EXERCISES | 1.1

Access end-of-section exercises online at www.webassign.net **WebAssign** ENHANCED

OBJECTIVES

1.2

- To determine whether a relation is a function
- To state the domains and ranges of certain functions
- To use function notation
- To perform operations with functions
- To find the composite of two functions

Functions

■ **| APPLICATION PREVIEW |**

The number of individual income tax returns filed electronically has increased in recent years. The relationship between the number of returns filed electronically and the number of years after 1995 can be described by the equation

$$y = 5.091x + 11.545$$

where y is in millions and x is the number of years past 1995 (*Source:* Internal Revenue Service). In this equation, y depends uniquely on x, so we say that y is a function of x. Understanding the mathematical meaning of the phrase *function of* and learning to interpret and apply such relationships are the goals of this section. (The function here will be discussed in Example 4.)

Relations and Functions

A **relation** is defined by a set of **ordered pairs** of real numbers. These ordered pairs may be determined from a table, a graph, an equation, or an inequality. For example, the solutions to the equation $y = 4x - 3$ are pairs of numbers (one for x and one for y), so $y = 4x - 3$ expresses a relation between x and y. We write the ordered pairs in this relation in the form (x, y) so that the first number is the x-value and the second is the y-value. Because we cannot list all of the ordered pairs that define this relation, we use the equation to give its definition. Similarly, the inequality $R \leq 5x$ expresses a relation between the two variables x and R.

Relation

> A **relation** is defined by a set of ordered pairs or by a rule that determines how the ordered pairs are found. The relation may also be defined by a table, a graph, an equation, or an inequality.

For example, the set of ordered pairs

$$\{(1, 3), (1, 6), (2, 6), (3, 9), (3, 12), (4, 12)\}$$

expresses a relation between the set of first components, $\{1, 2, 3, 4\}$, and the set of second components, $\{3, 6, 9, 12\}$. The set of first components is called the **domain** of the relation, and the set of second components is called the **range** of the relation. Figure 1.2(a) uses arrows to indicate how the inputs from the domain (the first components) are associated with the outputs in the range (the second components). Figure 1.2(b) shows another example of a relation. Because relations can also be defined by tables and graphs, Table 1.1 and Figure 1.3 are examples of relations.

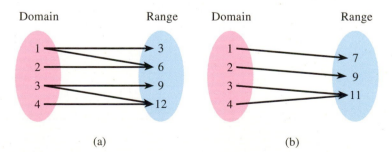

Figure 1.2   (a)   (b)

An equation frequently expresses how the second component (the output) is obtained from the first component (the input). For example, the equation

$$y = 4x - 3$$

expresses how the output y results from the input x. This equation expresses a special relation between x and y, because each value of x that is substituted into the equation results in exactly one value of y. If each value of x put into an equation results in one value of y, we say that the equation expresses y as a **function** of x.

Definition of a Function	A **function** is a relation between two sets such that to each element of the domain (input) there corresponds exactly one element of the range (output). A function may be defined by a set of ordered pairs, a table, a graph, or an equation.

When a function is defined, the variable that represents the numbers in the domain (input) is called the **independent variable** of the function, and the variable that represents the numbers in the range (output) is called the **dependent variable** (because its values depend on the values of the independent variable). The equation $y = 4x - 3$ defines y as a function of x, because only one value of y will result from each value of x that is substituted into the equation. Thus the equation defines a function in which x is the independent variable and y is the dependent variable.

■ | TABLE 1.1 | ■

Federal Income Tax

Income	Rate
$0–$8375	10%
$8376–$34,000	15%
$34,001–$82,400	25%
$82,401–$171,850	28%
$171,851–$373,650	33%
Over $373,650	35%

Source: Internal Revenue Service

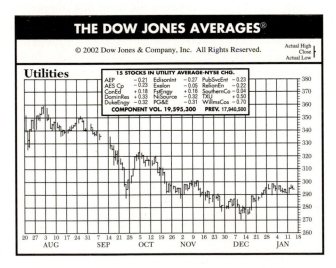

Figure 1.3

Source: Reprinted with permission of *The Wall Street Journal*, Copyright © 2002 Dow Jones & Company, Inc. All Rights Reserved Worldwide.

We can also apply this idea to a relation defined by a table or a graph. In Figure 1.2(b), because each input in the domain corresponds to exactly one output in the range, the relation is a function. Similarly, the data given in Table 1.1 (the tax brackets for U.S. income tax for single wage earners) represents the tax rate as a function of the income. Note in Table 1.1 that even though many different amounts of taxable income have the same tax rate, each amount of taxable income (input) corresponds to exactly one tax rate (output). On the other hand, the relation defined in Figure 1.3 is not a function because the graph representing the Dow Jones Utilities Average shows that for each day there are at least three different values—the actual high, the actual low, and the close. This particular figure also has historical interest because it shows a break in the graph when the New York Stock Exchange closed following the terrorist attacks of 9/11/2001.

EXAMPLE 1

Functions

Does $y^2 = 2x$ express y as a function of x?

Solution

No, because some values of x are associated with more than one value of y. In fact, there are two y-values for each $x > 0$. For example, if $x = 8$, then $y = 4$ or $y = -4$, two different y-values for the same x-value. The equation $y^2 = 2x$ expresses a relation between x and y, but y is not a function of x.

Graphs of Functions It is possible to picture geometrically the relations and functions that we have been discussing by sketching their graphs on a rectangular coordinate system. We construct a rectangular coordinate system by drawing two real number lines (called **coordinate axes**) that are perpendicular to each other and intersect at their origins (called the **origin** of the system).

The ordered pair (a, b) represents the point P that is located a units along the x-axis and b units along the y-axis (see Figure 1.4). Similarly, any point has a unique ordered pair that describes it.

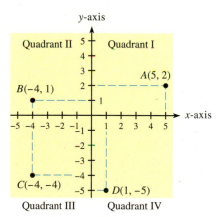

Figure 1.4

The values a and b in the ordered pair associated with the point P are called the **rectangular** (or **Cartesian**) **coordinates** of the point, where a is the **x-coordinate** (or **abscissa**), and b is the **y-coordinate** (or **ordinate**). The ordered pairs (a, b) and (c, d) are equal if and only if $a = c$ and $b = d$.

The **graph** of an equation that defines a function (or relation) is the picture that results when we plot the points whose coordinates (x, y) satisfy the equation. To sketch the graph, we plot enough points to suggest the shape of the graph and draw a smooth curve through the points. This is called the **point-plotting method** of sketching a graph.

EXAMPLE 2 **Graphing a Function**

Graph the function $y = 4x^2$.

Solution

We choose some sample values of x and find the corresponding values of y. Placing these in a table, we have sample points to plot. When we have enough to determine the shape of the graph, we connect the points to complete the graph. The table and graph are shown in Figure 1.5(a). ■

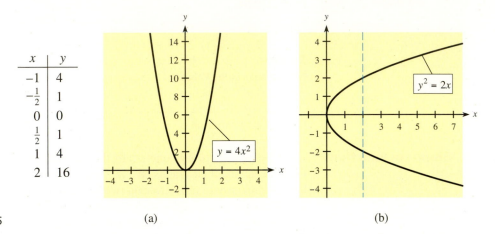

x	y
-1	4
$-\frac{1}{2}$	1
0	0
$\frac{1}{2}$	1
1	4
2	16

Figure 1.5   (a)   (b)

We can determine whether a relation is a function by inspecting its graph. If the relation is a function, then no one input (x-value) has two different outputs (y-values). This means that no two points on the graph will have the same first coordinate (component). Thus no two points of the graph will lie on the same vertical line.

Vertical-Line Test If no vertical line exists that intersects the graph at more than one point, then the graph is that of a function.

Performing this test on the graph of $y = 4x^2$ (Figure 1.5(a)), we easily see that this equation describes a function. The graph of $y^2 = 2x$ is shown in Figure 1.5(b), and we can see that the vertical-line test indicates that this is not a function (as we already saw in Example 1). For example, a vertical line at $x = 2$ intersects the curve at $(2, 2)$ and $(2, -2)$.

Function Notation We can use function notation to indicate that y is a function of x. The function is denoted by f, and we write $y = f(x)$. This is read "y is a function of x" or "y equals f of x." For specific values of x, $f(x)$ represents the values of the function (that is, outputs, or y-values) at those x-values. Thus if

$$f(x) = 3x^2 + 2x + 1$$

then
$$f(2) = 3(2)^2 + 2(2) + 1 = 17$$

and
$$f(-3) = 3(-3)^2 + 2(-3) + 1 = 22$$

Figure 1.6 shows the function notation $f(x)$ as (a) an operator on x and (b) a y-coordinate for a given x-value.

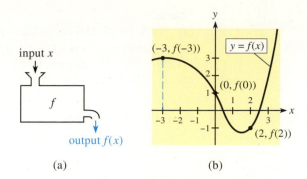

input x

output $f(x)$

Figure 1.6   (a)   (b)

Letters other than f may also be used to denote functions. For example, $y = g(x)$ or $y = h(x)$ may be used.

EXAMPLE 3   Evaluating Functions

If $y = f(x) = 2x^3 - 3x^2 + 1$, find the following.
(a) $f(3)$ (b) $f(-1)$ (c) $f(-a)$

Solution
(a) $f(3) = 2(3)^3 - 3(3)^2 + 1 = 2(27) - 3(9) + 1 = 28$
    Thus $y = 28$ when $x = 3$.
(b) $f(-1) = 2(-1)^3 - 3(-1)^2 + 1 = 2(-1) - 3(1) + 1 = -4$
    Thus $y = -4$ when $x = -1$.
(c) $f(-a) = 2(-a)^3 - 3(-a)^2 + 1 = -2a^3 - 3a^2 + 1$ ■

When a function fits a set of data exactly or approximately, we say that the function **models** the data. The model includes descriptions of all involved variables.

EXAMPLE 4   Electronic Income Tax Returns | APPLICATION PREVIEW |

The relationship between the number y of individual income tax returns filed electronically and the number of years after 1995 can be modeled by the function

$$y = f(x) = 5.091x + 11.545$$

where y is in millions and x is the number of years after 1995 (*Source:* Internal Revenue Service).

(a) Find $f(8)$.
(b) Write a sentence that explains the meaning of the result in part (a).

Solution
(a) $f(8) = 5.091(8) + 11.545 = 52.273$
(b) The statement $f(8) = 52.273$ means that in $1995 + 8 = 2003$, 52.273 million income tax returns were filed electronically.   ■

EXAMPLE 5   Mortgage Payment

| TABLE 1.2 |

$r(\%)$	$f(r)$
2.6	12
5.2	15
6.3	17
7.4	20
9	30

Table 1.2 shows the number of years that it will take a couple to pay off a $100,000 mortgage at several different interest rates if they pay $800 per month. If r denotes the rate and $f(r)$ denotes the number of years:

(a) What is $f(6.3)$ and what does it mean?
(b) If $f(r) = 30$, what is r?
(c) Does $2 \cdot f(2.6) = f(2 \cdot 2.6)$?

Solution

(a) $f(6.3) = 17$. This means that with a 6.3% interest rate, a couple can pay off the $100,000 mortgage in 17 years by paying $800 per month.

(b) The table indicates that $f(9) = 30$, so $r = 9$.

(c) $2 \cdot f(2.6) = 2 \cdot 12 = 24$ and $f(2 \cdot 2.6) = f(5.2) = 15$, so $2 \cdot f(2.6) \neq f(2 \cdot 2.6)$. ■

EXAMPLE 6 Function Notation

Given $f(x) = x^2 - 3x + 8$, find $\dfrac{f(x + h) - f(x)}{h}$ and simplify (if $h \neq 0$).

Solution

We find $f(x + h)$ by replacing each x in $f(x)$ with the expression $x + h$.

$$\frac{f(x + h) - f(x)}{h} = \frac{[(x + h)^2 - 3(x + h) + 8] - [x^2 - 3x + 8]}{h}$$

$$= \frac{[(x^2 + 2xh + h^2) - 3x - 3h + 8] - x^2 + 3x - 8}{h}$$

$$= \frac{x^2 + 2xh + h^2 - 3x - 3h + 8 - x^2 + 3x - 8}{h}$$

$$= \frac{2xh + h^2 - 3h}{h} = \frac{h(2x + h - 3)}{h} = 2x + h - 3 \quad ■$$

Domains and Ranges We will limit our discussion in this text to **real functions,** which are functions whose domains and ranges contain only real numbers. If the domain and range of a function are not specified, it is assumed that the domain consists of all real inputs (*x*-values) that result in real outputs (*y*-values), making the range a subset of the real numbers.

For the types of functions we are now studying, if the domain is unspecified, it will include all real numbers except

1. values that result in a denominator of 0, and
2. values that result in an even root of a negative number.

EXAMPLE 7 Domain and Range

Find the domain of each of the following functions; find the range for the functions in parts (a) and (b).

(a) $y = 4x^2$ (b) $y = \sqrt{4 - x}$ (c) $y = 1 + \dfrac{1}{x - 2}$

Solution

(a) There are no restrictions on the numbers substituted for x, so the domain consists of all real numbers. Because the square of any real number is nonnegative, $4x^2$ must be nonnegative. Thus the range is $y \geq 0$. The graph shown in Figure 1.7(a) illustrates our conclusions about the domain and range.

(b) We note the restriction that $4 - x$ cannot be negative. Thus the domain consists of only numbers less than or equal to 4. That is, the domain is the set of real numbers satisfying $x \leq 4$. Because $\sqrt{4 - x}$ is always nonnegative, the range is all $y \geq 0$. Figure 1.7(b) shows the graph of $y = \sqrt{4 - x}$. Note that the graph is located only where $x \leq 4$ and on or above the x-axis (where $y \geq 0$).

(c) $y = 1 + \dfrac{1}{x - 2}$ is undefined at $x = 2$ because $\dfrac{1}{0}$ is undefined. Hence, the domain consists of all real numbers except 2. Figure 1.7(c) shows the graph of $y = 1 + \dfrac{1}{x - 2}$.

The break where $x = 2$ indicates that $x = 2$ is not part of the domain. ■

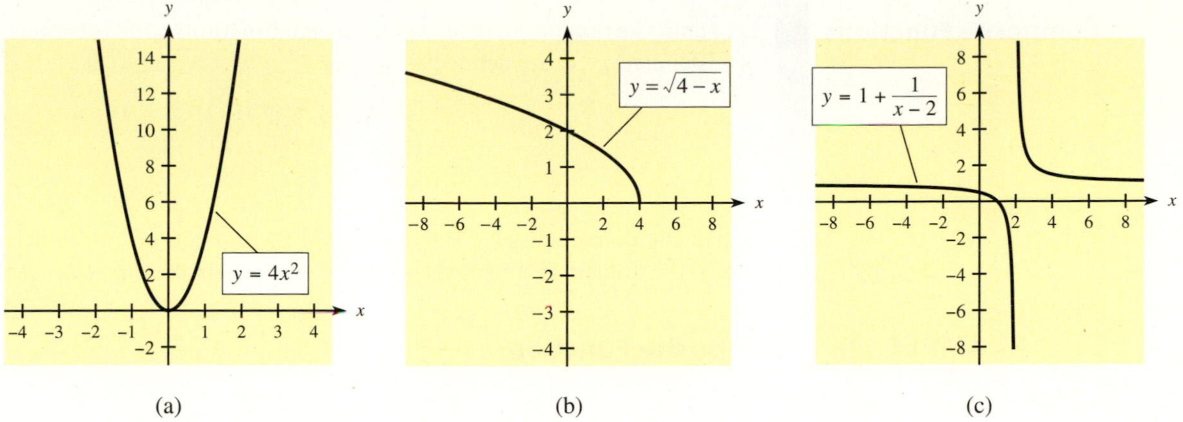

Figure 1.7

CHECKPOINT

1. If $y = f(x)$, the independent variable is _____ and the dependent variable is _____.
2. If $(1, 3)$ is on the graph of $y = f(x)$, then $f(1) = ?$
3. If $f(x) = 1 - x^3$, find $f(-2)$.
4. If $f(x) = 2x^2$, find $f(x + h)$.
5. If $f(x) = \dfrac{1}{x + 1}$, what is the domain of $f(x)$?

Operations with Functions

We can form new functions by performing algebraic operations with two or more functions. We define new functions that are the sum, difference, product, and quotient of two functions as follows.

Operations with Functions

Let f and g be functions of x, and define the following.

Sum $\quad\quad (f + g)(x) = f(x) + g(x)$
Difference $\quad (f - g)(x) = f(x) - g(x)$
Product $\quad (f \cdot g)(x) = f(x) \cdot g(x)$

Quotient $\quad \left(\dfrac{f}{g}\right)(x) = \dfrac{f(x)}{g(x)} \quad$ if $g(x) \neq 0$

EXAMPLE 8 **Operations with Functions**

If $f(x) = 3x + 2$ and $g(x) = x^2 - 3$, find the following functions.

(a) $(f + g)(x)$ (b) $(f - g)(x)$ (c) $(f \cdot g)(x)$ (d) $\left(\dfrac{f}{g}\right)(x)$

Solution

(a) $(f + g)(x) = f(x) + g(x) = (3x + 2) + (x^2 - 3) = x^2 + 3x - 1$
(b) $(f - g)(x) = f(x) - g(x) = (3x + 2) - (x^2 - 3) = -x^2 + 3x + 5$
(c) $(f \cdot g)(x) = f(x) \cdot g(x) = (3x + 2)(x^2 - 3) = 3x^3 + 2x^2 - 9x - 6$
(d) $\left(\dfrac{f}{g}\right)(x) = \dfrac{f(x)}{g(x)} = \dfrac{3x + 2}{x^2 - 3}$, if $x^2 - 3 \neq 0$ ∎

We now consider a new way to combine two functions. Just as we can substitute a number for the independent variable in a function, we can substitute a second function for the variable. This creates a new function called a **composite function.**

Composite Functions

Let f and g be functions. Then the **composite functions** g of f (denoted $g \circ f$) and f of g (denoted $f \circ g$) are defined as follows:

$$(g \circ f)(x) = g(f(x))$$
$$(f \circ g)(x) = f(g(x))$$

Note that the domain of $g \circ f$ is the subset of the domain of f for which $g \circ f$ is defined. Similarly, the domain of $f \circ g$ is the subset of the domain of g for which $f \circ g$ is defined.

EXAMPLE 9 Composite Functions

If $f(x) = 2x^3 + 1$ and $g(x) = x^2$, find the following.
(a) $(g \circ f)(x)$ (b) $(f \circ g)(x)$

Solution

(a) $(g \circ f)(x) = g(f(x))$
$= g(2x^3 + 1)$
$= (2x^3 + 1)^2 = 4x^6 + 4x^3 + 1$

(b) $(f \circ g)(x) = f(g(x))$
$= f(x^2)$
$= 2(x^2)^3 + 1$
$= 2x^6 + 1$ ■

Figure 1.8 illustrates both composite functions found in Example 9.

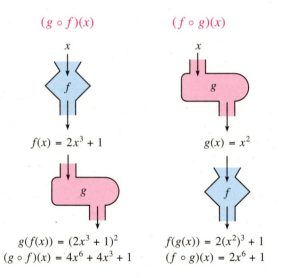

Figure 1.8

$(g \circ f)(x)$

$f(x) = 2x^3 + 1$

$g(f(x)) = (2x^3 + 1)^2$
$(g \circ f)(x) = 4x^6 + 4x^3 + 1$

$(f \circ g)(x)$

$g(x) = x^2$

$f(g(x)) = 2(x^2)^3 + 1$
$(f \circ g)(x) = 2x^6 + 1$

CHECKPOINT

6. If $f(x) = 1 - 2x$ and $g(x) = 3x^2$, find the following.
   (a) $(g - f)(x)$ (b) $(f \cdot g)(x)$ (c) $(f \circ g)(x)$ (d) $(g \circ f)(x)$
   (e) $(f \circ f)(x) = f(f(x))$

CHECKPOINT SOLUTIONS

1. Independent variable is x; dependent variable is y.
2. $f(1) = 3$
3. $f(-2) = 1 - (-2)^3 = 1 - (-8) = 9$
4. $f(x + h) = 2(x + h)^2$
   $= 2(x^2 + 2xh + h^2)$
5. The domain is all real numbers except $x = -1$, because $f(x)$ is undefined when $x = -1$.
6. (a) $(g - f)(x) = g(x) - f(x) = 3x^2 - (1 - 2x) = 3x^2 + 2x - 1$
   (b) $(f \cdot g)(x) = f(x) \cdot g(x) = (1 - 2x)(3x^2) = 3x^2 - 6x^3$
   (c) $(f \circ g)(x) = f(g(x)) = f(3x^2) = 1 - 2(3x^2) = 1 - 6x^2$
   (d) $(g \circ f)(x) = g(f(x)) = g(1 - 2x) = 3(1 - 2x)^2$
   (e) $(f \circ f)(x) = f(f(x)) = f(1 - 2x) = 1 - 2(1 - 2x) = 4x - 1$

| EXERCISES | 1.2

Access end-of-section exercises online at **www.webassign.net**   **WebAssign**

OBJECTIVES   1.3

- To find the intercepts of graphs
- To graph linear functions
- To find the slope of a line from its graph and from its equation
- To find the rate of change of a linear function
- To graph a line, given its slope and *y*-intercept or its slope and one point on the line
- To write the equation of a line, given information about its graph

Linear Functions

| APPLICATION PREVIEW |

The number of banks in the United States for selected years from 1980 to 2009 is given by

$$y = -380.961x + 18{,}483.167$$

where *x* is the number of years after 1980 (*Source:* Federal Deposit Insurance Corporation). What does this function tell about how the number of banks has changed per year during this period? (See Example 7.) In this section, we will find the slopes and intercepts of graphs of linear functions and apply them.

The function in the Application Preview is an example of a special function, called the **linear function,** defined as follows.

| **Linear Function** | A **linear function** is a function of the form
$$y = f(x) = ax + b$$
where *a* and *b* are constants. |

Intercepts Because the graph of a linear function is a line, only two points are necessary to determine its graph. It is frequently possible to use **intercepts** to graph a linear function. The point(s) where a graph intersects the *x*-axis are called the *x*-intercept points, and the *x*-coordinates of these points are the **x-intercepts.** Similarly, the points where a graph intersects the *y*-axis are the *y*-intercept points, and the *y*-coordinates of these points are the **y-intercepts.** Because any point on the *x*-axis has *y*-coordinate 0 and any point on the *y*-axis has *x*-coordinate 0, we find intercepts as follows.

| **Intercepts** | (a) To find the **y-intercept(s)** of the graph of an equation, set $x = 0$ in the equation and solve for *y*. *Note:* A function of *x* has at most one *y*-intercept.
(b) To find the **x-intercept(s)**, set $y = 0$ and solve for *x*. |

EXAMPLE 1   Intercepts

Find the intercepts and graph the following.
(a) $3x + y = 9$ (b) $x = 4y$

Solution

(a) To find the *y*-intercept, we set $x = 0$ and solve for *y*: $3(0) + y = 9$ gives $y = 9$, so the *y*-intercept is 9.

To find the *x*-intercept, we set $y = 0$ and solve for *x*: $3x + 0 = 9$ gives $x = 3$, so the *x*-intercept is 3. Using the intercepts gives the graph, shown in Figure 1.13 (next page).

(b) Letting $x = 0$ gives $y = 0$, and letting $y = 0$ gives $x = 0$, so the only intercept of the graph of $x = 4y$ is at the point $(0, 0)$. A second point is needed to graph the line. Hence, if we let $y = 1$ in $x = 4y$, we get $x = 4$ and have a second point $(4, 1)$ on the graph. It is wise to plot a third point as a check. The graph is shown in Figure 1.14 (next page). ■

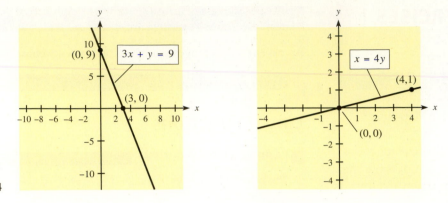

Figures 1.13 & 1.14

Note that the equation graphed in Figure 1.13 can be rewritten as

$$y = 9 - 3x \quad \text{or} \quad f(x) = 9 - 3x$$

We see in Figure 1.13 that the x-intercept $(3, 0)$ is the point where the function value is zero. The x-coordinate of such a point is called a **zero of the function.** Thus we see that the *x-intercepts of a function are the same as its zeros.*

EXAMPLE 2   ## Depreciation

A business property is purchased for \$122,880 and depreciated over a period of 10 years. Its value y is related to the number of months of service x by the equation

$$4096x + 4y = 491{,}520$$

Find the x-intercept and the y-intercept and use them to sketch the graph of the equation.

Solution

$$x\text{-intercept: } y = 0 \text{ gives } 4096x = 491{,}520$$
$$x = 120$$

Thus 120 is the x-intercept.

$$y\text{-intercept: } x = 0 \text{ gives } 4y = 491{,}520$$
$$y = 122{,}880$$

Thus 122,880 is the y-intercept. The graph is shown in Figure 1.15. Note that the units on the x- and y-axes are different and that the y-intercept corresponds to the value of the property 0 months after purchase. That is, the y-intercept gives the purchase price. The x-intercept corresponds to the number of months that have passed before the value is 0; that is, the property is fully depreciated after 120 months, or 10 years. Note that only positive values for x and y make sense in this application, so only the Quadrant I portion of the graph is shown. ■

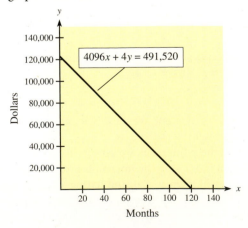

Figure 1.15

Despite the ease of using intercepts to graph linear equations, this method is not always the best. For example, vertical lines, horizontal lines, or lines that pass through the origin may have a single intercept, and if a line has both intercepts very close to the origin, using the intercepts may lead to an inaccurate graph.

Rate of Change; Slope of a Line

Note that in Figure 1.15, as the graph moves from the y-intercept point $(0, 122{,}880)$ to the x-intercept point $(120, 0)$, the y-value on the line changes $-122{,}880$ units (from $122{,}880$ to 0), whereas the x-value changes 120 units (from 0 to 120). Thus the **rate of change** of the value of the business property is

$$\frac{-122{,}880}{120} = -1024 \text{ dollars per month}$$

This means that each month the value of the property changes by -1024 dollars, or the value decreases by \$1024 per month. This **rate of change** of a linear function is called the **slope** of the line that is its graph (see Figure 1.15). For the graph of a linear function, the ratio of the change in y to the corresponding change in x measures the slope of the line. For any nonvertical line, the slope can be found by using any two points on the line, as follows.

Slope of a Line	If a nonvertical line passes through the points $P_1(x_1, y_1)$ and $P_2(x_2, y_2)$, its **slope,** denoted by m, is found by using $$m = \frac{y_2 - y_1}{x_2 - x_1} = \frac{\Delta y}{\Delta x}$$ where Δy, read "delta y," means "change in y" and Δx means "change in x." The slope of a vertical line is undefined. 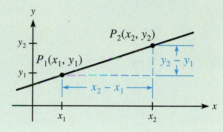

Note that for a given line, the slope is the same regardless of which two points are used in the calculation; this is because corresponding sides of similar triangles are in proportion.

EXAMPLE   3   Slopes

Find the slope of
(a) line ℓ_1, passing through $(-2, 1)$ and $(4, 3)$
(b) line ℓ_2, passing through $(3, 0)$ and $(4, -3)$

Solution

(a) $m = \dfrac{3 - 1}{4 - (-2)} = \dfrac{2}{6} = \dfrac{1}{3}$ or, equivalently, $m = \dfrac{1 - 3}{-2 - 4} = \dfrac{-2}{-6} = \dfrac{1}{3}$

This means that a point 3 units to the right and 1 unit up from any point on the line is also on the line. Line ℓ_1 is shown in Figure 1.16.

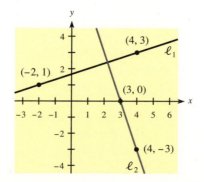

Figure 1.16

(b) $m = \dfrac{0 - (-3)}{3 - 4} = \dfrac{3}{-1} = -3$

This means that a point 1 unit to the right and 3 units down from any point on the line is also on the line. Line ℓ_2 is also shown in Figure 1.16. ∎

From the previous discussion, we see that the slope describes the direction of a line as follows.

Orientation of a Line and Its Slope

1. The slope is *positive* if the line slopes upward toward the right. The function is increasing.

$$m = \frac{\Delta y}{\Delta x} > 0$$

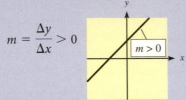

2. The slope is *negative* if the line slopes downward toward the right. The function is decreasing.

$$m = \frac{\Delta y}{\Delta x} < 0$$

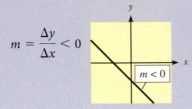

3. The slope of a *horizontal line* is 0, because $\Delta y = 0$. The function is constant.

$$m = \frac{\Delta y}{\Delta x} = 0$$

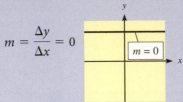

4. The slope of a *vertical line* is *undefined*, because $\Delta x = 0$.

$$m = \frac{\Delta y}{\Delta x} \text{ is undefined.}$$

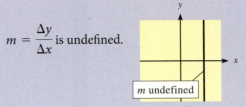

Two distinct nonvertical lines that have the same slope are parallel, and conversely, two nonvertical parallel lines have the same slope.

Parallel Lines

Two distinct nonvertical lines are *parallel* if and only if their slopes are *equal*.

In general, two lines are perpendicular if they intersect at right angles. However, the appearance of this perpendicularity relationship can be obscured when we graph lines, unless the axes have the same scale. Note that the lines ℓ_1 and ℓ_2 in Figure 1.16 appear to be perpendicular. However, the same lines graphed with different scales on the axes will not look perpendicular. To avoid being misled by graphs with different scales, we use slopes to tell us when lines are perpendicular. Note that the slope of ℓ_1, $\frac{1}{3}$, is the negative reciprocal of the slope of ℓ_2, -3. In fact, as with lines ℓ_1 and ℓ_2, any two nonvertical lines that are perpendicular have slopes that are negative reciprocals of each other.

Slopes of Perpendicular Lines

A line ℓ_1 with slope m, where $m \neq 0$, is *perpendicular* to line ℓ_2 if and only if the slope of ℓ_2 is $-1/m$. (The slopes are *negative reciprocals*.)

Because the slope of a vertical line is undefined, we cannot use slope in discussing parallel and perpendicular relations that involve vertical lines. Two vertical lines are parallel, and any horizontal line is perpendicular to any vertical line.

CHECKPOINT

1. Find the slope of the line through $(4, 6)$ and $(28, -6)$.
2. If a line has slope $m = 0$, then the line is _____. If a line has an undefined slope, then the line is _____.

3. Suppose that line 1 has slope $m_1 = 5$ and line 2 has slope m_2.
   (a) If line 1 is perpendicular to line 2, find m_2.
   (b) If line 1 is parallel to line 2, find m_2.

Writing Equations of Lines

If the slope of a line is m, then the slope between a fixed point (x_1, y_1) and any other point (x, y) on the line is also m. That is,

$$m = \frac{y - y_1}{x - x_1}$$

Solving for $y - y_1$ gives the point-slope form of the equation of a line.

Point-Slope Form

The equation of the line passing through the point (x_1, y_1) and with slope m can be written in the **point-slope form**

$$y - y_1 = m(x - x_1)$$

EXAMPLE 4   Equations of Lines

Write the equation for each line that passes through $(1, -2)$ and has

(a) slope $\frac{2}{3}$ (b) undefined slope   (c) point $(2, 3)$ also on the line

Solution

(a) Here $m = \frac{2}{3}$, $x_1 = 1$, and $y_1 = -2$. An equation of the line is

$$y - (-2) = \frac{2}{3}(x - 1)$$

$$y + 2 = \frac{2}{3}x - \frac{2}{3}$$

$$y = \frac{2}{3}x - \frac{8}{3}$$

This equation also may be written in **general form** as $2x - 3y - 8 = 0$. Figure 1.17 shows the graph of this line; the point $(1, -2)$ and the slope are highlighted.

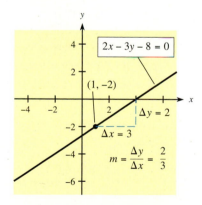

Figure 1.17

(b) Because m is undefined, we cannot use the point-slope form. This line is vertical, so every point on it has x-coordinate 1. Thus the equation is $x = 1$. Note that $x = 1$ is not a function.

(c) First use $(1, -2)$ and $(2, 3)$ to find the slope.

$$m = \frac{3 - (-2)}{2 - 1} = 5$$

Using $m = 5$ and the point $(1, -2)$ (the other point could also be used) gives

$$y - (-2) = 5(x - 1) \quad \text{or} \quad y = 5x - 7 \qquad \blacksquare$$

The graph of $x = 1$ (from Example 4(b)) is a vertical line, as shown in Figure 1.18(a); the graph of $y = 1$ has slope 0, and its graph is a horizontal line, as shown in Figure 1.18(b).

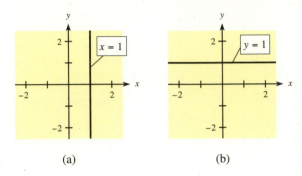

Figure 1.18   (a)   (b)

In general, **vertical lines** have undefined slope and the equation form $x = a$, where a is the x-coordinate of each point on the line. **Horizontal lines** have $m = 0$ and the equation form $y = b$, where b is the y-coordinate of each point on the line.

EXAMPLE 5   Pricing

U.S. Census Bureau data indicate that the average price p of digital television sets can be expressed as a linear function of the number of sets sold N (in thousands). In addition, as N increased by 1000, p dropped by \$10.40, and when 6485 (thousand) sets were sold, the average price per set was \$504.39. Write the equation of the line determined by this information.

Solution
We see that price p is a function of the number of sets N (in thousands), so the slope is given by

$$m = \frac{\text{change in } p}{\text{change in } N} = \frac{-10.40}{1000} = -0.0104$$

A point on the line is $(N_1, p_1) = (6485, 504.39)$. We use the point-slope form adapted to the variables N and p.

$$p - p_1 = m(N - N_1)$$
$$p - 504.39 = -0.0104(N - 6485)$$
$$p - 504.39 = -0.0104N + 67.444$$
$$p = -0.0104N + 571.834 \qquad \blacksquare$$

The point-slope form, with the y-intercept point $(0, b)$, can be used to derive a special form for the equation of a line.

$$y - b = m(x - 0)$$
$$y = mx + b$$

Slope-Intercept Form

The **slope-intercept form** of the equation of a line with slope m and y-intercept b is

$$y = mx + b$$

Note that if a linear equation has the form $y = mx + b$, then the coefficient of x is the slope and the constant term is the y-intercept.

EXAMPLE 6 **Writing the Equation of a Line**

Write the equation of the line with slope $\frac{1}{2}$ and y-intercept -3.

Solution

Substituting $m = \frac{1}{2}$ and $b = -3$ in the equation $y = mx + b$ gives $y = \frac{1}{2}x + (-3)$ or $y = \frac{1}{2}x - 3$. ∎

EXAMPLE 7 **Banks | APPLICATION PREVIEW |**

The number of banks in the United States for selected years from 1980 to 2009 is given by

$$y = -380.961x + 18{,}483.167$$

where x is the number of years after 1980 (*Source:* Federal Deposit Insurance Corporation).

(a) Find the slope and the y-intercept of this function.
(b) What does the y-intercept tell us about the banks?
(c) Interpret the slope as a rate of change.
(d) Why is the number of banks decreasing?

Solution

(a) The slope is $m = -380.961$ and the y-intercept is $b = 18{,}483.167$.
(b) Because x is the number of years after 1980, $x = 0$ represents 1980 and the y-intercept tells us that there were approximately 18,483 banks in 1980.
(c) The slope is -380.961, which tells us that the number of banks was decreasing at a rate of approximately 381 banks per year during this period.
(d) The decrease in number is due to consolidation of banks, with larger banks acquiring smaller banks, and the 2008 financial crisis. ∎

When a linear equation does not appear in slope-intercept form (and does not have the form $x = a$), it can be put into slope-intercept form by solving the equation for y.

EXAMPLE 8 **Slope-Intercept Form**

(a) Find the slope and y-intercept of the line whose equation is $x + 2y = 8$.
(b) Use this information to graph the equation.

Solution

(a) To put the equation in slope-intercept form, we must solve it for y.

$$2y = -x + 8 \quad \text{or} \quad y = -\frac{1}{2}x + 4$$

Thus the slope is $-\frac{1}{2}$ and the y-intercept is 4.
(b) First we plot the y-intercept point $(0, 4)$. Because the slope is $-\frac{1}{2} = \frac{-1}{2}$, moving 2 units to the right and down 1 unit from $(0, 4)$ gives the point $(2, 3)$ on the line. A third point (for a check) is plotted at $(4, 2)$. The graph is shown in Figure 1.19. ∎

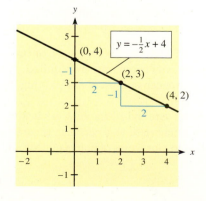

Figure 1.19

It is also possible to graph a straight line if we know its slope and any point on the line; we simply plot the point that is given and then use the slope to plot other points. The following summarizes the forms of equations of lines.

Forms of Linear Equations	
	General form: $ax + by + c = 0$
	Point-slope form: $y - y_1 = m(x - x_1)$
	Slope-intercept form: $y = mx + b$
	Vertical line: $x = a$
	Horizontal line: $y = b$

CHECKPOINT

4. Write the equation of the line that has slope $-\frac{3}{4}$ and passes through $(4, -6)$.
5. What are the slope and y-intercept of the graph of $x = -4y + 1$?

CHECKPOINT SOLUTIONS

1. $m = \dfrac{y_2 - y_1}{x_2 - x_1} = \dfrac{-6 - (6)}{28 - (4)} = \dfrac{-12}{24} = -\dfrac{1}{2}$

2. If $m = 0$, then the line is horizontal. If m is undefined, then the line is vertical.

3. (a) $m_2 = \dfrac{-1}{m_1} = -\dfrac{1}{5}$ (b) $m_2 = m_1 = 5$

4. Use $y - y_1 = m(x - x_1)$. Hence $y - (-6) = -\frac{3}{4}(x - 4)$, $y + 6 = -\frac{3}{4}x + 3$, or $y = -\frac{3}{4}x - 3$.

5. If $x = -4y + 1$, then the slope-intercept form is $y = mx + b$ or $y = -\frac{1}{4}x + \frac{1}{4}$. Hence $m = -\frac{1}{4}$ and the y-intercept is $\frac{1}{4}$.

| EXERCISES | 1.3

Access end-of-section exercises online at **www.webassign.net**

OBJECTIVES 1.4

- To use a graphing utility to graph equations in the standard viewing window
- To use a graphing utility and a specified range to graph equations
- To use a graphing utility to evaluate functions, to find intercepts, and to find zeros of a function
- To solve linear equations with a graphing utility

Graphs and Graphing Utilities

| APPLICATION PREVIEW |

Suppose that for a certain city the cost C of obtaining drinking water with p percent impurities (by volume) is given by

$$C = \frac{120{,}000}{p} - 1200$$

where $0 < p \le 100$. A graph of this equation would illustrate how the purity of water is related to the cost of obtaining it. Because this equation is not linear, graphing it would require many more than two points. Such a graph could be efficiently obtained with a graphing utility. (See Example 1.)

In Section 1.3, "Linear Functions," we saw that a linear equation could easily be graphed by plotting points because only two points are required. When we want the graph of an equation that is not linear, we can still use point plotting, but we must have enough points to sketch an accurate graph.

Some computer software and all graphing calculators have **graphing utilities** (also called *graphics utilities*) that can be used to generate an accurate graph. All graphing utilities use the point-plotting method to plot scores of points quickly and thereby graph an equation.

The computer monitor or calculator screen consists of a fine grid that looks like a piece of graph paper. In this grid, each tiny area is called a *pixel*. Essentially, a pixel corresponds to a point on a piece of graph paper, and as a graphing utility plots points, the pixels corresponding to the points are lighted and then connected, revealing the graph. The graph is shown in a *viewing window* or *viewing rectangle*. The values that define the viewing window or calculator range can be set individually or by using ZOOM keys. The important values are

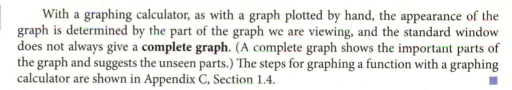

x-min: the smallest value on the *x*-axis
(the leftmost *x*-value in the window)

x-max: the largest value on the *x*-axis
(the rightmost *x*-value in the window)

y-min: the smallest value on the *y*-axis
(the lowest *y*-value in the window)

y-max: the largest value on the *y*-axis
(the highest *y*-value in the window)

Most graphing utilities have a *standard viewing window* that gives a window with *x*-values and *y*-values between −10 and 10.

$$x\text{-min: } -10 \qquad y\text{-min: } -10$$
$$x\text{-max: } \ \ 10 \qquad y\text{-max: } \ \ 10$$

The graph of $y = \frac{1}{3}x^3 - x^2 - 3x + 2$ with the standard viewing window is shown in Figure 1.20.

Figure 1.20

Calculator Note

With a graphing calculator, as with a graph plotted by hand, the appearance of the graph is determined by the part of the graph we are viewing, and the standard window does not always give a **complete graph**. (A complete graph shows the important parts of the graph and suggests the unseen parts.) The steps for graphing a function with a graphing calculator are shown in Appendix C, Section 1.4. ◼

EXAMPLE 1 Water Purity | APPLICATION PREVIEW |

For a certain city, the cost *C* of obtaining drinking water with *p* percent impurities (by volume) is given by

$$C = \frac{120{,}000}{p} - 1200$$

The equation for *C* requires that $p \neq 0$, and because *p* is the percent impurities, we know that $0 < p \leq 100$. Use the restriction on *p* and a graphing calculator to obtain an accurate graph of the equation.

Solution

To use a graphing calculator, we identify *C* with *y* and *p* with *x*. Thus we enter

$$y = \frac{120{,}000}{x} - 1200$$

Because $0 < p \leq 100$, we set the *x*-range to *include* these values, with the realization that only the portion of the graph above these values applies to our equation and that $p = 0$ (that is, $x = 0$) is excluded from the domain because it makes the denominator 0. With this *x*-range we can determine a *y*-, or *C*-, range from the equation or by tracing; see Figure 1.21(a) on the next page. We see that when *p* is near 0 (the value excluded from the domain), the *C*-coordinates of the points are very large, indicating that water free of impurities is very costly. However, Figure 1.21(a) does not accurately show what happens for large *p*-values. Figure 1.21(b) on the next page shows another view of the equation with the *C*-range (that is, the *y*-range) from −500 to 3500. We see the *p*-intercept is $p = 100$, which indicates that water containing 100% impurities costs nothing ($C = 0$). Note that $p > 100$ has no meaning. Why? ◼

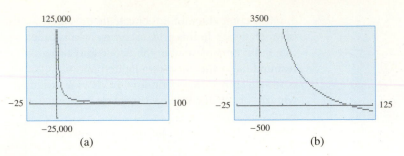

Figure 1.21   Two views of $C = \dfrac{120{,}000}{p} - 1200$

CHECKPOINT

Use a graphing calculator and the standard viewing window to graph the following.

1. $x^2 + 4y = 0$ 2. $y = \dfrac{4(x+1)^2}{x^2 + 1}$

We can evaluate a function $y = f(x)$ with a graphing calculator by using VALUE or TABLE to find the function values corresponding to values of the independent variable. We can also use TRACE to evaluate or approximate function values.

EXAMPLE 2   Women in the Work Force

The number y (in millions) of women in the work force is modeled by the function

$$y = -0.005x^2 + 1.38x + 0.07$$

where x is the number of years past 1940 (*Source:* Bureau of Labor Statistics).

(a) Graph this function from $x = 0$ to $x = 85$.
(b) Find the value of y when $x = 60$. Explain what this means.
(c) Use the model to predict the number of women in the work force in 2015.

Solution
(a) The graph is found by entering the equation with $x = 0$ to $x = 85$ in the window. Using TRACE for x-values between 0 and 85 helps determine that y-values between 0 and 85 give a complete graph. The graph, using this window, is shown in Figure 1.22(a).
(b) By using TRACE or TABLE, we can find $f(60) = 64.87$ (see Figure 1.22(b) and Figure 1.22(c)). This means that there were approximately 64.87 million, or 64,870,000, women in the work force in $1940 + 60 = 2000$.
(c) The year 2015 corresponds to $x = 75$ and $f(75) = 75.445$ (see Figure 1.22(c)). Thus in 2015, approximately 75,445,000 women are predicted to be in the work force. ■

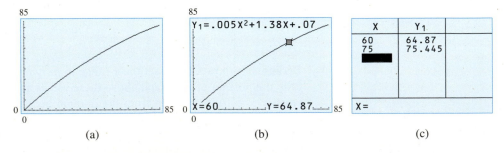

Figure 1.22   (a)   (b)   (c)

 Spreadsheet Note   The ability to represent quantities or their interrelationships algebraically is one of the keys to using spreadsheets. For details regarding spreadsheets, see Appendix D and the Online Excel Guide that accompanies this text. ■

Each cell in a spreadsheet has an address based on its row and column (see Table 1.3). These cell addresses can act like variables in an algebraic expression, and the "fill

| TABLE 1.3 |

	A	B
1	x	f(x) = 6x −3
2	−5	−33
3	−2	−15
4	−1	−9
5	0	−3
6	1	3
7	3	15
8	5	27

down" or "fill across" capabilities update this cell referencing while maintaining algebraic relationships.

We can use Excel to find the outputs of a function for given inputs. For details on entering data in a spreadsheet and evaluating functions with Excel, see Appendix D, Section 1.4. Table 1.3 shows a spreadsheet for the function $f(x) = 6x - 3$ for the input set $\{-5, -2, -1, 0, 1, 3, 5\}$, with these inputs listed as entries in the first column (column A). If -5 is in cell A2, then $f(-5)$ can be found in cell B2 by typing $= 6*A2 - 3$ in cell B2. By using the fill-down capability, we can obtain all the function values shown in column B. Excel can also graph this function (see Figure 1.23). (See Appendix D, Section 1.4, and the Online Excel Guide for details.)

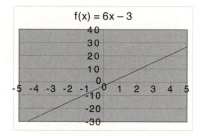

Figure 1.23

Graphical Solutions of Equations

We can use graphing utilities to find the x-intercepts of graphs of functions. Because an x-intercept is an x-value that makes the function equal to 0, the x-intercept is also a zero of the function. Thus if an equation is written in the form $0 = f(x)$, the x-intercept of $y = f(x)$ is a solution of the equation. This method of solving an equation is called the **x-intercept method.**

Calculator Note

Solving an Equation Using the x-Intercept Method with a Graphing Calculator

1. Rewrite the equation to be solved with 0 (and nothing else) on one side of the equation.
2. Enter the nonzero side of the equation found in the previous step in the equation editor of your graphing calculator, and graph the equation in an appropriate viewing window. Be certain that you can see the graph cross the horizontal axis.
3. Find or approximate the x-intercept by inspection with TRACE or by using the graphical solver that is often called ZERO or ROOT. The x-intercept is the value of x that makes the equation equal to zero, so it is the solution to the equation. The value of x found by this method is often displayed as a decimal approximation of the exact solution rather than as the exact solution.
4. Verify the solution in the original equation.

See Appendix C, Section 1.4, for more details for using the x-intercept method with a graphing calculator.

EXAMPLE 3 ### Solving an Equation with a Graphing Calculator

Graphically solve $5x + \dfrac{1}{2} = 7x - 8$ for x.

Solution

To solve this equation, we rewrite the equation with 0 on one side.

$$0 = 7x - 8 - \left(5x + \frac{1}{2}\right)$$

Graphing $y = 7x - 8 - (5x + \frac{1}{2})$ and finding the x-intercept gives x approximately equal to $4.25 = \frac{17}{4}$ (see Figure 1.24 on the next page). Checking $x = \frac{17}{4}$ in the original equation, or solving the equation analytically, shows that $x = \frac{17}{4}$ is the exact solution.

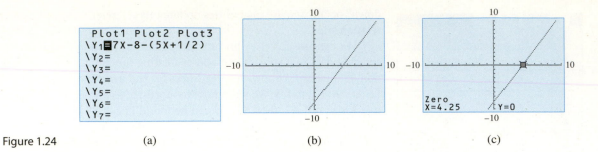

Figure 1.24   (a)   (b)   (c)

Spreadsheet Note

We can use Excel to solve the equation $0 = 5x - 8$ by finding the value of x that makes $f(x) = 5x - 8$ equal to 0—that is, by finding the zero of $f(x) = 5x - 8$. To find the zero of $f(x)$ with Excel, we enter the formula for $f(x)$ in the worksheet and use Goal Seek.

	A	B
1	x	f(x) = 5x – 8
2	1	= 5*A2 – 8

Under the goal set menu, setting cell B2 to the value 0 and choosing A2 as the changing cell gives $x = 1.6$, which is the solution to the equation. (See Appendix D, Section 1.4, and the Online Excel Guide for details.)

	A	B
1	x	f(x) = 5x – 8
2	1.6	0

CHECKPOINT SOLUTIONS

1. First solve $x^2 + 4y = 0$ for y.

    $$4y = -x^2$$

    $$y = \frac{-x^2}{4}$$

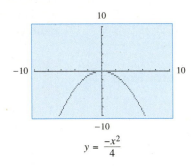

$$y = \frac{-x^2}{4}$$

2. The equation $y = \dfrac{4(x + 1)^2}{x^2 + 1}$ can be entered directly, but some care with parentheses is needed. Enter

    $$y_1 = (4(x + 1)^2)/(x^2 + 1)$$

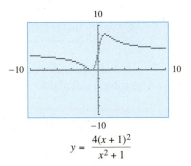

$$y = \frac{4(x + 1)^2}{x^2 + 1}$$

| EXERCISES | 1.4

OBJECTIVES   1.5

- To solve systems of linear equations in two variables by graphing
- To solve systems of linear equations by substitution
- To solve systems of linear equations by elimination
- To solve systems of three linear equations in three variables

Solutions of Systems of Linear Equations

▌| APPLICATION PREVIEW |▌

Suppose that a person has $200,000 invested, part at 9% and part at 8%, and that the yearly income from the two investments is $17,200. If x represents the amount invested at 9% and y represents the amount invested at 8%, then to find how much is invested at each rate we must find the values of x and y that satisfy both

$$x + y = 200{,}000 \quad \text{and} \quad 0.09x + 0.08y = 17{,}200$$

(See Example 3.)
Methods of solving systems of linear equations are discussed in this section.

Graphical Solution   In the previous sections, we graphed linear equations in two variables and observed that the graphs are straight lines. Each point on the graph represents an ordered pair of values (x, y) that satisfies the equation, so a point of intersection of two (or more) lines represents a solution to both (or all) the equations of those lines.

The equations are referred to as a **system of equations,** and the ordered pairs (x, y) that satisfy all the equations are the **solutions** (or **simultaneous solutions**) of the system. We can use graphing to find the solution of a system of equations.

EXAMPLE   1   Graphical Solution of a System

Use graphing to find the solution of the system

$$\begin{cases} 4x + 3y = 11 \\ 2x - 5y = -1 \end{cases}$$

Solution

The graphs of the two equations intersect (meet) at the point $(2, 1)$. (See Figure 1.25.) The solution of the system is $x = 2$, $y = 1$. Note that these values satisfy both equations. ■

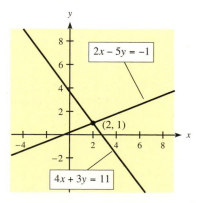

Figure 1.25

Two distinct nonparallel lines:
one solution

If the graphs of two equations are parallel lines, they have no point in common, and thus the system has no solution. Such a system of equations is called **inconsistent.** For example,

$$\begin{cases} 4x + 3y = 4 \\ 8x + 6y = 18 \end{cases}$$

is an **inconsistent system** (see Figure 1.26(a)).

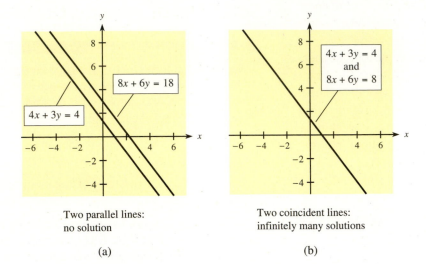

Two parallel lines: Two coincident lines:
no solution   infinitely many solutions

Figure 1.26   (a)   (b)

It is also possible that two equations describe the same line. When this happens, the equations are equivalent, and values that satisfy either equation are solutions of the system. For example,

$$\begin{cases} 4x + 3y = 4 \\ 8x + 6y = 8 \end{cases}$$

is called a **dependent system** because all points that satisfy one equation also satisfy the other (see Figure 1.26(b)).

Figures 1.25, 1.26(a), and 1.26(b) represent the three possibilities that can occur when we are solving a system of two linear equations in two variables.

Solution by Substitution   Graphical solution methods may yield only approximate solutions to some systems. Exact solutions can be found using algebraic methods, which are based on the fact that equivalent systems result when any of the following operations is performed.

Equivalent Systems

Equivalent systems result when

1. One expression is replaced by an equivalent expression.
2. Two equations are interchanged.
3. A multiple of one equation is added to another equation.
4. An equation is multiplied by a nonzero constant.

The **substitution method** is based on operation (1).

Substitution Method for Solving Systems

Procedure	Example
To solve a system of two equations in two variables by substitution:	Solve the system $\begin{cases} 2x + 3y = 4 \\ x - 2y = 3 \end{cases}$
1. Solve one of the equations for one of the variables in terms of the other.	1. Solving $x - 2y = 3$ for x gives $x = 2y + 3$.
2. Substitute this expression into the other equation to give one equation in one unknown.	2. Replacing x by $2y + 3$ in $2x + 3y = 4$ gives $2(2y + 3) + 3y = 4$.
3. Solve this linear equation for the unknown.	3. $4y + 6 + 3y = 4$ $$7y = -2 \Rightarrow y = -\frac{2}{7}$$
4. Substitute this solution into the equation in Step 1 or into one of the original equations to solve for the other variable.	4. $x = 2\left(-\dfrac{2}{7}\right) + 3 \Rightarrow x = \dfrac{17}{7}$
5. Check the solution by substituting for x and y in both original equations.	5. $2\left(\dfrac{17}{7}\right) + 3\left(-\dfrac{2}{7}\right) = 4 \checkmark$ $$\frac{17}{7} - 2\left(-\frac{2}{7}\right) = 3 \checkmark$$

EXAMPLE 2   Solution by Substitution

Solve the system

$$\begin{cases} 4x + 5y = 18 \quad (1) \\ 3x - 9y = -12 \quad (2) \end{cases}$$

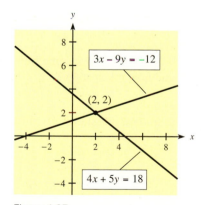

Figure 1.27

Solution

1. $x = \dfrac{9y - 12}{3} = 3y - 4$ Solve for x in equation (2).

2. $4(3y - 4) + 5y = 18$ Substitute for x in equation (1).

3. $12y - 16 + 5y = 18$ Solve for y.
$$17y = 34$$
$$y = 2$$

4. $x = 3(2) - 4$ Use $y = 2$ to find x.
$$x = 2$$

5. $4(2) + 5(2) = 18$ and $3(2) - 9(2) = -12$ $\checkmark$ Check.

Thus the solution is $x = 2$, $y = 2$. This means that when the two equations are graphed simultaneously, their point of intersection is $(2, 2)$. See Figure 1.27. ◼

Solution by Elimination   We can also eliminate one of the variables in a system by the **elimination method,** which uses addition or subtraction of equations.

Elimination Method for Solving Systems

Procedure	Example

To solve a system of two equations in two variables by the elimination method:

Solve the system $\begin{cases} 2x - 5y = 4 & (1) \\ x + 2y = 3 & (2) \end{cases}$

1. If necessary, multiply one or both equations by a non-zero number that will make the coefficients of one of the variables identical, except perhaps for signs.

1. Multiply equation (2) by -2.

$$2x - 5y = 4$$
$$-2x - 4y = -6$$

2. Add or subtract the equations to eliminate one of the variables.

2. Adding gives $0x - 9y = -2$

3. Solve for the variable in the resulting equation.

3. $y = \dfrac{2}{9}$

4. Substitute the solution into one of the original equations and solve for the other variable.

4. $2x - 5\left(\dfrac{2}{9}\right) = 4$

$$2x = 4 + \dfrac{10}{9} = \dfrac{36}{9} + \dfrac{10}{9}$$

$$2x = \dfrac{46}{9} \quad \text{so} \quad x = \dfrac{23}{9}$$

5. Check the solutions in both original equations.

5. $2\left(\dfrac{23}{9}\right) - 5\left(\dfrac{2}{9}\right) = 4$ ✔

$$\dfrac{23}{9} + 2\left(\dfrac{2}{9}\right) = 3 ✔$$

EXAMPLE 3   Investment Mix | APPLICATION PREVIEW |

A person has $200,000 invested, part at 9% and part at 8%. If the total yearly income from the two investments is $17,200, how much is invested at 9% and how much at 8%?

Solution

If x represents the amount invested at 9% and y represents the amount invested at 8%, then $x + y$ is the total investment.

$$x + y = 200,000 \qquad (1)$$

and $0.09x + 0.08y$ is the total income earned.

$$0.09x + 0.08y = 17,200 \qquad (2)$$

We solve these equations as follows:

$$
\begin{array}{lll}
-8x - 8y = -1,600,000 & (3) & \text{Multiply equation (1) by } -8. \\
\underline{9x + 8y = 1,720,000} & (4) & \text{Multiply equation (2) by 100.} \\
x = 120,000 & & \text{Add (3) and (4).}
\end{array}
$$

We find y by using $x = 120,000$ in equation (1).

$$120,000 + y = 200,000 \quad \text{gives} \quad y = 80,000$$

Thus $120,000 is invested at 9%, and $80,000 is invested at 8%.

As a check, we note that equation (1) is satisfied and

$$0.09(120{,}000) + 0.08(80{,}000) = 10{,}800 + 6400 = 17{,}200 \checkmark \quad \blacksquare$$

EXAMPLE 4   Solution by Elimination

Solve the systems:

(a) $\begin{cases} 4x + 3y = 4 \\ 8x + 6y = 18 \end{cases}$ (b) $\begin{cases} 4x + 3y = 4 \\ 8x + 6y = 8 \end{cases}$

Solution

(a) $\begin{cases} 4x + 3y = 4 \\ 8x + 6y = 18 \end{cases}$ Multiply by –2 to get: $\quad -8x - 6y = -8$

Leave as is, which gives: $\quad \underline{8x + 6y = 18}$

Add the equations to get: $\quad 0x + 0y = 10$

$$0 = 10$$

The system is solved when $0 = 10$. This is impossible, so there are no solutions of the system. The equations are inconsistent. Their graphs are parallel lines; see Figure 1.26(a) earlier in this section.

(b) $\begin{cases} 4x + 3y = 4 \\ 8x + 6y = 8 \end{cases}$ Multiply by –2 to get: $\quad -8x - 6y = -8$

Leave as is, which gives: $\quad \underline{8x + 6y = 8}$

Add the equations to get: $\quad 0x + 0y = 0$

$$0 = 0$$

This is an identity, so the two equations share infinitely many solutions. The equations are dependent. Their graphs coincide, and each point on this graph represents a solution of the system; see Figure 1.26(b) earlier in this section. $\blacksquare$

EXAMPLE 5   Medicine Concentrations

A nurse has two solutions that contain different concentrations of a certain medication. One is a 12.5% concentration and the other is a 5% concentration. How many cubic centimeters of each should she mix to obtain 20 cubic centimeters of an 8% concentration?

Solution

Let x equal the number of cubic centimeters of the 12.5% solution, and let y equal the number of cubic centimeters of the 5% solution. The total amount of substance is

$$x + y = 20$$

and the total amount of medication is

$$0.125x + 0.05y = (0.08)(20) = 1.6$$

Solving this pair of equations simultaneously gives

$$50x + 50y = 1000$$
$$\underline{-125x - 50y = -1600}$$
$$-75x = -600$$
$$x = 8$$
$$8 + y = 20, \text{ so } y = 12$$

Thus 8 cubic centimeters of a 12.5% concentration and 12 cubic centimeters of a 5% concentration yield 20 cubic centimeters of an 8% concentration.

Checking, we see that $8 + 12 = 20$ and

$$0.125(8) + 0.05(12) = 1 + 0.6 = 1.6 \checkmark \quad \blacksquare$$

CHECKPOINT

1. Solve by substitution: $\begin{cases} 3x - 4y = -24 \\ x + y = -1 \end{cases}$

2. Solve by elimination: $\begin{cases} 2x + 3y = 5 \\ 3x + 5y = -25 \end{cases}$

3. In Problems 1 and 2, the solution method is given.
   (a) In each case, explain why you think that method is appropriate.
   (b) In each case, would the other method work as well? Explain.

Calculator Note

Solutions of systems of equations, if they exist, can be found by using INTERSECT. INTERSECT will give the point of intersection exactly or approximately to a large number of significant digits. Details of solving systems of linear equations are shown in Appendix C, Section 1.5. Figure 1.28 shows that the solution to the following system is $x = 2$, $y = 3$.

$\begin{cases} 3x + 2y = 12 \\ 4x - 3y = -1 \end{cases}$

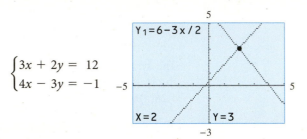

Figure 1.28

Spreadsheet Note

The details of how to use Goal Seek with Excel to solve a system of equations are shown in Appendix D, Section 1.5. The following spreadsheets show the input formulas and solution $(x = 2, y = 3)$ for the system solved in Figure 1.28.

	A	B	C	D
1	x	$= 6 - 1.5x$	$= 1/3 + 4x/3$	$= y1 - y2$
2	1	$= 6 - 1.5*A2$	$= 1/3 + 4*A2/3$	$= B2 - C2$

	A	B	C	D
1	x	$= 6 - 1.5x$	$= 1/3 + 4x/3$	$= y1 - y2$
2	2	3	3	0

EXAMPLE 6   College Enrollment

Suppose the percent of males who enrolled in college within 12 months of high school graduation is given by

$$y = -0.126x + 55.72$$

and the percent of females who enrolled in college within 12 months of high school graduation is given by

$$y = 0.73x + 39.7$$

where x is the number of years past 1960 (*Source: Statistical Abstract of the United States*). Graphically find the year these models indicate that the percent of females will equal the percent of males.

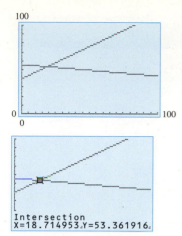

Figure 1.29

Solution

Entering the two functions in a graphing utility and setting the window with x-min $= 0$, x-max $= 100$, y-min $= 0$, and y-max $= 100$ gives the graph in Figure 1.29(a). Using INTERSECT (see Figure 1.29(b)) gives a point of intersection of these two graphs at approximately $x = 18.71$, $y = 53.36$. Thus the percent of females will reach the percent of males in 1979. The models indicate that in 1979, the male percent was 53.33% and the female percent was 53.57%. ■

Solving a system of equations by graphing, whether by hand or with a graphing utility, is limited by two factors. (1) It may be difficult to determine a viewing window that contains the point of intersection, and (2) the solution may be only approximate. With *some* systems of equations, the only practical method may be graphical approximation.

However, systems of *linear* equations can be consistently and accurately solved with algebraic methods. Computer algebra systems, some software packages (including spreadsheets), and some graphing calculators have the capability of solving systems of linear equations.

Three Equations in Three Variables

If a, b, c, and d represent constants, then

$$ax + by + cz = d$$

is a first-degree (linear) equation in three variables. When equations of this form are graphed in a three-dimensional coordinate system, their graphs are planes. Two different planes may intersect in a line (like two walls) or may not intersect at all (like a floor and ceiling). Three different planes may intersect in a single point (as when two walls meet the ceiling), may intersect in a line (as in a paddle wheel), or may not have a common intersection. (See Figures 1.30, 1.31, and 1.32.) Thus three linear equations in three variables may have a unique solution, infinitely many solutions, or no solution. For example, the solution of the system

$$\begin{cases} 3x + 2y + z = 6 \\ x - y - z = 0 \\ x + y - z = 4 \end{cases}$$

is $x = 1$, $y = 2$, $z = -1$, because these three values satisfy all three equations, and these are the only values that satisfy them. In this section, we will discuss only systems of three linear equations in three variables that have unique solutions. Additional systems will be discussed in Section 3.3, "Gauss-Jordan Elimination: Solving Systems of Equations."

We can solve three equations in three variables using a systematic procedure called the **left-to-right elimination method.**

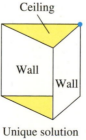

Ceiling

Wall

Wall

Unique solution

Figure 1.30

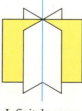

Infinitely many solutions

Figure 1.31

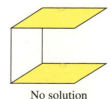

No solution

Figure 1.32

Left-to-Right Elimination Method

Procedure	Example

To solve a system of three equations in three variables by the left-to-right elimination method:

Solve: $\begin{cases} 2x + 4y + 5z = 4 \\ x - 2y - 3z = 5 \\ x + 3y + 4z = 1 \end{cases}$

1. If necessary, interchange two equations or use multiplication to make the coefficient of the first variable in equation (1) a factor of the other first variable coefficients.

1. Interchange the first two equations.

$$\begin{array}{rl} x - 2y - 3z = 5 & (1) \\ 2x + 4y + 5z = 4 & (2) \\ x + 3y + 4z = 1 & (3) \end{array}$$

2. Add multiples of the first equation to each of the following equations so that the coefficients of the first variable in the second and third equations become zero.

2. Add $(-2) \times$ equation (1) to equation (2) and add $(-1) \times$ equation (1) to equation (3).

$$\begin{array}{rl} x - 2y - 3z = 5 & (1) \\ 0x + 8y + 11z = -6 & (2) \\ 0x + 5y + 7z = -4 & (3) \end{array}$$

3. Add a multiple of the second equation to the third equation so that the coefficient of the second variable in the third equation becomes zero.

3. Add $\left(-\frac{5}{8}\right) \times$ equation (2) to equation (3).

$$\begin{array}{rl} x - 2y - 3z = 5 & (1) \\ 8y + 11z = -6 & (2) \\ 0y + \dfrac{1}{8}z = -\dfrac{2}{8} & (3) \end{array}$$

4. Solve the third equation and *back substitute* from the bottom to find the remaining variables.

4. $z = -2$ from equation (3)

$y = \dfrac{1}{8}(-6 - 11z) = 2$ from equation (2)

$x = 5 + 2y + 3z = 3$ from equation (1)

so $x = 3, y = 2, z = -2$

EXAMPLE 7   Left-to-Right Elimination

Solve: $\begin{cases} x + 2y + 3z = 6 & (1) \\ 2x + 3y + 2z = 6 & (2) \\ -x + y + z = 4 & (3) \end{cases}$

Solution

Using equation (1) to eliminate x from the other equations gives the equivalent system

$$\begin{cases} x + 2y + 3z = 6 & (1) \\ -y - 4z = -6 & (2) \quad \text{(−2) × equation (1) added to equation (2)} \\ 3y + 4z = 10 & (3) \quad \text{Equation (1) added to equation (3)} \end{cases}$$

Using equation (2) to eliminate y from equation (3) gives

$$\begin{cases} x + 2y + 3z = 6 & (1) \\ -y - 4z = -6 & (2) \\ -8z = -8 & (3) \quad \text{(3) × equation (2) added to equation (3)} \end{cases}$$

In a system of equations such as this, the first variable appearing in each equation is called the **lead variable.** Solving for each lead variable gives

$$x = 6 - 2y - 3z$$
$$y = 6 - 4z$$
$$z = 1$$

and using **back substitution** from the bottom gives

$$z = 1$$
$$y = 6 - 4 = 2$$
$$x = 6 - 4 - 3 = -1$$

Hence the solution is $x = -1, y = 2, z = 1$.   ■

Although other methods for solving systems of equations in three variables may be useful, the left-to-right elimination method is important because it is systematic and can easily be extended to larger systems and to systems solved with matrices. (See Section 3.3, "Gauss-Jordan Elimination: Solving Systems of Equations.")

CHECKPOINT

4. Use left-to-right elimination to solve.

$$\begin{cases} x - y - z = & 0 & (1) \\ y - 2z = -18 & (2) \\ x + y + z = & 6 & (3) \end{cases}$$

CHECKPOINT SOLUTIONS

1. $x + y = -1$ means $x = -y - 1$, so $3x - 4y = -24$ becomes

$$3(-y - 1) - 4y = -24$$
$$-3y - 3 - 4y = -24$$
$$-7y = -21$$
$$y = 3$$

Hence $x = -3 - 1 = -4$, and the solution is $x = -4, y = 3$.

2. Multiply the first equation by (-3) and the second by (2). Add the resulting equations to eliminate the x-variable and solve for y.

$$-6x - 9y = -15$$
$$\underline{6x + 10y = -50}$$
$$y = -65$$

Hence $2x + 3(-65) = 5$ or $2x - 195 = 5$, so $2x = 200$. Thus $x = 100$ and $y = -65$.

3. (a) Substitution works well in Problem 1 because it is easy to solve for x (or for y). Substitution would not work well in Problem 2 because solving for x (or for y) would introduce fractions.

    (b) Elimination would work well in Problem 1 if we multiplied the first equation by 4. As stated in (a), substitution would not be as easy in Problem 2.

4.
$$\begin{cases} x - y - z = & 0 & (1) \\ y - 2z = -18 & (2) \\ x + y + z = & 6 & (3) \end{cases}$$

Add $(-1) \times$ equation (1) to equation (3).

$$\begin{cases} x - y - z = & 0 & (1) \\ y - 2z = -18 & (2) \\ 2y + 2z = & 6 & (3) \end{cases}$$

Add $(-2) \times$ equation (2) to equation (3).

$$\begin{cases} x - y - z = & 0 \\ y - 2z = -18 \\ 6z = & 42 \end{cases}$$

Solve the equations for x, y, and z.

$$x = y + z$$
$$y = 2z - 18$$
$$z = 7$$

Back substitution gives $y = 14 - 18 = -4$ and $x = -4 + 7 = 3$. Thus the solution is $x = 3, y = -4, z = 7$.

| EXERCISES | 1.5

Access end-of-section exercises online at **www.webassign.net**

OBJECTIVES 1.6

- To formulate and evaluate total cost, total revenue, and profit functions
- To find marginal cost, revenue, and profit, given linear total cost, total revenue, and profit functions
- To write the equations of linear total cost, total revenue, and profit functions by using information given about the functions
- To find break-even points
- To evaluate and graph supply and demand functions
- To find market equilibrium

Applications of Functions in Business and Economics

| APPLICATION PREVIEW |

Suppose a firm manufactures MP3 players and sells them for $50 each, with costs incurred in the production and sale equal to $200,000 plus $10 for each unit produced and sold. Forming the total cost, total revenue, and profit as functions of the quantity x that is produced and sold (see Example 1) is called the theory of the firm. We will also discuss market analysis, in which supply and demand are found as functions of price, and market equilibrium is found.

Total Cost, Total Revenue, and Profit

The **profit** a firm makes on its product is the difference between the amount it receives from sales (its revenue) and its cost. If x units are produced and sold, we can write

$$P(x) = R(x) - C(x)$$

where

$$P(x) = \text{profit from sale of } x \text{ units}$$
$$R(x) = \text{total revenue from sale of } x \text{ units}$$
$$C(x) = \text{total cost of production and sale of } x \text{ units*}$$

In general, **revenue** is found by using the equation

$$\text{Revenue} = (\text{price per unit})(\text{number of units})$$

The **cost** is composed of two parts: fixed costs and variable costs. **Fixed costs** (FC), such as depreciation, rent, utilities, and so on, remain constant regardless of the number of units produced. **Variable costs** (VC) are those directly related to the number of units produced. Thus the cost is found by using the equation

$$\text{Cost} = \text{variable costs} + \text{fixed costs}$$

EXAMPLE 1 Cost, Revenue, and Profit | APPLICATION PREVIEW |

Suppose a firm manufactures MP3 players and sells them for $50 each. The costs incurred in the production and sale of the MP3 players are $200,000 plus $10 for each player produced and sold. Write the profit function for the production and sale of x players.

*The symbols generally used in economics for total cost, total revenue, and profit are TC, TR, and π, respectively. In order to avoid confusion, especially with the use of π as a variable, we do not use these symbols.

Solution

The total revenue for x MP3 players is $50x$, so the total revenue function is $R(x) = 50x$. The fixed costs are \$200,000, so the total cost for x players is $10x + 200,000$. Hence, $C(x) = 10x + 200,000$. The profit function is given by $P(x) = R(x) - C(x)$. Thus,

$$P(x) = 50x - (10x + 200,000)$$
$$P(x) = 40x - 200,000$$

Figure 1.33 shows the graphs of $R(x)$, $C(x)$, and $P(x)$. ■

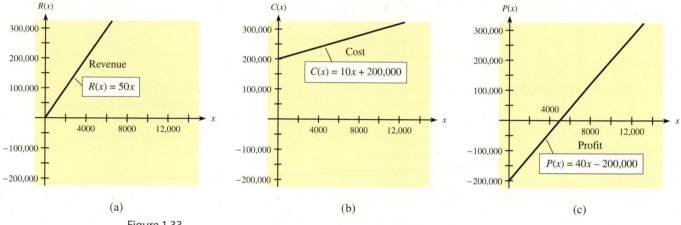

(a)   (b)   (c)

Figure 1.33

By observing the intercepts on the graphs in Figure 1.33, we note the following.

Revenue: 0 units produce 0 revenue; $R(0) = 0$.
Cost: 0 units' costs equal fixed costs = \$200,000; $C(0) = FC = 200,000$.
Profit: 0 units yield a loss equal to fixed costs = \$200,000;
$P(0) = -FC = -200,000$.
5000 units result in a profit of \$0 (no loss or gain); $P(5000) = 0$.

Marginals   In Example 1, both the total revenue function and the total cost function are linear, so their difference, the profit function, is also linear. The slope of the profit function represents the rate of change in profit with respect to the number of units produced and sold. This is called the **marginal profit** $(\overline{MP})$ for the product. Thus the marginal profit for the MP3 players in Example 1 is \$40. Similarly, the **marginal cost** $(\overline{MC})$ for this product is \$10 (the slope of the cost function), and the **marginal revenue** $(\overline{MR})$ is \$50 (the slope of the revenue function).

EXAMPLE 2   Marginal Cost

Suppose that the cost (in dollars) for a product is $C = 21.75x + 4890$. What is the marginal cost for this product, and what does it mean?

Solution

The equation has the form $C = mx + b$, so the slope is 21.75. Thus the marginal cost is $\overline{MC} = 21.75$ dollars per unit.

Because the marginal cost is the slope of the cost line, production of each additional unit will cost \$21.75 more, at any level of production. ■

Note that when total cost functions are linear, the marginal cost is the same as the variable cost per unit. This is not the case if the functions are not linear, as we shall see later.

CHECKPOINT

1. Suppose that when a company produces its product, fixed costs are $12,500 and variable cost per item is $75.
   (a) Write the total cost function if x represents the number of units.
   (b) Are fixed costs equal to $C(0)$?
2. Suppose the company in Problem 1 sells its product for $175 per item.
   (a) Write the total revenue function.
   (b) Find $R(100)$ and give its meaning.
3. (a) Give the formula for profit in terms of revenue and cost.
   (b) Find the profit function for the company in Problems 1 and 2.

Break-Even Analysis   We can solve the equations for total revenue and total cost simultaneously to find the point where cost and revenue are equal. This point is called the **break-even point.** On the graph of these functions, we use x to represent the quantity produced and y to represent the dollar value of revenue *and* cost. The point where the total revenue line crosses the total cost line is the break-even point.

EXAMPLE 3   **Break-Even**

A manufacturer sells a product for $10 per unit. The manufacturer's variable costs are $2.50 per unit and the cost of 100 units is $1450. How many units must the manufacturer produce each month to break even?

Solution
The total revenue for x units of the product is $10x$, so the equation for total revenue is $R = R(x) = 10x$. The equation for total cost is

$$C - 1450 = 2.50(x - 100) \qquad \text{or} \qquad C = 2.50x + 1200$$

We find the break-even point by solving the two equations simultaneously ($R = C$ at the break-even point). By substitution,

$$10x = 2.50x + 1200$$
$$7.5x = 1200 \quad \text{so} \quad x = 160$$

Thus the manufacturer will break even if 160 units are produced per month. The manufacturer will make a profit if more than 160 units are produced. Figure 1.34 shows that for $x < 160$, $R(x) < C(x)$ (resulting in a loss) and that for $x > 160$, $R(x) > C(x)$ (resulting in a profit).

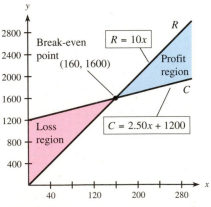

Figure 1.34

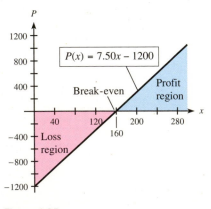

Figure 1.35

The profit function for Example 3 is given by

$$P(x) = R(x) - C(x) = 10x - (2.50x + 1200) \quad \text{or} \quad P(x) = 7.50x - 1200$$

We can find the point where the profit is zero (the break-even point) by setting $P(x) = 0$ and solving for x.

$$0 = 7.50x - 1200 \Rightarrow 1200 = 7.50x \Rightarrow x = 160$$

Note that this is the same break-even quantity that we found by solving the total revenue and total cost equations simultaneously (see Figure 1.35).

CHECKPOINT

4. Identify two ways in which break-even can be found.

Supply, Demand, and Market Equilibrium

Economists and managers also use points of intersection to determine market equilibrium. **Market equilibrium** occurs when the quantity of a commodity demanded is equal to the quantity supplied.

Demand by consumers for a commodity is related to the price of the commodity. The **law of demand** states that the quantity demanded will increase as price decreases and that the quantity demanded will decrease as price increases. The **law of supply** states that the quantity supplied for sale will increase as the price of a product increases. Note that although quantity demanded and quantity supplied are both functions of price p, economists traditionally graph these with price on the vertical axis. We will follow this tradition (see Figure 1.36).

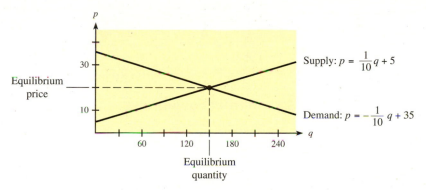

Figure 1.36

If the supply and demand curves for a commodity are graphed on the same coordinate system, with the same units, market equilibrium occurs at the point where the curves intersect. The price at that point is the **equilibrium price,** and the quantity at that point is the **equilibrium quantity.**

For the supply and demand functions shown in Figure 1.36, we see that the curves intersect at the point (150, 20). This means that when the price is $20, consumers are willing to purchase the same number of units (150) that producers are willing to supply.

In general, the equilibrium price and the equilibrium quantity must both be positive for the market equilibrium to have meaning.

We can find the market equilibrium by graphing the supply and demand functions on the same coordinate system and observing their point of intersection. As we have seen, finding the point(s) common to the graphs of two (or more) functions is called **solving a system of equations** or **solving simultaneously.**

EXAMPLE **4**   **Market Equilibrium**

Find the market equilibrium point for the following supply and demand functions.

$$\text{Demand:} \quad p = -3q + 36$$
$$\text{Supply:} \quad p = 4q + 1$$

Solution

At market equilibrium, the demand price equals the supply price. Thus,

$$-3q + 36 = 4q + 1$$
$$35 = 7q$$
$$q = 5$$
$$p = 21$$

The equilibrium point is (5, 21).

Checking, we see that

$$21 = -3(5) + 36 ✔ \quad \text{and} \quad 21 = 4(5) + 1 ✔$$ ■

Spreadsheet Note We can use Goal Seek with Excel to find the market equilibrium for given supply and demand functions, as we did to find solutions of systems in Section 1.5. The following spreadsheets show the Excel setup and solution for the market equilibrium problem of Example 4. ■

	A	B	C	D
1	q	p: demand	p: supply	demand − supply
2	1	$=-3*A2 + 36$	$=4*A2 + 1$	$=B2 - C2$

	A	B	C	D
1	q	p: demand	p: supply	demand − supply
2	5	21	21	0

EXAMPLE 5   **Market Equilibrium**

A group of wholesalers will buy 50 dryers per month if the price is $200 and 30 per month if the price is $300. The manufacturer is willing to supply 20 if the price is $210 and 30 if the price is $230. Assuming that the resulting supply and demand functions are linear, find the equilibrium point for the market.

Solution

Representing price by p and quantity by q, we can write the equations for the supply and demand functions as follows.

Demand function:

$$m = \frac{300 - 200}{30 - 50} = -5$$
$$p - 200 = -5(q - 50)$$
$$p = -5q + 450$$

Supply function:

$$m = \frac{230 - 210}{30 - 20} = 2$$
$$p - 230 = 2(q - 30)$$
$$p = 2q + 170$$

Because the prices are equal at market equilibrium, we have

$$-5q + 450 = 2q + 170$$
$$280 = 7q$$
$$q = 40$$
$$p = 250$$

The equilibrium point is (40, 250). See Figure 1.37 for the graphs of these functions. ■

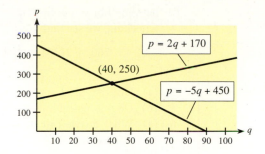

Figure 1.37

Supply and Demand with Taxation

Suppose a supplier is taxed $K per unit sold, and the tax is passed on to the consumer by adding $K to the selling price of the product. If the original supply function $p = f(q)$ gives the supply price per unit, then passing the tax on gives a new supply function, $p = f(q) + K$. Because the value of the product is not changed by the tax, the demand function is unchanged. Figure 1.38 shows the effect that this has on market equilibrium.

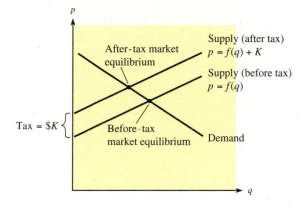

Figure 1.38

Note that the new market equilibrium point is the point of intersection of the original demand function and the new (after-tax) supply function.

EXAMPLE **6**   **Taxation**

In Example 5 the supply and demand functions for dryers were given as follows.

$$\text{Supply:} \quad p = 2q + 170$$
$$\text{Demand:} \quad p = -5q + 450$$

The equilibrium point was $q = 40$, $p = \$250$. If the wholesaler is taxed $14 per unit sold, what is the new equilibrium point?

Solution

The $14 tax per unit is passed on by the wholesaler, so the new supply function is

$$p = 2q + 170 + 14$$

and the demand function is unchanged. Thus we solve the system

$$\begin{cases} p = 2q + 184 \\ p = -5q + 450 \end{cases}$$
$$2q + 184 = -5q + 450$$
$$7q = 266$$
$$q = 38$$
$$p = 2(38) + 184 = 260$$

The new equilibrium point is $q = 38$, $p = \$260$.

Checking, we see that

$$260 = 2(38) + 184 \checkmark \qquad \text{and} \qquad 260 = -5(38) + 450 \checkmark \qquad \blacksquare$$

CHECKPOINT	5. (a) Does a typical linear demand function have positive slope or negative slope? Why? (b) Does a typical linear supply function have positive slope or negative slope? Why? 6. (a) What do we call the point of intersection of a supply function and a demand function? (b) If supply is given by $p = 0.1q + 20$ and demand is given by $p = 130 - 0.1q$, find the market equilibrium point.

CHECKPOINT SOLUTIONS

1. (a) $C(x) = 75x + 12{,}500$
   (b) Yes. $C(0) = 12{,}500 =$ Fixed costs. In fact, fixed costs are defined to be $C(0)$.
2. (a) $R(x) = 175x$
   (b) $R(100) = 175(100) = \$17{,}500$, which means that revenue is $17,500 when 100 units are sold.
3. (a) Profit = Revenue − Cost or $P(x) = R(x) - C(x)$
   (b) $P(x) = 175x - (75x + 12{,}500)$
   $$= 175x - 75x - 12{,}500 = 100x - 12{,}500$$
4. Break-even occurs where revenue equals cost $[R(x) = C(x)]$, or where profit is zero $[P(x) = 0]$.
5. (a) Negative slope, because demand falls as price increases.
   (b) Positive slope, because supply increases as price increases.
6. (a) Market equilibrium
   (b) Market equilibrium occurs when $0.1q + 20 = 130 - 0.1q$
   $$0.2q = 110 \text{ or } q = 550$$
   and $p = 0.1(550) + 20 = 75$.

| EXERCISES | 1.6

Access end-of-section exercises online at **www.webassign.net**

	KEY TERMS AND FORMULAS	
Section	**Key Terms**	**Formulas**

1.1	Equation, members; variable; solution
	 Identities; conditional equations
	 Properties of equality
	 Linear equation in one variable
	 Fractional equation
	 Linear equation in two variables
	 Aligning the data
	 Linear inequalities
	     Properties
	     Solutions	

1.2	Relation
	 Function
	     Vertical-line test
	     Domain, range	

Section	Key Terms	Formulas
	Coordinate system	
	Ordered pair, origin, x-axis, y-axis	
	Graph	
	Function notation	
	Operations with functions	
	Composite functions	$(f \circ g)(x) = f(g(x))$
1.3	Linear function	$y = ax + b$
	Intercepts	
	x-intercept (zero of a function)	where $y = 0$
	y-intercept	where $x = 0$
	Slope of a line	$m = \dfrac{y_2 - y_1}{x_2 - x_1}$
	Rate of change	
	Parallel lines	$m_1 = m_2$
	Perpendicular lines	$m_2 = -1/m_1$
	Point-slope form	$y - y_1 = m(x - x_1)$
	Slope-intercept form	$y = mx + b$
	Vertical line	$x = a$
	Horizontal line	$y = b$
1.4	Graphing utilities	
	Standard viewing window	
	Range	
	Evaluating functions	
	x-intercept	
	Zeros of functions	
	Solutions of equations	
	x-intercept method	
1.5	Systems of linear equations	
	Solutions	
	Graphical solutions	
	Equivalent systems	
	Substitution method	
	Elimination method	
	Left-to-right elimination method	
	Lead variable	
	Back substitution	
1.6	Cost and revenue functions	$C(x)$ and $R(x)$
	Profit functions	$P(x) = R(x) - C(x)$
	Marginal profit	Slope of linear profit function
	Marginal cost	Slope of linear cost function
	Marginal revenue	Slope of linear revenue function
	Break-even point	$C(x) = R(x)$　or　$P(x) = 0$
	Supply and demand functions	
	Market equilibrium	

REVIEW EXERCISES

Solve the equations in Problems 1–6.

1. $3x - 8 = 23$

2. $2x - 8 = 3x + 5$

3. $\dfrac{6x + 3}{6} = \dfrac{5(x - 2)}{9}$

4. $2x + \dfrac{1}{2} = \dfrac{x}{2} + \dfrac{1}{3}$

5. $\dfrac{6}{3x - 5} = \dfrac{6}{2x + 3}$

6. $\dfrac{2x + 5}{x + 7} = \dfrac{1}{3} + \dfrac{x - 11}{2x + 14}$

7. Solve for y: $3(y - 2) = -2(x + 5)$

8. Solve $3x - 9 \le 4(3 - x)$ and graph the solution.

9. Solve $\frac{2}{5}x \le x + 4$ and graph the solution.

10. Solve $5x + 1 \ge \frac{2}{3}(x - 6)$ and graph the solution.

11. If $p = 3q^3$, is p a function of q?

12. If $y^2 = 9x$, is y a function of x?

13. If $R = \sqrt[3]{x + 4}$, is R a function of x?

14. What are the domain and range of the function $y = \sqrt{9 - x}$?

15. If $f(x) = x^2 + 4x + 5$, find the following.

   (a) $f(-3)$ (b) $f(4)$ (c) $f\left(\dfrac{1}{2}\right)$

16. If $g(x) = x^2 + 1/x$, find the following.

   (a) $g(-1)$ (b) $g\left(\dfrac{1}{2}\right)$ (c) $g(0.1)$

17. If $f(x) = 9x - x^2$, find $\dfrac{f(x + h) - f(x)}{h}$ and simplify.

18. Does the graph in Figure 1.41 represent y as a function of x?

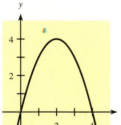

Figure 1.41

19. Does the graph in Figure 1.42 represent y as a function of x?

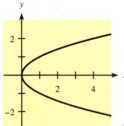

Figure 1.42

20. For the function f graphed in Figure 1.41, what is $f(2)$?

21. For the function f graphed in Figure 1.41, for what values of x does $f(x) = 0$?

22. The following table defines y as a function of x, denoted $y = f(x)$.

x	-2	-1	0	1	3	4
y	8	2	-3	4	2	7

Use the table to complete the following.
   (a) Identify the domain and range of $y = f(x)$.
   (b) Find $f(4)$.
   (c) Find all x-values for which $f(x) = 2$.
   (d) Graph $y = f(x)$.
   (e) Does the table define x as a function of y? Explain.

23. If $f(x) = 3x + 5$ and $g(x) = x^2$, find
   (a) $(f + g)(x)$ (b) $(f / g)(x)$
   (c) $f(g(x))$ (d) $(f \circ f)(x)$

In Problems 24–26, find the intercepts and graph.

24. $5x + 2y = 10$

25. $6x + 5y = 9$

26. $x = -2$

In Problems 27 and 28, find the slope of the line that passes through each pair of points.

27. $(2, -1)$ and $(-1, -4)$

28. $(-3.8, -7.16)$ and $(-3.8, 1.16)$

In Problems 29 and 30, find the slope and y-intercept of each line.

29. $2x + 5y = 10$

30. $x = -\dfrac{3}{4}y + \dfrac{3}{2}$

In Problems 31–37, write the equation of each line described.

31. Slope 4 and y-intercept 2

32. Slope $-\frac{1}{2}$ and y-intercept 3

33. Through $(-2, 1)$ with slope $\frac{2}{5}$

34. Through $(-2, 7)$ and $(6, -4)$

35. Through $(-1, 8)$ and $(-1, -1)$

36. Through $(1, 6)$ and parallel to $y = 4x - 6$

37. Through $(-1, 2)$ and perpendicular to $3x + 4y = 12$

In Problems 38 and 39, graph each equation with a graphing utility and the standard viewing window.

38. $x^2 + y - 2x - 3 = 0$

39. $y = \dfrac{x^3 - 27x + 54}{15}$

In Problems 40 and 41, use a graphing utility and
  (a) graph each equation in the viewing window given.
  (b) graph each equation in the standard viewing window.
  (c) explain how the two views differ and why.
40. $y = (x + 6)(x - 3)(x - 15)$ with x-min $= -15$, x-max $= 25$; y-min $= -700$, y-max $= 500$
41. $y = x^2 - x - 42$ with x-min $= -15$, x-max $= 15$; y-min $= -50$, y-max $= 50$

42. What is the domain of $y = \dfrac{\sqrt{x + 3}}{x}$? Check with a graphing utility.
43. Use a graphing utility and the x-intercept method to approximate the solution of $7x - 2 = 0$.

In Problems 44–50, solve each system of equations.

44. $\begin{cases} 4x - 2y = 6 \\ 3x + 3y = 9 \end{cases}$ 45. $\begin{cases} 2x + y = 19 \\ x - 2y = 12 \end{cases}$

46. $\begin{cases} 3x + 2y = 5 \\ 2x - 3y = 12 \end{cases}$ 47. $\begin{cases} 6x + 3y = 1 \\ \quad\quad y = -2x + 1 \end{cases}$

48. $\begin{cases} 4x - 3y = 253 \\ 13x + 2y = -12 \end{cases}$

49. $\begin{cases} x + 2y + 3z = 5 \\ \quad\quad y + 11z = 21 \\ \quad\quad 5y + 9z = 13 \end{cases}$ 50. $\begin{cases} x + y - z = 12 \\ \quad\quad 2y - 3z = -7 \\ 3x + 3y - 7z = 0 \end{cases}$

APPLICATIONS

51. *Endangered animals* The number of species of endangered animals, A, can be described by
$$A = 9.78x + 167.90$$
where x is the number of years past 1980 (*Source:* U.S. Fish and Wildlife Service).
   (a) To what year does $x = 17$ correspond?
   (b) What x-value corresponds to the year 2007?
   (c) If this equation remains valid, in what year will the number of species of endangered animals reach 461?

52. *Course grades* In a certain course, grades are based on three tests worth 100 points each, three quizzes worth 50 points each, and a final exam worth 200 points. A student has test grades of 91, 82, and 88, and quiz grades of 50, 42, and 42. What is the lowest percent the student can get on the final and still earn an A (90% or more of the total points) in the course?

53. *Cost analysis* The owner of a small construction business needs a new truck. He can buy a diesel truck for $38,000 and it will cost him $0.24 per mile to operate. He can buy a gas engine truck for $35,600 and it will cost him $0.30 per mile to operate. Find the number of miles he must drive before the costs are equal. If he normally keeps a truck for 5 years, which is the better buy?

54. *Heart disease risk* The Multiple Risk Factor Intervention Trial (MRFIT) used data from 356,222 men aged 35 to 57 to investigate the relationship between serum cholesterol and coronary heart disease (CHD) risk. Figure 1.43 shows the graph of the relationship of CHD risk and cholesterol, where a risk of 1 is assigned to 200 mg/dl of serum cholesterol and where the CHD risk is 4 times as high when serum cholesterol is 300 mg/dl.

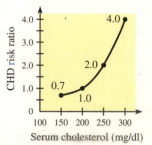

Figure 1.43

(a) Does this graph indicate that the CHD risk is a function of the serum cholesterol?
(b) Is the relationship a linear function?
(c) If CHD risk is a function f of serum cholesterol, what is $f(300)$?

55. *Mortgage loans* When a couple purchases a home, one of the first questions they face deals with the relationship between the amount borrowed and the monthly payment. In particular, if a bank offers 25-year loans at 7% interest, then the data in the following table would apply.

Amount Borrowed	Monthly Payment
$40,000	$282.72
50,000	353.39
60,000	424.07
70,000	494.75
80,000	565.44
90,000	636.11
100,000	706.78

Source: Comprehensive Mortgage Payment Tables, Publication No. 492, Financial Publishing Co., Boston

Assume that the monthly payment P is a function of the amount borrowed A (in thousands) and is denoted by $P = f(A)$, and answer the following.
(a) Find $f(80)$.
(b) Write a sentence that explains the meaning of $f(70) = 494.75$.

56. *Profit* Suppose that the profit from the production and sale of x units of a product is given by
$$P(x) = 330x - 0.05x^2 - 5000$$

In addition, suppose that for a certain month, the number of units produced on day t of the month is

$$x = q(t) = 100 + 10t$$

(a) Find $(P \circ q)(t)$ to express the profit as a function of the day of the month.

(b) Find the number of units produced, and the profit, on the fifteenth day of the month.

57. **Fish species growth** For many species of fish, the weight W is a function of the length L that can be expressed by

$$W = W(L) = kL^3 \quad k = \text{constant}$$

Suppose that for a particular species $k = 0.03$ and that for this species, the length (in centimeters) is a function of the number of years t the fish has been alive and that this function is given by

$$L = L(t) = 65 - 0.1(t - 25)^2 \quad 0 \le t \le 25$$

Find $(W \circ L)(t)$ in order to express W as a function of the age t of the fish.

58. **Distance to a thunderstorm** The distance d (in miles) to a thunderstorm is given by

$$d = \frac{t}{4.8}$$

where t is the number of seconds that elapse between seeing the lightning and hearing the thunder.

(a) Graph this function for $0 \le t \le 20$.

(b) The point $(9.6, 2)$ satisfies the equation. Explain its meaning.

59. **Body-heat loss** Body-heat loss due to convection depends on a number of factors. If H_c is body-heat loss due to convection, A_c is the exposed surface area of the body, $T_s - T_a$ is skin temperature minus air temperature, and K_c is the convection coefficient (determined by air velocity and so on), then we have

$$H_c = K_c A_c (T_s - T_a)$$

When $K_c = 1$, $A_c = 1$, and $T_s = 90$, the equation is

$$H_c = 90 - T_a$$

Sketch the graph.

60. **Profit** A company charting its profits notices that the relationship between the number of units sold, x, and the profit, P, is linear.

(a) If 200 units sold results in $3100 profit and 250 units sold results in $6000 profit, write the profit function for this company.

(b) Interpret the slope from part (a) as a rate of change.

61. **Health care costs** The average annual cost per consumer for health care can be modeled by

$$A = 427x + 4541$$

where x is the number of years from 2000 and projected to 2018 (*Source*: U.S. Census Bureau and U.S. Centers for Medicare and Medicaid Services).

(a) Is A a linear function of x?

(b) Find the slope and A-intercept of this function.

(c) Write a sentence that interprets the A-intercept.

(d) Write a sentence that interprets the slope of this function as a rate of change.

62. **Temperature** Write the equation of the linear relationship between temperature in Celsius (C) and Fahrenheit (F) if water freezes at 0°C and 32°F and boils at 100°C and 212°F.

63. **Photosynthesis** The amount y of photosynthesis that takes place in a certain plant depends on the intensity x of the light present, according to

$$y = 120x^2 - 20x^3 \quad \text{for } x \ge 0$$

(a) Graph this function with a graphing utility. (Use y-min $= -100$ and y-max $= 700$.)

(b) The model is valid only when $f(x) \ge 0$ (that is, on or above the x-axis). For what x-values is this true?

64. **Flow rates of water** The speed at which water travels in a pipe can be measured by directing the flow through an elbow and measuring the height to which it spurts out the top. If the elbow height is 10 cm, the equation relating the height h (in centimeters) of the water above the elbow and its velocity v (in centimeters per second) is given by

$$v^2 = 1960(h + 10)$$

(a) Solve this equation for h and graph the result, using the velocity as the independent variable.

(b) If the velocity is 210 cm/sec, what is the height of the water above the elbow?

65. **Investment mix** A retired couple have $150,000 to invest and want to earn $15,000 per year in interest. The safer investment yields 9.5%, but they can supplement their earnings by investing some of their money at 11%. How much should they invest at each rate to earn $15,000 per year?

66. **Botany** A botanist has a 20% solution and a 70% solution of an insecticide. How much of each must be used to make 4.0 liters of a 35% solution?

67. **Supply and demand** A certain product has supply and demand functions $p = 4q + 5$ and $p = -2q + 81$, respectively.

(a) If the price is $53, how many units are supplied and how many are demanded?

(b) Does this give a shortfall or a surplus?

(c) Is the price likely to increase from $53 or decrease from it?

68. **Market analysis** Of the equations $p + 6q = 420$ and $p = 6q + 60$, one is the supply function for a product and one is the demand function for that product.

(a) Graph these equations on the same set of axes.

(b) Label the supply function and the demand function.

(c) Find the market equilibrium point.

69. *Cost, revenue, and profit* The total cost and total revenue for a certain product are given by

$$C(x) = 38.80x + 4500$$
$$R(x) = 61.30x$$

(a) Find the marginal cost.
(b) Find the marginal revenue.
(c) Find the marginal profit.
(d) Find the number of units required to break even.

70. *Cost, revenue, and profit* A certain commodity has the following costs for a period.

Fixed cost:	$1500
Variable cost per unit:	$22

The commodity is sold for $52 per unit.
(a) What is the total cost function?
(b) What is the total revenue function?
(c) What is the profit function?
(d) What is the marginal cost?
(e) What is the marginal revenue?
(f) What is the marginal profit?
(g) What is the break-even quantity?

71. *Market analysis* The supply function and the demand function for a product are linear and are determined by the tables that follow. Find the quantity and price that will give market equilibrium.

Supply Function		Demand Function	
Price	Quantity	Price	Quantity
100	200	200	200
200	400	100	400
300	600	0	600

72. *Market analysis* Suppose that for a certain product the supply and demand functions prior to any taxation are

$$\text{Supply:} \quad p = \frac{q}{10} + 8$$
$$\text{Demand:} \quad 10p + q = 1500$$

If a tax of $2 per item is levied on the supplier and is passed on to the consumer as a price increase, find the market equilibrium after the tax is levied.

---

1 CHAPTER TEST

In Problems 1–4, solve the equations.
1. $10 - 2(2x - 9) - 4(6 + x) = 52$

2. $4x - 3 = \dfrac{x}{2} + 6$

3. $\dfrac{3}{x} + 4 = \dfrac{4x}{x + 1}$ 4. $\dfrac{3x - 1}{4x - 9} = \dfrac{5}{7}$

5. For $f(x) = 7 + 5x - 2x^2$, find and simplify

$$\frac{f(x + h) - f(x)}{h}.$$

6. Solve $1 + \dfrac{2}{3}t \le 3t + 22$ and graph the solution.

In Problems 7 and 8, find the intercepts and graph the functions.
7. $5x - 6y = 30$ 8. $7x + 5y = 21$
9. Consider the function $f(x) = \sqrt{4x + 16}$.
(a) Find the domain and range.
(b) Find $f(3)$.
(c) Find the y-coordinate on the graph of $y = f(x)$ when $x = 5$.
10. Write the equation of the line passing through $(-1, 2)$ and $(3, -4)$. Write your answer in slope-intercept form.
11. Find the slope and the y-intercept of the graph of $5x + 4y = 15$.
12. Write the equation of the line through $(-3, -1)$ that
(a) has undefined slope
(b) is perpendicular to $x = 4y - 8$.

13. Which of the following relations [(a), (b), (c)] are functions? Explain.

(a)

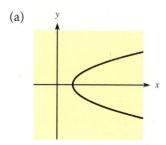

(b)

x	y
1	3
-1	3
4	2

(c) $y = \pm\sqrt{x^2 - 1}$

14. Graph $y = x^2 - 12x$
(a) in the standard window and
(b) in $x\text{-min} = -10$, $x\text{-max} = 30$; $y\text{-min} = -40$, $y\text{-max} = 40$.

15. Solve the system

$$\begin{cases} 3x + 2y = -2 \\ 4x + 5y = 2 \end{cases}$$

16. Given $f(x) = 5x^2 - 3x$ and $g(x) = x + 1$, find
(a) $(fg)(x)$
(b) $g(g(x))$
(c) $(f \circ g)(x)$

17. The total cost function for a product is $C(x) = 30x + 1200$, and the total revenue is $R(x) = 38x$, where x is the number of units produced and sold.
    (a) Find the marginal cost.
    (b) Find the profit function.
    (c) Find the number of units that gives the break-even point.
    (d) Find the marginal profit and explain what it means.

18. The selling price for each item of a product is $50, and the total cost is given by $C(x) = 10x + 18,000$, where x is the number of items.
    (a) Write the revenue function.
    (b) Find $C(100)$ and write a sentence that explains its meaning.
    (c) Find the number of units that gives the break-even point.

19. The supply function for a product is $p = 5q + 1500$ and the demand function is $p = -3q + 3100$. Find the quantity and price that give market equilibrium.

20. A building is depreciated by its owner, with the value y of the building after x months given by $y = 720,000 - 2000x$.
    (a) Find the y-intercept of the graph of this function, and explain what it means.
    (b) Find the slope of the graph and tell what it means.

21. An airline has 360 seats on a plane for one of its flights. If 90% of the people making reservations actually buy a ticket, how many reservations should the airline accept to be confident that it will sell 360 tickets?

22. Amanda plans to invest $20,000, part of it at a 9% interest rate and part of it in a safer fund that pays 6%. How much should be invested in each fund to yield an annual return of $1560?

I. Hospital Administration

Southwest Hospital has an operating room used only for eye surgery. The annual cost of rent, heat, and electricity for the operating room and its equipment is $360,000, and the annual salaries of the people who staff this room total $540,000.

Each surgery performed requires the use of $760 worth of medical supplies and drugs. To promote goodwill, every patient receives a bouquet of flowers the day after surgery. In addition, one-quarter of the patients require dark glasses, which the hospital provides free of charge. It costs the hospital $30 for each bouquet of flowers and $40 for each pair of glasses.

The hospital receives a payment of $2000 for each eye operation performed.

1. Identify the revenue per case and the annual fixed and variable costs for running the operating room.
2. How many eye operations must the hospital perform each year in order to break even?
3. Southwest Hospital currently averages 70 eye operations per month. One of the nurses has just learned about a machine that would reduce by $100 per patient the amount of medical supplies needed. It can be leased for $100,000 annually. Keeping in mind the financial cost and benefits, advise the hospital on whether it should lease this machine.
4. An advertising agency has proposed to the hospital's president that she spend $20,000 per month on television and radio advertising to persuade people that Southwest Hospital is the best place to have any eye surgery performed. Advertising account executives estimate that such publicity would increase business by 40 operations per month. If they are correct and if this increase is not big enough to affect fixed costs, what impact would this advertising have on the hospital's profits?
5. In case the advertising agency is being overly optimistic, how many extra operations per month are needed to cover the cost of the proposed ads?
6. If the ad campaign is approved and subsequently meets its projections, should the hospital review its decision about the machine discussed in Question 3?

II. Fundraising

At most colleges and universities, student organizations conduct fundraising activities, such as selling T-shirts or candy. If a club or organization decided to sell coupons for submarine sandwiches, it would want to find the best deal, based on the amount the club thought it could sell and how much profit it would make. In this project, you'll try to discover the best deal for this type of fundraiser conducted in your area.

1. Contact at least two different sub shops in your area from among local sub shops, a national chain, or a regional or national convenience store chain. From each contact, find the details of selling sub sandwich coupons as a fundraiser. In particular, determine the following for each sub shop, and present your findings in a chart.
   (a) The selling price for each coupon and its value to the coupon holder, including any expiration date
   (b) Your cost for each coupon sold and for each one returned
   (c) The total number of coupons provided
   (d) The duration of the sale
2. For each sub shop, determine a club's total revenue and total costs as linear functions of the number of coupons sold.
3. Form the profit function for each sub shop, and graph the profit functions together. For each function, determine the break-even point. Find the marginal profit for each function, and interpret its meaning.
4. The shop with the best deal will be the one whose coupons will provide the maximum profit for a club.
   (a) For each shop, determine a sales estimate that is based on location, local popularity, and customer value.
   (b) Use each estimate from part (a) with that shop's profit function to determine which shop gives the maximum profit.

Fully explain and support your claims; make and justify a recommendation.

CHAPTER

Quadratic and Other Special Functions

© Martin Strmiska/Alamy

In Chapter 1 we discussed functions in general and linear functions in particular. In this chapter we will discuss quadratic functions and their applications, and we will also discuss other types of functions, including identity, constant, power, absolute value, piecewise defined, and reciprocal functions. Graphs of polynomial and rational functions will also be introduced; they will be studied in detail in Chapter 10. We will pay particular attention to quadratic equations and to quadratic functions, and we will see that cost, revenue, profit, supply, and demand are sometimes modeled by quadratic functions. We also include, in an optional section on modeling, the creation of functions that approximately fit real data points.

The topics and some of the applications discussed in this chapter follow.

SECTIONS	APPLICATIONS
2.1 **Quadratic Equations** Factoring methods The quadratic formula	Social Security benefits, falling objects
2.2 **Quadratic Functions: Parabolas** Vertices of parabolas Zeros	Profit maximization, maximizing revenue
2.3 **Business Applications Using Quadratics**	Supply, demand, market equilibrium, break-even points, profit maximization
2.4 **Special Functions and Their Graphs** Basic functions Polynomial functions Rational functions Piecewise defined functions	Selling price, residential electric costs, average cost
2.5 **Modeling; Fitting Curves to Data with Graphing Utilities (optional)** Linear regression	Internet access, federal tax per capita, expected life span, consumer price index

Prerequisite Problem Type	For Section	Answer	Section for Review
Find $b^2 - 4ac$ if (a) $a = 1, b = 2, c = -2$ (b) $a = -2, b = 3, c = -1$ (c) $a = -3, b = -3, c = -2$	**2.1** **2.2** **2.3**	(a) 12 (b) 1 (c) -15	0.2 Signed numbers
(a) Factor $6x^2 - x - 2$. (b) Factor $6x^2 - 9x$.	**2.1** **2.2** **2.3**	(a) $(3x - 2)(2x + 1)$ (b) $3x(2x - 3)$	0.6 Factoring
Is $\dfrac{1 + \sqrt{-3}}{2}$ a real number?	**2.1**	No	0.4 Radicals
(a) Find the y-intercept of $y = x^2 - 6x + 8$. (b) Find the x-intercept of $y = 2x - 3$.	**2.2** **2.3**	(a) 8 (b) $\frac{3}{2}$	1.3 x- and y-intercepts
If $f(x) = -x^2 + 4x$, find $f(2)$. Find the domain of $$f(x) = \frac{12x + 8}{3x - 9}$$	**2.2** **2.3** **2.4**	4 All $x \neq 3$	1.2 Functions
Assume revenue is $R(x) = 500x - 2x^2$ and cost is $C(x) = 3600 + 100x + 2x^2$. (a) Find the profit function $P(x)$. (b) Find $P(50)$.	**2.3**	 (a) $P(x) = -3600 +$ $400x - 4x^2$ (b) $P(50) = 6400$	1.6 Cost, revenue, and profit

OBJECTIVES

2.1

- To solve quadratic equations with factoring methods
- To solve quadratic equations with the quadratic formula

Quadratic Equations

▮ | APPLICATION PREVIEW |

With fewer workers, future income from payroll taxes used to fund Social Security benefits won't keep pace with scheduled benefits. The function

$$B = -1.057t^2 + 8.259t + 74.071$$

describes the Social Security trust fund balance B, in billions of dollars, where t is the number of years past the year 2000 (*Source:* Social Security Administration). For planning purposes, it is important to know when the trust fund balance will be 0. That is, for what t-value does

$$0 = -1.057t^2 + 8.259t + 74.071?$$

(See Example 7 for the solution.)

In this section, we learn how to solve equations of this type, using factoring methods and using the quadratic formula.

Factoring Methods   A **quadratic equation** in one variable is an equation that can be written in the *general form*

$$ax^2 + bx + c = 0 \qquad (a \neq 0)$$

where a, b, and c represent constants. For example, the equations

$$3x^2 + 4x + 1 = 0 \qquad \text{and} \qquad 2x^2 + 1 = x^2 - x$$

are quadratic equations; the first of these is in general form, and the second can easily be rewritten in general form.

When we solve quadratic equations, we will be interested only in real number solutions and will consider two methods of solution: factoring and the quadratic formula. We will discuss solving quadratic equations by factoring first. (For a review of factoring, see Section 0.6, "Factoring.")

Solution by factoring is based on the following property of the real numbers.

Zero Product Property   For real numbers a and b, $ab = 0$ if and only if $a = 0$ or $b = 0$ or both.

Hence, to solve by factoring, we must first write the equation with zero on one side.

EXAMPLE **1**   **Solving Quadratic Equations**

Solve: (a) $6x^2 + 3x = 4x + 2$ (b) $6x^2 = 9x$

Solution

(a)
$$6x^2 + 3x = 4x + 2$$
$$6x^2 - x - 2 = 0 \qquad \text{Proper form for factoring}$$
$$(3x - 2)(2x + 1) = 0 \qquad \text{Factored}$$
$$3x - 2 = 0 \quad \text{or} \quad 2x + 1 = 0 \qquad \text{Factors equal to zero}$$
$$3x = 2 \qquad\qquad 2x = -1$$
$$x = \frac{2}{3} \qquad\qquad x = -\frac{1}{2} \qquad \text{Solutions}$$

We now check that these values are, in fact, solutions to our original equation.

$$6\left(\frac{2}{3}\right)^2 + 3\left(\frac{2}{3}\right) \stackrel{?}{=} 4\left(\frac{2}{3}\right) + 2 \qquad 6\left(-\frac{1}{2}\right)^2 + 3\left(-\frac{1}{2}\right) \stackrel{?}{=} 4\left(-\frac{1}{2}\right) + 2$$

$$\frac{14}{3} = \frac{14}{3} \; ✔ \qquad\qquad\qquad 0 = 0 \; ✔$$

(b)
$$6x^2 = 9x$$
$$6x^2 - 9x = 0$$
$$3x(2x - 3) = 0$$

$$3x = 0 \quad \text{or} \quad 2x - 3 = 0$$
$$x = 0 \qquad\qquad 2x = 3$$
$$x = \frac{3}{2}$$

Check: $6(0)^2 = 9(0)$ ✔ $\qquad 6\left(\frac{3}{2}\right)^2 = 9\left(\frac{3}{2}\right)$ ✔

Thus the solutions are $x = 0$ and $x = \frac{3}{2}$.

Note that in Example 1(b) it is tempting to divide both sides of the equation by x, but this is incorrect because it results in the loss of the solution $x = 0$. Never divide both sides of an equation by an expression containing the variable.

EXAMPLE 2 **Solving by Factoring**

Solve: (a) $(y - 3)(y + 2) = -4$ (b) $\dfrac{x + 1}{3x + 6} = \dfrac{3}{x} + \dfrac{2x + 6}{x(3x + 6)}$

Solution

(a) Note that the left side of the equation is factored, but the right member is not 0. Therefore, we must multiply the factors before we can rewrite the equation in general form.

$$(y - 3)(y + 2) = -4$$
$$y^2 - y - 6 = -4$$
$$y^2 - y - 2 = 0$$
$$(y - 2)(y + 1) = 0$$
$$y - 2 = 0 \quad \text{or} \quad y + 1 = 0$$
$$y = 2 \qquad\qquad y = -1$$

Check: $(2 - 3)(2 + 2) = -4$ ✔ $\qquad (-1 - 3)(-1 + 2) = -4$ ✔

(b) The LCD of all fractions is $x(3x + 6)$. Multiplying both sides of the equation by this LCD gives a quadratic equation that is equivalent to the original equation for $x \neq 0$ and $x \neq -2$. (The original equation is undefined for these values.)

$$\frac{x + 1}{3x + 6} = \frac{3}{x} + \frac{2x + 6}{x(3x + 6)}$$
$$x(3x + 6)\frac{x + 1}{3x + 6} = x(3x + 6)\left(\frac{3}{x} + \frac{2x + 6}{x(3x + 6)}\right) \qquad \text{Multiply both sides by } x(3x + 6).$$
$$x(x + 1) = 3(3x + 6) + (2x + 6)$$
$$x^2 + x = 9x + 18 + 2x + 6$$
$$x^2 - 10x - 24 = 0$$
$$(x - 12)(x + 2) = 0$$
$$x - 12 = 0 \text{ or } x + 2 = 0$$
$$x = 12 \text{ or } x = -2$$

Checking $x = 12$ and $x = -2$ in the original equation, we see that $x = -2$ makes the denominator equal to zero. Hence the only solution is $x = 12$. ✔

CHECKPOINT

1. The factoring method for solving a quadratic equation is based on the _____ product property. Hence, in order for us to solve a quadratic equation by factoring, one side of the equation must equal _____.
2. Solve the following equations by factoring.
   (a) $x^2 - 19x = 20$ (b) $2x^2 = 6x$

EXAMPLE 3   Falling Object

A tennis ball is thrown into a swimming pool from the top of a tall hotel. The height of the ball from the pool is modeled by $D(t) = -16t^2 - 4t + 300$ feet, where t is the time, in seconds, after the ball was thrown. How long after the ball was thrown was it 144 feet above the pool?

Both: Shutterstock.com

Solution

To find the number of seconds until the ball is 144 feet above the pool, we solve the equation

$$144 = -16t^2 - 4t + 300$$

for t. The solution follows.

$$144 = -16t^2 - 4t + 300$$
$$0 = -16t^2 - 4t + 156$$
$$0 = 4t^2 + t - 39$$
$$0 = (t - 3)(4t + 13)$$
$$t - 3 = 0 \text{ or } 4t + 13 = 0$$
$$t = 3 \text{ or } t = -13/4$$

This indicates that the ball will be 144 feet above the pool 3 seconds after it is thrown. The negative value for t has no meaning in this application.   ■

The Quadratic Formula   Factoring does not lend itself easily to solving quadratic equations such as

$$x^2 - 5 = 0$$

However, we can solve this equation by writing

$$x^2 = 5$$
$$x = \pm\sqrt{5}$$

In general, we can solve quadratic equations of the form $x^2 = C$ (no x-term) by taking the square root of both sides.

Square Root Property

The solution of $x^2 = C$ is

$$x = \pm\sqrt{C}$$

This property also can be used to solve equations such as those in the following example.

EXAMPLE 4   Square Root Method

Solve the following equations.
(a) $4x^2 = 5$ (b) $(3x - 4)^2 = 9$

Solution

We can use the square root property for both parts.
(a) $4x^2 = 5$ is equivalent to $x^2 = \frac{5}{4}$. Thus

$$x = \pm\sqrt{\frac{5}{4}} = \pm\frac{\sqrt{5}}{\sqrt{4}} = \pm\frac{\sqrt{5}}{2}$$

(b) $(3x - 4)^2 = 9$ is equivalent to $3x - 4 = \pm\sqrt{9}$. Thus

$3x - 4 = 3$	$3x - 4 = -3$
$3x = 7$	$3x = 1$
$x = \dfrac{7}{3}$	$x = \dfrac{1}{3}$

The solution of the general quadratic equation $ax^2 + bx + c = 0$, where $a \neq 0$, is called the **quadratic formula**. It can be derived by using the square root property, as follows:

$$ax^2 + bx + c = 0 \qquad \text{Standard form}$$
$$ax^2 + bx = -c \qquad \text{Subtract } c \text{ from both sides.}$$
$$x^2 + \frac{b}{a}x = -\frac{c}{a} \qquad \text{Divide both sides by } a.$$

We would like to make the left side of the last equation a perfect square of the form $(x + k)^2 = x^2 + 2kx + k^2$. If we let $2k = b/a$, then $k = b/(2a)$ and $k^2 = b^2/(4a^2)$. Hence we continue as follows:

$$x^2 + \frac{b}{a}x + \frac{b^2}{4a^2} = \frac{b^2}{4a^2} - \frac{c}{a} \qquad \text{Add } \frac{b^2}{4a^2} \text{ to both sides.}$$

$$\left(x + \frac{b}{2a}\right)^2 = \frac{b^2 - 4ac}{4a^2} \qquad \text{Simplify.}$$

$$x + \frac{b}{2a} = \pm\sqrt{\frac{b^2 - 4ac}{4a^2}} \qquad \text{Square root property}$$

$$x = \frac{-b}{2a} \pm \frac{\sqrt{b^2 - 4ac}}{2|a|} \qquad \text{Solve for } x.$$

Because $\pm 2|a|$ represents the same numbers as $\pm 2a$, we obtain the following.

Quadratic Formula

If $ax^2 + bx + c = 0$, where $a \neq 0$, then

$$x = \frac{-b \pm \sqrt{b^2 - 4ac}}{2a}$$

We may use the quadratic formula to solve all quadratic equations, but especially those in which factorization is difficult or impossible to see. The proper identification of values for a, b, and c to be substituted into the formula requires that the equation be in general form.

EXAMPLE 5   **Quadratic Formula**

Use the quadratic formula to solve $2x^2 - 3x - 6 = 0$ for x.

Solution

The equation is already in general form, with $a = 2$, $b = -3$, and $c = -6$. Hence

$$x = \frac{-b \pm \sqrt{b^2 - 4ac}}{2a} = \frac{-(-3) \pm \sqrt{(-3)^2 - 4(2)(-6)}}{2(2)}$$

$$= \frac{3 \pm \sqrt{9 + 48}}{4} = \frac{3 \pm \sqrt{57}}{4}$$

Thus the solutions are

$$x = \frac{3 + \sqrt{57}}{4} \quad \text{and} \quad x = \frac{3 - \sqrt{57}}{4}$$

EXAMPLE 6   **Nonreal Solution**

Using the quadratic formula, find the (real) solutions to $x^2 = x - 1$.

Solution

We must rewrite the equation in general form before we can determine the values of a, b, and c. We write $x^2 = x - 1$ as $x^2 - x + 1 = 0$. Note that $a = 1$, $b = -1$, and $c = 1$.

$$x = \frac{-(-1) \pm \sqrt{(-1)^2 - 4(1)(1)}}{2(1)} = \frac{1 \pm \sqrt{1 - 4}}{2} = \frac{1 \pm \sqrt{-3}}{2}$$

Because $\sqrt{-3}$ is not a real number, the values of x are not real. Hence there are no real solutions to the given equation.

In Example 6 there were no real solutions to the quadratic equation because the radicand of the quadratic formula was negative. In general, when solving a quadratic equation, we can use the sign of the radicand in the quadratic formula—that is, the sign of $b^2 - 4ac$—to determine how many real solutions there are. Thus we refer to $b^2 - 4ac$ as the **quadratic discriminant.**

Quadratic Discriminant

Given $ax^2 + bx + c = 0$ and $a \neq 0$,

If $b^2 - 4ac > 0$, the equation has two distinct real solutions.

If $b^2 - 4ac = 0$, the equation has exactly one real solution.

If $b^2 - 4ac < 0$, the equation has no real solutions.

The quadratic formula is especially useful when the coefficients of a quadratic equation are decimal values that make factorization impractical. This occurs in many applied problems.

EXAMPLE 7   **Social Security Trust Fund** | **APPLICATION PREVIEW** |

The function

$$B = -1.057t^2 + 8.259t + 74.071$$

describes the Social Security trust fund balance B, in billions of dollars, where t is the number of years past the year 2000 (*Source:* Social Security Administration). For planning purposes, it is important to know when the trust fund balance will be 0. To find when this occurs, solve

$$0 = -1.057t^2 + 8.259t + 74.071$$

Solution

We use the quadratic formula with

$$a = -1.057, \quad b = 8.259, \quad c = 74.071$$

Thus

$$t = \frac{-b \pm \sqrt{b^2 - 4ac}}{2a}$$

$$= \frac{-8.259 \pm \sqrt{(8.259)^2 - 4(-1.057)(74.071)}}{2(-1.057)}$$

$$\approx \frac{-8.259 \pm \sqrt{381.383}}{-2.114}$$

$$\approx \frac{-8.259 \pm 19.529}{-2.114}$$

We are interested only in the positive solution, so

$$t = \frac{-8.259 - 19.529}{-2.114} \approx 13.145$$

Therefore, the trust fund balance is projected to reach zero slightly more than 13 years after 2000, during the year 2014. ∎

Calculator Note   We can use graphing calculators to determine (or approximate) the solutions of quadratic equations. The solutions can be found by using commands (such as ZERO or SOLVER) or programs. TRACE can be used to approximate where the graph of the function $y = f(x)$ intersects the x-axis. This **x-intercept** is also the value of x that makes the function zero, so it is called a **zero of the function**. Because this value of x makes $f(x) = 0$, it is also a **solution** (or **root**) of the equation $0 = f(x)$. The steps in solving an equation with a graphing utility are shown in Appendix C, Section 2.1.

For example, we can find the solutions of the equation $0 = -7x^2 + 16x - 4$ by graphing $y = -7x^2 + 16x - 4$, as in Figures 2.1(a) and 2.1(b). The graphs show $y = 0$ when $x = 2$ and when $x = 0.2857$ (approximately).

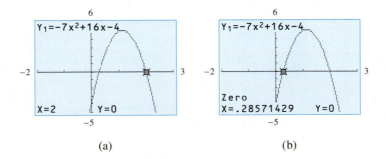

Figure 2.1   (a)   (b)

Solving $f(x) = 0$ algebraically gives solutions $x = 2$ and $x = \frac{2}{7} \approx 0.2857$. These values agree with those found graphically. ∎

Spreadsheet Note Excel can also be used to solve a quadratic equation. (See Appendix D, Section 2.1 for details.) For example, to solve the equation $0 = -7x^2 + 16x - 4$, we use its graph to approximate the x-intercepts (approximately 2 and 0).

The following screens show the input formula with values near 0 and 2, and the approximate solutions (in column A) after we enter 0 in column B and use Goal Seek. Goal Seek gives $-7x^2 + 16x - 4 = 0$ when $x = 2$ and when x is approximately 0.2857. ∎

	A	B
1	x	$-7x^2 + 16x - 4$
2	2	$= -7*A2^2 + 16*A2 - 4$
3	0	$= -7*A3^2 + 16*A3 - 4$

	A	B
1	x	$-7x^2 + 16x - 4$
2	2	0
3	0.285688	-0.00032

CHECKPOINT

3. The statement of the quadratic formula says that if _____ and $a \neq 0$, then $x = $ _____ .

4. Solve $2x^2 - 5x = 9$ using the quadratic formula.

CHECKPOINT SOLUTIONS

1. Zero; zero

2. (a) $x^2 - 19x - 20 = 0$
   $(x - 20)(x + 1) = 0$

$x - 20 = 0$	$x + 1 = 0$
$x = 20$	$x = -1$

   (b) $2x^2 - 6x = 0$
   $2x(x - 3) = 0$

$2x = 0$	$x - 3 = 0$
$x = 0$	$x = 3$

3. If $ax^2 + bx + c = 0$ with $a \neq 0$, then

$$x = \frac{-b \pm \sqrt{b^2 - 4ac}}{2a}$$

4. $2x^2 - 5x - 9 = 0$, so $a = 2, b = -5, c = -9$

$$x = \frac{-b \pm \sqrt{b^2 - 4ac}}{2a} = \frac{5 \pm \sqrt{25 - 4(2)(-9)}}{4}$$

$$= \frac{5 \pm \sqrt{25 + 72}}{4}$$

$$= \frac{5 \pm \sqrt{97}}{4}$$

$$x = \frac{5 + \sqrt{97}}{4} \approx 3.712 \quad \text{or} \quad x = \frac{5 - \sqrt{97}}{4} \approx -1.212$$

| EXERCISES | 2.1

Access end-of-section exercises online at **www.webassign.net** ENHANCED **Web**Assign

Quadratic Functions: Parabolas

| APPLICATION PREVIEW |

Because additional equipment, raw materials, and labor may cause variable costs of some products to increase dramatically as more units are produced, total cost functions are not always linear functions. For example, suppose that the total cost of producing a product is given by the equation

$$C = C(x) = 300x + 0.1x^2 + 1200$$

where x represents the number of units produced. That is, the cost of this product is represented by a second-degree function, or a quadratic function. We can find the cost of producing 10 units by evaluating

$$C(10) = 300(10) + 0.1(10)^2 + 1200 = 4210$$

If the revenue function for this product is

$$R = R(x) = 600x$$

then the profit is also a quadratic function:

$$P = R - C = 600x - (300x + 0.1x^2 + 1200)$$

$$P = -0.1x^2 + 300x - 1200$$

We can use this function to find how many units give maximum profit and what that profit is. (See Example 2.) In this section we will describe ways to find the maximum point or minimum point for a quadratic function.

Parabolas   In Chapter 1, we studied functions of the form $y = ax + b$, called linear (or first-degree) functions. We now turn our attention to **quadratic** (or second-degree) **functions.** The general equation of a quadratic function has the form

$$y = f(x) = ax^2 + bx + c$$

where a, b, and c are real numbers and $a \neq 0$.

The graph of a quadratic function,

$$y = ax^2 + bx + c \quad (a \neq 0)$$

has a distinctive shape called a **parabola.**

The basic function $y = x^2$ and a variation of it, $y = -\frac{1}{2}x^2$ are parabolas whose graphs are shown in Figure 2.2.

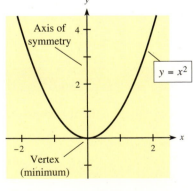

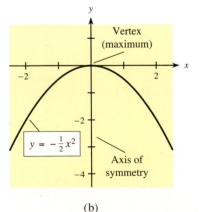

Figure 2.2   (a)   (b)

Vertex of a Parabola

As these examples illustrate, the graph of $y = ax^2$ is a parabola that opens upward if $a > 0$ and downward if $a < 0$. The **vertex**, where the parabola turns, is a **minimum point** if $a > 0$ and a **maximum point** if $a < 0$. The vertical line through the vertex of a parabola is called the **axis of symmetry** because one half of the graph is a reflection of the other half through this line.

The graph of $y = (x - 2)^2 - 1$ is the graph of $y = x^2$ shifted to a new location that is 2 units to the right and 1 unit down; its vertex is shifted from $(0, 0)$ to $(2, -1)$ and its axis of symmetry is shifted 2 units to the right. (See Figure 2.3(a).) The graph of $y = -\frac{1}{2}(x + 1)^2 + 2$ is the graph of $y = -\frac{1}{2}x^2$ shifted 1 unit to the left and 2 units up, with its vertex at $(-1, 2)$. (See Figure 2.3(b).)

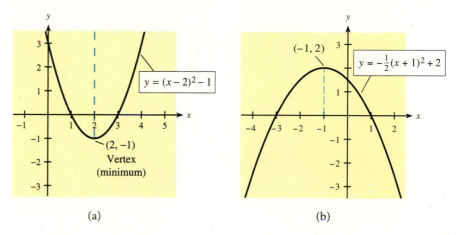

Figure 2.3   (a)   (b)

We can find the x-coordinate of the vertex of the graph of $y = ax^2 + bx + c$ by using the fact that the axis of symmetry of a parabola passes through the vertex. Regardless of the location of the vertex of $y = ax^2 + bx + c$ or the direction it opens, the y-intercept of the graph of $y = ax^2 + bx + c$ is $(0, c)$ and there is another point on the graph with y-coordinate c. See Figure 2.4.

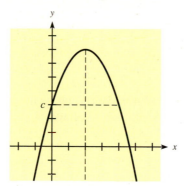

Figure 2.4

The x-coordinates of the points on this graph with y-coordinate c satisfy

$$c = ax^2 + bx + c$$

Solving this equation gives

$$0 = ax^2 + bx$$
$$0 = x(ax + b)$$
$$x = 0 \quad \text{or} \quad x = \frac{-b}{a}$$

The x-coordinate of the vertex is on the axis of symmetry, which is halfway from $x = 0$ to $x = \dfrac{-b}{a}$, at $x = \dfrac{-b}{2a}$. Thus we have the following.

Vertex of a Parabola

The quadratic function $y = f(x) = ax^2 + bx + c$ has its **vertex** at

$$\left(\frac{-b}{2a}, \; f\left(\frac{-b}{2a}\right) \right)$$

The optimum value (either maximum or minimum) of the function occurs at

$x = \dfrac{-b}{2a}$ and is $f\left(\dfrac{-b}{2a}\right)$. See Figure 2.5.

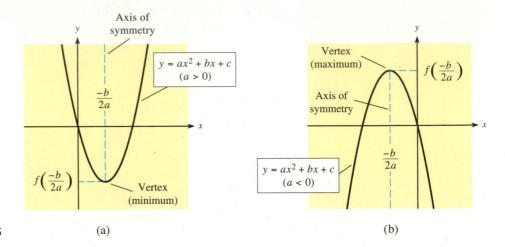

Figure 2.5   (a)   (b)

If we know the location of the vertex and the direction in which the parabola opens, we need very few other points to make a good sketch.

EXAMPLE   **1**   **Vertex and Graph of a Parabola**

Find the vertex and sketch the graph of

$$f(x) = 2x^2 - 4x + 4$$

Solution

Because $a = 2 > 0$, the graph of $f(x)$ opens upward and the vertex is the minimum point. We can calculate its coordinates as follows:

$$x = \frac{-b}{2a} = \frac{-(-4)}{2(2)} = 1$$

$$y = f(1) = 2$$

Thus the vertex is $(1, 2)$. Using x-values on either side of the vertex to plot additional points enables us to sketch the graph accurately. (See Figure 2.6.)   ■

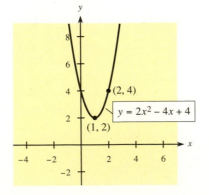

Figure 2.6

We can also use the coordinates of the vertex to find maximum or minimum values without a graph.

EXAMPLE 2   **Profit Maximization** | **APPLICATION PREVIEW** |

For the profit function

$$P(x) = -0.1x^2 + 300x - 1200$$

find the number of units that give maximum profit and find the maximum profit.

Solution

$P(x)$ is a quadratic function with $a < 0$. Thus the graph of $y = P(x)$ is a parabola that opens downward, so the vertex is a maximum point. The coordinates of the vertex are

$$x = \frac{-b}{2a} = \frac{-300}{2(-0.1)} = 1500$$

$$P = P(1500) = -0.1(1500)^2 + 300(1500) - 1200 = 223{,}800$$

Therefore, the maximum profit is $223,800 when 1500 units are sold.   ■

Calculator Note   We can also use a graphing calculator to graph the quadratic function in Example 2. Even with a graphing calculator, recognizing that the graph is a parabola and locating the vertex help us to set the viewing window (to include the vertex) and to know when we have a complete graph. Details of graphing the function of Example 2 are shown in Appendix C, Section 2.2. Figure 2.7 shows the graph. Note that the graph of this function has two x-intercepts.   ■

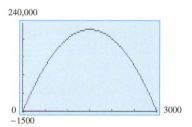

Figure 2.7

Zeros of Quadratic Functions   As we noted in Chapter 1, "Linear Equations and Functions," the x-intercepts of the graph of a function $y = f(x)$ are the values of x for which $f(x) = 0$, called the **zeros** of the function. As we saw in the previous section, the zeros of the quadratic function $y = f(x) = ax^2 + bx + c$ are the solutions of the quadratic equation $ax^2 + bx + c = 0$, which are given by the quadratic formula

$$x = \frac{-b \pm \sqrt{b^2 - 4ac}}{2a}$$

The information that is useful in graphing quadratic functions is summarized as follows.

Graphs of Quadratic Functions

Form: $y = f(x) = ax^2 + bx + c$
Graph: parabola

$a > 0$ parabola opens upward; vertex is a minimum point
$a < 0$ parabola opens downward; vertex is a maximum point

Coordinates of vertex: $x = \dfrac{-b}{2a}$, $y = f\left(\dfrac{-b}{2a}\right)$

Axis of symmetry equation: $x = \dfrac{-b}{2a}$

(continued)

x-intercepts or zeros (if real*):

$$x = \frac{-b + \sqrt{b^2 - 4ac}}{2a},$$

$$x = \frac{-b - \sqrt{b^2 - 4ac}}{2a}$$

y-intercept: Let $x = 0$; then $y = c$.

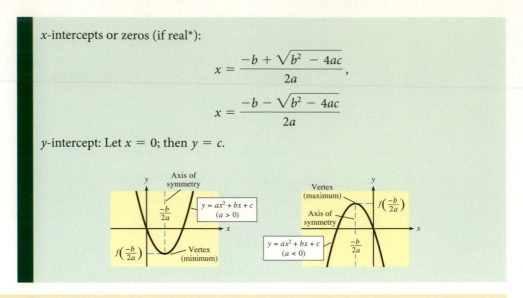

EXAMPLE 3   Graph of a Quadratic Function

For the function $y = 4x - x^2$, determine whether its vertex is a maximum point or a minimum point and find the coordinates of this point, find the zeros, if any exist, and sketch the graph.

Solution
The proper form is $y = -x^2 + 4x + 0$, so $a = -1$. Thus the parabola opens downward, and the vertex is the highest (maximum) point.

The vertex occurs at $x = \dfrac{-b}{2a} = \dfrac{-4}{2(-1)} = 2.$

The y-coordinate of the vertex is $f(2) = -(2)^2 + 4(2) = 4$.
The zeros of the function are solutions to

$$-x^2 + 4x = 0$$
$$x(-x + 4) = 0$$
$$\text{or } x = 0 \text{ and } x = 4$$

The graph of the function can be found by drawing a parabola with these three points (see Figure 2.8(a)), or by using a graphing calculator or a spreadsheet. Details of graphing quadratic (and other polynomial) functions with Excel are shown in Appendix D, Section 2.2 and the Online Excel Guide (see Figure 2.8(b)). ■

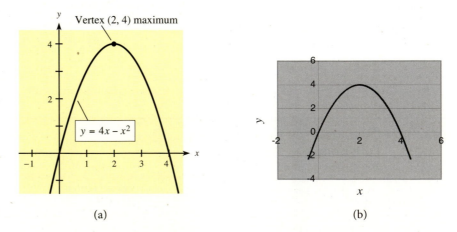

Figure 2.8   (a)   (b)

*If the zeros are not real, the graph does not cross the x-axis.

Example 3 shows how to use the solutions of $f(x) = 0$ to find the x-intercepts of a parabola if $y = f(x)$ is a quadratic function. Conversely, we can use the x-intercepts of $y = f(x)$, if they exist, to find or approximate the solutions of $f(x) = 0$. Exact integer solutions found in this way can also be used to determine factors of $f(x)$.

EXAMPLE 4 Graph Comparison

Figure 2.9 shows the graphs of two different quadratic functions. Use the figure to answer the following.

(a) Determine the vertex of each function.
(b) Determine the real solutions of $f_1(x) = 0$ and $f_2(x) = 0$.
(c) One of the graphs in Figure 2.9 is the graph of $y = 7 + 6x - x^2$, and one is the graph of $y = x^2 - 6x + 10$. Determine which is which, and why.

Solution
(a) For $y = f_1(x)$, the vertex is the maximum point, at $(3, 16)$.
     For $y = f_2(x)$, the vertex is the minimum point, at $(3, 1)$.
(b) For $y = f_1(x)$, the real solutions of $f_1(x) = 0$ are the zeros, or x-intercepts, at $x = -1$ and $x = 7$. For $y = f_2(x)$, the graph has no x-intercepts, so $f_2(x) = 0$ has no real solutions.
(c) Because the graph of $y = f_1(x)$ opens downward, the coefficient of x^2 must be negative. Hence Figure 2.9(a) shows the graph of $y = f_1(x) = 7 + 6x - x^2$. Similarly, the coefficient of x^2 in $y = f_2(x)$ must be positive, so Figure 2.9(b) shows the graph of $y = f_2(x) = x^2 - 6x + 10$. ∎

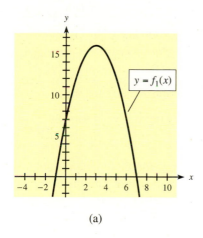

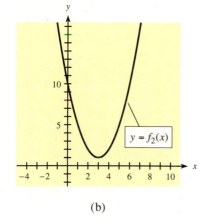

Figure 2.9   (a)   (b)

CHECKPOINT

1. Name the graph of a quadratic function.
2. (a) What is the x-coordinate of the vertex of $y = ax^2 + bx + c$?
    (b) For $y = 12x - \frac{1}{2}x^2$, what is the x-coordinate of the vertex? What is the y-coordinate of the vertex?
3. (a) How can you tell whether the vertex of $f(x) = ax^2 + bx + c$ is a maximum point or a minimum point?
    (b) In part 2(b), is the vertex a maximum point or a minimum point?
4. The zeros of a function correspond to what feature of its graph?

EXAMPLE 5 Maximizing Revenue

Ace Cruises offers a sunset cruise to a group of 50 people for a price of $30 per person, but it reduces the price per person by $0.50 for each additional person above the 50.

(a) Does reducing the price per person to get more people in the group give the cruise company more revenue?

(b) How many people will provide maximum revenue for the cruise company?

Solution

(a) The revenue to the company if 50 people are in the group and each pays $30 is $50(\$30) = \$1500$. The following table shows the revenue for the addition of people to the group. The table shows that as the group size begins to increase past 50 people, the revenue also increases. However, it also shows that increasing the group size too much (to 70 people) reduces revenue.

Increase in Group Size	Number of People	Decrease in Price	New Price ($)	Revenue ($)
0	50	0	30	$50(30) = 1500$
1	51	0.50(1)	29.50	$51(29.50) = 1504.50$
2	52	$0.50(2) = 1$	29	$52(29) = 1508$
3	53	$0.50(3) = 1.50$	28.50	$53(28.50) = 1510.50$
⋮	⋮	⋮	⋮	⋮
20	70	$0.50(20) = 10$	20	$70(20) = 1400$
⋮	⋮	⋮	⋮	⋮
x	$50 + x$	$0.50(x)$	$30 - 0.50x$	$(50 + x)(30 - 0.50x)$

(b) The last entry in the table shows the revenue for an increase of x people in the group.

$$R(x) = (50 + x)(30 - 0.50x)$$

Expanding this function gives a form from which we can find the vertex of its graph.

$$R(x) = 1500 + 5x - 0.50x^2$$

The vertex of the graph of this function is $x = \dfrac{-b}{2a} = \dfrac{-5}{2(-0.50)} = 5, R(5) = 1512.50$.

This means that the revenue will be maximized at $1512.50 when $50 + 5 = 55$ people are in the group, with each paying $30 - 0.50(5) = 27.50$ dollars.   ◼

Average Rate of Change   Because the graph of a quadratic function is not a line, the rate of change of the function is not constant. We can, however, find the **average rate of change** of a function between two input values if we know how much the function output values change between the two input values.

Average Rate of Change

The **average rate of change** of $f(x)$ with respect to x over the interval from $x = a$ to $x = b$ (where $a < b$) is calculated as

$$\text{Average rate of change} = \frac{\text{change of } f(x)}{\text{corresponding change in } x\text{-values}} = \frac{f(b) - f(a)}{b - a}$$

The average rate of change is also the slope of the segment (or secant line) joining the points $(a, f(a))$ and $(b, f(b))$.

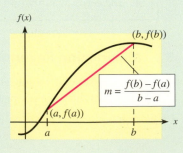

EXAMPLE **6** **Average Rate of Change of Revenue**

We found that the revenue for the cruise in Example 5 was defined by the quadratic function

$$R(x) = 1500 + 5x - 0.50x^2$$

where x is the increase in the group size beyond 50 people. What is the average rate of change of revenue if the group increases from 50 to 55 persons?

Solution

The average rate of change of revenue from 50 to 55 persons is the average rate of change of the function from $x = 0$ to $x = 5$.

$$\frac{R(5) - R(0)}{5 - 0} = \frac{1512.50 - 1500}{5} = 2.50 \text{ dollars per person}$$

This average rate of change is also the slope of the **secant line** connecting the points $(0, 1500)$ and $(5, 1512.50)$ on the graph of the function (see Figure 2.10). ■

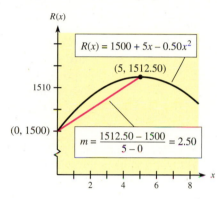

Figure 2.10

CHECKPOINT SOLUTIONS

1. Parabola

2. (a) $x = \dfrac{-b}{2a}$

   (b) $x = \dfrac{-12}{2(-1/2)} = \dfrac{-12}{(-1)} = 12$

   $y = 12(12) - \dfrac{1}{2}(12)^2 = 144 - 72 = 72$

3. (a) Maximum point if $a < 0$; minimum point if $a > 0$.

   (b) $(12, 72)$ is a maximum point because $a = -\frac{1}{2}$.

4. The x-intercepts

| EXERCISES | 2.2

Access end-of-section exercises online at **www.webassign.net**

OBJECTIVES

2.3

- To find break-even points by using quadratic cost and revenue functions
- To maximize quadratic revenue and profit functions
- To graph quadratic supply and demand functions
- To find market equilibrium by using quadratic supply and demand functions

Business Applications Using Quadratics

| APPLICATION PREVIEW |

Suppose that the supply function for a product is given by $2p - q - 38 = 0$ and the demand function is given by $p(q + 4) = 400$. Finding the market equilibrium involves solving a quadratic equation (see Example 3).

In this section, we graph quadratic supply and demand functions and find market equilibrium by solving supply and demand functions simultaneously using quadratic methods. We will also discuss quadratic revenue, cost, and profit functions, including break-even points and profit maximization.

Break-Even Points and Maximization

When we know the functions for total costs, $C(x)$, and total revenue, $R(x)$, we can find the break-even point by finding the quantity x that makes $C(x) = R(x)$.

For example, if $C(x) = 360 + 40x + 0.1x^2$ and $R(x) = 60x$, then setting $C(x) = R(x)$ we have

$$360 + 40x + 0.1x^2 = 60x$$
$$0.1x^2 - 20x + 360 = 0$$
$$x^2 - 200x + 3600 = 0$$
$$(x - 20)(x - 180) = 0$$
$$x = 20 \quad \text{or} \quad x = 180$$

Thus $C(x) = R(x)$ at $x = 20$ and at $x = 180$. If 20 items are produced and sold, $C(x)$ and $R(x)$ are both \$1200; if 180 items are sold, $C(x)$ and $R(x)$ are both \$10,800. Thus there are two break-even points. (See Figure 2.11.)

In a monopoly market, the revenue of a company is restricted by the demand for the product. In this case, the relationship between the price p of the product and the number of units sold x is described by the demand function $p = f(x)$, and the total revenue function for the product is given by

$$R = px = [f(x)]x$$

If, for example, the demand for a product is given by $p = 300 - x$, where x is the number of units sold and p is the price, then the revenue function for this product is the quadratic function

$$R = px = (300 - x)x = 300x - x^2$$

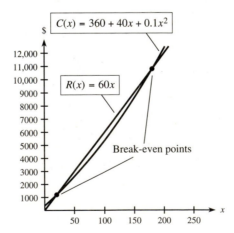

Figure 2.11

EXAMPLE   1   Break-Even Point

Suppose that in a monopoly market the total cost per week of producing a high-tech product is given by $C = 3600 + 100x + 2x^2$. Suppose further that the weekly demand function for this product is $p = 500 - 2x$. Find the number of units that will give the break-even point for the product.

Solution

The total cost function is $C(x) = 3600 + 100x + 2x^2$, and the total revenue function is $R(x) = px = (500 - 2x)x = 500x - 2x^2$.

Setting $C(x) = R(x)$ and solving for x gives

$$3600 + 100x + 2x^2 = 500x - 2x^2$$
$$4x^2 - 400x + 3600 = 0$$
$$x^2 - 100x + 900 = 0$$
$$(x - 90)(x - 10) = 0$$
$$x = 90 \quad \text{or} \quad x = 10$$

Does this mean the firm will break even at 10 units and at 90 units? Yes. Figure 2.12 shows the graphs of $C(x)$ and $R(x)$. From the graph we can observe that the firm makes a profit after $x = 10$ until $x = 90$, because $R(x) > C(x)$ in that interval. At $x = 90$, the profit is 0, and the firm loses money if it produces more than 90 units per week.   ■

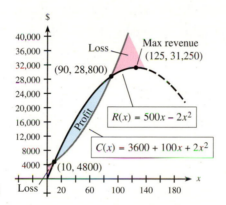

Figure 2.12

CHECKPOINT

1. The point of intersection of the revenue function and the cost function is called _____.
2. If $C(x) = 120x + 15,000$ and $R(x) = 370x - x^2$, finding the break-even points requires solution of what equation? Find the break-even points.

Note that for Example 1, the revenue function

$$R(x) = (500 - 2x)x = 500x - 2x^2$$

is a parabola that opens downward. Thus the vertex is the point at which revenue is maximum. We can locate this vertex by using the methods discussed in the previous section.

$$\text{Vertex: } x = \frac{-b}{2a} = \frac{-500}{2(-2)} = \frac{500}{4} = 125 \text{ (units)}$$

It is interesting to note that when $x = 125$, the firm achieves its maximum revenue of

$$R(125) = 500(125) - 2(125)^2 = 31,250 \text{ (dollars)}$$

but the costs when $x = 125$ are

$$C(125) = 3600 + 100(125) + 2(125)^2 = 47,350 \text{ (dollars)}$$

which results in a loss. This illustrates that maximizing revenue is not a good goal. We should seek to maximize profit. Figure 2.12 shows that maximum profit will occur where the distance between the revenue and cost curves (that is, $R(x) - C(x)$) is largest. This appears to be near $x = 50$, which is midway between the x-values of the break-even points. This is verified in the next example.

EXAMPLE 2   Profit Maximization

For the total cost function $C(x) = 3600 + 100x + 2x^2$ and the total revenue function $R(x) = 500x - 2x^2$ (from Example 1), find the number of units that maximizes profit and find the maximum profit.

Solution

Using Profit = Revenue − Cost, we can determine the profit function:

$$P(x) = (500x - 2x^2) - (3600 + 100x + 2x^2) = -3600 + 400x - 4x^2$$

This profit function is a parabola that opens downward, so the vertex will be the maximum point.

$$\text{Vertex: } x = \frac{-b}{2a} = \frac{-400}{2(-4)} = \frac{-400}{-8} = 50$$

Furthermore, when $x = 50$, we have

$$P(50) = -3600 + 400(50) - 4(50)^2 = 6400 \text{ (dollars)}$$

Thus, when 50 items are produced and sold, a maximum profit of \$6400 is made (see Figure 2.13).   ■

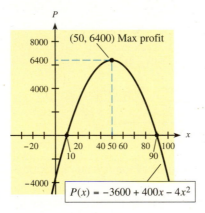

Figure 2.13

Figure 2.13 shows that the break-even points are at $x = 10$ and $x = 90$ and that the maximum profit occurs at the x-value midway between these x-values. This is reasonable because the graph of the profit function is a parabola, and the x-value of any parabola's vertex occurs midway between its x-intercepts.

It is important to note that the procedures for finding maximum revenue and profit in these examples depend on the fact that these functions are parabolas. For more general functions, procedures for finding maximum or minimum values are discussed in Chapter 10, "Applications of Derivatives."

Technology Note   Using methods discussed in Sections 2.1 and 2.2, we can use graphing utilities to locate maximum points, minimum points, and break-even points.   ■

Supply, Demand, and Market Equilibrium

The first-quadrant parts of parabolas or other quadratic equations are frequently used to represent supply and demand functions. For example, the first-quadrant part of $p = q^2 + q + 2$ (Figure 2.14(a)) may represent a supply curve, whereas the first-quadrant part of $q^2 + 2q + 6p - 23 = 0$ (Figure 2.14(b)) may represent a demand curve.

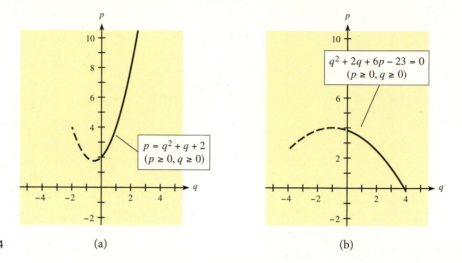

Figure 2.14   (a)   (b)

When quadratic equations are used to represent supply or demand curves, we can solve their equations simultaneously to find the market equilibrium as we did with linear supply and demand functions. As in Section 1.5, we can solve two equations in two variables by eliminating one variable and obtaining an equation in the other variable. When the functions are quadratic, the substitution method of solution is perhaps the best, and the resulting equation in one unknown will usually be quadratic.

EXAMPLE **3** **Market Equilibrium** | **APPLICATION PREVIEW** |

If the demand function for a commodity is given by $p(q + 4) = 400$ and the supply function is given by $2p - q - 38 = 0$, find the market equilibrium.

Solution

Solving the supply equation $2p - q - 38 = 0$ for p gives $p = \frac{1}{2}q + 19$. Substituting for p in $p(q + 4) = 400$ gives

$$\left(\frac{1}{2}q + 19\right)(q + 4) = 400$$

$$\frac{1}{2}q^2 + 21q + 76 = 400$$

$$\frac{1}{2}q^2 + 21q - 324 = 0$$

Multiplying both sides of the equation by 2 yields $q^2 + 42q - 648 = 0$. Factoring gives

$$(q - 12)(q + 54) = 0$$

$$q = 12 \quad \text{or} \quad q = -54$$

Thus the market equilibrium occurs when 12 items are sold, at a price of $p = \frac{1}{2}(12) + 19 = \$25$ each. The graphs of the demand and supply functions are shown in Figure 2.15 (on the next page). ∎

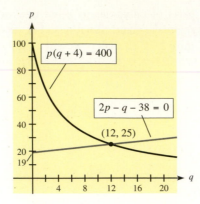

Figure 2.15

Calculator Note   Graphing calculators also can be used to sketch these graphs. A command such as INTERSECT could be used to determine points of intersection that give market equilibrium.

Figure 2.16(a) shows the graph of the supply and demand functions for the commodity in Example 3. Using the INTERSECT command gives the same market equilibrium point determined in Example 3 (see Figure 2.16(b)). ■

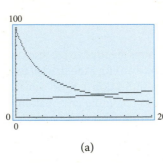
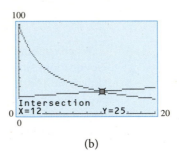

Figure 2.16   (a)   (b)

CHECKPOINT

3. The point of intersection of the supply and demand functions is called _____.
4. If the demand and supply functions for a product are

$$p + \frac{1}{10} q^2 = 1000 \quad \text{and} \quad p = \frac{1}{10} q + 10$$

respectively, finding the market equilibrium point requires solution of what equation? Find the market equilibrium.

CHECKPOINT SOLUTIONS

1. The break-even point
2. Solution of $C(x) = R(x)$. That is, solution of $120x + 15,000 = 370x - x^2$, or $x^2 - 250x + 15,000 = 0$.

$$(x - 100)(x - 150) = 0$$
$$x = 100 \quad \text{or} \quad x = 150$$

Thus the break-even points are when 100 units and 150 units are produced.
3. The market equilibrium point or market equilibrium
4. Solve $-\dfrac{1}{10} q^2 + 1000 = \dfrac{1}{10} q + 10$:

$$-q^2 + 10,000 = q + 100$$
$$0 = q^2 + q - 9900$$
$$0 = (q + 100)(q - 99)$$
$$q = -100 \text{ or } q = 99$$

Thus market equilibrium occurs when $q = 99$ and $p = 9.9 + 10 = 19.90$.

| EXERCISES | 2.3

Access end-of-section exercises online at **www.webassign.net**

OBJECTIVES   2.4

- To graph and apply basic functions, including constant and power functions
- To graph and apply polynomial and rational functions
- To graph and apply absolute value and piecewise defined functions

Special Functions and Their Graphs

| APPLICATION PREVIEW |

The average cost per item for a product is calculated by dividing the total cost by the number of items. Hence, if the total cost function for x units of a product is

$$C(x) = 900 + 3x + x^2$$

and if we denote the average cost function by $\overline{C}(x)$, then

$$\overline{C}(x) = \frac{C(x)}{x} = \frac{900 + 3x + x^2}{x}$$

This average cost function is a special kind of function, called a rational function. We can use this function to find the minimum average cost. (See Example 4.)

In this section, we discuss polynomial, rational, and other special functions.

Basic Functions   The special linear function

$$y = f(x) = x$$

is called the **identity function** (see Figure 2.17(a)), and a linear function defined by

$$y = f(x) = C \quad C \text{ a constant}$$

is called a **constant function.** Figure 2.17(b) shows the graph of the constant function $y = f(x) = 2$. (Note that the slope of the graph of any constant function is 0.)

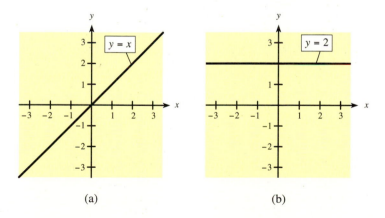

Figure 2.17   (a)   (b)

The functions of the form $y = ax^b$, where $b > 0$, are called **power functions.** Examples of power functions include $y = x^2$, $y = x^3$, $y = \sqrt{x} = x^{1/2}$, and $y = \sqrt[3]{x} = x^{1/3}$. (See Figure 2.18(a)–(d) on the next page.) The functions $y = \sqrt{x}$ and $y = \sqrt[3]{x}$ are also called **root functions,** and $y = x^2$ and $y = x^3$ are basic **polynomial functions.**

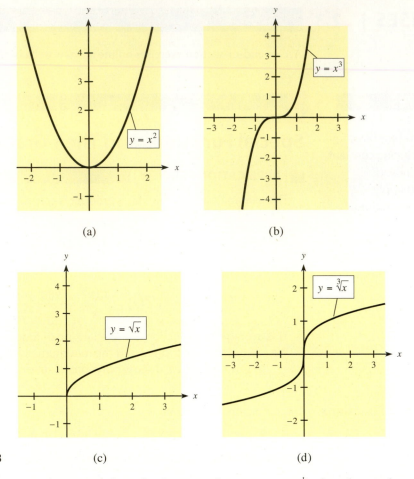

Figure 2.18   (c)   (d)

The general shape for the power function $y = ax^b$, where $b > 0$, depends on the value of b. Figure 2.19 shows the first-quadrant portions of typical graphs of $y = x^b$ for different values of b. Note how the direction in which the graph bends differs for $b > 1$ and for $0 < b < 1$. Getting accurate graphs of these functions requires plotting a number of points by hand or with a graphing utility. Our goal at this stage is to recognize the basic shapes of certain functions.

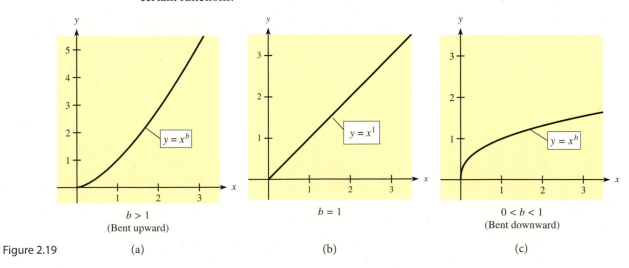

Figure 2.19   (a)   (b)   (c)

Shifts   In Section 2.2, "Quadratic Functions: Parabolas," we noted that the graph of $y = (x - h)^2 + k$ is a parabola that is shifted h units in the x-direction and k units in the y-direction. In general, we have the following.

Shifts of Graphs	The graph of $y = f(x - h) + k$ is the graph of $y = f(x)$ shifted h units in the x-direction and k units in the y-direction.

EXAMPLE 1   Shifted Graph

The graph of $y = x^3$ is shown in Figure 2.18(b) on the preceding page.

(a) Describe the graph of $y = x^3 - 3$ and graph this function.
(b) Describe the graph of $y = (x - 2)^3$ and graph this function.
(c) Describe the graph of $y = (x - 2)^3 - 3$ and graph this function.

Solution

(a) The graph of $y = x^3 - 3$ is the graph of $y = x^3$ shifted -3 units in the y-direction (down 3 units). The graph is shown in Figure 2.20(a).
(b) The graph of $y = (x - 2)^3$ is the graph of $y = x^3$ shifted 2 units in the x-direction (to the right 2 units). The graph is shown in Figure 2.20(b).
(c) The graph of $y = (x - 2)^3 - 3$ is the graph of $y = x^3$ shifted to the right 2 units and down 3 units. The graph is shown in Figure 2.20(c).   ∎

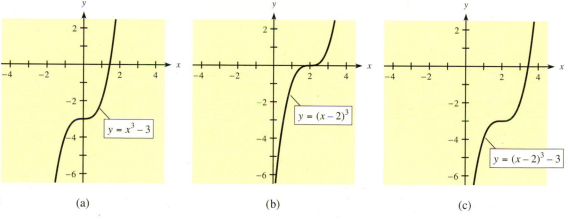

Figure 2.20   (a)   (b)   (c)

Polynomial Functions

A **polynomial function of degree n** has the form

$$y = a_n x^n + a_{n-1} x^{n-1} + \cdots + a_1 x + a_0$$

where $a_n \neq 0$, and n is an integer, $n \geq 0$.

A **linear function** is a polynomial function of degree 1, and a **quadratic function** is a polynomial function of degree 2.

Accurate graphing of a polynomial function of degree greater than 2 may require the methods of calculus; we will investigate these methods in Chapter 10, "Applications of Derivatives." For now, we will observe some characteristics of the graphs of polynomial functions of degrees 2, 3, and 4; these are summarized in Table 2.1 on the next page. Using this information and point plotting or using a graphing utility yields the graphs of these functions.

| TABLE 2.1 |

GRAPHS OF SOME POLYNOMIALS

	Degree 2	Degree 3	Degree 4
Turning points	1	0 or 2	1 or 3
x-intercepts	0, 1, or 2	1, 2, or 3	0, 1, 2, 3, or 4
Possible shapes	U or ∩ (a) (b)	(a) (b) (c) (d)	(a) (b) (c) (d) (e) (f)

Calculator Note The function $y = x^3 - 16x$ is a third-degree (cubic) polynomial function, so it has one of the four shapes shown in the "Degree 3" column in Table 2.1. Details of graphing this function are shown in Appendix C, Section 2.4 (see Figure 2.21). ■

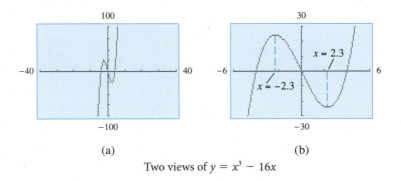

(a)   (b)

Figure 2.21   Two views of $y = x^3 - 16x$

EXAMPLE 2   Quartic Polynomial

Graph $y = x^4 - 2x^2$.

Solution

This is a degree 4 (quartic) polynomial function. Thus it has one or three turning points and has one of the six shapes in the "Degree 4" column of Table 2.1. We begin by graphing this function with a large viewing window to obtain the graph in Figure 2.22(a). Figure 2.22(b) shows the same graph with a smaller viewing window, near $x = 0$. We now see that the graph has shape (f) from Table 2.1, with turning points at $x = -1, x = 0$, and $x = 1$. ■

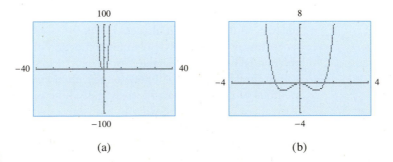

(a)   (b)

Figure 2.22

Spreadsheet Note It is also possible to graph polynomials with Excel. See Appendix D, Section 2.4 and the Online Excel Guide. ■

CHECKPOINT

1. All constant functions (such as $f(x) = 8$) have graphs that are _____.
2. Which of the following are polynomial functions?
   (a) $f(x) = x^3 - x + 4$
   (b) $f(x) = \dfrac{x + 1}{4x}$
   (c) $f(x) = 1 + \sqrt{x}$
   (d) $g(x) = \dfrac{1 + \sqrt{x}}{1 + x + \sqrt{x}}$
   (e) $h(x) = 5x$ Two views of $y = x^4 - 2x^2$
3. A third-degree polynomial can have at most _____ turning points.

Rational Functions

The function $f(x) = \dfrac{1}{x}$ is a special function called a **reciprocal function.** A table of sample values and the graph of $y = \dfrac{1}{x}$ are shown in Figure 2.23. Because division by 0 is not possible, $x = 0$ is not in the domain of the function.

x	y
−3	−0.33
−2	−0.5
−1	−1
−0.1	−10
−0.04	−25
−0.001	−1000
3	0.33
2	0.5
1	1
0.1	10
0.04	25
0.001	1000

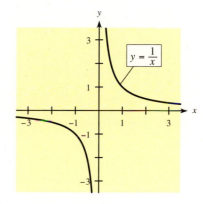

Figure 2.23

The reciprocal function $y = \dfrac{1}{x}$ is also a **rational function,** which is defined below.

Rational Function

A **rational function** is a function of the form

$$y = \frac{f(x)}{g(x)} \quad \text{with } g(x) \neq 0$$

where $f(x)$ and $g(x)$ are both polynomials. Its domain is the set of all real numbers for which $g(x) \neq 0$.

Asymptotes

The table of values and the graph in Figure 2.23 show that $|y|$ gets very large as x approaches 0, and that y is undefined at $x = 0$. In this case, we call the line $x = 0$ (that is, the y-axis) a **vertical asymptote.** On the graphs of polynomial functions, the turning points are usually the features of greatest interest. However, on graphs of rational functions, vertical asymptotes frequently are the most interesting features.

Rational functions sometimes have **horizontal asymptotes** as well as vertical asymptotes. Whenever the values of y approach some finite number b as $|x|$ becomes very large, we say that there is a horizontal asymptote at $y = b$.

Note that the graph of $y = \frac{1}{x}$ appears to get close to the x-axis as $|x|$ becomes large. Testing values of x for which $|x|$ is large, we see that y is close to 0. Thus, we say that the line $y = 0$ (or the x-axis) is a horizontal asymptote for the graph of $y = \frac{1}{x}$. The graph in the following example has both a vertical and a horizontal asymptote.

EXAMPLE 3   **Rational Function**

(a) Use values of x from -5 to 5 to develop a table of function values for the graph of

$$y = \frac{12x + 8}{3x - 9}$$

(b) Sketch the graph.

Solution

(a) Because $3x - 9 = 0$ when $x = 3$, it follows that $x = 3$ is not in the domain of this function. The values in the table indicate that the graph is approaching a vertical asymptote at $x = 3$.

x	-5	-4	-3	-2	-1	0	1
y	2.17	1.90	1.56	1.07	0.33	-0.89	-3.33

2	2.5	2.9	3.1	3.5	4	5
-10.7	-25.33	-142.67	150.67	33.33	18.67	11.33

To see if the function has a horizontal asymptote, we calculate y as $|x|$ becomes larger.

x	$-10{,}000$	-1000	1000	$10{,}000$
y	3.999	3.99	4.01	4.001

This table indicates that the graph is approaching $y = 4$ as $|x|$ increases, so we have a horizontal asymptote at $y = 4$.

(b) Using the information about vertical and horizontal asymptotes and plotting these points give the graph in Figure 2.24(a). ■

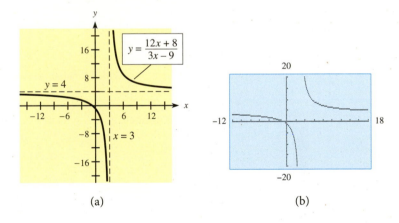

Figure 2.24   (a)   (b)

Calculator Note   When a graphing calculator is used to graph a discontinuous function, the discontinuity will be visible if the window is chosen carefully. For example, Figure 2.24(b) shows the graph of $y = \dfrac{12x + 8}{3x - 9}$ on a TI-84 calculator with the window $[-12, 18]$ by $[-20, 20]$. Note that a window centered at the x-value of the vertical asymptote ($x = 3$) shows the discontinuity. For details, see Appendix C, Section 2.4. ■

Spreadsheet Note   When graphing a function that is discontinuous at $x = a$ with Excel, we must leave the cell corresponding to $f(a)$ blank so that Excel will not connect the values. See Appendix D, Section 2.4 and the Online Excel Guide for details. ■

A general procedure for finding vertical and horizontal asymptotes of rational functions follows (see proof in Chapter 9) .

Asymptotes

Vertical Asymptote

The graph of the rational function $y = \dfrac{f(x)}{g(x)}$ has a vertical asymptote at $x = c$ if $g(c) = 0$ and $f(c) \neq 0$.

Horizontal Asymptote

Consider the rational function $y = \dfrac{f(x)}{g(x)} = \dfrac{a_n x^n + \cdots + a_1 x + a_0}{b_m x^m + \cdots + b_1 x + b_0}$.

1. If $n < m$ (that is, if the degree of the numerator is less than that of the denominator), a horizontal asymptote occurs at $y = 0$ (the x-axis).
2. If $n = m$ (that is, if the degree of the numerator equals that of the denominator), a horizontal asymptote occurs at $y = \dfrac{a_n}{b_m}$ (the ratio of the leading coefficients).
3. If $n > m$ (that is, if the degree of the numerator is greater than that of the denominator), there is no horizontal asymptote.

Recall that the graph of the function $y = \dfrac{12x + 8}{3x - 9}$ (shown in Figure 2.24) has a vertical asymptote at $x = 3$, and that $x = 3$ makes the denominator 0 but does not make the numerator 0. Observe also that this graph also has a horizontal asymptote at $y = 4$, and that the degree of the numerator is the same as that of the denominator with the ratio of the leading coefficients equal to $12/3 = 4$.

Determining the vertical and/or horizontal asymptotes of a rational function, if any exist, is very useful in determining a window for graphing. A rational function may or may not have turning points. The graph of the rational function in the following example has a vertical asymptote and a turning point.

EXAMPLE 4

Total Costs and Average Costs | APPLICATION PREVIEW |

Suppose the total cost function for x units of a product is given by

$$C(x) = 900 + 3x + x^2$$

Graph the average cost function for this product, and determine the minimum average cost.

Solution

The average cost per unit is

$$\overline{C}(x) = \frac{C(x)}{x} = \frac{900 + 3x + x^2}{x} \qquad \text{or} \qquad \overline{C}(x) = \frac{900}{x} + 3 + x$$

Because x represents the number of units produced, $x \geq 0$. Because $x = 0$ cannot be in the domain of the function, we choose a window with $x \geq 0$. Figure 2.25 shows the graph of $\overline{C}(x)$. The graph appears to have a minimum near $x = 30$. By plotting points or using MINIMUM on a graphing utility, we can verify that the minimum point occurs at $x = 30$ and $\overline{C} = 63$. Thus the minimum average cost is \$63 per unit when 30 units are produced. ■

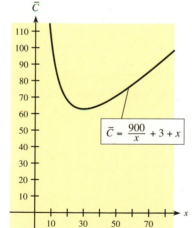

$\overline{C} = \dfrac{900}{x} + 3 + x$

Figure 2.25

CHECKPOINT

4. Given $f(x) = \dfrac{3x}{x - 4}$, decide whether the following are true or false.

   (a) $f(x)$ has a vertical asymptote at $x = 4$.
   (b) $f(x)$ has a horizontal asymptote at $x = 3$.

Piecewise Defined
Functions

Another special function comes from the definition of $|x|$. The **absolute value function** can be written as

$$f(x) = |x| \quad \text{or} \quad f(x) = \begin{cases} x & \text{if } x \geq 0 \\ -x & \text{if } x < 0 \end{cases}$$

Note that restrictions on the domain of the absolute value function specify different formulas for different parts of the domain. To graph $f(x) = |x|$, we graph the portion of the line $y = x$ for $x \geq 0$ (see Figure 2.26(a)) and the portion of the line $y = -x$ for $x < 0$ (see Figure 2.26(b)). When we put these pieces on the same graph (Figure 2.26(c)), they give us the graph of $f(x) = |x|$. Because the absolute value function is defined by two equations, we say it is a **piecewise defined function.**

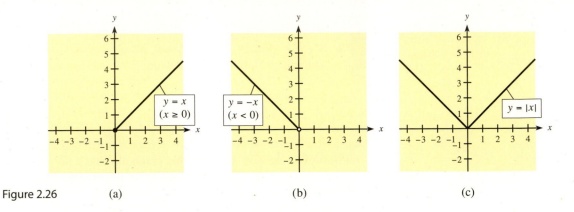

Figure 2.26   (a)   (b)   (c)

It is possible for the selling price S of a product to be defined as a piecewise function of the cost C of the product. For example, the selling price might be defined by two different equations on two different intervals, as follows:

$$S = f(C) = \begin{cases} 3C & \text{if } 0 \leq C \leq 20 \\ 1.5C + 30 & \text{if } C > 20 \end{cases}$$

When we write the equations in this way, the value of S depends on the value of C, so C is the independent variable and S is the dependent variable.

Note that the selling price of a product that costs \$15 would be $f(15) = 3(15) = 45$ (dollars) and that the selling price of a product that costs \$25 would be $f(25) = 1.5(25) + 30 = 67.50$ (dollars). Each of the two pieces of the graph of this function is a line and is easily graphed. It remains only to graph each in the proper interval. The graph is shown in Figure 2.27(a).

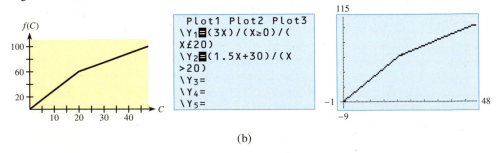

Figure 2.27   (b)

Calculator Note

Graphing calculators can be used to graph piecewise defined functions by entering each piece as a separate equation, along with the interval over which it is defined. The details of graphing piecewise defined functions with a graphing calculator are shown in Appendix C, Section 2.4. Figure 2.27(b) shows how the selling price function graphed in Figure 2.27(a) could be entered on a TI-84 calculator, using x to represent C and Figure 2.27(c) shows the graph. ∎

EXAMPLE 5   **Residential Electrical Costs**

The 2012 monthly charge in dollars for x kilowatt hours (kWh) of electricity used by a residential customer of Excelsior Electric Membership Corporation during the months of November through June is given by the function

$$C(x) = \begin{cases} 10 + 0.094x & \text{if } 0 \le x \le 100 \\ 19.40 + 0.075(x - 100) & \text{if } 100 < x \le 500 \\ 49.40 + 0.05(x - 500) & \text{if } x > 500 \end{cases}$$

(a) What is the monthly charge if 1100 kWh of electricity are consumed in a month?
(b) What is the monthly charge if 450 kWh are consumed in a month?

Solution

(a) We need to find $C(1100)$. Because $1100 > 500$, we use it in the third formula line.

$$C(1100) = 49.40 + 0.05(1100 - 500) = \$79.40$$

(b) We evaluate $C(450)$ by using the second formula line for $C(x)$.

$$C(450) = 19.40 + 0.075(450 - 100) = \$45.65$$

CHECKPOINT

5. If $f(x) = \begin{cases} 5 & \text{if } x \le 0 \\ 2x & \text{if } 0 < x < 5, \text{ find the following.} \\ x + 6 & \text{if } x \ge 5 \end{cases}$

(a) $f(-5)$ (b) $f(4)$ (c) $f(20)$

CHECKPOINT SOLUTIONS

1. Horizontal lines
2. Polynomial functions are (a) and (e).
3. Two
4. (a) True.
    (b) False. The horizontal asymptote is $y = 3$.
5. (a) $f(-5) = 5$. In fact, for any negative value of x, $f(x) = 5$.
    (b) $f(4) = 2(4) = 8$
    (c) $f(20) = 20 + 6 = 26$

| EXERCISES | 2.4

Access end-of-section exercises online at **www.webassign.net**   **WebAssign**

OBJECTIVES   2.5

- To graph data points in a scatter plot
- To determine the function type that will best model data
- To use a graphing utility to create an equation that models the data
- To graph the data points and model on the same graph

Modeling; Fitting Curves to Data with Graphing Utilities (optional)

■ | APPLICATION PREVIEW |

When the plot of a set of data points looks like a line, we can use a technique called linear regression (or the least-squares method) to find the equation of the line that is the best fit for the data points. The resulting equation (model) describes the data and gives a formula to plan for the future. We can use technology to fit linear and other functions to sets of data.

For selected years from 1970 to 2010, the data in Table 2.2 (on the next page) show the amount of tax paid per capita (per person). In this section we develop a model for these data (see Example 3).

Linear Regression

| TABLE 2.2 |

Year	Federal Tax per Capita
1970	$955
1975	1376
1980	2276
1985	3099
1990	4208
1995	5144
2000	7404
2005	7625
2010	7211

Source: Internal Revenue Service

Figure 2.28

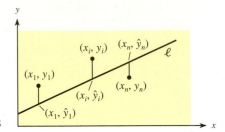

We saw in Section 1.3, "Linear Functions," that it is possible to write the equation of a straight line if we have two points on the line. Business firms frequently like to treat demand functions as though they were linear, even when they are not exactly linear. They do this because linear functions are much easier to handle than other functions. If a firm has more than two points describing the demand for its product, it is likely that the points will not all lie on the same straight line. However, by using a technique called **linear regression,** the firm can determine the "best line" that fits these points.

Suppose we have the points shown in Figure 2.28, and we seek the line that is the "best fit" for the points.

The formulas that give the best-fitting line for given data are developed in Chapter 14, "Functions of Two or More Variables," but graphing calculators, computer programs, and spreadsheets have built-in formulas or programs that give the equation of the best-fitting line for a set of data.

Modeling

Creating an equation, or **mathematical model,** for a set of data is a complex process that involves careful analysis of the data in accordance with a number of guidelines. Our approach to creating a model is much simplified and relies on a knowledge of the appearance of linear, quadratic, power, cubic, and quartic functions and the capabilities of computers and graphing calculators. We should note that the equations the computers and calculators provide are based on sophisticated formulas that are derived using calculus of several variables. These formulas provide the best fit for the data using the function type we choose, but we should keep the following limitations in mind.

1. The computer/calculator will give the best fit for whatever type of function we choose (even if the selected function is a bad fit for the points), so we must choose a function type carefully. To choose a function type, compare the plot of the data with the graphs of the functions discussed earlier in this chapter. The model will not fit the data if we choose a function type whose graph does not match the shape of the plotted data.
2. Some sets of data have no pattern, so they cannot be modeled by functions. Other data sets cannot be modeled by the functions we have studied.
3. Modeling provides a formula that relates the data. Even though a model may accurately fit a data set, it may not be a good predictor for values outside the data set.

To find a function that models a set of data, the first step is to plot the data and determine the shape of the curve that would best fit the data points. The plot of the data points is called a **scatter plot.** The scatter plot's shape determines the type of function that will be the best model for the data. Details of creating a scatter plot with a graphing calculator are shown in Appendix C, Section 2.5.

EXAMPLE 1 Curve Fitting

Graphs of three scatter plots are shown in Figure 2.29(a–c). Determine what type of function is your choice as the best-fitting curve for each scatter plot. If it is a polynomial function, state the degree.

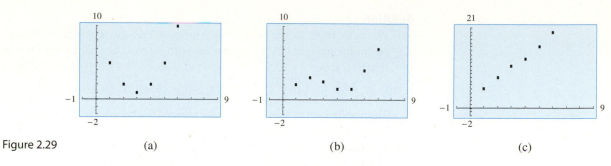

Figure 2.29   (a)   (b)   (c)

Solution

(a) It appears that a parabola will fit the data points in Figure 2.29(a), so the data points can be modeled by a second-degree (quadratic) function.

(b) The two "turns" in the scatter plot in Figure 2.29(b) suggest that it could be modeled by a cubic function.

(c) The data points in Figure 2.29(c) appear to lie (approximately) along a line, so the data can be modeled by a linear function.   ■

After the type of function that gives the best fit for the data is chosen, a graphing utility or spreadsheet can be used to develop the best-fitting equation for the function chosen. The following steps are used to find an equation that models data.

Modeling Data

1. Enter the data and use a graphing utility or spreadsheet to plot the data points. (This gives a scatter plot.)
2. Visually determine what type of function (including degree, if it is a polynomial function) would have a graph that best fits the data.
3. Use a graphing utility or spreadsheet to determine the equation of the chosen type that gives the best fit for the data.
4. To see how well the equation models the data, graph the equation and the data points on the same set of axes. If the graph of the equation does not fit the data points well, another type of function may model the data better.
5. After the model for a data set has been found, it can be rounded for reporting purposes. However, use the unrounded model in graphing and in calculations, unless otherwise instructed. Numerical answers found using a model should be rounded in a way that agrees with the context of the problem and with no more accuracy than the original output data.

The graphing calculator steps used to create a function that approximately fits a scatter plot are shown in Appendix C, Section 2.5.

EXAMPLE 2   U.S. Households with Internet Access

The following table gives the percent of U.S. households with Internet access in various years.

Year	1996	1997	1998	1999	2000	2001	2003	2008
Percent	8.5	14.3	26.2	28.6	41.5	50.5	52.4	78.0

Source: U.S. Census Bureau

(a) Use a graphing utility to create a scatter plot of the data, with x equal to the number of years after 1995.

(b) Use the utility to create an equation that models the data.

(c) Graph the function and the data on the same graph to see how well the function models ("fits") the data.

Solution

(a) When we use x as the number of years past 1995 (so $x = 0$ in 1995) and use y to represent the percent of U.S. households with Internet access, the points representing this function are

$$(1, 8.5) \quad (2, 14.3) \quad (3, 26.2) \quad (4, 28.6) \quad (5, 41.5) \quad (6, 50.5) \quad (8, 52.4) \quad (13, 78.0)$$

The scatter plot of these points is developed in Appendix C, Section 2.5, and shown in Figure 2.30(a).

(b) It appears that a line could be found that would fit close to the points in Figure 2.30(a). The points do not fit exactly on a line, because the percent of U.S. households with Internet access did not increase by exactly the same amount each year. The steps used to create the linear function that models these points are shown in Appendix C, Section 2.5. The linear equation that is the best fit for the points is

$$y = 5.79x + 7.13$$

(c) Figure 2.30(b) shows the graph of the function and the data on the same graph. Note that not all points fit on the graph of the equation, even though this is the line that is the best fit for the data.   ■

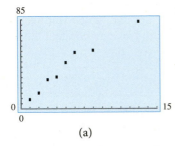

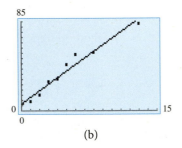

Figure 2.30   (a)   (b)

EXAMPLE   3   Federal Tax per Capita | APPLICATION PREVIEW |

The following table gives the amount of federal tax paid per capita (per person) for selected years from 1970 to 2010.

(a) Create a scatter plot of these data, with $x = 0$ representing 1970.
(b) Find a cubic model that fits the data points.
(c) Use the model to estimate the per capita tax for 2010.

Year	Federal Tax per Capita	Year	Federal Tax per Capita
1970	$955	1995	$5144
1975	1376	2000	7404
1980	2276	2005	7625
1985	3099	2010	7211
1990	4208		

Source: Internal Revenue Service

Solution

(a) The scatter plot is shown in Figure 2.31(a). The "turns" in the graph indicate that a cubic function would be a good fit for the data.

(b) The cubic function that is the best fit for the data is

$$y = -0.278x^3 + 16.2x^2 - 46.9x + 1110$$

The graph of the function and the scatter plot are shown in Figure 2.31(b).

(c) The value of y for $x = 40$ (shown in Figure 2.31(c)) predicts about \$7362 as the tax per capita in 2010. However, the downward turn in the graph suggests a decrease in future tax per capita and this is highly unlikely with current government debt. Thus, the model cannot be applied beyond 2010. ■

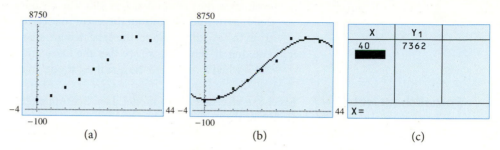

Figure 2.31 (a) (b) (c)

If it is not obvious what model will best fit a given set of data, several models can be developed and compared graphically with the data.

EXAMPLE 4 Expected Life Span

The expected life span of people in the United States depends on their year of birth.

Year	Life Span (years)	Year	Life Span (years)	Year	Life Span (years)	Year	Life Span (years)
1920	54.1	1960	69.7	1985	74.7	2005	77.9
1930	59.7	1970	70.8	1990	75.4	2010	78.1
1940	62.9	1975	72.6	1995	75.8	2015	78.9
1950	68.2	1980	73.7	2000	77.0	2020	79.5

Source: National Center for Health Statistics

(a) Create linear and quadratic models that give life span as a function of birth year with $x = 0$ representing 1900 and, by visual inspection, decide which model gives the better fit.
(b) Use both models to estimate the life span of a person born in the year 2000.
(c) Which model's prediction for the life span in 2015 seems better?

Solution

(a) The scatter plot for the data is shown in Figure 2.32(a). It appears that a linear function could be used to model the data. The linear equation that is the best fit for the data is

$$y = 0.233x + 53.71$$

The graph in Figure 2.32(b) shows how well the line fits the data points. The quadratic function that is the best fit for the data is

$$y = -0.00186x^2 + 0.498x + 46.0$$

Its graph is shown in Figure 2.32(c).

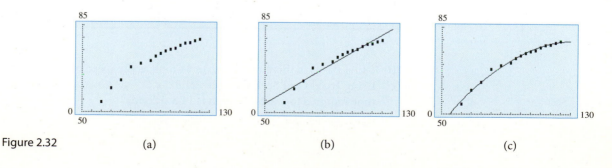

Figure 2.32 (a) (b) (c)

| TABLE 2.3 |

X	Y₁	Y₂
90	74.68	75.754
100	77.01	77.2
110	79.34	78.274
115	80.505	78.672
▬		

X=

The quadratic model appears to fit the data points better than the linear model. A table can be used to compare the models (see Table 2.3 with y_1 giving values from the linear model and y_2 giving values from the quadratic model).

(b) We can estimate the life span of people born in the year 2000 with either model by evaluating the functions at $x = 100$. The x-values in Table 2.3 represent the years 1990, 2000, 2010, and 2015. From the linear model the expected life span in 2000 is 77.0, and from the quadratic model the expected life span is 77.2.

(c) Looking at Table 2.3, we see that the linear model may be giving optimistic values in 2015, when $x = 115$, so the quadratic model may be better in the years after 2010. ■

Spreadsheet Note   Excel can also be used to model data. See Appendix D, Section 2.5 and the Online Excel Guide for details. ■

EXAMPLE 5   **Corvette Acceleration**

The following table shows the times that it takes a Corvette to reach speeds from 0 mph to 100 mph, in increments of 10 mph after 30 mph.

Time (sec)	Speed (mph)	Time (sec)	Speed (mph)
1.7	30	5.5	70
2.4	40	7.0	80
3.5	50	8.5	90
4.3	60	10.2	100

(a) Use a power function to model the data, and graph the points and the function to see how well the function fits the points.

(b) What does the model indicate the speed is 5 seconds after the car starts to move?

(c) Use the model to determine the number of seconds until the Corvette reaches 110 mph.

Solution

(a) Using x as the time and y as the speed, the scatter plot (found with Excel) is shown in Figure 2.33(a). An equation found using the power function $y = ax^b$ (graphed in Figure 2.33(b)) is

$$y = 21.875x^{0.6663}$$

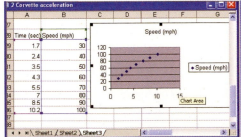
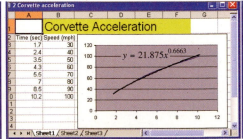

Figure 2.33

(b) Using $x = 5$ in this equation gives 64 mph as the speed of the Corvette.

(c) Using Excel Goal Seek with $y = 21.875x^{0.6663}$ set equal to 110 gives $x = 11.3$, so the time required to reach 110 mph is 11.3 seconds. ■

| EXERCISES | 2.5

Access end-of-section exercises online at **www.webassign.net**

KEY TERMS AND FORMULAS

Section	Key Terms	Formulas
2.1	Quadratic equation	$ax^2 + bx + c = 0 \quad (a \neq 0)$
	Square root property	$x^2 = C \Rightarrow x = \pm\sqrt{C}$
	Quadratic formula	$x = \dfrac{-b \pm \sqrt{b^2 - 4ac}}{2a}$
	Quadratic discriminant	$b^2 - 4ac$
2.2	Quadratic function; parabola	$y = f(x) = ax^2 + bx + c$
	Vertex of a parabola	$(-b/2a, f(-b/2a))$ Maximum point if $a < 0$ Minimum point if $a > 0$
	Axis of symmetry	$x = -b/2a$
	Zeros of a quadratic function	$x = \dfrac{-b \pm \sqrt{b^2 - 4ac}}{2a}$
2.3	Supply function Demand function Market equilibrium Cost, revenue, and profit functions Break-even points	 $P(x) = R(x) - C(x)$ $R(x) = C(x)$ or $P(x) = 0$
2.4	Basic functions	$f(x) = ax + b$ (linear) $f(x) = C, \; C = $ constant $f(x) = ax^b$ (power) $f(x) = x$ (identity) $f(x) = x^2, f(x) = x^3, f(x) = 1/x$
	Shifts of graphs	$f(x - h)$ shifts $f(x)$ h units horizontally $f(x) + k$ shifts $f(x)$ k units vertically
	Polynomial function of degree n	$f(x) = a_n x^n + a_{n-1}x^{n-1} + \cdots + a_1 x + a_0$ $a_n \neq 0, n \geq 0, n$ an integer
	Rational function	$f(x) = p(x)/q(x)$, where $p(x)$ and $q(x)$ are polynomials
	Vertical asymptote Horizontal asymptote	
	Absolute value function Piecewise defined functions	$f(x) = |x| = \begin{cases} x & \text{if } x \geq 0 \\ -x & \text{if } x < 0 \end{cases}$
2.5	Scatter plots Function types Mathematical model Fitting curves to data points Predicting from models	

REVIEW EXERCISES

In Problems 1–10, find the real solutions to each qua-dratic equation.

1. $3x^2 + 10x = 5x$
2. $4x - 3x^2 = 0$
3. $x^2 + 5x + 6 = 0$
4. $11 - 10x - 2x^2 = 0$
5. $(x - 1)(x + 3) = -8$
6. $4x^2 = 3$
7. $20x^2 + 3x = 20 - 15x^2$
8. $8x^2 + 8x = 1 - 8x^2$
9. $7 = 2.07x - 0.02x^2$
10. $46.3x - 117 - 0.5x^2 = 0$

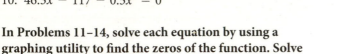

In Problems 11–14, solve each equation by using a graphing utility to find the zeros of the function. Solve the equation algebraically to check your results.

11. $4z^2 + 25 = 0$
12. $z(z + 6) = 27$
13. $3x^2 - 18x - 48 = 0$
14. $3x^2 - 6x - 9 = 0$
15. Solve $x^2 + ax + b = 0$ for x.
16. Solve $xr^2 - 4ar - x^2c = 0$ for r.

In Problems 17 and 18, approximate the real solutions to each quadratic equation to two decimal places.

17. $23.1 - 14.1x - 0.002x^2 = 0$
18. $1.03x^2 + 2.02x - 1.015 = 0$

For each function in Problems 19–24, find the vertex and determine if it is a maximum or minimum point, find the zeros if they exist, and sketch the graph.

19. $y = \frac{1}{2}x^2 + 2x$
20. $y = 4 + \frac{1}{4}x^2$
21. $y = 6 + x - x^2$
22. $y = x^2 - 4x + 5$
23. $y = x^2 + 6x + 9$
24. $y = 12x - 9 - 4x^2$

In Problems 25–30, use a graphing utility to graph each function. Use the vertex and zeros to determine an appropriate range. Be sure to label the maximum or minimum point.

25. $y = \frac{1}{3}x^2 - 3$
26. $y = \frac{1}{2}x^2 - 2$
27. $y = x^2 + 2x + 5$
28. $y = -10 + 7x - x^2$
29. $y = 20x - 0.1x^2$
30. $y = 50 - 1.5x + 0.01x^2$
31. Find the average rate of change of $f(x) = 100x - x^2$ from $x = 30$ to $x = 50$.
32. Find the average rate of change of

$$f(x) = x^2 - 30x + 22$$

over the interval $[10, 50]$.

In Problems 33–36, a graph is given. Use the graph to

(a) **locate the vertex,**

(b) **determine the zeros, and**

(c) **match the graph with one of the equations A, B, C, or D.**

A. $y = 7x - \frac{1}{2}x^2$

B. $y = \frac{1}{2}x^2 - x - 4$

C. $y = 8 - 2x - x^2$

D. $y = 49 - x^2$

33.
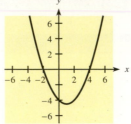

34.
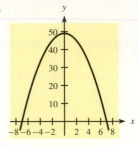

35.
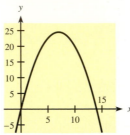

36.
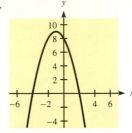

37. Sketch a graph of each of the following basic functions.
   (a) $f(x) = x^2$ (b) $f(x) = 1/x$ (c) $f(x) = x^{1/4}$

38. If $f(x) = \begin{cases} -x^2 & \text{if } x \le 0 \\ 1/x & \text{if } x > 0 \end{cases}$, find the following.

   (a) $f(0)$ (b) $f(0.0001)$
   (c) $f(-5)$ (d) $f(10)$

39. If $f(x) = \begin{cases} x & \text{if } x \le 1 \\ 3x - 2 & \text{if } x > 1 \end{cases}$, find the following.

   (a) $f(-2)$ (b) $f(0)$ (c) $f(1)$ (d) $f(2)$

In Problems 40 and 41, graph each function.

40. $f(x) = \begin{cases} x & \text{if } x \le 1 \\ 3x - 2 & \text{if } x > 1 \end{cases}$

41. (a) $f(x) = (x - 2)^2$ (b) $f(x) = (x + 1)^3$

In Problems 42 and 43, use a graphing utility to graph each function. Find any turning points.

42. $y = x^3 + 3x^2 - 9x$
43. $y = x^3 - 9x$

In Problems 44 and 45, graph each function. Find and identify any asymptotes.

44. $y = \dfrac{1}{x - 2}$
45. $y = \dfrac{2x - 1}{x + 3}$

46. *Modeling* Consider the data given in the table.
   (a) Make a scatter plot.
   (b) Fit a linear function to the data and comment on the fit.
   (c) Try other function types and find one that fits better than a linear function.

x	0	4	8	12	16	20	24
y	153	151	147	140	128	115	102

47. *Modeling* Consider the data given in the table.
    (a) Make a scatter plot.
    (b) Fit a linear function to the data and comment on the fit.
    (c) Try other function types and find one that fits better than a linear function.

x	3	5	10	15	20	25	30
y	35	45	60	70	80	87	95

APPLICATIONS

48. *Physics* A ball is thrown into the air from a height of 96 ft above the ground, and its height is given by $S = 96 + 32t - 16t^2$, where t is the time in seconds.
    (a) Find the values of t that make $S = 0$.
    (b) Do both of the values of t have meaning for this application?
    (c) When will the ball strike the ground?

49. *Profit* The profit for a product is given by $P(x) = 82x - 0.10x^2 - 1600$, where x is the number of units produced and sold. Break-even points will occur at values of x where $P(x) = 0$. How many units will give a break-even point for the product?

50. *Drug use* Data indicate that the percent of high school seniors who have tried hallucinogens can be described by the function

    $$f(t) = -0.037t^2 + 1.006t + 1.75$$

    where t is the number of years since 1990 (*Source: monitoringthefuture.org*).
    (a) For what years does the function estimate that the percent will be 8.1%?
    (b) For what year does the function estimate the percent will be a maximum? Find that maximum estimated percent.

51. *Maximum area* A rectangular lot is to be fenced in and then divided down the middle to create two identical fenced lots (see the figure). If 1200 ft of fence is to be used, the area of each lot is given by

    $$A = x\left(\frac{1200 - 3x}{4}\right)$$

    (a) Find the x-value that maximizes this area.
    (b) Find the maximum area.

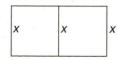

52. *Supply* Graph the first-quadrant portion of the supply function

    $$p = 2q^2 + 4q + 6$$

53. *Demand* Graph the first-quadrant portion of the demand function

    $$p = 18 - 3q - q^2$$

54. *Market equilibrium*
    (a) Suppose the supply function for a product is $p = 0.1q^2 + 1$ and the demand function is $p = 85 - 0.2q - 0.1q^2$. Sketch the first-quadrant portion of the graph of each function. Use the same set of axes for both and label the market equilibrium point.
    (b) Use algebraic methods to find the equilibrium price and quantity.

55. *Market equilibrium* The supply function for a product is given by $p = q^2 + 300$, and the demand is given by $p + q = 410$. Find the equilibrium quantity and price.

56. *Market equilibrium* If the demand function for a commodity is given by the equation $p^2 + 5q = 200$ and the supply function is given by $40 - p^2 + 3q = 0$, find the equilibrium quantity and price.

57. *Break-even points* If total costs for a product are given by $C(x) = 1760 + 8x + 0.6x^2$ and total revenues are given by $R(x) = 100x - 0.4x^2$, find the break-even quantities.

58. *Break-even points* If total costs for a commodity are given by $C(x) = 900 + 25x$ and total revenues are given by $R(x) = 100x - x^2$, find the break-even quantities.

59. *Maximizing revenue and profit* Find the maximum revenue and maximum profit for the functions described in Problem 58.

60. *Break-even and profit maximization* Given total profit $P(x) = 1.3x - 0.01x^2 - 30$, find maximum profit and the break-even quantities and sketch the graph.

61. *Maximum profit* Given $C(x) = 360 + 10x + 0.2x^2$ and $R(x) = 50x - 0.2x^2$, find the level of production that gives maximum profit and find the maximum profit.

62. *Break-even and profit maximization* A certain company has fixed costs of $15,000 for its product and variable costs given by $140 + 0.04x$ dollars per unit, where x is the total number of units. The selling price of the product is given by $300 - 0.06x$ dollars per unit.
    (a) Formulate the functions for total cost and total revenue.
    (b) Find the break-even quantities.
    (c) Find the level of sales that maximizes revenue.
    (d) Form the profit function and find the level of production and sales that maximizes profit.
    (e) Find the profit (or loss) at the production levels found in parts (c) and (d).

63. *Spread of AIDS* The function $H(t) = 8.81t^{0.476}$, where t is the number of years since 1990 and $H(t)$ is the number of world HIV infections (in millions), in year t, can be used as one means of predicting the spread of AIDS.
    (a) What type of function is this?
    (b) What does this function predict as the number of HIV cases in the year 2010?
    (c) Find $H(25)$ and write a sentence that explains its meaning.

64. *Photosynthesis* The amount y of photosynthesis that takes place in a certain plant depends on the intensity x of the light present, according to

$$y = 120x^2 - 20x^3 \quad \text{for } x \geq 0$$

(a) Graph this function with a graphing utility. (Use y-min $= -100$ and y-max $= 700$.)

(b) The model is valid only when $f(x) \geq 0$ (that is, on or above the x-axis). For what x-values is this true?

65. *Cost-benefit* Suppose the cost C, in dollars, of eliminating p percent of the pollution from the emissions of a factory is given by

$$C(p) = \frac{4800p}{100 - p}$$

(a) What type of function is this?

(b) Given that p represents the percent of pollution removed, what is the domain of $C(p)$?

(c) Find $C(0)$ and interpret its meaning.

(d) Find the cost of removing 99% of the pollution.

66. *Municipal water costs* The Borough Municipal Authority of Beaver, Pennsylvania, used the following function to determine charges for water.

$$C(x) = \begin{cases} 2.557x & \text{if } 0 \leq x \leq 100 \\ 255.70 + 2.04\,(x - 100) & \text{if } 100 < x \leq 1000 \\ 2091.07 + 1.689(x - 1000) & \text{if } x > 1000 \end{cases}$$

where $C(x)$ is the cost in dollars for x thousand gallons of water.

(a) Find the monthly charge for 12,000 gallons of water.

(b) Find the monthly charge for 825,000 gallons of water.

67. *Modeling (Subaru WRX)* The table shows the times that it takes a Subaru WRX to accelerate from 0 mph to speeds of 30 mph, 40 mph, etc., up to 90 mph, in increments of 10 mph.

Time (sec)	Speed (mph)	Time (sec)	Speed (mph)
1.6	30	7.8	70
2.7	40	10.2	80
4.0	50	12.9	90
5.6	60		

Source: Motor Trend

(a) Represent the times by x and the speeds by y, and model the function that is the best fit for the points.

(b) Graph the points and the function to see how well the function fits the points.

(c) What does the model indicate the speed is 5 seconds after the car starts to move?

(d) According to the model, in how many seconds will the car reach 79.3 mph?

68. *Modeling (public health care expenditures)* The total public expenditures (in billions of dollars) for national health care in the United States are given in the table for 2000 to 2007 with projections to 2018.

(a) Use x as the number of years past 2000 and find a quadratic model for these data.

(b) According to this model, when will public health care expenditures exceed $3 trillion (that is, $3000 billion)?

(c) How much confidence can we have in the projections beyond 2012? Explain.

Year	Health Care Expenditures (billions)	Year	Health Care Expenditures (billions)
2000	$597	2010	$1251
2001	662	2011	1332
2002	722	2012	1424
2003	779	2013	1527
2004	840	2014	1640
2005	899	2015	1768
2006	973	2016	1909
2007	1036	2017	2064
2008	1109	2018	2233
2009	1190		

Source: U.S Centers for Medicare & Medicaid Services

69. *Modeling (cohabiting households)* The table gives the number of cohabiting (without marriage) households (in thousands) for selected years.

(a) Find the power function that best fits the data, with x representing the number of years past 1950.

(b) Find a quadratic function that is the best fit for the data, with x equal to the number of years past 1950.

(c) Which model appears to fit the data better, or are they both good fits?

Year	Cohabiting Households (thousands)
1960	439
1970	523
1980	1589
1985	1983
1990	2856
1993	3510
1997	4130
2000	5457
2004	5841
2008	6214

Source: U.S. Census Bureau

2   CHAPTER TEST

1. Sketch a graph of each of the following functions.
   (a) $f(x) = x^4$
   (b) $g(x) = |x|$
   (c) $h(x) = -1$
   (d) $k(x) = \sqrt{x}$

2. The following figures show graphs of the power function $y = x^b$. Which is the graph for $b > 1$? Which is the graph for $0 < b < 1$?

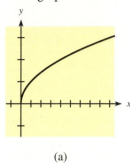

(a)

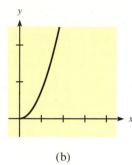

(b)

3. If $f(x) = ax^2 + bx + c$ and $a < 0$, sketch the shape of the graph of $y = f(x)$.

4. Graph:
   (a) $f(x) = (x + 1)^2 - 1$
   (b) $f(x) = (x - 2)^3 + 1$

5. Which of the following three graphs is the graph of $f(x) = x^3 - 4x^2$? Explain your choice.

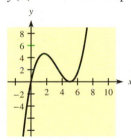

(a)

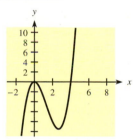

(b)

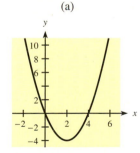

(c)

6. Let $f(x) = \begin{cases} 8x + 1/x & \text{if } x < 0 \\ 4 & \text{if } 0 \le x \le 2 \\ 6 - x & \text{if } x > 2 \end{cases}$

   Find (a) the y-coordinate of the point on the graph of $y = f(x)$ where $x = 16$; (b) $f(-2)$; (c) $f(13)$.

7. Sketch the graph of $g(x) = \begin{cases} x^2 & \text{if } x \le 1 \\ 4 - x & \text{if } x > 1 \end{cases}$

8. Find the vertex and zeros, if they exist, and sketch the graph of $f(x) = 21 - 4x - x^2$.

9. Solve $3x^2 + 2 = 7x$.

10. Solve $2x^2 + 6x = 9$.

11. Solve $\dfrac{1}{x} + 2x = \dfrac{1}{3} + \dfrac{x + 1}{x}$

12. Which of the following three graphs is that of $g(x) = \dfrac{3x - 12}{x + 2}$? Explain your choice.

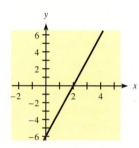

(a)

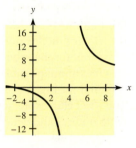

(b)

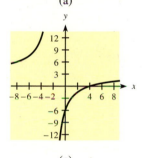

(c)

13. Find the horizontal and vertical asymptotes of the graph of
$$f(x) = \frac{8}{2x - 10}$$

14. Find the average rate of change of
$$P(x) = 92x - x^2 - 1760$$
over the interval from $x = 10$ to $x = 40$.

15. Choose the type of function that models each of the following graphs.

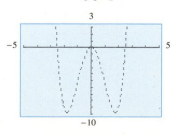

(a)

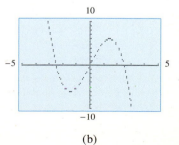

(b)

16. *Modeling*
    (a) Make a scatter plot, then develop a model, for the following data.
    (b) What does the model predict for $x = 40$?
    (c) When does the model predict $f(x) = 0$?

x	0	2	4	6	8	10	12	14	16	18	20
y	20.1	19.2	18.8	17.5	17.0	15.8	16	14.9	13.8	13.7	13.0

17. Suppose the supply and demand functions for a product are given by $6p - q = 180$ and $(p + 20)q = 30,000$, respectively. Find the equilibrium price and quantity.

18. Suppose a company's total cost for a product is given by $C(x) = 15,000 + 35x + 0.1x^2$ and the total revenue for the product is given by $R(x) = 285x - 0.9x^2$, where x is the number of units produced and sold.
    (a) Find the profit function.
    (b) Determine the number of units at which the profit for the product is maximized, and find the maximum possible profit.
    (c) Find the break-even point(s) for the product.

19. The wind chill expresses the combined effects of low temperatures and wind speeds as a single temperature reading. When the outside temperature is 0°F, the wind chill, WC, is a function of the wind speed s (in mph) and is approximated by the following function.*

$$WC = f(s) = \begin{cases} 0 & \text{if } 0 \le s \le 1.13 \\ 36.23 - 35.50\sqrt[6]{s} & \text{if } 1.13 < s < 55 \\ -33 & \text{if } s \ge 55 \end{cases}$$

*Source: National Weather Service

(a) Find $f(15)$ and write a sentence that explains its meaning.
(b) Find the wind chill when the wind speed is 48 mph.

20. The table gives the per capita out-of-pocket expenses for health care in the United States for selected years from 2000 and projected to 2018.

Year	Expenses	Year	Expenses
2000	$683.91	2010	$938.10
2002	733.40	2012	992.73
2004	802.32	2014	1070.10
2006	854.56	2016	1165.20
2008	917.46	2018	1274.63

Source: U.S. Centers for Medicare & Medicaid Services

(a) Plot the data, with x representing the number of years after 2000 and y representing the expenses.
(b) Find a linear and a cubic function model for the data.
(c) Use both models to estimate the per capita expenses in 2020.
(d) Can we be sure that the actual expenses for 2018 agree with the model's output for 2018?

I. An Inconvenient Truth

There are many functions mentioned in the movie *An Inconvenient Truth*. Read the questions below, then watch the movie and describe the functions in it as follows.

1. (a) Find a function in the movie that is described by a graph and answer the following questions for it.
   (b) What does the input describe? What units does it have?
   (c) What does the output describe? What units does it have?
   (d) How does Al Gore use this function to support his arguments on climate change?
2. (a) Find a second function in the movie and answer the following questions for it.
   (b) Is the function described by a graph, a table, an equation, or words?
   (c) What does the input describe? What units does it have?
   (d) What does the output describe? What units does it have?
   (e) How does Al Gore use this function to support his arguments on climate change?
   (f) Does Al Gore use extrapolation with this function to support his position? If so, do you think this extrapolation supports his position?
3. (a) For a function described in the movie, find a set of data that also describes the same or a similar function and use your function to answer the following questions.
   (b) What does the input describe? What units does it have?
   (c) What does the output describe? What units does it have?
   (d) Does your function support Al Gore's arguments on climate change?

II. Body Mass Index

Obesity is a risk factor for the development of several medical problems, including high blood pressure, high cholesterol, heart disease, and diabetes. Of course, whether a person is considered obese depends not only on weight but also on height. One way of comparing weights that accounts for height is the *body mass index (BMI)*. The table below gives the BMI for a variety of heights and weights for people. Medical professionals generally agree that a BMI of 30 or greater increases the risk of developing medical problems associated with obesity.

Describe how to assist a group of people in using this information. Some things you would want to include in your description follow.

A. How a person uses the table to determine his or her BMI.
B. How a person determines the weight that will put him or her at medical risk.
C. How a person whose weight or height is not in the table can determine whether his or her BMI is 30 or higher. To address this particular issue, develop a formula to find the weight that would give a BMI of 30 for a person of a given height, including how you would:
   1. Pick the points from the table that correspond to a BMI of 30 and create a table of these heights and weights. Change the units of height measurement to simplify the data.
   2. Determine what type of model seems to be the best fit for the data.
   3. Create a model for the data.
   4. Test the model with existing data.
   5. Explain how to use the model to test for obesity. In particular, use the model to find the obesity threshold for someone who is 5 feet tall and someone who is 6 feet 2 inches tall (heights that are not in the table).
   6. Use the Internet to find the definition of BMI, and, if necessary, convert it to a formula that uses inches and pounds.
   7. Make a table for heights from 61 inches to 73 inches that compares the threshold weights for a BMI of 30 from your model, from the definition of BMI, and from the table.

Body Mass Index for Specified Height (ft/in.) and Weight (lb)														
Height/Weight	120	130	140	150	160	170	180	190	200	210	220	230	240	250
5'1"	23	25	27	28	30	32	34	36	38	40	42	44	45	47
5'2"	22	24	26	27	29	31	33	35	37	38	40	42	44	46
5'3"	21	23	25	27	28	30	32	34	36	37	39	41	43	44
5'4"	21	22	24	26	28	29	31	33	34	36	38	40	41	43
5'5"	20	22	23	25	27	28	30	32	33	35	37	38	40	42
5'6"	19	21	23	24	26	27	29	31	32	34	36	37	39	40
5'7"	19	20	23	24	25	27	28	30	31	33	35	36	38	39
5'8"	18	20	21	23	24	26	27	29	30	32	34	35	37	38
5'9"	18	19	21	22	24	25	27	28	30	31	33	34	36	37
5'10"	17	19	20	22	23	24	25	27	28	29	31	33	35	36
5'11"	17	18	20	21	22	24	25	27	28	29	31	32	34	35
6'0"	16	18	19	20	22	23	24	26	27	29	30	31	33	34
6'1"	16	17	19	20	21	22	24	25	26	28	29	30	32	33

Source: Roche Pharmaceuticals

III. Operating Leverage and Business Risk

Once you have determined the break-even point for your product, you can use it to examine the effects of increasing or decreasing the role of fixed costs in your operating structure. The extent to which a business uses fixed costs (compared to variable costs) in its operations is referred to as **operating leverage.** The greater the use of operating leverage (fixed costs, often associated with fixed assets), the larger the increase in profit as sales rise, and the larger the increase in loss as sales fall (*Source:* **www.toolkit.cch.com/text/P06_7540.asp**). The higher the break-even quantity for your product, the greater your **business risk.** To see how operating leverage is related to business risk, complete the following.

1. If x is the number of units of a product that is sold and the price is $\$p$ per unit, write the function that gives the revenue R for the product.
2. (a) If the fixed cost for production of this product is $\$10,000$ and the variable cost is $\$100$ per unit, find the function that gives the total cost C for the product.
   (b) What type of function is this?
3. Write an equation containing only the variables p and x that describes when the break-even point occurs.
4. (a) Solve the equation from Question 3 for x, so that x is written as a function of p.
   (b) What type of function is this?
   (c) What is the domain of this function?
   (d) What is the domain of this function in the context of this application?
5. (a) Graph the function from Question 4 for the domain of the problem.
   (b) Does the function increase or decrease as p increases?
6. Suppose that the company has a monopoly for this product so that it can set the price as it chooses. What are the benefit and the danger to the company of pricing its product at $\$1100$ per unit?
7. If the company has a monopoly for this product so that it can set the price, what are the benefit and the danger to the company of pricing its product at $\$101$ per unit?
8. Suppose the company currently is labor-intensive, so that the monthly fixed cost is $\$10,000$ and the variable cost is $\$100$ per unit, but installing modern equipment will reduce the variable cost to $\$50$ per unit while increasing the monthly fixed cost to $\$30,000$. Also suppose the selling price per unit is $P = \$200$.
   (a) Which of these operations will have the highest operating leverage? Explain.
   (b) Which of these operations will have the highest business risk? Explain.
   (c) What is the relation between high operating leverage and risk? Discuss the benefit and danger of the second operating plan to justify your answer.

Matrices

© Deco/Alamy

A wide variety of application problems can be solved using matrices. A **matrix** (plural: matrices) is a rectangular array of numbers. In addition to storing data in a matrix and making comparisons of data, we can analyze data and make business decisions by defining the operations of addition, subtraction, scalar multiplication, and matrix multiplication. Matrices can be used to solve systems of linear equations, as we will see in this chapter. Matrices and matrix operations are the basis for computer spreadsheets such as Excel, which are used extensively in education, research, and business. Matrices are also useful in linear programming, which is discussed in Chapter 4.

The topics and applications discussed in this chapter include the following.

SECTIONS	APPLICATIONS
3.1 Matrices Operations with matrices	Balance of trade, supply charges, power to influence
3.2 Multiplication of Matrices	Material supply, encoding messages, venture capital
3.3 Gauss-Jordan Elimination: **Solving Systems of Equations** Systems with unique solutions Systems with nonunique solutions	Manufacturing, investment, traffic flow
3.4 Inverse of a Square Matrix; **Matrix Equations** Matrix equations Determinants	Decoding messages, venture capital, freight logistics
3.5 Applications of Matrices: **Leontief Input-Output** **Models**	Simple economy, outputs of an economy, production, open economy, closed economy, manufacturing

Prerequisite Problem Type	For Section	Answer	Section for Review
Evaluate: (a) $4-(-2)$ (b) $-4\left(\dfrac{2}{3}\right)+3$ (c) $(-5)(4)+(8)(1)+(2)(5)$ (d) Multiply each of the numbers $\quad 0,-3,-2,\,2$ by $-\dfrac{1}{3}$.	**3.1** **3.2** **3.3** **3.4** **3.5**	(a) 6 (b) $\dfrac{1}{3}$ (c) -2 (d) $0, 1, \dfrac{2}{3}, -\dfrac{2}{3}$	0.2 Signed numbers

Prerequisite Problem Type	For Section	Answer				Section for Review
Write the coefficients of x, y and z and the constant term in each equation: (a) $2x+5y+4z=4$ (b) $x-3y-2z=5$ (c) $3x+y=4$	**3.3** **3.4** **3.5**	*Coefficients*			*Const. term*	0.5 Algebraic expressions
		x	y	z		
		2	5	4	4	
		1	-3	-2	5	
		3	1	0	4	

Prerequisite Problem Type	For Section	Answer	Section for Review
Solve: (a) $x_1-2x_3=2$ for x_1 (b) $x_2+x_3=-3$ for x_2 (c) $450{,}000-3z\geq 0$ for z	**3.3** **3.4** **3.5**	(a) $x_1 = 2+2x_3$ (b) $x_2 = -3-x_3$ (c) $z \leq 150{,}000$	1.1 Linear equations and inequalities

Prerequisite Problem Type	For Section	Answer	Section for Review
Solve the system: $\begin{cases} x-0.25y = 11{,}750 \\ -0.20x+y = 10{,}000 \end{cases}$	**3.3** **3.4** **3.5**	$x = 15{,}000$ $y = 13{,}000$	1.5 Systems of linear equations

OBJECTIVES

3.1

- To organize and interpret data stored in matrices
- To add and subtract matrices
- To find the transpose of a matrix
- To multiply a matrix by a scalar (real number)

Matrices

| APPLICATION PREVIEW |

Data in tabular form can be stored in matrices, which are useful for solving an assortment of problems, including the following. Table 3.1 summarizes the dollar value (in millions) of 2006 U.S. exports and imports of cars, trucks, and automotive parts for selected countries. If the data in Table 3.1 are formed into two matrices, one for exports and one for imports, then the balance of trade with the selected countries for cars, trucks, and parts will be the difference of those two matrices.

| TABLE 3.1 |

	Exports			Imports		
	Cars	Trucks	Parts	Cars	Trucks	Parts
Canada	13,165	11,992	31,952	36,600	12,250	20,132
Japan	477	59	1755	43,522	986	15,704
Mexico	3451	1007	12,606	14,201	9559	25,217

Source: U.S. Bureau of Economic Analysis

As we noted in the introduction to this chapter, matrices can be used to store data and to perform operations with the data. The information in Table 3.1 can be organized into an export matrix *A* and an import matrix *B*, as follows.

$$A = \begin{bmatrix} 13{,}165 & 11{,}992 & 31{,}952 \\ 477 & 59 & 1755 \\ 3451 & 1007 & 12{,}606 \end{bmatrix} \quad B = \begin{bmatrix} 36{,}600 & 12{,}250 & 20{,}132 \\ 43{,}522 & 986 & 15{,}704 \\ 14{,}201 & 9559 & 25{,}217 \end{bmatrix}$$

We will discuss how to find the difference of these two and other pairs of matrices, how to find the sum of two matrices, and how to multiply a matrix by a real number. (The difference of these two matrices is found in Example 5.)

In the export matrix *A* in the Application Preview, the *rows* correspond to the countries and the *columns* correspond to items exported. The rows of a matrix are numbered from the top to the bottom, and the columns are numbered from left to right, as the following example matrix illustrates.

$$M = \begin{bmatrix} 20 & 25 & 22 & 18 \\ 27 & 40 & 31 & 25 \\ 16 & 16 & 19 & 16 \end{bmatrix} \begin{matrix} \text{Row 1} \\ \text{Row 2} \\ \text{Row 3} \end{matrix}$$

Column 1   Column 2   Column 3   Column 4

Matrices are classified in terms of the numbers of rows and columns they have. Matrix *M* has three rows and four columns, so we say this is a 3 × 4 (read "three by four") matrix.

The matrix

$$
A = \begin{bmatrix}
a_{11} & a_{12} & a_{13} & \cdots & a_{1n} \\
a_{21} & a_{22} & a_{23} & \cdots & a_{2n} \\
\vdots & & & & \vdots \\
a_{m1} & a_{m2} & a_{m3} & \cdots & a_{mn}
\end{bmatrix}
$$

has m rows and n columns, so it is an $m \times n$ matrix. When we designate A as an $m \times n$ matrix, we are indicating the size of the matrix. Two matrices are said to have the same **order** (be the same size) if they have the same number of rows and the same number of columns. For example, $C = \begin{bmatrix} 1 & 3 & 2 \\ 4 & 1 & 2 \end{bmatrix}$ and $D = \begin{bmatrix} 1 & 4 \\ 2 & 1 \\ 3 & 0 \end{bmatrix}$ do not have the same order. C is a 2×3 matrix and D is a 3×2 matrix.

The numbers in a matrix are called its **entries** or **elements.** Note that the subscripts on an entry in matrix A above correspond respectively to the row and column in which the entry is located. Thus a_{23} represents the entry in the second row and the third column, and we refer to it as the "two-three entry." In matrix B below, the entry denoted by b_{23} is 1.

Some matrices take special names because of their size. If the number of rows equals the number of columns, we say the matrix is a **square matrix.** Matrix B below is a 3×3 square matrix.

$$
B = \begin{bmatrix}
4 & 3 & 0 \\
0 & 0 & 1 \\
-4 & 0 & 0
\end{bmatrix}
$$

EXAMPLE 1 Population and Labor Force

Figure 3.1 gives the U.S. population and labor force (in millions, actual and projected) for the years 2008, 2012, and 2016.

Population and labor force

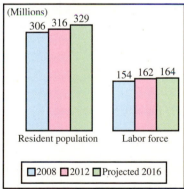

Figure 3.1

Source: U.S. Bureau of Labor Statistics and U.S. Census Bureau

A partial table of values from Figure 3.1 is shown. Complete the table that summarizes these data, and use it to create a 2×3 matrix describing the information.

	2008	2012	2016
Resident Population	306		
Labor Force		162	

Solution

Reading the data from Figure 3.1 and completing the table give

	2008	2012	2016
Resident Population	306	316	329
Labor Force	154	162	164

Writing these data in a matrix gives

$$\begin{bmatrix} 306 & 316 & 329 \\ 154 & 162 & 164 \end{bmatrix}$$

A matrix with one row, such as [9 5] or [3 2 1 6], is called a **row matrix**, and a matrix with one column, such as

$$\begin{bmatrix} 1 \\ 3 \\ 5 \end{bmatrix}$$

is called a **column matrix.** Row and column matrices are also called **vectors.**

Any matrix in which *every* entry is zero is called a **zero matrix;** examples include

$$\begin{bmatrix} 0 & 0 \\ 0 & 0 \end{bmatrix}, \quad \begin{bmatrix} 0 & 0 \\ 0 & 0 \\ 0 & 0 \end{bmatrix}, \quad \text{and} \quad \begin{bmatrix} 0 & 0 & 0 & 0 \\ 0 & 0 & 0 & 0 \\ 0 & 0 & 0 & 0 \end{bmatrix}$$

We define two matrices to be **equal** if they are of the same order and if each entry in one equals the corresponding entry in the other.

When the columns and rows of matrix A are interchanged to create a matrix B, and vice versa, we say that A and B are **transposes** of each other and write $A^T = B$ and $B^T = A$. We will see valuable uses for transposes of matrices in Chapter 4, "Inequalities and Linear Programming."

EXAMPLE 2   Matrices

(a) Which element of

$$A = \begin{bmatrix} 1 & 0 & 3 \\ 3 & 4 & 2 \\ 7 & 8 & 3 \end{bmatrix} \quad B = \begin{bmatrix} 1 & 0 & 3 \\ 3 & 4 & 2 \\ 7 & 8 & 4 \end{bmatrix}$$

is represented by a_{32}?

(b) Is A a square matrix?

(c) Find the transpose of matrix A.

(d) Does $A = B$?

Solution

(a) a_{32} represents the element in row 3 and column 2 of matrix A—that is, $a_{32} = 8$.

(b) Yes, it is a 3×3 (square) matrix.

(c)
$$A^T = \begin{bmatrix} 1 & 3 & 7 \\ 0 & 4 & 8 \\ 3 & 2 & 3 \end{bmatrix}$$

(d) No, $a_{33} \neq b_{33}$.

CHECKPOINT

1. (a) Do matrices A and B have the same order?

$$A = \begin{bmatrix} 1 & 2 & 3 \\ 4 & 5 & 6 \\ 7 & 8 & 9 \end{bmatrix} \quad \text{and} \quad B = \begin{bmatrix} 1 & 4 & 7 \\ 2 & 5 & 8 \\ 3 & 6 & 9 \end{bmatrix}$$

(b) Does matrix A equal matrix B?

(c) Does $B^T = A$?

Addition and Subtraction of Matrices

If two matrices have the same number of rows and columns, we can add the matrices by adding their corresponding entries.

Sum of Two Matrices

If matrix A and matrix B are of the same order and have elements a_{ij} and b_{ij}, respectively, then their **sum** $A + B$ is a matrix C whose elements are $c_{ij} = a_{ij} + b_{ij}$ for all i and j. That is,

$$A + B = \begin{bmatrix} a_{11} & a_{12} & \cdots & a_{1n} \\ a_{21} & a_{22} & \cdots & a_{2n} \\ \vdots & & & \vdots \\ a_{m1} & a_{m2} & \cdots & a_{mn} \end{bmatrix} + \begin{bmatrix} b_{11} & b_{12} & \cdots & b_{1n} \\ b_{21} & b_{22} & \cdots & b_{2n} \\ \vdots & & & \vdots \\ b_{m1} & b_{m2} & \cdots & b_{mn} \end{bmatrix}$$

$$= \begin{bmatrix} a_{11} + b_{11} & a_{12} + b_{12} & \cdots & a_{1n} + b_{1n} \\ a_{21} + b_{21} & a_{22} + b_{22} & \cdots & a_{2n} + b_{2n} \\ \vdots & & & \vdots \\ a_{m1} + b_{m1} & a_{m2} + b_{m2} & \cdots & a_{mn} + b_{mn} \end{bmatrix} = C$$

EXAMPLE 3 Matrix Sums

Find the sum of A and B if

$$A = \begin{bmatrix} 1 & 2 & 3 \\ 4 & -1 & -2 \end{bmatrix} \quad \text{and} \quad B = \begin{bmatrix} -1 & 2 & -3 \\ -2 & 0 & 1 \end{bmatrix}$$

Solution

$$A + B = \begin{bmatrix} 1 + (-1) & 2 + 2 & 3 + (-3) \\ 4 + (-2) & -1 + 0 & -2 + 1 \end{bmatrix} = \begin{bmatrix} 0 & 4 & 0 \\ 2 & -1 & -1 \end{bmatrix} \quad \blacksquare$$

Note that the sum of A and B in Example 3 could be found by adding the matrices in either order. That is, $A + B = B + A$. This is known as the **commutative law of addition** for matrices. We will see in the next section that multiplication of matrices is *not* commutative.

The matrix $-B$ is called the **negative** of the matrix B, and each element of $-B$ is the negative of the corresponding element of B. For example, if

$$B = \begin{bmatrix} -1 & 2 & -3 \\ -2 & 0 & 1 \end{bmatrix}, \quad \text{then} \quad -B = \begin{bmatrix} 1 & -2 & 3 \\ 2 & 0 & -1 \end{bmatrix}$$

Using the negative, we can define the difference $A - B$ (when A and B have the same order) by $A - B = A + (-B)$, or by subtracting corresponding elements.

EXAMPLE 4   Matrix Differences

For matrices A and B in Example 3, find $A - B$.

Solution

$A - B$ can be found by subtracting corresponding elements.

$$A - B = \begin{bmatrix} 1 - (-1) & 2 - 2 & 3 - (-3) \\ 4 - (-2) & -1 - 0 & -2 - 1 \end{bmatrix} = \begin{bmatrix} 2 & 0 & 6 \\ 6 & -1 & -3 \end{bmatrix}$$

CHECKPOINT

2. If $A = \begin{bmatrix} 1 & 2 & 0 \\ 3 & 1 & 2 \end{bmatrix}$, $B = \begin{bmatrix} 3 & 6 & 1 \\ 0 & 2 & 1 \end{bmatrix}$, and $C = \begin{bmatrix} 1 & 0 & 2 \\ 0 & 1 & 1 \end{bmatrix}$, find $A + B - C$.

3. (a) What matrix D must be added to matrix A so that their sum is matrix Z?

$$A = \begin{bmatrix} 1 & -2 & 3 \\ -5 & 1 & 2 \end{bmatrix} \text{ and } Z = \begin{bmatrix} 0 & 0 & 0 \\ 0 & 0 & 0 \end{bmatrix}$$

(b) Does $D = -A$?

EXAMPLE 5   Balance of Trade | APPLICATION PREVIEW |

Table 3.2 summarizes the dollar value (in millions) of 2006 U.S. exports and imports of cars, trucks, and automotive parts for selected countries. Write the matrix that describes the balance of trade with the selected countries for cars, trucks, and parts.

| TABLE 3.2 |

	Exports			Imports		
	Cars	Trucks	Parts	Cars	Trucks	Parts
Canada	13,165	11,992	31,952	36,600	12,250	20,132
Japan	477	59	1755	43,522	986	15,704
Mexico	3451	1007	12,606	14,201	9559	25,217

Source: U.S. Bureau of Economic Analysis

Solution

From Table 3.2 we represent the exports as matrix A and the imports as matrix B.

$$A = \begin{bmatrix} 13{,}165 & 11{,}992 & 31{,}952 \\ 477 & 59 & 1755 \\ 3451 & 1007 & 12{,}606 \end{bmatrix} \qquad B = \begin{bmatrix} 36{,}600 & 12{,}250 & 20{,}132 \\ 43{,}522 & 986 & 15{,}704 \\ 14{,}201 & 9559 & 25{,}217 \end{bmatrix}$$

The balance of trade we seek is given by the difference $A - B$, as follows.

$$A - B = \begin{bmatrix} 13{,}165 - 36{,}600 & 11{,}992 - 12{,}250 & 31{,}952 - 20{,}132 \\ 477 - 43{,}522 & 59 - 986 & 1755 - 15{,}704 \\ 3451 - 14{,}201 & 1007 - 9559 & 12{,}606 - 25{,}217 \end{bmatrix}$$

$$= \begin{bmatrix} -23{,}435 & -258 & 11{,}820 \\ -43{,}045 & -927 & -13{,}949 \\ -10{,}750 & -8552 & -12{,}611 \end{bmatrix}$$

Note that any negative entry indicates an unfavorable balance of trade for the item and with the country corresponding to that entry. Also note that the 1–3 entry is the only positive one and that it indicates a favorable trade balance, in automotive parts with Canada, worth $11,820,000,000. ■

Calculator Note   Graphing calculators have the capability to perform a number of operations on matrices. Steps for entering data in matrices and performing operations with matrices using a graphing calculator are shown in Appendix C, Section 3.1.

To compute $A + B$ and $A - B$ for

$$A = \begin{bmatrix} -1 & 3 & 2 \\ 1 & 0 & 2 \\ 0 & 1 & -1 \\ -4 & 2 & 3 \end{bmatrix} \text{ and } B = \begin{bmatrix} -3 & 2 & 1 \\ 0 & -2 & 2 \\ -1 & 6 & 8 \\ 3 & 2 & 4 \end{bmatrix}$$

We enter matrices in a graphing calculator (see Figure 3.2) and find the sum and difference of the two matrices by using the regular ⊞ and ⊟ keys (see Figure 3.3). ■

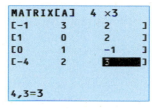

Figure 3.2

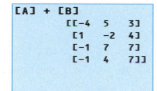

Figure 3.3

Scalar Multiplication   Consider the matrix

$$A = \begin{bmatrix} 3 & 2 \\ 1 & 1 \\ 2 & 0 \end{bmatrix}$$

Because $2A$ is $A + A$, we see that

$$2A = \begin{bmatrix} 3 & 2 \\ 1 & 1 \\ 2 & 0 \end{bmatrix} + \begin{bmatrix} 3 & 2 \\ 1 & 1 \\ 2 & 0 \end{bmatrix} = \begin{bmatrix} 6 & 4 \\ 2 & 2 \\ 4 & 0 \end{bmatrix}$$

Note that $2A$ could have been found by multiplying each entry of A by 2. In the same manner, it can be shown that $3A = A + A + A$. (Try it.)

We can define **scalar multiplication** as follows.

Scalar Multiplication

Multiplying a matrix by a real number (called a *scalar*) results in a matrix in which each entry of the original matrix is multiplied by the real number. Thus, if

$$A = \begin{bmatrix} a_{11} & a_{12} \\ a_{21} & a_{22} \end{bmatrix}, \text{ then } cA = \begin{bmatrix} ca_{11} & ca_{12} \\ ca_{21} & ca_{22} \end{bmatrix}$$

EXAMPLE 6   **Scalar Multiplication**

If

$$A = \begin{bmatrix} 4 & 1 & 4 & 0 \\ 2 & -7 & 3 & 6 \\ 0 & 0 & 2 & 5 \end{bmatrix}$$

find $5A$ and $-2A$.

Solution

$$5A = \begin{bmatrix} 5 \cdot 4 & 5 \cdot 1 & 5 \cdot 4 & 5 \cdot 0 \\ 5 \cdot 2 & 5(-7) & 5 \cdot 3 & 5 \cdot 6 \\ 5 \cdot 0 & 5 \cdot 0 & 5 \cdot 2 & 5 \cdot 5 \end{bmatrix} = \begin{bmatrix} 20 & 5 & 20 & 0 \\ 10 & -35 & 15 & 30 \\ 0 & 0 & 10 & 25 \end{bmatrix}$$

$$-2A = \begin{bmatrix} -2 \cdot 4 & -2 \cdot 1 & -2 \cdot 4 & -2 \cdot 0 \\ -2 \cdot 2 & -2(-7) & -2 \cdot 3 & -2 \cdot 6 \\ -2 \cdot 0 & -2 \cdot 0 & -2 \cdot 2 & -2 \cdot 5 \end{bmatrix} = \begin{bmatrix} -8 & -2 & -8 & 0 \\ -4 & 14 & -6 & -12 \\ 0 & 0 & -4 & -10 \end{bmatrix}$$

Figure 3.4 shows the matrix $-2A$ on a graphing calculator.   ■

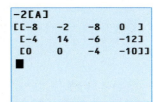

Figure 3.4

CHECKPOINT

4. (a) Find $3D - D$ if $D = \begin{bmatrix} 1 & 3 & 2 & 10 \\ 20 & 5 & 6 & 3 \end{bmatrix}$

   (b) Does $3D - D = 2D$?

EXAMPLE 7   **Supply Charges**

Suppose the purchase prices and delivery costs (per unit) for wood, siding, and roofing used in construction are given by Table 3.3. Then the table of unit costs may be represented by the matrix C.

■ | TABLE 3.3 |

	Wood	Siding	Roofing
Purchase	6	4	2
Delivery	1	1	0.5

$$C = \begin{bmatrix} 6 & 4 & 2 \\ 1 & 1 & 0.5 \end{bmatrix}$$

If the supplier announces a 10% increase on both purchase and delivery of these items, find the new unit cost matrix.

Solution

A 10% increase means that the new unit costs are the former costs plus 0.10 times the former cost. That is, the new costs are 1.10 times the former, so the new unit cost matrix is given by

$$1.10C = 1.10 \begin{bmatrix} 6 & 4 & 2 \\ 1 & 1 & 0.5 \end{bmatrix} = \begin{bmatrix} 6.60 & 4.40 & 2.20 \\ 1.10 & 1.10 & 0.55 \end{bmatrix}$$

■

EXAMPLE 8   **Power to Influence**

Suppose that in a government agency, paperwork is constantly flowing among offices according to the diagram.

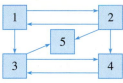

(a) Construct matrix A with elements

$$a_{ij} = \begin{cases} 1 & \text{if paperwork flows directly from } i \text{ to } j. \\ 0 & \text{if paperwork does not flow directly from } i \text{ to } j. \end{cases}$$

(b) Construct matrix B with elements

$$b_{ij} = \begin{cases} 1 & \text{if paperwork can flow from } i \text{ to } j \text{ through at most} \\ & \text{one intermediary, with } i \neq j. \\ 0 & \text{if this is not true.} \end{cases}$$

(c) The person in office i has the most power to influence others if the sum of the elements of row i in the matrix $A + B$ is the largest. What is the office number of this person?

Solution

(a) $A = \begin{bmatrix} 0 & 1 & 1 & 0 & 0 \\ 1 & 0 & 0 & 1 & 1 \\ 0 & 0 & 0 & 1 & 1 \\ 0 & 0 & 1 & 0 & 0 \\ 0 & 0 & 0 & 0 & 0 \end{bmatrix}$ (b) $B = \begin{bmatrix} 0 & 1 & 1 & 1 & 1 \\ 1 & 0 & 1 & 1 & 1 \\ 0 & 0 & 0 & 1 & 1 \\ 0 & 0 & 1 & 0 & 1 \\ 0 & 0 & 0 & 0 & 0 \end{bmatrix}$

(c) $A + B = \begin{bmatrix} 0 & 2 & 2 & 1 & 1 \\ 2 & 0 & 1 & 2 & 2 \\ 0 & 0 & 0 & 2 & 2 \\ 0 & 0 & 2 & 0 & 1 \\ 0 & 0 & 0 & 0 & 0 \end{bmatrix}$

Because the sum of the elements in row 2 is the largest (7), the person in office 2 has the most power to influence others. Because row 5 has the smallest sum (0), the person in office 5 has the least power.

■

Spreadsheet Note

Once data are entered in a spreadsheet, they can be stored in a matrix and then added to or subtracted from other matrices. See Appendix D, Section 3.1 and the Online Excel Guide for details.

■

CHECKPOINT SOLUTIONS

1. (a) Yes, A and B have the same order (size).
    (b) No. They have the same elements, but they are not equal because their corresponding elements are not equal.
    (c) Yes.

2. $\begin{bmatrix} 3 & 8 & -1 \\ 3 & 2 & 2 \end{bmatrix}$ 3. (a) $\begin{bmatrix} -1 & 2 & -3 \\ 5 & -1 & -2 \end{bmatrix}$ (b) Yes

4. (a) $3D - D = \begin{bmatrix} 3 & 9 & 6 & 30 \\ 60 & 15 & 18 & 9 \end{bmatrix} - \begin{bmatrix} 1 & 3 & 2 & 10 \\ 20 & 5 & 6 & 3 \end{bmatrix}$

$\qquad\qquad\quad = \begin{bmatrix} 2 & 6 & 4 & 20 \\ 40 & 10 & 12 & 6 \end{bmatrix}$

(b) Yes

| EXERCISES | 3.1

Access end-of-section exercises online at **www.webassign.net**   **Web**Assign

OBJECTIVE   3.2

• To multiply two matrices

Multiplication of Matrices

| APPLICATION PREVIEW |

Pentico Industries is a manufacturing company that has two divisions, located at Clarion and Brooks, each of which needs different amounts (units) of production materials as described by the following table.

	Steel	Wood	Plastic
Clarion	20	30	8
Brooks	22	25	15

These raw materials are supplied by Western Supply and Coastal Supply, with prices given in dollars per unit in the following table.

	Western	Coastal
Steel	300	290
Wood	100	90
Plastic	145	180

If one supplier must be chosen to provide all materials for either or both divisions of Pentico, company officials can decide which supplier to choose by constructing matrices from these tables and using matrix multiplication. (See Example 2.)

In this section we will discuss finding the product of two matrices.

Many computations that occur in business and manufacturing operations can be expressed and analyzed by using matrix multiplication. For example, suppose one store of an electronics retailer has 30 plasma TVs, 20 LCD TVs, and 10 home theater systems in its inventory. If the value of each plasma TV is \$800, that of each LCD TV is \$750, and that of each home theater system is \$500, then the value of this inventory is

$$30 \cdot 800 + 20 \cdot 750 + 10 \cdot 500 = \$44{,}000$$

If we write the value of each of the items in the *row matrix, A*, and the number of each of the items in the *column matrix, B*, then the value of this inventory may be represented by

$$AB = \begin{bmatrix} 800 & 750 & 500 \end{bmatrix} \begin{bmatrix} 30 \\ 20 \\ 10 \end{bmatrix} = \begin{bmatrix} 800 \cdot 30 + 750 \cdot 20 + 500 \cdot 10 \end{bmatrix} = \begin{bmatrix} \$44{,}000 \end{bmatrix}$$

This useful way of operating with a row matrix and a column matrix is called the **product** *AB*.

In general, we can multiply a row matrix *A* times a column matrix *B* if *A* and *B* have the same number of elements. The product *AB* is then

$$AB = \begin{bmatrix} a_1 & a_2 & a_3 & \cdots & a_n \end{bmatrix} \begin{bmatrix} b_1 \\ b_2 \\ b_3 \\ \vdots \\ b_n \end{bmatrix} = \begin{bmatrix} a_1b_1 + a_2b_2 + a_3b_3 + \cdots + a_nb_n \end{bmatrix}$$

Suppose the retailer has a second store with 40 plasma TVs, 25 LCD TVs, and 5 home theater systems. We can use matrix *C* to represent the inventories of the two stores.

$$C = \begin{matrix} & \text{Store I} & \text{Store II} & \\ & \begin{bmatrix} 30 & 40 \\ 20 & 25 \\ 10 & 5 \end{bmatrix} & \begin{matrix} \text{Plasma TVs} \\ \text{LCD TVs} \\ \text{Home theater systems} \end{matrix} \end{matrix}$$

If the values of the items are \$800, \$750, and \$500, respectively, we have already seen that the value of the inventory of store I is \$44,000. The value of the inventory of store II is

$$800 \cdot 40 + 750 \cdot 25 + 500 \cdot 5 = \$53,250$$

Because the value of the inventory of store I can be found by multiplying the row matrix *A* times the first column of matrix *C*, and the value of the inventory of store II can be found by multiplying matrix *A* times the second column of matrix *C*, we can write this result as

$$AC = \begin{bmatrix} 800 & 750 & 500 \end{bmatrix} \begin{bmatrix} 30 & 40 \\ 20 & 25 \\ 10 & 5 \end{bmatrix} = \begin{bmatrix} 44,000 & 53,250 \end{bmatrix}$$

The matrix we have represented as *AC* is called the **product** of the row matrix *A* and the 3×2 matrix *C*. We should note that matrix *A* is a 1×3 matrix, matrix *C* is a 3×2 matrix, and their product is a 1×2 matrix. In general, we can multiply an $m \times n$ matrix times an $n \times p$ matrix and the product will be an $m \times p$ matrix, as follows.

Product of Two Matrices

Given an $m \times n$ matrix *A* and an $n \times p$ matrix *B*, the **matrix product** *AB* is an $m \times p$ matrix *C*, with the *ij* entry of *C* given by the formula

$$c_{ij} = a_{i1}b_{1j} + a_{i2}b_{2j} + \cdots + a_{in}b_{nj}$$

which is illustrated in Figure 3.5.

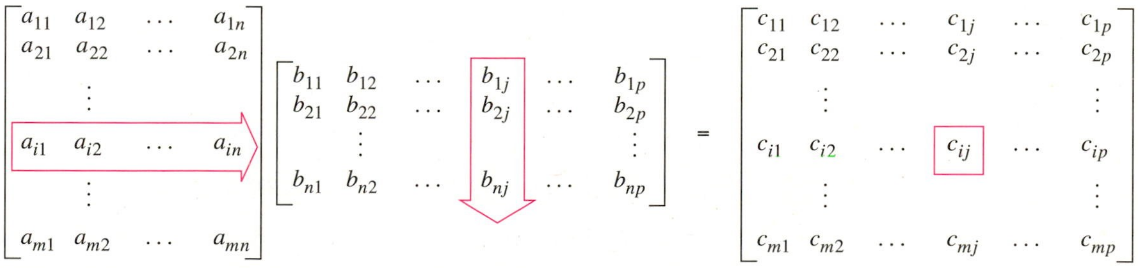

Figure 3.5

EXAMPLE 1 **Matrix Product**

(a) Find AB if

$$A = \begin{bmatrix} 3 & 4 \\ 2 & 5 \\ 6 & 10 \end{bmatrix} \text{ and } B = \begin{bmatrix} a & b & c & d \\ e & f & g & h \end{bmatrix}$$

(b) Find AB and BA if

$$A = \begin{bmatrix} 3 & 2 \\ 1 & 0 \end{bmatrix} \text{ and } B = \begin{bmatrix} 1 & 2 \\ 3 & 1 \end{bmatrix}$$

Solution

(a) A is a 3×2 matrix and B is a 2×4 matrix, so the number of columns of A equals the number of rows of B. Thus we can find the product AB, which is a 3×4 matrix, as follows:

$$AB = \begin{bmatrix} 3 & 4 \\ 2 & 5 \\ 6 & 10 \end{bmatrix} \begin{bmatrix} a & b & c & d \\ e & f & g & h \end{bmatrix}$$

$$= \begin{bmatrix} 3a + 4e & 3b + 4f & 3c + 4g & 3d + 4h \\ 2a + 5e & 2b + 5f & 2c + 5g & 2d + 5h \\ 6a + 10e & 6b + 10f & 6c + 10g & 6d + 10h \end{bmatrix}$$

This example shows that if $AB = C$, element c_{32} is found by multiplying each entry of A's *third* row by the corresponding entry of B's *second* column and then adding these products.

(b) Both A and B are 2×2 matrices, so both AB and BA are defined.

$$AB = \begin{bmatrix} 3 & 2 \\ 1 & 0 \end{bmatrix} \begin{bmatrix} 1 & 2 \\ 3 & 1 \end{bmatrix} = \begin{bmatrix} 3 \cdot 1 + 2 \cdot 3 & 3 \cdot 2 + 2 \cdot 1 \\ 1 \cdot 1 + 0 \cdot 3 & 1 \cdot 2 + 0 \cdot 1 \end{bmatrix} = \begin{bmatrix} 9 & 8 \\ 1 & 2 \end{bmatrix}$$

$$BA = \begin{bmatrix} 1 & 2 \\ 3 & 1 \end{bmatrix} \begin{bmatrix} 3 & 2 \\ 1 & 0 \end{bmatrix} = \begin{bmatrix} 1 \cdot 3 + 2 \cdot 1 & 1 \cdot 2 + 2 \cdot 0 \\ 3 \cdot 3 + 1 \cdot 1 & 3 \cdot 2 + 1 \cdot 0 \end{bmatrix} = \begin{bmatrix} 5 & 2 \\ 10 & 6 \end{bmatrix}$$

Note that for these matrices, the two products AB and BA are matrices of the same size, but they are not equal. That is, $BA \neq AB$. Thus we say that *matrix multiplication is not commutative.* ∎

CHECKPOINT

1. What is element c_{23} if $C = AB$ with

$$A = \begin{bmatrix} 1 & 2 & 0 \\ 2 & 1 & 3 \end{bmatrix} \text{ and } B = \begin{bmatrix} 1 & 0 & 2 \\ 3 & 1 & 1 \\ -2 & 1 & 0 \end{bmatrix}?$$

2. Find the product AB.

EXAMPLE 2 **Material Supply** | APPLICATION PREVIEW |

Pentico Industries must choose a supplier for the raw materials that it uses in its two manufacturing divisions at Clarion and Brooks. Each division uses different unit amounts of steel, wood, and plastic as shown in the table below.

	Steel	Wood	Plastic
Clarion	20	30	8
Brooks	22	25	15

The two supply companies being considered, Western and Coastal, can each supply all of these materials, but at different prices per unit, as described in the following table.

	Western	Coastal
Steel	300	290
Wood	100	90
Plastic	145	180

Use matrix multiplication to decide which supplier should be chosen to supply (a) the Clarion division and (b) the Brooks division.

Solution

The different amounts of products needed can be placed in matrix A and the prices charged by the suppliers in matrix B.

$$A = \begin{array}{c} \\ \\ \end{array}\overset{\text{Steel \ Wood \ Plastic}}{\begin{bmatrix} 20 & 30 & 8 \\ 22 & 25 & 15 \end{bmatrix}}\begin{array}{l} \text{Clarion} \\ \text{Brooks} \end{array} \qquad B = \overset{\text{Western \ Coastal}}{\begin{bmatrix} 300 & 290 \\ 100 & 90 \\ 145 & 180 \end{bmatrix}}\begin{array}{l} \text{Steel} \\ \text{Wood} \\ \text{Plastic} \end{array}$$

The price from each supplier for each division is found from the product AB.

$$AB = \begin{bmatrix} 20 & 30 & 8 \\ 22 & 25 & 15 \end{bmatrix}\begin{bmatrix} 300 & 290 \\ 100 & 90 \\ 145 & 180 \end{bmatrix}$$

$$= \begin{bmatrix} 20 \cdot 300 + 30 \cdot 100 + 8 \cdot 145 & 20 \cdot 290 + 30 \cdot 90 + 8 \cdot 180 \\ 22 \cdot 300 + 25 \cdot 100 + 15 \cdot 145 & 22 \cdot 290 + 25 \cdot 90 + 15 \cdot 180 \end{bmatrix}$$

$$= \overset{\text{Western \qquad Coastal}}{\begin{bmatrix} 10{,}160 & 9940 \\ 11{,}275 & 11{,}330 \end{bmatrix}}\begin{array}{l} \text{Clarion} \\ \text{Brooks} \end{array}$$

(a) The price for the Clarion division for supplies from Western is in the first row, first column and is \$10,160; the price for Clarion for supplies from Coastal is in the first row, second column and is \$9940. Thus the best price for Clarion is from Coastal.

(b) The price for the Brooks division for supplies from Western is in the second row, first column and is \$11,275; the price for Brooks for supplies from Coastal is in the second row, second column and is \$11,330. Thus the best price for Brooks is from Western. ■

An $n \times n$ (square) matrix (where n is any natural number) that has 1s down its diagonal and 0s everywhere else is called an **identity matrix**. The matrix

$$I = \begin{bmatrix} 1 & 0 & 0 \\ 0 & 1 & 0 \\ 0 & 0 & 1 \end{bmatrix}$$

is a 3×3 identity matrix. The matrix I is called an identity matrix because for any 3×3 matrix A, $AI = IA = A$. That is, if I is multiplied by a 3×3 matrix A, the product matrix is A. Note that when one of two square matrices being multiplied is an identity matrix, the product is *commutative*.

EXAMPLE 3   **Identity Matrix**

(a) Write the 2×2 identity matrix.

(b) Given $A = \begin{bmatrix} 4 & -7 \\ 13 & 2 \end{bmatrix}$, show that $AI = IA = A$.

Solution

(a) We denote the 2 × 2 identity matrix by

$$I = \begin{bmatrix} 1 & 0 \\ 0 & 1 \end{bmatrix}$$

(b) $AI = \begin{bmatrix} 4 & -7 \\ 13 & 2 \end{bmatrix} \begin{bmatrix} 1 & 0 \\ 0 & 1 \end{bmatrix} = \begin{bmatrix} 4+0 & 0-7 \\ 13+0 & 0+2 \end{bmatrix} = \begin{bmatrix} 4 & -7 \\ 13 & 2 \end{bmatrix}$

$$IA = \begin{bmatrix} 1 & 0 \\ 0 & 1 \end{bmatrix} \begin{bmatrix} 4 & -7 \\ 13 & 2 \end{bmatrix} = \begin{bmatrix} 4 & -7 \\ 13 & 2 \end{bmatrix}$$

Therefore, $AI = IA = A$.

EXAMPLE 4 **Encoding Messages**

Messages can be encoded through the use of a code and an encoding matrix. For example, given the code

a	b	c	d	e	f	g	h	i	j	k	l
1	2	3	4	5	6	7	8	9	10	11	12

m	n	o	p	q	r	s	t	u	v	w	x
13	14	15	16	17	18	19	20	21	22	23	24

y	z	blank
25	26	27

and the encoding matrix

$$A = \begin{bmatrix} 3 & 5 \\ 4 & 6 \end{bmatrix}$$

we can encode a message by separating it into number pairs (because A is a 2 × 2 matrix) and then multiplying each pair by A. Use this code and matrix to encode the message "good job."

Solution

The phrase "good job" is written

g	o	o	d	blank	j	o	b
7	15	15	4	27	10	15	2

and is encoded by writing each pair of numbers as a column matrix and multiplying each matrix by matrix A. Performing this multiplication gives the encoded message.

$$\begin{bmatrix} 3 & 5 \\ 4 & 6 \end{bmatrix} \begin{bmatrix} 7 \\ 15 \end{bmatrix} = \begin{bmatrix} 96 \\ 118 \end{bmatrix}$$

$$\begin{bmatrix} 3 & 5 \\ 4 & 6 \end{bmatrix} \begin{bmatrix} 15 \\ 4 \end{bmatrix} = \begin{bmatrix} 65 \\ 84 \end{bmatrix}$$

$$\begin{bmatrix} 3 & 5 \\ 4 & 6 \end{bmatrix} \begin{bmatrix} 27 \\ 10 \end{bmatrix} = \begin{bmatrix} 131 \\ 168 \end{bmatrix}$$

$$\begin{bmatrix} 3 & 5 \\ 4 & 6 \end{bmatrix} \begin{bmatrix} 15 \\ 2 \end{bmatrix} = \begin{bmatrix} 55 \\ 72 \end{bmatrix}$$

Thus the code sent is 96, 118, 65, 84, 131, 168, 55, 72. Notice that the letter "o" occurred three times in the original message but was encoded differently each time. We can also get the same code to send by putting the pairs of numbers as columns in one 2 × 4 matrix and multiplying by A.

$$\begin{bmatrix} 3 & 5 \\ 4 & 6 \end{bmatrix} \begin{bmatrix} 7 & 15 & 27 & 15 \\ 15 & 4 & 10 & 2 \end{bmatrix} = \begin{bmatrix} 96 & 65 & 131 & 55 \\ 118 & 84 & 168 & 72 \end{bmatrix}$$

In this case, the code is sent column by column.

Note in Example 4 that if we used a different encoding matrix, then the same message would be sent with a different sequence of code. We will discuss the methods of decoding messages in Section 3.4, "Inverse of a Square Matrix; Matrix Equations."

CHECKPOINT

3. (a) Compute AB if $A = \begin{bmatrix} 2 & 4 \\ 3 & 6 \end{bmatrix}$ and $B = \begin{bmatrix} 2 & -6 \\ -1 & 3 \end{bmatrix}$.

    (b) Can the product of two matrices be a zero matrix even if neither matrix is a zero matrix?

    (c) Does BA result in a zero matrix?

Calculator Note   Graphing calculators can be used to perform matrix multiplications. For example, if

$$A = \begin{bmatrix} 2 & 3 & 4 & 5 \\ 3 & 1 & 3 & 5 \\ 3 & -4 & 3 & 1 \end{bmatrix} \quad B = \begin{bmatrix} 2 & 2 & 5 \\ 3 & 5 & 7 \\ 1 & 1 & 2 \\ 5 & 4 & 3 \end{bmatrix} \quad C = \begin{bmatrix} 0.5 & 3.5 & 2.4 \\ 2 & 7 & 0.8 \\ 5.5 & 3.5 & 7 \\ 2.5 & 4 & 3 \end{bmatrix}$$

then Figure 3.6 shows the matrix products AB and BA and Figure 3.7 shows that the product BC is not defined. Details of finding matrix products with graphing calculators are shown in Appendix C, Section 3.2.

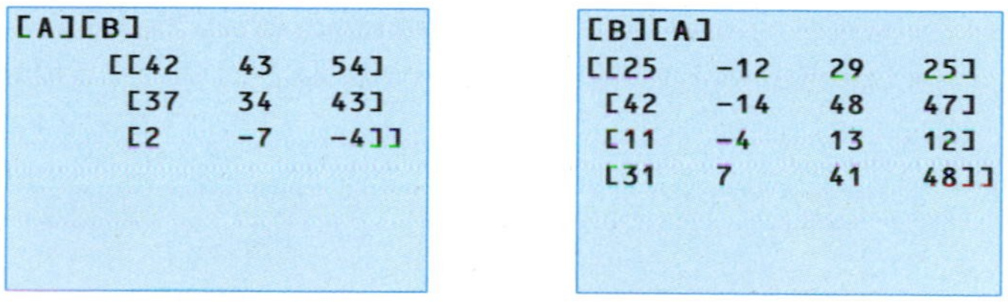

Figure 3.6

```
[A][B]
    [[42   43   54]
     [37   34   43]
     [2    -7   -4]]
```
(a)

```
[B][A]
    [[25  -12   29   25]
     [42  -14   48   47]
     [11   -4   13   12]
     [31    7   41   48]]
```
(b)

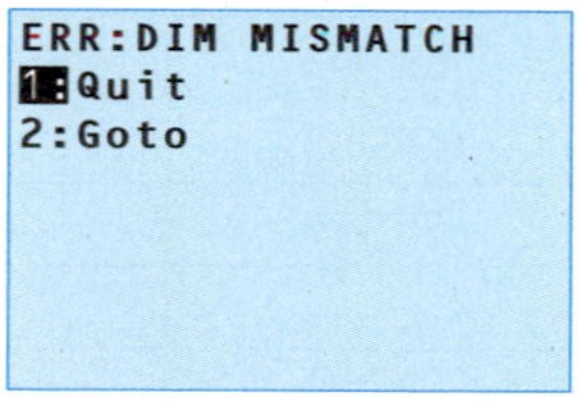

Figure 3.7

```
ERR:DIM MISMATCH
1:Quit
2:Goto
```

Spreadsheet Note   Like graphing calculators and software programs, Excel can be used to multiply two matrices.

EXAMPLE 5   Venture Capital

Suppose that a bank has three main sources of income—business loans, auto loans, and home mortgages—and that it draws funds from these sources for venture capital used to provide start-up funds for new businesses. Suppose the income from these sources for each of 3 years is given in Table 3.4, and the bank uses 45% of its income from business loans, 20% of its income from auto loans, and 30% of its income from home mortgages to get its venture capital funds. Write a matrix product that gives the venture capital for these years, and find the available venture capital in each of the 3 years.

| TABLE 3.4 |

INCOME FROM LOANS

Year	Business	Auto	Home
2010	63,300	20,024	51,820
2011	48,305	15,817	63,722
2012	55,110	18,621	64,105

Solution

The matrix that describes the sources of income for the 3 years is

$$\begin{bmatrix} 63{,}300 & 20{,}024 & 51{,}820 \\ 48{,}305 & 15{,}817 & 63{,}722 \\ 55{,}110 & 18{,}621 & 64{,}105 \end{bmatrix}$$

and the matrix that describes the percent of each that is used for venture capital is

$$\begin{bmatrix} 0.45 \\ 0.20 \\ 0.30 \end{bmatrix}$$

We can find the product of these matrices with Excel by using the MMULT command as shown in Table 3.5 (see Appendix D, Section 3.2 and the Online Excel Guide for details).

| TABLE 3.5 |

	A	B	C	D
		= MMULT(B3:C5,B11:B13)		
1		Income from Loans		
2		Business	Auto	Home
3	Matrix A	63,300	20,024	51,820
4		48,305	15,817	63,722
5		55,110	18,621	64,105
6				
7	Matrix B	0.45		
8		0.20		
9		0.30		
10				
11	Product AxB	48,035.80		
12		44,017.25		
13		47,755.20		

Thus the available venture capital for each of the 3 years is as follows.

$$
\begin{array}{ll}
2010: & \$48,035,80 \\
2011: & \$44,017.25 \\
2012: & \$47,755.20
\end{array}
$$

CHECKPOINT SOLUTIONS

1. $2 \cdot 2 + 1 \cdot 1 + 3 \cdot 0 = 5$

2. $\begin{bmatrix} 7 & 2 & 4 \\ -1 & 4 & 5 \end{bmatrix}$

3. (a) $AB = \begin{bmatrix} 0 & 0 \\ 0 & 0 \end{bmatrix}$ (b) Yes. See (a).   (c) No. $BA = \begin{bmatrix} -14 & -28 \\ 7 & 14 \end{bmatrix}$

| EXERCISES | 3.2

Access end-of-section exercises online at **www.webassign.net** WebAssign

OBJECTIVES   **3.3**

- To use matrices to solve systems of linear equations with unique solutions
- To use matrices to solve systems of linear equations with nonunique solutions

Gauss-Jordan Elimination: Solving Systems of Equations

| APPLICATION PREVIEW |

The Walters Manufacturing Company makes three types of metal storage sheds. The company has three departments: stamping, painting, and packaging. The following table gives the number of hours each division requires for each shed.

Department	Shed		
	Type I	Type II	Type III
Stamping	2	3	4
Painting	1	2	1
Packaging	1	1	2

By using the information in the table and by solving a system of equations, we can determine how many of each type of shed can be produced if the stamping department has 3200 hours available, the painting department has 1700 hours, and the packaging department has 1300 hours. The set of equations used to solve this problem (in Example 2) is

$$
\begin{cases}
2x_1 + 3x_2 + 4x_3 = 3200 \\
x_1 + 2x_2 + x_3 = 1700 \\
x_1 + x_2 + 2x_3 = 1300
\end{cases}
$$

In this section, we will solve such systems by using matrices and the Gauss-Jordan elimination method.

Systems with Unique Solutions

In solving systems with the left-to-right elimination method used in Example 8 in Section 1.5, "Solutions of Systems of Linear Equations," we operated on the coefficients of the variables x, y, and z and on the constants. If we keep the coefficients of the variables x, y, and z in distinctive columns, we do not need to write the equations. In solving a system of linear equations with matrices, we first write the coefficients and constants from the system in the **augmented matrix.** For example, the augmented matrix associated with

$$\begin{cases} x + 2y + 3z = 6 \\ x \quad\quad - z = 0 \\ x - y - z = -4 \end{cases} \text{ is } \left[\begin{array}{ccc|c} 1 & 2 & 3 & 6 \\ 1 & 0 & -1 & 0 \\ 1 & -1 & -1 & -4 \end{array}\right]$$

In the augmented matrix, the numbers on the left side of the solid line form the **coefficient matrix,** with each column containing the coefficients of a variable (0 represents any missing variable). The column on the right side of the line (called the *augment*) contains the constants. Each row of the matrix gives the corresponding coefficients of an equation.

CHECKPOINT

1. (a) Write the augmented matrix for the following system of linear equations.

$$\begin{cases} 3x + 2y - z = 3 \\ x - y + 2z = 4 \\ 2x + 3y - z = 3 \end{cases}$$

(b) Write the coefficient matrix for the system.

We can use matrices to solve systems of linear equations by performing the same operations on the rows of a matrix to reduce it as we do on equations in a linear system. The three different operations we can use to reduce the matrix are called **elementary row operations** and are similar to the operations with equations that result in equivalent systems. These operations are

1. Interchange two rows.
2. Add a multiple of one row to another row.
3. Multiply a row by a nonzero constant.

When a new matrix results from one or more of these elementary row operations being performed on a matrix, the new matrix is said to be **equivalent** to the original because both these matrices represent equivalent systems of equations. Thus, if it can be shown that the augmented matrix

$$\left[\begin{array}{ccc|c} 1 & 2 & 3 & 6 \\ 1 & 0 & -1 & 0 \\ 1 & -1 & -1 & -4 \end{array}\right] \text{ is equivalent to } \left[\begin{array}{ccc|c} 1 & 0 & 0 & -0.5 \\ 0 & 1 & 0 & 4 \\ 0 & 0 & 1 & -0.5 \end{array}\right]$$

then the systems corresponding to these matrices have the same solution. That is,

$$\begin{cases} x + 2y + 3z = 6 \\ x \quad\quad - z = 0 \\ x - y - z = -4 \end{cases} \text{ has the same solution as } \begin{cases} x = -0.5 \\ y = 4 \\ z = -0.5 \end{cases}$$

Thus, if a system of linear equations has a unique solution, we can solve the system by reducing the associated matrix to one whose coefficient matrix is the identity matrix and then "reading" the values of the variables that give the solution.

The process that we use to solve a system of equations with matrices (called the **elimination method** or **Gauss-Jordan elimination method**) is a systematic procedure that uses row operations to attempt to reduce the coefficient matrix to an identity matrix.

For example, the system

$$\begin{cases} 2x + 5y + 4z = 4 \\ x + 4y + 3z = 1 \\ x - 3y - 2z = 5 \end{cases}$$

can be represented by the augmented matrix

$$\begin{bmatrix} 2 & 5 & 4 & | & 4 \\ 1 & 4 & 3 & | & 1 \\ 1 & -3 & -2 & | & 5 \end{bmatrix}$$

The following procedure (using the augmented matrix above) is used to reduce the augmented matrix for a system of equations to an equivalent matrix from which the solutions to the system may be found.

Gauss-Jordan Elimination Method

Goal (for Each Step)	Row Operation	Equivalent Matrix
1. Get a 1 in row 1, column 1.	Interchange row 1 and row 2.	$\begin{bmatrix} 1 & 4 & 3 & \| & 1 \\ 2 & 5 & 4 & \| & 4 \\ 1 & -3 & -2 & \| & 5 \end{bmatrix}$
2. Add multiples of row 1 *only* to the other rows to get zeros as the other entries of column 1.	Add -2 times row 1 to row 2; put the result in row 2. Add -1 times row 1 to row 3; put the result in row 3.	$\begin{bmatrix} 1 & 4 & 3 & \| & 1 \\ 0 & -3 & -2 & \| & 2 \\ 0 & -7 & -5 & \| & 4 \end{bmatrix}$
3. Use rows below row 1 to get a 1 in row 2, column 2.	Multiply row 2 by $-\frac{1}{3}$; put the result in row 2.	$\begin{bmatrix} 1 & 4 & 3 & \| & 1 \\ 0 & 1 & \frac{2}{3} & \| & -\frac{2}{3} \\ 0 & -7 & -5 & \| & 4 \end{bmatrix}$
4. Add multiples of row 2 *only* to the other rows to get zeros as the other entries in column 2.	Add -4 times row 2 to row 1; put the result in row 1. Add 7 times row 2 to row 3; put the result in row 3.	$\begin{bmatrix} 1 & 0 & \frac{1}{3} & \| & \frac{11}{3} \\ 0 & 1 & \frac{2}{3} & \| & -\frac{2}{3} \\ 0 & 0 & -\frac{1}{3} & \| & -\frac{2}{3} \end{bmatrix}$
5. Use rows below row 2 to get a 1 in row 3, column 3.	Multiply row 3 by -3; put the result in row 3.	$\begin{bmatrix} 1 & 0 & \frac{1}{3} & \| & \frac{11}{3} \\ 0 & 1 & \frac{2}{3} & \| & -\frac{2}{3} \\ 0 & 0 & 1 & \| & 2 \end{bmatrix}$
6. Add multiples of row 3 *only* to the other rows to get zeros as the other entries in column 3.	Add $-\frac{1}{3}$ times row 3 to row 1; put the result in row 1; Add $-\frac{2}{3}$ times row 3 to row 2; put the result in row 2.	$\begin{bmatrix} 1 & 0 & 0 & \| & 3 \\ 0 & 1 & 0 & \| & -2 \\ 0 & 0 & 1 & \| & 2 \end{bmatrix}$
7. Repeat the process until it cannot be continued.	All rows have been used. The matrix is in reduced form.	

The previous display shows a series of step-by-step goals for reducing a matrix and shows a series of row operations that reduces this 3×3 matrix. The reduction procedure and row operations apply regardless of the size of the system.

In the Gauss-Jordan method box, the systems of equations corresponding to the initial and the final augmented matrices are

$$\begin{array}{ccc} \text{Initial} & & \text{Final} \\ 2x + 5y + 4z = 4 & & x + 0y + 0z = 3 \\ x + 4y + 3z = 1 \quad \text{and} & & 0x + y + 0z = -2 \\ x - 3y - 2z = 5 & & 0x + 0y + z = 2 \end{array}$$

Because the augmented matrices for these systems are equivalent, both systems have the same solution, $x = 3$, $y = -2$, and $z = 2$.

When a system of linear equations has a unique solution, the coefficient part of the reduced augmented matrix will be an identity matrix. Note that for this to occur, the coefficient matrix must be square; that is, the number of equations must equal the number of variables.

EXAMPLE 1   Using the Gauss-Jordan Method

Solve the system

$$
\begin{aligned}
x_1 + x_2 + x_3 + 2x_4 &= 6 \\
x_1 + 2x_2 \quad\quad + x_4 &= -2 \\
x_1 + x_2 + 3x_3 - 2x_4 &= 12 \\
x_1 + x_2 - 4x_3 + 5x_4 &= -16
\end{aligned}
$$

Solution

First note that there is a variable missing in the second equation. When we form the augmented matrix, we must insert a zero in this place so that each column represents the coefficients of one variable.

To solve this system, we must reduce the augmented matrix.

$$
\left[\begin{array}{cccc|c}
1 & 1 & 1 & 2 & 6 \\
1 & 2 & 0 & 1 & -2 \\
1 & 1 & 3 & -2 & 12 \\
1 & 1 & -4 & 5 & -16
\end{array}\right]
$$

The reduction process follows. The entry in row 1, column 1 is 1. To get zeros in the first column, add -1 times row 1 to row 2, and put the result in row 2; add -1 times row 1 to row 3 and put the result in row 3; and add -1 times row 1 to row 4 and put the result in row 4. These steps follow.

$$
\left[\begin{array}{cccc|c}
1 & 1 & 1 & 2 & 6 \\
1 & 2 & 0 & 1 & -2 \\
1 & 1 & 3 & -2 & 12 \\
1 & 1 & -4 & 5 & -16
\end{array}\right]
\begin{array}{c}
(-1)R_1 + R_2 \rightarrow R_2 \\
(-1)R_1 + R_3 \rightarrow R_3 \\
(-1)R_1 + R_4 \rightarrow R_4 \\
\longrightarrow
\end{array}
\left[\begin{array}{cccc|c}
1 & 1 & 1 & 2 & 6 \\
0 & 1 & -1 & -1 & -8 \\
0 & 0 & 2 & -4 & 6 \\
0 & 0 & -5 & 3 & -22
\end{array}\right]
$$

The entry in row 2, column 2 is 1. To get zeros in the second column, add -1 times row 2 to row 1, and put the result in row 1.

$$
\begin{array}{c}
(-1)R_2 + R_1 \rightarrow R_1 \\
\longrightarrow
\end{array}
\left[\begin{array}{cccc|c}
1 & 0 & 2 & 3 & 14 \\
0 & 1 & -1 & -1 & -8 \\
0 & 0 & 2 & -4 & 6 \\
0 & 0 & -5 & 3 & -22
\end{array}\right]
$$

The entry in row 3, column 3 is 2. To get a 1 in this position, multiply row 3 by $\frac{1}{2}$.

$$
\begin{array}{c}
\frac{1}{2}R_3 \rightarrow R_3 \\
\longrightarrow
\end{array}
\left[\begin{array}{cccc|c}
1 & 0 & 2 & 3 & 14 \\
0 & 1 & -1 & -1 & -8 \\
0 & 0 & 1 & -2 & 3 \\
0 & 0 & -5 & 3 & -22
\end{array}\right]
$$

Using row 3, add -2 times row 3 to row 1 and put the result in row 1; add row 3 to row 2 and put the result in row 2; and add 5 times row 3 to row 4 and put the result in row 4.

$$
\begin{array}{c}
(-2)R_3 + R_1 \rightarrow R_1 \\
R_3 + R_2 \rightarrow R_2 \\
5R_3 + R_4 \rightarrow R_4 \\
\longrightarrow
\end{array}
\left[\begin{array}{cccc|c}
1 & 0 & 0 & 7 & 8 \\
0 & 1 & 0 & -3 & -5 \\
0 & 0 & 1 & -2 & 3 \\
0 & 0 & 0 & -7 & -7
\end{array}\right]
$$

Multiplying $-\frac{1}{7}$ times row 4 and putting the result in row 4 gives a 1 in row 4, column 4.

$$\xrightarrow{-\frac{1}{7}R_4 \rightarrow R_4} \left[\begin{array}{cccc|c} 1 & 0 & 0 & 7 & 8 \\ 0 & 1 & 0 & -3 & -5 \\ 0 & 0 & 1 & -2 & 3 \\ 0 & 0 & 0 & 1 & 1 \end{array}\right]$$

Adding appropriate multiples of row 4 to the other rows gives

$$\begin{array}{c} (-7)R_4 + R_1 \rightarrow R_1 \\ 3R_4 + R_2 \rightarrow R_2 \\ \hline 2R_4 + R_3 \rightarrow R_3 \end{array} \left[\begin{array}{cccc|c} 1 & 0 & 0 & 0 & 1 \\ 0 & 1 & 0 & 0 & -2 \\ 0 & 0 & 1 & 0 & 5 \\ 0 & 0 & 0 & 1 & 1 \end{array}\right]$$

This corresponds to the system

$$\begin{aligned} x_1 & & & = 1 \\ & x_2 & & = -2 \\ & & x_3 & = 5 \\ & & x_4 & = 1 \end{aligned}$$

Thus the solution is $x_1 = 1$, $x_2 = -2$, $x_3 = 5$, and $x_4 = 1$. ■

Calculator Note Graphing calculators are especially useful in solving systems of equations because they can be used to perform the row reductions necessary to obtain a reduced matrix. On some calculators, a direct command will give a reduced form for the augmented matrix. Figure 3.8(a) shows the augmented matrix for the system of equations in Example 1, and Figure 3.8(b) shows the reduced form of the augmented matrix (which is the same as the reduced matrix found in Example 1). The steps used to solve a system of linear equations by finding the reduced matrix are shown in Appendix C, Section 3.3. ■

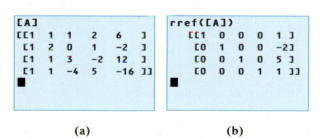

Figure 3.8   (a)   (b)

CHECKPOINT   2. Solve the system.

$$\begin{cases} 3x + 2y - z = 3 \\ x - y + 2z = 4 \\ 2x + 3y - z = 3 \end{cases}$$

EXAMPLE 2   **Manufacturing** | APPLICATION PREVIEW |

The Walters Manufacturing Company needs to know how best to use the time available within its three manufacturing departments in the construction and packaging of the three types of metal storage sheds. Each one must be stamped, painted, and packaged. Table 3.6 on the next page shows the number of hours required for the processing of each type of shed. Using the information in the table, determine how many of each type of shed can be produced if the stamping department has 3200 hours available, the painting department has 1700 hours, and the packaging department has 1300 hours.

▌| TABLE 3.6 | ▬▬▬▬▬▬▬▬▬▬

	Shed		
Department	Type I	Type II	Type III
Stamping	2	3	4
Painting	1	2	1
Packaging	1	1	2

Solution

If we let x_1 be the number of type I sheds, x_2 be the number of type II sheds, and x_3 be the number of type III sheds, we can develop the following equations.

$$\begin{aligned}
\text{Stamping hours:} \quad & 2x_1 + 3x_2 + 4x_3 = 3200 \\
\text{Painting hours:} \quad & 1x_1 + 2x_2 + 1x_3 = 1700 \\
\text{Packaging hours:} \quad & 1x_1 + 1x_2 + 2x_3 = 1300
\end{aligned}$$

The augmented matrix for this system of equations is

$$\left[\begin{array}{ccc|c}
2 & 3 & 4 & 3200 \\
1 & 2 & 1 & 1700 \\
1 & 1 & 2 & 1300
\end{array}\right]$$

Reducing this augmented matrix proceeds as follows:

$$\left[\begin{array}{ccc|c}
2 & 3 & 4 & 3200 \\
1 & 2 & 1 & 1700 \\
1 & 1 & 2 & 1300
\end{array}\right]
\xrightarrow{\text{Switch } R_1 \text{ and } R_3}
\left[\begin{array}{ccc|c}
1 & 1 & 2 & 1300 \\
1 & 2 & 1 & 1700 \\
2 & 3 & 4 & 3200
\end{array}\right]$$

$$\xrightarrow[\substack{(-1)R_1 + R_2 \rightarrow R_2 \\ (-2)R_1 + R_3 \rightarrow R_3}]{}
\left[\begin{array}{ccc|c}
1 & 1 & 2 & 1300 \\
0 & 1 & -1 & 400 \\
0 & 1 & 0 & 600
\end{array}\right]$$

$$\xrightarrow[\substack{(-1)R_2 + R_1 \rightarrow R_1 \\ (-1)R_2 + R_3 \rightarrow R_3}]{}
\left[\begin{array}{ccc|c}
1 & 0 & 3 & 900 \\
0 & 1 & -1 & 400 \\
0 & 0 & 1 & 200
\end{array}\right]
\xrightarrow[\substack{(-3)R_3 + R_1 \rightarrow R_1 \\ R_3 + R_2 \rightarrow R_2}]{}
\left[\begin{array}{ccc|c}
1 & 0 & 0 & 300 \\
0 & 1 & 0 & 600 \\
0 & 0 & 1 & 200
\end{array}\right]$$

Thus the solution to the system is

$$x_1 = 300 \qquad x_2 = 600 \qquad x_3 = 200$$

The company should make 300 type I, 600 type II, and 200 type III sheds. ■

Systems with Nonunique Solutions

All the systems considered so far have unique solutions, but it is also possible for a system of linear equations to have an infinite number of solutions or no solution at all. Although coefficient matrices for systems with an infinite number of solutions or no solution will not reduce to identity matrices, row operations can be used to obtain a reduced form from which the solutions, if they exist, can be determined.

A matrix is said to be in **reduced form** when it is in the following form:

1. The first nonzero element in each row is 1.
2. Every column containing a first nonzero element for some row has zeros everywhere else.
3. The first nonzero element of each row is to the right of the first nonzero element of every row above it.
4. All rows containing zeros are grouped together below the rows containing nonzero entries.

The following matrices are in reduced form because they satisfy these conditions.

$$\left[\begin{array}{cc|c} 1 & 0 & 4 \\ 0 & 1 & 2 \\ 0 & 0 & 0 \end{array}\right] \qquad \left[\begin{array}{ccc|c} 1 & 0 & 0 & 0 \\ 0 & 1 & 0 & 3 \\ 0 & 0 & 1 & 1 \\ 0 & 0 & 0 & 0 \end{array}\right] \qquad \left[\begin{array}{ccc|c} 1 & 4 & 0 & 1 \\ 0 & 0 & 1 & 2 \\ 0 & 0 & 0 & 0 \end{array}\right]$$

The following matrices are *not* in reduced form.

$$\left[\begin{array}{cc|c} 1 & ① & 0 \\ 0 & 1 & 3 \\ 0 & 0 & 0 \end{array}\right] \qquad \left[\begin{array}{cc|c} 1 & 0 & 1 \\ 0 & ② & 1 \\ 0 & 0 & 0 \end{array}\right] \qquad \left[\begin{array}{cc|c} 0 & 1 & 0 \\ 1 & 0 & 0 \\ 0 & 0 & 1 \end{array}\right]$$

In the first two matrices, the circled element must be changed to obtain a reduced form. Can you see what row operations would transform each of these matrices into reduced form? The third matrix does not satisfy condition 3.

We can solve a system of linear equations by using row operations on the augmented matrix until the coefficient matrix is transformed to an equivalent matrix in reduced form. We begin by following the goals outlined in the Gauss-Jordan elimination method and in the Example 1 discussion.

EXAMPLE 3   Matrix Solution of a System

Solve the system

$$\begin{cases} x_1 + x_2 - x_3 + x_4 = 3 \\ x_2 + x_3 + x_4 = 1 \\ x_1 - 2x_3 + x_4 = 6 \\ 2x_1 - x_2 - 5x_3 - 3x_4 = -5 \end{cases}$$

Solution

Attempting to reduce the coefficient matrix of the associated augmented matrix to an identity matrix requires the following.

$$\left[\begin{array}{cccc|c} 1 & 1 & -1 & 1 & 3 \\ 0 & 1 & 1 & 1 & 1 \\ 1 & 0 & -2 & 1 & 6 \\ 2 & -1 & -5 & -3 & -5 \end{array}\right] \begin{array}{l} (-1)R_1 + R_3 \to R_3 \\ \underrightarrow{(-2)R_1 + R_4 \to R_4} \end{array} \left[\begin{array}{cccc|c} 1 & 1 & -1 & 1 & 3 \\ 0 & 1 & 1 & 1 & 1 \\ 0 & -1 & -1 & 0 & 3 \\ 0 & -3 & -3 & -5 & -11 \end{array}\right]$$

$$\begin{array}{l} (-1)R_2 + R_1 \to R_1 \\ R_2 + R_3 \to R_3 \\ \underrightarrow{3R_2 + R_4 \to R_4} \end{array} \left[\begin{array}{cccc|c} 1 & 0 & -2 & 0 & 2 \\ 0 & 1 & 1 & 1 & 1 \\ 0 & 0 & 0 & 1 & 4 \\ 0 & 0 & 0 & -2 & -8 \end{array}\right]$$

The entry in row 3, column 3 cannot be made 1 using rows *below* row 2. Moving to column 4, the entry in row 3, column 4 is a 1. Using row 3 gives

$$\begin{array}{l} (-1)R_3 + R_2 \to R_2 \\ \underrightarrow{2R_3 + R_4 \to R_4} \end{array} \left[\begin{array}{cccc|c} 1 & 0 & -2 & 0 & 2 \\ 0 & 1 & 1 & 0 & -3 \\ 0 & 0 & 0 & 1 & 4 \\ 0 & 0 & 0 & 0 & 0 \end{array}\right]$$

The augmented matrix is now reduced; this matrix corresponds to the system

$$\begin{cases} x_1 - 2x_3 = 2 \\ x_2 + x_3 = -3 \\ x_4 = 4 \\ 0 = 0 \end{cases}$$

If we solve each of the equations for the leading variable (the variable corresponding to the first 1 in each row of the reduced form of a matrix) and let any non-leading variables equal any real number (x_3 in this case), we obtain the **general solution** of the system.

$$\begin{aligned}
x_1 &= 2 + 2x_3 \\
x_2 &= -3 - x_3 \\
x_4 &= 4 \\
x_3 &= \text{any real number}
\end{aligned}$$

The general solution gives the values of x_1 and x_2 dependent on the value of x_3, so we can get many different solutions of the system by specifying different values of x_3. For example, if $x_3 = 1$, then $x_1 = 4, x_2 = -4, x_3 = 1$, and $x_4 = 4$ is a solution of the system; if we let $x_3 = -2$, then $x_1 = -2, x_2 = -1, x_3 = -2$, and $x_4 = 4$ is another solution. ■

We have seen two different possibilities for solutions of systems of linear equations. In Examples 1 and 2, there was only one solution, whereas in Example 3 we saw that there were infinitely many solutions, one for each possible value of x_3. A third possibility exists: that the system has no solution. Recall that these possibilities also existed for two equations in two unknowns, as discussed in Chapter 1, "Linear Equations and Functions."

EXAMPLE 4 A System with No Solution

Solve the system

$$\begin{cases}
x + 2y - z = 3 \\
3x + y = 4 \\
2x - y + z = 2
\end{cases}$$

Solution

The augmented matrix for this system is reduced as follows:

$$\begin{bmatrix} 1 & 2 & -1 & | & 3 \\ 3 & 1 & 0 & | & 4 \\ 2 & -1 & 1 & | & 2 \end{bmatrix} \xrightarrow[(-2)R_1 + R_3 \to R_3]{(-3)R_1 + R_2 \to R_2} \begin{bmatrix} 1 & 2 & -1 & | & 3 \\ 0 & -5 & 3 & | & -5 \\ 0 & -5 & 3 & | & -4 \end{bmatrix}$$

$$\xrightarrow{-\frac{1}{5}R_2 \to R_2} \begin{bmatrix} 1 & 2 & -1 & | & 3 \\ 0 & 1 & -\frac{3}{5} & | & 1 \\ 0 & -5 & 3 & | & -4 \end{bmatrix}$$

$$\xrightarrow[(5)R_2 + R_3 \to R_3]{(-2)R_2 + R_1 \to R_1} \begin{bmatrix} 1 & 0 & \frac{1}{5} & | & 1 \\ 0 & 1 & -\frac{3}{5} & | & 1 \\ 0 & 0 & 0 & | & 1 \end{bmatrix}$$

The system of equations corresponding to the reduced matrix has $0 = 1$ as the third equation. This is clearly an impossibility, so there is no solution. ■

Now that we have seen examples of each of the different solution possibilities for n equations in n variables, let us consider what we have learned. In all cases the solution procedure began by setting up and then reducing the augmented matrix. Once the matrix is reduced, we can write the corresponding reduced system of equations. The solutions, if they exist, are easily found from this reduced system. We summarize the possibilities in Table 3.7.

| TABLE 3.7

UNIQUE AND NONUNIQUE SOLUTIONS

Reduced Form of Augmented Matrix for n Equations in n Variables	Solution to System
1. Coefficient array is an identity matrix.	Unique solution (see Examples 1 and 2).
2. Coefficient array is *not* an identity matrix with either:	
(a) A row of 0s in the coefficient array with a nonzero entry in the augment.	(a) No solution (see Example 4).
(b) Or otherwise.	(b) Infinitely many solutions. Solve for lead variables in terms of non-lead variables; non-lead variables can equal any real number (see Example 3).

Calculator Note

Whether or not a unique solution exists, we can use a graphing calculator to reduce an augmented matrix and then find any solution that exists (or that no solution exists). For example, following is a system of equations along with the calculator screens for its augmented matrix and the reduced form of its augmented matrix.

$$\begin{cases} x + 2y - 2z = 4 \\ 2x + 5y \quad\quad = 3 \\ x + \ y - 6z = 9 \end{cases}$$

```
[A]
    [[1  2  -2  4]
     [2  5   0  3]
     [1  1  -6  9]]
```

```
rref([A])
    [[1  0  -10  14]
     [0  1   4   -5]
     [0  0   0    0]]
```

From the reduced matrix we obtain the following equations and general solution.

$$x - 10z = \ 14 \qquad\qquad x = 10z + 14$$
$$y + \ 4z = -5 \qquad\qquad y = -4z - 5$$
$$\qquad\qquad\qquad\qquad\qquad z = \ \text{any real number}$$

Most calculators have a command (like rref([A]) above) that gives the reduced form of a matrix. See Appendix C, Section 3.3, for details. ∎

CHECKPOINT

For each system of equations, the reduced form of the augmented matrix is given. Give the solution to the system, if it exists.

3. $\begin{cases} 2x + 3y + 2z = 180 \\ -x + 2y + 4z = 180 \\ 2x + 6y + 5z = 270 \end{cases} \qquad \begin{bmatrix} 1 & 0 & 0 & | & 100 \\ 0 & 1 & 0 & | & -80 \\ 0 & 0 & 1 & | & 110 \end{bmatrix}$

4. $\begin{cases} x - 2y + 9z = 12 \\ 2x - \ y + 3z = 18 \\ x + \ y - 6z = \ 6 \end{cases} \qquad \begin{bmatrix} 1 & 0 & -1 & | & 8 \\ 0 & 1 & -5 & | & -2 \\ 0 & 0 & 0 & | & 0 \end{bmatrix}$

5. $\begin{cases} x - 4y + 3z = 4 \\ 2x - 2y + \ z = 6 \\ x + 2y - 2z = 4 \end{cases} \qquad \begin{bmatrix} 1 & 0 & -\frac{1}{3} & | & 0 \\ 0 & 1 & -\frac{5}{6} & | & 0 \\ 0 & 0 & 0 & | & 1 \end{bmatrix}$

Nonsquare Systems   Finally, we consider systems with fewer equations than variables. Because the coefficient matrix for such a system is not square, it cannot reduce to an identity matrix. Hence these systems of equations cannot have a unique solution but may have no solution or infinitely many solutions. Nevertheless, they are solved in the same manner as those considered so far but with the following two possible results.

1. If the reduced augmented matrix contains a row of 0s in the coefficient matrix with a nonzero number in the augment, the system has no solution.
2. Otherwise, the system has infinitely many solutions.

EXAMPLE 5   Investment

A trust account manager has \$500,000 to be invested in three different accounts. The accounts pay annual interest rates of 8%, 10%, and 14%, and the goal is to earn \$49,000 a year. To accomplish this, assume that x dollars is invested at 8%, y dollars at 10%, and z dollars at 14%. Find how much should be invested in each account to satisfy the conditions.

KULISH VICTORIIA/Shutterstock.com

Solution
The sum of the three investments is \$500,000, so we have the equation

$$x + y + z = 500{,}000$$

The interest earned from the 8% investment is $0.08x$, the amount earned from the 10% investment is $0.10y$, and the amount earned from the 14% investment is $0.14z$, so the total amount earned from the investments is given by the equation

$$0.08x + 0.10y + 0.14z = 49{,}000$$

These two equations represent all the given information, so the system is

$$\begin{cases} x + y + z = 500{,}000 \\ 0.08x + 0.10y + 0.14z = 49{,}000 \end{cases}$$

We solve this system using matrices, as follows:

$$\begin{bmatrix} 1 & 1 & 1 & | & 500{,}000 \\ 0.08 & 0.10 & 0.14 & | & 49{,}000 \end{bmatrix} \xrightarrow{-0.08R_1 + R_2 \to R_2} \begin{bmatrix} 1 & 1 & 1 & | & 500{,}000 \\ 0 & 0.02 & 0.06 & | & 9{,}000 \end{bmatrix}$$

$$\xrightarrow{50R_2 \to R_2} \begin{bmatrix} 1 & 1 & 1 & | & 500{,}000 \\ 0 & 1 & 3 & | & 450{,}000 \end{bmatrix} \xrightarrow{-R_2 + R_1 \to R_1} \begin{bmatrix} 1 & 0 & -2 & | & 50{,}000 \\ 0 & 1 & 3 & | & 450{,}000 \end{bmatrix}$$

This matrix represents the equivalent system

$$\begin{cases} x - 2z = 50{,}000 \\ y + 3z = 450{,}000 \end{cases} \quad \text{so} \quad \begin{aligned} x &= 50{,}000 + 2z \\ y &= 450{,}000 - 3z \\ z &= \text{any real number} \end{aligned}$$

Because a negative amount cannot be invested, x, y, and z must all be nonnegative. Note that $x \geq 0$ when $z \geq 0$ and that $y \geq 0$ when $450{,}000 - 3z \geq 0$, or when $z \leq 150{,}000$. Thus the amounts invested are

$$\begin{aligned} \$z \text{ at } 14\%, \quad &\text{where} \quad 0 \leq z \leq 150{,}000 \\ \$x \text{ at } 8\%, \quad &\text{where} \quad x = 50{,}000 + 2z \\ \$y \text{ at } 10\%, \quad &\text{where} \quad y = 450{,}000 - 3z \end{aligned}$$

There are many possible investment plans. One example would be with $z = \$100{,}000$, $x = \$250{,}000$, and $y = \$150{,}000$. Observe that the values for x and y depend on z, and once z is chosen, x will be between \$50,000 and \$350,000, and y will be between \$0 and \$450,000. ■

CHECKPOINT SOLUTIONS

1. (a) $\begin{bmatrix} 3 & 2 & -1 & | & 3 \\ 1 & -1 & 2 & | & 4 \\ 2 & 3 & -1 & | & 3 \end{bmatrix}$ (b) $\begin{bmatrix} 3 & 2 & -1 \\ 1 & -1 & 2 \\ 2 & 3 & -1 \end{bmatrix}$

2. $\begin{bmatrix} 3 & 2 & -1 & | & 3 \\ 1 & -1 & 2 & | & 4 \\ 2 & 3 & -1 & | & 3 \end{bmatrix}$ $\xrightarrow[\substack{\text{Then } -3R_1 + R_2 \to R_2 \\ -2R_1 + R_3 \to R_3}]{\text{Switch } R_1 \text{ and } R_2}$ $\begin{bmatrix} 1 & -1 & 2 & | & 4 \\ 0 & 5 & -7 & | & -9 \\ 0 & 5 & -5 & | & -5 \end{bmatrix}$

$\xrightarrow[\substack{R_2 + R_1 \to R_1 \\ -5R_2 + R_3 \to R_3}]{\substack{\text{Switch } R_2 \text{ and } R_3 \\ \text{Then } \frac{1}{5}R_2 \to R_2}}$ $\begin{bmatrix} 1 & 0 & 1 & | & 3 \\ 0 & 1 & -1 & | & -1 \\ 0 & 0 & -2 & | & -4 \end{bmatrix}$

$\xrightarrow[\substack{\text{Then } R_3 + R_2 \to R_2 \\ -R_3 + R_1 \to R_1}]{-\frac{1}{2}R_3 \to R_3}$ $\begin{bmatrix} 1 & 0 & 0 & | & 1 \\ 0 & 1 & 0 & | & 1 \\ 0 & 0 & 1 & | & 2 \end{bmatrix}$ Thus $x = 1$, $y = 1$, $z = 2$.

3. $x = 100$, $y = -80$, $z = 110$

4. $x - z = 8$, $y - 5z = -2 \Rightarrow x = z + 8$, $y = 5z - 2$, $z = $ any real number

5. No solution

| EXERCISES | 3.3

Access end-of-section exercises online at **www.webassign.net**

OBJECTIVES

3.4

- To find the inverse of a square matrix
- To use inverse matrices to solve systems of linear equations
- To find determinants of certain matrices

Inverse of a Square Matrix; Matrix Equations

| APPLICATION PREVIEW |

Enigma Machine →

markhiggins/Shutterstock.com

During World War II, military commanders sent instructions that used simple substitution codes. These codes were easily broken, so they were further coded by using a coding matrix and matrix multiplication. The resulting coded messages, when received, could be unscrambled with a decoding matrix that was the inverse matrix of the coding matrix. In Section 3.2, "Multiplication of Matrices," we encoded messages by using the following code:

a	b	c	d	e	f	g	h	i	j	k	l
1	2	3	4	5	6	7	8	9	10	11	12

m	n	o	p	q	r	s	t	u	v	w	x
13	14	15	16	17	18	19	20	21	22	23	24

y	z	blank
25	26	27

and the encoding matrix

$$A = \begin{bmatrix} 3 & 5 \\ 4 & 6 \end{bmatrix}$$

which converts the numbers to the coded message (in pairs of numbers). To decode a message encoded by this matrix A, we find the inverse of matrix A, which is denoted A^{-1}, and multiply A^{-1} by the coded message. (See Example 4.) In this section, we will discuss how to find and apply the inverse of a given square matrix.

Inverse Matrices

If the product of A and B is the identity matrix, I, we say that B is the inverse of A (and A is the inverse of B). The matrix B is called the **inverse matrix** of A, denoted A^{-1}.

Inverse Matrices

Two square matrices, A and B, are called **inverses** of each other if

$$AB = I \quad \text{and} \quad BA = I$$

In this case, $B = A^{-1}$ and $A = B^{-1}$.

EXAMPLE | **1** | **Inverse Matrices**

Is B the inverse of A if $A = \begin{bmatrix} 1 & 2 \\ 1 & 1 \end{bmatrix}$ and $B = \begin{bmatrix} -1 & 2 \\ 1 & -1 \end{bmatrix}$?

Solution

$$AB = \begin{bmatrix} 1 & 2 \\ 1 & 1 \end{bmatrix} \cdot \begin{bmatrix} -1 & 2 \\ 1 & -1 \end{bmatrix} = \begin{bmatrix} 1 & 0 \\ 0 & 1 \end{bmatrix} = I$$

$$BA = \begin{bmatrix} -1 & 2 \\ 1 & -1 \end{bmatrix} \cdot \begin{bmatrix} 1 & 2 \\ 1 & 1 \end{bmatrix} = \begin{bmatrix} 1 & 0 \\ 0 & 1 \end{bmatrix} = I$$

Thus B is the inverse of A, or $B = A^{-1}$. We also say A is the inverse of B, or $A = B^{-1}$. Thus $(A^{-1})^{-1} = A$. ∎

CHECKPOINT

1. Are the matrices A and B inverses if

$$A = \begin{bmatrix} 1 & 1 & -1 \\ 1 & -1 & 3 \\ 4 & 2 & 1 \end{bmatrix} \text{ and } B = \begin{bmatrix} 3.5 & 1.5 & -1 \\ -5.5 & -2.5 & 2 \\ -3 & -1 & 1 \end{bmatrix}?$$

We have seen how to use elementary row operations on augmented matrices to solve systems of linear equations. We can also find the inverse of a matrix by using elementary row operations.

If the inverse exists for a square matrix A, we find A^{-1} as follows.

Finding the Inverse of a Square Matrix

Procedure

To find the inverse of the square matrix A:

1. Form the augmented matrix $[A \mid I]$, where A is the $n \times n$ matrix and I is the $n \times n$ identity matrix.

2. Perform elementary row operations on $[A \mid I]$ until we have an augmented matrix of the form $[I \mid B]$—that is, until the matrix A on the left is transformed into the identity matrix.

   If A has no inverse, the reduction process on $[A \mid I]$ will yield a row of zeros in the left half of the augmented matrix.

3. The matrix B (on the right) is the inverse of matrix A.

Example

Find the inverse of matrix $A = \begin{bmatrix} 1 & 2 \\ 1 & 1 \end{bmatrix}$.

1. $\left[\begin{array}{cc|cc} 1 & 2 & 1 & 0 \\ 1 & 1 & 0 & 1 \end{array}\right]$

2. $\left[\begin{array}{cc|cc} 1 & 2 & 1 & 0 \\ 1 & 1 & 0 & 1 \end{array}\right]$

   $\xrightarrow{-R_1 + R_2 \to R_2} \left[\begin{array}{cc|cc} 1 & 2 & 1 & 0 \\ 0 & -1 & -1 & 1 \end{array}\right]$

   $\xrightarrow[-R_2 \to R_2]{2R_2 + R_1 \to R_1} \left[\begin{array}{cc|cc} 1 & 0 & -1 & 2 \\ 0 & 1 & 1 & -1 \end{array}\right]$

3. The inverse of A is $B = \begin{bmatrix} -1 & 2 \\ 1 & -1 \end{bmatrix}$.

   We verified this in Example 1.

EXAMPLE 2 Finding an Inverse Matrix

Find the inverse of matrix A.

$$A = \begin{bmatrix} 2 & 5 & 4 \\ 1 & 4 & 3 \\ 1 & -3 & -2 \end{bmatrix}$$

Solution

To find the inverse of matrix A, we reduce A in $[A \mid I]$ (the matrix A augmented with the identity matrix I). When the left side becomes an identity matrix, the inverse is in the augment.

$$\left[\begin{array}{ccc|ccc} 2 & 5 & 4 & 1 & 0 & 0 \\ 1 & 4 & 3 & 0 & 1 & 0 \\ 1 & -3 & -2 & 0 & 0 & 1 \end{array}\right] \xrightarrow{\text{Switch } R_1 \text{ and } R_2} \left[\begin{array}{ccc|ccc} 1 & 4 & 3 & 0 & 1 & 0 \\ 2 & 5 & 4 & 1 & 0 & 0 \\ 1 & -3 & -2 & 0 & 0 & 1 \end{array}\right]$$

$$\xrightarrow[{-R_1 + R_3 \to R_3}]{-2R_1 + R_2 \to R_2} \left[\begin{array}{ccc|ccc} 1 & 4 & 3 & 0 & 1 & 0 \\ 0 & -3 & -2 & 1 & -2 & 0 \\ 0 & -7 & -5 & 0 & -1 & 1 \end{array}\right]$$

$$\xrightarrow{-\frac{1}{3}R_2 \to R_2} \left[\begin{array}{ccc|ccc} 1 & 4 & 3 & 0 & 1 & 0 \\ 0 & 1 & \frac{2}{3} & -\frac{1}{3} & \frac{2}{3} & 0 \\ 0 & -7 & -5 & 0 & -1 & 1 \end{array}\right]$$

$$\xrightarrow[{7R_2 + R_3 \to R_3}]{-4R_2 + R_1 \to R_1} \left[\begin{array}{ccc|ccc} 1 & 0 & \frac{1}{3} & \frac{4}{3} & -\frac{5}{3} & 0 \\ 0 & 1 & \frac{2}{3} & -\frac{1}{3} & \frac{2}{3} & 0 \\ 0 & 0 & -\frac{1}{3} & -\frac{7}{3} & \frac{11}{3} & 1 \end{array}\right]$$

$$\xrightarrow[{-3R_3 \to R_3}]{\substack{R_3 + R_1 \to R_1 \\ 2R_3 + R_2 \to R_2}} \left[\begin{array}{ccc|ccc} 1 & 0 & 0 & -1 & 2 & 1 \\ 0 & 1 & 0 & -5 & 8 & 2 \\ 0 & 0 & 1 & 7 & -11 & -3 \end{array}\right]$$

The inverse we seek is

$$A^{-1} = \begin{bmatrix} -1 & 2 & 1 \\ -5 & 8 & 2 \\ 7 & -11 & -3 \end{bmatrix}$$

Calculator Note To see how a graphing calculator can be used to find the inverse of a matrix, see Appendix C, Section 3.4. For example, the inverse of the matrix

$$\begin{bmatrix} 1 & 2 & 4 \\ 3 & 1 & 3 \\ 5 & 2 & 4 \end{bmatrix}$$

is found by entering this matrix, using the inverse key or command, and converting the entries to fractions.

```
[A]⁻¹▸Frac
[[-1/4    0      1/4  ]
 [3/8    -2      9/8  ]
 [1/8     1     -5/8]]
■
```

so $A^{-1} = \begin{bmatrix} -\frac{1}{4} & 0 & \frac{1}{4} \\ \frac{3}{8} & -2 & \frac{9}{8} \\ \frac{1}{8} & 1 & -\frac{5}{8} \end{bmatrix}$

CHECKPOINT

2. Find the inverse of the matrix

$$A = \begin{bmatrix} 1 & 2 & 1 \\ 0 & 1 & 1 \\ 1 & 2 & 2 \end{bmatrix}$$

The technique of reducing $[A \mid I]$ can be used to find A^{-1} or to find that A^{-1} doesn't exist for any square matrix A. However, a special formula can be used to find the inverse of a 2×2 matrix, if the inverse exists.

Inverse of a 2 × 2 Matrix

If $A = \begin{bmatrix} a & b \\ c & d \end{bmatrix}$, then $A^{-1} = \dfrac{1}{ad - bc} \begin{bmatrix} d & -b \\ -c & a \end{bmatrix}$, provided $ad - bc \neq 0$.
If $ad - bc = 0$, then A^{-1} does not exist.

This result can be verified by direct calculation or by reducing $[A \mid I]$.

EXAMPLE 3 **Inverse of a 2 × 2 Matrix**

Find the inverse, if it exists, for each of the following.

(a) $A = \begin{bmatrix} 3 & 7 \\ 2 & 5 \end{bmatrix}$ (b) $B = \begin{bmatrix} 2 & -1 \\ -4 & 2 \end{bmatrix}$

Solution

(a) For A, $ad - bc = (3)(5) - (2)(7) = 1 \neq 0$, so A^{-1} exists.

$$A^{-1} = \frac{1}{1} \begin{bmatrix} 5 & -7 \\ -2 & 3 \end{bmatrix} = \begin{bmatrix} 5 & -7 \\ -2 & 3 \end{bmatrix}$$

(b) For B, $ad - bc = (2)(2) - (-4)(-1) = 4 - 4 = 0$, so B^{-1} does not exist.

EXAMPLE 4 **Decoding Messages** **| APPLICATION PREVIEW |**

In the Application Preview we recalled how to encode messages with

a	b	c	d	e	f	g	h	i	j	k	l
1	2	3	4	5	6	7	8	9	10	11	12

m	n	o	p	q	r	s	t	u	v	w	x
13	14	15	16	17	18	19	20	21	22	23	24

y	z	blank
25	26	27

and with the encoding matrix

$$A = \begin{bmatrix} 3 & 5 \\ 4 & 6 \end{bmatrix}$$

which converts the numbers to the coded message (in pairs of numbers).

To decode any message encoded by A, we must find the inverse of A and multiply A^{-1} by the coded message. Use the inverse of A to decode the coded message 96, 118, 65, 84, 131, 168, 55, 72.

Solution
We first find the inverse of matrix A.

$$A^{-1} = \frac{1}{18 - 20} \begin{bmatrix} 6 & -5 \\ -4 & 3 \end{bmatrix} = -\frac{1}{2} \begin{bmatrix} 6 & -5 \\ -4 & 3 \end{bmatrix} = \begin{bmatrix} -3 & 2.5 \\ 2 & -1.5 \end{bmatrix}$$

We can write the coded message numbers (in pairs) as columns of a matrix B and then find the product $A^{-1}B$.

$$A^{-1}B = \begin{bmatrix} -3 & 2.5 \\ 2 & -1.5 \end{bmatrix} \begin{bmatrix} 96 & 65 & 131 & 55 \\ 118 & 84 & 168 & 72 \end{bmatrix} = \begin{bmatrix} 7 & 15 & 27 & 15 \\ 15 & 4 & 10 & 2 \end{bmatrix}$$

Reading the numbers (down the respective columns) gives the result 7, 15, 15, 4, 27, 10, 15, 2, which is the message "good job." ■

Spreadsheet Note We can also find the inverse of a square matrix with Excel, if the inverse exists (See Appendix D, Section 3.4 and the Online Excel Guide). The following spreadsheet shows a matrix A and its inverse, A^{-1}.

=minverse(B1:D3)			
	A	**B**	**C**	**D**
1	Matrix A	−1	0	1
2		1	4	−3
3		1	−2	1
4				
5	Inverse A	0.5	0.5	1
6		1	0.5	0.5
7		1.5	0.5	1

Warning: Roundoff error can occur when technology is used to find the inverse of a matrix. ■

Matrix Equations   The system $\begin{cases} 2x + 5y + 4z = 4 \\ x + 4y + 3z = 1 \\ x - 3y - 2z = 5 \end{cases}$ can be written as either of the following matrix equations.

$$\begin{bmatrix} 2x + 5y + 4z \\ x + 4y + 3z \\ x - 3y - 2z \end{bmatrix} = \begin{bmatrix} 4 \\ 1 \\ 5 \end{bmatrix} \quad \text{or} \quad \begin{bmatrix} 2 & 5 & 4 \\ 1 & 4 & 3 \\ 1 & -3 & -2 \end{bmatrix} \begin{bmatrix} x \\ y \\ z \end{bmatrix} = \begin{bmatrix} 4 \\ 1 \\ 5 \end{bmatrix}$$

We can see that this second form of a matrix equation is equivalent to the first if we carry out the multiplication. With the system written in this second form, we could also solve the system by multiplying both sides of the equation by the inverse of the coefficient matrix.

EXAMPLE 5   ### Solution with Inverse Matrices

Use an inverse matrix to solve the system:

$$\begin{cases} 2x + 5y + 4z = 4 \\ x + 4y + 3z = 1 \\ x - 3y - 2z = 5 \end{cases}$$

Solution

To solve the system, we multiply both sides of the associated matrix equation, written above, by the inverse of the coefficient matrix, which we found in Example 2.

$$\begin{bmatrix} -1 & 2 & 1 \\ -5 & 8 & 2 \\ 7 & -11 & -3 \end{bmatrix} \begin{bmatrix} 2 & 5 & 4 \\ 1 & 4 & 3 \\ 1 & -3 & -2 \end{bmatrix} \begin{bmatrix} x \\ y \\ z \end{bmatrix} = \begin{bmatrix} -1 & 2 & 1 \\ -5 & 8 & 2 \\ 7 & -11 & -3 \end{bmatrix} \begin{bmatrix} 4 \\ 1 \\ 5 \end{bmatrix}$$

Note that we must be careful to multiply both sides *from the left* because matrix multiplication is not commutative. If we carry out the multiplications, we obtain

$$\begin{bmatrix} 1 & 0 & 0 \\ 0 & 1 & 0 \\ 0 & 0 & 1 \end{bmatrix} \begin{bmatrix} x \\ y \\ z \end{bmatrix} = \begin{bmatrix} 3 \\ -2 \\ 2 \end{bmatrix}, \quad \text{or} \quad \begin{bmatrix} x \\ y \\ z \end{bmatrix} = \begin{bmatrix} 3 \\ -2 \\ 2 \end{bmatrix}$$

which yields $x = 3, y = -2, z = 2$, the same solution as that found in the previous section.   ■

Just as we wrote the system of three equations as a matrix equation of the form $AX = B$, this can be done in general. If A is an $n \times n$ matrix and if B and X are $n \times 1$ matrices, then

$$AX = B$$

is a **matrix equation.**

If the inverse of a matrix A exists, then we can use that inverse to solve the matrix equation for the matrix X. The general solution method follows.

$$AX = B$$

Multiplying both sides of the equation (from the left) by A^{-1} gives

$$A^{-1}(AX) = A^{-1}B$$
$$(A^{-1}A)X = A^{-1}B$$
$$IX = A^{-1}B$$
$$X = A^{-1}B$$

Thus inverse matrices can be used to solve systems of equations. Unfortunately, this method will work only if the solution to the system is unique. In fact, a system $AX = B$ has a unique solution if and only if A^{-1} exists. If the inverse of the coefficient matrix exists, the solution method above can be used to solve the system.

CHECKPOINT   3.   Use inverse matrices to solve the system.

$$\begin{cases} 7x - 5y = 12 \\ 2x - 3y = 6 \end{cases}$$

Calculator Note   Because we often can easily find the inverse of a matrix with a graphing calculator, it is easy to use a graphing calculator to solve a system of linear equations by using the inverse of the coefficient matrix.   ■

 EXAMPLE 6   ### Freight Logistics with Technology

On-Time Freight carries three classes of cargo, I, II, and III, and sets their delivery rates according to the class. The following table gives the details of each class and On-Time's capacity limits.

	Class I	Class II	Class III	Limits
Volume (cubic feet)	25	20	50	15,000
Weight (pounds)	25	50	100	27,500
Insured value (dollars)	250	50	500	92,250

(a) Create a system of linear equations that could be used to find the number of units of each cargo class that should be hauled. Then use the inverse of the coefficient matrix to solve the system.

(b) Use the inverse from part (a) to find how to change the hauling decision if an additional 300 cubic feet of volume becomes available.

(c) What is the connection between column 1 of the inverse matrix used in parts (a) and (b) and the answer found in part (b)?

Solution

(a) Let x = the number of units of class I cargo

y = the number of units of class II cargo

z = the number of units of class III cargo

Then we can formulate the following system of equations:

$$25x + 20y + 50z = 15{,}000$$
$$25x + 50y + 100z = 27{,}500$$
$$250x + 50y + 500z = 92{,}250$$

The matrix equation for this system is $AX = B$, as follows:

$$AX = \begin{bmatrix} 25 & 20 & 50 \\ 25 & 50 & 100 \\ 250 & 50 & 500 \end{bmatrix} \begin{bmatrix} x \\ y \\ z \end{bmatrix} = \begin{bmatrix} 15{,}000 \\ 27{,}500 \\ 92{,}250 \end{bmatrix} = B$$

We use technology to find the inverse of the coefficient matrix, A. If we multiply both sides (from the left) by A^{-1}, we get $X = A^{-1}B$. (See Figure 3.9.)

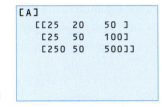
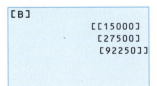
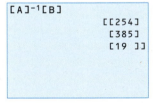

Figure 3.9

Thus the solution is $x = 254$ units of class I cargo, $y = 385$ units of class II cargo, and $z = 19$ units of class III cargo.

(b) If an additional 300 cubic feet of cargo space becomes available, this changes only the first entry of B. The new B and the new solution, $A^{-1}B$, are as follows, along with $300\,A^{-1}$. (See Figure 3.10.)

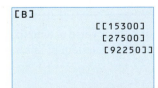

Figure 3.10

(c) This new solution is the original $X = A^{-1}B$ plus 300 times column 1 of A^{-1}.

$$[\text{original } X] + 300[\text{column 1 of } A^{-1}] = \begin{bmatrix} 254 \\ 385 \\ 19 \end{bmatrix} + \begin{bmatrix} 32 \\ 20 \\ -18 \end{bmatrix} = \begin{bmatrix} 286 \\ 405 \\ 1 \end{bmatrix}$$

The effects of similar changes in weight or insured value also could be found by using the appropriate column of A^{-1}. Thus we see that the inverse is useful not only in solving the original problem, but also in adjusting that solution to changes in resource amounts. ■

Spreadsheet Note   We can also use Excel to solve a system of linear equations, if a unique solution exists. We can solve such a system by finding the inverse of the coefficient matrix and multiplying this inverse times the matrix containing the constants. The spreadsheet that follows shows the solution of Example 5 found with Excel. See also Appendix D, Section 3.4 and the Online Excel Guide.

	A	B	C	D
1	Matrix A	2	5	4
2		1	4	3
3		1	−3	−2
4				
5	Inverse A	−1	2	1
6		−5	8	2
7		7	−11	−3
8				
9	Matrix B	4		
10		1		
11		5		
12				
13	Solution X	3		
14		−2		
15		2		

■

EXAMPLE 7 **Venture Capital**

Suppose a bank draws its venture capital funds from three main sources of income—business loans, auto loans, and home mortgages. The income from these sources for each of 3 years is given in Table 3.8, and the venture capital for these years is given in Table 3.9. If the bank uses a fixed percent of its income from each of the business loans (x), auto loans (y), and home mortgages (z) to get its venture capital funds, find the percent of income used for each of these sources of income.

| TABLE 3.8 |

INCOME FROM LOANS

	Business	Auto	Home
2010	68,210	23,324	57,234
2011	43,455	19,335	66,345
2012	58,672	15,654	69,334

| TABLE 3.9 |

	Venture Capital
2010	53,448.74
2011	43,447.50
2012	50,582.88

Solution

The matrix equation describing this problem is $AX = V$, or

$$\begin{bmatrix} 68,210 & 23,324 & 57,234 \\ 43,455 & 19,335 & 66,345 \\ 58,672 & 15,654 & 69,334 \end{bmatrix} \begin{bmatrix} x \\ y \\ z \end{bmatrix} = \begin{bmatrix} 53,448.74 \\ 43,447.50 \\ 50,582.88 \end{bmatrix}$$

Finding the inverse of matrix A and multiplying both sides of the matrix equation (from the left) by A^{-1} give the solution $X = A^{-1}V$:

$$\begin{bmatrix} x \\ y \\ z \end{bmatrix} = \begin{bmatrix} 0.47 \\ 0.23 \\ 0.28 \end{bmatrix}$$

Thus 47%, 23%, and 28% come from business, auto, and home loans, respectively. ■

Determinants Recall that the inverse of $A = \begin{bmatrix} a & b \\ c & d \end{bmatrix}$ can be found with the formula

$$A^{-1} = \frac{1}{ad - bc} \begin{bmatrix} d & -b \\ -c & a \end{bmatrix}$$

if $ad - bc \neq 0$.

The value $ad - bc$ used in finding the inverse of this 2×2 matrix is used so frequently that it is given a special name, the **determinant** of A (denoted $det\ A$ or $|A|$).

Determinant of a 2 × 2 Matrix

$$det \begin{bmatrix} a & b \\ c & d \end{bmatrix} = \begin{vmatrix} a & b \\ c & d \end{vmatrix} = ad - bc$$

EXAMPLE 8 Determinant of a 2 × 2 Matrix

Find $det\ A$ if $A = \begin{bmatrix} 2 & 4 \\ 3 & -4 \end{bmatrix}$.

Solution

$$det\ A = (2)(-4) - (3)(4) = -8 - 12 = -20$$ ■

Calculator Note   There are formulas for finding the determinants of square matrices of orders larger than 2×2, but for the matrices used in this text, graphing calculators make it easy to find determinants. (See Appendix C, Section 3.4.) ■

We have already seen that the inverse of a 2×2 matrix A does not exist if and only if $det\ A = 0$. In general, the inverse of an $n \times n$ matrix B does not exist if $det\ B = 0$. This also means that if the coefficient matrix of a system of linear equations has a determinant equal to 0, the system does not have a unique solution.

EXAMPLE 9 Solution with Technology

Consider the following matrix equation $AX = B$.

$$\begin{bmatrix} 2 & 3 & 0 & 1 & 3 \\ 1 & 0 & 2 & 3 & 1 \\ 0 & 1 & 0 & 2 & 3 \\ 1 & 0 & 2 & 2 & 1 \\ 1 & 0 & 0 & 0 & 3 \end{bmatrix} \begin{bmatrix} x_1 \\ x_2 \\ x_3 \\ x_4 \\ x_5 \end{bmatrix} = \begin{bmatrix} 16 \\ 2 \\ 9 \\ 3 \\ 10 \end{bmatrix}$$

(a) Use a calculator to find the determinant of A.

(b) Use the result from part (a) to determine whether $AX = B$ has a unique solution.

Solution

(a) Entering matrix A and using the determinant command gives det $A = 24$.

(b) Because A is the coefficient matrix for the system and det $A = 24$, the system has a unique solution. Using the inverse matrix gives the solution $x_1 = 1$, $x_2 = 2$, $x_3 = 0.5$, $x_4 = -1$, $x_5 = 3$. ■

CHECKPOINT SOLUTIONS

1. Yes, because both AB and BA give the 3×3 identity matrix.

2. $$\left[\begin{array}{ccc|ccc} 1 & 2 & 1 & 1 & 0 & 0 \\ 0 & 1 & 1 & 0 & 1 & 0 \\ 1 & 2 & 2 & 0 & 0 & 1 \end{array}\right] \xrightarrow{-R_1 + R_3 \to R_3} \left[\begin{array}{ccc|ccc} 1 & 2 & 1 & 1 & 0 & 0 \\ 0 & 1 & 1 & 0 & 1 & 0 \\ 0 & 0 & 1 & -1 & 0 & 1 \end{array}\right]$$

$$\xrightarrow{-2R_2 + R_1 \to R_1} \left[\begin{array}{ccc|ccc} 1 & 0 & -1 & 1 & -2 & 0 \\ 0 & 1 & 1 & 0 & 1 & 0 \\ 0 & 0 & 1 & -1 & 0 & 1 \end{array}\right]$$

$$\xrightarrow[-R_3 + R_2 \to R_2]{R_3 + R_1 \to R_1} \left[\begin{array}{ccc|ccc} 1 & 0 & 0 & 0 & -2 & 1 \\ 0 & 1 & 0 & 1 & 1 & -1 \\ 0 & 0 & 1 & -1 & 0 & 1 \end{array}\right]$$

The inverse is $A^{-1} = \begin{bmatrix} 0 & -2 & 1 \\ 1 & 1 & -1 \\ -1 & 0 & 1 \end{bmatrix}$.

3. This system can be written

$$\begin{bmatrix} 7 & -5 \\ 2 & -3 \end{bmatrix} \begin{bmatrix} x \\ y \end{bmatrix} = \begin{bmatrix} 12 \\ 6 \end{bmatrix}$$

and for the coefficient matrix A, $ad - bc = -21 - (-10) = -11$. Thus A^{-1} exists,

$$A^{-1} = -\frac{1}{11} \begin{bmatrix} -3 & 5 \\ -2 & 7 \end{bmatrix} = \begin{bmatrix} \frac{3}{11} & -\frac{5}{11} \\ \frac{2}{11} & -\frac{7}{11} \end{bmatrix}$$

so multiplying by A^{-1} *from the left* gives the solution.

$$\begin{bmatrix} \frac{3}{11} & -\frac{5}{11} \\ \frac{2}{11} & -\frac{7}{11} \end{bmatrix} \begin{bmatrix} 7 & -5 \\ 2 & -3 \end{bmatrix} \begin{bmatrix} x \\ y \end{bmatrix} = \begin{bmatrix} \frac{3}{11} & -\frac{5}{11} \\ \frac{2}{11} & -\frac{7}{11} \end{bmatrix} \begin{bmatrix} 12 \\ 6 \end{bmatrix}$$

$$\begin{bmatrix} 1 & 0 \\ 0 & 1 \end{bmatrix} \begin{bmatrix} x \\ y \end{bmatrix} = \begin{bmatrix} \frac{6}{11} \\ -\frac{18}{11} \end{bmatrix}$$

Thus the solution is $x = \frac{6}{11}$, $y = -\frac{18}{11}$.

| EXERCISES | 3.4

Access end-of-section exercises online at **www.webassign.net**   **WebAssign**

OBJECTIVES

3.5

- To interpret Leontief technology matrices
- To use Leontief models to solve input-output problems

Applications of Matrices: Leontief Input-Output Models

| APPLICATION PREVIEW |

Suppose we consider a simple economy as being based on three commodities: agricultural products, manufactured goods, and fuels. Suppose further that production of 10 units of agricultural products requires 5 units of agricultural products, 2 units of manufactured goods, and 1 unit of fuels; that production of 10 units of manufactured goods requires 1 unit of agricultural products, 5 units of manufactured products, and 3 units of fuels; and that production of 10 units of fuels requires 1 unit of agricultural products, 3 units of manufactured goods, and 4 units of fuels.

Table 3.10 summarizes this information in terms of production of 1 unit. The first column represents the units of agricultural products, manufactured goods, and fuels, respectively, that are needed to produce 1 unit of agricultural products. Column 2 represents the units required to produce 1 unit of manufactured goods, and column 3 represents the units required to produce 1 unit of fuels.

| TABLE 3.10 |

| | Outputs | | |
Inputs	Agricultural Products	Manufactured Goods	Fuels
Agricultural products	0.5	0.1	0.1
Manufactured goods	0.2	0.5	0.3
Fuels	0.1	0.3	0.4

From Table 3.10 we can form matrix A, which is called a technology matrix or a Leontief matrix.

$$A = \begin{bmatrix} 0.5 & 0.1 & 0.1 \\ 0.2 & 0.5 & 0.3 \\ 0.1 & 0.3 & 0.4 \end{bmatrix}$$

The Leontief input-output model is named for Wassily Leontief. The model Leontief developed was useful in predicting the effects on the economy of price changes or shifts in government spending.*

Leontief's work divided the economy into 500 sectors, which, in a subsequent article in *Scientific American* (October 1951), were reduced to a more manageable 42 departments of production. We can examine the workings of input-output analysis with a very simplified view of the economy. (We will do this in Example 2 with the economy described above.)

In this section, we will interpret Leontief technology matrices for simple economies and will solve input-output problems for these economies by using Leontief models.

---

*Leontief's work dealt with a massive analysis of the American economy. He was awarded a Nobel Prize in economics in 1973 for his study.

In Table 3.10, we note that the sum of the units of agricultural products, manufactured goods, and fuels required to produce 1 unit of fuels (column 3) does not add up to 1 unit. This is because not all commodities or industries are represented in the model. In particular, it is customary to omit labor from models of this type.

EXAMPLE 1   Leontief Model

Use Table 3.10 to answer the following questions.

(a) How many units of agricultural products and of fuels are required to produce 100 units of manufactured goods?
(b) Production of which commodity is least dependent on the other two?
(c) If fuel costs rise, which two industries will be most affected?

Solution

(a) Referring to column 2, manufactured goods, we see that 1 unit requires 0.1 unit of agricultural products and 0.3 unit of fuels. Thus 100 units of manufactured goods require 10 units of agricultural products and 30 units of fuels.
(b) Looking down the columns, we see that 1 unit of agricultural products requires 0.3 unit of the other two commodities; 1 unit of manufactured goods requires 0.4 unit of the other two; and 1 unit of fuels requires 0.4 unit of the other two. Thus production of agricultural products is least dependent on the others.
(c) A rise in the cost of fuels would most affect those industries that use the larger amounts of fuels. One unit of agricultural products requires 0.1 unit of fuels, whereas a unit of manufactured goods requires 0.3 unit, and a unit of fuels requires 0.4 of its own units. Thus manufacturing and the fuel industry would be most affected by a cost increase in fuels. ■

CHECKPOINT

The following technology matrix for a simple economy describes the relationship of certain industries to each other in the production of 1 unit of product.

	Ag	Mfg	Fuels	Util
Agriculture	0.32	0.06	0.09	0.06
Manfacturing	0.11	0.41	0.19	0.31
Fuels	0.31	0.14	0.18	0.20
Utilities	0.18	0.26	0.21	0.25

1. Which industry is most dependent on its own goods for its operation?
2. How many units of fuels are required to produce 100 units of utilities?

Open Leontief Model   For a simplified model of the economy, such as that described with Table 3.10, not all information is contained within the technology matrix. In particular, each industry has a gross production. The **gross production matrix** for the economy can be represented by the column matrix

$$X = \begin{bmatrix} x_1 \\ x_2 \\ x_3 \end{bmatrix}$$

where x_1 is the gross production of agricultural products, x_2 is the gross production of manufactured goods, and x_3 is the gross production of fuels.

The amounts of the gross productions used within the economy by the various industries are given by AX. Those units of gross production not used by these industries are called **final demands** or **surpluses** and may be considered as being available for consumers, the government, or export. If we place these surpluses in a column matrix D, then they can be represented by the equation

$$X - AX = D, \quad \text{or} \quad (I - A)X = D$$

where I is an identity matrix of appropriate size. This matrix equation is called the **technological equation** for an **open Leontief model.** This is called an open model because some of the goods from the economy are "open," or available to those outside the economy.

Open Leontief Model

The technological equation for an open Leontief model is

$$(I - A)X = D$$

EXAMPLE 2 Gross Outputs | APPLICATION PREVIEW |

Technology matrix A represents a simple economy with an agricultural industry, a manufacturing industry, and a fuels industry (as introduced in Table 3.10). If we wish to have surpluses of 85 units of agricultural products, 65 units of manufactured goods, and 0 units of fuels, what should the gross outputs be?

$$A = \begin{bmatrix} 0.5 & 0.1 & 0.1 \\ 0.2 & 0.5 & 0.3 \\ 0.1 & 0.3 & 0.4 \end{bmatrix}$$

Solution

Let X be the matrix of gross outputs for the industries and let D be the column matrix of each industry's surpluses. Then the technological equation is $(I - A)X = D$. We begin by finding the matrix $I - A$.

$$I - A = \begin{bmatrix} 1 & 0 & 0 \\ 0 & 1 & 0 \\ 0 & 0 & 1 \end{bmatrix} - \begin{bmatrix} 0.5 & 0.1 & 0.1 \\ 0.2 & 0.5 & 0.3 \\ 0.1 & 0.3 & 0.4 \end{bmatrix} = \begin{bmatrix} 0.5 & -0.1 & -0.1 \\ -0.2 & 0.5 & -0.3 \\ -0.1 & -0.3 & 0.6 \end{bmatrix}$$

Hence we must solve the matrix equation

$$\begin{bmatrix} 0.5 & -0.1 & -0.1 \\ -0.2 & 0.5 & -0.3 \\ -0.1 & -0.3 & 0.6 \end{bmatrix} \begin{bmatrix} x_1 \\ x_2 \\ x_3 \end{bmatrix} = \begin{bmatrix} 85 \\ 65 \\ 0 \end{bmatrix}$$

The augmented matrix is

$$\begin{bmatrix} 0.5 & -0.1 & -0.1 & | & 85 \\ -0.2 & 0.5 & -0.3 & | & 65 \\ -0.1 & -0.3 & 0.6 & | & 0 \end{bmatrix}$$

If we reduce by the methods of Section 3.3, "Gauss-Jordan Elimination: Solving Systems of Equations," we obtain

$$\begin{bmatrix} 1 & 0 & 0 & | & 300 \\ 0 & 1 & 0 & | & 400 \\ 0 & 0 & 1 & | & 250 \end{bmatrix}$$

so the gross outputs for the industries are

$$
\begin{array}{lll}
\text{Agriculture:} & x_1 = 300 \\
\text{Manufacturing:} & x_2 = 400 \\
\text{Fuels:} & x_3 = 250
\end{array}
$$

In general, the technological equation for the open Leontief model can be solved by using the inverse of $I - A$. That is,

$$(I - A)X = D \qquad \text{has the solution} \qquad X = (I - A)^{-1}D$$

Technology Note If we have the technology matrix A and the surplus matrix D, we can solve the technological equation for an open Leontief model with a calculator or computer by defining matrices I, A, and D and evaluating $X = (I - A)^{-1}D$. Figure 3.11 shows the solution of Example 2 using a graphing calculator. ∎

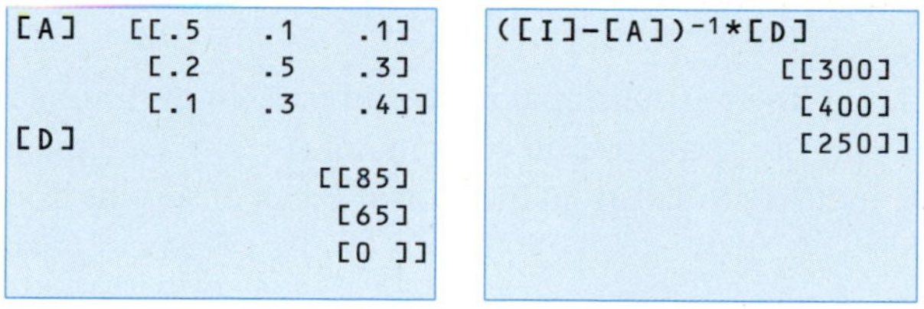

```
[A]   [[.5   .1   .1]        ([I]-[A])⁻¹*[D]
       [.2   .5   .3]                  [[300]
       [.1   .3   .4]]                  [400]
[D]                                     [250]]

       [[85]
        [65]
        [0 ]]
```

Figure 3.11

EXAMPLE **3**   **Production**

A primitive economy with a lumber industry and a power industry has the following technology matrix.

$$\begin{array}{cc} & \text{Lumber} \quad \text{Power} \\ A = & \begin{bmatrix} 0.1 & 0.2 \\ 0.2 & 0.4 \end{bmatrix} \begin{array}{l} \text{Lumber} \\ \text{Power} \end{array} \end{array}$$

(a) If surpluses of 30 units of lumber and 70 units of power are desired, find the gross production of each industry.
(b) If an additional 5 units of surplus power were required, how much additional output would be required from each industry?

Solution

(a) Let x be the gross production of lumber, and let y be the gross production of power. Then

$$I - A = \begin{bmatrix} 0.9 & -0.2 \\ -0.2 & 0.6 \end{bmatrix}, \quad X = \begin{bmatrix} x \\ y \end{bmatrix}, \quad \text{and} \quad D = \begin{bmatrix} 30 \\ 70 \end{bmatrix}$$

We must solve the matrix equation $(I - A)X = D$ for X, and we can do this by using

$$X = (I - A)^{-1}D$$

We can find $(I - A)^{-1}$ with the formula from Section 3.4: "Inverse of a Square Matrix; Matrix Equations" or with a graphing calculator.

$$(I - A)^{-1} = \frac{1}{0.54 - 0.04}\begin{bmatrix} 0.6 & 0.2 \\ 0.2 & 0.9 \end{bmatrix} = 2\begin{bmatrix} 0.6 & 0.2 \\ 0.2 & 0.9 \end{bmatrix} = \begin{bmatrix} 1.2 & 0.4 \\ 0.4 & 1.8 \end{bmatrix}$$

Then we can find the gross production of each industry:

$$X = (I - A)^{-1}D = \begin{bmatrix} 1.2 & 0.4 \\ 0.4 & 1.8 \end{bmatrix}\begin{bmatrix} 30 \\ 70 \end{bmatrix} = \begin{bmatrix} 64 \\ 138 \end{bmatrix}$$

Hence the gross productions are

$$\text{Lumber: } x = 64$$
$$\text{Power: } y = 138$$

(b) If 5 more units of surplus power were required, then the new gross productions would be

$$(I - A)^{-1} \begin{bmatrix} 30 \\ 70 + 5 \end{bmatrix} = \begin{bmatrix} 1.2 & 0.4 \\ 0.4 & 1.8 \end{bmatrix} \begin{bmatrix} 30 \\ 75 \end{bmatrix} = \begin{bmatrix} 66 \\ 147 \end{bmatrix}$$

Thus, for 5 additional units of power, the increased production required from each industry would be

$$\begin{bmatrix} \text{Lumber} \\ \text{Power} \end{bmatrix} = \begin{bmatrix} 66 \\ 147 \end{bmatrix} - \begin{bmatrix} 64 \\ 138 \end{bmatrix} = \begin{bmatrix} 2 \\ 9 \end{bmatrix} \quad \text{which is} \quad 5 \begin{bmatrix} \text{Col 2 of} \\ (I - A)^{-1} \end{bmatrix} = 5 \begin{bmatrix} 0.4 \\ 1.8 \end{bmatrix} = \begin{bmatrix} 2 \\ 9 \end{bmatrix} \quad \blacksquare$$

The relationship observed in part (b) of Example 3 is true in general about $(I - A)^{-1}$. Namely, the entries in column j of $(I - A)^{-1}$ are the increased amounts needed from each industry to satisfy an increase of 1 unit of surplus from industry j.

EXAMPLE 4   Open Economy

The economy of a developing nation is based on agricultural products, steel, and coal. An output of 1 ton of agricultural products requires inputs of 0.1 ton of agricultural products, 0.02 ton of steel, and 0.05 ton of coal. An output of 1 ton of steel requires inputs of 0.01 ton of agricultural products, 0.13 ton of steel, and 0.18 ton of coal. An output of 1 ton of coal requires inputs of 0.01 ton of agricultural products, 0.20 ton of steel, and 0.05 ton of coal. Write the technology matrix for this economy. Find the necessary gross productions to provide surpluses of 2350 tons of agricultural products, 4552 tons of steel, and 911 tons of coal.

Solution

The technology matrix for the economy is

$$\begin{array}{ccc} \text{Ag} & \text{Steel} & \text{Coal} \end{array}$$
$$\begin{bmatrix} 0.1 & 0.01 & 0.01 \\ 0.02 & 0.13 & 0.20 \\ 0.05 & 0.18 & 0.05 \end{bmatrix} \begin{array}{l} \text{Ag} \\ \text{Steel} \\ \text{Coal} \end{array}$$

The technological equation for this economy is $(I - A)X = D$, or

$$\left(\begin{bmatrix} 1 & 0 & 0 \\ 0 & 1 & 0 \\ 0 & 0 & 1 \end{bmatrix} - \begin{bmatrix} 0.1 & 0.01 & 0.01 \\ 0.02 & 0.13 & 0.20 \\ 0.05 & 0.18 & 0.05 \end{bmatrix} \right) X = \begin{bmatrix} 2350 \\ 4552 \\ 911 \end{bmatrix}$$

Thus the solution is

```
[A] [[.1    .01   .01]          ([I]-[A])⁻¹*[D]
     [.02   .13   .2 ]                   [[2700]
     [.05   .18   .05]]                   [5800]
[D]                                       [2200]]
        [[2350]         ■
         [4552]
         [911 ]]
```

Thus the necessary gross productions of the industries are

$$X = \begin{bmatrix} 2700 \\ 5800 \\ 2200 \end{bmatrix} \begin{array}{l} \text{Ag} \\ \text{Steel} \\ \text{Coal} \end{array}$$

  ■

CHECKPOINT

3. Suppose a primitive economy has a wood products industry and a minerals industry, with technology matrix A (open Leontief model).

$$A = \begin{array}{c} \\ \\ \end{array} \begin{array}{cc} \text{W} & \text{M} \\ \left[\begin{array}{cc} 0.12 & 0.11 \\ 0.09 & 0.15 \end{array} \right] & \begin{array}{c} \text{Wood products} \\ \text{Minerals} \end{array} \end{array}$$

If surpluses of 1738 units of wood products and 1332 units of minerals are desired, find the gross production of each industry.

Closed Leontief Model   The model we have considered is referred to as an open Leontief model because not all inputs and outputs were incorporated within the technology matrix. If a model is developed in which all inputs and outputs are used within the system, then such a model is called a **closed Leontief model.** In such a model, labor (and perhaps other factors) must be included. Labor is included by considering a new industry, households, which produces labor. When such a closed model is developed, all outputs are used within the system, and the sum of the entries in each column equals 1. In this case, there is no surplus, so $D = 0$, and we have the following:

Closed Leontief Model

The technological equation for a closed Leontief model is

$$(I - A)X = 0$$

where 0 is a zero column matrix.

For closed models, the technological equation does *not* have a unique solution, so the matrix $(I - A)^{-1}$ does *not* exist and hence cannot be used to find the solution.

EXAMPLE 5   **Closed Economy**

The following closed Leontief model with technology matrix A might describe the economy of the entire country, with x_1 equal to the government budget, x_2 the value of industrial output (profit-making organizations), x_3 the budget of nonprofit organizations, and x_4 the households budget.

$$A = \begin{array}{c} \\ \\ \end{array} \begin{array}{cccc} \text{G} & \text{I} & \text{N} & \text{H} \\ \left[\begin{array}{cccc} 0.4 & 0.2 & 0.1 & 0.3 \\ 0.2 & 0.2 & 0.2 & 0.1 \\ 0.2 & 0 & 0.2 & 0.1 \\ 0.2 & 0.6 & 0.5 & 0.5 \end{array} \right] & \begin{array}{c} \text{Government} \\ \text{Industry} \\ \text{Nonprofit} \\ \text{Households} \end{array} \end{array}$$

Find the total budgets (or outputs) x_1, x_2, x_3, and x_4.

Solution

This is a closed Leontief model; therefore, we solve the technological equation $(I - A)X = 0$.

$$\left(\left[\begin{array}{cccc} 1 & 0 & 0 & 0 \\ 0 & 1 & 0 & 0 \\ 0 & 0 & 1 & 0 \\ 0 & 0 & 0 & 1 \end{array} \right] - \left[\begin{array}{cccc} 0.4 & 0.2 & 0.1 & 0.3 \\ 0.2 & 0.2 & 0.2 & 0.1 \\ 0.2 & 0 & 0.2 & 0.1 \\ 0.2 & 0.6 & 0.5 & 0.5 \end{array} \right] \right) X = \left[\begin{array}{c} 0 \\ 0 \\ 0 \\ 0 \end{array} \right]$$

or

$$\left[\begin{array}{cccc} 0.6 & -0.2 & -0.1 & -0.3 \\ -0.2 & 0.8 & -0.2 & -0.1 \\ -0.2 & 0 & 0.8 & -0.1 \\ -0.2 & -0.6 & -0.5 & 0.5 \end{array} \right] \left[\begin{array}{c} x_1 \\ x_2 \\ x_3 \\ x_4 \end{array} \right] = \left[\begin{array}{c} 0 \\ 0 \\ 0 \\ 0 \end{array} \right]$$

The augmented matrix for this system is

$$\begin{bmatrix} 0.6 & -0.2 & -0.1 & -0.3 & 0 \\ -0.2 & 0.8 & -0.2 & -0.1 & 0 \\ -0.2 & 0 & 0.8 & -0.1 & 0 \\ -0.2 & -0.6 & -0.5 & 0.5 & 0 \end{bmatrix}$$

Reducing this matrix with row operations, we obtain

$$\begin{bmatrix} 1 & 0 & 0 & -\frac{55}{82} & 0 \\ 0 & 1 & 0 & -\frac{15}{41} & 0 \\ 0 & 0 & 1 & -\frac{12}{41} & 0 \\ 0 & 0 & 0 & 0 & 0 \end{bmatrix}$$

See Figure 3.12 for this reduction with a graphic calculator.

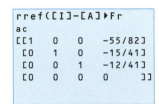

```
rref([I]-[A]▶Fr
ac
[[1   0   0   -55/82]
 [0   1   0   -15/41]
 [0   0   1   -12/41]
 [0   0   0    0    ]]
```

Figure 3.12

The system of equations that corresponds to this augmented matrix is

$$x_1 - \tfrac{55}{82}x_4 = 0 \qquad\qquad x_1 = \tfrac{55}{82}x_4$$

$$x_2 - \tfrac{15}{41}x_4 = 0 \quad \text{or} \quad x_2 = \tfrac{15}{41}x_4$$

$$x_3 - \tfrac{12}{41}x_4 = 0 \qquad\qquad x_3 = \tfrac{12}{41}x_4$$

$$x_4 = \text{households budget}$$

Note that the economy satisfies the given equation if the government budget is $\frac{55}{82}$ times the households budget, if the value of industrial output is $\frac{15}{41}$ times the households budget, and if the budget of nonprofit organizations is $\frac{12}{41}$ times the households budget. The dependency here is expected because industrial output is limited by labor supply. ■

CHECKPOINT

4. For a closed Leontief model:
   (a) What is the sum of the column entries of the technology matrix?
   (b) State the technological equation.

We have examined two types of input-output models as they pertain to the economy as a whole. Since Leontief's original work with the economy, various other applications of input-output models have been developed. Consider the following parts-listing problem.

EXAMPLE 6   Manufacturing

A storage shed consists of 4 walls and a roof. The walls and roof are made from stamped aluminum panels and are reinforced with 4 braces, each of which is held with 6 bolts. The final assembly joins the walls to each other and to the roof, using a total of 20 more bolts. The parts listing for these sheds can be described by the following matrix.

$$Q = \begin{bmatrix} & \text{S} & \text{W} & \text{R} & \text{P} & \text{Br} & \text{Bo} \\ 0 & 0 & 0 & 0 & 0 & 0 \\ 4 & 0 & 0 & 0 & 0 & 0 \\ 1 & 0 & 0 & 0 & 0 & 0 \\ 0 & 1 & 1 & 0 & 0 & 0 \\ 0 & 4 & 4 & 0 & 0 & 0 \\ 20 & 24 & 24 & 0 & 0 & 0 \end{bmatrix} \begin{matrix} \text{Sheds} \\ \text{Walls} \\ \text{Roofs} \\ \text{Panels} \\ \text{Braces} \\ \text{Bolts} \end{matrix}$$

This matrix is a type of input-output matrix, with the rows representing inputs that are used directly to produce the items that head the columns. We note that the entries here are quantities used rather than proportions used. Also, the columns containing all zeros indicate primary assembly items because they are not made from other parts.

Thus to produce a shed requires 4 walls, 1 roof, and 20 bolts, and to produce a roof requires 1 aluminum panel, 4 braces, and 24 bolts.

Suppose that an order is received for 4 completed sheds and spare parts including 2 walls, 1 roof, 8 braces, and 24 bolts. How many of each of the assembly items are required to fill the order?

Solution

If matrix D represents the order and matrix X represents the gross production, we have

$$D = \begin{bmatrix} 4 \\ 2 \\ 1 \\ 0 \\ 8 \\ 24 \end{bmatrix} \begin{matrix} \text{Sheds} \\ \text{Walls} \\ \text{Roofs} \\ \text{Panels} \\ \text{Braces} \\ \text{Bolts} \end{matrix} \qquad X = \begin{bmatrix} x_1 \\ x_2 \\ x_3 \\ x_4 \\ x_5 \\ x_6 \end{bmatrix} = \begin{bmatrix} \text{total sheds required} \\ \text{total walls required} \\ \text{total roofs required} \\ \text{total panels required} \\ \text{total braces required} \\ \text{total bolts required} \end{bmatrix}$$

Then X must satisfy $X - QX = D$, which is the technological equation for an open Leontief model.

We can find X by solving the system $(I - Q)X = D$, which is represented by the augmented matrix

$$\begin{bmatrix} 1 & 0 & 0 & 0 & 0 & 0 & 4 \\ -4 & 1 & 0 & 0 & 0 & 0 & 2 \\ -1 & 0 & 1 & 0 & 0 & 0 & 1 \\ 0 & -1 & -1 & 1 & 0 & 0 & 0 \\ 0 & -4 & -4 & 0 & 1 & 0 & 8 \\ -20 & -24 & -24 & 0 & 0 & 1 & 24 \end{bmatrix}$$

Although this matrix is quite large, it is easily reduced and yields

$$\begin{bmatrix} 1 & 0 & 0 & 0 & 0 & 0 & 4 \\ 0 & 1 & 0 & 0 & 0 & 0 & 18 \\ 0 & 0 & 1 & 0 & 0 & 0 & 5 \\ 0 & 0 & 0 & 1 & 0 & 0 & 23 \\ 0 & 0 & 0 & 0 & 1 & 0 & 100 \\ 0 & 0 & 0 & 0 & 0 & 1 & 656 \end{bmatrix}$$

This matrix tells us that the order will have 4 complete sheds. The 18 walls from the matrix include the 16 from the complete sheds, and the 5 roofs include the 4 from the complete sheds. The 23 panels, 100 braces, and 656 bolts are the total number of *primary assembly items* (bolts, braces, and panels) required to fill the order. ■

The inverse of $I - Q$ exists, so it could be used directly or with technology to find the solution to the parts-listing problem of Example 6.

CHECKPOINT SOLUTIONS

1. Manufacturing
2. Under the "Util" column and across from the "Fuels" row is 0.20, indicating that producing 1 unit of utilities requires 0.20 unit of fuels. Thus producing 100 units of utilities requires 20 units of fuels.
3. Let x = the gross production of wood products
   y = the gross production of minerals

We must solve $(I - A)X = D$.

$$\left(\begin{bmatrix} 1 & 0 \\ 0 & 1 \end{bmatrix} - \begin{bmatrix} 0.12 & 0.11 \\ 0.09 & 0.15 \end{bmatrix} \right) \begin{bmatrix} x \\ y \end{bmatrix} = \begin{bmatrix} 1738 \\ 1332 \end{bmatrix}$$

$$\begin{bmatrix} 0.88 & -0.11 \\ -0.09 & 0.85 \end{bmatrix} \begin{bmatrix} x \\ y \end{bmatrix} = \begin{bmatrix} 1738 \\ 1332 \end{bmatrix}$$

$$\begin{bmatrix} 0.88 & -0.11 \\ -0.09 & 0.85 \end{bmatrix}^{-1} = \frac{1}{(0.88)(0.85) - (0.09)(0.11)} \begin{bmatrix} 0.85 & 0.11 \\ 0.09 & 0.88 \end{bmatrix}$$

Hence

$$\begin{bmatrix} x \\ y \end{bmatrix} = \frac{1}{0.7381} \begin{bmatrix} 0.85 & 0.11 \\ 0.09 & 0.88 \end{bmatrix} \begin{bmatrix} 1738 \\ 1332 \end{bmatrix} = \begin{bmatrix} 2200 \\ 1800 \end{bmatrix}$$

4. (a) The sum of the column entries is 1.
    (b) $(I - A)X = 0$.

| EXERCISES | 3.5

Access end-of-section exercises online at **www.webassign.net**   **WebAssign**

KEY TERMS AND FORMULAS		
Section	**Key Terms**	**Formulas**
3.1	Matrices	
	Entries	
	Order	
	Square, row, column, and zero matrices	
	Transpose	A^T
	Sum of two matrices	
	Negative of a matrix	
	Scalar multiplication	
3.2	Matrix product	
	Identity matrix	2×2 is $\begin{bmatrix} 1 & 0 \\ 0 & 1 \end{bmatrix}$
3.3	Augmented matrix	
	Coefficient matrix	
	Elementary row operations	
	Solving systems of equations	
	Gauss-Jordan elimination method	
	Nonunique solutions	
	Reduced form	
3.4	Inverse matrices	
	Matrix equations	$AX = B$
	Determinants	$\begin{vmatrix} a & b \\ c & d \end{vmatrix} = ad - bc$

Section	Key Terms	Formulas
3.5	Technology matrix	
	Leontief input-output model	
	Gross production matrix	
	Final demands (surpluses)	
	Leontief model	
	Technological equation	
	Open	$(I - A)X = D$
	Closed	$(I - A)X = 0$

REVIEW EXERCISES

Use the matrices below as needed to complete Problems 1–25.

$$A = \begin{bmatrix} 4 & 4 & 2 & -5 \\ 6 & 3 & -1 & 0 \\ 0 & 0 & -3 & 5 \end{bmatrix} \quad B = \begin{bmatrix} 2 & -5 & -11 & 8 \\ 4 & 0 & 0 & 4 \\ -2 & -2 & 1 & 9 \end{bmatrix}$$

$$C = \begin{bmatrix} 4 & -2 \\ 5 & 0 \\ 6 & 0 \\ 1 & 3 \end{bmatrix} \quad D = \begin{bmatrix} 3 & 5 \\ 1 & 2 \end{bmatrix} \quad E = \begin{bmatrix} 1 & 1 \\ 1 & 1 \\ 4 & 6 \\ 0 & 5 \end{bmatrix}$$

$$F = \begin{bmatrix} -1 & 6 \\ 4 & 11 \end{bmatrix} \quad G = \begin{bmatrix} 2 & -5 \\ -1 & 3 \end{bmatrix} \quad I = \begin{bmatrix} 1 & 0 \\ 0 & 1 \end{bmatrix}$$

1. Find a_{12} in matrix A.
2. Find b_{23} in matrix B.
3. Which of the matrices, if any, are 3×4?
4. Which of the matrices, if any, are 2×4?
5. Which matrices are square?
6. Write the negative of matrix B.
7. If a matrix is added to its negative, what kind of matrix results?
8. Two matrices can be added if they have the same _____.

Perform the indicated operations in Problems 9–25.

9. $A + B$
10. $C - E$
11. $D^T - I$
12. $3C$
13. $4I$
14. $-2F$
15. $4D - 3I$
16. $F + 2D$
17. $3A - 5B$
18. AC
19. CD
20. DF
21. FD
22. FI
23. IF
24. DG^T
25. $(DG)F$

In Problems 26–28, the reduced matrix for a system of equations is given. (a) Identify the type of solution for the system (unique solution, no solution, or infinitely many solutions). Explain your reasoning. (b) For each system that has a solution, find it. If the system has infinitely many solutions, find two different specific solutions.

26. $\begin{bmatrix} 1 & 0 & -2 & | & 6 \\ 0 & 1 & 3 & | & 7 \\ 0 & 0 & 0 & | & 0 \end{bmatrix}$

27. $\begin{bmatrix} 1 & 0 & -4 & | & 0 \\ 0 & 1 & 3 & | & 0 \\ 0 & 0 & 0 & | & 1 \end{bmatrix}$

28. $\begin{bmatrix} 1 & 0 & 0 & | & 0 \\ 0 & 1 & 0 & | & -10 \\ 0 & 0 & 1 & | & 14 \end{bmatrix}$

29. Given the following augmented matrix representing a system of linear equations, find the solution.

$$\begin{bmatrix} 1 & 1 & 2 & | & 5 \\ 4 & 0 & 1 & | & 5 \\ 2 & 1 & 1 & | & 5 \end{bmatrix}$$

In Problems 30–36, solve each system using matrices.

30. $\begin{cases} x - 2y = 4 \\ -3x + 10y = 24 \end{cases}$

31. $\begin{cases} x + y + z = 4 \\ 3x + 4y - z = -1 \\ 2x - y + 3z = 3 \end{cases}$

32. $\begin{cases} -x + y + z = 3 \\ 3x - z = 1 \\ 2x - 3y - 4z = -2 \end{cases}$

33. $\begin{cases} x + y - 2z = 5 \\ 3x + 2y + 5z = 10 \\ -2x - 3y + 15z = 2 \end{cases}$

34. $\begin{cases} x - y = 3 \\ x + y + 4z = 1 \\ 2x - 3y - 2z = 7 \end{cases}$

35. $\begin{cases} x - 3y + z = 4 \\ 2x - 5y - z = 6 \end{cases}$

36. $\begin{cases} x_1 + x_2 + x_3 + x_4 = 3 \\ x_1 - 2x_2 + x_3 - 4x_4 = -5 \\ x_1 - x_3 + x_4 = 0 \\ x_2 + x_3 + x_4 = 2 \end{cases}$

37. Are D and G inverse matrices if

$$D = \begin{bmatrix} 3 & 5 \\ 1 & 2 \end{bmatrix} \quad \text{and} \quad G = \begin{bmatrix} 2 & -5 \\ -1 & 3 \end{bmatrix}?$$

In Problems 38–40, find the inverse of each matrix.

38. $\begin{bmatrix} 7 & -1 \\ -10 & 2 \end{bmatrix}$

39. $\begin{bmatrix} 1 & 0 & 2 \\ 3 & 4 & -1 \\ 1 & 1 & 0 \end{bmatrix}$

40. $\begin{bmatrix} 3 & 3 & 2 \\ -1 & 4 & 2 \\ 2 & 5 & 3 \end{bmatrix}$

In Problems 41–43, solve each system of equations by using inverse matrices.

41. $\begin{cases} x \qquad + 2z = 5 \\ 3x + 4y - z = 2 \\ x + y \qquad = -3 \end{cases}$ (See Problem 39.)

42. $\begin{cases} 3x + 3y + 2z = 1 \\ -x + 4y + 2z = -10 \\ 2x + 5y + 3z = -6 \end{cases}$ (See Problem 40.)

43. $\begin{cases} x + 3y + z = 0 \\ x + 4y + 3z = 2 \\ 2x - y - 11z = -12 \end{cases}$

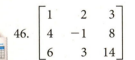

44. Does the following matrix have an inverse?

$$\begin{bmatrix} 1 & 2 & 4 \\ 2 & 0 & -4 \\ 1 & 0 & -2 \end{bmatrix}$$

In Problems 45 and 46:

(a) **Find the determinant of the matrix.**

(b) **Use it to decide whether the matrix has an inverse.**

45. $\begin{bmatrix} 4 & 4 \\ -2 & 2 \end{bmatrix}$

46. $\begin{bmatrix} 1 & 2 & 3 \\ 4 & -1 & 8 \\ 6 & 3 & 14 \end{bmatrix}$

APPLICATIONS

The Burr Cabinet Company manufactures bookcases and filing cabinets at two plants, A and B. Matrix M gives the production for the two plants during June, and matrix N gives the production for July. Use them in Problems 47–49.

$$M = \begin{matrix} & \begin{matrix} A & \ \ B \end{matrix} & \\ \begin{bmatrix} 150 & 80 \\ 280 & 300 \end{bmatrix} & \begin{matrix} \text{Bookcases} \\ \text{Files} \end{matrix} \end{matrix} \qquad N = \begin{matrix} \begin{matrix} A & \ \ B \end{matrix} \\ \begin{bmatrix} 100 & 60 \\ 200 & 400 \end{bmatrix} \end{matrix}$$

$$S = \begin{bmatrix} 120 & 80 \\ 180 & 300 \end{bmatrix} \qquad P = \begin{bmatrix} 1000 & 800 \\ 600 & 1200 \end{bmatrix}$$

47. *Production* Write the matrix that represents total production at the two plants for the 2 months.

48. *Production* If matrix P represents the inventories at the plants at the beginning of June and matrix S represents shipments from the plants during June, write the matrix that represents the inventories at the end of June.

49. *Production* If the company sells its bookcases to wholesalers for $100 and its filing cabinets for $120, for which month was the value of production higher: (a) at plant A? (b) at plant B?

A small church choir is made up of men and women who wear choir robes in the sizes shown in matrix A.

$$A = \begin{matrix} & \begin{matrix} \text{Men} & \text{Women} \end{matrix} & \\ \begin{bmatrix} 1 & 14 \\ 12 & 10 \\ 8 & 3 \end{bmatrix} & \begin{matrix} \text{Small} \\ \text{Medium} \\ \text{Large} \end{matrix} \end{matrix}$$

$$B = \begin{matrix} & \begin{matrix} \text{S} & \ \text{M} & \ \text{L} \end{matrix} & \\ \begin{bmatrix} 25 & 40 & 45 \\ 10 & 10 & 10 \end{bmatrix} & \begin{matrix} \text{Robes} \\ \text{Hoods} \end{matrix} \end{matrix}$$

Matrix B contains the prices (in dollars) of new robes and hoods according to size. Use these matrices in Problems 50 and 51.

50. *Cost* Find the product BA, and label the rows and columns to show what each entry represents.

51. *Cost* To find a matrix that gives the cost of new robes and the cost of new hoods, find

$$BA \begin{bmatrix} 1 \\ 1 \end{bmatrix}$$

52. *Manufacturing* Two departments of a firm, A and B, need different amounts of the same products. The following table gives the amounts of the products needed by the two departments.

	Steel	Plastic	Wood
Department A	30	20	10
Department B	20	10	20

These three products are supplied by two suppliers, Ace and Kink, with the unit prices given in the following table.

	Ace	Kink
Steel	300	280
Plastic	150	100
Wood	150	200

(a) Use matrix multiplication to find how much these two orders will cost at the two suppliers. The result should be a 2 × 2 matrix.

(b) From which supplier should each department make its purchase?

53. *Investment* Over the period of time 2002–2011, Maura's actual return (per dollar of stock) was 0.013469 per month, Ward's actual return was 0.013543 per month, and Goldsmith's actual return was 0.006504. Under certain assumptions, including that whatever underlying process generated the past set of observations will continue to hold in the present and near future, we can

estimate the expected return for a stock portfolio containing these assets in some combination.

Suppose we have a portfolio that has 20% of our total wealth in Maura, 30% in Ward, and 50% in Goldsmith stock. Use the following steps to estimate the portfolio's historical return, which can be used to estimate the future return.

(a) Write a 1×3 matrix that defines the decimal part of our total wealth in Maura, in Ward, and in Goldsmith.

(b) Write a 3×1 matrix that gives the monthly returns from each of these companies.

(c) Find the historical return of the portfolio by computing the product of these two matrices.

(d) What is the expected monthly return of the stock portfolio?

54. *Investment* A woman has $50,000 to invest. She has decided to invest all of it by purchasing some shares of stock in each of three companies: a fast-food chain that sells for $50 per share and has an expected growth of 11.5% per year, a software company that sells for $20 per share and has an expected growth of 15% per year, and a pharmaceutical company that sells for $80 per share and has an expected growth of 10% per year. She plans to buy twice as many shares of stock in the fast-food chain as in the pharmaceutical company. If her goal is 12% growth per year, how many shares of each stock should she buy?

55. *Nutrition* A biologist is growing three different types of slugs (types A, B, and C) in the same laboratory environment. Each day, the slugs are given a nutrient mixture that contains three different ingredients (I, II, and III). Each type A slug requires 1 unit of I, 3 units of II, and 1 unit of III per day. Each type B slug requires 1 unit of I, 4 units of II, and 2 units of III per day. Each type C slug requires 2 units of I, 10 units of II, and 6 units of III per day.

(a) If the daily mixture contains 2000 units of I, 8000 units of II, and 4000 units of III, find the number of slugs of each type that can be supported.

(b) Is it possible to support 500 type A slugs? If so, how many of the other types are there?

(c) What is the maximum number of type A slugs possible? How many of the other types are there in this case?

56. *Transportation* An airline company has three types of aircraft that carry three types of cargo. The payload of each type is summarized in the table below.

Units Carried	Plane Type		
	Passenger	Transport	Jumbo
First-class mail	100	100	100
Passengers	150	20	350
Air freight	20	65	35

(a) Suppose that on a given day the airline must move 1100 units of first-class mail, 460 units of air freight, and 1930 passengers. How many aircraft of each type should be scheduled? Use inverse matrices.

(b) How should the schedule from part (a) be adjusted to accommodate 730 more passengers?

(c) What column of the inverse matrix used in part (a) can be used to answer part (b)?

57. *Economy models* An economy has a shipping industry and an agricultural industry with technology matrix A.

$$A = \begin{bmatrix} 0.1 & 0.2 \\ 0.2 & 0.4 \end{bmatrix} \begin{matrix} \text{Shipping} \\ \text{Agriculture} \end{matrix}$$

(with column headers S   A)

(a) Surpluses of 4720 tons of shipping output and 40 tons of agricultural output are desired. Find the gross production of each industry.

(b) Find the additional production needed from each industry for 1 more unit of agricultural surplus.

58. *Economy models* A simple economy has a shoe industry and a cattle industry. Each unit of shoe output requires inputs of 0.1 unit of shoes and 0.2 unit of cattle products. Each unit of cattle products output requires inputs of 0.1 unit of shoes and 0.05 unit of cattle products.

(a) Write the technology matrix for this simple economy.

(b) If surpluses of 850 units of shoes and 275 units of cattle products are desired, find the gross production of each industry.

59. *Economy models* A look at the industrial sector of an economy can be simplified to include three industries: the mining industry, the manufacturing industry, and the fuels industry. The technology matrix for this sector of the economy is given by

$$A = \begin{bmatrix} 0.4 & 0.2 & 0.2 \\ 0.2 & 0.4 & 0.2 \\ 0.1 & 0.2 & 0.2 \end{bmatrix} \begin{matrix} \text{Mining} \\ \text{Manufacturing} \\ \text{Fuels} \end{matrix}$$

(with column headers M   Mf   F)

Find the gross production of each industry if surpluses of 72 units of mined goods, 40 units of manufactured goods, and 220 units of fuels are desired.

60. *Economy models* Suppose a closed Leontief model for a nation's economy has the following technology matrix.

$$A = \begin{bmatrix} 0.4 & 0.2 & 0.2 & 0.2 \\ 0.2 & 0.4 & 0.1 & 0.2 \\ 0.2 & 0.1 & 0.3 & 0.1 \\ 0.2 & 0.3 & 0.4 & 0.5 \end{bmatrix} \begin{matrix} \text{Government} \\ \text{Agriculture} \\ \text{Manufacturing} \\ \text{Households} \end{matrix}$$

(with column headers G   A   M   H)

Find the gross production of each industry.

3   CHAPTER TEST

In Problems 1–6, perform the indicated matrix operations with the following matrices.

$$A = \begin{bmatrix} 1 & -2 \\ 3 & 2 \\ 4 & 1 \end{bmatrix} \quad B = \begin{bmatrix} 2 & -2 & 1 \\ 3 & 1 & 5 \end{bmatrix}$$

$$C = \begin{bmatrix} 3 & -4 & -1 \\ 2 & 2 & -1 \end{bmatrix} \quad D = \begin{bmatrix} 1 & 2 & 4 \\ 3 & 5 & 41 \\ 3 & 2 & 3 \end{bmatrix}$$

1. $A^T + B$ 2. $B - C$ 3. CD
4. DA 5. BA 6. ABD

7. Find the inverse of the matrix $\begin{bmatrix} 1 & 3 \\ 2 & 4 \end{bmatrix}$.

8. Find the inverse of the matrix

$$\begin{bmatrix} 1 & 2 & 4 \\ 1 & 2 & 2 \\ 1 & 1 & 4 \end{bmatrix}$$

9. If $AX = B$ and

$$A^{-1} = \begin{bmatrix} 1 & 2 & 0 \\ 3 & 1 & 2 \\ 4 & 1 & 1 \end{bmatrix} \quad \text{and} \quad B = \begin{bmatrix} 3 \\ 1 \\ 2 \end{bmatrix}$$

find the matrix X.

10. Solve

$$\begin{cases} x - y + 2z = 4 \\ x + 4y + z = 4 \\ 2x + 2y + 4z = 10 \end{cases}$$

11. Solve

$$\begin{cases} x - y + 2z = 4 \\ x + 4y + z = 4 \\ 2x + 3y + 3z = 8 \end{cases}$$

12. Solve

$$\begin{cases} x - y + 3z = 4 \\ x + 5y + 2z = 3 \\ 2x + 4y + 5z = 8 \end{cases}$$

13. Use an inverse matrix to solve

$$\begin{cases} x + y + 2z = 4 \\ x + 2y + z + w = 4 \\ 2x + 5y + 4z + 2w = 10 \\ 2y + z + 2w = 0 \end{cases}$$

14. Solve

$$\begin{cases} x - y + 2z - w = 4 \\ x + 4y + z + w = 4 \\ 2x + 2y + 4z + 2w = 10 \\ - y + z + 2w = 2 \end{cases}$$

15. Suppose that the solution of an investment problem involving a system of linear equations is given by

$$B = 75{,}000 - 3H \quad \text{and} \quad E = 20{,}000 + 2H$$

where B represents the dollars invested in Barton Bank stocks, H is the dollars invested in Heath Healthcare stocks, and E is the dollars invested in Electronics Depot stocks.

(a) If $10,000 is invested in the Heath Healthcare stocks, how much is invested in the other two stocks?

(b) What is the dollar range that could be invested in the Heath Healthcare stocks?

(c) What is the minimum amount that could be invested in the Electronics Depot stocks? How much is invested in the other two stocks in this case?

16. In an ecological model, matrix A gives the fractions of several types of plants consumed by the herbivores in the ecosystem. Similarly, matrix B gives the fraction of each type of herbivore that is consumed by the carnivores in the system. Each row of matrix A refers to a type of plant, and each column refers to a kind of herbivore. Similarly, each row of matrix B is a kind of herbivore and each column is a kind of carnivore.

$$A = \begin{matrix} & \text{Herbivores} & \\ \begin{bmatrix} 0.2 & 0.1 & 0.4 \\ 0.3 & 0.3 & 0.1 \\ 0.2 & 0.1 & 0.1 \\ 0.1 & 0.2 & 0.5 \\ 0.4 & 0.4 & 0 \end{bmatrix} \end{matrix} \text{Plants}$$

$$B = \begin{matrix} & \text{Carnivores} & \\ \begin{bmatrix} 0.1 & 0.1 & 0.2 \\ 0.2 & 0 & 0.4 \\ 0.1 & 0.5 & 0.1 \end{bmatrix} \end{matrix} \text{Herbivores}$$

The matrix AB gives the fraction of each plant that is consumed by each carnivore.

(a) Find AB.

(b) What fraction of plant type 1 (row 1) is consumed by each carnivore?

(c) Identify the plant type that each carnivore consumes the most.

17. A furniture manufacturer produces four styles of chairs and has orders for 1000 of style A, 4000 of style B, 2000 of style C, and 1000 of style D. The following table lists the numbers of units of raw materials the manufacturer needs for each style. Suppose wood costs $5 per unit, nylon costs $3 per unit, velvet costs $4 per unit, and springs cost $4 per unit.

	Wood	Nylon	Velvet	Springs
Style A	10	5	0	0
Style B	5	0	20	10
Style C	5	20	0	10
Style D	5	10	10	10

(a) Form a 1×4 matrix containing the number of units of each style ordered.

(b) Use matrix multiplication to determine the total number of units of each raw material needed.

(c) Write a 4×1 matrix representing the costs of the raw materials.

(d) Use matrix multiplication to find the 1×1 matrix that represents the total investment to fill the orders.

(e) Use matrix multiplication to determine the cost of materials for each style of chair.

18. Complete parts (a) and (b) by using the code discussed in Sections 3.2 and 3.4 (where a is 1, ..., z is 26, and blank is 27), and the coding matrix is

$$A = \begin{bmatrix} 8 & 5 \\ 3 & 2 \end{bmatrix}$$

(a) Encode the message "Less is more," the motto of architect Mies van der Rohe.

(b) Decode the message 138, 54, 140, 53, 255, 99, 141, 54, 201, 76, 287, 111.

19. A young couple with a $120,000 inheritance wants to invest all this money. They plan to diversify their investments, choosing some of each of the following types of stock.

Type	Cost/Share	Expected Total Growth/Share
Growth	$ 30	$ 4.60
Blue-chip	100	11.00
Utility	50	5.00

Their strategy is to have the total investment in growth stocks equal to the sum of the other investments, and their goal is a 13% growth for their investment. How many shares of each type of stock should they purchase?

20. Suppose an economy has two industries, agriculture and minerals, and the economy has the technology matrix

$$A = \begin{bmatrix} & \text{Ag} & \text{M} \\ & 0.4 & 0.2 \\ & 0.1 & 0.3 \end{bmatrix} \begin{matrix} \text{Ag} \\ \text{M} \end{matrix}$$

(a) If surpluses of 100 agriculture units and 140 minerals units are desired, find the gross production of each industry.

(b) How many additional units must each industry produce to have 4 more units of agricultural surplus?

(c) How many additional units must each industry produce for 1 more unit of minerals surplus?

21. Suppose the technology matrix for a closed model of a simple economy is given by matrix A. Find the gross productions of the industries.

$$A = \begin{bmatrix} \text{P} & \text{NP} & \text{H} \\ 0.4 & 0.3 & 0.4 \\ 0.2 & 0.4 & 0.2 \\ 0.4 & 0.3 & 0.4 \end{bmatrix} \begin{matrix} \text{Profit} \\ \text{Nonprofit} \\ \text{Households} \end{matrix}$$

Use the following information in Problems 22 and 23. The national economy of Swiziland has four products: agricultural products, machinery, fuel, and steel. Producing 1 unit of agricultural products requires 0.2 unit of agricultural products, 0.3 unit of machinery, 0.2 unit of fuel, and 0.1 unit of steel. Producing 1 unit of machinery requires 0.1 unit of agricultural products, 0.2 unit of machinery, 0.2 unit of fuel, and 0.4 unit of steel. Producing 1 unit of fuel requires 0.1 unit of agricultural products, 0.2 unit of machinery, 0.3 unit of fuel, and 0.2 unit of steel. Producing 1 unit of steel requires 0.1 unit of agricultural products, 0.2 unit of machinery, 0.3 unit of fuel, and 0.2 unit of steel.

22. Create the technology matrix for this economy.

23. Determine how many units of each product will give surpluses of 1700 units of agriculture products, 1900 units of machinery, 900 units of fuel, and 300 units of steel.

24. A simple closed economy has the technology matrix A. Find the total outputs of the four sectors of the economy.

$$A = \begin{bmatrix} \text{Ag} & \text{S} & \text{F} & \text{H} \\ 0.25 & 0.25 & 0.3 & 0.1 \\ 0.3 & 0.25 & 0.2 & 0.4 \\ 0.15 & 0.2 & 0.1 & 0.3 \\ 0.3 & 0.3 & 0.4 & 0.2 \end{bmatrix} \begin{matrix} \text{Agriculture} \\ \text{Steel} \\ \text{Fuel} \\ \text{Households} \end{matrix}$$

I. Taxation

Federal income tax allows a deduction for any state income tax paid during the year. In addition, the state of Alabama allows a deduction from its state income tax for any federal income tax paid during the year. The federal corporate income tax rate is equivalent to a flat rate of 34% if the taxable federal income is between $335,000 and $10,000,000, and the Alabama rate is 5% of the taxable state income.*

Both the Alabama and the federal taxable income for a corporation are $1,000,000 *before* either tax is paid. Because each tax is deductible on the other return, the taxable income will differ for the state and federal taxes. One procedure often used by tax accountants to find the tax due in this and similar situations is called *iteration* and is described by the first five steps below.[†]

1. Make an estimate of the federal taxes due by assuming no state tax is due, deduct this estimated federal tax due from the state taxable income, and calculate an estimate of the state taxes due on the basis of this assumption.
2. Deduct the estimated state tax due as computed in (1) from the federal taxable income, and calculate a new estimate of the federal tax due under this assumption.
3. Deduct the new estimated federal income tax due as computed in (2) from the state taxable income, and calculate a new estimate of the state taxes due on the basis of these calculations.
4. Repeat the process in (2) and (3) to get a better estimate of the taxes due to both governments.
5. Continue this process until the federal tax changes from one repetition to another by less than $1. What federal tax is due? What state tax is due?

To find the tax due each government directly, we can solve a matrix equation.

6. Create two linear equations that describe this taxation situation, convert the system to a matrix equation, and solve the system to see exactly what tax is due to each government.
7. What is the effective rate that this corporation pays to each government?

---

*Source: Federal Income Tax Tables and Alabama Income Tax Instructions.
[†]Performing these steps with an electronic spreadsheet is described by Kenneth H. Johnson in "A Simplified Calculation of Mutually Dependent Federal and State Tax Liabilities," in *The Journal of Taxation*, December 1994.

II. Company Profits after Bonuses and Taxes

A company earns $800,000 profit before paying bonuses and taxes. Suppose that a bonus of 2% is paid to each employee after all taxes are paid, that state taxes of 6% are paid on the profit after the bonuses are paid, and that federal taxes are 34% of the profit that remains after bonuses and state taxes are paid.

1. How much profit remains after all taxes and bonuses are paid? Give your answer to the nearest dollar.
2. If the employee bonuses are increased to 3%, employees would receive 50% more bonus money. How much more profit would this cost the company than the 2% bonuses?
3. Use a calculator or spreadsheet to determine the loss of profit associated with bonuses of 4%, 5%, etc., up to 10%, and make a management decision about what bonuses to give in order to balance employee morale with loss of profit.
4. Use technology and the data points found above to create an equation that models the company's profit after expenses as a function of the employee bonus percent.

4

Inequalities and Linear Programming

Weldon Schloneger/Shutterstock.com

Most companies seek to maximize profits subject to the limitations imposed by product demand and available resources (such as raw materials and labor) or to minimize production costs subject to the need to fill customer orders. If the relationships among the various resources, production requirements, costs, and profits are all linear, then these activities may be planned (or programmed) in the best possible (optimal) way by using **linear programming.** Because linear programming provides the best possible solution to problems involving allocation of limited resources among various activities, its impact has been tremendous.

In this chapter, we introduce and discuss this important method. The chapter topics and applications include the following.

SECTIONS	APPLICATIONS
4.1 Linear Inequalities in Two Variables	Land management, manufacturing constraints
4.2 Linear Programming: Graphical Methods	Profit, revenue, production costs
4.3 The Simplex Method: Maximization **Nonunique solutions** **Shadow prices**	Profit, manufacturing
4.4 The Simplex Method: Duality and Minimization	Purchasing, nutrition
4.5 The Simplex Method with Mixed Constraints	Production scheduling

Prerequisite Problem Type	For Section	Answer	Section for Review
(a) Solve $3x - 2y \geq 4$ for y. (b) Solve $2x + 4y = 60$ for y.	4.1	(a) $y \leq \frac{3}{2}x - 2$ (b) $y = 15 - \frac{1}{2}x$	1.1 Linear equations and inequalities
Graph the equation $y = \frac{3}{2}x - 2$.	4.1	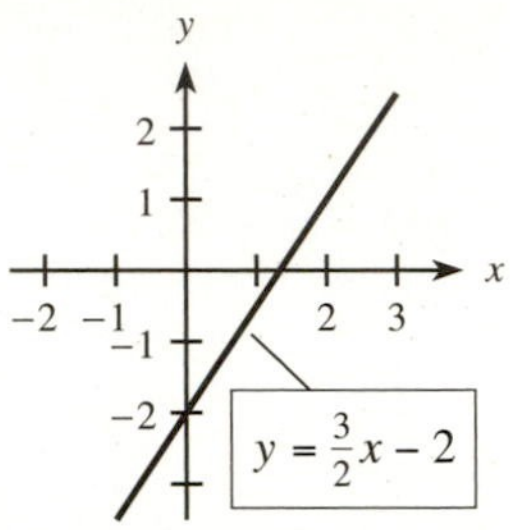	1.3 Graphing linear equations
Solve the systems: (a) $\begin{cases} x + 2y = 10 \\ 2x + y = 14 \end{cases}$ (b) $\begin{cases} x + 0.5y = 16 \\ x + y = 24 \end{cases}$	4.1 4.2	(a) $x = 6, \ y = 2$ (b) $x = 8, \ y = 16$	1.5 Systems of linear equations
Write the system in an augmented matrix: $\begin{cases} x + 2y + s_1 = 10 \\ 2x + y + s_2 = 14 \\ -2x - 3y + f = 0 \end{cases}$	4.3 4.4 4.5	$\begin{bmatrix} 1 & 2 & 1 & 0 & 0 & \vert & 10 \\ 2 & 1 & 0 & 1 & 0 & \vert & 14 \\ -2 & -3 & 0 & 0 & 1 & \vert & 0 \end{bmatrix}$	3.3 Gauss-Jordan elimination
Write a matrix equivalent to matrix A with the entry in row 1, column 2 equal to 1 and with all other entries in column 2 equal to 0. First multiply row 1 by $\frac{1}{2}$. $A = \begin{bmatrix} 1 & 2 & 1 & 0 & 0 & \vert & 10 \\ 2 & 1 & 0 & 1 & 0 & \vert & 14 \\ -2 & -3 & 0 & 0 & 1 & \vert & 0 \end{bmatrix}$	4.3 4.4 4.5	$\begin{bmatrix} \frac{1}{2} & 1 & \frac{1}{2} & 0 & 0 & \vert & 5 \\ \frac{3}{2} & 0 & -\frac{1}{2} & 1 & 0 & \vert & 9 \\ -\frac{1}{2} & 0 & \frac{3}{2} & 0 & 1 & \vert & 15 \end{bmatrix}$	3.3 Gauss-Jordan elimination

The graph shows the line $y = \frac{3}{2}x - 2$.

OBJECTIVES

4.1

- To graph linear inequalities in two variables
- To solve systems of linear inequalities in two variables

Linear Inequalities in Two Variables

| APPLICATION PREVIEW |

CDF Appliances has assembly plants in Atlanta and Fort Worth, where the company produces a variety of kitchen appliances, including a 12-cup coffee maker and a cappuccino machine. At the Atlanta plant, 160 of the coffee makers and 200 of the cappuccino machines can be assembled each hour. At the Fort Worth plant, 800 of the coffee makers and 200 of the cappuccino machines can be assembled each hour. CDF Appliances expects orders for at least 64,000 of the coffee makers and at least 40,000 of the cappuccino machines. At each plant the number of assembly hours available for these two appliances is limited (constrained) by each plant's capacity and the need to fill the orders. Finding the assembly hours for each plant that satisfy all these constraints at the same time is called solving the system of inequalities. (See Example 4.)

One Linear Inequality in Two Variables

Before we look at systems of inequalities, we will discuss solutions of one inequality in two variables, such as $y < x$. The solutions of this inequality are the ordered pairs (x, y) that satisfy the inequality. Thus $(1, 0)$, $(3, 2)$, $(0, -1)$, and $(-2, -5)$ are solutions of $y < x$, but $(3, 7)$, $(-4, -3)$, and $(2, 2)$ are not.

The graph of $y < x$ consists of all points in which the y-coordinate is less than the x-coordinate. The graph of the region $y < x$ can be found by graphing the line $y = x$ (as a dashed line, because the given inequality does not include $y = x$). This line separates the xy-plane into two **half-planes**, $y < x$ and $y > x$. We can determine which half-plane is the solution region by selecting as a **test point** any point *not on the line*; let's choose $(2, 0)$. Because the coordinates of this test point satisfy the inequality $y < x$, the half-plane containing this point is the solution region for $y < x$. (See Figure 4.1.) If the coordinates of the test point do not satisfy the inequality, then the other half-plane is the solution region. For example, say we had chosen $(0, 4)$ as our test point. Its coordinates do not satisfy $y < x$, so the half-plane that does *not* contain $(0, 4)$ is the solution region. (Note that we get the same region.)

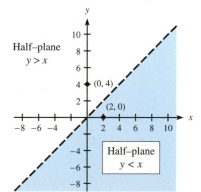

Figure 4.1

EXAMPLE 1 **Graphing an Inequality**

Graph the inequality $4x - 2y \leq 6$.

Solution

First we graph the line $4x - 2y = 6$, or (equivalently) $y = 2x - 3$, as a solid line, because points lying on the line satisfy the given inequality. Next we pick a test point that is not on the line. If we use $(0, 0)$, we see that its coordinates satisfy $4x - 2y \leq 6$—that is, $y \geq 2x - 3$. Hence the **solution region** is the line $y = 2x - 3$ and the half-plane that contains the test point $(0, 0)$. See Figure 4.2.   ■

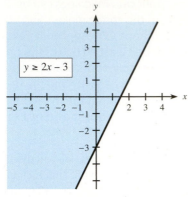

Figure 4.2

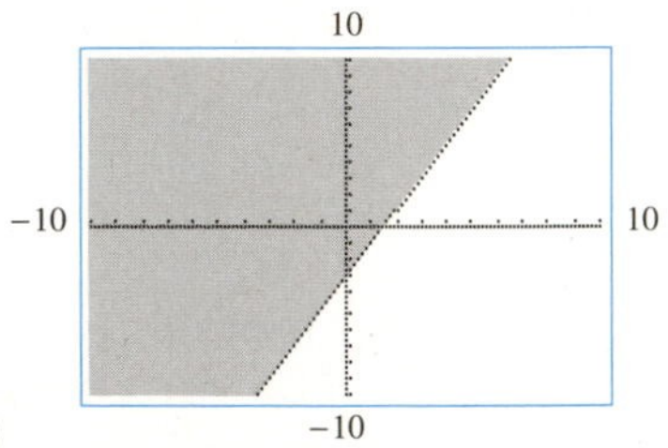

Figure 4.3

Calculator Note Graphing calculators can also be used to shade the solution region of an inequality. Details are shown in Appendix C, Section 4.1. Figure 4.3 shows a graphing calculator window for the solution region for $y \geq 2x - 3$.   ■

Systems of Linear Inequalities

If we have two or more inequalities in two variables, we can find the solutions that satisfy all the inequalities. We call the inequalities a **system of inequalities,** and the solution of the system can be found by finding the intersection of the solution sets of all the inequalities.

The solution set of the system of inequalities can be found by graphing the inequalities on the same set of axes and noting their points of intersection.

EXAMPLE 2 **Graphical Solution of a System of Inequalities**

Graph the solution of the system

$$\begin{cases} 3x - 2y \geq 4 \\ x + y - 3 > 0 \end{cases}$$

Solution

Begin by graphing the equations $3x - 2y = 4$ and $x + y = 3$ (from $x + y - 3 = 0$) by the intercept method: Find y when $x = 0$ and find x when $y = 0$.

$3x - 2y = 4$		$x + y = 3$	
x	y	x	y
0	-2	0	3
4/3	0	3	0

We graph $3x - 2y = 4$ as a solid line and $x + y = 3$ as a dashed line (see Figure 4.4(a)). We use any point not on either line as a test point; let's use $(0, 0)$. Note that the coordinates of $(0, 0)$ do not satisfy either $3x - 2y \geq 4$ or $x + y - 3 > 0$. Thus the solution region for each individual inequality is the half-plane that does not contain the point $(0, 0)$. Figure 4.4(b) indicates the half-plane solution for each inequality with arrows pointing from the line into the desired half-plane (away from the test point). The points that satisfy both of these inequalities lie in the intersection of the two individual solution regions, shown in Figure 4.4(c). This solution region is the graph of the solution of this system of inequalities.

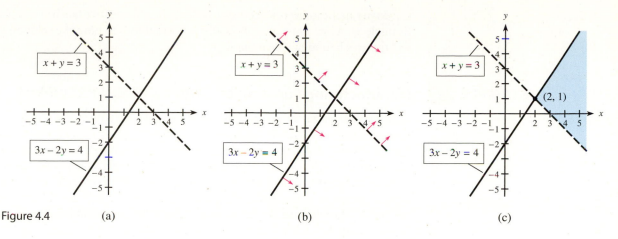

Figure 4.4   (a)   (b)   (c)

The point $(2, 1)$ in Figure 4.4(c), where the two regions form a "corner," can be found by solving the *equations* $3x - 2y = 4$ and $x + y = 3$ simultaneously. ◼

EXAMPLE 3   ## Graphical Solution of a System of Inequalities

Graph the solution of the system

$$\begin{cases} x + 2y \le 10 \\ 2x + \ y \le 14 \\ x \ge 0, y \ge 0 \end{cases}$$

Solution

The two inequalities $x \ge 0$ and $y \ge 0$ restrict the solution to Quadrant I (and the axes bounding Quadrant I).

We seek points in the first quadrant (on or above $y = 0$ and on or to the right of $x = 0$) that satisfy $x + 2y \le 10$ *and* $2x + y \le 14$. Note that the test point $(0, 0)$ satisfies both of these inequalities. Thus the solution region lies in Quadrant I and toward the origin from the lines $x + 2y = 10$ and $2x + y = 14$. (See Figure 4.5.) We can observe from the graph that the points $(0, 0)$, $(7, 0)$, and $(0, 5)$ are corners of the solution region. The corner $(6, 2)$ is found by solving the *equations* $y = 5 - \frac{1}{2}x$ and $y = 14 - 2x$ simultaneously as follows:

$$5 - \frac{1}{2}x = 14 - 2x$$

$$\frac{3}{2}x = 9$$

$$x = 6$$

$$y = 2$$

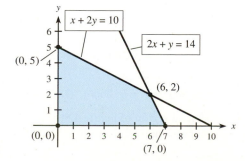

Figure 4.5

We will see that the corners of the solution region are important in solving linear programming problems. ◼

Many applications restrict the variables to be nonnegative (such as $x \geq 0$ and $y \geq 0$ in Example 3). As we noted, the effect of this restriction is to limit the solution to Quadrant I and the axes bounding Quadrant I.

EXAMPLE 4   Manufacturing Constraints | APPLICATION PREVIEW |

CDF Appliances has assembly plants in Atlanta and Fort Worth, where the company produces a variety of kitchen appliances, including a 12-cup coffee maker and a cappuccino machine. At the Atlanta plant, 160 of the coffee makers and 200 of the cappuccino machines can be assembled each hour. At the Fort Worth plant, 800 of the coffee makers and 200 of the cappuccino machines can be assembled each hour. CDF Appliances expects orders for at least 64,000 of the coffee makers and at least 40,000 of the cappuccino machines. At each plant, the number of assembly hours available for these two appliances is constrained by each plant's capacity and the need to fill the orders. Write the system of inequalities that describes these assembly plant constraints, and graph the solution region for the system.

Solution

Let x be the number of assembly hours at the Atlanta plant, and let y be the number of assembly hours at the Fort Worth plant. The production capabilities of each facility and the anticipated orders are summarized in the following table.

	Atlanta	Fort Worth	Needed
Coffee maker	160/hr	800/hr	At least 64,000
Cappuccino machine	200/hr	200/hr	At least 40,000

This table and the fact that both x and y must be nonnegative give the following constraints.

$$160x + 800y \geq 64{,}000$$
$$200x + 200y \geq 40{,}000$$
$$x \geq 0, y \geq 0$$

The solution is restricted to Quadrant I and the axes that bound Quadrant I. The equation $160x + 800y = 64{,}000$ can be graphed with x- and y-intercepts:

$$x\text{-intercept: } y = 0 \text{ gives } 160x = 64{,}000 \text{ or } x = 400$$
$$y\text{-intercept: } x = 0 \text{ gives } 800y = 64{,}000 \text{ or } y = 80$$

The test point $(0, 0)$ does not satisfy $160x + 800y \geq 64{,}000$, so the region lies above the line (see Figure 4.6(a)). In like manner, the region satisfying $200x + 200y \geq 40{,}000$ lies above the line $200x + 200y = 40{,}000$, and the solution region for this system is shown in Figure 4.6(b).

The corner $(150, 50)$ is found by solving simultaneously as follows.

$$\begin{cases} 160x + 800y = 64{,}000 \\ 200x + 200y = 40{,}000 \end{cases} \text{ is equivalent to } \begin{cases} x + 5y = 400 & (1) \\ x + y = 200 & (2) \end{cases}$$

Finding equation (1) minus equation (2) gives one equation in one variable and allows us to complete the solution.

$$4y = 200, \text{ so } y = 200/4 = 50 \text{ and } x = 200 - y = 150$$

Any point in the shaded region of Figure 4.6(b) represents a possible number of assembly hours at each plant that would fill the orders.

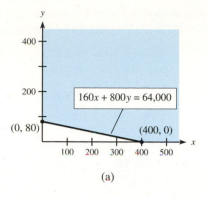
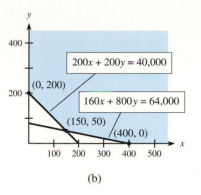

Figure 4.6   (a)   (b)

Michal Bednarek/Shutterstock.com; Barbro Bergfeldt/Shutterstock.com; Fedorov Oleksiy/Shutterstock.com; Thorsten Schuh/Shutterstock.com

EXAMPLE 5   Land Management

A farm co-op has 6000 acres available to plant with corn and soybeans. Each acre of corn requires 9 gallons of fertilizer/herbicide and 3/4 hour of labor to harvest. Each acre of soybeans requires 3 gallons of fertilizer/herbicide and 1 hour of labor to harvest. The co-op has available at most 40,500 gallons of fertilizer/herbicide and at most 5250 hours of labor for harvesting. The number of acres of each crop is limited (constrained) by the available resources: land, fertilizer/herbicide, and labor for harvesting. Write the system of inequalities that describes the constraints and graph the solution region for the system.

Solution

If x represents the number of acres of corn and y represents the number of acres of soybeans, then both x and y must be nonnegative and we have the following constraints.

Land acres:	$x + y \leq 6{,}000$ (1)
Fertilizer/herbicide gallons:	$9x + 3y \leq 40{,}500$ (2)
Labor hours:	$\dfrac{3}{4}x + y \leq 5{,}250$ (3)
	$x \geq 0, y \geq 0$

The intercepts of the lines associated with the inequalities are

$(6000, 0)$ and $(0, 6000)$ for the line $x + y = 6000$ (from Inequality (1))
$(4500, 0)$ and $(0, 13{,}500)$ for the line $9x + 3y = 40{,}500$ (from Inequality (2))

$(7000, 0)$ and $(0, 5250)$ for the line $\dfrac{3}{4}x + y = 5250$ (from Inequality (3))

The solution region is shaded in Figure 4.7, with three of the corners at $(0, 0)$, $(4500, 0)$, and $(0, 5250)$. The corner $(3750, 2250)$ is found by solving simultaneously.

$$\begin{cases} 9x + 3y = 40{,}500 & (4) \\ x + y = 6{,}000 & (5) \end{cases}$$

$$\begin{cases} 9x + 3y = 40{,}500 & (4) \\ -(3x + 3y = 18{,}000) & [3 \times \text{Eq. } (5)] \end{cases}$$

$$6x \quad\quad = 22{,}500$$

$$x = \frac{22{,}500}{6} = 3750$$

$$y = 6000 - x = 2250$$

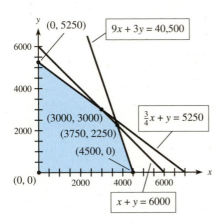

Figure 4.7

The corner $(3000, 3000)$ is found similarly by solving simultaneously with $x + y = 6000$ and $\frac{3}{4}x + y = 5250$.

Any point in the shaded region of Figure 4.7 represents a possible number of acres of corn and of soybeans that the co-op could plant, treat with fertilizer/herbicide, and harvest. ■

CHECKPOINT

1. Graph the region determined by the inequalities

$$2x + 3y \leq 12$$
$$4x + 2y \leq 16$$
$$x \geq 0, y \geq 0$$

2. Determine the corners of the region.

Calculator Note

Graphing calculators can also be used in the graphical solution of a system of inequalities. See Appendix C, Section 4.1 and Example 6 for details. ■

EXAMPLE 6   Finding a Solution Region with a Graphing Calculator

Use a graphing calculator to find the following.

(a) Find the region determined by the inequalities below.

$$5x + 2y \leq 54$$
$$2x + 4y \leq 60$$
$$x \geq 0, y \geq 0$$

(b) Find the corners of this region.

Solution

(a) We first write the inequalities as equations, solved for y.

$$5x + 2y = 54 \Rightarrow y = 27 - \frac{5}{2}x$$

$$2x + 4y = 60 \Rightarrow y = 15 - \frac{1}{2}x$$

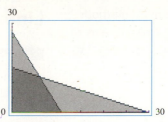

Figure 4.8

Graphing these equations with a graphing calculator and using shading shows the region satisfying the inequalities (see Figure 4.8).

(b) Finding the x- and y-intercepts and using the command INTERSECT with pairs of lines that form the borders of this region will give the points where the boundaries intersect. These points, $(0, 0)$, $(0, 15)$, $(6, 12)$, and $(10.8, 0)$, can also be found algebraically. These points are the corners of the solution region. These corners will be important to us in the graphical solutions of linear programming problems in the next section. ■

CHECKPOINT SOLUTIONS

1.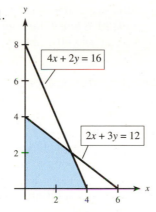

2. The graph shows corners at $(0, 0)$, $(0, 4)$, and $(4, 0)$. A corner also occurs where $2x + 3y = 12$ and $4x + 2y = 16$ intersect. The point of intersection is $x = 3$, $y = 2$, so $(3, 2)$ is a corner.

| EXERCISES | 4.1

Access end-of-section exercises online at **www.webassign.net**

- To use graphical methods to find the optimal value of a linear function subject to constraints

Linear Programming: Graphical Methods

| APPLICATION PREVIEW |

Many practical problems in business and economics seek to optimize some function (such as cost or profit) subject to complex relationships among capital, raw materials, labor, and so forth. Consider the following example.

A farm co-op has 6000 acres available to plant with corn and soybeans. Each acre of corn requires 9 gallons of fertilizer/herbicide and 3/4 hour of labor to harvest. Each acre of soybeans requires 3 gallons of fertilizer/herbicide and 1 hour of labor to harvest. The co-op has available at most 40,500 gallons of fertilizer/herbicide and at most 5250 hours of labor for harvesting. If the profits per acre are $60 for corn and $40 for soybeans, how many acres of each crop should the co-op plant to maximize their profit? What is the maximum profit?

The farm co-op's maximum profit can be found by using a mathematical technique called linear programming. (See Example 1.) Linear programming can be used to solve problems such as this if the limits on the variables (called constraints) can be expressed as linear inequalities and if the function that is to be maximized or minimized (called the objective function) is a linear function.

Feasible Regions and Solutions

Linear programming is widely used by businesses for problems that involve many variables (sometimes more than 100). In this section we begin our study of this important technique by considering problems involving two variables. With two variables we can use graphical methods to help solve the problem. The constraints form a system of linear inequalities in two variables that we can solve by graphing. The solution of the system of constraint inequalities determines a region, any point of which may yield the *optimal* (maximum or minimum) value for the objective function.* Hence any point in the region determined by the constraints is called a **feasible solution,** and the region itself is called the **feasible region.** In a linear programming problem, we seek the feasible solution that maximizes (or minimizes) the objective function.

EXAMPLE **1**   **Maximizing Profit** | APPLICATION PREVIEW |

A farm co-op has 6000 acres available to plant with corn and soybeans. Each acre of corn requires 9 gallons of fertilizer/herbicide and 3/4 hour of labor to harvest. Each acre of soybeans requires 3 gallons of fertilizer/herbicide and 1 hour of labor to harvest. The co-op has available at most 40,500 gallons of fertilizer/herbicide and at most 5250 hours of labor for harvesting. If the profits per acre are $60 for corn and $40 for soybeans, how many acres of each crop should the co-op plant to maximize their profit? What is the maximum profit?

Solution

We begin by letting x equal the number of acres of corn and y equal the number of acres of soybeans. Because the profit P is $60 for each of the x acres of corn and $40 for each of the y acres of soybeans, the function that describes the co-op's profit is given by

$$\text{Profit} = P = 60x + 40y \quad (\text{in dollars})$$

The constraints for this farm co-op application were developed in Example 5 of Section 4.1.

---

*The region determined by the constraints must be *convex* for the optimal to exist. A convex region is one such that for any two points in the region, the segment joining those points lies entirely within the region. We restrict our discussion to convex regions.

Combining the profit equation and the constraints found earlier gives the linear programming problem

$$\text{Maximize Profit} \quad P = 60x + 40y$$
$$\text{Subject to:} \quad x + y \le 6{,}000$$
$$9x + 3y \le 40{,}500$$
$$\frac{3}{4}x + y \le 5{,}250$$
$$x \ge 0, y \ge 0$$

The solution of the system of inequalities, or constraints, forms the feasible region shaded in Figure 4.9(a). *Any* point inside the shaded region or on its boundary is a feasible (possible) solution of the problem. For example, point A (1000, 2000) is in the feasible region, and at this point the profit is $P = 60(1000) + 40(2000) = 140{,}000$ dollars.

To find the maximum value of $P = 60x + 40y$, we cannot possibly evaluate P at every point in the feasible region. However, many points in the feasible region may correspond to the same value of P. For example, at point A in Figure 4.9(a), the value of P is 140,000 and the profit function becomes

$$60x + 40y = 140{,}000 \qquad \text{or} \qquad y = \frac{140{,}000 - 60x}{40} = 3500 - \frac{3}{2}x$$

The graph of this function is a line with slope $m = -3/2$ and y-intercept 3500. Many points in the feasible region lie on this line (see Figure 4.9(b)), and their coordinates all result in a profit $P = 140{,}000$. Any point in the feasible region that results in profit P satisfies

$$P = 60x + 40y \qquad \text{or} \qquad y = \frac{P - 60x}{40} = \frac{P}{40} - \frac{3}{2}x$$

In this form we can see that different P-values change the y-intercept for the line but the slope is always $m = -3/2$, so the lines for different P-values are parallel. Figure 4.9(b) shows the feasible region and the lines representing the objective function for $P = 140{,}000$, $P = 315{,}000$, and $P = 440{,}000$. Note that the line corresponding to $P = 315{,}000$ intersects the feasible region at the corner point (3750, 2250). Values of P less than 315,000 give lines that pass through the feasible region, but represent a profit less than \$315,000 (such as the line for $P = 140{,}000$ through point A). Similarly, values of P greater than 315,000 give lines that miss the feasible region, and hence cannot be solutions of the problem.

Thus, the farm co-op's maximum profit, subject to the constraints (i.e., the solution of the co-op's linear programming problem), is $P = \$315{,}000$ when $x = 3750$ and $y = 2250$. That is, when the co-op plants 3750 acres of corn and 2250 acres of soybeans, it achieves maximum profit of \$315,000. ■

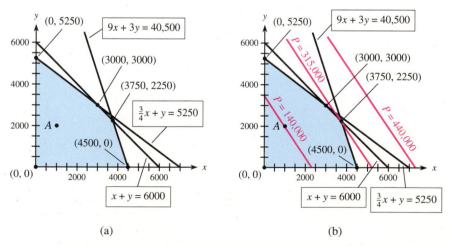

Figure 4.9   (a)   (b)

Notice in Example 1 that the objective function was maximized at one of the "corners" (vertices) of the feasible region. This is not a coincidence; it turns out that a corner will always lie on the objective function line corresponding to the optimal value.

Solving Graphically The feasible region in Figure 4.9 is an example of a **closed and bounded region** because it is entirely enclosed by, and includes, the lines associated with the constraints.

Solutions of Linear Programming Problems

1. When the feasible region for a linear programming problem is closed and bounded, the objective function has a maximum value and a minimum value.

2. When the feasible region is not closed and bounded, the objective function may have a maximum only, a minimum only, or no solution.

3. If a linear programming problem has a solution, then the optimal (maximum or minimum) value of an objective function occurs at a corner of the feasible region determined by the constraints.

4. If the objective function has its optimal value at two corners, then it also has that optimal value at any point on the line (boundary) connecting those two corners.

Thus, for a closed and bounded region, we can find the maximum or minimum value of the objective function by evaluating the function at each of the corners of the feasible region formed by the solution of the constraint inequalities. If the feasible region is not closed and bounded, we must check to make sure the objective function has an optimal value.

The steps involved in solving a linear programming problem are as follows.

Linear Programming (Graphical Method)

Procedure	Example
To find the optimal value of a function subject to constraints:	Find the maximum and minimum values of $C = 2x + 3y$ subject to the constraints $$\begin{cases} x + 2y \leq 10 \\ 2x + y \leq 14 \\ x \geq 0, y \geq 0 \end{cases}$$
1. Write the objective function and constraint inequalities from the problem.	1. Objective function: $C = 2x + 3y$ Constraints: $x + 2y \leq 10$ $2x + y \leq 14$ $x \geq 0, y \geq 0$

2. Graph the solution of the constraint system.
  (a) If the feasible region is closed and bounded, proceed to Step 3.
  (b) If the region is not closed and bounded, check whether an optimal value exists. If not, state this. If so, proceed to Step 3.
	2. The constraint region is closed and bounded. See Figure 4.10.
	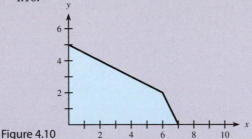
	Figure 4.10

3. Find the corners of the resulting feasible region. This may require simultaneous solution of two or more pairs of boundary equations.	3. Corners are $(0, 0)$, $(0, 5)$, $(6, 2)$, $(7, 0)$.
4. Evaluate the objective function at each corner of the feasible region determined by the constraints.	4. At $(0, 0)$, $C = 2x + 3y = 2(0) + 3(0) = 0$ At $(0, 5)$, $C = 2x + 3y = 2(0) + 3(5) = 15$ At $(6, 2)$, $C = 2x + 3y = 2(6) + 3(2) = 18$ At $(7, 0)$, $C = 2x + 3y = 2(7) + 3(0) = 14$
5. If two corners give the optimal value of the objective function, then all points on the boundary line joining these two corners also optimize the function.	5. The function is maximized at $x = 6$, $y = 2$. The maximum value is $C = 18$. The function is minimized at $x = 0$, $y = 0$. The minimum value is $C = 0$.

EXAMPLE 2 Maximizing Revenue

STANDARD   PLUSH

Chairco manufactures two types of chairs, standard and plush. Standard chairs require 2 hours to construct and finish, and plush chairs require 3 hours to construct and finish. Upholstering takes 1 hour for standard chairs and 3 hours for plush chairs. There are 240 hours per day available for construction and finishing, and 150 hours per day are available for upholstering. If the revenue for standard chairs is $89 and for plush chairs is $133.50, how many of each type should be produced each day to maximize revenue?

Solution

Let x be the number of standard chairs produced each day, and let y be the number of plush chairs produced. Then the daily revenue function is given by $R = 89x + 133.5y$. There are constraints for construction and finishing (no more than 240 hours/day) and for upholstering (no more than 150 hours/day). Thus we have the following.

$$\text{Construction/finishing constraint: } 2x + 3y \le 240$$
$$\text{Upholstering constraint: } x + 3y \le 150$$

Because all quantities must be nonnegative, we also have the constraints $x \ge 0$ and $y \ge 0$. Thus we seek to solve the following problem.

$$\text{Maximize } R = 89x + 133.5y \text{ subject to}$$
$$2x + 3y \le 240$$
$$x + 3y \le 150$$
$$x \ge 0, y \ge 0$$

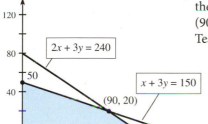

The feasible set is the closed and bounded region shaded in Figure 4.11. The corners of the feasible region are $(0, 0)$, $(120, 0)$, $(0, 50)$, and $(90, 20)$. All of these are obvious except $(90, 20)$, which can be found by solving $2x + 3y = 240$ and $x + 3y = 150$ simultaneously. Testing the objective function at the corners gives the following.

$$\text{At } (0, 0), \quad R = 89x + 133.5y = 89(0) + 133.5(0) = 0$$
$$\text{At } (120, 0), \ R = 89(120) + 133.5(0) = 10{,}680$$
$$\text{At } (0, 50), \quad R = 89(0) + 133.5(50) = 6675$$
$$\text{At } (90, 20), \ R = 89(90) + 133.5(20) = 10{,}680$$

Maximum at two corners

Figure 4.11

Thus the maximum revenue of $10,680 occurs at either the point $(120, 0)$ or the point $(90, 20)$. This means that the revenue function will be maximized not only at these two corner points but also at any point on the segment joining them. Since the number of each type of chair must be an integer, Chairco has maximum revenue of $10,680 at any pair of integer values along the segment joining $(120, 0)$ and $(90, 20)$. For example, the point $(105, 10)$ is on this segment, and the revenue at this point is also $10,680:

$$89x + 133.5y = 89(105) + 133.5(10) = 10{,}680$$

CHECKPOINT

1. Find the maximum and minimum values of the objective function $f = 4x + 3y$ on the shaded region in the figure, determined by the following constraints.

$$\begin{cases} 2x + 3y \le 12 \\ 4x - 2y \le 8 \\ x \ge 0, y \ge 0 \end{cases}$$

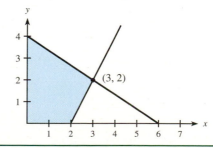

The examples so far have involved closed and bounded regions. Similar procedures apply for an unbounded region, but optimal solutions are no longer guaranteed.

EXAMPLE 3   Minimization

Find the maximum and minimum values (if they exist) of $C = x + y$ subject to the constraints

$$3x + 2y \geq 12$$
$$x + 3y \geq 11$$
$$x \geq 0, y \geq 0$$

Solution

The graph of the constraint system is shown in Figure 4.12(a). Note that the feasible region is not closed and bounded, so we must check whether optimal values exist. This check is done by graphing $C = x + y$ for selected values of C and noting the trend. Figure 4.12(b) shows the solution region with graphs of $C = x + y$ for $C = 3$, $C = 5$, and $C = 8$. Note that the objective function has a minimum but no maximum.

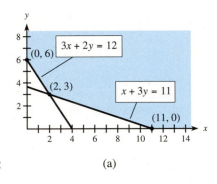

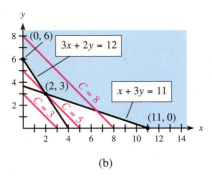

Figure 4.12   (a)   (b)

The corners $(0, 6)$ and $(11, 0)$ can be identified from the graph. The third corner, $(2, 3)$, can be found by solving the equations $3x + 2y = 12$ and $x + 3y = 11$ simultaneously.

$$
\begin{aligned}
3x + 2y &= 12 \\
-3x - 9y &= -33 \\
\hline
-7y &= -21 \quad \Rightarrow y = 3 \quad \text{and} \quad x = 2
\end{aligned}
$$

Examining the value of C at each corner point, we have

$$
\begin{aligned}
\text{At } (0, 6), \quad C = x + y &= 0 + 6 = 6 \\
\text{At } (11, 0), \quad C = x + y &= 11 + 0 = 11 \\
\text{At } (2, 3), \quad C = x + y &= 2 + 3 = 5
\end{aligned}
$$

Thus, we conclude the following:

Minimum value of $C = x + y$ is 5 at $(2, 3)$.
Maximum value of $C = x + y$ does not exist.

Note that C can be made arbitrarily large in the feasible region.   ∎

CHECKPOINT

2. Find the maximum and minimum values (if they exist) of the objective function $g = 3x + 4y$ subject to the following constraints.

$$x + 2y \geq 12, \quad x \geq 0$$
$$3x + 4y \geq 30, \quad y \geq 2$$

EXAMPLE 4   **Minimizing Production Costs**

Two chemical plants, one at Macon and one at Jonesboro, produce three types of fertilizer, low phosphorus (LP), medium phosphorus (MP), and high phosphorus (HP). At each plant, the fertilizer is produced in a single production run, so the three types are produced in fixed proportions. The Macon plant produces 1 ton of LP, 2 tons of MP, and 3 tons of HP in a single operation, and it charges $600 for what is produced in one operation, whereas one operation of the Jonesboro plant produces 1 ton of LP, 5 tons of MP, and 1 ton of HP, and it charges $1000 for what it produces in one operation. If a customer needs 100 tons of LP, 260 tons of MP, and 180 tons of HP, how many production runs should be ordered from each plant to minimize costs?

Solution

If x represents the number of operations requested from the Macon plant and y represents the number of operations requested from the Jonesboro plant, then we seek to minimize cost

$$C = 600x + 1000y$$

The following table summarizes production capabilities and requirements.

	Macon Plant	Jonesboro Plant	Requirements
Units of LP	1	1	100
Units of MP	2	5	260
Units of HP	3	1	180

Using the number of operations requested and the fact that requirements must be met or exceeded, we can formulate the following constraints.

$$\begin{cases} x + y \geq 100 \\ 2x + 5y \geq 260 \\ 3x + y \geq 180 \\ x \geq 0, y \geq 0 \end{cases}$$

Graphing this system gives the feasible set shown in Figure 4.13. The objective function has a minimum even though the feasible set is not closed and bounded. The corners are $(0, 180)$, $(40, 60)$, $(80, 20)$, and $(130, 0)$, where $(40, 60)$ is obtained by solving $x + y = 100$ and $3x + y = 180$ simultaneously, and where $(80, 20)$ is obtained by solving $x + y = 100$ and $2x + 5y = 260$ simultaneously.

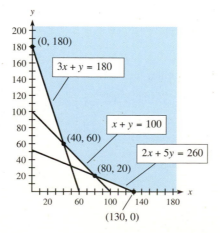

Figure 4.13

Evaluating $C = 600x + 1000y$ at each corner, we obtain

$$
\begin{aligned}
&\text{At } (0, 180), &C &= 600(0) + 1000(180) = 180{,}000 \\
&\text{At } (40, 60), &C &= 600(40) + 1000(60) = 84{,}000 \\
&\text{At } (80, 20), &C &= 600(80) + 1000(20) = 68{,}000 \\
&\text{At } (130, 0), &C &= 600(130) + 1000(0) = 78{,}000
\end{aligned}
$$

Thus, by placing orders requiring 80 production runs from the Macon plant and 20 production runs from the Jonesboro plant, the customer's needs will be satisfied at a minimum cost of $68,000. ■

EXAMPLE 5   Maximization Subject to Constraints with Graphing Calculators

Use a graphing calculator to maximize $f = 5x + 11y$ subject to the constraints

$$
\begin{aligned}
5x + 2y &\le 54 \\
2x + 4y &\le 60 \\
x \ge 0, \, y &\ge 0
\end{aligned}
$$

Solution
The feasible region (with dark shading) is shown in Figure 4.14. The boundaries intersect at $(0, 0)$, $(0, 15)$, $(6, 12)$ and $(10.8, 0)$. These corners can be found with INTERSECT and/or TRACE. (See Appendix C, Section 4.2, for details.) These corners can also be found algebraically. Testing the objective function at each corner gives the values of f shown below.

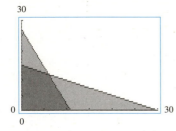

Figure 4.14

$$
\begin{aligned}
&\text{At } (0, 0), &f &= 5(0) + 11(0) = 0 \\
&\text{At } (0, 15), &f &= 5(0) + 11(15) = 165 \\
&\text{At } (6, 12), &f &= 5(6) + 11(12) = 162 \\
&\text{At } (10.8, 0), &f &= 5(10.8) + 11(0) = 54
\end{aligned}
$$

The maximum value is $f = 165$ at $x = 0, y = 15$. ■

CHECKPOINT SOLUTIONS

1. The values of $f = 4x + 3y$ at the corners are found as follows.

$$
\begin{aligned}
&\text{At } (0, 0), &f &= 4(0) + 3(0) = 0 \\
&\text{At } (2, 0), &f &= 4(2) + 3(0) = 8 \\
&\text{At } (3, 2), &f &= 4(3) + 3(2) = 18 \\
&\text{At } (0, 4), &f &= 4(0) + 3(4) = 12
\end{aligned}
$$

The maximum value of f is 18 at $x = 3, y = 2$, and the minimum value is $f = 0$ at $x = 0, y = 0$.

2. The graph of the feasible region is shown in Figure 4.15. The values of $g = 3x + 4y$ at the corners are found as follows:

$$
\begin{aligned}
&\text{At } (0, 7.5), &g &= 3(0) + 4(7.5) = 30 \\
&\text{At } (6, 3), &g &= 3(6) + 4(3) = 30 \\
&\text{At } (8, 2), &g &= 3(8) + 4(2) = 32
\end{aligned}
$$

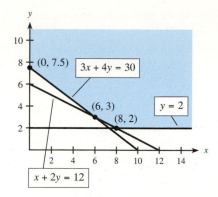

Figure 4.15

The minimum value of g is 30 at both $(0, 7.5)$ and $(6, 3)$. Thus any point on the border joining $(0, 7.5)$ and $(6, 3)$ will give the minimum value 30. For example, $(2, 6)$ is on this border and gives the value $6 + 24 = 30$. The maximum value of g does not exist; g can be made arbitrarily large on this feasible region.

| EXERCISES | 4.2

Access end-of-section exercises online at **www.webassign.net**

OBJECTIVE   4.3

- To use the simplex method to maximize functions subject to constraints

The Simplex Method: Maximization

| APPLICATION PREVIEW |

The Solar Technology Company manufactures three different types of hand calculators and classifies them as scientific, business, and graphing according to their calculating capabilities. The three types have production requirements given by the following table.

	Scientific	Business	Graphing
Electronic circuit components	5	7	10
Assembly time (hours)	1	3	4
Cases	1	1	1

The firm has a monthly limit of 90,000 circuit components, 30,000 hours of labor, and 9000 cases. If the profit is $6 for each scientific, $13 for each business, and $20 for each graphing calculator, the number of each that should be produced to yield maximum profit and the amount of that maximum possible profit can be found by using the simplex method of linear programming. (See Example 3.)

The Simplex Method   The graphical method for solving linear programming problems is practical only when there are two variables. The method discussed in this section, the **simplex method,** is especially useful if there are more than two variables in a problem. This method gives a systematic way of moving from one feasible corner of the convex region to another in such a way that the value of the objective function increases until an optimal value is reached or it is discovered that no solution exists.

The simplex method was developed by George Dantzig in 1947. One of the earliest applications of the method was to the scheduling problem that arose in connection with the Berlin airlift, begun in 1948. Since then, linear programming and the simplex method have been used to solve optimization problems in a wide variety of businesses. In fact, in a recent survey of Fortune 500 firms, 85% of the respondents indicated that they use linear programming.

In 1984, N. Karmarkar, a researcher at Bell Labs, developed a new method (that uses interior points rather than corner points) for solving linear programming problems. When the number of variables is very large, Karmarkar's method is more efficient than the simplex method. However, the simplex method is just as effective for many applications and is still widely used.

In discussing the simplex method, we initially restrict ourselves to **standard maximization problems,** which satisfy the following conditions.

1. The objective function is to be maximized.
2. All variables are nonnegative.
3. The constraints are of the form

$$a_1x_1 + a_2x_2 + \cdots + a_nx_n \leq b$$

where $b > 0$.

These conditions may seem to restrict the types of problems unduly, but in applied situations where the objective function is to be maximized, the constraints often satisfy conditions 2 and 3.

Before outlining a procedure for the simplex method, let us develop the procedure and investigate its rationale by seeing how the simplex method compares with the graphical method.

EXAMPLE | **1** | **Maximizing Profit**

A farm co-op has 6000 acres available to plant with corn and soybeans. Each acre of corn requires 9 gallons of fertilizer/herbicide and 3/4 hour of labor to harvest. Each acre of soybeans requires 3 gallons of fertilizer/herbicide and 1 hour of labor to harvest. The co-op has available at most 40,500 gallons of fertilizer/herbicide and at most 5250 hours of labor for harvesting. If the profits per acre are $60 for corn and $40 for soybeans, how many acres of each crop should the co-op plant in order to maximize their profit? What is the maximum profit?

Solution

This application was solved graphically in Example 1 in Section 4.2. As we saw there, with $x =$ the number of acres of corn and $y =$ the number of acres of soybeans, the farm co-op linear programming problem is

Maximize profit $P = 60x + 40y$

subject to $x + y \leq 6{,}000$ (Land constraint)

$9x + 3y \leq 40{,}500$ (Fertilizer/herbicide constraint)

$\dfrac{3}{4}x + y \leq 5{,}250$ (Labor hours constraint)

Note that $x \geq 0$ and $y \geq 0$. Because variables must be nonnegative in the simplex method, we shall no longer state these conditions.

The simplex method uses matrix methods on systems of equations, so we convert each constraint inequality to an equation by using a new variable, called a **slack variable.** We can think of each slack variable as representing the amount of the constraint left unused. In the land constraint, $x + y$ represents the total number of acres planted, and this number cannot exceed the 6000 acres available. Thus, if we let s_1 represent the unused (unplanted) acres of land, then $s_1 \geq 0$ and

$$x + y \leq 6000 \quad \text{becomes} \quad x + y + s_1 = 6000$$

Similarly, if s_2 represents the unused gallons of fertilizer/herbicide, then $s_2 \geq 0$ and

$$9x + 3y \leq 40{,}500 \quad \text{becomes} \quad 9x + 3y + s_2 = 40{,}500$$

And if s_3 represents the unused hours of labor, then $s_3 \geq 0$ and

$$\frac{3}{4}x + y \leq 5250 \quad \text{becomes} \quad \frac{3}{4}x + y + s_3 = 5250$$

Notice in each constraint that the expressions in x and y represent the amount of that resource that is used and that the slack variable represents the amount not used.

Previously, when we used matrices with systems of equations we arranged the equations with all the variables on the left side of the equation and the constants on the right. If we write the objective function in this form

$$P = 60x + 40y \quad \text{becomes} \quad -60x - 40y + P = 0$$

and then the following system of equations describes this problem.

$$\begin{cases} x & + & y & + & s_1 & & & & & & & = & 6{,}000 \\ 9x & + & 3y & & & + & s_2 & & & & & = & 40{,}500 \\ \frac{3}{4}x & + & y & & & & & + & s_3 & & & = & 5{,}250 \\ -60x & - & 40y & & & & & & & + & P & = & 0 \end{cases}$$

We seek to maximize P in the last equation subject to the constraints in the first three equations.

We can place this system of equations in a matrix called a **simplex matrix** or a **simplex tableau.** Note that the objective function is in the last (bottom) row of this matrix with all variables on the left side and that the rows above the function row correspond to the constraints.

$$A = \begin{bmatrix} x & y & s_1 & s_2 & s_3 & P & \\ 1 & 1 & 1 & 0 & 0 & 0 & 6{,}000 \\ 9 & 3 & 0 & 1 & 0 & 0 & 40{,}500 \\ \frac{3}{4} & 1 & 0 & 0 & 1 & 0 & 5{,}250 \\ -60 & -40 & 0 & 0 & 0 & 1 & 0 \end{bmatrix} \begin{array}{l} \text{Land constraint} \\ \text{Fertilizer/herbicide constraint} \\ \text{Labor hours constraint} \\ \text{Objective function (maximum profit)} \end{array}$$

From the graph in Figure 4.16, we know that $(0, 0)$ is a feasible solution giving a value of 0 for P. Because the origin is always a feasible solution of the type of linear programming problems we are considering, the simplex method begins there and systematically moves to new corners while increasing the value of P.

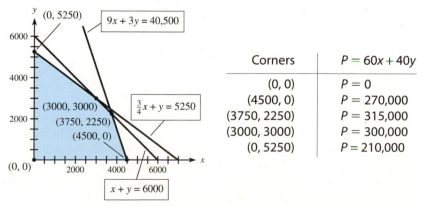

Corners	$P = 60x + 40y$
$(0, 0)$	$P = 0$
$(4500, 0)$	$P = 270{,}000$
$(3750, 2250)$	$P = 315{,}000$
$(3000, 3000)$	$P = 300{,}000$
$(0, 5250)$	$P = 210{,}000$

Figure 4.16

Note that if $x = 0$ and $y = 0$, then all 6000 acres of land are unused ($s_1 = 6000$), all 40,500 gallons of fertilizer/herbicide are unused ($s_2 = 40,500$), all 5250 labor hours are unused ($s_3 = 5250$), and profit is $P = 0$. From rows 1, 2, 3, and 4 of matrix A, we can read these values of s_1, s_2, s_3, and P, respectively.

Recall that the last row still corresponds to the objective function $P = 60x + 40y$. If we seek to improve (that is, increase the value of) $P = 60x + 40y$ by changing the value of only one of the variables, then because the coefficient of x is larger, the rate of increase is greatest by increasing x.

In general, we can find the variable that will improve the objective function most rapidly by looking for the *most negative* value in the last row (the objective function row) of the simplex matrix. The column that contains this value is called the **pivot column.**

The simplex matrix A is shown again below with the x-column indicated as the pivot column.

$$A = \begin{bmatrix} x & y & s_1 & s_2 & s_3 & P & \\ \boxed{1} & 1 & 1 & 0 & 0 & 0 & 6{,}000 \\ \boxed{9} & 3 & 0 & 1 & 0 & 0 & 40{,}500 \\ \frac{3}{4} & 1 & 0 & 0 & 1 & 0 & 5{,}250 \\ \boxed{-60} & -40 & 0 & 0 & 0 & 1 & 0 \end{bmatrix}$$

$6000 \div 1 = 6000$

$40{,}500 \div 9 = 4500$ smallest

$5250 \div \left(\frac{3}{4}\right) = 5250 \left(\frac{4}{3}\right) = 7000$

↑— Most negative entry; pivot column

The amount by which x can be increased is limited by the constraining equations (rows 1, 2, and 3 of the simplex matrix). As the graph in Figure 4.16 shows, the constraints limit x to no larger than $x = 4500$. Note that when the positive coefficients in the pivot column (the x-column) of the simplex matrix are divided into the constants in the augment, the *smallest* quotient is $40{,}500 \div 9 = 4500$. This smallest quotient identifies the **pivot row.** The **pivot entry** (or **pivot**) is the entry in both the pivot column and the pivot row. The 9 (circled) in row 2, column 1 of matrix A is the pivot entry. Notice that the smallest quotient, 4500, represents the amount that x can be increased (while $y = 0$), and the point (4500, 0) corresponds to a corner adjoining (0, 0) on the constraint region (see Figure 4.16).

To continue this solution method, we use row operations with the pivot row of the simplex matrix to simulate moving to another corner of the constraint region. We multiply the second row by $\frac{1}{9}$ to convert the pivot entry to 1, and then add multiples of row 2 to each of the remaining rows to make all elements of the pivot column, except the pivot element, equal to 0.

$$A = \begin{bmatrix} x & y & s_1 & s_2 & s_3 & P & \\ 1 & 1 & 1 & 0 & 0 & 0 & 6{,}000 \\ \boxed{9} & 3 & 0 & 1 & 0 & 0 & 40{,}500 \\ \frac{3}{4} & 1 & 0 & 0 & 1 & 0 & 5{,}250 \\ -60 & -40 & 0 & 0 & 0 & 1 & 0 \end{bmatrix} \xrightarrow{\frac{1}{9}R_2 \to R_2} \begin{bmatrix} x & y & s_1 & s_2 & s_3 & P & \\ 1 & 1 & 1 & 0 & 0 & 0 & 6{,}000 \\ \boxed{1} & \frac{1}{3} & 0 & \frac{1}{9} & 0 & 0 & 4{,}500 \\ \frac{3}{4} & 1 & 0 & 0 & 1 & 0 & 5{,}250 \\ -60 & -40 & 0 & 0 & 0 & 1 & 0 \end{bmatrix}$$

$$\begin{array}{c} -R_2 + R_1 \to R_1 \\ -\frac{3}{4}R_2 + R_3 \to R_3 \\ \hline 60R_2 + R_4 \to R_4 \end{array} \begin{bmatrix} x & y & s_1 & s_2 & s_3 & P & \\ 0 & \frac{2}{3} & 1 & -\frac{1}{9} & 0 & 0 & 1{,}500 \\ 1 & \frac{1}{3} & 0 & \frac{1}{9} & 0 & 0 & 4{,}500 \\ 0 & \frac{3}{4} & 0 & -\frac{1}{12} & 1 & 0 & 1{,}875 \\ 0 & -20 & 0 & \frac{20}{3} & 0 & 1 & 270{,}000 \end{bmatrix} = B$$

This "transforms" the simplex matrix A into the simplex matrix B. Recall that this transformation was based on holding $y = 0$ and increasing x as much as possible so the resulting point would be feasible—that is, moving from (0, 0) to (4500, 0) on the graph in Figure 4.16. At the point (4500, 0) the profit is $P = 60x + 40y = 60(4500) + 40(0) = 270{,}000$, and the resources used (and not used, or slack) are as follows.

Land: $x + y + s_1 = 6000 \Rightarrow 4500 + 0 + s_1 = 6000 \Rightarrow s_1 = 1500$

Fertilizer/herbicide: $9x + 3y + s_2 = 40{,}500 \Rightarrow 40{,}500 + 0 + s_2 = 40{,}500 \Rightarrow s_2 = 0$

Labor hours: $\frac{3}{4}x + y + s_3 = 5250 \Rightarrow 3375 + 0 + s_3 = 5250 \Rightarrow s_3 = 1875$

Let's see how these values $(x = 4500, y = 0, s_1 = 1500, s_2 = 0, s_3 = 1875,$ and $P = 270{,}000)$ can be found in matrix B.

Notice that the columns of B that correspond to the variables x, s_1, and s_3 and to the objective function P are all different, but each contains a single entry of 1 and the other entries are 0. Variables with columns of this type (but *not* the objective function) are called **basic variables.** Similarly, the columns for variables y and s_2 do not have this special form and these columns correspond to **nonbasic variables.** Notice also that the bottom row of B represents

$$-20y + \frac{20}{3}s_2 + P = 270{,}000, \quad \text{so} \quad P = 270{,}000 + 20y - \frac{20}{3}s_2$$

depends on y and s_2 (the nonbasic variables). If these nonbasic variables equal 0 (that is, $y = 0$ and $s_2 = 0$), then rows 1, 2, 3, and 4 of matrix B allow us to read the values of s_1, x, s_3, and P, respectively.

$$s_1 = 1500, \quad x = 4500, \quad s_3 = 1875, \quad P = 270{,}000$$

These solution values correspond to the feasible solution at the corner $(4500, 0)$ in Figure 4.16 and are called a **basic feasible solution.**

We also see that the value of $P = 270{,}000 + 20y - \frac{20}{3}s_2$ can be increased by increasing the value of y while holding the value of s_2 at 0. In matrix B, we can see this fact by observing that the most negative number in the last row (objective function row) occurs in the y-column. This means the y-column is the new pivot column.

The amount of increase in y is limited by the constraints. In matrix B we can discover the limitations on y by dividing the coefficient of y (if positive) into the constant term in the augment of each row and choosing the smallest nonnegative quotient. This identifies the pivot row, and hence the pivot entry, circled in matrix B.

$$B = \begin{bmatrix} & x & y & s_1 & s_2 & s_3 & P & \\ 0 & \tfrac{2}{3} & 1 & -\tfrac{1}{9} & 0 & 0 & 1{,}500 \\ 1 & \tfrac{1}{3} & 0 & \tfrac{1}{9} & 0 & 0 & 4{,}500 \\ 0 & \tfrac{3}{4} & 0 & -\tfrac{1}{12} & 1 & 0 & 1{,}875 \\ 0 & -20 & 0 & \tfrac{20}{3} & 0 & 1 & 270{,}000 \end{bmatrix}$$

$1500 \div \left(\tfrac{2}{3}\right) = 1500\left(\tfrac{3}{2}\right) = 2250$ smallest

$4500 \div \left(\tfrac{1}{3}\right) = 4500\left(\tfrac{3}{1}\right) = 13{,}500$

$1875 \div \left(\tfrac{3}{4}\right) = 1875\left(\tfrac{4}{3}\right) = 2500$

↑— Largest negative; pivot column

The smallest quotient is 2250; note that if we move to a corner adjoining $(4500, 0)$ on the feasible region in Figure 4.16, the largest y-value is 2250 [at the point $(3750, 2250)$].

Let us transform the simplex matrix B by multiplying row 1 by $\frac{3}{2}$ to convert the pivot entry to 1, and then adding multiples of row 1 to the other rows to make all other entries in the y-column (pivot column) equal to 0. The new simplex matrix is C.

$$B = \begin{bmatrix} & x & y & s_1 & s_2 & s_3 & P & \\ 0 & \tfrac{2}{3} & 1 & -\tfrac{1}{9} & 0 & 0 & 1{,}500 \\ 1 & \tfrac{1}{3} & 0 & \tfrac{1}{9} & 0 & 0 & 4{,}500 \\ 0 & \tfrac{3}{4} & 0 & -\tfrac{1}{12} & 1 & 0 & 1{,}875 \\ 0 & -20 & 0 & \tfrac{20}{3} & 0 & 1 & 270{,}000 \end{bmatrix}$$

$\xrightarrow{\frac{3}{2}R_1 \to R_1}$

$$\begin{bmatrix} & x & y & s_1 & s_2 & s_3 & P & \\ 0 & 1 & \tfrac{3}{2} & -\tfrac{1}{6} & 0 & 0 & 2{,}250 \\ 1 & \tfrac{1}{3} & 0 & \tfrac{1}{9} & 0 & 0 & 4{,}500 \\ 0 & \tfrac{3}{4} & 0 & -\tfrac{1}{12} & 1 & 0 & 1{,}875 \\ 0 & -20 & 0 & \tfrac{20}{3} & 0 & 1 & 270{,}000 \end{bmatrix}$$

$\begin{array}{l} -\tfrac{1}{3}R_1 + R_2 \to R_2 \\ -\tfrac{3}{4}R_1 + R_3 \to R_3 \\ 20R_1 + R_4 \to R_4 \end{array}$ $\xrightarrow{\hspace{2cm}}$

$$\begin{bmatrix} & x & y & s_1 & s_2 & s_3 & P & \\ 0 & 1 & \tfrac{3}{2} & -\tfrac{1}{6} & 0 & 0 & 2{,}250 \\ 1 & 0 & -\tfrac{1}{2} & \tfrac{1}{6} & 0 & 0 & 3{,}750 \\ 0 & 0 & -\tfrac{9}{8} & \tfrac{1}{24} & 1 & 0 & 187.5 \\ 0 & 0 & 30 & \tfrac{10}{3} & 0 & 1 & 315{,}000 \end{bmatrix} = C$$

In this new simplex matrix C, the basic variables (corresponding to the columns with a single entry of 1) are x, y, and s_3, the nonbasic variables are s_1 and s_2, and the last row (the objective function row) corresponds to

$$30s_1 + \frac{10}{3}s_2 + P = 315{,}000 \quad \text{or} \quad P = 315{,}000 - 30s_1 - \frac{10}{3}s_2$$

From the objective function we see that any increase in s_1 or s_2 will cause a *decrease* in P. Observing this or noting that all entries in the last row are nonnegative tells us that we have found the optimal value of the objective function and it occurs when $s_1 = 0$ and $s_2 = 0$. Using this, rows 1, 2, 3, and 4 of matrix C give the values for y, x, s_3, and P, respectively.

$$y = 2250, \quad x = 3750, \quad s_3 = 187.5, \quad P = 315{,}000, \quad \text{with} \quad s_1 = 0 \quad \text{and} \quad s_2 = 0$$

This corresponds to the corner $(3750, 2250)$ on the graph and to the maximum value of $P = 315{,}000$ at this corner (see Figure 4.16).

Just as we found graphically, these values mean that the co-op should plant $x = 3750$ acres of corn and $y = 2250$ acres of soybeans for a maximum profit of \$315,000. And this planting scheme results in $s_1 = 0$ unused acres of land, $s_2 = 0$ unused gallons of fertilizer/herbicide, and $s_3 = 187.5$ unused hours of labor. ◾

As we review the simplex solution in Example 1, we observe that each of the basic feasible solutions corresponded to a corner (x, y) of the feasible region and that two of the five variables x, y, s_1, s_2, and s_3 were equal to 0 at each feasible solution.

At $(0, 0)$: x and y were equal to 0.
At $(4500, 0)$: y and s_2 were equal to 0.
At $(3750, 2250)$: s_1 and s_2 were equal to 0.

See Figure 4.17 for an illustration of how the simplex method moved us from corner to corner until the optimal solution was obtained.

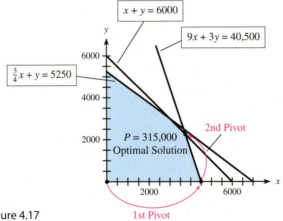

Pivoting Summary

1st Pivot:
from: $(0, 0)$ where $P = 0$
to: $(4500, 0)$ where $P = 270{,}000$

2nd Pivot:
from: $(4500, 0)$ where $P = 270{,}000$
to: $(3750, 2250)$ where $P = 315{,}000$
The Optimal Solution

Figure 4.17

Recall from Example 1 that at each step in the simplex method we can identify the *basic variables* directly from the simplex matrix by noting that their columns contain a *single entry of 1* with all other entries being zero. (Check this.) The remaining columns identify the *nonbasic variables*, and at each feasible solution (that is, at each step in the simplex method) the nonbasic variables were set equal to zero. With the nonbasic variables equal to zero, the values of the basic variables and the objective function can be read from the rows of the simplex matrix.

The Simplex Method: Tasks and Procedure

The simplex method involves a series of decisions and operations using matrices. It involves three major tasks.

Simplex Method Tasks

Task A: Setting up the matrix for the simplex method.
Task B: Determining necessary operations and implementing those operations to reach a solution.
Task C: Reading the solution from the simplex matrix.

From Example 1, it should be clear that the pivot selection and row operations of Task B are designed to move the system to the corner that achieves the most rapid increase in the function and at the same time is sensitive to the nonnegative limitations on the variables. The procedure that combines these three tasks to maximize a function is given next.

Linear Programming (Simplex Method)

Procedure

To use the simplex method to solve linear programming problems:

1. Use a different slack variable to write each constraint inequality as an equation, with positive constants on the right side.

2. Set up the simplex matrix. It contains the system of constraint equations with the objective function in the last row with all variables on the left and coefficient 1 for f.

3. Find the pivot entry:
   (a) The pivot column has the most negative number in the last row. (If a tie occurs, use either column.)

   (b) The positive coefficient in the pivot column that gives the smallest nonnegative quotient when divided into the constant in the constraint rows of the augment is the pivot entry. If there is a tie, either coefficient may be chosen. If there are no positive coefficients in the pivot column, no solution exists. Stop.

4. Use row operations with only the row containing the pivot entry to make the pivot entry a 1 and all other entries in the pivot column zeros. This process is called **pivoting**. This makes the variable corresponding to the pivot column a basic variable.

Example

Maximize $f = 2x + 3y$ subject to
$$x + 2y \le 10$$
$$2x + y \le 14$$

1. $x + 2y \le 10$ becomes $x + 2y + s_1 = 10$.
   $2x + y \le 14$ becomes $2x + y + s_2 = 14$.
   Maximize f in $-2x - 3y + f = 0$.

2.
$$
\begin{array}{ccccc}
x & y & s_1 & s_2 & f \\
\end{array}
$$
$$
\left[
\begin{array}{ccccc|c}
1 & 2 & 1 & 0 & 0 & 10 \\
2 & 1 & 0 & 1 & 0 & 14 \\
\hline
-2 & -3 & 0 & 0 & 1 & 0
\end{array}
\right]
$$

3. (a)
$$
\left[
\begin{array}{ccccc|c}
1 & \boxed{2} & 1 & 0 & 0 & 10 \\
2 & 1 & 0 & 1 & 0 & 14 \\
\hline
-2 & \boxed{-3} & 0 & 0 & 1 & 0
\end{array}
\right]
$$
   ↑— Most negative entry; pivot column

   (b) Row 1 quotient: $\dfrac{10}{2} = 5$

   Row 2 quotient: $\dfrac{14}{1} = 14$

   Therefore, pivot row is row 1.

$$
\left[
\begin{array}{ccccc|c}
1 & ② & 1 & 0 & 0 & 10 \\
2 & 1 & 0 & 1 & 0 & 14 \\
\hline
-2 & -3 & 0 & 0 & 1 & 0
\end{array}
\right]
$$
   Pivot entry is 2, circled.

4. Row operations:
   Multiply row 1 by $\frac{1}{2}$.
   Add -1 times new row 1 to row 2.
   Add 3 times new row 1 to row 3.

   Result:
$$
\left[
\begin{array}{ccccc|c}
\frac{1}{2} & 1 & \frac{1}{2} & 0 & 0 & 5 \\
\frac{3}{2} & 0 & -\frac{1}{2} & 1 & 0 & 9 \\
\hline
-\frac{1}{2} & 0 & \frac{3}{2} & 0 & 1 & 15
\end{array}
\right]
$$
   Indicators

(continued)

Linear Programming (Simplex Method) *(continued)*

Procedure	Example

5. The numerical entries in the last row are the indicators of what to do next.

   (a) If there is a negative indicator, return to Step 3.

   (b) If all indicators are positive or zero, an optimum value has been obtained for the objective function.

   (c) In a complete solution, the variables that have positive entries in the last row are nonbasic variables.

5. $-\frac{1}{2}$ is negative, so we identify a new pivot column and reduce again.

$$\begin{bmatrix} \frac{1}{2} & 1 & \frac{1}{2} & 0 & 0 & 5 \\ \frac{3}{2} & 0 & -\frac{1}{2} & 1 & 0 & 9 \\ -\frac{1}{2} & 0 & \frac{3}{2} & 0 & 1 & 15 \end{bmatrix}$$

$5 \div \frac{1}{2} = 10$
$9 \div \frac{3}{2} = 6^*$

Pivot column   *6 is smaller quotient.

The new pivot entry is $\frac{3}{2}$ (circled). Now reduce, using the new pivot row, to obtain:

$$\begin{bmatrix} 0 & 1 & \frac{2}{3} & -\frac{1}{3} & 0 & 2 \\ 1 & 0 & -\frac{1}{3} & \frac{2}{3} & 0 & 6 \\ 0 & 0 & \frac{4}{3} & \frac{1}{3} & 1 & 18 \end{bmatrix}$$

Indicators all 0 or positive, so the solution is complete.
Nonbasic variables: s_1 and s_2. Basic variables: x and y.

6. After setting the nonbasic variables equal to 0, read the values of the basic variables and the objective function from the rows of the matrix.

6. f is maximized at 18 when $y = 2$, $x = 6$, $s_1 = 0$, and $s_2 = 0$. (This same example was solved graphically in the Procedure/Example in Section 4.2.)

As we look back over our work, we can observe the following:

1. Steps 1 and 2 complete Task A (setting up the simplex matrix).
2. Steps 3–5 complete Task B (pivoting to obtain the final matrix).
3. Step 6 completes Task C (reading the optimal solution).

EXAMPLE 2   Simplex Method Tasks

Complete Tasks A, B, and C to find the maximum value of $f = 4x + 3y$ subject to

$$x + 2y \le 8$$
$$2x + y \le 10$$

Solution

Task A: Introducing the slack variables gives the system of equations

$$\begin{cases} x + 2y + s_1 & = 8 \\ 2x + y & + s_2 & = 10 \\ -4x - 3y & + f = 0 \end{cases}$$

Writing this in a matrix with the objective function in the last row gives

$$A = \begin{bmatrix} x & y & s_1 & s_2 & f & \\ 1 & 2 & 1 & 0 & 0 & 8 \\ 2 & 1 & 0 & 1 & 0 & 10 \\ -4 & -3 & 0 & 0 & 1 & 0 \end{bmatrix}$$

$8/1 = 8$
$10/2 = 5^*$

Most negative   *Smallest quotient

Note that in this initial simplex matrix A, variables x and y are nonbasic (so set to 0) and this corresponds to the basic feasible solution $x = 0$, $y = 0$ and $s_1 = 8$, $s_2 = 10$, $f = 0$.

Task B: The most negative entry in the last row is -4 (indicated by the arrow), so the pivot column is column 1. The smallest quotient formed by dividing the positive coefficients from column 1 into the constants in the augment is 5 (indicated by the asterisk), so row 2 is the pivot row. Thus the pivot entry is in row 2, column 1 (circled in matrix A).

Next use row operations with row 2 only to make the pivot entry equal 1 and all other entries in column 1 equal zero. The row operations are

(a) Multiply row 2 by $\frac{1}{2}$ to get a 1 as the new pivot entry.
(b) Add -1 times the new row 2 to row 1 to get a 0 in row 1 above the pivot entry.
(c) Add 4 times the new row 2 to row 3 to get a 0 in row 3 below the pivot entry.

This gives the simplex matrix B, which has nonbasic variables $y = 0$ and $s_2 = 0$ and corresponds to the basic feasible solution $y = 0$, $s_2 = 0$ and $x = 5$, $s_1 = 3$, $f = 20$.

$$B = \begin{array}{c} \\ \\ \\ \end{array} \begin{array}{ccccc} x & y & s_1 & s_2 & f \\ \end{array}$$

$$B = \left[\begin{array}{ccccc|c} 0 & \tfrac{3}{2} & 1 & -\tfrac{1}{2} & 0 & 3 \\ 1 & \tfrac{1}{2} & 0 & \tfrac{1}{2} & 0 & 5 \\ 0 & -1 & 0 & 2 & 1 & 20 \end{array}\right] \begin{array}{l} 3 \div \tfrac{3}{2} = 2^* \\ \\ 5 \div \tfrac{1}{2} = 10 \end{array}$$

↑
Most negative   *Smallest quotient

The last row contains a negative entry, so the solution is not optimal and we locate another pivot column to continue. The pivot column is indicated with an arrow, and the smallest quotient is indicated with an asterisk. The new pivot entry is circled.

The row operations using this pivot row are

(a) Multiply row 1 by $\frac{2}{3}$ to get a 1 as the new pivot element.
(b) Add $-\frac{1}{2}$ times the new row 1 to row 2 to get a 0 in row 2 below the pivot element.
(c) Add the new row 1 to row 3 to get a 0 in row 3 below the pivot element.

This gives

$$C = \begin{array}{ccccc} x & y & s_1 & s_2 & f \\ \end{array}$$

$$C = \left[\begin{array}{ccccc|c} 0 & 1 & \tfrac{2}{3} & -\tfrac{1}{3} & 0 & 2 \\ 1 & 0 & -\tfrac{1}{3} & \tfrac{2}{3} & 0 & 4 \\ 0 & 0 & \tfrac{2}{3} & \tfrac{5}{3} & 1 & 22 \end{array}\right]$$

Because there are no negative indicators in the last row, the solution is complete.

Task C: s_1 and s_2 are nonbasic variables, so they equal zero. Thus f is maximized at 22 when $y = 2$, $x = 4$, $s_1 = 0$, and $s_2 = 0$. ∎

EXAMPLE 3   Manufacturing | APPLICATION PREVIEW |

The Solar Technology Company manufactures three different types of hand calculators and classifies them as scientific, business, and graphing according to their calculating capabilities. The three types have production requirements given by the following table.

	Scientific	Business	Graphing
Electronic circuit components	5	7	10
Assembly time (hours)	1	3	4
Cases	1	1	1

The firm has a monthly limit of 90,000 circuit components, 30,000 hours of labor, and 9000 cases. If the profit is $6 for each scientific, $13 for each business, and $20 for each graphing calculator, how many of each should be produced to yield the maximum profit? What is the maximum profit?

Solution

Let x_1 be the number of scientific calculators produced, x_2 the number of business calculators produced, and x_3 the number of graphing calculators produced. Then the problem is to maximize the profit $f = 6x_1 + 13x_2 + 20x_3$ subject to

Inequalities	Equations with Slack Variables
$5x_1 + 7x_2 + 10x_3 \leq 90{,}000$	$5x_1 + 7x_2 + 10x_3 + s_1 \qquad\quad = 90{,}000$
$x_1 + 3x_2 + 4x_3 \leq 30{,}000$	$x_1 + 3x_2 + 4x_3 + \quad s_2 \quad\;\; = 30{,}000$
$x_1 + x_2 + x_3 \leq 9000$	$x_1 + x_2 + x_3 + \qquad\quad s_3 = 9000$

The slack variables can be interpreted as follows: s_1 is the number of unused circuit components, s_2 is the number of unused hours of labor, and s_3 is the number of unused calculator cases. The simplex matrix, with the first pivot entry circled, is shown below.

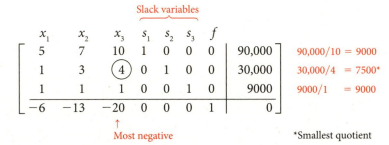

We now use row operations to change the pivot entry to 1 and create zeros elsewhere in the pivot column. The row operations are

1. Multiply row 2 by $\frac{1}{4}$ to convert the pivot entry to 1.
2. Using the new row 2,
   (a) Add -10 times new row 2 to row 1.
   (b) Add -1 times new row 2 to row 3.
   (c) Add 20 times new row 2 to row 4.

The result is the second simplex matrix.

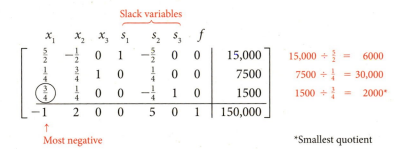

We check the last row indicators. Because there is an entry that is negative, we repeat the simplex process. Again the pivot entry is circled, and we apply row operations similar to those above. What are the row operations this time? The resulting matrix is

$$
\begin{array}{ccccccc}
x_1 & x_2 & x_3 & s_1 & s_2 & s_3 & f \\
\end{array}
$$

$$
\left[\begin{array}{ccccccc|c}
0 & -\frac{4}{3} & 0 & 1 & -\frac{5}{3} & -\frac{10}{3} & 0 & 10{,}000 \\
0 & \frac{2}{3} & 1 & 0 & \frac{1}{3} & -\frac{1}{3} & 0 & 7000 \\
1 & \frac{1}{3} & 0 & 0 & -\frac{1}{3} & \frac{4}{3} & 0 & 2000 \\
0 & \frac{7}{3} & 0 & 0 & \frac{14}{3} & \frac{4}{3} & 1 & 152{,}000
\end{array}\right]
$$

Checking the last row, we see that all the entries in this row are 0 or positive, so the solution is complete. From the final simplex matrix we see that the nonbasic variables are x_2, s_2, and s_3, and with these equal to zero we determine all the solution values:

$$x_2 = 0, s_2 = 0, s_3 = 0 \quad \text{and} \quad x_1 = 2000, x_3 = 7000, s_1 = 10,000, f = 152,000$$

Thus the numbers of calculators that should be produced are

$$x_1 = 2000 \text{ scientific calculators}$$
$$x_2 = 0 \text{ business calculators}$$
$$x_3 = 7000 \text{ graphing calculators}$$

in order to obtain a maximum profit of \$152,000 for the month. Note also that the slack variables tell us that with this optimal solution there were $s_1 = 10,000$ unused circuit components, $s_2 = 0$ unused labor hours, and $s_3 = 0$ unused calculator cases. ■

CHECKPOINT

1. Write the following constraints as equations by using slack variables.

$$3x_1 + x_2 + x_3 \le 9$$
$$2x_1 + x_2 + 3x_3 \le 8$$
$$2x_1 + x_2 \qquad \le 5$$

2. Set up the simplex matrix to maximize $f = 2x_1 + 5x_2 + x_3$ subject to the constraints discussed in Problem 1.

Using the simplex matrix from Problem 2, perform the following.

3. Find the first pivot entry.
4. Use row operations to make the variable of the first pivot entry basic.
5. Find the second pivot entry and make a second variable basic.
6. Find the solution.

Spreadsheet Note

As our discussion indicates, the simplex method is quite complicated, even when the number of variables is relatively small (2 or 3). For this reason, computer software packages are often used to solve linear programming problems. The steps for solving linear programming problems with Excel are shown in Appendix D, Section 4.3, and the Online Excel Guide.

The following Excel screen shows entries for beginning to solve the linear programming problem that we solved "by hand" with the simplex method in Example 3.

	A	B	C
1	Variables		
2			
3	# scientific calculators (x)	0	
4	# business calculators (y)	0	
5	# graphing calculators (z)	0	
6			
7	Objective		
8			
9	Maximize profit	= 6*B3 + 13*B4 + 20*B5	
10			
11	Constraints		
12		Amount used	Maximum
13	Circuit components	= 5*B3 + 7*B4 + 10*B5	90000
14	Labor hours	= B3 + 3*B4 + 4*B5	30000
15	Cases	= B3 + B4 + B5	9000

The screen below gives the solution. Step-by-step instructions are found in Appendix D, Section 4.3, and the Online Excel Guide.

	A	B	C
1	Variables		
2			
3	# scientific calculators (x)	2000	
4	# business calculators (y)	0	
5	# graphing calculators (z)	7000	
6			
7	Objective		
8			
9	Maximize profit	152000	
10			
11	Constraints		
12		Amount used	Maximum
13	Circuit components	80000	90000
14	Labor hours	30000	30000
15	Cases	9000	9000

This screen indicates that the profit is maximized at $152,000 when 2000 scientific and 7000 graphing calculators are produced. Note that under "Constraints," the difference between the columns labeled "Maximum" and "Amount used" gives the amount of each resource that was unused—that is, the value of each slack variable: $s_1 = 10{,}000$, $s_2 = 0$, and $s_3 = 0$. Note that this is the same solution found in Example 3. ◼

Nonunique Solutions: Multiple Solutions and No Solution

As we've seen with graphical methods, a linear programming problem can have multiple solutions when the optimal value for the objective function occurs at two adjacent corners of the feasible region. Consider the following problem.

Maximize $f = 2x + y$ subject to

$$x + \frac{1}{2}y \le 16$$
$$x + y \le 24$$
$$x \ge 0, \, y \ge 0$$

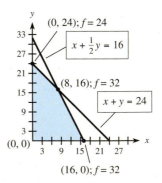

Figure 4.18

From Figure 4.18, we can see that the maximum value of f is $f = 32$ at the two corners $(8, 16)$ and $(16, 0)$. This means that $f = 32$ at any point on the segment joining those corners. Thus this problem has infinitely many solutions. Let's examine how we can discover these multiple solutions with the simplex method.

The simplex matrix for this problem (with the slack variables introduced) is

$$\begin{array}{c} \begin{array}{ccccc} x & y & s_1 & s_2 & f \end{array} \\ \left[\begin{array}{ccccc|c} \boxed{1} & \frac{1}{2} & 1 & 0 & 0 & 16 \\ 1 & 1 & 0 & 1 & 0 & 24 \\ \hline -2 & -1 & 0 & 0 & 1 & 0 \end{array}\right] \end{array}$$

The most negative value in the last row is -2, so the x-column is the pivot column. The smallest quotient occurs in row 1; the pivot entry (circled) is 1. Making x a basic variable gives the following transformed simplex matrix.

$$\begin{array}{c} \begin{array}{ccccc} x & y & s_1 & s_2 & f \end{array} \\ \left[\begin{array}{ccccc|c} 1 & \frac{1}{2} & 1 & 0 & 0 & 16 \\ 0 & \frac{1}{2} & -1 & 1 & 0 & 8 \\ \hline 0 & 0 & 2 & 0 & 1 & 32 \end{array}\right] \end{array}$$

This simplex matrix has no negative indicators, so the optimal value of f has been found ($f = 32$ when $x = 16$, $y = 0$). Note that one of the nonbasic variables (y) has a zero indicator. When this occurs, we can use the column with the 0 indicator as our pivot column and often obtain a new solution that has the same value for the objective function. In this final tableau, if we used the y-column as the pivot column, then we could find the second solution (at $x = 8$, $y = 16$) that gives the same optimal value of f. You will be asked to do this in Problem 39 of the exercises for this section.

Multiple Solutions

When the simplex matrix for the optimal value of f has a nonbasic variable with a zero indicator in its column, there may be multiple solutions giving the *same* optimal value for f. We can discover whether another solution exists by using the column of that nonbasic variable as the pivot column.

Spreadsheet Note Note that with Excel, multiple solutions may be found by using different initial values for the variables when the problem is entered into Excel. ■

No Solution It is also possible for a linear programming problem to have an unbounded solution (and thus no maximum value for f).

EXAMPLE **4** ## Simplex Method When No Maximum Exists

Maximize $f = 2x_1 + x_2 + 3x_3$ subject to

$$x_1 - 3x_2 + x_3 \leq 3$$
$$x_1 - 6x_2 + 2x_3 \leq 6$$

if a maximum exists.

Solution

The simplex matrix for this problem (after introduction of slack variables) is

$$\begin{array}{cccccc} x_1 & x_2 & x_3 & s_1 & s_2 & f \\ \left[\begin{array}{ccc|ccc|c} 1 & -3 & \boxed{1} & 1 & 0 & 0 & 3 \\ 1 & -6 & 2 & 0 & 1 & 0 & 6 \\ \hline -2 & -1 & -3 & 0 & 0 & 1 & 0 \end{array}\right] \end{array}$$

We see that the x_3 column has the most negative number, so we use that column as the pivot column.

We see that the smallest quotient is 3. Both coefficients give this quotient, so we can use either coefficient as the pivot entry. Using the element in row 1, we can make x_3 basic. The new simplex matrix is

$$\begin{array}{cccccc} x_1 & x_2 & x_3 & s_1 & s_2 & f \\ \left[\begin{array}{ccc|ccc|c} 1 & -3 & 1 & 1 & 0 & 0 & 3 \\ -1 & 0 & 0 & -2 & 1 & 0 & 0 \\ \hline 1 & -10 & 0 & 3 & 0 & 1 & 9 \end{array}\right] \end{array}$$

The most negative value in the last row of this matrix is -10, so the pivot column is the x_2-column. But there are no positive coefficients in the x_2-column; what does this mean? Even if we could pivot using the x_2-column, the variables x_1 and s_1 would remain equal to 0. And the relationships among the remaining variables x_2, x_3, and s_2 would be the same.

$$-3x_2 + x_3 = 3 \quad \text{or} \quad x_3 = 3 + 3x_2$$
$$s_2 = 0$$

As x_2 is increased, x_3 also increases because all variables are nonnegative. This means that no matter how much we increase x_2, all other variables remain nonnegative. Furthermore, as we increase x_2 (and hence x_3), the value of f also increases. That is, there is no maximum value for f. ∎

This example indicates that when a linear programming problem has an unbounded solution (and thus no maximum value for f), this is identified by the following condition in the simplex method.

No Solution

If, after the pivot column has been found, there are no positive coefficients in that column, no maximum solution exists.

Some difficulties can arise in using the simplex method to solve linear programming problems. These "difficulties" are discussed in more advanced courses, and we shall not encounter them in this text.

Shadow Prices   Finally, we note that one of the most powerful aspects of the simplex method is that it not only solves the given problem but also tells how to adapt the given solution if the situation changes. In particular, from the final simplex matrix (final tableau), the solution to the problem can be read, and there is information about the amount of change in the objective function if 1 additional unit of a resource becomes available (this change is called the **shadow price** of the resource) and how this additional resource changes the optimal solution values.

For example, if one more acre of land became available in the Example 1 application (for a total of 6001 acres), it can be shown that if the co-op planted 3749.5 acres of corn and 2251.5 acres of soybeans, the profit would increase by \$30. That is, the shadow price of land is \$30/acre (or the marginal profit value of the land is \$30/acre). The following simplex matrix C is the final one from Example 1 (where x = acres of corn, y = acres of soybeans, and s_1 = the slack variable for the land constraint). The bottom entry in the s_1-column of C is the shadow price for land. Furthermore, if the s_1-column is added to the last column in C, then matrix D results. The solution obtained from D gives the new planting scheme and the new maximum profit if one more acre of land is planted (x = 3749.5 acres of corn and y = 2251.5 acres of soybeans for a new maximum profit of P = \$315,030).

$$C = \begin{bmatrix} x & y & s_1 & s_2 & s_3 & P & \\ 0 & 1 & \frac{3}{2} & -\frac{1}{6} & 0 & 0 & 2{,}250 \\ 1 & 0 & -\frac{1}{2} & \frac{1}{6} & 0 & 0 & 3{,}750 \\ 0 & 0 & -\frac{9}{8} & \frac{1}{24} & 1 & 0 & 187.5 \\ 0 & 0 & 30 & \frac{10}{3} & 0 & 1 & 315{,}000 \end{bmatrix}$$

$$D = \begin{bmatrix} x & y & s_1 & s_2 & s_3 & P & \\ 0 & 1 & \frac{3}{2} & -\frac{1}{6} & 0 & 0 & 2{,}251.5 \\ 1 & 0 & -\frac{1}{2} & \frac{1}{6} & 0 & 0 & 3{,}749.5 \\ 0 & 0 & -\frac{9}{8} & \frac{1}{24} & 1 & 0 & 186\frac{3}{8} \\ 0 & 0 & 30 & \frac{10}{3} & 0 & 1 & 315{,}030 \end{bmatrix}$$

Additional material on shadow prices appears in the second Extended Application/Group Project at the end of this chapter.

CHECKPOINT SOLUTIONS

1. $3x_1 + x_2 + x_3 + s_1 \qquad\qquad = 9$
   $2x_1 + x_2 + 3x_3 \qquad + s_2 \qquad = 8$
   $2x_1 + x_2 \qquad\qquad\qquad + s_3 = 5$

2. $$\begin{bmatrix} 3 & 1 & 1 & 1 & 0 & 0 & 0 & 9 \\ 2 & 1 & 3 & 0 & 1 & 0 & 0 & 8 \\ 2 & 1 & 0 & 0 & 0 & 1 & 0 & 5 \\ -2 & -5 & -1 & 0 & 0 & 0 & 1 & 0 \end{bmatrix}$$

3. The first pivot entry is the 1 in row 3, column 2.

4. $$\begin{bmatrix} 1 & 0 & 1 & 1 & 0 & -1 & 0 & 4 \\ 0 & 0 & 3 & 0 & 1 & -1 & 0 & 3 \\ 2 & 1 & 0 & 0 & 0 & 1 & 0 & 5 \\ 8 & 0 & -1 & 0 & 0 & 5 & 1 & 25 \end{bmatrix}$$

5. The second pivot entry is the 3 in row 2, column 3.

$$\begin{bmatrix} 1 & 0 & 0 & 1 & -\frac{1}{3} & -\frac{2}{3} & 0 & 3 \\ 0 & 0 & 1 & 0 & \frac{1}{3} & -\frac{1}{3} & 0 & 1 \\ 2 & 1 & 0 & 0 & 0 & 1 & 0 & 5 \\ 8 & 0 & 0 & 0 & \frac{1}{3} & \frac{14}{3} & 1 & 26 \end{bmatrix}$$

6. All indicators are 0 or positive, so f is maximized at 26 when $x_1 = 0$, $x_2 = 5$, and $x_3 = 1$. Note also that $s_1 = 3$, $s_2 = 0$, and $s_3 = 0$.

| EXERCISES | 4.3

Access end-of-section exercises online at **www.webassign.net**

- To formulate the dual for minimization problems
- To solve minimization problems using the simplex method on the dual

The Simplex Method: Duality and Minimization

| APPLICATION PREVIEW |

A beef producer is considering two different types of feed. Each feed contains some or all of the necessary ingredients for fattening beef. Brand 1 feed costs 20 cents per pound and brand 2 costs 30 cents per pound. Table 4.1 contains all the relevant data about nutrition and cost of each brand and the minimum requirements per unit of beef. The producer would like to determine how much of each brand to buy in order to satisfy the nutritional requirements for Ingredients A and B at minimum cost. (See Example 3.)

| TABLE 4.1 |

	Brand 1	Brand 2	Minimum Requirement
Ingredient A	3 units/lb	5 units/lb	40 units
Ingredient B	4 units/lb	3 units/lb	46 units
Cost per pound	20¢	30¢	

In this section, we will use the simplex method to solve minimization problems such as this one.

Dual Problems We have solved both maximization and minimization problems using graphical methods, but the simplex method, as discussed in the previous section, applies only to maximization problems. Now let us turn our attention to extending the use of the simplex method to solve minimization problems. Such problems might arise when a company seeks to minimize its production costs yet fill customers' orders or purchase items necessary for production.

As with maximization problems, we shall consider only **standard minimization problems,** in which the constraints satisfy the following conditions.

1. All variables are nonnegative.
2. The constraints are of the form

$$a_1y_1 + a_2y_2 + \cdots + a_ny_n \geq b \quad \text{where } b \text{ is positive.}$$

Because the simplex method specifically seeks to increase the objective function, it does not apply to minimization problems. However, associated with each minimization problem there is a related maximization problem called the **dual problem** or **dual.**

EXAMPLE 1   A Minimization Problem and Its Dual

Given the problem

$$\text{Minimize } g = 11y_1 + 7y_2$$
$$\text{subject to} \quad y_1 + 2y_2 \geq 10$$
$$3y_1 + y_2 \geq 15$$

find the dual maximization problem.

Solution

First, we write a matrix A that has a form similar to the simplex matrix, but without slack variables and with positive coefficients for the variables in the last row (and with the function name in the augment).

The matrix for the dual problem is formed by interchanging the rows and columns of matrix A. Matrix B is the result of this procedure, and recall that B is the **transpose** of A.

Minimization   Dual

$$A = \begin{bmatrix} 1 & 2 & | & 10 \\ 3 & 1 & | & 15 \\ \hline 11 & 7 & | & g \end{bmatrix} \qquad B = \begin{bmatrix} 1 & 3 & | & 11 \\ 2 & 1 & | & 7 \\ \hline 10 & 15 & | & g \end{bmatrix}$$

Note that row 1 of A becomes column 1 of B, row 2 of A becomes column 2 of B, and row 3 of A becomes column 3 of B.

In the same way that we formed matrix A, we can now "convert back" from matrix B to the maximization problem that is the dual of the original minimization problem. We state this dual problem, using different letters to emphasize that this is a different problem from the original.

$$\text{Maximize } f = 10x_1 + 15x_2$$
$$\text{subject to} \quad x_1 + 3x_2 \le 11$$
$$2x_1 + x_2 \le 7$$

Next let's see how the solutions of these problems are related. The following box shows the solution of the Example 1 minimization problem and its dual maximization problem.

Comparison of a Minimization Problem and Its Dual

Maximization Problem—Simplex Method

Minimization Problem—Graphical Method

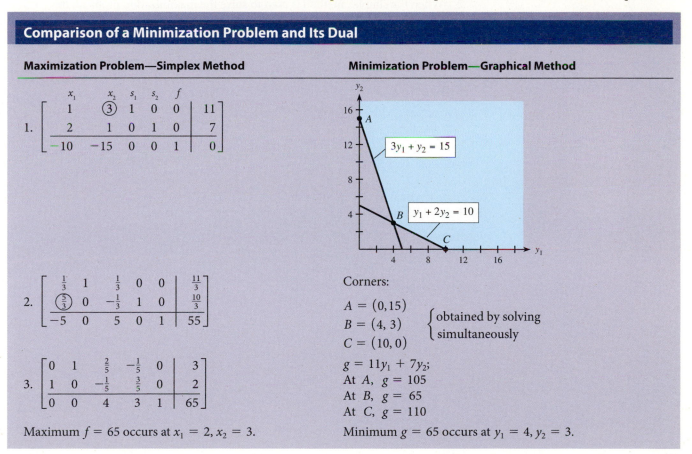

1.
$$\begin{bmatrix} x_1 & x_2 & s_1 & s_2 & f & \\ 1 & ③ & 1 & 0 & 0 & | & 11 \\ 2 & 1 & 0 & 1 & 0 & | & 7 \\ -10 & -15 & 0 & 0 & 1 & | & 0 \end{bmatrix}$$

2.
$$\begin{bmatrix} \frac{1}{3} & 1 & \frac{1}{3} & 0 & 0 & | & \frac{11}{3} \\ ⑤/3 & 0 & -\frac{1}{3} & 1 & 0 & | & \frac{10}{3} \\ -5 & 0 & 5 & 0 & 1 & | & 55 \end{bmatrix}$$

3.
$$\begin{bmatrix} 0 & 1 & \frac{2}{5} & -\frac{1}{5} & 0 & | & 3 \\ 1 & 0 & -\frac{1}{5} & \frac{3}{5} & 0 & | & 2 \\ 0 & 0 & 4 & 3 & 1 & | & 65 \end{bmatrix}$$

Maximum $f = 65$ occurs at $x_1 = 2, x_2 = 3$.

Corners:

$A = (0, 15)$
$B = (4, 3)$ } obtained by solving simultaneously
$C = (10, 0)$

$g = 11y_1 + 7y_2;$
At A, $g = 105$
At B, $g = 65$
At C, $g = 110$

Minimum $g = 65$ occurs at $y_1 = 4, y_2 = 3$.

In comparing the solutions of the minimization problem and its dual, the first thing to note is that the maximum value of f in the maximization problem and the minimum value of g in the minimization problem are the same. Furthermore, looking at the final simplex matrix, we can also find the values of y_1 and y_2 that give the minimum value of g by looking at the last (bottom) entries in the slack variable columns of the matrix.

$$
\begin{array}{ccccccc}
x_1 & y_2 & s_1 & s_2 & f & \\
\left[\begin{array}{ccccc|c}
0 & 1 & \frac{2}{5} & -\frac{1}{5} & 0 & 3 \\
1 & 0 & -\frac{1}{5} & \frac{3}{5} & 0 & 2 \\
0 & 0 & 4 & 3 & 1 & 65
\end{array}\right]
\end{array}
$$

These values give the values of y_1 and y_2 in the minimization problem.

Duality and Solving   From Example 1 we see that the given minimization problem and its dual maximization problem are very closely related. Furthermore, when problems exhibit this relationship, the simplex method can be used to solve them both. This fact was proved by John von Neumann, and it is summarized by the Principle of Duality.

Principle of Duality

1. When a standard minimization problem and its dual have a solution, the maximum value of the function to be maximized is the same value as the minimum value of the function to be minimized.
2. When the simplex method is used to solve the maximization problem, the values of the variables that solve the corresponding minimization problem are *the last entries in the columns corresponding to the slack variables.*

EXAMPLE 2   **Minimization with the Simplex Method**

Given the problem

$$
\begin{array}{rl}
\text{Minimize } & 18y_1 + 12y_2 = g \\
\text{subject to } & 2y_1 + y_2 \geq 8 \\
& 6y_1 + 6y_2 \geq 36
\end{array}
$$

(a) State the dual of this minimization problem.
(b) Solve this minimization problem by first solving the dual maximization problem.

Solution
(a) The dual of this minimization problem will be a maximization problem. To form the dual, first write the matrix A (without slack variables and with positive coefficients for the variables in the last row) for the given problem.

$$
A = \left[\begin{array}{cc|c}
2 & 1 & 8 \\
6 & 6 & 36 \\
18 & 12 & g
\end{array}\right]
$$

Next, transpose matrix A to yield matrix B.

$$
B = \left[\begin{array}{cc|c}
2 & 6 & 18 \\
1 & 6 & 12 \\
8 & 36 & g
\end{array}\right]
$$

Then write the dual maximization problem, renaming the variables and the function.

Dual problem
$$
\begin{array}{rl}
\text{Maximize } & f = 8x_1 + 36x_2 \\
\text{subject to } & 2x_1 + 6x_2 \leq 18 \\
& x_1 + 6x_2 \leq 12
\end{array}
$$

(b) To solve the given minimization problem, we use the simplex method on its dual maximization problem. The complete simplex matrix for each step is given, and each pivot entry is circled.

1. $\begin{bmatrix} 2 & 6 & 1 & 0 & 0 & | & 18 \\ 1 & ⑥ & 0 & 1 & 0 & | & 12 \\ -8 & -36 & 0 & 0 & 1 & | & 0 \end{bmatrix}$ 2. $\begin{bmatrix} ① & 0 & 1 & -1 & 0 & | & 6 \\ \frac{1}{6} & 1 & 0 & \frac{1}{6} & 0 & | & 2 \\ -2 & 0 & 0 & 6 & 1 & | & 72 \end{bmatrix}$

3. $\begin{array}{ccccc} x_1 & x_2 & s_1 & s_2 & f \end{array}$
$\begin{bmatrix} 1 & 0 & 1 & -1 & 0 & | & 6 \\ 0 & 1 & -\frac{1}{6} & \frac{2}{6} & 0 & | & 1 \\ 0 & 0 & 2 & 4 & 1 & | & 84 \end{bmatrix}$

The solution of the maximization problem is $f = 84$ (at $x_1 = 6$, $x_2 = 1$). By the Principle of Duality, $g = 84$ is the minimum value of the objective function $g = 18y_1 + 12y_2$, and the values that give this minimum are the last entries in the slack variable columns, $y_1 = 2$ and $y_2 = 4$.   ■

CHECKPOINT

Perform the following steps to begin the process of finding the minimum value of $g = y_1 + 4y_2$, subject to the following constraints.

$$2y_1 + 4y_2 \geq 18$$
$$y_1 + 5y_2 \geq 15$$

1. Form the matrix associated with the minimization problem.
2. Find the transpose of this matrix.
3. State the dual problem by using the matrix from Problem 2.
4. Write the simplex matrix for the dual of this problem.
5. The dual of this minimization problem has the following solution matrix. What is the solution of the minimization problem?

$$\begin{bmatrix} 1 & 0 & \frac{5}{6} & -\frac{1}{6} & 0 & | & \frac{1}{6} \\ 0 & 1 & -\frac{2}{3} & \frac{1}{3} & 0 & | & \frac{2}{3} \\ 0 & 0 & 5 & 2 & 1 & | & 13 \end{bmatrix}$$

EXAMPLE 3   Purchasing | APPLICATION PREVIEW |

A beef producer is considering two different types of feed. Each feed contains some or all of the necessary ingredients for fattening beef. Brand 1 feed costs 20 cents per pound and brand 2 costs 30 cents per pound. How much of each brand should the producer buy in order to satisfy the nutritional requirements for ingredients A and B at minimum cost? Table 4.2 contains all the relevant data about nutrition and cost of each brand and the minimum requirements per unit of beef.

| TABLE 4.2 |

	Brand 1	Brand 2	Minimum Requirement
Ingredient A	3 units/lb	5 units/lb	40 units
Ingredient B	4 units/lb	3 units/lb	46 units
Cost per pound	20¢	30¢	

Solution

Let y_1 be the number of pounds of brand 1, and let y_2 be the number of pounds of brand 2. Then we can formulate the problem as follows:

$$\text{Minimize cost } C = 20y_1 + 30y_2$$
$$\text{subject to} \quad 3y_1 + 5y_2 \geq 40$$
$$4y_1 + 3y_2 \geq 46$$

The original linear programming problem is usually called the **primal problem.** If we write this in an augmented matrix, without the slack variables, then we can use the transpose to find the dual problem.

Primal   Dual (with function renamed)

$$\begin{bmatrix} 3 & 5 & | & 40 \\ 4 & 3 & | & 46 \\ \hline 20 & 30 & | & C \end{bmatrix} \qquad \begin{bmatrix} 3 & 4 & | & 20 \\ 5 & 3 & | & 30 \\ \hline 40 & 46 & | & f \end{bmatrix}$$

Thus the dual maximization problem is

$$\text{Maximize } f = 40x_1 + 46x_2$$
$$\text{subject to} \quad 3x_1 + 4x_2 \leq 20$$
$$5x_1 + 3x_2 \leq 30$$

The simplex matrix for this maximization problem, with slack variables included, is

$$1. \quad \begin{bmatrix} 3 & ④ & 1 & 0 & 0 & | & 20 \\ 5 & 3 & 0 & 1 & 0 & | & 30 \\ \hline -40 & -46 & 0 & 0 & 1 & | & 0 \end{bmatrix}$$

Solving the problem gives

$$2. \quad \begin{bmatrix} \frac{3}{4} & 1 & \frac{1}{4} & 0 & 0 & | & 5 \\ ⑪/④ & 0 & -\frac{3}{4} & 1 & 0 & | & 15 \\ \hline -\frac{11}{2} & 0 & \frac{23}{2} & 0 & 1 & | & 230 \end{bmatrix} \qquad 3. \quad \begin{bmatrix} 0 & 1 & \frac{5}{11} & -\frac{3}{11} & 0 & | & \frac{10}{11} \\ 1 & 0 & -\frac{3}{11} & \frac{4}{11} & 0 & | & \frac{60}{11} \\ \hline 0 & 0 & 10 & 2 & 1 & | & 260 \end{bmatrix}$$

The solution to this problem is

$$y_1 = 10 \text{ lb of brand 1}$$
$$y_2 = 2 \text{ lb of brand 2}$$
$$\text{Minimum cost} = 260 \text{ cents or \$2.60 per unit of beef} \quad ■$$

In a more extensive study of linear programming, the duality relationship that we have used in this section as a means of solving minimization problems can be shown to have more properties than we have mentioned.

Spreadsheet Note When we use Excel, minimization problems are solved in a manner similar to maximization problems. Because of the constraints, the Solver for this problem has different directions on the inequality symbols; Solver is also set to find the minimum (min) rather than the maximum (max) value of the objective function. (For details on using the Excel Solver, see Appendix D, Section 4.3, and the Online Excel Guide.)

The solution to the minimization problem in Example 3, found with Solver and shown below, is that cost is minimized at 260 cents when 10 lb of brand 1 and 2 lb of brand 2 are used. This is the same solution we found "by hand" with the simplex method.   ■

	A	B	C
1	Variables		
2			
3	Pounds Brand 1	10	
4	Pounds Brand 2	2	
5			
6	Objective		
7			
8	Minimize cost	260	
9			
10	Constraints		
11		Amount in feeds	Required
12	Ingredient A	40	40
13	Ingredient B	46	46

CHECKPOINT SOLUTIONS

1. $\begin{bmatrix} 2 & 4 & | & 18 \\ 1 & 5 & | & 15 \\ 1 & 4 & | & g \end{bmatrix}$ 2. $\begin{bmatrix} 2 & 1 & | & 1 \\ 4 & 5 & | & 4 \\ 18 & 15 & | & g \end{bmatrix}$

3. Find the maximum value of $f = 18x_1 + 15x_2$ subject to the constraints

$$2x_1 + x_2 \le 1$$
$$4x_1 + 5x_2 \le 4$$

4. The simplex matrix for this dual problem is

$$\begin{bmatrix} 2 & 1 & 1 & 0 & 0 & | & 1 \\ 4 & 5 & 0 & 1 & 0 & | & 4 \\ -18 & -15 & 0 & 0 & 1 & | & 0 \end{bmatrix}$$

5. The minimum value is $g = 13$ when $y_1 = 5$ and $y_2 = 2$.

| EXERCISES | 4.4

Access end-of-section exercises online at **www.webassign.net**   **WebAssign**

OBJECTIVES   4.5

- To solve maximization problems with mixed constraints
- To solve minimization problems with mixed constraints

The Simplex Method with Mixed Constraints

| APPLICATION PREVIEW |

The Laser Company manufactures two models of home theater systems, the Star and the Allstar, at two plants, located in Ashville and in Cleveland. The maximum daily output at the Ashville plant is 900, and the profits there are $200 per unit of the Allstar and $100 per unit of the Star. The maximum daily output at the Cleveland plant is 800, and the profits there are $210 per unit of the Allstar and $80 per unit of the Star. In addition, restrictions at the Ashville plant mean that the number of units of the Star model cannot exceed 100 more than the number of units of the Allstar model produced. If the company gets a rush order for 800 units of the Allstar model and 600 units of the Star model, finding the number of units of each model that should be produced at each location to fill the order and obtain the maximum profit is a mixed constraint linear programming problem (see Example 3).

Mixed Constraints and Maximization

In the previous two sections, we used the simplex method to solve two different types of standard linear programming problems. These types are summarized as follows.

	Maximization Problems	Minimization Problems
Function	Maximized	Minimized
Variables	Nonnegative	Nonnegative
Constraints	$a_1x_1 + a_2x_2 + \cdots$ $+ a_nx_n \le b\ (b \ge 0)$	$c_1y_1 + c_2y_2 + \cdots + c_ny_n \ge d\ (d \ge 0)$
Simplex method	Applied directly	Applied to dual

In this section we will show how to apply the simplex method to mixed constraint linear programming problems that do not exactly fit either of these types. The term *mixed constraint* is used because some inequalities describing the constraints contain "≤" and some contain "≥" signs.

We begin by considering maximization problems with constraints that have a form different from those noted in the preceding table. This can happen in one of the following ways.

1. If a constraint has the form $a_1 x_1 + a_2 x_2 + \cdots + a_n x_n \leq b$, where $b < 0$, such as with
    $$x - 2y \leq -8$$
2. If a constraint has the form $a_1 x_1 + a_2 x_2 + \cdots + a_n x_n \geq b$, where $b \geq 0$, such as with
    $$2x - 3y \geq 6$$

Note that in the latter case, multiplying both sides of the inequality by (-1) changes it as follows:

$$a_1 x_1 + a_2 x_2 + \cdots + a_n x_n \geq b \text{ becomes } -a_1 x_1 - a_2 x_2 - \cdots - a_n x_n \leq -b$$

That is,

$$2x - 3y \geq 6 \text{ becomes } -2x + 3y \leq -6$$

Thus we see that the two possibilities for mixed constraints are actually one: constraints that have the form $a_1 x_1 + a_2 x_2 + \cdots + a_n x_n \leq b$, where b is any constant. We call these **less than or equal to constraints** and denote them as **≤ constraints.**

EXAMPLE 1   Maximization with Mixed Constraints

Maximize $f = x + 2y$ subject to

$$\begin{aligned} x + y &\leq 13 \\ 2x - 3y &\geq 6 \\ x \geq 0, y &\geq 0 \end{aligned}$$

Solution

We begin by expressing all constraints as ≤ constraints. In this case, we multiply $2x - 3y \geq 6$ by (-1) to obtain $-2x + 3y \leq -6$. We can now use slack variables to write the inequalities as equations and form the simplex matrix.

$$
\begin{aligned}
x + y + s_1 &= 13 \\
-2x + 3y + s_2 &= -6 \quad \text{gives} \\
-x - 2y + f &= 0
\end{aligned}
\qquad
\begin{array}{ccccc|c}
x_1 & y_2 & s_1 & s_2 & f & \\
1 & 1 & 1 & 0 & 0 & 13 \\
-2 & 3 & 0 & 1 & 0 & -6 \\
\hline
-1 & -2 & 0 & 0 & 1 & 0
\end{array}
$$

The -6 in the upper portion of the last column is a problem. If we compute values for the variables x, y, s_1, and s_2 that are associated with this matrix, we get $s_2 = -6$. This violates the condition of the simplex method that all variables be nonnegative. In order to use the simplex method, we must change the sign of any negative entry that appears in the upper portion of the last column (above the line separating the objective function from the constraints). If the problem has a solution, there will always be another negative entry in the same row as this negative entry, but in a different column. We choose this column as the pivot column, because pivoting with it will give a positive entry in the last column.

$$
\begin{array}{ccccc|c}
x & y & s_1 & s_2 & f & \\
1 & 1 & 1 & 0 & 0 & 13 \\
-2 & 3 & 0 & 1 & 0 & -6 \\
\hline
-1 & -2 & 0 & 0 & 1 & 0
\end{array}
$$

← Negative in last column

↑
Pivot column (negative entry in the same row as the negative entry in the last column)

Once we have identified the pivot column, we choose the pivot entry by forming *all* quotients and choosing the entry that gives the *smallest positive quotient*. Often this will mean pivoting by a negative number, which is the case in this problem.

$$
\begin{array}{ccccc}
x & y & s_1 & s_2 & f \\
\end{array}
$$

$$
\left[
\begin{array}{ccccc|c}
1 & 1 & 1 & 0 & 0 & 13 \\
\boxed{-2} & 3 & 0 & 1 & 0 & -6 \\
\hline
-1 & -2 & 0 & 0 & 1 & 0
\end{array}
\right]
\qquad
\begin{array}{l}
13/1 = 13 \\
(-6)/(-2) = 3* \\
\\
\end{array}
$$

↑
Pivot entry circled

*Smallest positive quotient

Pivoting with this entry gives the following matrix.

$$
\begin{array}{ccccc}
x & y & s_1 & s_2 & f \\
\end{array}
$$

$$
\left[
\begin{array}{ccccc|c}
0 & \boxed{\frac{5}{2}} & 1 & \frac{1}{2} & 0 & 10 \\
1 & -\frac{3}{2} & 0 & -\frac{1}{2} & 0 & 3 \\
\hline
0 & -\frac{7}{2} & 0 & -\frac{1}{2} & 1 & 3
\end{array}
\right]
\Big\} \text{No negatives}
$$

↑
Most negative

This new matrix has no negatives in the upper portion of the last column, so we proceed with the simplex method as we have used it previously. The new pivot column is found from the most negative entry in the last row (indicated), and the pivot is found from the smallest positive quotient (circled). Using this pivot completes the solution.

$$
\begin{array}{ccccc}
x & y & s_1 & s_2 & f \\
\end{array}
$$

$$
\left[
\begin{array}{ccccc|c}
0 & 1 & \frac{2}{5} & \frac{1}{5} & 0 & 4 \\
1 & 0 & \frac{3}{5} & -\frac{1}{5} & 0 & 9 \\
\hline
0 & 0 & \frac{7}{5} & \frac{1}{5} & 1 & 17
\end{array}
\right]
$$

We see that when $x = 9$ and $y = 4$, then the maximum value of $f = x + 2y$ is 17. ■

CHECKPOINT

1. Write the simplex matrix to maximize $f = 4x + y$ subject to the constraints

$$
\begin{aligned}
x + 2y &\leq 12 \\
3x - 4y &\geq 6 \\
x \geq 0, y &\geq 0
\end{aligned}
$$

2. Using the matrix from Problem 1, find the maximum value of $f = 4x + y$ subject to the constraints.

Mixed Constraints and Minimization

If mixed constraints occur in a minimization problem, the simplex method does not apply to the dual problem. However, the same techniques used in Example 1 can be slightly modified and applied to minimization problems with mixed constraints. If our objective is to minimize f, then we alter the problem so as to maximize $-f$ subject to "≤ constraints" and then proceed as in Example 1.

EXAMPLE 2   Minimization with Mixed Constraints

Minimize the function $f = 3x + 4y$ subject to the constraints

$$
\begin{aligned}
x + y &\geq 20 \\
x + 2y &\geq 25 \\
-5x + y &\leq 4 \\
x \geq 0, y &\geq 0
\end{aligned}
$$

Solution

Because of the mixed constraints, we seek to maximize $-f = -3x - 4y$ subject to

$$\begin{aligned}
-x - \;\;y &\le -20 \\
-x - 2y &\le -25 \\
-5x + \;\;y &\le \;\;\;4
\end{aligned}$$

The simplex matrix for this problem is

$$
\begin{array}{cccccc}
x & y & s_1 & s_2 & s_3 & -f
\end{array}
$$

$$
\left[\begin{array}{cccccc|c}
-1 & -1 & 1 & 0 & 0 & 0 & -20 \\
-1 & -2 & 0 & 1 & 0 & 0 & -25 \\
-5 & 1 & 0 & 0 & 1 & 0 & 4 \\
\hline
3 & 4 & 0 & 0 & 0 & 1 & 0
\end{array}\right]
\quad \leftarrow \text{Negatives}
$$

In this case, there are two negative entries in the upper portion of the last column. When this happens, we can start with either one; we choose -20 (row 1). In row 1 (to the left of -20), we find negative entries in columns 1 and 2. We can choose our pivot column as either of these columns; we choose column 1. Once this choice is made, the pivot is determined from the quotients.

$$
\begin{array}{cccccc}
x & y & s_1 & s_2 & s_3 & -f
\end{array}
$$

$$
\left[\begin{array}{cccccc|c}
\boxed{-1} & -1 & 1 & 0 & 0 & 0 & -20 \\
-1 & -2 & 0 & 1 & 0 & 0 & -25 \\
-5 & 1 & 0 & 0 & 1 & 0 & 4 \\
\hline
3 & 4 & 0 & 0 & 0 & 1 & 0
\end{array}\right]
\quad
\begin{array}{l}
-20/(-1) = 20^* \\
-25/(-1) = 25 \\
4/(-5) = -4/5
\end{array}
$$

$\uparrow$
Pivot column (pivot entry circled)   *Smallest positive quotient

Pivoting with this entry gives the following matrix.

$$
\begin{array}{cccccc}
x & y & s_1 & s_2 & s_3 & -f
\end{array}
$$

$$
\left[\begin{array}{cccccc|c}
1 & 1 & -1 & 0 & 0 & 0 & 20 \\
0 & \boxed{-1} & -1 & 1 & 0 & 0 & -5 \\
0 & 6 & -5 & 0 & 1 & 0 & 104 \\
\hline
0 & 1 & 3 & 0 & 0 & 1 & -60
\end{array}\right]
\quad \leftarrow \text{Negative}
$$

$\uparrow$
Pivot column

Now there is only one negative in the upper portion of the last column. Looking for negatives in this same row, we see that we can choose either column 2 or column 3 for our pivot column. We choose column 2, and the pivot is circled. Pivoting gives the following matrix.

$$
\begin{array}{cccccc}
x & y & s_1 & s_2 & s_3 & -f
\end{array}
$$

$$
\left[\begin{array}{cccccc|c}
1 & 0 & -2 & 1 & 0 & 0 & 15 \\
0 & 1 & 1 & -1 & 0 & 0 & 5 \\
0 & 0 & -11 & 6 & 1 & 0 & 74 \\
\hline
0 & 0 & 2 & 1 & 0 & 1 & -65
\end{array}\right]
$$

At this point we have all positives in the upper portion of the last column, so the simplex method can be applied. (The negative value in the lower portion is acceptable because $-f$ is our function.) Looking at the indicators, we see that the solution is complete.

From the matrix we have $x = 15$, $y = 5$, and $-f = -65$. Thus the solution is $x = 15$ and $y = 5$, which gives the minimum value for the function $f = 3x + 4y = 65$. ■

A careful review of these examples allows us to summarize the key procedural steps for solving linear programming problems with mixed constraints.

> ### Summary: Simplex Method for Mixed Constraints
>
> 1. If the problem is to minimize f, then maximize $-f$.
> 2. a. Make all constraints $\leq$ constraints by multiplying both sides of any $\geq$ constraints by (-1).
>   b. Use slack variables and form the simplex matrix.
> 3. In the simplex matrix, scan the *upper portion* of the last column for any negative entries.
>   a. If there are no negative entries, apply the simplex method.
>   b. If there are negative entries, go to Step 4.
> 4. When there is a negative value in the upper portion of the last column, proceed as follows:
>   a. Select any negative entry in the same row and use its column as the pivot column.
>   b. In the pivot column, compute all quotients for the entries above the line and determine the pivot from the *smallest positive quotient*.
>   c. After completing the pivot operations, return to Step 3.

EXAMPLE 3   Production Scheduling for Maximum Profit | APPLICATION PREVIEW |

The Laser Company manufactures two models of home theater systems, the Star and the Allstar, at two plants, located in Ashville and in Cleveland. The maximum daily output at the Ashville plant is 900, and the profits there are $200 per unit of the Allstar and $100 per unit of the Star. The maximum daily output at the Cleveland plant is 800, and the profits there are $210 per unit of the Allstar and $80 per unit of the Star. In addition, restrictions at the Ashville plant mean that the number of units of the Star model cannot exceed 100 more than the number of units of the Allstar model produced. If the company gets a rush order for 800 units of the Allstar model and 600 units of the Star model, how many units of each model should be produced at each location to fill the order and obtain the maximum profit?

Solution

The following table identifies the variables x and y and relates other important facts in the problem. In particular, we see that only variables x and y are required.

	Ashville Plant	Total Needed	Cleveland Plant
Allstar	x produced (profit = $200 each)	800	$800 - x$ produced (profit = $210 each)
Star	y produced (profit = $100 each)	600	$600 - y$ produced (profit = $80 each)
Capacity	900	–	800

We wish to maximize profit, and from the table we see that total profit is given by

$$P = 200x + 100y + 210(800 - x) + 80(600 - y)$$

or

$$P = 216{,}000 - 10x + 20y$$

We can read capacity constraints from the table.

$$x + y \leq 900 \quad (\text{Ashville})$$
$$(800 - x) + (600 - y) \leq 800 \quad \text{or} \quad x + y \geq 600 \quad (\text{Cleveland})$$

An additional Ashville plant constraint from the statement of the problem is

$$y \leq x + 100 \quad \text{or} \quad -x + y \leq 100$$

Thus our problem is

$$\text{Maximize } P = 216{,}000 - 10x + 20y$$
$$\text{subject to} \quad x + y \le 900$$
$$x + y \ge 600$$
$$-x + y \le 100$$
$$x \ge 0, y \ge 0$$

We must express $x + y \ge 600$ as a $\le$ constraint; multiplying both sides by (-1) gives $-x - y \le -600$. The simplex matrix is

$$
\begin{array}{cccccc}
x & y & s_1 & s_2 & s_3 & P \\
\end{array}
$$

$$
\left[
\begin{array}{cccccc|c}
1 & 1 & 1 & 0 & 0 & 0 & 900 \\
\boxed{-1} & -1 & 0 & 1 & 0 & 0 & -600 \\
-1 & 1 & 0 & 0 & 1 & 0 & 100 \\
\hline
10 & -20 & 0 & 0 & 0 & 1 & 216{,}000
\end{array}
\right]
\quad \leftarrow \text{Negative}
$$

$\uparrow$
Pivot column

The negative in the upper portion of the last column is indicated. In row 2 we find negatives in both column 1 and column 2, so either of these may be our pivot column. Our choice is indicated, and the pivot entry is circled. The matrix that results from the pivot operations follows.

$$
\begin{array}{cccccc}
x & y & s_1 & s_2 & s_3 & P \\
\end{array}
$$

$$
\left[
\begin{array}{cccccc|c}
0 & 0 & 1 & 1 & 0 & 0 & 300 \\
1 & 1 & 0 & -1 & 0 & 0 & 600 \\
0 & \boxed{2} & 0 & -1 & 1 & 0 & 700 \\
\hline
0 & -30 & 0 & 10 & 0 & 1 & 210{,}000
\end{array}
\right]
$$

$\uparrow$
Pivot column

No entry in the upper portion of the last column is negative, so we can proceed with the simplex method. From the indicators we see that column 2 is the pivot column (indicated above), and the pivot entry is circled. The simplex matrix that results from pivoting follows.

$$
\begin{array}{cccccc}
x & y & s_1 & s_2 & s_3 & P \\
\end{array}
$$

$$
\left[
\begin{array}{cccccc|c}
0 & 0 & 1 & \boxed{1} & 0 & 0 & 300 \\
1 & 0 & 0 & -\frac{1}{2} & -\frac{1}{2} & 0 & 250 \\
0 & 1 & 0 & -\frac{1}{2} & \frac{1}{2} & 0 & 350 \\
\hline
0 & 0 & 0 & -5 & 15 & 1 & 220{,}500
\end{array}
\right]
$$

$\uparrow$
Pivot column

The new pivot column is indicated, and the pivot entry is circled. The pivot operation yields the following matrix.

$$
\begin{array}{cccccc}
x & y & s_1 & s_2 & s_3 & P \\
\end{array}
$$

$$
\left[
\begin{array}{cccccc|c}
0 & 0 & 1 & 1 & 0 & 0 & 300 \\
1 & 0 & \frac{1}{2} & 0 & -\frac{1}{2} & 0 & 400 \\
0 & 1 & \frac{1}{2} & 0 & \frac{1}{2} & 0 & 500 \\
\hline
0 & 0 & 5 & 0 & 15 & 1 & 222{,}000
\end{array}
\right]
$$

This matrix shows that the solution is complete. We see that $x = 400$ (so $800 - x = 400$), $y = 500$ (so $600 - y = 100$), and $P = \$222{,}000$.

Thus the company should operate the Ashville plant at capacity and produce 400 units of the Allstar model and 500 units of the Star model. The remainder of the order, 400 units of the Allstar model and 100 units of the Star model, should be produced at the Cleveland plant. This production schedule gives maximum profit $P = \$222{,}000$. ■

Spreadsheet Note

When using Excel to solve a linear programming problem with mixed constraints, it is not necessary to convert the form of the inequalities. They can simply be entered in their "mixed" form, and the objective function can be maximized or minimized with Solver. ■

CHECKPOINT SOLUTIONS

1. $\begin{bmatrix} 1 & 2 & 1 & 0 & 0 & | & 12 \\ \boxed{-3} & 4 & 0 & 1 & 0 & | & -6 \\ -4 & -1 & 0 & 0 & 1 & | & 0 \end{bmatrix}$

2. $\begin{bmatrix} 0 & \boxed{\frac{10}{3}} & 1 & \frac{1}{3} & 0 & | & 10 \\ 1 & -\frac{4}{3} & 0 & -\frac{1}{3} & 0 & | & 2 \\ 0 & -\frac{19}{3} & 0 & -\frac{4}{3} & 1 & | & 8 \end{bmatrix} \rightarrow \begin{bmatrix} 0 & 1 & \frac{3}{10} & \boxed{\frac{1}{10}} & 0 & | & 3 \\ 1 & 0 & \frac{2}{5} & -\frac{1}{5} & 0 & | & 6 \\ 0 & 0 & \frac{19}{10} & -\frac{7}{10} & 1 & | & 27 \end{bmatrix}$

$\rightarrow \begin{bmatrix} 0 & 10 & 3 & 1 & 0 & | & 30 \\ 1 & 2 & 1 & 0 & 0 & | & 12 \\ 0 & 7 & 4 & 0 & 1 & | & 48 \end{bmatrix}$

The maximum value is $f = 48$ at $x = 12$, $y = 0$, $s_1 = 0$, and $s_2 = 30$.

| EXERCISES | 4.5

Access end-of-section exercises online at **www.webassign.net**

KEY TERMS AND FORMULAS

Section	Key Terms
4.1	Linear inequalities in two variables
	Graphs
	Systems
	Solution region
4.2	Linear programming
	Objective function
	Constraints
	Optimum values
	Feasible region
	Graphical solution
4.3	Simplex method (maximization)
	Standard maximization problem
	Slack variables
	Simplex matrix
	Pivot entry
	Basic and nonbasic variables
	Basic feasible solution
	Pivoting
	Nonunique solutions
	Shadow prices

Section	Key Terms
4.4	Simplex method (minimization) Standard minimization problem Dual problem Transpose Primal problem
4.5	Simplex method with mixed constraints Creating $\leq$ constraints

REVIEW EXERCISES

In Problems 1–4, graph the solution set of each inequality or system of inequalities.

1. $2x + 3y \leq 12$

2. $4x + 5y > 100$

3. $\begin{cases} x + 2y \leq 20 \\ 3x + 10y \leq 80 \\ x \geq 0,\ y \geq 0 \end{cases}$

4. $\begin{cases} 3x + y \geq 4 \\ x + y \geq 2 \\ -x + y \leq 4 \\ x \qquad \leq 5 \end{cases}$

In Problems 5–8, a function and the graph of a feasible region are given. In each case, find both the maximum and minimum values of the function, if they exist, and the point at which each occurs.

5. $f = -x + 3y$

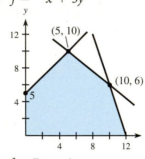

6. $f = 6x + 4y$

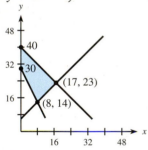

7. $f = 7x - 6y$

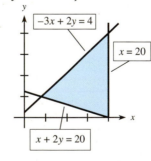

8. $f = 9x + 10y$

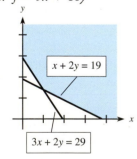

In Problems 9–14, solve the linear programming problems using graphical methods. Restrict $x \geq 0$ and $y \geq 0$.

9. Maximize $f = 5x + 6y$ subject to
$$x + 3y \leq 24$$
$$4x + 3y \leq 42$$
$$2x + y \leq 20$$

10. Maximize $f = x + 4y$ subject to
$$7x + 3y \leq 105$$
$$2x + 5y \leq 59$$
$$x + 7y \leq 70$$

11. Minimize $g = 5x + 3y$ subject to
$$3x + y \geq 12$$
$$x + y \geq 6$$
$$x + 6y \geq 11$$

12. Minimize $g = x + 5y$ subject to
$$8x + y \geq 85$$
$$x + y \geq 50$$
$$x + 4y \geq 80$$
$$x + 10y \geq 104$$

13. Maximize $f = 5x + 2y$ subject to
$$y \leq 20$$
$$2x + y \leq 32$$
$$-x + 2y \geq 4$$

14. Minimize $f = x + 4y$ subject to
$$y \leq 30$$
$$3x + 2y \geq 75$$
$$-3x + 5y \geq 30$$

In Problems 15–18, use the simplex method to solve the linear programming problems. Assume all variables are nonnegative.

15. Maximize $f = 7x + 12y$ subject to the conditions in Problem 10.

16. Maximize $f = 3x + 4y$ subject to
$$x + 4y \leq 160$$
$$x + 2y \leq 100$$
$$4x + 3y \leq 300$$

17. Maximize $f = 3x + 8y$ subject to the conditions in Problem 16.

18. Maximize $f = 3x + 2y$ subject to
$$x + 2y \leq 48$$
$$x + y \leq 30$$
$$2x + y \leq 50$$
$$x + 10y \leq 200$$

**Problems 19 and 20 have nonunique solutions. If there is
no solution, indicate this; if there are multiple solutions,
find two different solutions. Use the simplex method,
with $x \geq 0$, $y \geq 0$.**

19. Maximize $f = 4x + 4y$ subject to

$$x + 5y \leq 500$$
$$x + 2y \leq 230$$
$$x + y \leq 160$$

20. Maximize $f = 2x + 5y$ subject to

$$-4x + y \leq 40$$
$$x - 7y \leq 70$$

**In Problems 21–24, form the dual and use the simplex
method to solve the minimization problem. Assume all
variables are nonnegative.**

21. Minimize $g = 7y_1 + 6y_2$ subject to

$$5y_1 + 2y_2 \geq 16$$
$$3y_1 + 7y_2 \geq 27$$

22. Minimize $g = 3y_1 + 4y_2$ subject to

$$3y_1 + y_2 \geq 8$$
$$y_1 + y_2 \geq 6$$
$$2y_1 + 5y_2 \geq 18$$

23. Minimize $g = 2y_1 + y_2$ subject to the conditions in
Problem 22.

24. Minimize $g = 12y_1 + 11y_2$ subject to

$$y_1 + y_2 \geq 100$$
$$2y_1 + y_2 \geq 140$$
$$6y_1 + 5y_2 \geq 580$$

**Problems 25 and 26 involve mixed constraints.
Solve each with the simplex method. Assume all
variables are nonnegative.**

25. Maximize $f = 3x + 5y$ subject to

$$x + y \geq 19$$
$$-x + y \geq 1$$
$$-x + 10y \leq 190$$

26. Maximize $f = 4x + 6y$ subject to

$$2x + 5y \leq 37$$
$$5x - y \leq 34$$
$$-x + 2y \geq 4$$

**In Problems 27–34, use the simplex method. Assume all
variables are nonnegative.**

27. Maximize $f = 39x + 5y + 30z$ subject to

$$x + z \leq 7$$
$$3x + 5y \leq 30$$
$$3x + y \leq 18$$

28. Minimize $g = 12y_1 + 5y_2 + 2y_3$ subject to

$$y_1 + 2y_2 + y_3 \geq 60$$
$$12y_1 + 4y_2 + 3y_3 \geq 120$$
$$2y_1 + 3y_2 + y_3 \geq 80$$

29. Minimize $g = 25y_1 + 10y_2 + 4y_3$ subject to

$$7.5y_1 + 4.5y_2 + 2y_3 \geq 650$$
$$6.5y_1 + 3y_2 + 1.5y_3 \geq 400$$
$$y_1 + 1.5y_2 + 0.5y_3 \geq 200$$

30. Minimize $f = 10x + 3y$ subject to

$$-x + 10y \geq 5$$
$$4x + y \geq 62$$
$$x + y \leq 50$$

31. Minimize $f = 4x + 3y$ subject to

$$-x + y \geq 1$$
$$x + y \leq 45$$
$$10x + y \geq 45$$

32. Maximize $f = 88x_1 + 86x_2 + 100x_3 + 100x_4$ subject to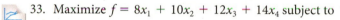

$$3x_1 + 2x_2 + 2x_3 + 5x_4 \leq 200$$
$$2x_1 + 2x_2 + 4x_3 + 5x_4 \leq 100$$
$$x_1 + x_2 + x_3 + x_4 \leq 200$$
$$x_1 \leq 40$$

33. Maximize $f = 8x_1 + 10x_2 + 12x_3 + 14x_4$ subject to

$$6x_1 + 3x_2 + 2x_3 + x_4 \geq 350$$
$$3x_1 + 2x_2 + 5x_3 + 6x_4 \leq 300$$
$$8x_1 + 3x_2 + 2x_3 + x_4 \leq 400$$
$$x_1 + x_2 + x_3 + x_4 \leq 100$$

34. Minimize $g = 10x_1 + 9x_2 + 12x_3 + 8x_4$ subject to

$$45x_1 + 58.5x_3 \geq 4680$$
$$36x_2 + 31.5x_4 \geq 4230$$
$$x_1 + x_2 \leq 100$$
$$x_3 +   x_4 \leq 100$$

APPLICATIONS

35. *Manufacturing* A company manufactures backyard
swing sets of two different sizes. The larger set requires
5 hours of labor to complete, the smaller set requires
2 hours, and there are 700 hours of labor available each
week. The packaging department can package at most
185 swing sets per week. If the profit is \$100 on each
larger set and \$50 on each smaller set, how many of
each should be produced to yield the maximum profit?
What is the maximum profit? Use graphical methods.

36. *Production* A company produces two different grades
of steel, A and B, at two different factories, 1 and 2. The
following table summarizes the production capabilities
of the factories, the cost per day, and the number of
units of each grade of steel that is required to fill orders.

	Factory 1	Factory 2	Required
Grade A steel	1 unit	2 units	80 units
Grade B steel	3 units	2 units	140 units
Cost per day	$5000	$6000	

How many days should each factory operate in order to fill the orders at minimum cost? What is the minimum cost? Use graphical methods.

Use the simplex method to solve Problems 37–45.

37. *Production* A small industry produces two items, I and II. It operates at capacity and makes a profit of $6 on each item I and $4 on each item II. The following table gives the hours required to produce each item and the hours available per day.

	I	II	Hours Available
Assembly	2 hours	1 hour	100
Packaging & inspection	1 hour	1 hour	60

Find the number of items that should be produced each day to maximize profits, and find the maximum daily profit.

38. *Production* Pinocchio Crafts makes two types of wooden crafts: Jacob's ladders and locomotive engines. The manufacture of these crafts requires both carpentry and finishing. Each Jacob's ladder requires 1 hour of finishing and $\frac{1}{2}$ hour of carpentry. Each locomotive engine requires 1 hour of finishing and 1 hour of carpentry. Pinocchio Crafts can obtain all the necessary raw materials, but only 120 finishing hours and 75 carpentry hours per week are available. Also, demand for Jacob's ladders is limited to at most 100 per week. If Pinocchio Crafts makes a profit of $3 on each Jacob's ladder and $5 on each locomotive engine, how many of each should it produce each week to maximize profits? What is the maximum profit?

39. *Profit* At its Jacksonville factory, Nolmaur Electronics manufactures 4 models of TV sets: LCD models in 27-, 32-, and 42-in. sizes and a 42-in. plasma model. The manufacturing and testing hours required for each model and available at the factory each week, as well as each model's profit, are shown in the following table.

	27-in. LCD	32-in. LCD	42-in. LCD	42-in. Plasma	Available Hours
Manufacturing (hr)	8	10	12	15	1870
Testing (hr)	2	4	4	4	530
Profit	$80	$120	$160	$200	

In addition, the supplier of the amplifier units can provide at most 200 units per week with at most 100 of these for both types of 42-in. models. The weekly demand for the 32-in. sets is at most 120. Nolmaur

wants to determine the number of each type of set that should be produced each week to obtain maximum profit.
(a) Carefully identify the variables for Nolmaur's linear programming problem.
(b) Carefully state Nolmaur's linear programming problem.
(c) Solve this linear programming problem to determine Nolmaur's manufacturing plan and maximum profit.

40. *Nutrition* A nutritionist wants to find the least expensive combination of two foods that meet minimum daily vitamin requirements, which are 5 units of A and 30 units of B. Each ounce of food I provides 2 units of A and 1 unit of B, and each ounce of food II provides 10 units of A and 10 units of B. If food I costs 30 cents per ounce and food II costs 20 cents per ounce, find the number of ounces of each food that will provide the required vitamins and minimize the cost.

41. *Nutrition* A laboratory wishes to purchase two different feeds, A and B, for its animals. The following table summarizes the nutritional contents of the feeds, the required amounts of each ingredient, and the cost of each type of feed.

	Feed A	Feed B	Required
Carbohydrates	1 unit/lb	4 units/lb	40 units
Protein	2 units/lb	1 unit/lb	80 units
Cost	14¢/lb	16¢/lb	

How many pounds of each type of feed should the laboratory buy in order to satisfy its needs at minimum cost?

42. *Production* A company makes three products, I, II, and III, at three different factories. At factory A, it can make 10 units of each product per day. At factory B, it can make 20 units of II and 20 units of III per day. At factory C, it can make 20 units of I, 20 units of II, and 10 units of III per day. The company has orders for 200 units of I, 500 units of II, and 300 units of III. If the daily costs are $200 at A, $300 at B, and $500 at C, find the number of days that each factory should operate in order to fill the company's orders at minimum cost. Find the minimum cost.

43. *Profit* A company makes pancake mix and cake mix. Each pound of pancake mix uses 0.6 lb of flour and 0.1 lb of shortening. Each pound of cake mix uses 0.4 lb of flour, 0.1 lb of shortening, and 0.4 lb of sugar. Suppliers can deliver at most 6000 lb of flour, at least 500 lb of shortening, and at most 1200 lb of sugar. If the profit per pound is $0.35 for pancake mix and $0.25 for cake mix, how many pounds of each mix should be made to earn maximum profit? What is the maximum profit?

44. *Manufacturing* A company manufactures desks and computer tables at plants in Texas and Louisiana. At the Texas plant, production costs are $12 for each desk and $20 for each computer table, and the plant can produce at most 120 units per day. At the Louisiana plant, costs

are $14 for each desk and $19 per computer table, and the plant can produce at most 150 units per day. The company gets a rush order for 130 desks and 130 computer tables at a time when the Texas plant is further limited by the fact that the number of computer tables it produces must be at least 10 more than the number of desks. How should production be scheduled at each location in order to fill the order at minimum cost? What is the minimum cost?

 45. *Cost* Armstrong Industries makes two different grades of steel at its plants in Midland and Donora. Weekly demand is at least 500 tons for grade 1 steel and at least 450 tons for grade 2. Due to differences in the equipment and the labor force, production times and costs per ton for each grade of steel differ slightly at each location, as shown in the following table.

| | Grade 1 | | Grade 2 | | Furnace |
	Min/ Ton	Cost/ Ton	Min/ Ton	Cost/ Ton	Time/Week (in Minutes)
Midland	40	$100	42	$120	19,460
Donora	44	$110	45	$90	21,380

Find the number of tons of each grade of steel that should be made each week at each plant so that Armstrong meets demand at minimum cost. Find the minimum cost.

---

4 CHAPTER TEST

1. Maximize $f = 3x + 5y$ subject to the constraints determined by the shaded region shown in the figure.

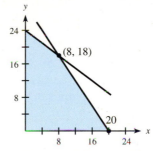

2. Following are three simplex matrices in various stages of the simplex method solution of a maximization problem.
   (a) Identify the matrix for which multiple solutions exist. Circle the pivot and specify all row operations with that pivot that are needed to find a second solution.
   (b) Identify the matrix for which no solution exists. Explain how you can tell there is no solution.
   (c) For the remaining matrix, find the basic feasible solution. Use the variables $x_1, x_2, \ldots$, the slack variables $s_1, s_2, \ldots$, and the objective function f, and tell whether these values give the optimal solution. If they do not, identify the next pivot.

$$A = \begin{bmatrix} 2 & 0 & -4 & 1 & -2 & 0 & 0 & 6 \\ -1 & 1 & 0 & 0 & 1 & 0 & 0 & 3 \\ 4 & 0 & -6 & 0 & -1 & 1 & 0 & 10 \\ -3 & 0 & -8 & 0 & 4 & 0 & 1 & 50 \end{bmatrix}$$

$$B = \begin{bmatrix} 0 & 1 & -2 & \frac{1}{4} & 0 & \frac{1}{5} & 0 & 12 \\ 0 & 0 & 4 & -\frac{3}{4} & 1 & -\frac{2}{5} & 0 & 20 \\ 1 & 0 & 3 & -\frac{5}{4} & 0 & \frac{4}{5} & 0 & 40 \\ 0 & 0 & 4 & 6 & 0 & -\frac{1}{5} & 1 & 170 \end{bmatrix}$$

$$C = \begin{bmatrix} 1 & 2 & 0 & 1 & 0 & -\frac{3}{2} & 0 & 40 \\ 0 & 1 & 0 & -2 & 1 & \frac{1}{2} & 0 & 15 \\ 0 & 3 & 1 & -1 & 0 & \frac{1}{4} & 0 & 60 \\ 0 & 0 & 0 & 4 & 0 & 6 & 1 & 220 \end{bmatrix}$$

3. Graph the solution region for each of the following.
   (a) $3x - 5y \le 30$ (b) $x > 2$ (c) $\begin{cases} 5y \ge 2x \\ x + y > 7 \\ 2y \le 5x \end{cases}$

4. Formulate the dual maximization problem associated with the following.
   Minimize $g = 2y_1 + 3y_2 + 5y_3$ subject to
   $$3y_1 + 5y_2 + y_3 \ge 100$$
   $$4y_1 + 6y_2 + 3y_3 \ge 120$$
   $$y_1 \ge 0, y_2 \ge 0, y_3 \ge 0$$

5. Find the maximum and minimum values of $f = 5x + 2y$ (if they exist) subject to the constraints determined by the shaded region in the figure.

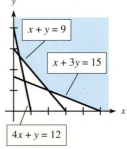

6. Use graphical methods to minimize $g = 3x + y$ subject to $5x - 4y \le 40, x + y \ge 80, 2x - y \ge 4, x \ge 0,$ and $y \ge 0$.

7. Express the following linear programming problem as a maximization problem with $\le$ constraints. Assume $x \ge 0, y \ge 0$.

Minimize $g = 7x + 3y$ subject to

$$-x + 4y \geq 4$$
$$x - y \leq 5$$
$$2x + 3y \leq 30$$

8. The final simplex matrix for a problem is as follows. Give the solutions to *both* the minimization problem and the dual maximization problem. Carefully label each solution.

$$\begin{bmatrix} 0 & 1 & -0.5 & 3 & -0.7 & 0 & 0 & | & 15 \\ 1 & 0 & -0.8 & -2 & -0.5 & 0 & 0 & | & 17 \\ 0 & 0 & -0.6 & -1 & -2.4 & 1 & 0 & | & 11 \\ 0 & 0 & 12 & 4 & 18 & 0 & 1 & | & 658 \end{bmatrix}$$

9. Use the simplex method to maximize $f = 70x + 5y$ subject to $x + 1.5y \leq 150$, $x + 0.5y \leq 90$, $x \geq 0$, and $y \geq 0$.

10. Maximize $f = 60x + 48y + 36z$ subject to

$$8x + 6y + z \leq 108$$
$$4x + 2y + 1.5z \leq 50$$
$$2x + 1.5y + 0.5z \leq 40$$
$$z \leq 16$$
$$x \geq 0, \ y \geq 0, \ z \geq 0$$

11. River Brewery is a microbrewery that produces an amber lager and an ale. Producing a barrel of lager requires 3 lb of corn and 2 lb of hops. Producing a barrel of ale requires 2 lb each of corn and hops. Profits are $35 from each barrel of lager and $30 from each barrel of ale. Suppliers can provide at most 1200 lb of corn and at most 1000 lb of hops per month. Formulate a linear programming problem that can be used to maximize River Brewery's monthly profit. Then solve it with the simplex method.

12. A marketing research group conducting a telephone survey must contact at least 150 wives and 120 hus-bands. It costs $3 to make a daytime call and (because of higher labor costs) $4 to make an evening call. On average, daytime calls reach wives 30% of the time, husbands 10% of the time, and neither of these 60% of the time, whereas evening calls reach wives 30% of the time, husbands 30% of the time, and neither of these 40% of the time. Staffing considerations mean that daytime calls must be less than or equal to half of the total calls made. Formulate a linear programming problem that can be used to minimize the cost of completing the survey. Then solve it with any method from this chapter.

13. Lawn Rich, Inc. makes four different lawn care products that have fixed percents of phosphate, nitrogen, potash, and other minerals. The following table shows the percent of each of the four ingredients in each of Lawn Rich's products and the profit per ton for each product.

	Product 1	Product 2	Product 3	Product 4
% Phosphate	60	30	20	10
% Nitrogen	20	20	20	20
% Potash	9	6	15	18
% Other minerals	8	3	2	1
Profit/ton	$40	$50	$60	$70

A lawn care company buys in bulk from Lawn Rich and then blends the products for its own purposes. For the blend it desires, the lawn care company wants its Lawn Rich order to contain at least 35 tons of phosphate, at most 20 tons of nitrogen, at most 9 tons of potash, and at most 4 tons of other minerals. How many tons of each product should Lawn Rich produce to fulfill the customer's needs and maximize its own profit? What is the maximum profit?

I. Transportation

The Kimble Firefighting Equipment Company has two assembly plants: plant A in Atlanta and plant B in Buffalo. It has two distribution centers: center I in Pittsburgh and center II in Savannah. The company can produce at most 800 alarm valves per week at plant A and at most 1000 per week at plant B. Center I must have at least 900 alarm valves per week, and center II must have at least 600 alarm valves per week. The costs per unit for transportation from the plants to the centers follow.

Plant A to center I	$2
Plant A to center II	3
Plant B to center I	4
Plant B to center II	1

What weekly shipping plan will meet the market demands and minimize the total cost of shipping the valves from the assembly plants to the distribution centers? What is the minimum transportation cost? Should the company eliminate the route with the most expensive unit shipment cost?

1. To assist the company in making these decisions, use linear programming with the following variables.

   y_1 = the number of units shipped from plant A to center I
   y_2 = the number of units shipped from plant A to center II
   y_3 = the number of units shipped from plant B to center I
   y_4 = the number of units shipped from plant B to center II

2. Form the constraint inequalities that give the limits for the plants and distribution centers, and form the total cost function that is to be minimized.
3. Solve the mixed constraint problem for the minimum transportation cost.
4. Determine whether any shipment route is unnecessary.

II. Slack Variables and Shadow Prices

Part A: With linear programming problems, one set of interesting questions deals with how the optimal solution changes if the constraint quantities change. Surprisingly, answers to these questions can be found in the final simplex tableau. In this investigation, we restrict ourselves to standard maximization problems and begin by reexamining the farm co-op application discussed in Example 1 in Section 4.3.

Farm Co-op Problem Statement

Maximize profit $P = 60x + 40y$ subject to

$$
\begin{aligned}
x + y &\leq 6000 && \text{(Land constraint, in acres)} \\
9x + 3y &\leq 40{,}500 && \text{(Fertilizer/herbicide constraint, in gallons)} \\
\tfrac{3}{4}x + y &\leq 5250 && \text{(Labor constraint, in hours)}
\end{aligned}
$$

Initial Tableau

$$
A = \left[\begin{array}{ccccccc|c}
x & y & s_1 & s_2 & s_3 & P & \\
1 & 1 & 1 & 0 & 0 & 0 & 6{,}000 \\
9 & 3 & 0 & 1 & 0 & 0 & 40{,}500 \\
\tfrac{3}{4} & 1 & 0 & 0 & 1 & 0 & 5{,}250 \\
\hline
-60 & -40 & 0 & 0 & 0 & 1 & 0
\end{array}\right]
$$

Final Tableau

$$
C = \left[\begin{array}{ccccccc|c}
x & y & s_1 & s_2 & s_3 & P & \\
0 & 1 & \tfrac{3}{2} & -\tfrac{1}{6} & 0 & 0 & 2{,}250 \\
1 & 0 & -\tfrac{1}{2} & \tfrac{1}{6} & 0 & 0 & 3{,}750 \\
0 & 0 & -\tfrac{9}{8} & \tfrac{1}{24} & 1 & 0 & 187.5 \\
\hline
0 & 0 & 30 & \tfrac{10}{3} & 0 & 1 & 315{,}000
\end{array}\right]
$$

Notice in the initial tableau that the columns for the slack variables and the objective function form an identity matrix (here a 4×4 matrix), and the row operations of the simplex method transform this identity matrix into another matrix in the same columns of the final tableau. Call this matrix S.

1. (a) Identify S and calculate SA, where A is the initial tableau above. What do you notice when you compare the result of SA with the final tableau, C?

   (b) Look at the Solar Technology manufacturing application in Example 3 in Section 4.3 and identify the corresponding matrix S there. Then find S times the initial tableau and compare your result with the final tableau. What do you notice?

2. (a) In the farm co-op application above, if the number of gallons of fertilizer/herbicide were increased by h to $40{,}500 + h$, then the row 2 entry in the last column of A would become $40{,}500 + h$. Call this new initial tableau A_h. Find the effect this change in fertilizer/herbicide would have on the optimal solution by calculating SA_h. Describe the differences between C (the final tableau of the original co-op problem) and SA_h.

   (b) Repeat part (a) if the number of acres of land is changed by k.

   (c) Use the results of parts (a) and (b) to complete the following general statement, A.
   Statement A: If constraint i has slack variable s_i and the quantity of constraint i is changed by h, then the new optimal solution can be found by using

   $h \cdot$ (the ____column) plus the ____ column as the new ____ column

(d) Based on your results so far (and similar calculations with other examples, if needed), decide whether the following statement is true or false.

"The process described in statement A for determining the new optimal solution applies for any increment of change in any constraint or combination of constraints as long as the last column of the new final tableau has *all* nonnegative entries."

(e) Apply these results to find the new optimal solution to the farm co-op application if the co-op obtained 16 additional acres of land and 48 more gallons of fertilizer/herbicide.

3. (a) In part 2(a), how much is the objective function changed in the new optimal solution?

(b) If $h = 1$ in part 2(a), how much would the objective function change and where in the final tableau is the value that gives this amount of change?

(c) Recall that the shadow price for a constraint measures the change in the objective function if one additional unit of the resource for that constraint becomes available. What is the shadow price for fertilizer/herbicide, and what does it tell us?

(d) Find the shadow prices for acres of land and labor hours and tell what each means.

(e) Complete the following general statement.

Statement B: The shadow price for constraint i is found in the final tableau as the _____ entry in the column for _____.

(f) From this example (and others if you check), notice that if a slack variable is nonzero, then the shadow price of the associated constraint will equal zero. Why do you think this is true?

Part B: Each week the Plush Seating Co. makes stuffed chairs (x) and love seats (y) by using hours of carpentry time (s_1), hours of assembly time (s_2), and hours of upholstery time (s_3) in order to maximize profit P, in dollars. Plush Seating's final tableau is shown below; use it and the results from part A to answer the questions that follow.

$$
\begin{array}{ccccccc|c}
x & y & s_1 & s_2 & s_3 & P & \\
\hline
1 & 0 & \frac{1}{3} & -1 & 0 & 0 & 28 \\
0 & 1 & -\frac{1}{3} & 2 & 0 & 0 & 8 \\
0 & 0 & \frac{4}{3} & -10 & 1 & 0 & 10 \\
\hline
0 & 0 & \frac{8}{3} & 64 & 0 & 1 & 2816
\end{array}
$$

1. Producing how many chairs and love seats will yield a maximum profit? Find the maximum profit.

2. How many hours of assembly time are left unused?

3. How many hours of upholstery time are left unused?

4. How much will profit increase if an extra hour of carpentry time becomes available? What is this amount of increase called?

5. How much will profit increase if an extra hour of upholstery time becomes available?

6. If 1 more hour of labor were available, where should Plush Seating use it? Explain why.

7. If each chair takes 2 hours of upholstery time and each love seat takes 6 hours, how many hours of upholstery time are available each week?

8. If Plush Seating could hire a part-timer to do 6 hours of carpentry and 1 hour of assembly, determine the new optimal solution.

9. Suppose that each worker can do any of the three jobs, and, rather than hiring a part-timer, managers want to reassign the extra upholstery hours. If they use 9 of those hours in carpentry and the other 1 in assembly, find the new optimal solution.

10. Find a way to reassign any number of the upholstery hours in a way different from that in Question 9, so that profit is improved even more than in Question 9.

5

Exponential and Logarithmic Functions

Monkey Business Images/Shutterstock.com

In this chapter we study exponential and logarithmic functions, which provide models for many applications that at first seem remote and unrelated. In our study of these functions, we will examine their descriptions, their properties, their graphs, and the special inverse relationship between these two functions. We will see how exponential and logarithmic functions are applied to some of the concerns of social scientists, business managers, and life scientists.

The chapter topics and applications include the following.

SECTIONS	APPLICATIONS
5.1 Exponential Functions Graphs Modeling	Investments, purchasing power and inflation, consumer price index
5.2 Logarithmic Functions Graphs and modeling Properties Change of base	Doubling time, market share, life expectancy, Richter scale
5.3 Exponential Equations Growth and decay Gompertz curves and logistic functions	Sales decay, demand, total revenue, organizational growth, Blu-ray player sales

Prerequisite Problem Type	For Section	Answer	Section for Review
Write the following with positive exponents: (a) x^{-3} (b) $\dfrac{1}{x^{-2}}$ (c) $\sqrt{x}$	**5.1** **5.2** **5.3**	(a) $\dfrac{1}{x^3}$ (b) x^2 (c) $x^{1/2}$	0.3, 0.4 Exponents and radicals
Simplify: (a) 2^0 (b) $x^0 \ (x \neq 0)$ (c) $49^{1/2}$ (d) 10^{-2}	**5.1** **5.2** **5.3**	(a) 1 (b) 1 (c) 7 (d) $\dfrac{1}{100}$	0.3, 0.4 Exponents
Answer true or false: (a) $\left(\dfrac{1}{2}\right)^x = 2^{-x}$ (b) $\sqrt{50} = 50^{1/2}$ (c) If $8 = 2^y$, then $y = 4$. (d) If $x^3 = 8$, then $x = 2$.	**5.1** **5.2**	(a) True (b) True (c) False; $y = 3$ (d) True	0.3, 0.4 Exponents and radicals
(a) If $f(x) = 2^{-2x}$, what is $f(-2)$? (b) If $f(x) = 2^{-2x}$, what is $f(1)$? (c) If $f(t) = (1 + 0.02)^t$, what is $f(0)$? (d) If $f(t) = 100(0.03)^{0.02t}$, what is $f(0)$? (e) If $f(x) = \dfrac{9.46}{1 + 53.08e^{-1.28x}}$, what is $f(5)$?	**5.1** **5.2** **5.3**	(a) 16 (b) $\dfrac{1}{4}$ (c) 1 (d) 3 (e) ≈ 8.693	1.2 Function notation

OBJECTIVES

5.1

- To graph exponential functions
- To evaluate exponential functions
- To model with exponential functions

Exponential Functions

If $10,000 is invested at 6%, compounded monthly, then the future value S of the investment after x years is given by the exponential function

$$S = 10,000(1.005)^{12x}$$

(We will find the future value of the investment after 5 years in Example 2.)
In this section we will evaluate and graph exponential functions, and we will model data with exponential functions.

Exponential Functions and Graphs

Suppose a culture of bacteria has the characteristic that each minute, every microorganism splits into two new organisms. We can describe the number of bacteria in the culture as a function of time. That is, if we begin the culture with one microorganism, we know that after 1 minute we will have two organisms, after 2 minutes, 4, and so on. Table 5.1 gives a few of the values that describe this growth. If x represents the number of minutes that have passed and y represents the number of organisms, the points (x, y) lie on the graph of the function with equation

$$y = 2^x$$

The equation $y = 2^x$ is an example of a special group of functions called **exponential functions**, defined as follows.

| TABLE 5.1 |

Minutes Passed	Number of Organisms
0	1
1	2
2	4
3	8
4	16

Exponential Functions

If a is a real number with $a > 0$ and $a \neq 1$, then the function

$$f(x) = a^x$$

is an **exponential function** with base a.

A table of some values satisfying $y = 2^x$ and the graph of this function are given in Figure 5.1. This function is said to model the growth of the number of organisms in the previous discussion, even though some points on the graph do not correspond to a time and a number of organisms. For example, time x could not be negative, and the number of organisms y could not be fractional.

x	$y = 2^x$
-3	$2^{-3} = \frac{1}{8}$
-2	$2^{-2} = \frac{1}{4}$
-1	$2^{-1} = \frac{1}{2}$
0	$2^0 = 1$
1	$2^1 = 2$
2	$2^2 = 4$
3	$2^3 = 8$

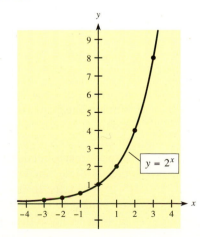

Figure 5.1

We defined rational powers of x in terms of radicals in Chapter 0, so 2^x makes sense for any rational power x. It can also be shown that the laws of exponents apply for irrational numbers. We will assume that if we graphed $y = 2^x$ for irrational values of x, those points would lie on the curve in Figure 5.1. Thus, in general, we can graph an exponential function by plotting easily calculated points, such as those in the table in Figure 5.1, and drawing a smooth curve through the points.

EXAMPLE 1 Graphing an Exponential Function

Graph $y = 10^x$.

Solution

A table of values and the graph are given in Figure 5.2.

x	y
-3	$10^{-3} = 1/1000$
-2	$10^{-2} = 1/100$
-1	$10^{-1} = 1/10$
0	$10^0 = 1$
1	$10^1 = 10$
2	$10^2 = 100$
3	$10^3 = 1000$

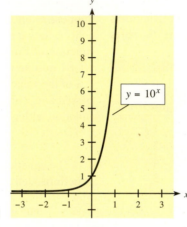

Figure 5.2

Calculator Note We also could have used a graphing calculator to graph $y = 2^x$ and $y = 10^x$. Figure 5.3 shows graphs of $f(x) = 1.5^x$, $f(x) = 4^x$, and $f(x) = 20^x$ that were obtained using a graphing calculator. Note the viewing window for each graph in the figure. See Appendix C, Section 5.1, for details.

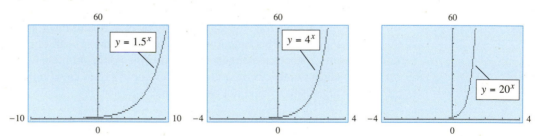

Figure 5.3

Note that the three graphs in Figure 5.3 and the graphs of $y = 2^x$ and $y = 10^x$ are similar. The graphs of $y = 2^x$ and $y = 10^x$ clearly approach, but do not touch, the negative x-axis. However, the graphs in Figure 5.3 appear as if they might eventually merge with the negative x-axis. By adjusting the viewing window, however, we would see that these graphs also approach, but do not touch, the negative x-axis. That is, the negative x-axis is an asymptote for these functions.

In fact, the shapes of the graphs of functions of the form $y = f(x) = a^x$, with $a > 1$, are similar to those in Figures 5.1, 5.2, and 5.3. Exponential functions of this type, and more generally of the form $f(x) = C(a^x)$, where $C > 0$ and $a > 1$, are called **exponential growth functions** because they are used to model growth in diverse applications. Their graphs have the basic shape shown in the following box.

Graphs of Exponential Growth Functions

Function: $y = f(x) = C(a^x)$ $(C > 0, a > 1)$
y-intercept: $(0, C)$
Graph shape:

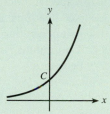

Domain: All real numbers;   Range: $y > 0$.
Asymptote: The x-axis (negative half)

One exponential growth model concerns money invested at compound interest. Often we seek to evaluate rather than graph these functions.

EXAMPLE **2** **Future Value of an Investment** | APPLICATION PREVIEW |

If \$10,000 is invested at 6%, compounded monthly, then the future value of the investment S after x years is given by

$$S = 10,000(1.005)^{12x}$$

Find the future value of the investment after (a) 5 years and (b) 30 years.

Solution
These future values can be found with a calculator.

(a) $S = 10,000(1.005)^{12(5)} = 10,000(1.005)^{60} = \$13,488.50$ (nearest cent)
(b) $S = 10,000(1.005)^{12(30)} = 10,000(1.005)^{360} = \$60,225.75$ (nearest cent)

Note that the amount after 30 years is significantly more than the amount after 5 years, a result consistent with exponential growth models. ■

iStockphoto.com/Martin McElligott

1. Can any value of x give a negative value for y if $y = a^x$ and $a > 1$?
2. If $a > 1$, what asymptote does the graph of $y = a^x$ approach?

A special function that occurs frequently in economics and biology is $y = e^x$, where e is a fixed irrational number (approximately $2.71828\ldots$). We will see how e arises when we discuss interest that is compounded continuously, and we will formally define e in Section 9.2, "Continuous Functions; Limits at Infinity."

Because $e > 1$, the graph of $y = e^x$ will have the same basic shape as other growth exponentials. We can calculate the y-coordinate for points on the graph of this function with a calculator. A table of some values (with y-values rounded to two decimal places) and the graph are shown in Figure 5.4.

x	$y = e^x$
-2	0.14
-1	0.37
0	1.00
1	2.72
2	7.39

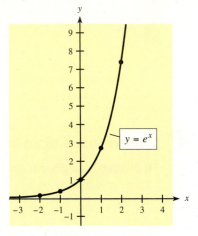

Figure 5.4

Exponentials whose bases are between 0 and 1, such as $y = \left(\frac{1}{2}\right)^x$, have graphs different from those of the exponentials just discussed. Using the properties of exponents, we have

$$y = \left(\frac{1}{2}\right)^x = (2^{-1})^x = 2^{-x}$$

By using this technique, all exponentials of the form $y = b^x$, where $0 < b < 1$, can be rewritten in the form $y = a^{-x}$, where $a = \frac{1}{b} > 1$. Thus, graphs of equations of the form $y = a^{-x}$, where $a > 1$, and of the form $y = b^x$, where $0 < b < 1$, will have the same shape.

EXAMPLE 3 Graphing an Exponential Function

Graph $y = 2^{-x}$.

Solution

A table of values and the graph are given in Figure 5.5.

x	$y = 2^{-x}$
-3	$2^3 = 8$
-2	$2^2 = 4$
-1	$2^1 = 2$
0	$2^0 = 1$
1	$2^{-1} = \frac{1}{2}$
2	$2^{-2} = \frac{1}{4}$
3	$2^{-3} = \frac{1}{8}$

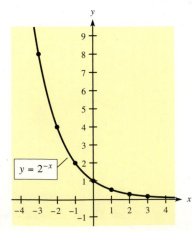

Figure 5.5

EXAMPLE 4   **Graphing an Exponential Function with Base e**

Graph $y = e^{-2x}$.

Solution

Using a calculator to find the values of powers of e (to 2 decimal places), we get the graph shown in Figure 5.6. ∎

x	$y = e^{-2x}$
-3	$e^6 = 403.43$
-2	$e^4 = 54.60$
-1	$e^2 = 7.39$
0	$e^0 = 1.00$
1	$e^{-2} = 0.14$
2	$e^{-4} = 0.02$

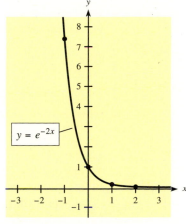

Figure 5.6

We can also use a graphing utility to graph exponential functions such as those in Examples 3 and 4. Additional examples of functions of this type would yield graphs with the same shape. **Exponential decay functions** have the form $y = a^{-x}$, where $a > 1$, or, more generally, the form $y = C(a^{-x})$, where $C > 0$ and $a > 1$. They model decay for various phenomena, and their graphs have the characteristics and shape shown in the following box.

Graphs of Exponential Decay Functions

Function: $y = f(x) = C(a^{-x})$ $(C > 0, a > 1)$ or
$y = f(x) = C(b^x)$ $(C > 0, 0 < b < 1)$

y-intercept: $(0, C)$

Graph shape:

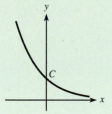

Domain: All real numbers;   Range: $y > 0$.
Asymptote: The x-axis (positive half)

Exponential decay functions can be used to describe various physical phenomena, such as the number of atoms of a radioactive element (radioactive decay), the lingering effect of an advertising campaign on new sales after the campaign ends, and the effect of inflation on the purchasing power of a fixed income.

EXAMPLE 5   Purchasing Power and Inflation

The purchasing power P of a fixed income of \$30,000 per year (such as a pension) after t years of 4% inflation can be modeled by

$$P = 30,000e^{-0.04t}$$

Find the purchasing power after (a) 5 years and (b) 20 years.

Solution
We can use a calculator to answer both parts.

(a) $P = 30,000e^{-0.04(5)} = 30,000e^{-0.2} = \$24,562$ (nearest dollar)

(b) $P = 30,000e^{-0.04(20)} = 30,000e^{-0.8} = \$13,480$ (nearest dollar)

Note that the impact of inflation over time significantly erodes purchasing power and provides some insight into the plight of elderly people who live on fixed incomes.   ■

Exponential functions with base e often arise in natural ways. As we will see in Section 6.2, the growth of money that is compounded continuously is given by $S = Pe^{rt}$, where P is the original principal, r is the annual interest rate, and t is the time in years. Certain populations (of insects, for example) grow exponentially, and the number of individuals can be closely approximated by the equation $y = P_0e^{ht}$, where P_0 is the original population size, h is a constant that depends on the type of population, and y is the population size at any instant t. Also, the amount y of a radioactive substance is modeled by the exponential decay equation $y = y_0e^{-kt}$, where y_0 is the original amount and k is a constant that depends on the radioactive substance.

There are other important exponential functions that use base e but whose graphs are different from those we have discussed. For example, the standard normal probability curve (often referred to as a bell-shaped curve) is the graph of an exponential function with base e (see Figure 5.7). Later in this chapter we will study other exponential functions that model growth but whose graphs are also different from those discussed previously.

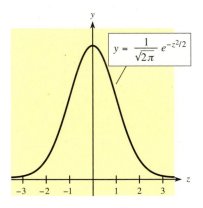

Figure 5.7   Standard normal probability curve

CHECKPOINT

3. True or false: The graph of $y = \left(\dfrac{1}{a}\right)^x$, with $a > 1$, is the same as the graph of $y = a^{-x}$.

4. True or false: The graph of $y = a^{-x}$, with $a > 1$, approaches the positive x-axis as an asymptote.

5. True or false: The graph of $y = 2^{-x}$ is the graph of $y = 2^x$ reflected about the y-axis.

Calculator Note

Graphing calculators allow us to investigate variations of the growth and decay exponentials. For example, we can examine the similarities and differences between the graphs of

$y = f(x) = a^x$ and those of $y = mf(x), y = f(kx)$, $y = f(x + h)$, and $y = f(x) + C$ for constants m, k, h, and C. We also can explore these similarities and differences for $y = f(x) = a^{-x}$. Figure 5.8 shows graphs of $y = f(x) = e^x$ and $y = mf(x) = me^x$ for $m = -6, -3, 2,$ and 10.

From the graphs, we see that when $m > 0$, the shape of the graph is still that of a growth exponential. When $m < 0$, the shape is that of a growth exponential turned upside down, or reflected through the x-axis (making the range $y < 0$). In all cases, the asymptote is unchanged and is the negative x-axis. In each case, the y-intercept is $(0, m)$.

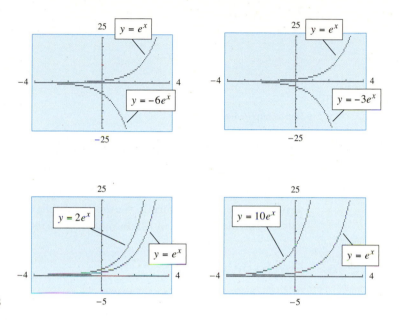

Figure 5.8

Because the outputs of exponential functions grow very rapidly, it is sometimes hard to find an appropriate window in which to show the graph of an exponential function on a graphing calculator. One way to find appropriate values of y-min and y-max when x-min and x-max are given is to use TABLE to find values of y for selected values of x. See Appendix C, Section 5.1 for details.

Figure 5.9(a) shows a table of values of $y = 1000e^{0.12x}$ that correspond to certain values of x, and Figure 5.9(b) shows the graph of this function in a window that uses the table values to set y-max equal to 3500, with y-min $= 0$.   ■

Figure 5.9

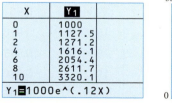
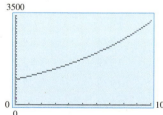

Modeling with Exponential Functions   Many types of data can be modeled using an exponential growth function. Figure 5.10(a) on the next page shows a graph of amounts of carbon in the atmosphere due to emissions from the burning of fossil fuels. With curve-fitting tools available with some computer software or on graphing calculators, we can develop an equation that models, or approximates, these data. The model and its graph are shown in Figure 5.10(b) on the next page.

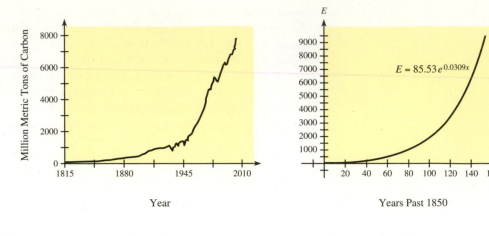

Figure 5.10

(a)

(b)

Source: http://cdiac.ornl.gov/

Calculator Note   The model in Figure 5.10(b) is an exponential growth function. As we have noted, functions of this type have the general form $y = C(a^x)$ with $a > 1$, and exponential decay functions have the general form $y = C(a^{-x})$ with $a > 1$ [or $y = C(b^x)$ with $0 < b < 1$]. Note that when technology is applied to model exponentials (either growth or decay), the model is written in an equivalent form $y = a*b^x$. The details of modeling exponential functions with a graphing calculator are shown in Appendix C, Section 5.1.   ■

EXAMPLE   6   Purchasing Power

The purchasing power of $1 is calculated by assuming that the consumer price index (CPI) was $1 during 1983. It is used to show the values of goods that could be purchased for $1 in 1983. Table 5.2 gives the purchasing power of a 1983 dollar for the years 1968–2008.

(a) Use these data, with $x = 0$ in 1960, to find an exponential function that models the decay of the dollar.

(b) What does this model predict as the purchasing power of $1 in 2016?

(c) When will the purchasing power of a dollar fall below $0.20?

| TABLE 5.2 |

Year	Purchasing Power of $1	Year	Purchasing Power of $1	Year	Purchasing Power of $1
1968	2.873	1982	1.035	1996	0.637
1969	2.726	1983	1.003	1997	0.623
1970	2.574	1984	0.961	1998	0.613
1971	2.466	1985	0.928	1999	0.600
1972	2.391	1986	0.913	2000	0.581
1973	2.251	1987	0.880	2001	0.565
1974	2.029	1988	0.846	2002	0.556
1975	1.859	1989	0.807	2003	0.543
1976	1.757	1990	0.766	2004	0.529
1977	1.649	1991	0.734	2005	0.512
1978	1.532	1992	0.713	2006	0.500
1979	1.380	1993	0.692	2007	0.482
1980	1.215	1994	0.675	2008	0.464
1981	1.098	1995	0.656		

Source: U.S. Bureau of Labor Statistics

Solution

(a) From a scatter plot of the points, with $x = 0$ in 1960, it appears that an exponential decay model is appropriate (see Figure 5.11(a)). Using the function of the form $a*b^x$, the data can be modeled by the equation $y = 3.443(0.9556^x)$ with $x = 0$ in 1960. Figure 5.11(b) shows how the graph of the equation fits the data.

To write this equation with a base greater than 1, we can write 0.9556 as $1/(0.9556)^{-1} = 1.0465^{-1}$. Thus the model also could be written as

$$y = 3.443(1.0465^{-x})$$

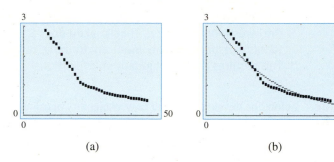

Figure 5.11   (a)   (b)

(b) Recall that we use an unrounded model in our calculations and graphs, unless otherwise instructed. For the year 2016, we use $x = 56$. Evaluating the function for $x = 56$ gives the purchasing power of $0.270.

(c) Evaluation from the graph or a table shows that the purchasing power is below $0.20 when x is 63, or in the year 2023, if the model remains valid. ■

Spreadsheet Note

A spreadsheet also can be used to create a scatter plot of the data in Table 5.2 and to develop an exponential model. Figure 5.12(a) shows an Excel scatter plot of the Table 5.2 data, and Figure 5.12(b) shows the scatter plot with Excel's base e exponential model. (Procedural details for using Excel are in Appendix D, Section 5.1, and in the Online Excel Guide.) Note that the model $y = 3.443(0.9556^x)$ is equivalent to $y = 3.443e^{-0.0454x}$ because $e^{-0.0454} \approx 0.9556$. ■

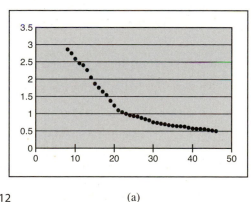

Figure 5.12   (a)   (b)

CHECKPOINT SOLUTIONS

1. No, all values of y are positive.
2. The left side of the x-axis ($y = 0$)
3. True
4. True
5. True

| **EXERCISES** | 5.1

Access end-of-section exercises online at **www.webassign.net**

Logarithmic Functions and Their Properties

| APPLICATION PREVIEW |

If P dollars is invested at an annual rate r, compounded continuously, then the future value of the investment after t years is given by

$$S = Pe^{rt}$$

A common question with investments such as this is "How long will it be before the investment doubles?" That is, when does $S = 2P$? (See Example 3.) The answer to this question gives an important formula, called a "doubling-time" formula. This requires the use of logarithmic functions.

Logarithmic Functions and Graphs

Before the development and easy availability of calculators and computers, certain arithmetic computations, such as $(1.37)^{13}$ and $\sqrt[16]{3.09}$, were difficult to perform. The computations could be performed relatively easily using **logarithms,** which were developed in the 17th century by John Napier, or by using a slide rule, which is based on logarithms. The use of logarithms as a computing technique has all but disappeared today, but the study of **logarithmic functions** is still very important because of the many applications of these functions.

For example, let us again consider the culture of bacteria described at the beginning of the previous section. If we know that the culture is begun with one microorganism and that each minute every microorganism present splits into two new ones, then we can find the number of minutes it takes until there are 1024 organisms by solving

$$1024 = 2^y$$

The solution of this equation may be written in the form

$$y = \log_2 1024$$

which is read "y equals the logarithm of 1024 to the base 2."

In general, we may express the equation $x = a^y \ (a > 0, a \neq 1)$ in the form $y = f(x)$ by defining a **logarithmic function.**

Logarithmic Function

For $a > 0$ and $a \neq 1$, the **logarithmic function**

$$y = \log_a x \ \text{(logarithmic form)}$$

has domain $x > 0$, base a, and is defined by

$$a^y = x \ \text{(exponential form)}$$

| TABLE 5.3 |

Logarithmic Form	Exponential Form
$\log_{10} 100 = 2$	$10^2 = 100$
$\log_{10} 0.1 = -1$	$10^{-1} = 0.1$
$\log_2 x = y$	$2^y = x$
$\log_a 1 = 0 \ (a > 0)$	$a^0 = 1$
$\log_a a = 1 \ (a > 0)$	$a^1 = a$

From the definition, we know that $\log_3 81 = 4$ because $3^4 = 81$. In this case the logarithm, 4, is the exponent to which we have to raise the base 3 to obtain 81. In general, if $y = \log_a x$, then y is the exponent to which the base a must be raised to obtain x.

The a is called the **base** in both $\log_a x = y$ and $a^y = x$, and y is the *logarithm* in $\log_a x = y$ and the *exponent* in $a^y = x$. Thus **a logarithm is an exponent.**

Table 5.3 shows some logarithmic equations and their equivalent exponential forms.

EXAMPLE 1 Logarithms and Exponential Forms

(a) Write $64 = 4^3$ in logarithmic form.
(b) Write $\log_4\left(\frac{1}{64}\right) = -3$ in exponential form.
(c) If $4 = \log_2 x$, find x.
(d) Evaluate: $\log_2 8$
(e) Evaluate: $\log_5\left(\frac{1}{25}\right)$

Solution

(a) In $64 = 4^3$, the base is 4 and the exponent (or logarithm) is 3. Thus $64 = 4^3$ is equivalent to $3 = \log_4 64$.
(b) In $\log_4\left(\frac{1}{64}\right) = -3$, the base is 4 and the logarithm (or exponent) is -3. Thus $\log_4\left(\frac{1}{64}\right) = -3$ is equivalent to $4^{-3} = \frac{1}{64}$.
(c) If $4 = \log_2 x$, then $2^4 = x$ and $x = 16$.
(d) If $y = \log_2 8$, then $8 = 2^y$. Because $2^3 = 8$, $\log_2 8 = 3$.
(e) If $y = \log_5\left(\frac{1}{25}\right)$, then $\frac{1}{25} = 5^y$. Because $5^{-2} = \frac{1}{25}$, $\log_5\left(\frac{1}{25}\right) = -2$. ∎

EXAMPLE 2 Graphing a Logarithmic Function

Graph $y = \log_2 x$.

Solution

We may graph $y = \log_2 x$ by graphing $x = 2^y$. The table of values (found by substituting values of y and calculating x) and the graph are shown in Figure 5.13. ∎

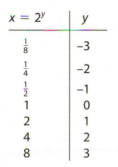

$x = 2^y$	y
$\frac{1}{8}$	-3
$\frac{1}{4}$	-2
$\frac{1}{2}$	-1
1	0
2	1
4	2
8	3

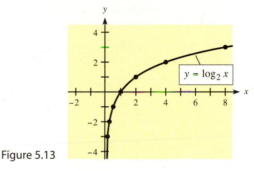

Figure 5.13

From the definition of logarithms, we see that every logarithm has a base. Most applications of logarithms involve logarithms to the base 10 (called **common logarithms**) or logarithms to the base e (called **natural logarithms**). In fact, logarithms to the base 10 and to the base e are the only ones that have function keys on calculators. Thus it is important to be familiar with their names and designations.

Common and Natural Logarithms

Common logarithms:	$\log x$	means	$\log_{10} x$.
Natural logarithms:	$\ln x$	means	$\log_e x$.

Values of common and natural logarithmic functions are usually found with a calculator. For example, a calculator gives $\log 2 \approx 0.301$ and $\ln 2 \approx 0.693$.

EXAMPLE 3 Doubling Time for an Investment | APPLICATION PREVIEW |

If \$$P$ is invested for t years at interest rate r, compounded continuously, then the future value of the investment is given by $S = Pe^{rt}$. The doubling time for this investment can be found by solving for t in $S = Pe^{rt}$ when $S = 2P$. That is, we must solve $2P = Pe^{rt}$, or (equivalently) $2 = e^{rt}$. Express this equation in logarithmic form and then solve for t to find the doubling-time formula.

Solution

In logarithmic form, $2 = e^{rt}$ is equivalent to $\log_e 2 = rt$. Solving for t gives the doubling-time formula

$$t = \frac{\log_e 2}{r} = \frac{\ln 2}{r}$$

Note that if the interest rate is $r = 10\%$, compounded continuously, the time required for the investment to double is

$$t = \frac{\ln 2}{0.10} \approx \frac{0.693}{0.10} = \frac{69.3}{10} \approx 6.93$$

In general we can approximate the doubling time for an investment at $r\%$, compounded continuously, with $\frac{70}{r}$. (In economics, this is called the **Rule of 70.**) ■

EXAMPLE 4   **Market Share**

Suppose that after a company introduces a new product, the number of months m before its market share is s percent can be modeled by

$$m = 20 \ln \left(\frac{40}{40 - s} \right)$$

When will its product have a 35% share of the market?

Solution

A 35% market share means $s = 35$. Hence

$$m = 20 \ln \left(\frac{40}{40 - s} \right) = 20 \ln \left(\frac{40}{40 - 35} \right) = 20 \ln \left(\frac{40}{5} \right) = 20 \ln 8 \approx 41.6$$

Thus, the market share will be 35% after about 41.6 months. ■

EXAMPLE 5   **Graphing the Natural Logarithm**

Graph $y = \ln x$.

Solution

We can graph $y = \ln x$ by evaluating $y = \ln x$ for $x > 0$ (including some values $0 < x < 1$) with a calculator. The graph is shown in Figure 5.14. ■

x	$y = \ln x$
0.05	-3.000
0.10	-2.303
0.50	-0.693
1	0.000
2	0.693
3	1.099
5	1.609
10	2.303

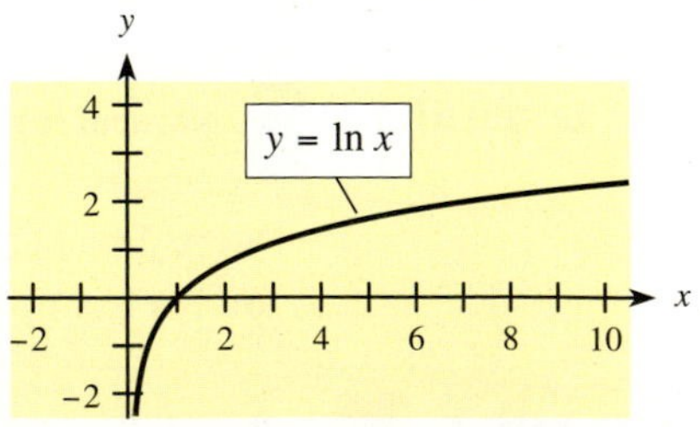

Figure 5.14

Note that because $\ln x$ is a standard function on a graphing utility, such a utility could be used to obtain its graph. Note also that the graphs of $y = \log_2 x$ (Figure 5.13) and $y = \ln x$ (Figure 5.14) are very similar. The shapes of graphs of equations of the form $y = \log_a x$ with $a > 1$ are similar to these two graphs.

| **Graphs of Logarithmic Functions** | Equation: $y = \log_a x \quad (a > 1)$
x-intercept: $(1, 0)$
Domain: All positive reals
Range: All reals
Asymptote: y-axis (negative half) | 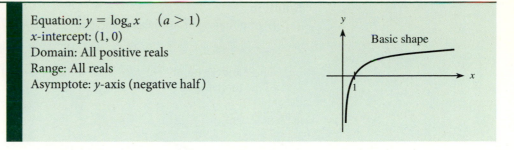 |

CHECKPOINT

1. What asymptote does the graph of $y = \log_a x$ approach when $a > 1$?
2. For $a > 1$, does the equation $y = \log_a x$ represent the same function as the equation $x = a^y$?
3. For what values of x is $y = \log_a x$, $a > 0$, $a \neq 1$, defined?

Modeling with Logarithmic Functions

The basic shape of a logarithmic function is important for two reasons. First, when we graph a logarithmic function, we know that the graph should have this shape. Second, when data points have this basic shape, they suggest a logarithmic model.

 EXAMPLE 6   **Life Expectancy**

The expected life span of people in the United States depends on their year of birth (see the table and scatter plot, with $x = 0$ in 1900, in Figure 5.15). In Section 2.5, "Modeling; Fitting Curves to Data with Graphing Utilities," we modeled life span with a linear function and with a quadratic function. However, the scatter plot suggests that the best model may be logarithmic.

(a) Use technology to find a logarithmic equation that models the data.
(b) The National Center for Health Statistics projects the expected life span for people born in 2000 to be 77.0 and that for those born in 2010 to be 78.1. Use your model to project the life spans for those years.
(c) Could the expected life spans for 2015 and 2020, as given in the table, have been found with the model found in part (a)?

Life Span		**Life Span**	
Year	**(years)**	**Year**	**(years)**
1920	54.1	1994	75.7
1930	59.7	1996	76.1
1940	62.9	1998	76.7
1950	68.2	1999	76.7
1960	69.7	2000	77.0
1970	70.8	2001	77.2
1975	72.6	2002	77.3
1980	73.7	2003	77.5
1987	75.0	2004	77.8
1988	74.9	2005	77.9
1989	75.2	2010	78.1
1990	75.4	2015	78.9
1992	75.8	2020	79.5

Source: National Center for Health Statistics

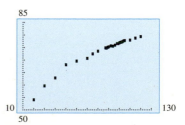

Figure 5.15

Solution

(a) A graphing calculator gives the logarithmic model

$$L(x) = 11.027 + 14.304 \ln x$$

where x is the number of years since 1900. Figure 5.16 shows the scatter plot and graph of the model.

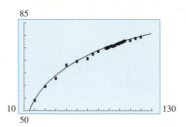

Figure 5.16

(b) For the year 2000, the value of x is 100, so the projected life span is $L(100) = 76.9$. For 2010, the model gives 78.3 as the expected life span. These calculations closely approximate the projections from the National Center for Health Statistics.

(c) Yes, the values from the model match perfectly. ■

Properties of Logarithms   The definition of the logarithmic function and the previous examples suggest a special relationship between the logarithmic function $y = \log_a x$ and the exponential function $y = a^x$ ($a > 0$, $a \neq 1$). Because we can write $y = \log_a x$ in exponential form as $x = a^y$, we see that the connection between

$$y = \log_a x \quad \text{and} \quad y = a^x$$

is that x and y have been interchanged from one function to the other. This is true for the functional description and hence for the ordered pairs or coordinates that satisfy these functions. This is illustrated in Table 5.4 for the functions $y = \log x$ ($y = \log_{10} x$) and $y = 10^x$.

| TABLE 5.4 |

$y = \log_{10} x = \log x$		$y = 10^x$	
Coordinates	Justification	Coordinates	Justification
$(1000, 3)$	$3 = \log 1000$	$(3, 1000)$	$1000 = 10^3$
$(100, 2)$	$2 = \log 100$	$(2, 100)$	$100 = 10^2$
$\left(\frac{1}{10}, -1\right)$	$-1 = \log \frac{1}{10}$	$\left(-1, \frac{1}{10}\right)$	$\frac{1}{10} = 10^{-1}$

In general we say that $y = f(x)$ and $y = g(x)$ are **inverse functions** if, whenever the pair (a, b) satisfies $y = f(x)$, the pair (b, a) satisfies $y = g(x)$. Furthermore, because the values of the x- and y-coordinates are interchanged for inverse functions, their graphs are reflections of each other about the line $y = x$.

Thus for $a > 0$ and $a \neq 1$, the logarithmic function $y = \log_a x$ (also written $x = a^y$) and the exponential function $y = a^x$ are inverse functions.

The logarithmic function $y = \log x$ is the inverse of the exponential function $y = 10^x$. Thus the graphs of $y = \log x$ and $y = 10^x$ are reflections of each other about the line $y = x$. Some values of x and y for these functions are given in Table 5.4, and their graphs are shown in Figure 5.17.

This inverse relationship is a consequence of the definition of the logarithmic function. We can use this definition to discover several other properties of logarithms.

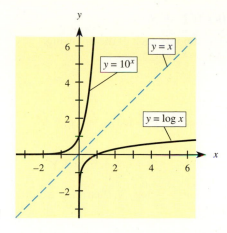

Figure 5.17

Because logarithms are exponents, the properties of logarithms can be derived from the properties of exponents. (The properties of exponents are discussed in Chapter 0, "Algebraic Concepts.") The following properties of logarithms are useful in simplifying expressions that contain logarithms.

Logarithmic Property I	If $a > 0$, $a \neq 1$, then $\log_a a^x = x$, for any real number x.

To prove this result, note that the exponential form of $y = \log_a a^x$ is $a^y = a^x$, so $y = x$. That is, $\log_a a^x = x$.

EXAMPLE 7   Logarithmic Property I

Use Property I to simplify each of the following.
(a) $\log_4 4^3$ (b) $\ln e^x$

Solution
(a) $\log_4 4^3 = 3$
Check: The exponential form is $4^3 = 4^3$. ✔

(b) $\ln e^x = \log_e e^x = x$
Check: The exponential form is $e^x = e^x$. ✔   ■

We note that two special cases of Property I are used frequently; these are when $x = 1$ and $x = 0$.

Special Cases of Logarithmic Property I	Because $a^1 = a$, we have $\log_a a = 1$. Because $a^0 = 1$, we have $\log_a 1 = 0$.

The logarithmic form of $y = a^{\log_a x}$ is $\log_a y = \log_a x$, so $y = x$. This means that $a^{\log_a x} = x$ and proves Property II.

Logarithmic Property II	If $a > 0$, $a \neq 1$, then $a^{\log_a x} = x$, for any positive real number x.

EXAMPLE 8 **Logarithmic Property II**

Use Property II to simplify each of the following.
(a) $2^{\log_2 4}$ (b) $e^{\ln x}$

Solution

(a) $2^{\log_2 4} = 4$
Check: The logarithmic form is $\log_2 4 = \log_2 4$. ✔

(b) $e^{\ln x} = x$
Check: The logarithmic form is $\log_e x = \ln x$. ✔

■

If $u = \log_a M$ and $v = \log_a N$, then their exponential forms are $a^u = M$ and $a^v = N$, respectively. Thus

$$\log_a (MN) = \log_a (a^u \cdot a^v) = \log_a (a^{u+v}) = u + v = \log_a M + \log_a N$$

and Property III is established.

Logarithmic Property III	If $a > 0$, $a \neq 1$, and M and N are positive real numbers, then $$\log_a (MN) = \log_a M + \log_a N$$

EXAMPLE 9 **Logarithmic Property III**

(a) Find $\log_2 (4 \cdot 16)$, if $\log_2 4 = 2$ and $\log_2 16 = 4$.
(b) Find $\ln 77$, if $\ln 7 = 1.9459$ and $\ln 11 = 2.3979$ (to 4 decimal places).

Solution

(a) $\log_2 (4 \cdot 16) = \log_2 4 + \log_2 16 = 2 + 4 = 6$
Check: $\log_2 (4 \cdot 16) = \log_2 (64) = 6$, because $2^6 = 64$. ✔

(b) $\ln 77 = \ln (7 \cdot 11) = \ln 7 + \ln 11 = 1.9459 + 2.3979 = 4.3438$
Check: Using a calculator, we can see that $e^{4.3438} \approx 77$. ✔

■

Logarithmic Property IV	If $a > 0$, $a \neq 1$, and M and N are positive real numbers, then $$\log_a (M/N) = \log_a M - \log_a N$$

The proof of this property is left to the student.

EXAMPLE 10 **Logarithmic Property IV**

(a) Evaluate $\log_3 \left(\frac{9}{27}\right)$.
(b) Find $\log_{10} \left(\frac{16}{5}\right)$, if $\log_{10} 16 = 1.2041$ and $\log_{10} 5 = 0.6990$ (to 4 decimal places).

Solution

(a) $\log_3 \left(\frac{9}{27}\right) = \log_3 9 - \log_3 27 = 2 - 3 = -1$
Check: $\log_3 \left(\frac{1}{3}\right) = -1$ because $3^{-1} = \frac{1}{3}$. ✔

(b) $\log_{10} \left(\frac{16}{5}\right) = \log_{10} 16 - \log_{10} 5 = 1.2041 - 0.6990 = 0.5051$
Check: Using a calculator, we can see that $10^{0.5051} \approx 3.2 = \frac{16}{5}$. ✔

■

Logarithmic Property V	If $a > 0$, $a \neq 1$, M is a positive real number, and N is any real number, then $$\log_a (M^N) = N \log_a M$$

The proof of this property is left to the student.

EXAMPLE **11**   **Logarithmic Property V**

(a) Simplify $\log_3 (9^2)$.

(b) Simplify $\ln 8^{-4}$, if $\ln 8 = 2.0794$ (to 4 decimal places).

Solution

(a) $\log_3 (9^2) = 2 \log_3 9 = 2 \cdot 2 = 4$

(b) $\ln 8^{-4} = -4 \ln 8 = -4(2.0794) = -8.3176$ ■

CHECKPOINT

4. Simplify:

   (a) $6^{\log_6 x}$ (b) $\log_7 7^3$ (c) $\ln \left(\dfrac{1}{e^2} \right)$ (d) $\log 1$

5. If $\log_7 20 = 1.5395$ and $\log_7 11 = 1.2323$, find the following.

   (a) $\log_7 220 = \log_7 (20 \cdot 11)$ (b) $\log_7 \left(\dfrac{11}{20} \right)$ (c) $\log_7 20^4$

Change of Base   With a calculator, we can directly evaluate only those logarithms with base 10 or base e. Also, logarithms with base 10 or base e are the only ones that are standard functions on a graphing calculator. Thus, if we had a way to express a logarithmic function with any base in terms of a logarithm with base 10 or base e, we would be able to evaluate the original logarithmic function and graph it with a graphing calculator.

In general, if we use the properties of logarithms, we can write

$$y = \log_b x \quad \text{in the form} \quad b^y = x$$

If we take the base a logarithm of both sides, we have

$$\log_a b^y = \log_a x$$
$$y \log_a b = \log_a x$$
$$y = \frac{\log_a x}{\log_a b}$$

This gives us the **change-of-base formula** from base b to base a.

Change-of-Base Formulas

If $a \neq 1, b \neq 1, a > 0, b > 0$, then

$$\log_b x = \frac{\log_a x}{\log_a b}$$

For calculation purposes, we can convert logarithms to base e or base 10.

Base e: $\log_b x = \dfrac{\ln x}{\ln b}$ Base 10: $\log_b x = \dfrac{\log x}{\log b}$

EXAMPLE **12**   **Change-of-Base Formula**

Evaluate $\log_7 15$ by using a change-of-base formula.

Solution

$$\log_7 15 = \frac{\ln 15}{\ln 7} = \frac{2.70805}{1.94591} = 1.39166 \ (\text{approximately})$$

As a check, a calculator shows that $7^{1.39166} \approx 15$. ✔   ■

Calculator Note

The change-of-base formula expands the capability of graphing calculators so they can be used to evaluate logarithms and to graph logarithmic functions to any base. See Appendix C, Section 5.2, for details.

Figure 5.18(a) shows the graph of $y = \log_2 x = \dfrac{\ln x}{\ln 2}$. Note that it is exactly the same as the graph in Figure 5.13 in Example 2.

Figure 5.18(b) shows the graph of $y = \log_7 x = \dfrac{\ln x}{\ln 7}$. ∎

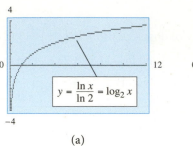

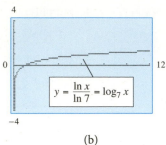

Figure 5.18   (a)   (b)

Spreadsheet Note

The graph of a logarithm with any base b can be created with Excel by entering the formula " $= \log(x, b)$." See Appendix D, Section 5.2, and the Online Excel Guide for details. ∎

Natural logarithms, $y = \ln x$ (and the inverse exponentials with base e), have many practical applications, some of which are considered in the next section. Common logarithms, $y = \log x$, were widely used for computation before computers and calculators became popular. They also have several applications to scaling variables, where the purpose is to reduce the scale of variation when a natural physical variable covers a wide range.

EXAMPLE 13   **Richter Scale**

The Richter scale is used to measure the intensity of an earthquake. The magnitude on the Richter scale of an earthquake of intensity I is given by

$$R = \log(I/I_0)$$

where I_0 is a certain minimum intensity used for comparison.

(a) Find R if I is 15,800,000 times as great as I_0.
(b) As of 2010, the 1964 Alaskan earthquake was the most violent U.S. earthquake and the second strongest ever; it measured 9.2 on the Richter scale. Find the intensity of the 1964 Alaskan earthquake.

Solution

(a) If $I = 15,800,000 I_0$, then $I/I_0 = 15,800,000$. Hence

$$R = \log(15,800,000)$$
$$= 7.2 \text{ (approximated to 1 decimal place)}$$

(b) For $R = 9.2$, it follows that

$$9.2 = \log(I/I_0)$$

Rewriting this in exponential form gives

$$10^{9.2} = I/I_0$$

and from a calculator we obtain

$$I/I_0 = 1{,}580{,}000{,}000 \text{ (approximately)}$$

Thus the intensity is 1,580,000,000 times I_0.

Note that a Richter scale measurement that is 2 units larger means that the intensity is $10^2 = 100$ *times* greater. ∎

CHECKPOINT SOLUTIONS

1. The (negative) y-axis
2. Yes
3. $x > 0$ is the domain of $\log_a x$.
4. (a) x, by Property II
   (b) 3, by Property I
   (c) $\ln\left(\dfrac{1}{e^2}\right) = \ln\left(e^{-2}\right) = -2$, by Property I
   (d) 0, because $\log_a 1 = 0$ for any base a.
5. (a) By Property III,
   $$\log_7 (20 \cdot 11) = \log_7 20 + \log_7 11 = 1.5395 + 1.2323 = 2.7718$$
   (b) By Property IV,
   $$\log_7 \left(\tfrac{11}{20}\right) = \log_7 11 - \log_7 20 = 1.2323 - 1.5395 = -0.3072$$
   (c) By Property V,
   $$\log_7 20^4 = 4\log_7 20 = 4(1.5395) = 6.1580$$

| EXERCISES | 5.2

Access end-of-section exercises online at **www.webassign.net**

OBJECTIVES   5.3

- To solve exponential equations
- To solve exponential growth or decay equations when sufficient data are known
- To solve exponential equations representing demand, supply, total revenue, or total cost when sufficient data are known
- To model and solve problems involving Gompertz curves and logistic functions

Solutions of Exponential Equations: Applications of Exponential and Logarithmic Functions

| APPLICATION PREVIEW |

A company's sales decline following an advertising campaign so that the number of sales is $S = 2000(2^{-0.1x})$, where x is the number of days after the campaign ends. When should the company begin a new campaign? In other words, how many days will elapse before sales drop below 350 (see Example 3)?

In this section, we discuss applications of exponential and logarithmic functions to the solutions of problems involving exponential growth and decay, and supply and demand. With exponential decay, we consider various applications, including decay of radioactive materials and decay of sales following an advertising campaign. With exponential growth, we focus on population growth and growth that can be modeled by Gompertz curves and logistic functions. Finally, we consider applications of demand functions that can be modeled by exponential functions.

Solving Exponential Equations Using Logarithmic Properties

Questions such as the one in the Application Preview can be answered by solving an equation in which the variable appears in the exponent, called an **exponential equation.** Logarithmic properties are useful in converting an equation of this type to one that can be solved easily. In this section we limit our discussion and applications to equations involving a single exponential.

Solving Exponential Equations

Procedure	Example
To solve an exponential equation by using properties of logarithms:	Solve: $4(25^{2x}) = 312{,}500$
1. Isolate the exponential by rewriting the equation with a base raised to a power on one side.	1. $\dfrac{4(25^{2x})}{4} = \dfrac{312{,}500}{4}$ $25^{2x} = 78{,}125$
2. Take the logarithm (either base e or base 10) of both sides.	2. We choose base e, although base 10 would work similarly. $\ln(25^{2x}) = \ln(78{,}125)$
3. Use a property of logarithms to remove the variable from the exponent.	3. Using logarithmic Property V gives $2x \ln 25 = \ln(78{,}125)$
4. Solve for the variable.	4. Dividing both sides by $2 \ln 25$ gives $x = \dfrac{\ln(78{,}125)}{2 \ln 25} = 1.75$

In Step 2 we noted that using a logarithm base 10 would work similarly, and it gives the same result (try it). Unless the exponential in the equation has base e or base 10, the choice of a logarithm in Step 2 makes no difference. However, if the exponential in the equation has base e, then choosing a base e logarithm is slightly easier (see Examples 2 and 4), and similarly for choosing a base 10 logarithm when the equation has a base 10 exponential.

EXAMPLE 1 Solving an Exponential Equation

Solve the equation $6(4^{3x-2}) = 120$.

Solution

We first divide both sides by 6 to isolate the exponential.

$$4^{3x-2} = 20$$

Taking the logarithm, base 10, of both sides leads to the solution.

$$\log 4^{3x-2} = \log 20$$

$$(3x - 2) \log 4 = \log 20$$

$$3x - 2 = \frac{\log 20}{\log 4}$$

$$x = \frac{1}{3}\left(\frac{\log 20}{\log 4} + 2\right) \approx 1.387$$

An alternate method of solving the equation is to write $4^{3x-2} = 20$ in logarithmic form.

$$\log_4 20 = 3x - 2$$

$$x = \frac{\log_4 20 + 2}{3} \approx 1.387$$

The change-of-base formula can be used to verify that these solutions are the same. ■

Growth and Decay   Recall that **exponential decay models** are those that can be described by a function of the following form.

Decay Models	$f(x) = C(a^{-x})$ with $a > 1$ and $C > 0$ or $f(x) = C(b^x)$ with $0 < b < 1$ and $C > 0$

Some applications that use exponential decay models are radioactive decay, demand curves, sales decay, and blood pressure in the aorta, illustrated in the following example.

EXAMPLE 2   Pressure in the Aorta

Medical research has shown that over short periods of time when the valves to the aorta of a normal adult close, the pressure in the aorta is a function of time and can be modeled by the equation

$$P = 95e^{-0.491t}$$

where t is in seconds. How long will it be before the pressure reaches 80?

Solution
Setting $P = 80$ and solving for t will give us the length of time before the pressure reaches 80.

$$80 = 95e^{-0.491t}$$

To solve this equation for t, we first isolate the exponential containing t by dividing both sides by 95.

$$\frac{80}{95} = e^{-0.491t}$$

Because e is the base of the exponential, we take the logarithm, base e, of both sides, then use logarithmic Property I to simplify the right-hand side of the expression.

$$\ln\left(\frac{80}{95}\right) = \ln e^{-0.491t} = -0.491t$$

Evaluating the left-hand side with a calculator gives $-0.172 = -0.491t$. Thus,

$$\frac{-0.172}{-0.491} = t \quad \text{so} \quad t = 0.35 \text{ second} \quad \text{(approximately)} \qquad ■$$

EXAMPLE 3   Sales Decay | APPLICATION PREVIEW |

A company finds that its daily sales begin to fall after the end of an advertising campaign, and the decline is such that the number of sales is $S = 2000(2^{-0.1x})$, where x is the number of days after the campaign ends.

(a) How many sales will be made 10 days after the end of the campaign?
(b) If the company does not want sales to drop below 350 per day, when should it start a new campaign?

Solution
(a) If $x = 10$, sales are given by $S = 2000(2^{-1}) = 1000$.

(b) Setting $S = 350$ and solving for x will give us the number of days after the end of the campaign when sales will reach 350.

$$350 = 2000(2^{-0.1x})$$

$$\frac{350}{2000} = 2^{-0.1x} \qquad \text{so} \qquad 0.175 = 2^{-0.1x}$$

With the base 2 exponential isolated, we take logarithms of both sides. This time we choose base 10 logarithms and complete the solution as follows.

$$\log 0.175 = \log(2^{-0.1x})$$

$$\log 0.175 = (-0.1x)(\log 2) \qquad \text{\textcolor{red}{Property V}}$$

$$\frac{\log 0.175}{\log 2} = -0.1x$$

$$-2.515 \approx -0.1x \qquad \text{so} \qquad x \approx 25.15$$

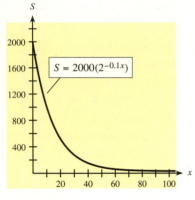

Figure 5.19

Thus sales will be 350 at day 25.15, and fall below 350 during the 26th day after the end of the campaign. If a new campaign isn't begun on or before the 26th day, sales will drop below 350. (See Figure 5.19.) ■

In business, economics, biology, and the social sciences, **exponential growth models** describe the growth of money, bacteria, or population.

Growth Models | $f(x) = C(a^x)$ with $a > 1$ and $C > 0$

The function $y = P_0 e^{ht}$ can be used to model population growth for humans, insects, or bacteria.

EXAMPLE 4 **Population**

The population of a certain city was 30,000 in 1990 and 40,500 in 2000. If the formula $P = P_0 e^{ht}$ applies to the growth of the city's population, what should the population be in the year 2020?

Solution

We can first use the data from 1990 ($t = 0$) and 2000 ($t = 10$) to find the value of h in the formula. Letting $P_0 = 30,000$ and $P = 40,500$, we get

$$40,500 = 30,000e^{h(10)}$$

$$1.35 = e^{10h} \qquad \text{\textcolor{red}{Isolate the exponential.}}$$

Taking the natural logarithms of both sides and using logarithmic Property I give

$$\ln 1.35 = \ln e^{10h} = 10h$$

$$0.3001 = 10h \qquad \text{so} \qquad h \approx 0.0300$$

Thus the formula for this population is $P = P_0 e^{0.03t}$. To predict the population for the year 2020, we set the most recent data point (for 2000) as $t = 0$, $P_0 = 40,500$, and find P for $t = 20$. This gives

$$P = 40,500e^{0.03(20)} = 40,500e^{0.6} = 40,500(1.8221) \approx 73,795 \qquad ■$$

CHECKPOINT 1. Suppose the sales of a product, in dollars, are given by $S = 1000e^{-0.07x}$, where x is the number of days after the end of an advertising campaign.
(a) What are sales 2 days after the end of the campaign?
(b) How long will it be before sales are $300?

Economic and Management Applications

Sometimes cost, revenue, demand, and supply may also be modeled by exponential or logarithmic equations. For example, suppose the demand for a product is given by $p = 30(3^{-q/2})$, where q is the number of thousands of units demanded at a price of p dollars per unit. Then the graph of the demand curve is as given in Figure 5.20.

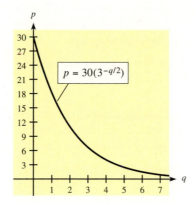

Figure 5.20

EXAMPLE 5 Demand

Suppose the demand function for q thousand units of a certain commodity is given by

$$p = 30(3^{-q/2})$$

(a) At what price per unit will the demand equal 4000 units?
(b) How many units, to the nearest thousand units, will be demanded if the price is $17.32?

Solution
(a) If 4000 units are demanded, then $q = 4$ and

$$p = 30(3^{-4/2}) = 30(0.1111) \approx 3.33 \text{ dollars}$$

(b) If $p = 17.32$, then $17.32 = 30(3^{-q/2})$

$$0.5773 = 3^{-q/2} \qquad \text{\color{red}Isolate the exponential.}$$

$$\ln 0.5773 = \ln 3^{-q/2} = -\frac{q}{2}\ln 3$$

$$\frac{-2\ln 0.5773}{\ln 3} = q \qquad \text{so} \qquad 1.000 \approx q$$

The number of units demanded would be approximately 1000 units. ∎

CHECKPOINT 2. Suppose the monthly demand for a product is given by $p = 400e^{-0.003x}$, where p is the price in dollars and x is the number of units. How many units will be demanded when the price is $100?

EXAMPLE 6   Total Revenue

Suppose the demand function for a commodity is given by $p = 100e^{-x/10}$, where p is the price per unit when x units are sold.

(a) What is the total revenue function for the commodity?
(b) What would be the total revenue if 30 units were demanded and supplied?

Solution

(a) The total revenue can be computed by multiplying the quantity sold and the price per unit. The demand function gives the price per unit when x units are sold, so the total revenue for x units is $R(x) = x \cdot p = x(100e^{-x/10})$. Thus the total revenue function is $R(x) = 100xe^{-x/10}$.

(b) If 30 units are sold, the total revenue is

$$R(30) = 100(30)e^{-30/10} \approx 100(30)(0.0498) = 149.40 \text{ (dollars)} \quad \blacksquare$$

Gompertz Curves and Logistic Functions

Gompertz Curves

One family of curves that has been used to describe human growth and development, the growth of organisms in a limited environment, and the growth of many types of organizations is the family of **Gompertz curves**. These curves are graphs of equations of the form

$$N = Ca^{R^t}$$

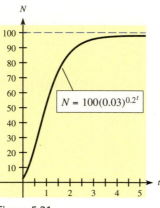

where t represents the time, $R\ (0 < R < 1)$ is a constant that depends on the population, a represents the proportion of initial growth, C is the maximum possible number of individuals, and N is the number of individuals at a given time t.

For example, the equation $N = 100(0.03)^{0.2^t}$ could be used to predict the size of a deer herd introduced on a small island. Here the maximum number of deer C would be 100, the proportion of the initial growth a is 0.03, and R is 0.2. For this example, t represents time, measured in decades. The graph of this equation is given in Figure 5.21.

$$N = 100(0.03)^{0.2^t}$$

Figure 5.21

EXAMPLE 7   Organizational Growth

A hospital administrator predicts that the growth in the number of hospital employees will follow the Gompertz equation

$$N = 2000(0.6)^{0.5^t}$$

where t represents the number of years after the opening of a new facility.

(a) What is the number of employees when the facility opens?
(b) How many employees are predicted after 1 year of operation?
(c) Graph the curve.
(d) What is the maximum value of N that the curve will approach?

Solution

(a) The facility opens when $t = 0$, so $N = 2000(0.6)^{0.5^0} = 2000(0.6)^1 = 1200$.
(b) In 1 year, $t = 1$, so $N = 2000(0.6)^{0.5} = 2000\sqrt{0.6} \approx 1549$.
(c) The graph is shown in Figure 5.22.
(d) From the graph we can see that as larger values of t are substituted in the function, the values of N approach, but never reach, 2000. We say that the line $N = 2000$ (dashed) is an **asymptote** for this curve and that 2000 is the maximum possible value. $\quad \blacksquare$

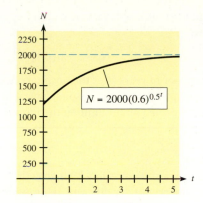

Figure 5.22

EXAMPLE 8 **Wildlife Populations**

The Gompertz equation

$$N = 100(0.03)^{0.2^t}$$

predicts the size of a deer herd on a small island t decades from now. During what year will the deer population reach or exceed 70?

Solution
We solve the equation with $N = 70$.

$$70 = 100(0.03)^{0.2^t}$$

$$0.7 = 0.03^{0.2^t} \qquad \text{Isolate the exponential.}$$

$$\ln 0.7 = \ln 0.03^{0.2^t} = 0.2^t \ln 0.03$$

$$\frac{\ln 0.7}{\ln 0.03} = 0.2^t \qquad \text{Again, isolate the exponential.}$$

$$\ln\left(\frac{\ln 0.7}{\ln 0.03}\right) = t \ln 0.2 \qquad \text{Take the logarithm of both sides.}$$

$$\ln(0.10172) \approx t(-1.6094)$$

$$\frac{-2.2855}{-1.6094} \approx t \qquad \text{so} \qquad t \approx 1.42 \text{ decades}$$

The population will exceed 70 in just over 14 years, or during the 15th year. ■

Calculator Note   Some graphing calculators have a SOLVER feature or program that can be used to solve an equation. Solution with such a feature or program may not be successful unless a reasonable guess at the solution (which can be seen from a graph) is chosen to start the process used to solve the equation. Figure 5.23 shows the equation from Example 8 and its solution with the SOLVER feature of a graphing calculator. ■

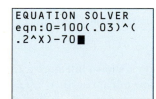

Figure 5.23

CHECKPOINT

3. Suppose the number of employees at a new regional hospital is predicted by the Gompertz curve

$$N = 3500(0.1)^{0.5^t}$$

where t is the number of years after the hospital opens.
(a) How many employees did the hospital have when it opened?
(b) What is the expected upper limit on the number of employees?

Logistic Functions

Gompertz curves describe situations in which growth is limited. There are other equations that model this phenomenon under different assumptions. Two examples are

(a) $y = c(1 - e^{-ax})$, $a > 0$ See Figure 5.24(a).

(b) $y = \dfrac{A}{1 + ce^{-ax}}$, $a > 0$ See Figure 5.24(b).

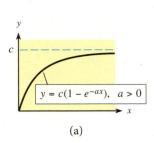
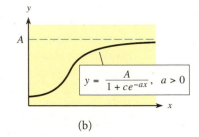

Figure 5.24   (a)   (b)

These equations have many applications. In general, both (a) and (b) can be used to model learning, sales of new products, and population growth, and (b) can be used to describe the spread of epidemics. The equation

$$y = \frac{A}{1 + ce^{-ax}}   a > 0$$

is called a **logistic function,** and the graph in Figure 5.24(b) is called a **logistic curve.**

EXAMPLE 9   **Blu-ray Players**

In recent years, Blu-ray players have been one of the most frequently purchased pieces of home electronics equipment. One company's revenues from the sales of Blu-ray players from 2006 to 2011 can be modeled by the logistic function

$$y = \frac{9.46}{1 + 53.08e^{-1.28x}}$$

where x is the number of years past 2005 and y is in millions of dollars.

(a) Graph this function.
(b) Use the function to estimate the sales revenue for 2011.

Solution

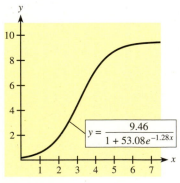

Figure 5.25

(a) The graph of this function is shown in Figure 5.25. Note that the graph is an S-shaped curve with a relatively slow initial rise, then a steep climb, followed by a leveling off. It is reasonable that sales would eventually level off. Very seldom will exponential growth continue indefinitely, so logistic functions often better represent many types of sales and organizational growth.

(b) For the 2011 estimate, use $x = 6$ in the function.

$$y = \frac{9.46}{1 + 53.08e^{-1.28(6)}} = \frac{9.46}{1 + 53.08e^{-7.68}} \approx 9.234$$

Thus the model estimates that in 2011, the sales revenue from Blu-ray players for this company was about $9.234 million. ■

CHECKPOINT SOLUTIONS

1. (a) $S = 1000e^{-0.07(2)} = 1000e^{-0.14} \approx \$869.36$ Nearest cent

   (b) $300 = 1000e^{-0.07x}$

   $0.3 = e^{-0.07x}$

   $\ln(0.3) = -0.07x$

   $\dfrac{\ln(0.3)}{-0.07} = x \approx 17$ Nearest day

2. $100 = 400e^{-0.003x}$

   $0.25 = e^{-0.003x}$

   $\ln(0.25) = -0.003x$

   $\dfrac{\ln(0.25)}{-0.003} = x \approx 462$ Nearest unit

3. (a) $3500(0.1) = 350$

   (b) 3500

| EXERCISES | 5.3

Access end-of-section exercises online at **www.webassign.net**

KEY TERMS
AND FORMULAS

Section	Key Terms	Formulas
5.1	Exponential functions	
	Growth functions	$f(x) = a^x \quad (a > 1)$
		$f(x) = C(a^x) \quad (C > 0, a > 1)$
	Decay functions	$f(x) = C(a^{-x}) \quad (C > 0, a > 1)$
		$f(x) = C(b^x) \quad (C > 0, 0 < b < 1)$
	e	$e \approx 2.71828$
5.2	Logarithmic function	$y = \log_a x$, defined by $x = a^y$
	Common logarithm	$\log x = \log_{10} x$
	Natural logarithm	$\ln x = \log_e x$
	Inverse functions	
	Logarithmic Properties I–V	$\log_a a^x = x;$
		$a^{\log_a x} = x;$
		$\log_a(MN) = \log_a M + \log_a N;$
		$\log_a(M/N) = \log_a M - \log_a N;$
		$\log_a(M^N) = N(\log_a M)$

Section	Key Terms	Formulas
	Change-of-base formulas	$\log_b x = \dfrac{\log_a x}{\log_a b}$
		$\log_b x = \dfrac{\ln x}{\ln b}$
5.3	Solving exponential equations	Isolate the exponential; take logarithm of both sides.
	Exponential growth and decay	
	Decay models	$f(x) = Ca^{-x} \quad (a > 1, C > 0)$
	Growth models	$f(x) = Ca^{x} \quad (a > 1, C > 0)$
	Gompertz curves	$N = Ca^{R^t}$
	Logistic functions	$y = \dfrac{A}{1 + ce^{-ax}} \quad (a > 0)$

REVIEW EXERCISES

1. Write each statement in logarithmic form.
   (a) $2^x = y$ (b) $3^y = 2x$
2. Write each statement in exponential form.
   (a) $\log_7\left(\dfrac{1}{49}\right) = -2$ (b) $\log_4 x = -1$

Graph the following functions.

3. $y = e^x$
4. $y = e^{-x}$
5. $y = \log_2 x$
6. $y = 2^x$
7. $y = \frac{1}{2}(4^x)$
8. $y = \ln x$
9. $y = \log_4 x$
10. $y = 3^{-2x}$
11. $y = \log x$
12. $y = 10(2^{-x})$

In Problems 13–20 evaluate each logarithm without using a calculator. In Problems 13–17, check with the change-of-base formula.

13. $\log_5 1$
14. $\log_2 8$
15. $\log_{25} 5$
16. $\log_3\left(\dfrac{1}{3}\right)$
17. $\log_3 3^8$
18. $\ln e$
19. $e^{\ln 5}$
20. $10^{\log 3.15}$

In Problems 21–24, if $\log_a x = 1.2$ and $\log_a y = 3.9$, find each of the following by using the properties of logarithms.

21. $\log_a\left(\dfrac{x}{y}\right)$
22. $\log_a \sqrt{x}$
23. $\log_a(xy)$
24. $\log_a(y^4)$

In Problems 25 and 26, use the properties of logarithms to write each expression as the sum or difference of two logarithmic functions containing no exponents.

25. $\log(yz)$
26. $\ln \sqrt{\dfrac{x+1}{x}}$

27. Is it true that $\ln x + \ln y = \ln(x + y)$ for all positive values of x and y?
28. If $f(x) = \ln x$, find $f(e^{-2})$.
29. If $f(x) = 2^x + \log(7x - 4)$, find $f(2)$.
30. If $f(x) = e^x + \ln(x + 1)$, find $f(0)$.
31. If $f(x) = \ln(3e^x - 5)$, find $f(\ln 2)$.

In Problems 32 and 33, use a change-of-base formula to evaluate each logarithm.

32. $\log_9 2158$
33. $\log_{12}(0.0195)$

In Problems 34 and 35, rewrite each logarithm by using a change-of-base formula, then graph the function with a graphing utility.

34. $y = \log_{\sqrt{3}} x$
35. $f(x) = \log_{11}(2x - 5)$

In Problems 36–42, solve each equation.

36. $6^{4x} = 46{,}656$
37. $8000 = 250(1.07)^x$
38. $11{,}000 = 45{,}000e^{-0.05x}$
39. $312 = 300 + 300e^{-0.08x}$
40. $\log_3(4x - 5) = 3$
41. $\ln(2x - 1) - \ln 3 = \ln 9$
42. $3 + 2\log_2 x = \log_2(x + 3) + 5$

APPLICATIONS

43. *Medicare spending* The graph in the figure on the following page shows the projected federal spending for Medicare as a percent of the gross domestic product. If these expenditures were modeled as a function of time,

which of a growth exponential, a decay exponential, or a logarithm would be the best model? Justify your answer.

A Consuming Problem

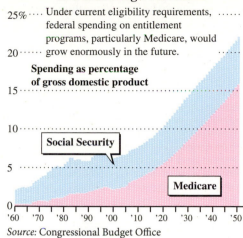

25% ···· Under current eligibility requirements, ······
federal spending on entitlement
programs, particularly Medicare, would
20 ······ grow enormously in the future.

**Spending as percentage
of gross domestic product**

Social Security

Medicare

'60 '70 '80 '90 '00 '10 '20 '30 '40 '50

Source: Congressional Budget Office

44. *Students per computer* The following figure shows the history of the number of students per computer in U.S. public schools. If this number of students were modeled by a function, do you think the best model would be linear, quadratic, exponential, or logarithmic? Explain.

Students per Computer in U.S. Public Schools

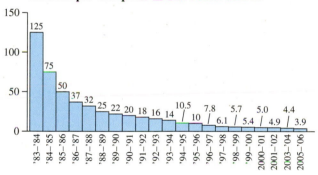

Source: MCH Strategic Data, Sweet Springs, MO. Reprinted by permission.

45. *Inflation* The purchasing power P of a $60,000 pension after t years of 3% annual inflation is modeled by

$$P(t) = 60{,}000(0.97)^t$$

(a) What is the purchasing power after 20 years?
(b) Graph this function for $t = 0$ to $t = 25$ with a graphing utility.

46. *Modeling Consumer credit* The data in the following table show the total consumer credit (in billions of dollars) that was outstanding in the United States for various years from 1980 to 2008.

(a) Find an exponential function that models these data. Use $x = 0$ in 1975.
(b) What does the model predict for the total consumer credit outstanding in 2015?

(c) In what year did the total consumer credit reach $5300 billion, according to the model?

Year	Amount (in billions)
1980	$349.4
1985	$593.2
1990	$789.1
1995	$1095.8
2000	$1560.2
2001	$1666.9
2002	$1725.7
2003	$2119.9
2004	$2232.3
2005	$2323.4
2006	$2415.0
2007	$2551.9
2008	$2592.1

Source: Federal Reserve

47. *Poverty threshold* The average poverty threshold for 1990–2008 for a single individual can be modeled by

$$y = -3130.3 + 4056.8 \ln x$$

where x is the number of years past 1980 and y is the annual income in dollars (*Source:* U.S. Bureau of the Census).

(a) What does the model predict as the poverty threshold in 2015?
(b) Graph this function for $x = 5$ to $x = 40$ with a graphing utility.

48. *Modeling Percent of paved roads* The following table shows various years from 1960 to 2008 and, for those years, the percents of U.S. roads and streets that were paved.

(a) Find a logarithmic equation that models these data. Use $x = 0$ in 1950.
(b) What percent of paved roads does the model predict for 2015?
(c) When does the model predict that 70% of U.S. roads and streets will be paved?

Year	Percent	Year	Percent
1960	34.7	1995	60.8
1965	39.4	2000	63.4
1970	44.5	2002	64.8
1975	48.3	2004	64.5
1980	53.7	2006	65.2
1985	54.7	2008	67.4
1990	58.4		

Source: U.S. Department of Transportation

49. **Stellar magnitude** The stellar magnitude M of a star is related to its brightness B as seen from earth according to

$$M = -\frac{5}{2} \log (B/B_0)$$

where B_0 is a standard level of brightness (the brightness of the star Vega).
(a) Find the magnitude of Venus if its brightness is 36.3 times B_0.
(b) Find the brightness (as a multiple of B_0) of the North Star if its magnitude is 2.1.
(c) If the faintest stars have magnitude 6, find their brightness (as a multiple of B_0).
(d) Is a star with magnitude -1.0 brighter than a star with magnitude $+1.0$?

50. **Sales decay** The sales decay for a product is given by $S = 50{,}000e^{-0.1x}$, where S is the weekly sales (in dollars) and x is the number of weeks that have passed since the end of an advertising campaign.
(a) What will sales be 6 weeks after the end of the campaign?
(b) How many weeks will pass before sales drop below $15,000?

51. **Sales decay** The sales decay for a product is given by $S = 50{,}000e^{-0.6x}$, where S is the monthly sales (in dollars) and x is the number of months that have passed since the end of an advertising campaign. What will sales be 6 months after the end of the campaign?

52. **Compound interest** If $1000 is invested at 12%, compounded monthly, the future value S at any time t (in years) is given by

$$S = 1000(1.01)^{12t}$$

How long will it take for the amount to double?

53. **Compound interest** If $5000 is invested at 13.5%, compounded continuously, then the future value S at any time t (in years) is given by $S = 5000e^{0.135t}$.
(a) What is the amount after 9 months?
(b) How long will it be before the investment doubles?

54. **World population** With data from the International Data Base of the U.S. Bureau of the Census, the

accompanying figure shows a scatter plot of actual or projected world population (in billions) in 5-year increments from 1950 to 2050 ($x = 0$ in 1945). Which of the following functions would be the best model: exponential growth, exponential decay, logarithmic, or logistic? Explain.

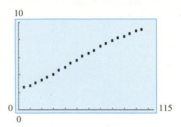

55. **Advertising and sales** Because of a new advertising campaign, a company predicts that sales will increase and that the yearly sales will be given by the equation

$$N = 10{,}000(0.3)^{0.5^t}$$

where t represents the number of years after the start of the campaign.
(a) What are the sales when the campaign begins?
(b) What are the predicted sales for the third year?
(c) What are the maximum predicted sales?

56. **Spread of a disease** The spread of a highly contagious virus in a high school can be described by the logistic function

$$y = \frac{5000}{1 + 1000e^{-0.8x}}$$

where x is the number of days after the virus is identified in the school and y is the total number of people who are infected by the virus in the first x days.
(a) Graph the function for $0 \le x \le 15$.
(b) How many students had the virus when it was first discovered?
(c) What is the total number infected by the virus during the first 15 days?
(d) In how many days will the total number infected reach 3744?

5   CHAPTER TEST

In Problems 1–4, graph the functions.
1. $y = 5^x$ 2. $y = 3^{-x}$
3. $y = \log_5 x$ 4. $y = \ln x$

In Problems 5–8, use technology to graph the functions.
5. $y = 3^{0.5x}$ 6. $y = \ln(0.5x)$
7. $y = e^{2x}$ 8. $y = \log_7 x$

In Problems 9–12, use a calculator to give a decimal approximation of the numbers to three decimal places.
9. e^4 10. $3^{-2.1}$
11. $\ln 4$ 12. $\log 21$
13. Use the definition of a logarithmic function to write $\log_7 x = 3.1$ in exponential form. Then find x to three decimal places.

14. Write $3^{2x} = 27$ in logarithmic form.

In Problems 15 and 16, solve each equation for x to 3 decimal places.

15. $3 + 6e^{-2x} = 7$ 16. $8 - \log_3(5x - 13) = 5$

In Problems 17–20, simplify the expressions, using properties or definitions of logarithms.

17. $\log_2 8$ 18. $e^{\ln x^4}$ 19. $\log_7 7^3$ 20. $\ln e^{x^2}$

21. Write $\ln(M \cdot N)$ as a sum involving M and N.

22. Write $\ln\left(\dfrac{x^3 - 1}{x + 2}\right)$ as a difference involving two binomials.

23. Write $\log_4(x^3 + 1)$ as a base e logarithm using a change-of-base formula.

24. Solve for x: $47{,}500 = 1500(1.015)^{6x}$

25. The graph in the following figure shows the number of active workers who will be (or have been) supporting each Social Security beneficiary. What type of function might be an appropriate model for this situation?

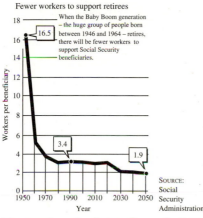

Fewer workers to support retirees

When the Baby Boom generation – the huge group of people born between 1946 and 1964 – retires, there will be fewer workers to support Social Security beneficiaries.

SOURCE: Social Security Administration

26. The graph in the figure shows the history of U.S. cellular telephone subscribership for selected years from 1986 to 2008. If these data were modeled by a smooth curve with years on the horizontal axis, what type of function might be appropriate? Explain.

U.S. Cellular Telephone Subscribership, 1986–2008

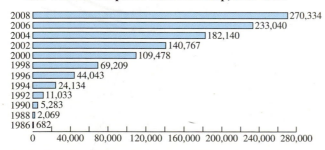

Source: The CTIA Semi-Annual Wireless Industry Indices. Used with the permission of CTIA—The Wireless Association:*

*Data may slightly from other sources.

27. The total national expenditures (in billions of dollars) for health services and supplies from 2000 and projected to 2018 can be modeled with the equation

$$E = 1327e^{0.062t}$$

where t is the number of years past 2000 (*Source:* U.S. Centers for Medicare & Medicaid Services).

(a) Find the predicted total expenditures for 2015.

(b) According to the model, how long will it take for the 2015 total expenditures to double?

28. A company plans to phase out one model of its product and replace it with a new model. An advertising campaign for the product being replaced just ended, and typically after such a campaign, monthly sales volume S (in dollars) decays according to

$$S = 22{,}000e^{-0.35t}$$

where t is in months. When the monthly sales volume for this product reaches \$2500, the company plans to discontinue production and launch the new model. How long will it be until this happens?

29. The total U.S. personal income I (in billions of dollars) from 1988 and projected to 2018 can be modeled by

$$I = \frac{70{,}290}{1 + 18.26e^{-0.058t}}$$

where t is the number of years past 1985 (*Source:* U.S. Department of Labor).

(a) What does the model predict for the total U.S. personal income in 2015?

(b) When will the total U.S. personal income reach \$25,000 billion?

30. *Modeling* The following table shows the receipts (in billions of dollars) for world tourism for selected years 1990–2007.

(a) From a scatter plot of the data (with $x = 0$ in 1980), indicate whether an exponential, logarithmic, or logistic model is most appropriate. Explain.

(b) Find the best-fitting model of the type found in part (a).

(c) What does the model predict receipts for world tourism will be in 2010?

Year	Receipts (in billions)	Year	Receipts (in billions)
1990	\$264	2000	\$473
1992	\$317	2002	\$474
1994	\$356	2003	\$524
1996	\$439	2004	\$623
1998	\$445	2007	\$735

Source: World Tourism Organization

I. Starbucks Stores (Modeling)

We all know that the number of Starbucks stores increased rapidly during 1992–2009. To see how rapidly, observe the following table, which gives the number of U.S. stores and the total number of stores during this period.

Year	Number of U.S. Stores	Total Stores
1992	113	127
1993	163	183
1994	264	300
1995	430	483
1996	663	746
1997	974	1121
1998	1321	1568
1999	1657	2028
2000	2119	2674
2001	2925	3817
2002	3756	5104
2003	4453	6193
2004	5452	7567
2005	7353	10,241
2006	8896	12,440
2007	10,684	15,011
2008	11,567	16,680
2009	11,128	16,635

Source: Starbucks.com

To investigate how the number of stores is likely to increase in the future and how the total number of stores compares with the number of U.S. stores, complete the following.

1. Create a scatter plot of the points $(x, f(x))$, with x equal to the number of the years past 1990 and $f(x)$ equal to the number of U.S. stores in the designated year.
2. Find an exponential function that models these data. Rewrite the function with base e.
3. How accurately does this model estimate the number of U.S. stores in 2000 and in 2006?
4. Find a logistic function that models the data. Does this model estimate the number of U.S. stores in 2000 and in 2006 better than the exponential model?
5. Graph the exponential model and the logistic model on the same axes as the scatter plot to determine whether the exponential or the logarithmic model is the better fit.
6. Find the exponential and the logistic functions that model the total number of Starbucks stores as a function of the number of years past 1992.
7. Which of the two models found in Question 6 is the better fit for the data?
8. Create a new column of data containing the ratio of total stores to U.S. stores by dividing the total number of stores by the number of U.S. stores in each row of the table.
9. Create a scatter plot with the number of years past 1990 as the inputs and the ratios found in Question 8 as the outputs.
10. Discuss the growth of the ratio of total stores to U.S. stores during 1992–2009. At what average rate is this ratio growing during 1992–2009?
11. Check the Internet to see whether the growth models developed previously remain accurate into the next decade.

II. Agricultural Business Management

A commercial vegetable and fruit grower carefully observes the relationship between the amount of fertilizer used on a certain variety of pumpkin and the extra revenue made from the sales of the resulting pumpkin crop. These observations are recorded in the following table.

Amount of fertilizer (pounds/acre)	0	250	500	750	1000	1250	1500	1750	2000
Extra revenue earned (dollars/acre)	0	96	145	172	185	192	196	198	199

The fertilizer costs 14¢ per pound. What would you advise the grower is the most profitable amount of fertilizer to use?

Check your advice by answering the following questions.

1. Is each pound of fertilizer equally effective?
2. Estimate the maximum amount of extra revenue that can be earned per acre by observing the data.
3. Graph the data points giving the extra revenue made from improved pumpkin sales as a function of the amount of fertilizer used. What does the graph tell you about the growth of the extra revenue as a function of the amount of fertilizer used?
4. Which of the following equations best models the graph you obtained in Question 3?

$$\text{(a) } N = Ca^{R^x} \qquad \text{(b) } y = \frac{A}{1 + ce^{-ax}} \qquad \text{(c) } y = c(1 - e^{-ax})$$

   Determine a specific equation that fits the observed data. That is, find the value of the constants in the model you have chosen.
5. Graph the total cost of the fertilizer as a function of the amount of fertilizer used on the graph you created for Question 3. Determine an equation that models this relationship. For what amount of fertilizer is the extra revenue earned from pumpkin sales offset by the cost of the fertilizer?
6. Use the graphs or the equations you have created to estimate the maximum extra profit that can be made and the amount of fertilizer per acre that would result in that extra profit.

HamsterMan/Shutterstock.com

Mathematics of Finance

Regardless of whether or not your career is in business, understanding how interest is computed on investments and loans is important to you as a consumer. The proliferation of personal finance and money management software attests to this importance. The goal of this chapter is to provide some understanding of the methods used to determine the interest and future value (principal plus interest) resulting from savings plans and the methods used in repayment of debts.

The topics and applications discussed in this chapter include the following.

Prerequisite Problem Type	For Section	Answer	Section for Review
What is $\dfrac{(-1)^n}{2n}$, if (a) $n = 1$? (b) $n = 2$? (c) $n = 3$?	6.1	(a) $-\frac{1}{2}$ (b) $\frac{1}{4}$ (c) $-\frac{1}{6}$	0.2 Signed numbers
(a) What is $(-2)^6$? (b) What is $\dfrac{\frac{1}{4}[1 - (\frac{1}{2})^6]}{1 - \frac{1}{2}}$?	6.2	(a) 64 (b) $\frac{63}{128}$	0.2 Signed numbers
If $f(x) = \dfrac{1}{2x}$, what is (a) $f(2)$? (b) $f(4)$?	6.1	(a) $\frac{1}{4}$ (b) $\frac{1}{8}$	1.2 Function notation
Evaluate: $e^{(0.08)(20)}$	6.2	4.95303	5.1 Exponential functions
Evaluate: (a) $\dfrac{1 - (1.05)^{-16}}{0.05}$ (b) $5000 + 5000\left[\dfrac{1-(1.059)^{-9}}{0.059}\right]$	6.3 6.4	(a) 10.83777 (b) 39,156.96575	0.3 Integral exponents
Solve: (a) $1100 = 1000 + 1000(0.058)t$ (b) $12,000 = P(1.03)^6$ (c) $850 = A_n\left[\dfrac{0.0065}{1 - (1.0065)^{-360}}\right]$	6.1–6.5	(a) $t \approx 1.72$ (b) $P \approx 10{,}049.81$ (c) $A_n \approx 118{,}076.79$	1.1 Linear equations
Solve: (a) $2 = (1.08)^n$ (b) $20{,}000 = 10{,}000e^{0.08t}$	6.2	(a) $n \approx 9.01$ (b) $t \approx 8.66$	5.3 Solutions of exponential equations

Simple Interest; Sequences

| APPLICATION PREVIEW |

Mary Spaulding purchased some Wind-Gen Electric stock for $6125.00. After 6 months, the stock had risen in value by $138.00 and had paid dividends totaling $144.14. One way to evaluate this investment and compare it with a bank savings plan is to find the simple interest rate that these gains represent. (See Example 3.)

In this section we begin our study of the mathematics of finance by considering simple interest. Simple interest forms the basis for all calculations involving interest that is paid on an investment or that is due on a loan. Also, we can evaluate short-term investments (such as in stocks or real estate) by calculating their simple interest rates.

Simple Interest   If a sum of money P (called the **principal**) is invested for a time period t (frequently in years) at an interest rate r per period, the **simple interest** is given by the following formula.

Simple Interest

The **simple interest** I is given by

$$I = Prt$$

where
$I =$ interest (in dollars)
$P =$ principal (in dollars)
$r =$ annual interest rate (written as a decimal)
$t =$ time (in years)*

Note that the time measurements for r and t must agree.

Simple interest is paid on investments involving time certificates issued by banks and on certain types of bonds, such as U.S. government bonds and municipal bonds. The interest for a given period is paid to the investor, and the principal remains the same.

If you borrow money from a friend or a relative, interest on your loan might be calculated with the simple interest formula. We'll consider some simple interest loans in this section, but interest on loans from banks and other lending institutions is calculated using methods discussed later.

EXAMPLE 1 Simple Interest

(a) If $8000 is invested for 2 years at an annual interest rate of 9%, how much interest will be received at the end of the 2-year period?
(b) If $4000 is borrowed for 39 weeks at an annual interest rate of 15%, how much interest is due at the end of the 39 weeks?

*Periods of time other than years can be used, as can other monetary systems.

Solution

(a) The interest is $I = Prt = \$8000(0.09)(2) = \$1440$.
(b) Use $I = Prt$ with $t = 39/52 = 0.75$ year. Thus

$$I = \$4000(0.15)(0.75) = \$450$$

Future Value

The **future amount of an investment,** or its **future value,** at the end of an interest period is the sum of the principal and the interest. Thus, in Example 1(a), the future value is

$$S = \$8000 + \$1440 = \$9440$$

Similarly, the **future amount of a loan,** or its **future value,** is the amount of money that must be repaid. In Example 1(b), the future value of the loan is the principal plus the interest, or

$$S = \$4000 + \$450 = \$4450$$

Future Value	If we use the letter S to denote the **future value** of an investment or a loan, then we have
	Future value of investment or loan: $S = P + I$
	where P is the principal (in dollars) and I is the interest (in dollars).

The principal P of a loan is also called the **face value** or the **present value** of the loan.

EXAMPLE 2 Future Value and Present Value

(a) If \$2000 is borrowed for one-half year at a simple interest rate of 12% per year, what is the future value of the loan at the end of the half-year?
(b) An investor wants to have \$20,000 in 9 months. If the best available interest rate is 6.05% per year, how much must be invested now to yield the desired amount?

Solution

(a) The interest for the half-year period is $I = \$2000(0.12)(0.5) = \$120$. Thus the future value of the investment for the period is

$$S = P + I = \$2000 + \$120 = \$2120$$

(b) We know that $S = P + I = P + Prt$. In this case, we must solve for P, the present value. Also, the time 9 months is $(9/12)$ of a year.

$$\$20,000 = P + P(0.0605)(9/12) = P + 0.045375P$$
$$\$20,000 = 1.045375P$$
$$\frac{\$20,000}{1.045375} = P \quad \text{so} \quad P = \$19,131.89$$

EXAMPLE 3 Return on an Investment | APPLICATION PREVIEW |

Mary Spaulding bought Wind-Gen Electric stock for \$6125.00. After 6 months, the value of her shares had risen by \$138.00 and dividends totaling \$144.14 had been paid. Find the simple interest rate she earned on this investment if she sold the stock at the end of the 6 months.

Solution

To find the simple interest rate that Mary earned on this investment, we find the rate that would yield an amount of simple interest equal to all of Mary's gains (that is, equal to the rise in the stock's price plus the dividends she received). Thus the principal is $6125.00, the time is 1/2 year, and the interest earned is the total of all gains (that is, interest $I = \$138.00 + \$144.14 = \$282.14$). Using these values in $I = Prt$ gives

$$\$282.14 = (\$6125)r(0.5) = \$3062.5r$$

$$r = \frac{\$282.14}{\$3062.5} \approx 0.092 = 9.2\%$$

Thus Mary's return was equivalent to an annual simple interest rate of about 9.2%. ◼

EXAMPLE 4   **Duration of an Investment**

If $1000 is invested at 5.8% simple interest, how long will it take to grow to $1100?

Solution

We use $P = \$1000$, $S = \$1100$, and $r = 0.058$ in $S = P + Prt$ and solve for t.

$$\$1100 = \$1000 + \$1000(0.058)t$$

$$\$100 = \$58t \qquad \text{so} \qquad \frac{\$100}{\$58} = t \ \text{ and } \ t \approx 1.72 \text{ years}$$ ◼

CHECKPOINT

1. What is the simple interest formula?
2. If $8000 is invested at 6% simple interest for 9 months, find the future value of the investment.
3. If a $2500 investment grows to $2875 in 15 months, what simple interest rate was earned?

Sequences

Let's look at the monthly future values of a $2000 investment that earns 1% simple interest for each of 5 months.

Month	Interest ($I = Prt$)	Future Value of the Investment
1	($2000)(0.01)(1) = $20	$2000 + $20 = $2020
2	($2000)(0.01)(1) = $20	$2020 + $20 = $2040
3	($2000)(0.01)(1) = $20	$2040 + $20 = $2060
4	($2000)(0.01)(1) = $20	$2060 + $20 = $2080
5	($2000)(0.01)(1) = $20	$2080 + $20 = $2100

These future values are outputs that result when the inputs are positive integers that correspond to the number of months of the investment. Outputs (such as these future values) that arise uniquely from positive integer inputs define a special type of function.

Sequence

A function whose domain is the set of positive integers is called a **sequence function**. The set of function outputs of a sequence function

$$f(1) = a_1, f(2) = a_2, \ldots, f(n) = a_n, \ldots$$

forms an ordered list called a **sequence**. The outputs $a_1, a_2, a_3, \ldots$ are called **terms** of the sequence, with a_1 the first term, a_2 the second term, and so on.

Because calculations involving interest often result from using positive integer inputs, sequences are the basis for most of the financial formulas derived in this chapter.

EXAMPLE 5   Terms of a Sequence

Write the first four terms of the sequence whose nth term is $a_n = (-1)^n/(2n)$.

Solution

The first four terms of the sequence are as follows:

$$a_1 = \frac{(-1)^1}{2(1)} = -\frac{1}{2} \qquad a_2 = \frac{(-1)^2}{2(2)} = \frac{1}{4}$$

$$a_3 = \frac{(-1)^3}{2(3)} = -\frac{1}{6} \qquad a_4 = \frac{(-1)^4}{2(4)} = \frac{1}{8}$$

We usually write these terms in the form $-\frac{1}{2}, \frac{1}{4}, -\frac{1}{6}, \frac{1}{8}$. ■

Arithmetic Sequences   The sequence 2020, 2040, 2060, 2080, 2100, ... can also be described in the following way:

$$a_1 = 2020, \quad a_n = a_{n-1} + 20 \quad \text{for } n > 1$$

This sequence is an example of a special kind of sequence called an **arithmetic sequence.** In such a sequence, each term after the first can be found by adding a constant to the preceding term. Thus we have the following definition.

Arithmetic Sequence

A sequence is called an **arithmetic sequence** (progression) if there exists a number d, called the **common difference,** such that

$$a_n = a_{n-1} + d \quad \text{for } n > 1$$

EXAMPLE 6   Arithmetic Sequences

Write the next three terms of each of the following arithmetic sequences.
(a) $1, 3, 5, \ldots$
(b) $9, 6, 3, \ldots$
(c) $\frac{1}{2}, \frac{5}{6}, \frac{7}{6}, \ldots$

Solution
(a) The common difference is 2, so the next three terms are 7, 9, 11.
(b) The common difference is -3, so the next three terms are $0, -3, -6$.
(c) The common difference is $\frac{1}{3}$, so the next three terms are $\frac{3}{2}, \frac{11}{6}, \frac{13}{6}$. ■

Because each term after the first term in an arithmetic sequence is obtained by adding d to the preceding term, the second term is $a_1 + d$, the third is $(a_1 + d) + d = a_1 + 2d, \ldots$, and the nth term is $a_1 + (n-1)d$. Thus we have the following formula.

nth Term of an Arithmetic Sequence

The **nth term of an arithmetic sequence** (progression) is given by

$$a_n = a_1 + (n-1)d$$

where a_1 is the first term and d is the common difference between successive terms.

EXAMPLE 7 nth Terms of Arithmetic Sequences

(a) Find the 11th term of the arithmetic sequence with first term 3 and common difference -2.
(b) If the first term of an arithmetic sequence is 4 and the 9th term is 20, find the 75th term.

Solution

(a) The 11th term is $a_{11} = 3 + (11 - 1)(-2) = -17$.

(b) Substituting the values $a_1 = 4$, $a_n = 20$, and $n = 9$ in $a_n = a_1 + (n - 1)d$ gives $20 = 4 + (9 - 1)d$. Solving this equation gives $d = 2$. Therefore, the 75th term is $a_{75} = 4 + (75 - 1)(2) = 152$. ■

Sum of an Arithmetic Sequence

Consider the arithmetic sequence with first term a_1, common difference d, and nth term a_n. The first n terms of an arithmetic sequence can be written in two ways, as follows.

$$\text{From } a_1 \text{ to } a_n: \quad a_1, a_1 + d, a_1 + 2d, a_1 + 3d, \ldots, a_1 + (n - 1)d$$
$$\text{From } a_n \text{ to } a_1: \quad a_n, a_n - d, a_n - 2d, a_n - 3d, \ldots, a_n - (n - 1)d$$

If we let s_n represent the sum of the first n terms of the sequence just described, then we have the following equivalent forms for the sum.

$$s_n = a_1 + (a_1 + d) + (a_1 + 2d) + \cdots + [a_1 + (n - 1)d] \qquad (1)$$
$$s_n = a_n + (a_n - d) + (a_n - 2d) + \cdots + [a_n - (n - 1)d] \qquad (2)$$

If we add equations (1) and (2) term by term, we obtain

$$2s_n = (a_1 + a_n) + (a_1 + a_n) + (a_1 + a_n) + \cdots + (a_1 + a_n)$$

in which $(a_1 + a_n)$ appears as a term n times. Thus,

$$2s_n = n(a_1 + a_n) \qquad \text{so} \qquad s_n = \frac{n}{2}(a_1 + a_n)$$

Sum of an Arithmetic Sequence

The **sum of the first n terms of an arithmetic sequence** is given by the formula

$$s_n = \frac{n}{2}(a_1 + a_n)$$

where a_1 is the first term of the sequence and a_n is the nth term.

EXAMPLE 8   Sums of Arithmetic Sequences

Find the sum of

(a) the first 10 terms of the arithmetic sequence with first term 2 and common difference 4

(b) the first 91 terms of the arithmetic sequence $\frac{1}{4}, \frac{7}{12}, \frac{11}{12}, \ldots$.

Solution

(a) We are given the values $n = 10$, $a_1 = 2$, and $d = 4$. Thus the 10th term is $a_{10} = 2 + (10 - 1)4 = 38$, and the sum of the first 10 terms is

$$s_{10} = \frac{10}{2}(2 + 38) = 200$$

(b) The first term is $\frac{1}{4}$ and the common difference is $\frac{1}{3}$. Therefore, the 91st term is $a_{91} = \frac{1}{4} + (91 - 1)\left(\frac{1}{3}\right) = 30\frac{1}{4} = \frac{121}{4}$. The sum of the first 91 terms is

$$s_{91} = \frac{91}{2}\left(\frac{1}{4} + \frac{121}{4}\right) = \frac{91}{2}\left(\frac{122}{4}\right) = \frac{91(61)}{4} = \left(\frac{5551}{4}\right) = 1387\frac{3}{4} \qquad ■$$

CHECKPOINT

4. Of the following sequences, identify the arithmetic sequences.

(a) $1, 4, 9, 16, \ldots$

(b) $1, 4, 7, 10, \ldots$

(c) $1, 2, 4, 8, \ldots$

5. Given the arithmetic sequence $-10, -6, -2, \ldots$, find

(a) the 51st term

(b) the sum of the first 51 terms.

Technology Note Finally, we should note that many graphing calculators, some graphics software packages, and spreadsheets have the capability of defining sequence functions and then operating on them by finding additional terms, graphing the terms, or summing a fixed number of terms.

CHECKPOINT SOLUTIONS

1. $I = Prt$

2. $S = P + I = 8000 + 8000(0.06)\left(\frac{9}{12}\right) = \$8360$

3. $S = P + I$, so

$$2875 = 2500 + 2500r\left(\frac{15}{12}\right)$$

$$375 = 3125r \quad \text{so} \quad r = \frac{375}{3125} = 0.12, \quad \text{or} \quad 12\%$$

4. Only (b) is an arithmetic sequence; the common difference is 3.

5. (a) The common difference is $d = 4$. The 51st term $(n = 51)$ is

$$a_{51} = -10 + (51 - 1)4 = -10 + 50(4) = -10 + 200 = 190$$

(b) For $a_1 = -10$ and $a_{51} = 190$, the sum of the first 51 terms is
$$s_{51} = \tfrac{51}{2}(-10 + 190) = 4590.$$

| EXERCISES | 6.1

Access end-of-section exercises online at **www.webassign.net**

OBJECTIVES 6.2

- To find the future value of a compound interest investment and the amount of interest earned when interest is compounded at regular intervals or continuously
- To find the annual percentage yield (APY), or the effective annual interest rate, of money invested at compound interest
- To find the time it takes for an investment to reach a specified amount
- To find specified terms, and sums of specified numbers of terms, of geometric sequences

Compound Interest; Geometric Sequences

| APPLICATION PREVIEW |

A common concern for young parents is how they will pay for their children's college educations. To address this, there are an increasing number of college savings plans, such as the so-called "529 plans," which are tax-deferred.

Suppose Jim and Eden established such a plan for their baby daughter Maura. If this account earns 9.8% compounded quarterly and if their goal is to have $200,000 by Maura's 18th birthday, what would be the impact of having $10,000 in the account by Maura's 1st birthday? (See Example 3.)

A compound interest investment (such as Jim and Eden's account) is one in which interest is paid into the account at regular intervals. In this section we consider investments of this type and develop formulas that enable us to determine the impact of making a $10,000 investment in Maura's college tuition account by her 1st birthday.

Compound Interest   In the previous section we discussed simple interest. A second method of paying interest is the **compound interest** method, where the interest for each period is added to the principal before interest is calculated for the next period. With compound interest, both the interest added and the principal earn interest for the next period. With this method, the principal grows as the interest is added to it. Think of it like building a snowman. You start with a snowball the size of your fist (your investment), and as you roll the snowball around the yard, the snowball picks up additional snow (interest). The more you roll it, the snowball itself (principal) keeps growing and as it does, the amount of snow it picks up along the way (interest) also grows. This method is used in investments such as savings accounts and U.S. government series EE bonds.

Pixel 4 Images/Shutterstock.com

An understanding of compound interest is important not only for people planning careers with financial institutions but also for anyone planning to invest money. To see how compound interest is computed, consider the following table, which tracks the annual growth of $20,000 invested for 3 years at 10% compounded annually. (Notice that the ending principal for each year becomes the beginning principal for the next year.)

Year	Beginning Principal $= P$	10% Annual Interest $= I$	Ending Principal $= P + I$
1	$20,000	0.10($20,000) = $2000	$22,000
2	$22,000	0.10($22,000) = $2200	$24,200
3	$24,200	0.10($24,200) = $2420	$26,620

Note that the future value for each year can be found by multiplying the beginning principal for that year by $1 + 0.10$, or 1.10. That is,

First year: $\qquad$ $20,000(1.10) = \$22,000$
Second year: $\qquad$ $[\$20,000(1.10)](1.10) = \$20,000(1.10)^2 = \$24,200$
Third year: $\qquad$ $[\$20,000(1.10)^2](1.10) = \$20,000(1.10)^3 = \$26,620$

This suggests that if we maintained this investment for n years, the future value at the end of this time would be $\$20,000(1.10)^n$. Thus we have the following general formula.

Future Value (Annual Compounding)

If $\$P$ is invested at an interest rate of r per year, compounded annually, the future value S at the end of the nth year is

$$S = P(1 + r)^n$$

EXAMPLE 1 **Annual Compounding**

If $3000 is invested for 4 years at 9% compounded annually, how much interest is earned?

Solution
The future value is

$$
\begin{aligned}
S &= \$3000(1 + 0.09)^4 \\
&= \$3000(1.4115816) \\
&= \$4234.7448 \\
&= \$4234.74, \text{ to the nearest cent}
\end{aligned}
$$

Because $3000 of this amount was the original investment, the interest earned is $4234.74 − $3000 = $1234.74. ■

Some accounts have the interest compounded semiannually, quarterly, monthly, or daily. Unless specifically stated otherwise, a stated interest rate, called the **nominal annual rate,** is the rate per year and is denoted by r. The interest rate *per period*, denoted by i, is the nominal rate divided by the number of interest periods per year. The interest periods are also called *conversion periods*, and the number of periods is denoted by n. Thus, if $100 is invested for 5 years at 6% compounded semiannually (twice a year), it has been invested for $n = 10$ periods (5 years × 2 periods per year) at $i = 3\%$ per period (6% per year ÷ 2 periods per year). The future value of an investment of this type is found using the following formula.

Future Value (Periodic Compounding)

If $P is invested for t years at a nominal interest rate r, compounded m times per year, then the total number of compounding periods is

$$n = mt$$

the interest rate per compounding period is

$$i = \frac{r}{m} \quad (\text{expressed as a decimal})$$

and the future value is

$$S = P(1 + i)^n = P\left(1 + \frac{r}{m}\right)^{mt}$$

EXAMPLE 2 **Periodic Compounding**

For each of the following investments, find the interest rate per period, i, and the number of compounding periods, n.

(a) 12% compounded monthly for 7 years
(b) 7.2% compounded quarterly for 11 quarters

Solution
(a) If the compounding is monthly and $r = 12\% = 0.12$, then $i = 0.12/12 = 0.01$. The number of compounding periods is $n = (7 \text{ yr})(12 \text{ periods/yr}) = 84$.
(b) $i = 0.072/4 = 0.018$, $n = 11$ (the number of quarters given) ■

Once we know i and n, we can calculate the future value from the formula with a calculator.

EXAMPLE 3   **Future Value**   **| APPLICATION PREVIEW |**

Jim and Eden want to have $200,000 in Maura's college fund on her 18th birthday, and they want to know the impact on this goal of having $10,000 invested at 9.8%, compounded quarterly, on her 1st birthday. To advise Jim and Eden regarding this, find

(a)   the future value of the $10,000 investment
(b)   the amount of compound interest that the investment earns
(c)   the impact this would have on their goal.

Solution

(a)   For this situation, $i = 0.098/4 = 0.0245$ and $n = 4(17) = 68$. Thus the future value of the $10,000 is given by

$$S = P(1 + i)^n = \$10,000(1 + 0.0245)^{68} = \$10,000(5.18577) = \$51,857.70$$

(b)   The amount of interest earned is $51,857.70 − $10,000 = $41,857.70.
(c)   Thus $10,000 invested by Maura's 1st birthday grows to an amount that is slightly more than 25% of their goal. This rather large early investment has a substantial impact on their goal.   ■

Technology Note

The steps for finding the future value of an investment with a calculator or with Excel are found in Appendices C and D, Section 6.2. See also the Online Excel Guide.   ■

We saw previously that compound interest calculations are based on those for simple interest, except that interest payments are added to the principal. In this way, interest is earned on both principal and previous interest payments. Let's examine the effect of this compounding by comparing compound interest and simple interest. If the investment in Example 3 had been at simple interest, the interest earned would have been $Prt = \$10,000(0.098)(17) = \$16,660$. This is almost $25,200 less than the amount of compound interest earned. And this difference would have been magnified over a longer period of time. Try reworking Example 3's investment over 30 years and compare the compound interest earned with the simple interest earned. This comparison begins to shed some light on why Albert Einstein characterized compound interest as "the most powerful force in the Universe."

EXAMPLE 4   **Present Value**

What amount must be invested now in order to have $12,000 after 3 years if money is worth 6% compounded semiannually?

Solution

We need to find the present value P, knowing that the future value is $S = \$12,000$. Use $i = 0.06/2 = 0.03$ and $n = 3(2) = 6$.

$$S = P(1 + i)^n$$

$$\$12,000 = P(1 + 0.03)^6 = P(1.03)^6 \approx P(1.1940523)$$

$$P = \frac{\$12,000}{1.1940523} = \$10,049.81, \text{ to the nearest cent}$$ ■

EXAMPLE 5   **Rate Earned**

As Figure 6.1 on the next page shows, three years after Google stock was first sold publicly, its share price had risen 650%. Google's 650% increase means that $10,000 invested in Google stock at its initial public offering (I.P.O.) was worth $65,000 three years later. What interest rate compounded annually does this represent? (For similar information and calculations regarding Microsoft's annual compounding performance, see Problem 33 in the exercises for this section.)

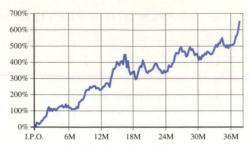

Figure 6.1

Source: Based on data from Google Finance

Solution

We use $P = \$10{,}000$, $S = \$65{,}000$, and $n = 3$ in the formula $S = P(1 + i)^n$, and solve for i.

$$\$65{,}000 = \$10{,}000(1 + i)^3$$
$$6.5 = (1 + i)^3$$

At this point we take the cube root (third root) of both sides (or, equivalently, raise both sides to the $1/3$ power).

$$6.5^{(1/3)} = [(1 + i)^3]^{(1/3)}$$
$$1.8663 \approx 1 + i \quad \text{so} \quad 0.8663 \approx i$$

Thus, this investment earned about 86.63% compounded annually.   ■

CHECKPOINT

1. If $5000 is invested at 6%, compounded quarterly, for 5 years, find
   (a) the number of compounding periods per year, m
   (b) the number of compounding periods for the investment, n
   (c) the interest rate for each compounding period, i
   (d) the future value of the investment.
2. Find the present value of an investment that is worth $12,000 after 5 years at 9% compounded monthly.

If we invest a sum of money, say $100, then the higher the interest rate, the greater the future value. Figure 6.2 shows a graphical comparison of the future values when $100 is invested at 5%, 8%, and 10%, all compounded annually over a period of 30 years. Note that higher interest rates yield consistently higher future values and have dramatically higher future values after 15–20 years. Note also that these graphs of the future values are growth exponentials.

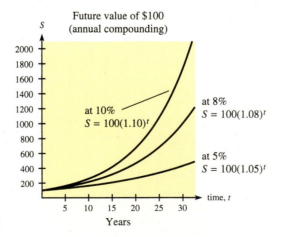

Figure 6.2

Spreadsheet Note

Spreadsheets and financial calculators can also be used to investigate and analyze compound interest investments, such as tracking the future value of an investment or comparing investments with different interest rates.

Table 6.1 shows a portion of a spreadsheet that tracks the monthly growth of two investments, both compounded monthly and both with $P = \$1000$, but one at 6% (giving $i = 0.005$) and the other at 6.9% (giving $i = 0.00575$). ■

| TABLE 6.1 |

**SECOND 12 MONTHS OF TWO INVESTMENTS
AT DIFFERENT INTEREST RATES**

	A	B	C
1	Month #	$ = 1000(1.005)^n	$ = 1000(1.00575)^n
2	22	$1115.97	$1134.44
3	23	$1121.55	$1140.96
4	24	$1127.16	$1147.52

Continuous Compounding

Because more frequent compounding means that interest is paid more often (and hence more interest on interest is earned), it would seem that the more frequently the interest is compounded, the larger the future value will become. To determine the interest that results from *continuous* compounding (compounding every instant), consider an investment of $1 for 1 year at a 100% interest rate. If the interest is compounded m times per year, the future value is given by

$$S = \left(1 + \frac{1}{m}\right)^m$$

Table 6.2 shows the future values that result as the number of compounding periods increases.

| TABLE 6.2 |

Compounded	Number of Periods per Year	Future Value of $1
Annually	1	$\left(1 + \frac{1}{1}\right)^1 = 2$
Monthly	12	$\left(1 + \frac{1}{12}\right)^{12} = 2.6130\ldots$
Daily	360 (business year)	$\left(1 + \frac{1}{360}\right)^{360} = 2.7145\ldots$
Hourly	8640	$\left(1 + \frac{1}{8640}\right)^{8640} = 2.71812\ldots$
Each minute	518,400	$\left(1 + \frac{1}{518,400}\right)^{518,400} = 2.71827\ldots$

Table 6.2 shows that as the number of periods per year increases, the future value increases, although not very rapidly. In fact, no matter how often the interest is compounded, the future value will never exceed $2.72. We say that as the number of periods increases, the future value approaches a limit, which is the number e:

$$e = 2.7182818\ldots.$$

We discussed the number e and the function $y = e^x$ in Chapter 5, "Exponential and Logarithmic Functions." The discussion here shows one way the number we call e may be derived. We will define e more formally later.

Future Value (Continuous Compounding)

In general, if $\$P$ is invested for t years at a nominal rate r compounded continuously, then the future value is given by the exponential function

$$S = Pe^{rt}$$

EXAMPLE 6   Continuous Compounding

(a) Find the future value if $1000 is invested for 20 years at 8%, compounded continuously.

(b) What amount must be invested at 6.5%, compounded continuously, so that it will be worth $25,000 after 8 years?

Solution

(a) The future value is

$$S = \$1000e^{(0.08)(20)} = \$1000e^{1.6}$$
$$= \$1000(4.95303) \quad (\text{because } e^{1.6} \approx 4.95303)$$
$$= \$4953.03$$

(b) Solve for the present value P in $\$25{,}000 = Pe^{(0.065)(8)}$.

$$\$25{,}000 = Pe^{(0.065)(8)} = Pe^{0.52} = P(1.68202765)$$
$$\frac{\$25{,}000}{1.68202765} = P \quad \text{so} \quad P = \$14{,}863.01, \text{ to the nearest cent}$$

EXAMPLE 7   Comparing Investments

How much more will you earn if you invest $1000 for 5 years at 8% compounded continuously instead of at 8% compounded quarterly?

Solution

If the interest is compounded continuously, the future value at the end of the 5 years is

$$S = \$1000e^{(0.08)(5)} = \$1000e^{0.4} \approx \$1000(1.49182) = \$1491.82$$

If the interest is compounded quarterly, the future value at the end of the 5 years is

$$S = \$1000(1.02)^{20} \approx \$1000(1.485947) = \$1485.95, \text{ to the nearest cent}$$

Thus the extra interest earned by compounding continuously is

$$\$1491.82 - \$1485.95 = \$5.87$$

Annual Percentage Yield

As Example 7 shows, when we invest money at a given compound interest rate, the method of compounding affects the amount of interest we earn. As a result, a rate of 8% can earn more than 8% interest if compounding is more frequent than annually.

For example, suppose $1 is invested for 1 year at 8%, compounded semiannually. Then $i = 0.08/2 = 0.04$, $n = 2$, the future value is $S = \$1(1.04)^2 = \$1.0816$, and the interest earned for the year is $\$1.0816 - \$1 = \$0.0816$. Note that this amount of interest represents an annual percentage yield of 8.16%, so we say that 8% compounded semiannually has an **annual percentage yield (APY)**, or **effective annual rate,** of 8.16%. Similarly, if $1 is invested at 8% compounded continuously, then the interest earned is $\$1(e^{0.08}) - \$1 = \$0.0833$, for an APY of 8.33%.

Banks acknowledge this difference between stated nominal interest rates and annual percentage yields by posting both rates for their investments. Note that the annual percentage yield is equivalent to the stated rate when compounding is annual. In general, the annual percentage yield equals I/P, or just I if $P = \$1$. Hence we can calculate the APY with the following formulas.

Annual Percentage Yield (APY)	Let r represent the annual (nominal) interest rate for an investment. Then the **annual percentage yield (APY)*** is found as follows.

Periodic Compounding. If m is the number of compounding periods per year, then $i = r/m$ is the interest rate per period, and

$$\text{APY} = \left(1 + \frac{r}{m}\right)^m - 1 = (1 + i)^m - 1 \quad \text{(as a decimal)}$$

Continuous Compounding

$$\text{APY} = e^r - 1 \quad \text{(as a decimal)}$$

Thus, although we cannot directly compare two nominal rates with different compounding periods, we can compare their corresponding APYs.

EXAMPLE 8 Comparing Yields

Suppose a young couple such as Jim and Eden from our Application Preview found three different investment companies that offered college savings plans: (a) one at 10% compounded annually, (b) another at 9.8% compounded quarterly, and (c) a third at 9.65% compounded continuously. Find the annual percentage yield (APY) for each of these three plans to discover which plan is best.

Solution
(a) For annual compounding, the stated rate is the APY. Thus, APY = 10%.
(b) Because the number of periods per year is $m = 4$ and the nominal rate is $r = 0.098$, the rate per period is $i = r/m = 0.098/4 = 0.0245$. Thus,

$$\text{APY} = (1 + 0.0245)^4 - 1 = 1.10166 - 1 = 0.10166 = 10.166\%$$

(c) For continuous compounding and a nominal rate of 9.65%, we have

$$\text{APY} = e^{0.0965} - 1 = 1.10131 - 1 = 0.10131 = 10.131\%$$

Hence we see that of these three choices, 9.8% compounded quarterly is best. Furthermore, even 9.65% compounded continuously has a higher APY than 10% compounded annually. ∎

CHECKPOINT

3. For each future value formula below, decide which is used for interest that is compounded periodically and which is used for interest that is compounded continuously.
   (a) $S = P(1 + i)^n$ (b) $S = Pe^{rt}$
4. If $5000 is invested at 6% compounded continuously for 5 years, find the future value of the investment.
5. Find the annual percentage yield of an investment that earns 7% compounded semi-annually.

EXAMPLE 9 Doubling Time

How long does it take an investment of $10,000 to double if it is invested at
(a) 8%, compounded annually
(b) 8%, compounded continuously?

*Note that the annual percentage yield is also called the **effective annual rate.**

Solution

(a) We solve for n in $\$20{,}000 = \$10{,}000(1 + 0.08)^n$.

$$2 = 1.08^n$$

Taking the logarithm, base e, of both sides of the equation gives

$$\ln 2 = \ln 1.08^n$$
$$\ln 2 = n \ln 1.08 \quad \text{\color{red}Logarithm Property V}$$
$$n = \frac{\ln 2}{\ln 1.08} \approx 9.0 \text{ (years)}$$

(b) Solve for t in $\$20{,}000 = \$10{,}000e^{0.08t}$.

$$2 = e^{0.08t} \quad \text{\color{red}Isolate the exponential.}$$
$$\ln 2 = \ln e^{0.08t} \quad \text{\color{red}Take "ln" of both sides.}$$
$$\ln 2 = 0.08t \quad \text{\color{red}Logarithm Property I}$$
$$t = \frac{\ln 2}{0.08} \approx 8.7 \text{ (years)}$$

Spreadsheet Note   To understand better the effect of compounding periods on an investment, we can use a spreadsheet to compare different compounding schemes. Table 6.3 shows a spreadsheet that tracks the growth at 5-year intervals of a $100 investment for quarterly compounding and continuous compounding (both at 10%).

| TABLE 6.3 |

COMPARISON OF TWO INVESTMENTS WITH DIFFERENT COMPOUNDING SCHEMES

	A	B	C
1	Year	$S=100(1.025)\wedge(4n)$	$S=100*\exp(.10n)$
2	5	163.86	164.87
3	10	268.51	271.83
4	15	439.98	448.17
5	20	720.96	738.91
6	25	1181.37	1218.25
7	30	1935.81	2008.55

Also, most spreadsheets (including Excel) and some calculators have built-in finance packages that can be used to solve the compound interest problems of this section and later sections of this chapter. See Appendix D, Section 6.2, and the Online Excel Guide for details.

CHECKPOINT   6. How long does it take $5000 to double if it is invested at 9% compounded monthly?

Geometric Sequences   If $\$P$ is invested at an interest rate of i per period, compounded at the end of each period, the future value at the end of each succeeding period is

$$P(1 + i), P(1 + i)^2, P(1 + i)^3, \ldots, P(1 + i)^n, \ldots$$

The future values for each of the succeeding periods form a sequence in which each term (after the first) is found by multiplying the previous term by the same number. Such a sequence is called a **geometric sequence.**

Geometric Sequence   A sequence is called a **geometric sequence** (progression) if there exists a number r, called the **common ratio,** such that

$$a_n = ra_{n-1} \quad \text{for } n > 1$$

Geometric sequences form the foundation for other applications involving compound interest.

EXAMPLE 10 Geometric Sequences

Write the next three terms of the following geometric sequences.
(a) $1, 3, 9, \ldots$ (b) $4, 2, 1, \ldots$ (c) $3, -6, 12, \ldots$

Solution
(a) The common ratio is 3, so the next three terms are 27, 81, 243.
(b) The common ratio is $\frac{1}{2}$, so the next three terms are $\frac{1}{2}, \frac{1}{4}, \frac{1}{8}$.
(c) The common ratio is -2, so the next three terms are $-24, 48, -96$. ■

Because each term after the first in a geometric sequence is obtained by multiplying the previous term by r, the second term is $a_1 r$, the third is $a_1 r^2$, etc., and the nth term is $a_1 r^{n-1}$. Thus we have the following formula.

---

nth Term of a Geometric Sequence

The ***nth term of a geometric sequence*** (progression) is given by

$$a_n = a_1 r^{n-1}$$

where a_1 is the first term of the sequence and r is the common ratio.

EXAMPLE 11 nth Term of a Geometric Sequence

Find the seventh term of the geometric sequence with first term 5 and common ratio -2.

Solution
The seventh term is $a_7 = 5(-2)^{7-1} = 5(64) = 320$. ■

EXAMPLE 12 Ball Rebounding

A ball is dropped from a height of 125 feet. If it rebounds $\frac{3}{5}$ of the height from which it falls every time it hits the ground, how high will it bounce after it strikes the ground for the fifth time?

Solution
The first rebound is $\frac{3}{5}(125) = 75$ feet; the second rebound is $\frac{3}{5}(75) = 45$ feet. The heights of the rebounds form a geometric sequence with first term 75 and common ratio $\frac{3}{5}$. Thus the fifth term is

$$a_5 = 75\left(\frac{3}{5}\right)^4 = 75\left(\frac{81}{625}\right) = \frac{243}{25} = 9\frac{18}{25} \text{ feet}$$ ■

Sum of a Geometric Sequence

Next we develop a formula for the sum of a geometric sequence, a formula that is important in our study of annuities. The sum of the first n terms of a geometric sequence is

$$s_n = a_1 + a_1 r + a_1 r^2 + \cdots + a_1 r^{n-1} \qquad (1)$$

If we multiply Equation (1) by r, we have

$$r s_n = a_1 r + a_1 r^2 + a_1 r^3 + \cdots + a_1 r^n \qquad (2)$$

Subtracting Equation (2) from Equation (1), we obtain

$$s_n - r s_n = a_1 + (a_1 r - a_1 r) + (a_1 r^2 - a_1 r^2) + \cdots + (a_1 r^{n-1} - a_1 r^{n-1}) - a_1 r^n$$

Thus

$$s_n(1 - r) = a_1 - a_1 r^n \qquad \text{so} \qquad s_n = \frac{a_1 - a_1 r^n}{1 - r} \quad \text{if } r \neq 1$$

This gives the following.

Sum of a Geometric Sequence	The **sum of the first n terms of the geometric sequence** with first term a_1 and common ratio r is

$$s_n = \frac{a_1(1 - r^n)}{1 - r} \quad \text{provided that } r \neq 1$$

EXAMPLE 13 Sums of Geometric Sequences

(a) Find the sum of the first five terms of the geometric progression with first term 4 and common ratio -3.

(b) Find the sum of the first six terms of the geometric sequence $\frac{1}{4}, \frac{1}{8}, \frac{1}{16}, \ldots$.

Solution

(a) We are given that $n = 5$, $a_1 = 4$, and $r = -3$. Thus

$$s_5 = \frac{4\left[1 - (-3)^5\right]}{1 - (-3)} = \frac{4\left[1 - (-243)\right]}{4} = 244$$

(b) We know that $n = 6$, $a_1 = \frac{1}{4}$, and $r = \frac{1}{2}$. Thus

$$s_6 = \frac{\frac{1}{4}\left[1 - \left(\frac{1}{2}\right)^6\right]}{1 - \frac{1}{2}} = \frac{\frac{1}{4}\left(1 - \frac{1}{64}\right)}{\frac{1}{2}} = \frac{1 - \frac{1}{64}}{2} = \frac{64 - 1}{128} = \frac{63}{128}$$

CHECKPOINT

7. Identify any geometric sequences among the following.
   (a) $1, 4, 9, 16, \ldots$ (b) $1, 4, 7, 10, \ldots$ (c) $1, 4, 16, 64, \ldots$

8. (a) Find the 40th term of the geometric sequence $8, 6, \frac{9}{2}, \ldots$.
   (b) Find the sum of the first 20 terms of the geometric sequence $2, 6, 18, \ldots$.

CHECKPOINT SOLUTIONS

1. (a) $m = 4$ (b) $n = m \cdot t = 4 \cdot 5 = 20$ (c) $i = \dfrac{r}{m} = \dfrac{0.06}{4} = 0.015$

   (d) $S = P(1 + i)^n = \$5000(1 + 0.015)^{20} \approx \$6734.28$, to the nearest cent

2. $i = 0.09/12 = 0.0075$ and $n = (5)(12) = 60$

   $$\$12,000 = P(1 + 0.0075)^{60} = P(1.5656810270)$$
   $$P = \$12,000/1.565681027 = \$7664.40, \text{ to the nearest cent}$$

3. (a) Periodic compounding
   (b) Continuous compounding

4. $S = Pe^{rt} = \$5000e^{(0.06)(5)} = \$5000e^{0.3} \approx \$6749.29$, to the nearest cent

5. $i = 0.07/2 = 0.035$ and $m = 2$

   $$\text{APY} = (1 + 0.035)^2 - 1 \approx 0.0712, \text{ or } 7.12\%$$

6. With $i = 0.09/12 = 0.0075$, we solve for n in

   $$\$10,000 = \$5000(1 + 0.0075)^n$$
   $$2 = (1.0075)^n$$
   $$\ln(2) = \ln(1.0075)^n = n \ln(1.0075)$$
   $$\frac{\ln(2)}{\ln(1.0075)} = n \quad \text{so} \quad n \approx 92.8 \text{ months}$$

7. Only sequence (c) is geometric with common ratio $r = 4$.

8. (a) $a_{40} = a_1 r^{40-1} = 8\left(\frac{3}{4}\right)^{39} \approx 0.0001$

   (b) $s_{20} = \dfrac{a_1(1 - r^{20})}{1 - r} = \dfrac{2(1 - 3^{20})}{1 - 3} = -(1 - 3^{20})$

   $$= 3^{20} - 1 = 3,486,784,400$$

| EXERCISES | 6.2

Access end-of-section exercises online at **www.webassign.net**

OBJECTIVES   6.3

- To compute the future values of ordinary annuities and annuities due
- To compute the payments required in order for ordinary annuities and annuities due to have specified future values
- To compute the payment required to establish a sinking fund
- To find how long it will take to reach a savings goal

Future Values of Annuities

| APPLICATION PREVIEW |

Twins graduate from college together and start their careers. Twin 1 invests $2000 at the end of each of 8 years in an account that earns 10%, compounded annually. After the initial 8 years, no additional contributions are made, but the investment continues to earn 10%, compounded annually. Twin 2 invests no money for 8 years but then contributes $2000 at the end of each year for a period of 36 years (to age 65) to an account that pays 10%, compounded annually. How much money does each twin have at age 65? (See Example 2 and Example 3.)

In this section we consider ordinary annuities and annuities due, and we develop a formula for each type of annuity that allows us to find its future value—that is, a formula for the value of the account after regular deposits have been made over a period of time.

Each twin's contributions form an annuity. An **annuity** is a financial plan characterized by regular payments. We can view an annuity as a savings plan in which the regular payments are contributions to the account, and then we can ask what the total value of the account will become (as in the Application Preview). Also, we can view an annuity as a payment plan (such as for retirement) in which regular payments are made from an account, often to an individual.

Ordinary Annuities   Most people save (or invest) money by depositing relatively small amounts at different times. If a depositor makes equal deposits at regular intervals, he or she is contributing to an annuity. The payments (deposits) may be made weekly, monthly, quarterly, yearly, or at any other interval of time. The sum of all payments plus all interest earned is called the **future amount of the annuity** or its **future value.**

In this text we will deal with annuities in which the payments begin and end on fixed dates, and we will deal first with annuities in which the payments are made at the end of each of the equal payment intervals. This type of annuity is called an **ordinary annuity** (and also an **annuity immediate**). The ordinary annuities we will consider have payment intervals that coincide with the compounding period of the interest.

Suppose you invested $100 at the end of each year for 5 years in an account that paid interest at 10%, compounded annually. How much money would you have in the account at the end of the 5 years?

Because you are making payments at the end of each period (year), this annuity is an ordinary annuity.

To find the future value of your annuity at the end of the 5 years, we compute the future value of each payment separately and add the amounts (see Figure 6.3). The $100 invested *at the end* of the first year will draw interest for 4 years, so it will amount to $100(1.10)^4$. Figure 6.3 shows similar calculations for years 2–4. The $100 invested at the end of the fifth year will draw no interest, so it will amount to $100.

Thus the future value of the annuity is given by

$$S = 100 + 100(1.10) + 100(1.10)^2 + 100(1.10)^3 + 100(1.10)^4$$

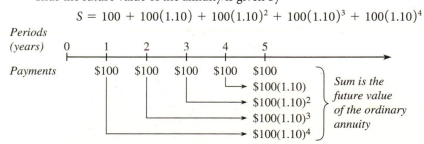

Figure 6.3

The terms of this sum are the first five values of the geometric sequence having $a_1 = 100$ and $r = 1.10$. Thus

$$S = \frac{100[1 - (1.10)^5]}{1 - 1.10} = \frac{100(-0.61051)}{-0.10} = 610.51$$

Thus the investment ($100 at the end of each year for 5 years, at 10%, compounded annually) would return $610.51.

Because every such annuity will take the same form, we can state that if a periodic payment R is made for n periods at an interest rate i *per period*, the **future amount of the annuity**, or its **future value**, will be given by

$$S = R \cdot \frac{1 - (1 + i)^n}{1 - (1 + i)} \quad \text{or} \quad S = R\left[\frac{(1 + i)^n - 1}{i}\right]$$

> **Future Value of an Ordinary Annuity**
>
> If R is deposited at the end of each period for n periods in an annuity that earns interest at a rate of i per period, the **future value of the annuity** will be
>
> $$S = R \cdot s_{\overline{n}|i} = R \cdot \left[\frac{(1 + i)^n - 1}{i}\right]$$
>
> where $s_{\overline{n}|i}$ is read "s, n angle i" and represents the future value of an ordinary annuity of $1 per period for n periods with an interest rate of i per period.

The value of $s_{\overline{n}|i}$ can be computed directly with a calculator or found in Table I in Appendix A.

EXAMPLE 1 Future Value

Richard Lloyd deposits $200 at the end of each quarter in an account that pays 4%, compounded quarterly. How much money will he have in his account in $2\frac{1}{4}$ years?

Solution
The number of periods is $n = (4)(2.25) = 9$, and the rate *per period* is $i = 0.04/4 = 0.01$. At the end of $2\frac{1}{4}$ years the future value of the annuity will be

$$S = \$200 \cdot s_{\overline{9}|0.01} = \$200\left[\frac{(1 + 0.01)^9 - 1}{0.01}\right]$$

$$= \$200(9.368527) = \$1873.71, \text{ to the nearest cent}$$

Note that Table I in Appendix A also gives $s_{\overline{9}|0.01} = 9.368527$. ∎

In the Application Preview, we described savings plans for twins. In the next two examples, we answer the questions posed in the Application Preview.

EXAMPLE 2 Future Value for Twin 2 | APPLICATION PREVIEW |

Twin 2 in the Application Preview invests $2000 at the end of each year for 36 years (until age 65) in an account that pays 10%, compounded annually. How much does twin 2 have at age 65?

Solution
This savings plan is an ordinary annuity with $i = 0.10$, $n = 36$, and $R = \$2000$. The future value (to the nearest dollar) is

$$S = R\left[\frac{(1 + i)^n - 1}{i}\right] = \$2000\left[\frac{(1.10)^{36} - 1}{0.10}\right] = \$598,254$$ ∎

EXAMPLE 3   **Future Value for Twin 1 | APPLICATION PREVIEW |**

Twin 1 invests $2000 at the end of each of 8 years in an account that earns 10%, compounded annually. After the initial 8 years, no additional contributions are made, but the investment continues to earn 10%, compounded annually, for 36 more years (until twin 1 is age 65). How much does twin 1 have at age 65?

Solution
We seek the future values of two different investments. The first is an ordinary annuity with $R = \$2000$, $n = 8$ periods, and $i = 0.10$. The second is a compound interest investment with $n = 36$ periods and $i = 0.10$ and whose principal (that is, its present value) is the future value of this twin's ordinary annuity.
    We first find the future value of the ordinary annuity.

$$S = R\left[\frac{(1 + i)^n - 1}{i}\right] = \$2000\left[\frac{(1 + 0.10)^8 - 1}{0.10}\right] \approx \$22{,}871.78$$

This amount is the principal of the compound interest investment. If no deposits or withdrawals were made for the next 36 years, the future value of this investment would be

$$S = P(1 + i)^n = \$22{,}871.78(1 + 0.10)^{36} = \$707{,}028.03, \text{ to the nearest cent}$$

Thus, at age 65, twin 1's investment is worth about $707,028.   ∎

Looking back at Examples 2 and 3, we can extract the following summary and see which twin was the wiser.

	Contributions	Account Value at Age 65
Twin 1	$2000/year for 8 years = $16,000	$707,028
Twin 2	$2000/year for 36 years = $72,000	$598,254

Note that twin 1 contributed $56,000 less than twin 2 but had $108,774 more at age 65. This illustrates the powerful effect that time and compounding have on investments.

Calculator Note   Steps for finding the future value of an ordinary annuity with a graphing calculator are found in Appendix C, Section 6.3.   ∎

Figure 6.4 shows a comparison (each graphed as a smooth curve) of the future values of two annuities that are invested at 6% compounded monthly. One annuity has monthly payments of $100, and one has payments of $125. Note the impact of a slightly larger contribution on the future value of the annuity after 5 years, 10 years, 15 years, and 20 years. Of course, a slightly higher interest rate also can have a substantial impact over time; try a graphical comparison yourself.

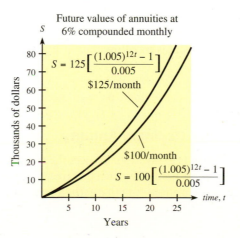

Figure 6.4

Spreadsheet Note

Excel can be used to find the future value of an ordinary annuity. See Appendix D, Section 6.3 or the Online Excel Guide. Excel also can be used to compare investments such as those compared graphically in Figure 6.4. Table 6.4 shows the future values at the end of every two years for the first 10 years of two ordinary annuities: one of $100 per month and a second of $125 per month and both at 6%, compounded monthly.

TABLE 6.4			
	A	**B**	**C**
1		Future Value for R = $100	Future Value for R = $125
2	End of Year #	at 6% compounded monthly	at 6% compounded monthly
3	2	2543.20	3178.99
4	4	5409.78	6762.23
5	6	8640.89	10,801.11
6	8	12,282.85	15,353.57
7	10	16,387.93	20,484.92

Sometimes we want to know how long it will take for an annuity to reach a desired future value.

EXAMPLE 4   **Time to Reach a Goal**

A small business invests $1000 at the end of each month in an account that earns 6% compounded monthly. How long will it take until the business has $100,000 toward the purchase of its own office building?

Solution
This is an ordinary annuity with $S = \$100,000$, $R = \$1000$, $i = 0.06/12 = 0.005$, and $n =$ the number of months. To answer the question of how long, solve for n.

Use $S = R\left[\dfrac{(1+i)^n - 1}{i}\right]$ and solve for n in $100,000 = 1000\left[\dfrac{(1+0.005)^n - 1}{0.005}\right]$.

This is an exponential equation. Hence, we isolate $(1.005)^n$, take the natural logarithm of both sides, and then solve for n as follows.

$$100,000 = \frac{1000}{0.005}[(1.005)^n - 1]$$
$$0.5 = (1.005)^n - 1$$
$$1.5 = (1.005)^n$$
$$\ln(1.5) = \ln[(1.005)^n] = n[\ln(1.005)]$$
$$n = \frac{\ln(1.5)}{\ln(1.005)} \approx 81.3$$

Because this investment is monthly, after 82 months the company will be able to purchase its own office building.

We can also find the periodic payment needed to obtain a specified future value.

EXAMPLE 5   **Payment for an Ordinary Annuity**

A young couple wants to save $50,000 over the next 5 years and then to use this amount as a down payment on a home. To reach this goal, how much money must they deposit at the end of each quarter in an account that earns interest at a rate of 5%, compounded quarterly?

Solution

This plan describes an ordinary annuity with a future value of $50,000 whose payment size, R, is to be determined. Quarterly compounding gives $n = 5(4) = 20$ and $i = 0.05/4 = 0.0125$.

$$S = R\left[\frac{(1 + i)^n - 1}{i}\right]$$

$$\$50,000 = R\left[\frac{(1 + 0.0125)^{20} - 1}{0.0125}\right] = R(22.56297854)$$

$$R = \frac{\$50,000}{22.56297854} = \$2216.02, \text{ to the nearest cent}$$

Sinking Funds

Just as the couple in Example 5 was saving for a future purchase, some borrowers, such as municipalities, may have a debt that must be paid in a single large sum on a specified future date. If these borrowers make periodic deposits that will produce that sum on a specified date, we say that they have established a **sinking fund.** If the deposits are all the same size and are made regularly, they form an ordinary annuity whose future value (on a specified date) is the amount of the debt. To find the size of these periodic deposits, we solve for R in the equation for the future value of an annuity:

$$S = R\left[\frac{(1 + i)^n - 1}{i}\right]$$

EXAMPLE 6   **Sinking Fund**

A company establishes a sinking fund to discharge a debt of $300,000 due in 5 years by making equal semiannual deposits, the first due in 6 months. If the deposits are placed in an account that pays 6%, compounded semiannually, what is the size of the deposits?

Solution

For this sinking fund, we want to find the payment size, R, given that the future value is $S = \$300,000$, $n = 2(5) = 10$, and $i = 0.06/2 = 0.03$. Thus we have

$$S = R\left[\frac{(1 + i)^n - 1}{i}\right]$$

$$\$300,000 = R\left[\frac{(1 + 0.03)^{10} - 1}{0.03}\right]$$

$$\$300,000 = R(11.463879)$$

$$\frac{\$300,000}{11.463879} = R \quad \text{so} \quad R = \$26,169.15$$

Thus the semiannual deposit is $26,169.15.

CHECKPOINT

1. Suppose that $500 is deposited at the end of every quarter for 6 years in an account that pays 8%, compounded quarterly.
   (a) What is the total number of payments (periods)?
   (b) What is the interest rate per period?
   (c) What formula is used to find the future value of the annuity?
   (d) Find the future value of the annuity.

2. A sinking fund of $100,000 is to be established with equal payments at the end of each half-year for 15 years. Find the amount of each payment if money is worth 10%, compounded semiannually.

Annuities Due

Deposits in savings accounts, rent payments, and insurance premiums are examples of **annuities due.** Unlike an ordinary annuity, an annuity due has the periodic payments made at the *beginning* of the period. The *term* of an annuity due is from the first payment to the end of one period after the last payment. Thus an annuity due draws interest for one period more than the ordinary annuity.

We can find the future value of an annuity due by treating each payment as though it were made at the *end* of the preceding period in an ordinary annuity. Then that amount, $Rs_{\overline{n}|i}$, remains in the account for one additional period (see Figure 6.5) and its future value is $Rs_{\overline{n}|i}(1 + i)$.

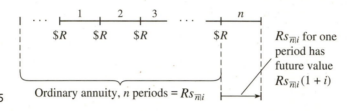

Figure 6.5

Thus the formula for the future value of an annuity due is as follows.

Future Value of an Annuity Due

$$S_{\text{due}} = Rs_{\overline{n}|i}(1 + i) = R\left[\frac{(1 + i)^n - 1}{i}\right](1 + i)$$

EXAMPLE 7 Future Value

Find the future value of an investment if $150 is deposited at the beginning of each month for 9 years and the interest rate is 7.2%, compounded monthly.

Solution

Because deposits are made at the *beginning* of each month, this is an annuity due with $R = \$150$, $n = 9(12) = 108$, and $i = 0.072/12 = 0.006$.

$$S_{\text{due}} = R\left[\frac{(1 + i)^n - 1}{i}\right](1 + i) = \$150\left[\frac{(1 + 0.006)^{108} - 1}{0.006}\right](1 + 0.006)$$

$$= \$150(151.3359308)(1.006) = \$22,836.59, \text{ to the nearest cent}$$

We can also use the formula for the future value of an annuity due to determine the payment size required to reach an investment goal.

EXAMPLE 8 Required Payment

Suppose a company wants to have $450,000 after $2\frac{1}{2}$ years to modernize its production equipment. How much of each previous quarter's profits should be deposited at the beginning of the current quarter to reach this goal, if the company's investment earns 6.8%, compounded quarterly?

Solution

We seek the payment size, R, for an annuity due with $S_{due} = \$450,000$, $n = (2.5)(4) = 10$, and $i = 0.068/4 = 0.017$.

$$S_{due} = R\left[\frac{(1+i)^n - 1}{i}\right](1+i)$$

$$\$450,000 = R\left[\frac{(1+0.017)^{10} - 1}{0.017}\right](1+0.017)$$

$$\$450,000 = R(10.80073308)(1.017) = (10.98434554)R$$

$$R = \frac{\$450,000}{10.98434554} = \$40,967.39, \text{ to the nearest cent}$$

Thus, the company needs to deposit about $40,967 at the beginning of each quarter for the next $2\frac{1}{2}$ years to reach its goal.   ■

CHECKPOINT

3. Suppose $100 is deposited at the beginning of each month for 3 years in an account that pays 6%, compounded monthly.
    (a) What is the total number of payments (or periods)?
    (b) What is the interest rate per period?
    (c) What formula is used to find the future value of the annuity?
    (d) Find the future value.

CHECKPOINT SOLUTIONS

1. (a) $n = 4(6) = 24$ periods   (b) $i = 0.08/4 = 0.02$ per period

    (c) $S = R\left[\dfrac{(1+i)^n - 1}{i}\right]$ for an ordinary annuity

    (d) $S = \$500\left[\dfrac{(1+0.02)^{24} - 1}{0.02}\right] = \$15,210.93$, to the nearest cent

2. Use $n = 2(15) = 30$, $i = 0.10/2 = 0.05$, and $S = \$100,000$ in the formula

    $$S = R\left[\frac{(1+i)^n - 1}{i}\right]$$

    $$\$100,000 = R\left[\frac{(1+0.05)^{30} - 1}{0.05}\right] = R(66.4388475)$$

    $$\frac{\$100,000}{66.4388475} = R \text{ so } R = \$1505.14, \text{ to the nearest cent}$$

3. (a) $n = 3(12) = 36$ periods   (b) $i = 0.06/12 = 0.005$ per period

    (c) $S_{due} = R\left[\dfrac{(1+i)^n - 1}{i}\right](1+i)$ for an annuity due

    (d) $S_{due} = \$100\left[\dfrac{(1+0.005)^{36} - 1}{0.005}\right](1+0.005) = \$3953.28$, to the nearest cent

| EXERCISES | 6.3

Access end-of-section exercises online at **www.webassign.net**

OBJECTIVES 6.4

- To compute the present values of ordinary annuities, annuities due, and deferred annuities
- To compute the payments from various annuities
- To find how long an annuity will last
- To apply present values to bond pricing

Present Values of Annuities

▌| APPLICATION PREVIEW |▐

If you wanted to receive, at retirement, $1000 at the end of each month for 16 years, what lump sum would you need to invest in an annuity that paid 9%, compounded monthly? (See Example 2.) We call this lump sum the present value of the annuity. Note that the annuity in this case is an account from which a person receives equal periodic payments (withdrawals).

In this section, we will find present values of annuities, compute payments from annuities, and find how long annuities will last.

We have discussed how contributing to an annuity program will result in a sum of money, and we have called that sum the future value of the annuity. Just as the term *annuity* is used to describe an account in which a person makes equal periodic payments (deposits), this term is also used to describe an account from which a person receives equal periodic payments (withdrawals). That is, if you invest a lump sum of money in an account today, so that at regular intervals you will receive a fixed sum of money, you have established an annuity. The single sum of money required to purchase an annuity that will provide these payments at regular intervals is the **present value** of the annuity.

Ordinary Annuities

Suppose we wish to invest a lump sum of money (denoted by A_n) in an annuity that earns interest at rate i per period in order to receive (withdraw) payments of size R from this account at the end of each of n periods (after which time the account balance will be 0). Recall that receiving payments at the end of each period means that this is an ordinary annuity.

To find a formula for A_n, we can find the present value of each future payment and then add these present values (see Figure 6.6).

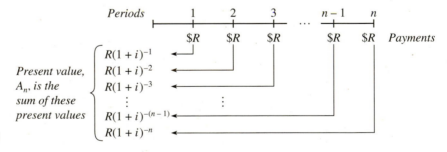

Figure 6.6

Figure 6.6 shows that we can express A_n as follows.

$$A_n = R(1 + i)^{-1} + R(1 + i)^{-2} + R(1 + i)^{-3} + \cdots + R(1 + i)^{-(n-1)} + R(1 + i)^{-n} \quad (1)$$

Multiplying both sides of equation (1) by $(1 + i)$ gives

$$(1 + i)A_n = R + R(1 + i)^{-1} + R(1 + i)^{-2} + \cdots + R(1 + i)^{-(n-2)} + R(1 + i)^{-(n-1)} \quad (2)$$

If we subtract Equation (1) from Equation (2), we obtain $iA_n = R - R(1 + i)^{-n}$. Then solving for A_n gives the following.

Present Value of an Ordinary Annuity	If a payment of $\$R$ is to be made at the end of each period for n periods from an account that earns interest at a rate of i per period, then the account is an **ordinary annuity,** and the **present value** is $$A_n = R \cdot a_{\overline{n}|i} = R \cdot \left[\frac{1 - (1 + i)^{-n}}{i}\right]$$ where $a_{\overline{n}|i}$ represents the present value of an ordinary annuity of $\$1$ per period for n periods, with an interest rate of i per period.

The values of $a_{\overline{n}|i}$ can be computed directly with a calculator or found in Table II in Appendix A.

EXAMPLE 1   Present Value

What is the present value of an annuity of $\$1500$ payable at the end of each 6-month period for 2 years if money is worth 8%, compounded semiannually?

Solution
We are given that $R = \$1500$ and $i = 0.08/2 = 0.04$. Because a payment is made twice a year for 2 years, the number of periods is $n = (2)(2) = 4$. Thus,

$$A_n = R \cdot a_{\overline{4}|0.04} = \$1500\left[\frac{1 - (1 + 0.04)^{-4}}{0.04}\right] = \$1500(3.629895) \approx \$5444.84$$

If we had used Table II in Appendix A, we would have seen that $a_{\overline{4}|0.04}$, the present value of an ordinary annuity of $\$1$ per period for 4 periods with an interest rate of $0.04 = 4\%$ per period, is 3.629895, and we would have obtained the same present value, $\$5444.84$. ∎

EXAMPLE 2   Present Value | APPLICATION PREVIEW |

Find the lump sum that one must invest in an annuity in order to receive $\$1000$ at the end of each month for the next 16 years, if the annuity pays 9%, compounded monthly.

Solution
The sum we seek is the present value of an ordinary annuity, A_n, with $R = \$1000$, $i = 0.09/12 = 0.0075$, and $n = (16)(12) = 192$.

$$A_n = R\left[\frac{1 - (1 + i)^{-n}}{i}\right]$$

$$= \$1000\left[\frac{1 - (1.0075)^{-192}}{0.0075}\right] = \$1000(101.5727689) = \$101,572.77$$

Thus the required lump sum, to the nearest dollar, is $\$101,573$. ∎

Technology Note   The steps for finding the present value of an ordinary annuity with a calculator and with Excel are found in Appendices C and D, Section 6.4. See also the Online Excel Guide. ■

It is important to note that all annuities involve both periodic payments and a lump sum of money. It is whether this lump sum is in the present or in the future that distinguishes problems that use formulas for present values of annuities from those that use formulas for future values of annuities. In Example 2, the lump sum was needed now (in the present) to generate the $1000 payments, so we used the present value formula.

Figure 6.7 shows a graph that compares the present value of an annuity of $1000 per year at an interest rate of 6%, compounded annually, with the same annuity at an interest rate of 10%, compounded annually. Note the impact that the higher interest rate has on the present value needed to establish such an annuity for a longer period of time. Specifically, at 6% interest, a present value of more than $12,500 is needed to generate 25 years of $1000 payments, but at an interest rate of 10%, the necessary present value is less than $10,000.

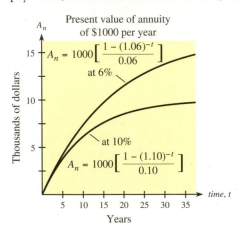

Figure 6.7

Spreadsheet Note   The analysis shown graphically in Figure 6.7 can also be done with a spreadsheet, as Table 6.5 shows for years 5, 10, 15, 20, and 25. ■

| TABLE 6.5 |

PRESENT VALUE GIVING R = $1000 FOR N YEARS

	A	B	C
1	Year number	at 6% compounded annually	at 10% compounded annually
2	5	$4212.36	$3790.79
3	10	$7360.09	$6144.57
4	15	$9712.25	$7606.08
5	20	$11469.92	$8513.56
6	25	$12783.36	$9077.04

Our graphical and spreadsheet comparisons in Figure 6.7 and Table 6.5 and those from other sections emphasize the truth of the saying "Time is money." We have seen that, given enough time, relatively small differences in contributions or in interest rates can result in substantial differences in amounts (both present and future values).

EXAMPLE 3   **Payments from an Annuity**

Suppose that a couple plans to set up an ordinary annuity with a $100,000 inheritance they received. What is the size of the quarterly payments they will receive for the next 6 years (while their children are in college) if the account pays 7%, compounded quarterly?

Solution

The $100,000 is the amount the couple has now, so it is the present value of an ordinary annuity whose payment size, R, we seek. Using present value $A_n = \$100,000$, $n = 6(4) = 24$, and $i = 0.07/4 = 0.0175$, we solve for R.

$$A_n = R\left[\frac{1 - (1 + i)^{-n}}{i}\right]$$

$$\$100,000 = R\left[\frac{1 - (1 + 0.0175)^{-24}}{0.0175}\right]$$

$$\$100,000 = R(19.46068565)$$

$$R = \frac{\$100,000}{19.46068565} \approx \$5138.57$$

CHECKPOINT

1. Suppose an annuity pays $2000 at the end of each 3-month period for $3\frac{1}{2}$ years and money is worth 4%, compounded quarterly.
    (a) What is the total number of periods?
    (b) What is the interest rate per period?
    (c) What formula is used to find the present value of the annuity?
    (d) Find the present value.
2. An inheritance of $400,000 will provide how much at the end of each year for the next 20 years, if money is worth 7%, compounded annually?

EXAMPLE 4   **Number of Payments from an Annuity**

An inheritance of $250,000 is invested at 9%, compounded monthly. If $2500 is withdrawn at the end of each month, how long will it be until the account balance is $0?

Solution

The regular withdrawals form an ordinary annuity with present value $A_n = \$250,000$, payment $R = \$2500$, $i = 0.09/12 = 0.0075$, and $n = $ the number of months.

Use $A_n = R\left[\dfrac{1 - (1 + i)^{-n}}{i}\right]$ and solve for n in $250,000 = 2500\left[\dfrac{1 - (1.0075)^{-n}}{0.0075}\right]$.

This is an exponential equation, and to solve it we isolate $(1.0075)^{-n}$ and then take the natural logarithm of both sides.

$$\frac{250,000(0.0075)}{2500} = 1 - (1.0075)^{-n}$$

$$0.75 = 1 - (1.0075)^{-n}$$

$$(1.0075)^{-n} = 0.25$$

$$\ln\left[(1.0075)^{-n}\right] = \ln(0.25) \quad \Rightarrow \quad -n\left[\ln(1.0075)\right] = \ln(0.25)$$

$$-n = \frac{\ln(0.25)}{\ln(1.0075)} \approx -185.5, \quad \text{so} \quad n \approx 185.5$$

Thus, the account balance will be $0 in 186 months.

Bond Pricing

Bonds represent a relatively safe investment similar to a bank certificate of deposit (CD), but unlike CDs (and like stocks), bonds can be traded. And, as is also true of stocks, the trading or market price of a bond may fluctuate.

Most commonly, bonds are issued by the government, corporations, or municipalities for periods of 10 years or longer. Bonds actually constitute a loan in which the issuer of the bond is the borrower, the bond holders (or purchasers) are the lenders, and the interest payments to the bond holders are called **coupons**. In the simplest case, a bond's issue price, or **par value,** is the same as its **maturity value,** and the coupons are paid semiannually.

For example, if a corporation plans to issue $5000 bonds at 6% semiannually, then the maturity value is $5000, the **coupon rate** is 6%, and each semiannual coupon payment will be

$$(\$5000)(0.06/2) = (\$5000)(0.03) = \$150$$

The coupons are paid at the end of every 6 months and constitute an ordinary annuity. At maturity, the bond holder receives the final coupon payment of $150 plus the $5000 maturity value of the bond.

Because the amount of the coupon is fixed by the coupon rate for the entire term of the bond, the market price of the bond is strongly influenced by current interest rates. If a $5000 bond pays a 6% coupon rate, but market interest rates are higher than that, an investor will typically invest in the bond only if the bond's price makes its rate of return comparable to the market rate. The rate of return that the investor requires in order to buy the bond is called the **yield rate.**

EXAMPLE 5   Bond Pricing

Suppose a 15-year corporate bond has a maturity value of $10,000 and coupons at 5% paid semiannually. If an investor wants to earn a yield of 7.2% compounded semiannually, what should he or she pay for this bond?

Solution
Each semiannual coupon payment is

$$(\$10{,}000)(0.05/2) = (\$10{,}000)(0.025) = \$250$$

Since the desired rate of return is 7.2% semiannually and the bond is for 15 years, we set $i = 0.072/2 = 0.036$ and $n = (15)(2) = 30$.

Then the market price of this bond is the sum of the present values found in (1) and (2) below.

(1) the principal (or present value) of a compound interest investment at $i = 0.036$ for $n = 30$ periods with future value $S = \$10{,}000$

$$S = P(1 + i)^n$$
$$\$10{,}000 = P(1 + 0.036)^{30} = P(1.036)^{30}$$
$$P = \frac{\$10{,}000}{(1.036)^{30}} \approx \$3461.05 \quad (\text{nearest cent})$$

(2) the present value of the ordinary annuity formed by the coupon payments of $R = \$250$ at $i = 0.036$ for $n = 30$ periods

$$A_n = R\left[\frac{1 - (1 + i)^{-n}}{i}\right]$$

$$A_n = \$250\left[\frac{1 - (1.036)^{-30}}{0.036}\right] \approx \$4540.94 \quad (\text{nearest cent})$$

Thus, to earn the desired 7.2% yield, the market price an investor should pay for this bond is

$$\text{Price} = \$3461.05 + \$4540.94 = \$8001.99$$

Note that in Example 5 the bond's market price is less than its maturity value. In this case the bond is said to be **selling at a discount.** This has to be the case in order for the yield rate to exceed the coupon rate. Similarly, if the yield rate is lower than the coupon rate, then the market price of the bond will exceed its maturity value. When this happens, the bond is said to be **selling at a premium.** In general, the market price of a bond moves in the opposite direction from current yield rates.

Annuities Due

Recall that an annuity due is one in which payments are made at the beginning of each period. This means that the present value of an annuity due of n payments (denoted $A_{(n,\text{due})}$) of R at interest rate i per period can be viewed as an initial payment of R plus the payment program for an ordinary annuity of $n - 1$ payments of R at interest rate i per period (see Figure 6.8).

Present Value of an Annuity Due

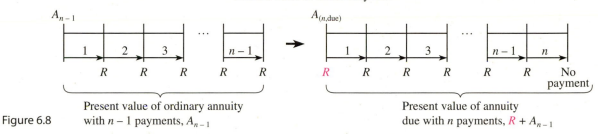

Figure 6.8

From Figure 6.8, we have

$$A_{(n,\text{due})} = R + A_{n-1} = R + R\left[\frac{1 - (1 + i)^{-(n-1)}}{i}\right]$$

$$= R\left[1 + \frac{1 - (1 + i)^{-(n-1)}}{i}\right] = R\left[\frac{i}{i} + \frac{1 - (1 + i)^{-(n-1)}}{i}\right]$$

$$= R\left[\frac{(1 + i) - (1 + i)^{-(n-1)}}{i}\right] = R(1 + i)\left[\frac{1 - (1 + i)^{-n}}{i}\right]$$

Thus we have the following formula for the present value of an annuity due.

Present Value of an Annuity Due

If a payment of R is to be made at the beginning of each period for n periods from an account that earns interest rate i per period, then the account is an **annuity due,** and its **present value** is given by

$$A_{(n,\text{due})} = R\left[\frac{1 - (1 + i)^{-n}}{i}\right](1 + i) = Ra_{\overline{n}|i}(1 + i)$$

where $a_{\overline{n}|i}$ denotes the present value of an ordinary annuity of $1 per period for n periods at interest rate i per period.

EXAMPLE 6   Lottery Prize

A lottery prize worth $1,200,000 is awarded in payments of $10,000 at the beginning of each month for 10 years. Suppose money is worth 7.8%, compounded monthly. What is the *real* value of the prize?

Solution

The *real* value of this prize is its present value when it is awarded. That is, it is the present value of an annuity due with $R = \$10,000$, $i = 0.078/12 = 0.0065$, and $n = 12(10) = 120$.

Thus

$$A_{(120,\text{due})} = \$10,000 \left[\frac{1 - (1 + 0.0065)^{-120}}{0.0065} \right] (1 + 0.0065)$$

$$= \$10,000(83.1439199)(1.0065) = \$836,843.55, \text{ to the nearest cent.}$$

This means the lottery operator needs this amount to generate the 120 monthly payments of $10,000 each. ∎

EXAMPLE 7   Court Settlement Payments

Suppose that a court settlement results in a $750,000 award. If this is invested at 9%, compounded semiannually, how much will it provide at the beginning of each half-year for a period of 7 years?

Solution
Because payments are made at the beginning of each half-year, this is an annuity due. We seek the payment size, R, and use the present value $A_{(n,\text{due})} = \$750,000$, $n = 2(7) = 14$, and $i = 0.09/2 = 0.045$.

$$A_{(n,\text{due})} = R \left[\frac{1 - (1 + i)^{-n}}{i} \right] (1 + i)$$

$$\$750,000 = R \left[\frac{1 - (1 + 0.045)^{-14}}{0.045} \right] (1 + 0.045)$$

$$\$750,000 \approx R(10.22282528)(1.045) \approx R(10.682852)$$

$$R = \frac{\$750,000}{10.682852} = \$70,205.97, \text{ to the nearest cent} \quad ∎$$

CHECKPOINT

3. What lump sum will be needed to generate payments of $5000 at the beginning of each quarter for a period of 5 years if money is worth 7%, compounded quarterly?

Deferred Annuities

A **deferred annuity** is one in which the first payment is made not at the beginning or end of the first period, but at some later date. An annuity that is deferred for k periods and then has payments of R per period at the end of each of the next n periods is an ordinary deferred annuity and can be illustrated by Figure 6.9.

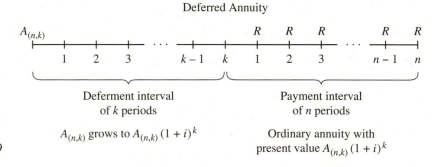

Deferred Annuity

Figure 6.9

We now consider how to find the present value of such a deferred annuity when the interest rate is i per period. If payment is deferred for k periods, then the present value deposited now, denoted by $A_{(n,k)}$, is a compound interest investment for these k periods, and its future value is $A_{(n,k)}(1 + i)^k$. This amount then becomes the present value of the ordinary annuity for the next n periods. From Figure 6.9, we see that we have two equivalent expressions for the amount at the beginning of the first payment period.

$$A_{(n,k)}(1 + i)^k = R \left[\frac{1 - (1 + i)^{-n}}{i} \right]$$

Multiplying both sides by $(1 + i)^{-k}$ gives the following.

Present Value of a Deferred Annuity	The **present value of a deferred annuity** of R per period for n periods, deferred for k periods with interest rate i per period, is given by $$A_{(n,k)} = R\left[\frac{1 - (1 + i)^{-n}}{i}\right](1 + i)^{-k} = Ra_{\overline{n}|i}(1 + i)^{-k}$$

EXAMPLE 8   Present Value of a Deferred Annuity

A deferred annuity is purchased that will pay \$10,000 per quarter for 15 years after being deferred for 5 years. If money is worth 6% compounded quarterly, what is the present value of this annuity?

Solution

We use $R = \$10,000$, $n = 4(15) = 60$, $k = 4(5) = 20$, and $i = 0.06/4 = 0.015$ in the formula for the present value of a deferred annuity.

$$A_{(60,20)} = R\left[\frac{1 - (1 + i)^{-60}}{i}\right](1 + i)^{-20}$$

$$= \$10,000\left[\frac{1 - (1.015)^{-60}}{0.015}\right](1.015)^{-20}$$

$$= \$292,386.85, \text{ to the nearest cent} \qquad \blacksquare$$

EXAMPLE 9   Lottery Prize Payments

Suppose a lottery prize of \$50,000 is invested by a couple for future use as their child's college fund. The family plans to use the money as 8 semiannual payments at the end of each 6-month period after payments are deferred for 10 years. How much would each payment be if the money can be invested at 8.6% compounded semiannually?

Solution

We seek the payment R for a deferred annuity with $n = 8$ payment periods, deferred for $k = 2(10) = 20$ periods, $i = 0.086/2 = 0.043$, and $A_{(n,k)} = \$50,000$.

$$A_{(n,k)} = R\left[\frac{1 - (1 + i)^{-n}}{i}\right](1 + i)^{-k}$$

$$\$50,000 = R\left[\frac{1 - (1.043)^{-8}}{0.043}\right](1.043)^{-20} = R(6.650118184)(0.4308378316)$$

$$\$50,000 = R(2.865122499)$$

$$R = \frac{\$50,000}{2.865122499} = \$17,451.26, \text{ to the nearest cent} \qquad \blacksquare$$

In Example 9, note the effect of the deferral time. The family receives $8(\$17,451.26) = \$139,610.08$ from the original \$50,000 investment.

CHECKPOINT

4. Suppose an annuity at 6% compounded semiannually will pay $5000 at the end of each 6-month period for 5 years with the first payment deferred for 10 years.
   (a) What is the number of payment periods and the number of deferral periods?
   (b) What is the interest rate per period?
   (c) What formula is used to find the present value of this annuity?
   (d) Find the present value of this annuity.

CHECKPOINT SOLUTIONS

1. (a) $n = 3.5(4) = 14$
   (b) $i = 0.04/4 = 0.01$
   (c) $A_n = R\left[\dfrac{1 - (1 + i)^{-n}}{i}\right]$ for an ordinary annuity
   (d) $A_n = \$2000\left[\dfrac{1 - (1.01)^{-14}}{0.01}\right] = \$26,007.41$, to the nearest cent

2. We seek R for an ordinary annuity with $A_n = \$400,000$, $n = 20$, and $i = 0.07$.

$$A_n = R\left[\frac{1 - (1 + i)^{-n}}{i}\right]$$

$$\$400,000 = R\left[\frac{1 - (1.07)^{-20}}{0.07}\right] = R(10.59401425)$$

$$R = \frac{\$400,000}{10.59401425} = \$37,757.17, \text{ to the nearest cent}$$

3. With payments at the beginning of each quarter, we seek the present value $A_{(n,\text{due})}$ with $n = 4(5) = 20$, $i = 0.07/4 = 0.0175$, and $R = \$5000$. Hence,

$$A_{(20,\text{due})} = R\left[\frac{1 - (1 + i)^{-n}}{i}\right](1 + i)$$

$$= \$5000\left[\frac{1 - (1 + 0.0175)^{-20}}{0.0175}\right](1 + 0.0175)$$

$$= \$5000(16.752881)(1.0175) = \$85,230.28, \text{ to the nearest cent}$$

4. (a) $n = 5(2) = 10$ payment periods, and $k = 10(2) = 20$ deferral periods
   (b) $0.06/2 = 0.03$
   (c) $A_{(n,k)} = R\left[\dfrac{1 - (1 + i)^{-n}}{i}\right](1 + i)^{-k}$ for a deferred annuity
   (d) $A_{(n,k)} = \$5000\left[\dfrac{1 - (1.03)^{-10}}{0.03}\right](1.03)^{-20} = \$23,614.83$, to the nearest cent

| **EXERCISES** | **6.4** |

Access end-of-section exercises online at **www.webassign.net**

OBJECTIVES   **6.5**

- To find the regular payments required to amortize a debt
- To find the amount that can be borrowed for a specified payment
- To develop an amortization schedule
- To find the unpaid balance of a loan
- To find the effect of paying an extra amount

Loans and Amortization

▌| APPLICATION PREVIEW |

Chuckie and Angelica have been married for 4 years and are ready to buy their first home. They have saved $30,000 for a down payment, and their budget can accommodate a monthly mortgage payment of $1200.00. They want to know in what price range they can purchase a home if the rate for a 30-year loan is 7.8% compounded monthly. (See Example 3.)

When businesses, individuals, or families (such as Chuckie and Angelica) borrow money to make a major purchase, four questions commonly arise:

1. What will the payments be?
2. How much can be borrowed and still fit a budget?
3. What is the payoff amount of the loan before the final payment is due?
4. What is the total of all payments needed to pay off the loan?

In this section we discuss the most common way that consumers and businesses discharge debts—regular payments of fixed size made on a loan (called amortization)—and determine ways to answer Chuckie and Angelica's question (Question 2) and the other three questions posed above.

Amortization   Just as we invest money to earn interest, banks and lending institutions lend money and collect interest for its use. Federal law now requires that the full cost of any loan and the **true annual percentage rate (APR)** be disclosed with the loan. Because of this law, loans now are usually paid off by a series of partial payments with interest charged on the unpaid balance at the end of each payment period.

This type of loan is usually repaid by making all payments (including principal and interest) of equal size. This process of repaying the loan is called **amortization.**

When a bank makes a loan of this type, it is purchasing from the borrower an ordinary annuity that pays a fixed return each payment period. The lump sum the bank gives to the borrower (the principal of the loan) is the present value of the ordinary annuity, and each payment the bank receives from the borrower is a payment from the annuity. Hence, to find the size of these periodic payments, we solve for R in the formula for the present value of an ordinary annuity,

$$A_n = R\left[\frac{1 - (1 + i)^{-n}}{i}\right]$$

which yields the following algebraically equivalent formula.

Amortization Formula   If the debt of A_n, with interest at a rate of i per period, is amortized by n equal periodic payments (each payment being made at the end of a period), the size of each payment is

$$R = A_n \cdot \left[\frac{i}{1 - (1 + i)^{-n}}\right]$$

EXAMPLE 1 **Payments to Amortize a Debt**

A debt of $1000 with interest at 16%, compounded quarterly, is to be amortized by 20 quarterly payments (all the same size) over the next 5 years. What will the size of these payments be?

Solution

The amortization of this loan is an ordinary annuity with present value $1000. Therefore, $A_n = \$1000$, $n = 20$, and $i = 0.16/4 = 0.04$. Thus we have

$$R = A_n \left[\frac{i}{1 - (1 + i)^{-n}} \right] = \$1000 \left[\frac{0.04}{1 - (1.04)^{-20}} \right]$$

$$= \$1000(0.07358175) = \$73.58, \text{ to the nearest cent}$$

EXAMPLE 2 **Buying a Home**

A man buys a house for $200,000. He makes a $50,000 down payment and agrees to amortize the rest of the debt with quarterly payments over the next 10 years. If the interest on the debt is 12%, compounded quarterly, find (a) the size of the quarterly payments, (b) the total amount of the payments, and (c) the total amount of interest paid.

Solution

(a) We know that $A_n = \$200,000 - \$50,000 = \$150,000$, $n = 4(10) = 40$, and $i = 0.12/4 = 0.03$. Thus, for the quarterly payment, we have

$$R = \$150,000 \left[\frac{0.03}{1 - (1.03)^{-40}} \right]$$

$$= (\$150,000)(0.043262378) = \$6489.36, \text{ to the nearest cent}$$

(b) The man made 40 payments of $6489.36, so his payments totaled

$$(40)(\$6489.36) = \$259,574.40$$

plus the $50,000 down payment, or $309,574.40.

(c) Of the $309,574.40 paid, $200,000 was for payment of the house. The remaining $109,574.40 was the total amount of interest paid.

Technology Note The steps for finding the payment size to amortize a debt with a calculator or Excel are found in Appendices C and D, Section 6.5. See also the Online Excel Guide.

EXAMPLE 3 **Affordable Home** | APPLICATION PREVIEW |

Chuckie and Angelica have $30,000 for a down payment, and their budget can accommodate a monthly mortgage payment of $1200.00. What is the most expensive home they can buy if they can borrow money for 30 years at 7.8%, compounded monthly?

Solution

We seek the amount that Chuckie and Angelica can borrow, or A_n, knowing that $R = \$1200$, $n = 30(12) = 360$, and $i = 0.078/12 = 0.0065$. We can use these values in the amortization formula and solve for A_n (or use the formula for the present value of an ordinary annuity).

$$R = A_n \left[\frac{i}{1 - (1 + i)^{-n}} \right]$$

$$\$1200 = A_n \left[\frac{0.0065}{1 - (1.0065)^{-360}} \right]$$

$$A_n = \$1200 \left[\frac{1 - (1.0065)^{-360}}{0.0065} \right]$$

$$A_n = \$1200(138.9138739) = \$166,696.65, \text{ to the nearest cent}$$

Thus, if they borrow $166,697 (to the nearest dollar) and put down $30,000, the most expensive home they can buy would cost $166,697 + $30,000 = $196,697.   ■

CHECKPOINT

1. A debt of $25,000 is to be amortized with equal quarterly payments over 6 years, and money is worth 7%, compounded quarterly.
   (a) Find the total number of payments.
   (b) Find the interest rate per period.
   (c) Write the formula used to find the size of each payment.
   (d) Find the amount of each payment.
2. A new college graduate determines that she can afford a car payment of $400 per month. If the auto manufacturer is offering a special 2.1% financing rate, compounded monthly for 5 years, how much can she borrow and still have a $400 monthly payment?

Amortization Schedule

We can construct an **amortization schedule** that summarizes all the information regarding the amortization of a loan.

For example, a loan of $10,000 with interest at 10% could be repaid in 5 equal annual payments of size

$$R = \$10,000 \left[\frac{0.10}{1 - (1 + 0.10)^{-5}} \right] = \$10,000(0.263797) = \$2637.97$$

Each time this $2637.97 payment is made, some is used to pay the interest on the unpaid balance, and some is used to reduce the principal. For the first payment, the unpaid balance is $10,000, so the interest payment is 10% of $10,000, or $1000. The remaining $1637.97 is applied to the principal. Hence, after this first payment, the unpaid balance is $10,000 − $1637.97 = $8362.03.

For the second payment of $2637.97, the amount used for interest is 10% of $8362.03, or $836.20; the remainder, $1801.77, is used to reduce the principal.

This information for these two payments and for the remaining payments is summarized in the following amortization schedule.

Period	Payment	Interest	Balance Reduction	Unpaid Balance
				10000.00
1	2637.97	1000.00	1637.97	8362.03
2	2637.97	836.20	1801.77	6560.26
3	2637.97	656.03	1981.94	4578.32
4	2637.97	457.83	2180.14	2398.18
5	2638.00	239.82	2398.18	0.00
Total	13189.88	3189.88	10000.00	

Note that the last payment was increased by 3¢ so that the balance was reduced to $0 at the end of the 5 years. Such an adjustment is normal in amortizing a loan.

Spreadsheet Note

The preceding amortization schedule involves only a few payments. For loans involving more payments, a spreadsheet program is an excellent tool to develop an amortization schedule. The spreadsheet requires the user to understand the interrelationships among the columns. The accompanying spreadsheet output in Table 6.6 shows the amortization schedule for the first, sixth, twelfth, eighteenth, and twenty-fourth monthly payments on a $100,000 loan amortized for 30 years at 6%, compounded monthly. It is interesting to observe how little the unpaid balance changes during these first 2 years—and, consequently, how much of each payment is devoted to paying interest due rather than to reducing the debt. In fact, the column labeled "Total Interest" shows a running total of how much has been paid toward interest. We see that after 12 payments of $599.55, a total of $7194.60 has been paid, and $5966.59 of this has been interest payments. ■

TABLE 6.6						
	A	**B**	**C**	**D**	**E**	**F**
1	Payment Number	Payment Amount	Interest	Balance Reduction	Unpaid Balance	Total Interest
2	0	0	0	0	$100,000.00	$0.00
3	1	$599.55	$500.00	$99.55	$99,900.45	$500.00
4	6	$599.55	$497.49	$102.06	$99,395.18	$2,992.48
5	12	$599.55	$494.39	$105.16	$98,771.99	$5,966.59
6	18	$599.55	$491.19	$108.36	$98,129.87	$8,921.77
7	24	$599.55	$487.90	$111.65	$97,468.25	$11,857.45

Unpaid Balance of a Loan

Many people who borrow money, such as for a car or a home, do not pay on the loan for its entire term. Rather, they pay off the loan early by making a final lump sum payment. The unpaid balance found in the amortization schedule is the "payoff amount" of the loan and represents the lump sum payment that must be made to complete payment on the loan. When the number of payments is large, we may wish to find the unpaid balance of a loan without constructing an amortization schedule.

Recall that calculations for amortization of a debt are based on the present value formula for an ordinary annuity. Because of this, the **unpaid balance of a loan** (also called the **payoff amount** and the **outstanding principal of the loan**) is the present value needed to generate all the remaining payments.

Unpaid Balance or Payoff Amount of a Loan

For a loan of n payments of R per period at interest rate i per period, the **unpaid balance**, or **payoff amount**, after k payments have been made is the present value of an ordinary annuity with $n - k$ payments. That is, with $n - k$ payments remaining,

$$\text{Unpaid balance} = A_{n-k} = R\left[\frac{1 - (1 + i)^{-(n-k)}}{i}\right]$$

EXAMPLE 4   Unpaid Balance

In Example 2, we found that the monthly payment for a loan of $150,000 at 12%, compounded quarterly, for 10 years is $6489.36 (to the nearest cent). Find the unpaid balance immediately after the 15th payment.

Solution

The unpaid balance after the 15th payment is the present value of the annuity with $40 - 15 = 25$ payments remaining. Thus we use $R = \$6489.36$, $i = 0.03$, and $n - k = 25$ in the formula for the unpaid balance of a loan.

$$A_{n-k} = R\left[\frac{1 - (1 + i)^{-(n-k)}}{i}\right] = \$6489.36\left[\frac{1 - (1.03)^{-25}}{0.03}\right]$$

$$= (\$6489.36)\,(17.4131477) = \$113,000.18, \text{ to the nearest cent} \quad \blacksquare$$

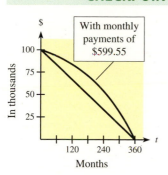

Figure 6.10

CHECKPOINT

3. A 42-month auto loan has monthly payments of \$411.35. If the interest rate is 8.1%, compounded monthly, find the unpaid balance immediately after the 24th payment.

The curve in Figure 6.10 shows the unpaid balance of a \$100,000 loan at 6%, compounded monthly, for 30 years (with monthly payments of \$599.55, as in Table 6.6). In the figure, the straight line represents how the unpaid balance would decrease if each payment diminished the debt by the same amount. Note how the curve decreases slowly at first and more rapidly as the unpaid balance nears zero. The reason for this behavior is that when the unpaid balance is large, much of each payment is devoted to interest. Similarly, when the debt decreases, more of each payment goes toward the principal. An interesting question then is what would be the effect of paying an extra amount (directly toward the principal) with each payment.

EXAMPLE 5   **Effect of Paying an Extra Amount**

Consider the loan in Table 6.6: \$100,000 borrowed at 6% compounded monthly for 30 years, with monthly payments of \$599.55. As Table 6.6 shows, the unpaid balance after 24 monthly payments is \$97,468.25. Suppose that from this point the borrower decides to pay \$650 per month.

(a) How many more payments must be made?
(b) How much would this save over the life of the loan?

Solution

(a) Use $A_n = R\left[\dfrac{1 - (1 + i)^{-n}}{i}\right]$ with $A_n = \$97,468.25$, $R = \$650$, $i = 0.06/12 = 0.005$,

and solve for n (by first isolating the exponential $(1.005)^{-n}$).

$$97,468.25 = 650\left[\frac{1 - (1.005)^{-n}}{0.005}\right] \tag{1}$$

$$\frac{(97,468.25)(0.005)}{650} = 1 - (1.005)^{-n}$$

$$(1.005)^{-n} = 1 - \frac{(97,468.25)(0.005)}{650}$$

$$\ln\left[(1.005)^{-n}\right] = \ln\,(0.2502442308)$$

$$-n\left[\ln\,(1.005)\right] = \ln\,(0.2502442308)$$

$$n = \frac{\ln\,(0.2502442308)}{-\ln\,(1.005)} \approx 277.8$$

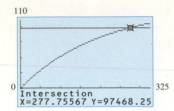

Figure 6.11

(b) For the purposes of determining the savings, assume $n = 277.8$ payments of $650 were made. Thus the total paid with each method is as follows.

Paying Extra: $(24 \text{ payments of } \$599.55) + (277.8 \text{ payments of } \$650) = \$194,959.20$
Paying $599.55: $(360 \text{ payments of } \$599.55) = \$215,838.00$

Hence by increasing the payments slightly, the borrower saved

$$\$215,838.00 - \$194,959.20 = \$20,878.80$$

Calculator Note

Notice in Example 5 that we could have solved Equation (1) graphically by finding where $y_1 = 97,468.25$ and $y_2 = 650\left[\dfrac{1 - (1.005)^{-x}}{0.005}\right]$ intersect (see Figure 6.11).

CHECKPOINT SOLUTIONS

1. (a) $n = 6(4) = 24$ payments
   (b) $i = 0.07/4 = 0.0175$ per period
   (c) $R = A_n\left[\dfrac{i}{1 - (1 + i)^{-n}}\right]$
   (d) $R = \$25,000\left[\dfrac{0.0175}{1 - (1.0175)^{-24}}\right] = \$1284.64$, to the nearest cent

2. Use $R = A_n\left[\dfrac{i}{1 - (1 + i)^{-n}}\right]$ with $i = 0.021/12 = 0.00175$, $n = 60$, and $R = \$400$.

   $$\$400 = A_n\left[\dfrac{0.00175}{1 - (1.00175)^{-60}}\right] = A_n(0.0175715421)$$

   $$A_n = \dfrac{\$400}{0.0175715421} \approx \$22,764.08, \text{ to the nearest cent}$$

3. Use $R = \$411.35$, $i = 0.081/12 = 0.00675$, and $n - k = 42 - 24 = 18$.

   $$A_{n-k} = R\left[\dfrac{1 - (1 + i)^{-(n-k)}}{i}\right] = \$411.35\left[\dfrac{1 - (1.00675)^{-18}}{0.00675}\right]$$
   $$= \$6950.13, \text{ to the nearest cent}$$

| **EXERCISES** | **6.5**

Access end-of-section exercises online at **www.webassign.net**

KEY TERMS AND FORMULAS

Section	Key Terms	Formulas
6.1	Simple interest	$I = Prt$
	Future value of a simple interest investment or loan	$S = P + I$
	Sequence function	
	Arithmetic sequence	$a_n = a_{n-1} + d \quad (n > 1)$
	Common difference	d
	nth term of	$a_n = a_1 + (n - 1)d$
	Sum of first n terms	$s_n = \dfrac{n}{2}(a_1 + a_n)$
6.2	Future value of compound interest investment	
	Periodic compounding	$S = P(1 + i)^n = P\left(1 + \dfrac{r}{m}\right)^{mt}$
	Continuous compounding	$S = Pe^{rt}$
	Annual percentage yield	
	Periodic compounding	$APY = \left(1 + \dfrac{r}{m}\right)^m - 1$
		$= (1 + i)^m - 1$
	Continuous compounding	$APY = e^r - 1$
	Geometric sequence	$a_n = ra_{n-1} \quad (n > 1)$
	Common ratio	r
	nth term of	$a_n = a_1 r^{n-1}$
	Sum of first n terms	$s_n = \dfrac{a_1(1 - r^n)}{1 - r} \quad (\text{if } r \neq 1)$
6.3	Future values of annuities	
	Ordinary annuity	$S = R\left[\dfrac{(1 + i)^n - 1}{i}\right]$
	Required payment into an ordinary annuity	
	Sinking fund	
	Annuity due	$S_{\text{due}} = R\left[\dfrac{(1 + i)^n - 1}{i}\right](1 + i)$
6.4	Present values of annuities	
	Ordinary annuity	$A_n = R\left[\dfrac{1 - (1 + i)^{-n}}{i}\right]$
	Bond pricing	
	Annuity due	$A_{(n,\text{due})} = R\left[\dfrac{1 - (1 + i)^{-n}}{i}\right](1 + i)$
	Deferred annuity	$A_{(n,k)} = R\left[\dfrac{1 - (1 + i)^{-n}}{i}\right](1 + i)^{-k}$
6.5	True annual percentage rate (APR)	
	Amortization (Loan amount is the same as present value of an ordinary annuity)	$R = A_n\left[\dfrac{i}{1 - (1 + i)^{-n}}\right]$
	Unpaid balance ($n - k$ payments left)	$A_{n-k} = R\left[\dfrac{1 - (1 + i)^{-(n-k)}}{i}\right]$
	Amortization schedule	

REVIEW EXERCISES

1. Find the first 4 terms of the sequence with nth term

$$a_n = \frac{1}{n^2}$$

2. Identify any arithmetic sequences and find the common differences.
   (a) $12, 7, 2, -3, \ldots$ (b) $1, 3, 6, 10, \ldots$
   (c) $\frac{1}{6}, \frac{1}{3}, \frac{1}{2}, \frac{2}{3}, \ldots$

3. Find the 80th term of the arithmetic sequence with first term -2 and common difference 3.

4. Find the 36th term of the arithmetic sequence with third term 10 and eighth term 25.

5. Find the sum of the first 60 terms of the arithmetic sequence $\frac{1}{3}, \frac{1}{2}, \frac{2}{3}, \ldots$

6. Identify any geometric sequences and their common ratios.
   (a) $\frac{1}{4}, 2, 16, 128, \ldots$ (b) $16, -12, 9, -\frac{27}{4}, \ldots$
   (c) $4, 16, 36, 64, \ldots$

7. Find the fourth term of the geometric sequence with first term 64 and eighth term $\frac{1}{2}$.

8. Find the sum of the first 16 terms of the geometric sequence $\frac{1}{9}, \frac{1}{3}, 1, \ldots$.

APPLICATIONS

Finance **Identify which of the following formulas applies to each situation in Problems 9–14.**

$$I = Prt \qquad S = P(1 + i)^n \qquad S = Pe^{rt}$$

$$S = R\left[\frac{(1 + i)^n - 1}{i}\right] \qquad A_n = R\left[\frac{1 - (1 + i)^{-n}}{i}\right]$$

9. The future value of a series of payments, with interest compounded when the payments are made

10. The simple interest earned on an investment

11. The present value of a series of payments, with interest compounded when the payments are made

12. The future value of an investment that earns interest compounded periodically

13. The regular payment size needed to amortize a debt, when interest is compounded when the payments are made

14. The future value of an investment that earns interest compounded continuously

15. *Loans* If $8000 is borrowed at 12% simple interest for 3 years, what is the future value of the loan at the end of the 3 years?

16. *Loan rate* Mary Toy borrowed $2000 from her parents and repaid them $2100 after 9 months. What simple interest rate did she pay?

17. *Tuition* How much summer earnings must a college student deposit on August 31 in order to have $3000 for tuition and fees on December 31 of the same year, if the investment earns 6% simple interest?

18. *Contest payments* Suppose the winner of a contest receives $10 on the first day of the month, $20 on the second day, $30 on the third day, and so on for a 30-day month. What is the total won?

19. *Salaries* Suppose you are offered two identical jobs: one paying a starting salary of $40,000 with yearly raises of $2000 and one paying a starting salary of $36,000 with yearly raises of $2500. Which job will pay more money over a 10-year period?

20. *Investments* An investment is made at 8%, compounded quarterly, for 10 years.
    (a) Find the number of periods.
    (b) Find the interest rate per period.

21. *Future value* Write the formula for the future value of a compound interest investment, if interest is compounded
    (a) periodically   (b) continuously.

22. *Comparing rates* Which compounding method, at 6%, earns more money?
    (a) semiannually   (b) monthly

23. *Interest* If $1000 is invested for 4 years at 8%, compounded quarterly, how much interest will be earned?

24. *Savings goal* How much must one invest now in order to have $18,000 in 4 years if the investment earns 5.4%, compounded monthly?

25. *Future value* What is the future value if $1000 is invested for 6 years at 8%, compounded continuously?

26. *College fund* A couple received an inheritance and plan to invest some of it for their grandchild's college education. How much must they invest if they would like the fund to have $100,000 after 15 years, and their investment earns 10.31%, compounded continuously?

27. *Investments* If $15,000 is invested at 6%, compounded quarterly, how long will it be before it grows to $25,000?

28. *Investment rates* (a) If an initial investment of $35,000 grows to $257,000 in 15 years, what annual interest rate, continuously compounded, was earned? (b) What is the annual percentage yield on this investment?

29. *Comparing yields* Find the annual percentage yield equivalent to a nominal rate of 7.2% (a) compounded quarterly and (b) compounded continuously.

30. *Chess legend* Legend has it that when the king of Persia wanted to reward the inventor of chess with whatever he desired, the inventor asked for one grain of wheat on the first square of the chessboard, with the number of grains doubled on each square thereafter for the remaining 63 squares. Find the number of grains on the 64th square. (The sum of all the grains of wheat on all the squares would cover Alaska more than 3 inches deep with wheat!)

31. *Chess legend* If, in Problem 30, the king granted the inventor's wish for *half* a chessboard, how many grains would the inventor receive?

32. *Annuity* Find the future value of an ordinary annuity of $800 paid at the end of every 6-month period for 10 years, if it earns interest at 12%, compounded semiannually.

33. *Sinking fund* How much would have to be invested at the end of each year at 6%, compounded annually, to pay off a debt of $80,000 in 10 years?

34. *Annuity* Find the future value of an annuity due of $800 paid at the beginning of every 6-month period for 10 years, if it earns interest at 12%, compounded semiannually.

35. *Construction fund* A company wants to have $250,000 available in $4\frac{1}{2}$ years for new construction. How much must be deposited at the beginning of each quarter to reach this goal if the investment earns 10.2%, compounded quarterly?

36. *Time to reach a goal* If $1200 is deposited at the end of each quarter in an account that earns 7.2%, compounded quarterly, how long will it be until the account is worth $60,000?

37. *Annuity* What lump sum would have to be invested at 9%, compounded semiannually, to provide an annuity of $10,000 at the end of each half-year for 10 years?

38. *Cruise fund* A couple wish to set up an annuity that will provide 6 monthly payments of $3000 while they take an extended cruise. How much of a $15,000 inheritance must be set aside if they plan to leave in 5 years and want the first payment before they leave? Assume that money is worth 7.8%, compounded monthly.

39. *Powerball lottery* Winners of lotteries receive the jackpot distributed over a period of years, usually 20 or 25 years. The winners of the Powerball lottery on July 29, 1998, elected to take a one-time cash payout rather than receive the $295.7 million jackpot in 25 annual payments beginning on the date the lottery was won.
    (a) How much money would the winners have received at the beginning of each of the 25 years?
    (b) If the value of money was 5.91%, compounded annually, what one-time payout did they receive in lieu of the annual payments?

40. *Payment from an annuity* A recent college graduate's gift from her grandparents is $20,000. How much will this provide at the end of each month for the next 12 months while the graduate travels? Assume that money is worth 6.6%, compounded monthly.

41. *Payment from an annuity* An IRA of $250,000 is rolled into an annuity that pays a retired couple at the beginning of each quarter for the next 20 years. If the annuity earns 6.2%, compounded quarterly, how much will the couple receive each quarter?

42. *Duration of an annuity* A retirement account that earns 6.8%, compounded semiannually, contains $488,000. How long can $40,000 be withdrawn at the end of each half-year until the account balance is $0?

43. *Amortization* A debt of $1000 with interest at 12%, compounded monthly, is amortized by 12 monthly payments (of equal size). What is the size of each payment?

44. *Loan pay-off* A debt of $8000 is amortized with eight semiannual payments of $1288.29 each. If money is worth 12%, compounded semiannually, find the unpaid balance after five payments have been made.

45. *Cash value* A woman paid $90,000 down for a house and agreed to pay 18 quarterly payments of $4500 each. If money is worth 4%, compounded quarterly, how much would the house have cost if she had paid cash?

46. *Amortization schedule* Complete the next two lines of the amortization schedule for a $100,000 loan for 30 years at 7.5%, compounded monthly, with monthly payments of $699.22.

Payment Number	Payment Amount	Interest	Balance Reduction	Unpaid Balance
56	$699.22	$594.67	$104.55	$95,042.20

MISCELLANEOUS FINANCIAL PROBLEMS

47. If an initial investment of $2500 grows to $38,000 in 18 years, what annual interest rate, compounded annually, did this investment earn?

48. How much must be deposited at the end of each month in an account that earns 8.4%, compounded monthly, if the goal is to have $40,000 after 10 years?

49. What is the future value if $1000 is invested for 6 years at 8% (a) simple interest and (b) compounded semiannually?

50. Quarterly payments of $500 are deposited in an account that pays 8%, compounded quarterly. How much will have accrued in the account at the end of 4 years if each payment is made at the end of each quarter?

51. If $8000 is invested at 7%, compounded continuously, how long will it be before it grows to $22,000?

52. An investment broker bought some stock at $87.89 per share and sold it after 3 months for $105.34 per share. What was the annual simple interest rate earned on this transaction?

53. Suppose that a 30-year municipal bond has a maturity value of $5000 and a coupon rate of 8%, with coupons paid semiannually. Find the market price of the bond if the current yield rate is 10% compounded semiannually. Is this bond selling at a discount or at a premium?

54. A $5000 Mauranol, Inc. corporate bond matures in 10 years and has a coupon rate of 7.2% paid semiannually.
    (a) If the yield rate is 8% compounded semiannually, find the bond's market price.
    (b) Three years later the yield rate is 5.8%. If the bond holder wishes to sell at this time, what is the bond's current market price?
    (c) If you bought a $5000 Mauranol, Inc. bond at the price in part (a), collected 3 years of coupons, and sold at the price in part (b), what amount would you have earned on your original investment?
    (d) Suppose you made a single investment with principal equal to the amount paid in part (a), and after 3 years you received a single lump sum return equal to the amount from part (c). What interest rate compounded semiannually would you have earned?

55. Kevin Patrick paid off the loan he took out to buy his car, but once the loan was paid, he continued to deposit $400 on the first of each month in an account that paid 5.4%, compounded monthly. After four years of making these deposits, Kevin was ready to buy a new car. How much did he have in the account?

56. A young couple receive an inheritance of $72,000 that they want to set aside for a college fund for their two children. How much will this provide at the end of each half-year for a period of 9 years if it is deferred for 11 years and can be invested at 7.3%, compounded semiannually?

57. A bank is trying to decide whether to advertise some new 18-month certificates of deposit (CDs) at 6.52%, compounded quarterly, or at 6.48%, compounded continuously. Which rate is a better investment for the consumer who buys such a CD? Which rate is better for the bank?

58. A divorce settlement of $40,000 is paid in $1000 payments at the end of each of 40 months. What is the present value of this settlement if money is worth 12%, compounded monthly?

59. If $8000 is invested at 12%, compounded continuously, for 3 years, what is the total interest earned at the end of the 3 years?

60. A couple borrowed $184,000 to buy a condominium. Their loan was for 25 years and money is worth 6%, compounded monthly.
    (a) Find their monthly payment size.
    (b) Find the total amount they would pay over 25 years.
    (c) Find the total interest they would pay over 25 years.
    (d) Find the unpaid balance after 7 years.

61. Suppose a salesman invests his $12,500 bonus in a fund that earns 10.8%, compounded monthly. Suppose also that he makes contributions of $150 at the end of each month to this fund.
    (a) Find the future value after $12\frac{1}{2}$ years.
    (b) If after the $12\frac{1}{2}$ years, the fund is used to set up an annuity, how much will it pay at the end of each month for the next 10 years?

62. Three years from now, a couple plan to spend 4 months traveling in China, Japan, and Southeast Asia. When they take their trip, they would like to withdraw $10,000 at the beginning of each month to cover their expenses for that month. Starting now, how much must they deposit at the beginning of each month for the next 3 years so that the account will provide the money they want while they are traveling? Assume that such an account pays 6.6%, compounded monthly.

63. At age 22, Aruam Sdlonyer receives a $4000 IRA from her parents. At age 30, she decides that at age 67 she'd like to have a retirement fund that would pay $20,000 at the end of each month for 20 years. Suppose all investments earn 8.4%, compounded monthly. How much does Aruam need to deposit at the end of each month from ages 30 to 67 to realize her goal?

64. A company borrows $2.6 million for 15 years at 5.6%, compounded quarterly. After 2 years of regular payments, the company's profits are such that management feels they can increase the quarterly payments to $70,000 each. How long will it take to pay off the loan, and how much interest will be saved?

6 CHAPTER TEST

1. If $8000 is invested at 7%, compounded continuously, how long will it be before it grows to $47,000?

2. How much must you put aside at the end of each half-year if, in 6 years, you want to have $12,000, and the current interest rate is 6.2%, compounded semiannually?

3. To buy a municipal bond that matures to $10,000 in 9 months, you must pay $9510. What simple interest rate is earned?

4. A debt of $280,000 is amortized with 40 equal semiannual payments of $14,357.78. If interest is 8.2%, compounded semiannually, find the unpaid balance of the debt after 25 payments have been made.

5. If an investment of $1000 grew to $13,500 in 9 years, what interest rate, compounded annually, did this investment earn?

6. A couple borrowed $97,000 at 7.2%, compounded monthly, for 25 years to purchase a condominium.
   (a) Find their monthly payment.
   (b) Over the 25 years, how much interest will they pay?

7. If you borrow $2500 for 15 months at 4% simple interest, how much money must you repay after the 15 months?

8. Suppose you invest $100 at the end of each month in an account that earns 6.9%, compounded monthly. What is the future value of the account after $5\frac{1}{2}$ years?

9. Find the annual percentage yield (the effective annual rate) for an investment that earns 8.4%, compounded monthly.

10. An accountant wants to withdraw $3000 from an investment at the beginning of each quarter for the next 15 years. How much must be deposited originally if the investment earns 6%, compounded quarterly? Assume that after the 15 years, the balance is zero.

11. If $10,000 is invested at 7%, compounded continuously, what is the value of the investment after 20 years?

12. A collector made a $10,000 down payment for a classic car and agreed to make 18 quarterly payments of $1500 at the end of each quarter. If money is worth 8%, compounded quarterly, how much would the car have cost if the collector had paid cash?

13. What amount must you invest in an account that earns 6.8%, compounded quarterly, if you want to have $9500 after 5 years?

14. If $1500 is deposited at the end of each half-year in a retirement account that earns 8.2%, compounded semiannually, how long will it be before the account contains $500,000?

15. A couple would like to have $50,000 at the end of $4\frac{1}{2}$ years. They can invest $10,000 now and make additional contributions at the end of each quarter over the $4\frac{1}{2}$ years. If the investment earns 7.7%, compounded quarterly, what size payments will enable them to reach their goal?

16. A woman makes $3000 contributions at the end of each half-year to a retirement account for a period of 8 years. For the next 20 years, she makes no additional contributions and no withdrawals.
   (a) If the account earns 7.5%, compounded semiannually, find the value of the account after the 28 years.
   (b) If this account is used to set up an annuity that pays her an amount at the beginning of each 6-month period for the next 20 years, how much will each payment be?

17. Maxine deposited $400 at the beginning of each month for 15 years in an account that earned 6%, compounded monthly. Find the value of the account after the 15 years.

18. Grandparents plan to establish a college trust for their youngest grandchild. How much is needed now so the trust, which earns 6.4%, compounded quarterly, will provide $4000 at the end of each quarter for 4 years after being deferred for 10 years?

19. The following sequence is arithmetic: 298.8, 293.3, 287.8, 282.3, ...
   (a) If you were not told that this sequence was arithmetic, how could you tell that it was?
   (b) Find the 51st term of this sequence.
   (c) Find the sum of the first 51 terms.

20. A 400-milligram dose of heart medicine is taken daily. During each 24-hour period, the body eliminates 40% of this drug (so that 60% remains in the body). Thus

the amount of drug in the bloodstream just after the 31st dose is given by

$$400 + 400(0.6) + 400(0.6)^2 + 400(0.6)^3 + \ldots$$
$$+ 400(0.6)^{29} + 400(0.6)^{30}$$

Find the level of the drug in the bloodstream at this time.

21. Suppose that a 10-year corporate bond has a maturity value of $10,000 and a coupon rate of 10%, with coupons paid semiannually. Find the market price of the bond if the current yield rate is 6% compounded semiannually. Is this bond selling at a discount or at a premium?

22. If you took a home mortgage loan of $150,000 for 30 years at 8.4%, compounded monthly, your monthly payments would be $1142.76. Suppose after 2 years you had an additional $2000. Would you save more over the life of the loan by paying an extra $2000 with your 24th payment or by paying $1160 per month beginning with the 25th payment? To help you answer this question, complete the following.

(a) Find the unpaid balance after 24 payments (both including and not including the $2000 payment).

(b) Determine how long it takes to pay off the loan with the regular $1142.76 payments and the additional $2000 payment. Then find the total interest paid.

(c) Determine how long it takes to pay off the loan without the additional $2000 payment, but with monthly payments of $1160 from the 25th payment. Then find the total interest paid.

(d) Which payment method costs less?

I. Mail Solicitation Home Equity Loan: Is This a Good Deal?

Discover® offers a home equity line of credit that lowers the total of the monthly payments on several hypothetical loans by $366.12. (See the following table.)

The home equity loan payment is based on "10.49% APR annualized over a 10-year term." This means the loan is amortized at 10.49% with monthly payments.

(a) Using this interest rate and monthly payments for the 10 years, how much money will still be owed at the end of the 10 years?
(b) How much is paid during the 10 years?
(c) How much would the payment have to be to have the debt paid in the 10 years?
(d) How long would it take to pay off the debts under the original plan if interest were 15% on each loan?

| Loan Type | Current Monthly Payment Example | | Discover Home Equity | |
	Loan Amount	Monthly Payment	Loan Amount	Monthly Payment
Bank cards	$10,000	$186,000	PAID OFF	NONE
Auto loan	$12,500	$320.03	PAID OFF	NONE
Department store cards	$2,500	$78.63	PAID OFF	NONE
Discover home equity line of credit			$25,000	$218.54
Total	$25,000	$584.66	$25,000	$218.54

Source: Discover Loan Center mailer

II. Profit Reinvestment

T. C. Hardware Store wants to construct a new building at a second location. If construction could begin immediately, the cost would be $500,000. At this time, however, the owners do not have the required 20% down payment, so they plan to invest $2000 per month of their profits until they have the necessary amount. They can invest their money in an annuity account that pays 6%, compounded monthly, but they are concerned about the 3% average inflation rate in the construction industry. They would like you to give them some projections about how 3% inflation will affect the time required to accrue the down payment and the building's eventual cost. They would also like to know how their projected profits, after the new building is complete, will affect their schedule for paying off the mortgage loan.

Specifically, the owners would like you to prepare a report that answers the following questions.

1. If the 3% annual inflation rate is accurate, how long will it take to get the down payment? *Hint:* Assume inflation is compounded monthly. Then the required down payment will be 20% of

$$500,000\left(1 + \frac{0.03}{12}\right)^n$$

where n is in months. Thus you must find n such that this amount equals the future value of the owners' annuity. The solution to the resulting equation is difficult to obtain by algebraic methods but can be found with a graphing utility.

2. If the 3% annual inflation rate is accurate (compounded monthly), what will T. C. Hardware's projected construction costs be (to the nearest hundred dollars) when it has its down payment?

3. If T. C. Hardware borrows 80% of its construction costs and amortizes that amount at 7.8%, compounded monthly, for 15 years, what will its monthly payment be?

4. Finally, the owners believe that 2 years after this new building is begun (that is, after 2 years of payments on the loan), their profits will increase enough so that they'll be able to make double mortgage payments each month until the loan is paid. Under these assumptions, how long will it take T. C. Hardware to pay off the loan?

III. Purchasing a Home

Buying a home is the biggest single investment or purchase that most individuals make. This project is designed to give you some insight into the home-buying process and the associated costs.

Find all the costs associated with buying a home by making a 20% down payment and borrowing 80% of the cost of the home.

1. Decide how much your home will cost (and how much you'll have to put down and how much to mortgage). Call a realtor to determine the cost of a home in your area (this could be a typical starter home, an average home for the area, or your dream house). Ask the realtor to identify and estimate all the closing costs associated with buying your home. Closing costs include your down payment, attorney's fees, title fees, and so on.

2. Call at least three different banks to determine mortgage loan rates for 15 years, 25 years, and 30 years and the costs associated with obtaining your loan. These associated costs of a loan are paid at closing and include an application fee, an appraisal fee, points, and so on.

3. For each bank and each different loan duration, develop a summary that contains
   (a) an itemization and explanation of all closing costs and the total amount due at closing
   (b) the monthly payment
   (c) the total amount paid over the life of the loan
   (d) the amortization schedule for each loan (use a spreadsheet or financial software package).

4. For each duration (15 years, 25 years, and 30 years), identify which bank gives the best loan rate and explain why you think it is best. Be sure to consider what is best for someone who is likely to remain in the home for 7 years or less compared with someone who is likely to stay 20 years or more.

7

Mike Ehrmann/Getty Images

Introduction to Probability

An economist cannot predict exactly how the gross national product will change, a physician cannot determine exactly the cause of lung cancer, and a psychologist cannot determine the exact effect of environment on behavior. But each of these determinations can be made with varying probabilities. Thus, an understanding of the meaning and determination of the probabilities of events occurring is important to success in business, economics, the life sciences, and the social sciences. Probability was initially developed to solve gambling problems, but it is now the basis for solving problems in a wide variety of areas.

The topics and applications discussed in this chapter include the following.

SECTIONS		APPLICATIONS
7.1	**Probability; Odds**	Quality control, industrial safety
7.2	**Unions and Intersections of Events: One-Trial Experiments**	Politics, heart attack risk, property development
7.3	**Conditional Probability: The Product Rule**	Public opinion, delinquent accounts, inspections, identity theft
7.4	**Probability Trees and Bayes' Formula**	Quality control, medical tests
7.5	**Counting: Permutations and Combinations**	License plates, banquet seating, club officers, auto options, pizza choices
7.6	**Permutations, Combinations, and Probability**	Intelligence tests, quality control, jury selection, auto keys
7.7	**Markov Chains**	Credit card use, politics, educational attainment

Prerequisite Problem Type	For Section	Answer	Section for Review
(a) The sum of x, $\frac{1}{4}$, and $\frac{2}{3}$ is 1. What is x? (b) If the sum of x, $\frac{1}{3}$, and $\frac{1}{3}$ is 1, what is x?	7.1	(a) $x = \frac{1}{12}$ (b) $x = \frac{1}{3}$	1.1 Linear equations
If 12 numbers are weighted so that each of the 12 numbers has an equal weight and the sum of the weights is 1, what weight should be assigned to each number?	7.1	$\frac{1}{12}$	1.1 Linear equations
(a) List the set of integers that satisfies $2 \le x < 6$. (b) List the set of integers that satisfies $4 \le s \le 12$.	7.1	(a) $\{2, 3, 4, 5\}$ (b) $\{4, 5, 6, 7, 8, 9, 10, 11, 12\}$	1.1 Linear inequalities
For the sets $E = \{1, 2, 3, 4, 5\}$ and $F = \{2, 4, 6, 8\}$, find (a) $E \cup F$ (b) $E \cap F$	7.1 7.2	(a) $\{1, 2, 3, 4, 5, 6, 8\}$ (b) $\{2, 4\}$	0.1 Sets
Find the product: $[0.9 \quad 0.1]\begin{bmatrix} 0.8 & 0.2 \\ 0.3 & 0.7 \end{bmatrix}$	7.7	$[0.75 \quad 0.25]$	3.2 Matrix multiplication
Solve the system using matrices: $\begin{cases} 0.5V_1 + 0.4V_2 + 0.3V_3 = V_1 \\ 0.4V_1 + 0.5V_2 + 0.3V_3 = V_2 \\ 0.1V_1 + 0.1V_2 + 0.4V_3 = V_3 \end{cases}$	7.7	$V_1 = 3V_3$ $V_2 = 3V_3$ $V_3 =$ any real number	3.3 Gauss-Jordan elimination

OBJECTIVES **7.1**

- To compute the probability of the occurrence of an event
- To construct a sample space for a probability experiment
- To compute the odds that an event will occur
- To compute the empirical probability that an event will occur

Probability; Odds

■ | **APPLICATION PREVIEW** |

Before the General Standard Company can place a 3-year warranty on all of the faucets it produces, it needs to find the probability that each faucet it produces will function properly (not leak) for 3 years. (See Example 5.)

In this section we will use sample spaces to find the probability that an event will occur by using assumptions or knowledge of the entire population of data under consideration. When knowledge of the entire population is not available, we can compute the empirical probability that an event will occur by using data collected about the probability experiment. This method can be used to compute the probability that the General Standard faucets will work properly if information about a large sample of the faucets is gathered.

Probability; Sample Spaces

When we toss a coin, it can land in one of two ways, heads or tails. If the coin is a "fair" coin, these two possible **outcomes** have an equal chance of occurring, and we say the outcomes of this **probability experiment** are **equally likely.** If we seek the probability that an experiment has a certain result, that result is called an **event.**

Suppose that an experiment can have a total of n equally likely outcomes and that k of these outcomes would be considered successes. Then the probability of achieving a success in the experiment is k/n, the number of ways the experiment can result in a success divided by the total number of possible outcomes.

Probability of a Single Event

If an event E can happen in k ways out of a total of n equally likely possibilities, the probability of the occurrence of the event is denoted by

$$\Pr(E) = \frac{\text{Number of successes}}{\text{Number of possible outcomes}} = \frac{k}{n}$$

EXAMPLE 1 Probability

If we draw a ball from a bag containing 4 white balls and 6 black balls, what is the probability of

(a) getting a white ball?   (b) getting a black ball?   (c) not getting a white ball?

Solution

(a) A white ball can be drawn in 4 ways out of 10 equally likely possibilities.

$$\Pr(W) = \frac{4}{10} = \frac{2}{5}$$

(b) We can draw a black ball in 6 ways out of 10 possible outcomes.

$$\Pr(B) = \frac{6}{10} = \frac{3}{5}$$

(c) The probability of not getting a white ball is $\Pr(B)$, because not getting a white ball is the same as getting a black ball. Thus $\Pr(\text{not } W) = \Pr(B) = 3/5$.   ■

A set that contains all the possible outcomes of an experiment is called a **sample space.** For the experiment of randomly drawing a numbered card* from a box of cards numbered 1 through 12, a sample space is

$$S = \{1, 2, 3, 4, 5, 6, 7, 8, 9, 10, 11, 12\}$$

Each element of the sample space is called a **sample point,** and an **event** is a subset of the sample space. Each sample point in the sample space is assigned a nonnegative **probability measure** or **probability weight** such that the sum of the weights in the sample space is 1. The probability of an event is the sum of the weights of the sample points in the event's **subspace** (subset of S).

For example, the experiment "drawing a number at random from the numbers 1 through 12" has the sample space S given above and each element has a probability weight of $1/12$. The event "drawing an even number" in this experiment has the subspace $E = \{2, 4, 6, 8, 10, 12\}$ and the probability is

$$\text{Pr}(\text{even}) = \text{Pr}(E) = \frac{1}{12} + \frac{1}{12} + \frac{1}{12} + \frac{1}{12} + \frac{1}{12} + \frac{1}{12} = \frac{6}{12} = \frac{1}{2}$$

Some experiments have an infinite number of outcomes, but we will concern ourselves only with experiments having finite sample spaces. We can usually construct more than one sample space for an experiment. The sample space in which each sample point is equally likely is called an **equiprobable sample space.** Suppose the number of elements in a sample space is $n(S) = n$, with each element representing an equally likely outcome of a probability experiment. Then the weight (probability) assigned to each element is

$$\frac{1}{n(S)} = \frac{1}{n}$$

If an event E that is a subset of an equiprobable sample space contains $n(E) = k$ elements of S, then we can restate the probability of an event E in terms of the number of elements of E.

Probability of an Event	If an event E can occur in $n(E) = k$ ways out of $n(S) = n$ equally likely ways, then
	$$\text{Pr}(E) = \frac{n(E)}{n(S)} = \frac{k}{n}$$

EXAMPLE 2   Probability of Events

If a number is to be selected at random from the integers 1 through 12, what is the probability that it is

(a) divisible by 3?   (b) even and divisible by 3?   (c) even or divisible by 3?

Solution

The set $S = \{1, 2, 3, 4, 5, 6, 7, 8, 9, 10, 11, 12\}$ is an equiprobable sample space for this experiment (as mentioned previously).

(a) The set $E = \{3, 6, 9, 12\}$ contains the numbers that are divisible by 3. Thus

$$\text{Pr}(\text{divisible by 3}) = \frac{n(E)}{n(S)} = \frac{4}{12} = \frac{1}{3}$$

(b) The numbers 1 through 12 that are even *and* divisible by 3 are 6 and 12, so

$$\text{Pr}(\text{even and divisible by 3}) = \frac{2}{12} = \frac{1}{6}$$

---

*Selecting a card randomly means every card has an equal chance of being selected.

(c) The numbers that are even *or* divisible by 3 are 2, 3, 4, 6, 8, 9, 10, 12, so

$$\Pr(\text{even or divisible by } 3) = \frac{8}{12} = \frac{2}{3}$$ ∎

CHECKPOINT

1. If a coin is drawn from a box containing 4 gold, 10 silver, and 16 copper coins, all the same size, what is the probability of
   (a) getting a gold coin?
   (b) getting a copper coin?
   (c) not getting a copper coin?

Probability

If an event E is certain to occur, E contains all of the elements of the sample space, S. Hence the sum of the probability weights of E is the same as that of S, so

$$\Pr(E) = 1 \qquad \text{if } E \text{ is certain to occur}$$

If event E is impossible, $E = \varnothing$, so

$$\Pr(E) = 0 \qquad \text{if } E \text{ is impossible}$$

Because all probability weights of elements of a sample space are nonnegative,

$$0 \le \Pr(E) \le 1 \qquad \text{for any event } E$$

EXAMPLE 3 **Sample Spaces**

Suppose a coin is tossed 3 times.

(a) Construct an equiprobable sample space for the experiment.
(b) Find the probability of obtaining 0 heads.
(c) Find the probability of obtaining 2 heads.

Solution

(a) Perhaps the most obvious way to record the possibilities for this experiment is to list the number of heads that could result: {0, 1, 2, 3}. But the probability of obtaining 0 heads is different from the probability of obtaining 2 heads, so this sample space is not equiprobable. A sample space in which each outcome is equally likely is {HHH, HHT, HTH, THH, HTT, THT, TTH, TTT}, where HHT indicates that the first two tosses were heads and the third was a tail.

(b) Because there are 8 equally likely possible outcomes, $n(S) = 8$. Only one of the eight possible outcomes, $E = \{\text{TTT}\}$, gives 0 heads, so $n(E) = 1$. Thus

$$\Pr(0 \text{ heads}) = \frac{n(E)}{n(S)} = \frac{1}{8}$$

$$= \frac{1}{8}$$

(c) The event "two heads" is $F = \{HHT, HTH, THH\}$, so

$$\Pr(2 \text{ heads}) = \frac{n(F)}{n(S)} = \frac{3}{8}$$

To find the probability of obtaining a given sum when a pair of dice is rolled, we need to determine the outcomes. However, the sums $\{2, 3, 4, 5, 6, 7, 8, 9, 10, 11, 12\}$ are not equally likely; there is only one way to obtain a sum of 2, but there are several ways to obtain a sum of 6. If we distinguish between the two dice we are rolling, and we record all the possible outcomes for each die, we see that there are 36 possibilities, each of which is equally likely. Table 7.1 shows these possibilities, with the event "sum = 9" encircled.

| TABLE 7.1 |

| First Die | \multicolumn{6}{c}{Second Die} |
	1	2	3	4	5	6
1	(1, 1)	(1, 2)	(1, 3)	(1, 4)	(1, 5)	(1, 6)
2	(2, 1)	(2, 2)	(2, 3)	(2, 4)	(2, 5)	(2, 6)
3	(3, 1)	(3, 2)	(3, 3)	(3, 4)	(3, 5)	(3, 6)
4	(4, 1)	(4, 2)	(4, 3)	(4, 4)	(4, 5)	(4, 6)
5	(5, 1)	(5, 2)	(5, 3)	(5, 4)	(5, 5)	(5, 6)
6	(6, 1)	(6, 2)	(6, 3)	(6, 4)	(6, 5)	(6, 6)

This list of possible outcomes for finding the sum of two dice is an equiprobable sample space for the experiment. Because the 36 elements in the sample space are equally likely, we can find the probability that a given sum results by determining the number of ways this sum can occur and dividing that number by 36.

EXAMPLE 4   Rolling Dice

Use Table 7.1 to find the following probabilities if a pair of distinguishable dice is rolled.

(a) Pr(sum is 5)   (b) Pr(sum is 2)   (c) Pr(sum is 8)

Solution

(a) If E is the event "sum is 5," then $E = \{(4, 1), (3, 2), (2, 3), (1, 4)\}$.
Therefore, $\Pr(\text{sum is 5}) = \frac{n(E)}{n(S)} = \frac{4}{36} = \frac{1}{9}$.

(b) The sum 2 results only from $\{(1, 1)\}$. Thus $\Pr(\text{sum is 2}) = 1/36$.
(c) The sample points that give a sum of 8 are in the subspace

$$F = \{(6, 2), (5, 3), (4, 4), (3, 5), (2, 6)\}$$

Thus $\Pr(\text{sum is 8}) = \frac{n(F)}{n(S)} = \frac{5}{36}$.

CHECKPOINT

2. A ball is to be drawn from a bag containing balls numbered 1, 2, 3, 4, and 5. To find the probability that a ball with an even number is drawn, we can use the sample space $S_1 = \{\text{even, odd}\}$ or $S_2 = \{1, 2, 3, 4, 5\}$.
(a) Which of these sample spaces is an equiprobable sample space?
(b) What is the probability of drawing a ball that is even-numbered?

3. A coin is tossed and a die is rolled.
   (a) Write an equiprobable sample space listing the possible outcomes for this experiment.
   (b) Find the probability of getting a head on the coin and an even number on the die.
   (c) Find the probability of getting a tail on the coin and a number divisible by 3 on the die.

Empirical Probability

Up to this point, the probabilities assigned to events were determined *theoretically*, either by assumption (the probability of obtaining a head in one toss of a fair coin is 1/2) or by knowing the entire population under consideration (if 1000 people are in a room and 400 of them are males, the probability of picking a person at random and getting a male is $400/1000 = 2/5$). We can also determine probabilities **empirically,** where the assignment of the probability of an event is derived from our experiences or from data that have been gathered. Probabilities developed in this way are called **empirical probabilities.** An empirical probability is formed by conducting an experiment, or by observing a situation a number of times, and noting how many times a certain event occurs.

Empirical Probability of an Event

$$\text{Pr}(\text{Event}) = \frac{\text{Number of observed occurrences of the event}}{\text{Total number of trials or observations}}$$

In some cases empirical probability is the only way to assign a probability to an event. For example, life insurance premiums are based on the probability that an individual will live to a certain age. These probabilities are developed empirically and organized into mortality tables.

In other cases, empirical probabilities may have a theoretical framework. For example, for tossing a coin,

$$\text{Pr}(\text{Head}) = \text{Pr}(\text{Tail}) = \frac{1}{2}$$

gives the theoretical probability. If a coin was actually tossed 1000 times, it might come up heads 517 times and tails 483 times, giving empirical probabilities for these 1000 tosses of

$$\text{Pr}(\text{Head}) = \frac{517}{1000} = 0.517 \quad \text{and} \quad \text{Pr}(\text{Tail}) = \frac{483}{1000} = 0.483$$

On the other hand, if the coin was tossed only 4 times, it might come up heads 3 times and tails once, giving empirical probabilities vastly different from the theoretical ones. In general, empirical probabilities can differ from theoretical probabilities, but they are likely to reflect the theoretical probabilities when the number of observations is large. Moreover, when this is not the case, there may be reason to suspect that the theoretical model does not fit the event (for example, that a coin may be biased).

For example, suppose that 800 of the 500,000 microcomputer diskettes manufactured by a firm are defective. Then the 800 defective diskettes occurred in a very large number of diskettes manufactured, so the ratio of the number of defective diskettes to the total number of diskettes manufactured provides the *empirical* probability of a diskette picked at random being defective.

$$\text{Pr}(\text{defective}) = \frac{800}{500,000} = \frac{1}{625}$$

EXAMPLE 5   Quality Control | APPLICATION PREVIEW |

The General Standard Company can use information about a large sample of its faucets to compute the empirical probability that a faucet chosen at random will function properly (not leak) for 3 years. Of a sample of 12,316 faucets produced in 2007, 137 were found to fail within 3 years. Find the empirical probability that any faucet the company produces will work properly for 3 years.

Solution

If 137 of these faucets fail to work properly, then $12{,}316 - 137 = 12{,}179$ will work properly. Thus the probability that any faucet selected at random will function properly is

$$\Pr(\text{not leak}) = \frac{12{,}179}{12{,}316} \approx 0.989$$

Odds and Probability

We sometimes use **odds** to describe the likelihood that an event will occur. The odds in favor of an event E occurring and the odds against E occurring are found as follows.

Odds

If the probability that event E occurs is $\Pr(E) \neq 1$, then the odds that E will occur are

$$\text{Odds in favor of } E = \frac{\Pr(E)}{1 - \Pr(E)}$$

The odds that E will not occur, if $\Pr(E) \neq 0$, are

$$\text{Odds against } E = \frac{1 - \Pr(E)}{\Pr(E)}$$

The odds $\dfrac{a}{b}$ are stated as "a to b" or "$a{:}b$."

EXAMPLE 6   Odds

If the probability of drawing a queen from a deck of playing cards is $1/13$, what are the odds
(a) in favor of drawing a queen?
(b) against drawing a queen?

Solution

(a) $\dfrac{1/13}{1 - 1/13} = \dfrac{1/13}{12/13} = \dfrac{1}{12}$

The odds in favor of drawing a queen are 1 to 12, which we write as 1:12.

(b) $\dfrac{12/13}{1/13} = \dfrac{12}{1}$

The odds against drawing a queen are 12 to 1, denoted 12:1.

Odds and Probability

If the odds in favor of an event E occurring are $a{:}b$, the probability that E will occur is

$$\Pr(E) = \frac{a}{a + b}$$

If the odds against E are $b{:}a$, then

$$\Pr(E) = \frac{a}{a + b}$$

For example, according to funny2.com, the odds that an American adult does not want to live to age 120 are 3 to 2. Thus the probability that an adult American chosen at random does not want to live to age 120 is $3/(3 + 2) = 3/5$.

EXAMPLE 7   Robbery

The day after O. J. Simpson was arrested for armed robbery in September 2007, the odds against his being convicted were given in Las Vegas as 6 to 5 (*Source:* Fox News Network). What probability of conviction did this indicate?

Solution
The odds against the conviction are 6:5, so the probability of conviction is

$$\Pr(\text{convicted}) = \frac{5}{5 + 6} = \frac{5}{11}$$

| CHECKPOINT SOLUTIONS |

1. (a) The number of coins is 30 and the number of gold coins is 4, so $\Pr(\text{gold}) = 4/30 = 2/15$.
   (b) The number of copper coins is 16, so $\Pr(\text{copper}) = 16/30 = 8/15$.
   (c) The number of coins that are not copper is 14, so $\Pr(\text{not copper}) = 14/30 = 7/15$.
2. (a) $S_1 = \{\text{even, odd}\}$ is not an equiprobable sample space for this experiment, because there are not equal numbers of even and odd numbers.
      $S_2 = \{1, 2, 3, 4, 5\}$ is an equiprobable sample space.
   (b) $\Pr(\text{even}) = 2/5$
3. (a) $\{H1, H2, H3, H4, H5, H6, T1, T2, T3, T4, T5, T6\}$
   (b) $\Pr(\text{head and even number}) = 3/12 = 1/4$
   (c) $\Pr(\text{tail and number divisible by 3}) = 2/12 = 1/6$

| EXERCISES | 7.1

Access end-of-section exercises online at **www.webassign.net**

OBJECTIVES   7.2

- To find the probability of the intersection of two events
- To find the probability of the union of two events
- To find the probability of the complement of an event

Unions and Intersections of Events: One-Trial Experiments

| APPLICATION PREVIEW |

Suppose that the directors of the Chatham County School Board consist of 8 Democrats, 3 of whom are female, and 4 Republicans, 2 of whom are female. If a director is chosen at random to discuss an issue on television, we can find the probability that the person will be a Democrat or female by finding the probability of the union of the two events "a Democrat is chosen" and "a female is chosen." (See Example 4.) In this section we will discuss methods of finding the probabilities of the intersection and union of events by using sample spaces and formulas.

Unions, Intersections, and Complements of Events

As we stated in the previous section, events determine subsets of the sample space for a probability experiment. The **intersection, union,** and **complement** of events are defined as follows.

Intersection, Union, and Complement

If E and F are two events in a sample space S, then the **intersection of E and F** is

$$E \cap F = \{a : a \in E \ \text{and} \ a \in F\}$$

the **union of E and F** is

$$E \cup F = \{a : a \in E \ \text{or} \ a \in F\}$$

and the **complement of E** is

$$E' = \{a : a \in S \ \text{and} \ a \notin E\}$$

(continued)

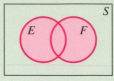 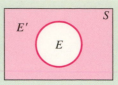

$$E \cap F \qquad\qquad E \cup F$$

Using these definitions, we can write the probabilities that certain events will occur as follows:

$$\Pr(E \text{ and } F \text{ both occur}) = \Pr(E \cap F)$$
$$\Pr(E \text{ or } F \text{ occurs}) = \Pr(E \cup F)$$
$$\Pr(E \text{ does } \textbf{not} \text{ occur}) = \Pr(E')$$

In Section 7.1, "Probability; Odds," we investigated probability questions by identifying the sample space elements for each described event. We now apply the properties just stated.

EXAMPLE   1   Unions and Intersections of Events

A card is drawn from a box containing 15 cards numbered 1 to 15. What is the probability that the card is

(a) even and divisible by 3?   (b) even or divisible by 3?   (c) not even?

Solution

The sample space S contains the 15 numbers.

$$S = \{1, 2, 3, 4, 5, 6, 7, 8, 9, 10, 11, 12, 13, 14, 15\}$$

If we let E represent "even-numbered" and D represent "number divisible by 3," we have

$$E = \{2, 4, 6, 8, 10, 12, 14\} \quad \text{and} \quad D = \{3, 6, 9, 12, 15\}$$

These sets are shown in the Venn diagram in Figure 7.1.

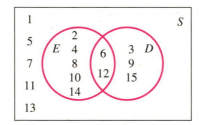

Figure 7.1

(a) The event "even and divisible by 3" is $E \cap D = \{6, 12\}$, so

$$\Pr(\text{even and divisible by 3}) = \Pr(E \cap D) = \frac{n(E \cap D)}{n(S)} = \frac{2}{15}$$

(b) The event "even or divisible by 3" is $E \cup D = \{2, 3, 4, 6, 8, 9, 10, 12, 14, 15\}$, so

$$\Pr(\text{even or divisible by 3}) = \Pr(E \cup D) = \frac{n(E \cup D)}{n(S)} = \frac{10}{15} = \frac{2}{3}$$

(c) The event "not even," or E', contains all elements of S not in E, so $E' = \{1, 3, 5, 7, 9, 11, 13, 15\}$, and the probability that the card is not even is

$$\Pr(\text{not } E) = \Pr(E') = \frac{n(E')}{n(S)} = \frac{8}{15}$$

CHECKPOINT

1. If a ball is drawn from a bag containing 4 red balls numbered 1, 2, 3, 4 and 3 white balls numbered 5, 6, 7, what is the probability that the ball is
    (a) red and even?   (b) white and even?

Probability of the Complement

Figure 7.2

Because the complement of an event contains all the elements of S *except* for those elements in E (see Figure 7.2), the number of elements in E' is $n(S) - n(E)$, so

$$Pr(E') = \frac{n(S) - n(E)}{n(S)} = 1 - \frac{n(E)}{n(S)} = 1 - Pr(E)$$

$$Pr(\text{not } E) = Pr(E') = 1 - Pr(E)$$

EXAMPLE 2 **Heart Attack Risk**

Suppose that a 50-year-old man has systolic blood pressure of 110 and that 60% of his cholesterol is HDL. The risk of heart attack in the next 10 years is given in Table 7.2. If a man satisfying these conditions is selected at random and has a cholesterol level of 200, what is the probability that he will not have a heart attack by age 60?

| TABLE 7.2 |

Total Cholesterol	Risk of Heart Attack
130	1%
150	1
175	2
200	2
250	3
300	5

Solution
If the man selected at random satisfies the given conditions and has a cholesterol level of 200, the probability that he will have a heart attack in the next 10 years (that is, by age 60) is $2\% = 0.02$. Thus the probability that he will not have a heart attack by age 60 is

$$Pr(\text{no attack}) = 1 - Pr(\text{attack}) = 1 - 0.02 = 0.98 \qquad \blacksquare$$

Inclusion-Exclusion Principle

The number of elements in the set $E \cup F$ is the number of elements in E plus the number of elements in F, *minus* the number of elements in $E \cap F$ [because this number was counted twice in $n(E) + n(F)$]. Thus,

$$n(E \cup F) = n(E) + n(F) - n(E \cap F)$$

Hence,

$$Pr(E \cup F) = \frac{n(E \cup F)}{n(S)} = \frac{n(E)}{n(S)} + \frac{n(F)}{n(S)} - \frac{n(E \cap F)}{n(S)}$$

$$= Pr(E) + Pr(F) - Pr(E \cap F)$$

This establishes what is called the **inclusion-exclusion principle.**

Inclusion-Exclusion Principle

If E and F are any two events, then the probability that one event or the other will occur, denoted $Pr(E \text{ or } F)$, is given by

$$Pr(E \text{ or } F) = Pr(E \cup F) = Pr(E) + Pr(F) - Pr(E \cap F)$$

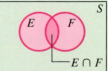

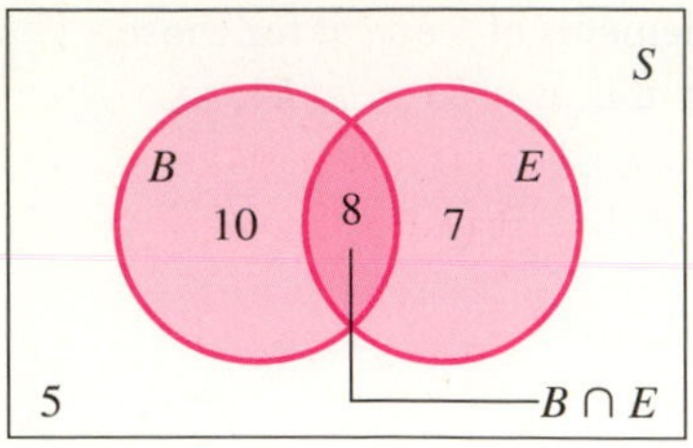

Figure 7.3

Suppose that of 30 people interviewed for a position, 18 had a business degree, 15 had previous experience, and 8 of those with experience also had a business degree. The Venn diagram in Figure 7.3 illustrates this situation.

Because 8 of the 30 people have both a business degree and experience,

$$\Pr(B \cap E) = \frac{8}{30}$$

We also see that

$$\Pr(B) = \frac{18}{30} \text{ and } \Pr(E) = \frac{15}{30}$$

Thus the probability of choosing a person at random who has a business degree or experience is

$$\Pr(B \cup E) = \Pr(B) + \Pr(E) - \Pr(B \cap E)$$
$$= \frac{18}{30} + \frac{15}{30} - \frac{8}{30} = \frac{25}{30} = \frac{5}{6}$$

EXAMPLE   3   Football

Suppose that in a given year, the probability that the Steelers will win their division is 1/6, the probability that the Jets will win their division is 1/12, and the probability that both will win their divisions is 1/24. What is the probability that one of the teams will win their division?

Solution

$\Pr(\text{one will win}) = \Pr(\text{Steelers or Jets will win})$

$$= \Pr(\text{Steelers win}) + \Pr(\text{Jets win}) - \Pr(\text{Steelers and Jets win})$$

$$= \frac{1}{6} + \frac{1}{12} - \frac{1}{24} = \frac{5}{24}$$

EXAMPLE   4   Politics | APPLICATION PREVIEW |

The directors of the Chatham County School Board consist of 8 Democrats, 3 of whom are female, and 4 Republicans, 2 of whom are female. If a director is chosen at random to discuss an issue on television, what is the probability that the person will be a Democrat or female?

Solution

There are 12 directors, of whom 8 are Democrats and 5 are female. But 3 of the directors are Democrats *and* females and cannot be counted twice. Thus

$$\Pr(D \text{ or } F) = \Pr(D \cup F) = \Pr(D) + \Pr(F) - \Pr(D \text{ and } F)$$

$$= \frac{8}{12} + \frac{5}{12} - \frac{3}{12} = \frac{10}{12} = \frac{5}{6}$$

CHECKPOINT

2. If a ball is drawn from a bag containing 4 red balls numbered 1, 2, 3, 4 and 3 white balls numbered 5, 6, 7, what is the probability that the ball is
   (a) red or even?   (b) white or even?

Mutually Exclusive Events We say that events E and F are **mutually exclusive** if and only if $E \cap F = \emptyset$. Thus

$$\Pr(E \cup F) = \Pr(E) + \Pr(F) - 0 = \Pr(E) + \Pr(F)$$

if and only if E and F are mutually exclusive.

Mutually Exclusive Events

If E and F are **mutually exclusive,** then $\Pr(E \cap F) = 0$, and

$$\Pr(E \text{ or } F) = \Pr(E \cup F) = \Pr(E) + \Pr(F)$$

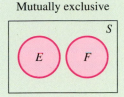

Mutually exclusive

EXAMPLE 5   Mutually Exclusive Events

Find the probability of obtaining a 6 or a 4 in one roll of a die.

Solution

Rolling a 6 and rolling a 4 on one roll of a die are mutually exclusive events. Let E be the event "rolling a 6" and F be the event "rolling a 4."

$$\Pr(E) = \Pr(\text{rolling } 6) = \frac{1}{6} \quad \text{and} \quad \Pr(F) = \Pr(\text{rolling } 4) = \frac{1}{6}$$

Then

$$\Pr(E \cup F) = \Pr(E) + \Pr(F) = \frac{1}{6} + \frac{1}{6} = \frac{1}{3}$$

EXAMPLE 6   Property Development

A firm is considering three possible locations for a new factory. The probability that site A will be selected is $1/3$ and the probability that site B will be selected is $1/5$. Only one location will be chosen.
(a) What is the probability that site A or site B will be chosen?
(b) What is the probability that neither site A nor site B will be chosen?

Solution

(a) The two events are mutually exclusive, so

$$\Pr(\text{site A or site B}) = \Pr(\text{site A} \cup \text{site B})$$

$$= \Pr(\text{site A}) + \Pr(\text{site B}) = \frac{1}{3} + \frac{1}{5} = \frac{8}{15}$$

(b) The probability that neither will be chosen is $1 - 8/15 = 7/15$.

The formula for the probability of the union of mutually exclusive events can be extended beyond two events.

$$\Pr(E_1 \cup E_2 \cup \ldots \cup E_n) = \Pr(E_1) + \Pr(E_2) + \cdots + \Pr(E_n)$$
for mutually exclusive events $E_1, E_2, \ldots, E_n$.

EXAMPLE   7   Dice

Find the probability of rolling two distinct dice and obtaining a sum greater than 9.

Solution

The event "a sum greater than 9" consists of the mutually exclusive events "sum = 10," "sum = 11," and "sum = 12," and these probabilities can be found in Table 7.1 in Section 7.1.

$$\text{Pr}(10 \text{ or } 11 \text{ or } 12) = \text{Pr}(10 \cup 11 \cup 12)$$

$$= \text{Pr}(10) + \text{Pr}(11) + \text{Pr}(12) = \frac{3}{36} + \frac{2}{36} + \frac{1}{36} = \frac{6}{36} = \frac{1}{6} \blacksquare$$

CHECKPOINT SOLUTIONS

1. (a) The sample space contains 7 balls: $S = \{R1, R2, R3, R4, W5, W6, W7\}$. The probability that a ball drawn at random is red and even is found by using the subspace that describes that event. Thus

$$\text{Pr}(\text{red and even}) = \text{Pr}(\text{red} \cap \text{even}) = \frac{2}{7}$$

   (b) The set $\{W6\}$ describes the event "white and even," so

$$\text{Pr}(\text{white and even}) = \text{Pr}(\text{white} \cap \text{even}) = \frac{1}{7}$$

2. (a) $\text{Pr}(\text{red or even}) = \text{Pr}(\text{red} \cup \text{even})$
$$= \text{Pr}(\text{red}) + \text{Pr}(\text{even}) - \text{Pr}(\text{red} \cap \text{even})$$
$$= \frac{4}{7} + \frac{3}{7} - \frac{2}{7} = \frac{5}{7}$$

   (b) $\text{Pr}(\text{white or even}) = \text{Pr}(\text{white} \cup \text{even})$
$$= \text{Pr}(\text{white}) + \text{Pr}(\text{even}) - \text{Pr}(\text{white} \cap \text{even})$$
$$= \frac{3}{7} + \frac{3}{7} - \frac{1}{7} = \frac{5}{7}$$

| EXERCISES | 7.2

Access end-of-section exercises online at **www.webassign.net**

OBJECTIVES   7.3

- To solve probability problems involving conditional probability
- To compute the probability that two or more independent events will occur
- To compute the probability that two or more dependent events will occur

Conditional Probability: The Product Rule

| APPLICATION PREVIEW |

Polk's Department Store has observed that 80% of its charge accounts have men's names on them and that 16% of the accounts with men's names on them have been delinquent at least once, whereas 5% of the accounts with women's names on them have been delinquent at least once. We can find the probability that an account selected at random is in a man's name and is delinquent by using the Product Rule. (See Example 5.) In this section we will use the Product Rule to find the probability that two events will occur, and we will solve problems involving conditional probability.

Conditional Probability   Each probability that we have computed has been relative to the sample space for the experiment. Sometimes information is given that reduces the sample space needed to solve a stated problem. For example, if a die is rolled, the probability that a 5 occurs, given that an odd number is rolled, is denoted by

$$\text{Pr}(5 \text{ rolled} \mid \text{odd number rolled})$$

The knowledge that an odd number occurs reduces the sample space from the numbers on a die, $S = \{1, 2, 3, 4, 5, 6\}$, to the sample space for odd numbers on a die, $S_1 = \{1, 3, 5\}$, so the event $E =$ "a 5 occurs" has probability

$$\text{Pr}(5 \mid \text{odd number}) = \frac{n(5 \text{ and odd number rolled})}{n(\text{odd number rolled})} = \frac{n(E \cap S_1)}{n(S_1)} = \frac{1}{3}$$

To find $\text{Pr}(A \mid B)$, we seek the probability that A occurs, given that B occurs. Figure 7.4 shows the original sample space S with the reduced sample space B shaded. Because all elements must be contained in B, we evaluate $\text{Pr}(A \mid B)$ by dividing the number of elements in $A \cap B$ by the number of elements in B.

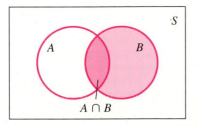

Figure 7.4

Conditional Probability

The **conditional probability** that A occurs, given that B occurs, is denoted by $\text{Pr}(A \mid B)$ and is given by

$$\text{Pr}(A \mid B) = \frac{n(A \cap B)}{n(B)}$$

EXAMPLE 1   Public Opinion

Suppose the table below summarizes the opinions of various groups in a survey on the issue of increased enforcement of immigration laws. What is the probability that a person selected at random from this group of people

(a) is nonwhite?
(b) is nonwhite and favors increased enforcement of immigration laws?
(c) favors increased enforcement of immigration laws, given that the person is nonwhite?

	Whites		Nonwhites		
Opinion	Republicans	Democrats	Republicans	Democrats	Totals
Favor	300	100	25	10	435
Oppose	100	250	25	190	565
Total	400	350	50	200	1000

Solution
(a) Of the 1000 people in the group, $50 + 200 = 250$ are nonwhite, so

$$\text{Pr}(\text{nonwhite}) = \frac{250}{1000} = \frac{1}{4}$$

(b) There are 1000 people in the group and $25 + 10 = 35$ who both are nonwhite and favor increased enforcement, so

$$\text{Pr}(\text{nonwhite and favor}) = \frac{35}{1000} = \frac{7}{200}$$

(c) There are $50 + 200 = 250$ nonwhite and $25 + 10 = 35$ people who both are non-white and favor increased enforcement, so

$$\text{Pr}(\text{favor} \mid \text{nonwhite}) = \frac{n(\text{favor and nonwhite})}{n(\text{nonwhite})} = \frac{35}{250} = \frac{7}{50}$$

Using the definition of $\text{Pr}(A \mid B)$ and dividing the numerator and denominator by $n(S)$, the number of elements in the sample space, gives

$$\text{Pr}(A \mid B) = \frac{n(A \cap B)}{n(B)} = \frac{\dfrac{n(A \cap B)}{n(S)}}{\dfrac{n(B)}{n(S)}} = \frac{\text{Pr}(A \cap B)}{\text{Pr}(B)}$$

Conditional Probability Formula

Let A and B be events in the sample space S with $\text{Pr}(B) > 0$. The **conditional probability** that event A occurs, given that event B occurs, is given by

$$\text{Pr}(A \mid B) = \frac{\text{Pr}(A \cap B)}{\text{Pr}(B)}$$

EXAMPLE 2 Conditional Probability

A red die and a green die are rolled. What is the probability that the sum rolled on the dice is 6, given that the sum is less than 7?

Solution
Using the conditional probability formula and the sample space in Table 7.1 in Section 7.1, we have

$$\text{Pr}(\text{sum is } 6 \mid \text{sum} < 7) = \frac{\text{Pr}(\text{sum is } 6 \cap \text{sum} < 7)}{\text{Pr}(\text{sum} < 7)}$$

Because $\text{Pr}(\text{sum is } 6 \cap \text{sum} < 7) = \text{Pr}(\text{sum is } 6) = 5/36$ the probability is

$$\text{Pr}(\text{sum is } 6 \mid \text{sum} < 7) = \frac{5/36}{1/36 + 2/36 + 3/36 + 4/36 + 5/36}$$

$$= \frac{5/36}{15/36} = \frac{1}{3}$$

CHECKPOINT

1. Suppose that one ball is drawn from a bag containing 4 red balls numbered 1, 2, 3, 4 and 3 white balls numbered 5, 6, 7. What is the probability that the ball is white, given that it is an even-numbered ball?

The Product Rule   We can solve the formula

$$\text{Pr}(A \mid B) = \frac{\text{Pr}(A \cap B)}{\text{Pr}(B)}$$

for $\text{Pr}(A \cap B)$ by multiplying both sides of the equation by $\text{Pr}(B)$. This gives

$$\text{Pr}(A \cap B) = \text{Pr}(B) \cdot \text{Pr}(A \mid B)$$

Similarly, we can solve

$$\Pr(B \mid A) = \frac{\Pr(B \cap A)}{\Pr(A)} = \frac{\Pr(A \cap B)}{\Pr(A)}$$

for $\Pr(A \cap B)$, getting

$$\Pr(A \cap B) = \Pr(A) \cdot \Pr(B \mid A)$$

Product Rule

If A and B are probability events, then the probability of the event "A and B" is $\Pr(A \text{ and } B)$, and it can be found by one or the other of these two formulas.

$$\Pr(A \text{ and } B) = \Pr(A \cap B) = \Pr(A) \cdot \Pr(B \mid A)$$

or

$$\Pr(A \text{ and } B) = \Pr(A \cap B) = \Pr(B) \cdot \Pr(A \mid B)$$

EXAMPLE **3**   **Cards**

Suppose that from a deck of 52 cards, two cards are drawn in succession, without replacement. Find the probability that both cards are kings.

Solution

We seek the probability that the first card will be a king and the probability that the second card will be a king given that the first card was a king. The probability that the first card is a king is

$$\Pr(1\text{st card is a king}) = \frac{4}{52}$$

and the probability that the second card is a king is

$$\Pr(2\text{nd king} \mid 1\text{st king}) = \frac{3}{51}$$

because one king has been removed from the deck on the first draw. Thus

$$\Pr(2 \text{ kings}) = \Pr(1\text{st king}) \cdot \Pr(2\text{d king} \mid 1\text{st king})$$

$$= \frac{4}{52} \cdot \frac{3}{51} = \frac{1}{221}$$

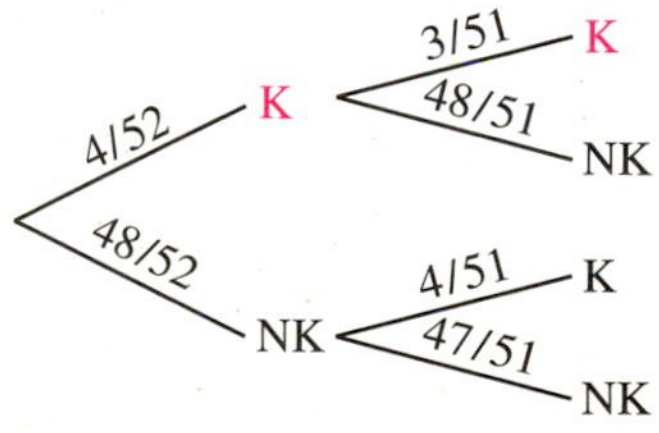

Figure 7.5

We can represent the outcomes in Example 3 with a **probability tree,** as shown in Figure 7.5. The outcome of the first draw is either a king or not a king; it is illustrated by the beginning two branches of the tree. The possible outcomes of the second trial can be treated as branches that emanate from each of the outcomes of the first trial. The tree shows that if a king occurs on the first trial, the outcome of the second trial is either a king or not a king, and the probabilities of these outcomes depend on the fact that a king occurred on the first trial. On the other hand, the probabilities of getting or not getting a king are different if the first draw does not yield a king, so the branches illustrating these possible outcomes have different probabilities.

There is only one "path" through the tree that illustrates getting kings on both draws, and the probability that both draws result in kings is the product of the probabilities on the branches of this path. Note that this gives $\frac{4}{52} \cdot \frac{3}{51} = \frac{1}{221}$, which is the same Product Rule probability found in Example 3.

In general, the connection between the Product Rule and a probability tree is as follows.

Probability Trees

The probability attached to *each branch* of a **probability tree** is the conditional probability that the specified event will occur, given that the events on the preceding branches have occurred.

When an event can be described by one path through a probability tree, the probability that that event will occur is the *product* of the probabilities on the branches *along the path* that represents the event.

EXAMPLE 4 Probability Tree

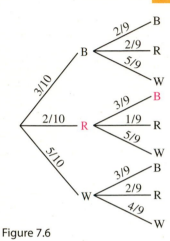

Figure 7.6

A box contains 3 black balls, 2 red balls, and 5 white balls. One ball is drawn, it is *not* replaced, and a second ball is drawn. Find the probability that the first ball is red and the second is black.

Solution

Let E_1 be "red ball first" and E_2 be "black ball second." The probability tree for this experiment is shown in Figure 7.6. The path representing the event whose probability we seek is the path through R (red ball) to B (black ball). The probability of this event is the product of the probabilities along this path.

$$\Pr(E_1 \cap E_2) = \Pr(E_1) \cdot \Pr(E_2 \,|\, E_1) = \frac{2}{10} \cdot \frac{3}{9} = \frac{1}{15}$$

CHECKPOINT

2. Two balls are drawn, without replacement, from the box described in Example 4.
   (a) Find the probability that both balls are white.
   (b) Find the probability that the first ball is white and the second is red.

EXAMPLE 5   Delinquent Accounts | APPLICATION PREVIEW |

Polk's Department Store has observed that 80% of its charge accounts have men's names on them and that 16% of the accounts with men's names on them have been delinquent at least once, while 5% of the accounts with women's names on them have been delinquent at least once. What is the probability that an account selected at random

(a) is in a man's name and is delinquent?
(b) is in a woman's name and is delinquent?

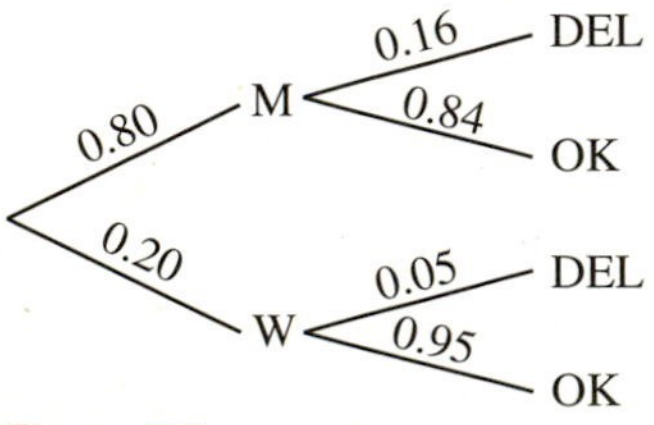

Solution

(a) Pr(man's account) = 0.80
    Pr(delinquent | man's) = 0.16
    Pr(man's and delinquent) = Pr(man's $\cap$ delinquent)
    $\qquad\qquad\qquad\qquad$ = Pr(man's) $\cdot$ Pr(delinquent | man's)
    $\qquad\qquad\qquad\qquad$ = (0.80)(0.16) = 0.128

Using the probability tree in Figure 7.7 gives the same probability.

Figure 7.7

(b) $\text{Pr(woman's account)} = 0.20$
$\text{Pr(delinquent | woman's)} = 0.05$

Using the probability tree in Figure 7.7 gives the probability

$$\text{Pr(woman's and delinquent)} = (0.20)(0.05) = 0.01 \qquad \blacksquare$$

EXAMPLE 6   Manufacturing Inspections

All products on an assembly line must pass two inspections. It has been determined that the probability that the first inspector will miss a defective item is 0.09. If a defective item gets past the first inspector, the probability that the second inspector will not detect it is 0.01. What is the probability that a defective item will not be rejected by either inspector? (All good items pass both inspections.)

Solution
The second inspector inspects only items passed by the first inspector. The probability that a defective item will pass both inspections is as follows.

$$\begin{aligned}
\text{Pr(defective pass both)} &= \text{Pr(pass first} \cap \text{pass second)} \\
&= \text{Pr(pass first)} \cdot \text{Pr(pass second | passed first)} \\
&= (0.09)(0.01) = 0.0009 \qquad \blacksquare
\end{aligned}$$

Independent Events   If a coin is tossed twice, the sample space is $S = \{\text{HH, HT, TH, TT}\}$, so

$$\text{Pr(2d is H | 1st is H)} = \frac{\text{Pr(1st is H} \cap \text{2d is H)}}{\text{Pr(1st is H)}} = \frac{\text{Pr(Both H)}}{\text{Pr(1st is H)}} = \frac{1/4}{1/2} = \frac{1}{2}$$

We can see that the probability that the second toss of a coin is a head, given that the first toss was a head, was $1/2$, which is the same as the probability that the second toss is a head, regardless of what happened on the first toss. This is because the results of the coin tosses are independent events. We can define **independent events** as follows.

Independent Events

The events A and B are **independent** if and only if
$$\text{Pr}(A \mid B) = \text{Pr}(A) \quad \text{and} \quad \text{Pr}(B \mid A) = \text{Pr}(B)$$
This means that the occurrence or nonoccurrence of one event does not affect the other.

If A and B are independent events, then this definition can be used to simplify the Product Rule:

$$\text{Pr}(A \cap B) = \text{Pr}(A) \cdot \text{Pr}(B \mid A) = \text{Pr}(A) \cdot \text{Pr}(B)$$

Thus we have the following.

Product Rule for Independent Events

If A and B are independent events, then
$$\text{Pr}(A \text{ and } B) = \text{Pr}(A \cap B) = \text{Pr}(A) \cdot \text{Pr}(B)$$

EXAMPLE 7   Independent Events

A die is rolled and a coin is tossed. Find the probability of getting a 4 on the die and a head on the coin.

Solution

Let E_1 be "4 on the die" and E_2 be "head on the coin." The events are independent because what occurs on the die does not affect what happens to the coin.

$$\Pr(E_1) = \Pr(4 \text{ on die}) = \frac{1}{6} \quad \text{and} \quad \Pr(E_2) = \Pr(\text{head on coin}) = \frac{1}{2}$$

Then

$$\Pr(E_1 \cap E_2) = \Pr(E_1) \cdot \Pr(E_2) = \frac{1}{6} \cdot \frac{1}{2} = \frac{1}{12} \qquad \blacksquare$$

The Product Rules can be expanded to include more than two events, as the next example shows.

EXAMPLE 8   Extending the Product Rules

A bag contains 3 red marbles, 4 white marbles, and 3 black marbles. Find the probability of getting a red marble on the first draw, a black marble on the second draw, and a white marble on the third draw (a) if the marbles are drawn with replacement, and (b) if the marbles are drawn without replacement.

Solution

(a) The marbles are replaced after each draw, so the contents are the same on each draw. Thus, the events are independent. Let E_1 be "red on first," let E_2 be "black on second," and let E_3 be "white on third." Then

$$\Pr(E_1 \cap E_2 \cap E_3) = \Pr(E_1) \cdot \Pr(E_2) \cdot \Pr(E_3)$$
$$= \frac{3}{10} \cdot \frac{3}{10} \cdot \frac{4}{10} = \frac{36}{1000} = \frac{9}{250}$$

(b) The marbles are not replaced, so the events are dependent.

$$\Pr(E_1 \cap E_2 \cap E_3) = \Pr(E_1) \cdot \Pr(E_2 | E_1) \cdot \Pr(E_3 | E_1 \text{ and } E_2)$$
$$= \frac{3}{10} \cdot \frac{3}{9} \cdot \frac{4}{8} = \frac{1}{20} \qquad \blacksquare$$

EXAMPLE 9   Identity Theft

Identity theft can occur when an unscrupulous individual learns someone else's Social Security number. Suppose a person learns the first five digits of your Social Security number. Calculate the probability that the person guesses the last four digits and thus gains access to your Social Security number.

Solution

The digits in your Social Security number can repeat, so the probability that each digit is guessed correctly is 1/10. Thus the probability the person guesses the last four digits is

$$\frac{1}{10} \cdot \frac{1}{10} \cdot \frac{1}{10} \cdot \frac{1}{10} = \frac{1}{10,000}$$

EXAMPLE 10 **Coin Tossing**

A coin is tossed 4 times. Find the probability that
(a) no heads occur.
(b) at least one head is obtained.

Solution

(a) Each toss is independent, and on each toss, the probability that a tail occurs is 1/2. Thus the probability of 0 heads (4 tails) in 4 tosses is

$$Pr(TTTT) = \frac{1}{2} \cdot \frac{1}{2} \cdot \frac{1}{2} \cdot \frac{1}{2} = \frac{1}{16}$$

(b) When a coin is tossed 4 times, if the event $E = \{0 \text{ heads}\}$, then

$$\text{"not } E\text{"} = E' = \{\text{at least one head}\}$$

Thus the probability that at least one head occurs is

$$Pr(\text{at least one head occurs}) = Pr(\text{not 0 heads})$$

$$= 1 - Pr(0 \text{ heads}) = 1 - \frac{1}{16} = \frac{15}{16}$$

In general, we can calculate the probability of at least one success in n trials as follows.

$$Pr(\text{at least 1 success in } n \text{ trials}) = 1 - Pr(0 \text{ successes})$$

CHECKPOINT SOLUTIONS

1. $Pr(\text{white} \mid \text{even-numbered}) = \dfrac{Pr(\text{white and even})}{Pr(\text{even})} = \dfrac{1/7}{3/7} = \dfrac{1}{3}$

2. (a) $Pr(\text{white on both draws}) = Pr(W_1 \cap W_2) = Pr(W_1) \cdot Pr(W_2 \mid W_1)$

    $$= \frac{5}{10} \cdot \frac{4}{9} = \frac{2}{9}$$

   (b) $Pr(\text{white on first and red on second}) = Pr(W_1 \cap R_2) = Pr(W_1) \cdot Pr(R_2 \mid W_1)$

    $$= \frac{5}{10} \cdot \frac{2}{9} = \frac{1}{9}$$

| EXERCISES | 7.3

Access end-of-section exercises online at **www.webassign.net**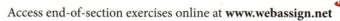

- To use probability trees to solve problems
- To use Bayes' formula to solve probability problems

Probability Trees and Bayes' Formula

| APPLICATION PREVIEW |

Suppose that a test for diagnosing a certain serious illness is successful in detecting the disease in 95% of all infected persons who are tested but that it incorrectly diagnoses 4% of all healthy people as having the serious disease. Suppose also that it incorrectly diagnoses 12% of all people having another minor disease as having the serious disease. If it is known that 2% of the population have the serious disease, 90% are healthy, and 8% have the minor disease, we can use a formula called Bayes' formula to find the probability that a person has the serious disease if the test indicates that he or she does. (See Example 3.) Such a Bayes problem can be solved with the formula or with the use of probability trees.

Probability Trees   Probability trees provide a systematic way to analyze probability experiments that have two or more trials or that use multiple paths within the tree. In this section, we will use trees to solve some additional probability problems and to solve **Bayes problems.** Recall that in a probability tree, we find the probability of an event as follows.

Probability Trees

The probability attached to each branch is the conditional probability that the specified event will occur, given that the events on the preceding branches have occurred.

1. For each event that can be described by a single path through the tree, the probability that that event

will occur is the product of the probabilities on the branches along the path that represents the event.

2. Because the paths represent mutually exclusive events, an event described by two or more paths through a probability tree has its probability found by adding the probabilities from the paths.

EXAMPLE   1   Probability Tree

A bag contains 5 red balls, 4 blue balls, and 3 white balls. Two balls are drawn, one after the other, without replacement. Draw a tree representing the experiment and find

(a) Pr(blue on first draw and white on second draw).
(b) Pr(white on both draws).
(c) Pr(drawing a blue ball and a white ball).
(d) Pr(second ball is red).

Solution
The tree is shown in Figure 7.8.

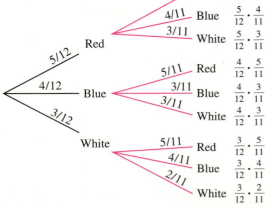

Figure 7.8   First draw   Second draw

(a) By multiplying the probabilities along the path that represents blue on the first draw and white on the second draw, we obtain

$$\Pr(\text{blue on the first draw and white on second draw}) = \frac{4}{12} \cdot \frac{3}{11} = \frac{1}{11}$$

(b) Multiplying the probabilities along the path that represents white on both draws gives

$$\Pr(\text{white on both draws}) = \frac{3}{12} \cdot \frac{2}{11} = \frac{1}{22}$$

(c) This result can occur by drawing a blue ball first and then a white ball *or* by drawing a white ball and then a blue ball. Thus either of two paths leads to this result. Because the results represented by these paths are mutually exclusive, the probability is found by *adding* the probabilities from the two paths. That is, letting B represent a blue ball and W represent a white ball, we have

$$\begin{aligned} \Pr(B \text{ and } W) &= \Pr(B \text{ first, then } W \text{ or } W \text{ first, then } B) \\ &= \Pr(B \text{ first, then } W) + \Pr(W \text{ first, then } B) \\ &= \frac{4}{12} \cdot \frac{3}{11} + \frac{3}{12} \cdot \frac{4}{11} = \frac{2}{11} \end{aligned}$$

(d) The second ball can be red after the first is red, blue, or white.

$$\begin{aligned} \Pr(\text{2d ball is red}) &= \Pr(R \text{ then } R, \text{ or } B \text{ then } R, \text{ or } W \text{ then } R) \\ &= \Pr(R \text{ then } R) + \Pr(B \text{ then } R) + \Pr(W \text{ then } R) \\ &= \frac{5}{12} \cdot \frac{4}{11} + \frac{4}{12} \cdot \frac{5}{11} + \frac{3}{12} \cdot \frac{5}{11} = \frac{5}{12} \end{aligned}$$

CHECKPOINT

1. Urn I contains 3 gold coins, urn II contains 1 gold coin and 2 silver coins, and urn III contains 1 gold coin and 1 silver coin. If an urn is selected at random and a coin is drawn from the urn, construct a probability tree and find the probability that a gold coin will be drawn.

EXAMPLE 2   **Quality Control**

A security system is manufactured with a "fail-safe" provision so that it functions properly if any two or more of its three main components, A, B, and C, are functioning properly. The probabilities that components A, B, and C are functioning properly are 0.95, 0.90, and 0.92, respectively. What is the probability that the system functions properly?

Solution

The probability tree in Figure 7.9 shows all of the possibilities for the components of the system. Each stage of the tree corresponds to a component, and G indicates that the component is good; N indicates it is not good. Four paths through the probability tree give at least two good (functioning) components, so there are four mutually exclusive possible outcomes. The probability that the system functions properly is

$$\Pr(\text{at least } 2G) = \Pr(GGG \cup GGN \cup GNG \cup NGG)$$
$$= (0.95)(0.90)(0.92) + (0.95)(0.90)(0.08) + (0.95)(0.10)(0.92) + (0.05)(0.90)(0.92)$$
$$= 0.9838$$

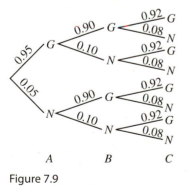

A   B   C

Figure 7.9

Bayes' Formula   Using a probability tree permits us to answer more difficult questions regarding conditional probability. For example, in the experiment described in Example 1, we can find the probability that the first ball is red, given that the second ball is blue. To find this probability, we use the formula for conditional probability, as follows.

$$\Pr(\text{1st is red} \mid \text{2d is blue}) = \frac{\Pr(\text{1st is red and 2d is blue})}{\Pr(\text{2d is blue})}$$

Figure 7.10 shows the probability tree with the possible outcomes for the experiment of Example 1. One path on the tree represents "1st is red and 2d is blue," and three paths (red-blue, blue-blue, white-blue) describe "2d is blue," so

$$Pr(\text{1st is red} \mid \text{2d is blue}) = \frac{\dfrac{5}{12} \cdot \dfrac{4}{11}}{\dfrac{5}{12} \cdot \dfrac{4}{11} + \dfrac{4}{12} \cdot \dfrac{3}{11} + \dfrac{3}{12} \cdot \dfrac{4}{11}}$$

$$= \frac{5/33}{5/33 + 1/11 + 1/11} = \frac{5}{11}$$

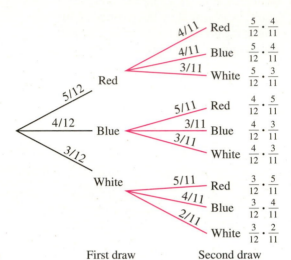

Figure 7.10   First draw   Second draw

This example illustrates a special type of problem, called a **Bayes problem.** In this type of problem, we know the result of the second stage of a two-stage experiment and wish to find the probability of a specified result in the first stage.

Suppose there are n possible outcomes in the first stage of the experiment, denoted $E_1, E_2, \ldots, E_n$, and m possible outcomes in the second stage, denoted $F_1, F_2, \ldots, F_m$ (see Figure 7.11). Then the probability that event E_1 occurs in the first stage, given that F_1 has occurred in the second stage, is

$$Pr(E_1 \mid F_1) = \frac{Pr(E_1 \cap F_1)}{Pr(F_1)} \tag{1}$$

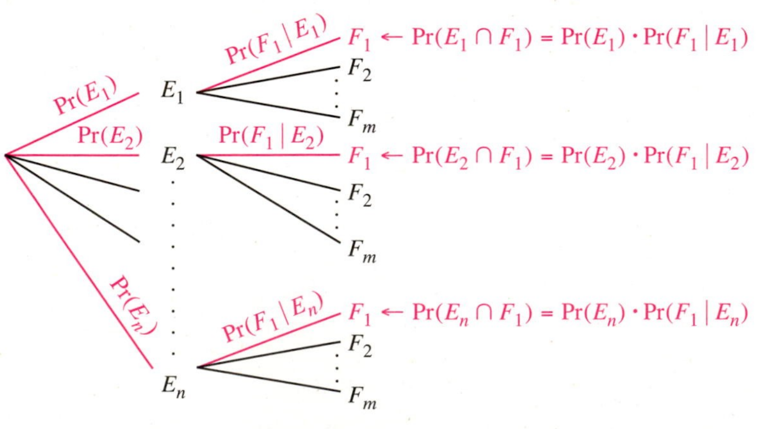

Figure 7.11

By looking at the probability tree in Figure 7.11, we see that

$$Pr(E_1 \cap F_1) = Pr(E_1) \cdot Pr(F_1 \mid E_1)$$

and that

$$Pr(F_1) = Pr(E_1 \cap F_1) + Pr(E_2 \cap F_1) + \cdots + Pr(E_n \cap F_1)$$

Using these facts, equation (1) becomes

$$\Pr(E_1 \mid F_1) = \frac{\Pr(E_1) \cdot \Pr(F_1 \mid E_1)}{\Pr(E_1 \cap F_1) + \Pr(E_2 \cap F_1) + \cdots + \Pr(E_n \cap F_1)} \qquad (2)$$

Using the fact that for any E_i, $\Pr(E_i$ and $F_1) = \Pr(E_i) \cdot \Pr(F_1 \mid E_i)$, we can rewrite equation (2) in a form called **Bayes' formula.**

Bayes' Formula	If a probability experiment has n possible outcomes in its first stage given by E_1, $E_2, \ldots, E_n$, and if F_1 is an event in the second stage, then the probability that event E_1 occurs in the first stage, given that F_1 has occurred in the second stage, is given by **Bayes' formula:** $$\Pr(E_1 \mid F_1) = \frac{\Pr(E_1) \cdot \Pr(F_1 \mid E_1)}{\Pr(E_1) \cdot \Pr(F_1 \mid E_1) + \Pr(E_2) \cdot \Pr(F_1 \mid E_2) + \cdots + \Pr(E_n) \cdot \Pr(F_1 \mid E_n)}$$

Note that Bayes problems can be solved either by this formula or by the use of a probability tree. Of the two methods, many students find using the tree easier because it is less abstract than the formula.

Bayes' Formula and Trees	$$\Pr(E_1 \mid F_1) = \frac{\text{Product of branch probabilities on path leading to } F_1 \text{ through } E_1}{\text{Sum of all branch products on paths leading to } F_1}$$

EXAMPLE 3   Medical Tests | APPLICATION PREVIEW |

Suppose a test for diagnosing a certain serious disease is successful in detecting the disease in 95% of all persons infected but that it incorrectly diagnoses 4% of all healthy people as having the serious disease. Suppose also that it incorrectly diagnoses 12% of all people having another minor disease as having the serious disease. It is known that 2% of the population have the serious disease, 90% of the population are healthy, and 8% have the minor disease. Find the probability that a person selected at random has the serious disease if the test indicates that he or she does. Use H to represent healthy, M to represent having the minor disease, and D to represent having the serious disease.

Solution
The tree that represents the health condition of a person chosen at random and the results of the test on that person are shown in Figure 7.12. (A test that indicates that a person has the disease is called a *positive* test.)

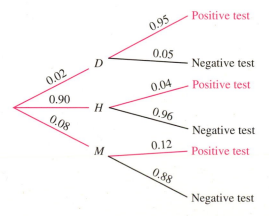

Figure 7.12

We seek $\Pr(D \mid \text{pos. test})$. Using the tree and the conditional probability formula gives

$$\Pr(D \mid \text{pos. test}) = \frac{\Pr(D \cap \text{pos. test})}{\Pr(\text{pos. test})}$$

$$= \frac{(0.02)(0.95)}{(0.90)(0.04) + (0.02)(0.95) + (0.08)(0.12)}$$

$$= \frac{0.0190}{0.0360 + 0.0190 + 0.0096}$$

$$\approx 0.2941$$

Using Bayes' formula directly gives the same result.

$$\Pr(D \mid \text{pos. test})$$

$$= \frac{\Pr(D) \cdot \Pr(\text{pos. test} \mid D)}{\Pr(H) \cdot \Pr(\text{pos. test} \mid H) + \Pr(D) \cdot \Pr(\text{pos. test} \mid D) + \Pr(M) \cdot \Pr(\text{pos. test} \mid M)}$$

$$= \frac{(0.02)(0.95)}{(0.90)(0.04) + (0.02)(0.95) + (0.08)(0.12)} \approx 0.2941 \qquad \blacksquare$$

CHECKPOINT

2. Use the information (and tree) from Example 3 to find the probability that a person chosen at random is healthy if the test is negative (a negative result indicates that the person does not have the disease).

Table 7.3 gives a summary of probability formulas and when they are used.

| TABLE 7.3

SUMMARY OF PROBABILITY FORMULAS

		One Trial	Two Trials	More Than Two Trials
$\Pr(A \text{ and } B)$	Independent	Sample space	$\Pr(A) \cdot \Pr(B)$	Product of probabilities
	Dependent	Sample space	$\Pr(A) \cdot \Pr(B \mid A)$	Product of conditional probabilities
$\Pr(A \text{ or } B)$	Mutually exclusive	$\Pr(A) + \Pr(B)$	Probability tree	Probability tree
	Not mutually exclusive	$\Pr(A) + \Pr(B) - \Pr(A \text{ and } B)$	Probability tree	Probability tree

CHECKPOINT SOLUTIONS

1. The probability tree for this probability experiment is given below. From the tree we have

$$\Pr(\text{gold}) = \Pr\big[(\text{I and } G) \text{ or } (\text{II and } G) \text{ or } (\text{III and } G)\big]$$

$$= \frac{1}{3} \cdot \frac{3}{3} + \frac{1}{3} \cdot \frac{1}{3} + \frac{1}{3} \cdot \frac{1}{2} = \frac{11}{18}$$

2. The probability that a person is healthy, given that the test is negative, is

$$\Pr(H \mid \text{neg. test}) = \frac{\Pr(H \cap \text{neg. test})}{\Pr(\text{neg. test})}$$

$$= \frac{(0.90)(0.96)}{(0.02)(0.05) + (0.90)(0.96) + (0.08)(0.88)} \approx 0.9237$$

| EXERCISES | 7.4

Access end-of-section exercises online at www.webassign.net

OBJECTIVES   7.5

- To use the Fundamental Counting Principle and permutations to solve counting problems
- To use combinations to solve counting problems

Counting: Permutations and Combinations

| APPLICATION PREVIEW |

During a national television advertising campaign, Little Caesar's Pizza stated that for $9.95, you could get 2 medium-sized pizzas, each with any of 0 to 5 toppings chosen from 11 that are available. The commercial asked the question, "How many different pairs of pizzas can you get?" (See Example 7.) Answering this question uses combinations, which we discuss in this section. Other counting techniques, including permutations, are also discussed in this section.

Fundamental Counting Principle

Suppose that you decide to dine at a restaurant that offers 3 appetizers, 8 entrees, and 6 desserts. How many different meals can you have? We can answer questions of this type using the **Fundamental Counting Principle.**

Fundamental Counting Principle

If there are $n(A)$ ways in which an event A can occur, and if there are $n(B)$ ways in which a second event B can occur after the first event has occurred, then the two events can occur in $n(A) \cdot n(B)$ ways.

The Fundamental Counting Principle can be extended to any number of events as long as they are independent. Thus the total number of possible meals at the restaurant mentioned above is

$$n(\text{appetizers}) \cdot n(\text{entrees}) \cdot n(\text{desserts}) = 3 \cdot 8 \cdot 6 = 144$$

EXAMPLE 1   License Plates

Pennsylvania offers a special "Save Wild Animals" license plate, which gives Pennsylvania zoos $15 for each plate sold. These plates have the two letters P and Z, followed by four numbers. When plates like these have all been purchased, additional plates will replace the final digit with a letter other than O. What is the maximum amount that the zoos could receive from the sale of the plates?

Solution

The following numbers represent the number of possible digits or letters in each of the spaces on the license plates as illustrated below.

$$\underline{1} \quad \underline{1} \quad \underline{10} \quad \underline{10} \quad \underline{10} \quad \underline{35}$$

Note that the last space can be a number from 0 to 9 or a letter other than O. By the Fundamental Counting Principle, the product of these numbers gives the number of license plates that can be made satisfying the given conditions.

$$1 \cdot 1 \cdot 10 \cdot 10 \cdot 10 \cdot 35 = 35,000$$

Thus 35,000 automobiles can be licensed with these plates, so the zoos could receive ($15)(35,000) = $525,000.

CHECKPOINT

1. If a state permits either a letter or a nonzero digit to be used in each of six places on its license plates, how many different plates can it issue?

EXAMPLE 2 **Head Table Seating**

Suppose a person planning a banquet cannot decide how to seat 6 honored guests at the head table. In how many arrangements can they be seated in the 6 chairs on one side of a table?

Solution

If we think of the 6 chairs as spaces, we can determine how many ways each space can be filled. The planner could place any one of the 6 people in the first space (say the first chair on the left). One of the 5 remaining persons could then be placed in the second space, one of the 4 remaining persons in the third space, and so on, as illustrated below.

$$\underline{6}\ \underline{5}\ \underline{4}\ \underline{3}\ \underline{2}\ \underline{1}$$

The total number of arrangements that can be made is the product of these numbers. That is, the total number of possible arrangements is

$$6 \cdot 5 \cdot 4 \cdot 3 \cdot 2 \cdot 1 = 720 \quad\blacksquare$$

Because special products such as $6 \cdot 5 \cdot 4 \cdot 3 \cdot 2 \cdot 1$ frequently occur in counting theory, we use special notation to denote them. We write 6! (read "6 **factorial**") to denote $6 \cdot 5 \cdot 4 \cdot 3 \cdot 2 \cdot 1$. Likewise, $4! = 4 \cdot 3 \cdot 2 \cdot 1$.

Factorial

For any positive integer n, we define **n factorial,** written $n!$, as

$$n! = n(n-1)(n-2)\cdots3 \cdot 2 \cdot 1$$

We define $0! = 1$.

Calculator Note   See Appendix C, Section 7.5, for details on evaluating factorials with a calculator. $\quad\blacksquare$

EXAMPLE 3 **Banquet Seating**

Suppose the planner in Example 2 knows that 8 people feel they should be at the head table, but only 6 spaces are available. How many arrangements can be made that place 6 of the 8 people at the head table?

Solution

There are again 6 spaces to fill, but any of 8 people can be placed in the first space, any of 7 people in the second space, and so on. Thus the total number of possible arrangements is

$$8 \cdot 7 \cdot 6 \cdot 5 \cdot 4 \cdot 3 = 20{,}160 \quad\blacksquare$$

Permutations The number of possible arrangements in Example 3 is called the number of **permutations** of 8 things taken 6 at a time, and it is denoted $_8P_6$. Note that $_8P_6$ gives the first 6 factors of 8!, so we can use factorial notation to write the product.

$$_8P_6 = \frac{8!}{2!} = \frac{8!}{(8-6)!}$$

Permutations

The number of possible ordered arrangements of r objects chosen from a set of n objects is called the number of **permutations** of n objects taken r at a time, and it equals

$$_nP_r = \frac{n!}{(n-r)!}$$

Note that $_nP_n = n!$.

EXAMPLE 4   **Club Officers**

In how many ways can a president, a vice president, a secretary, and a treasurer be selected from an organization with 20 members?

Solution

We seek the number of orders (arrangements) in which 4 people can be selected from a group of 20. This number is

$$_{20}P_4 = \frac{20!}{(20-4)!} = 116,280$$

Calculator Note   See Appendix C, Section 7.5, for details on evaluating permutations with a calculator.

CHECKPOINT   2. In a 5-question matching test, how many different answer sheets are possible if no answer can be used twice and there are
(a) 5 answers available for matching?
(b) 7 answers available for matching?

Combinations   When we talk about arrangements of people at a table or arrangements of digits on a license plate, order is important. We now consider counting problems in which order is not important.

To find the number of ways you can select 3 people from 5 without regard to order, we would first find the number of ways to select them *with order*, $_5P_3$, and then divide by the number of ways the 3 people could be ordered (3!). Thus 3 people can be selected from 5 (without regard for order) in $_5P_3/3! = 10$ ways. We say that the number of **combinations** of 5 things taken 3 at a time is 10.

Combinations   The number of ways in which *r* objects can be chosen from a set of *n* objects without regard to the order of selection is called the number of **combinations** of *n* objects taken *r* at a time, and it equals

$$_nC_r = \frac{_nP_r}{r!}$$

Note that the fundamental difference between permutations and combinations is that permutations are used when order is a factor in the selection, and combinations are used when it is not. Because $_nP_r = n!/(n-r)!$, we have

$$_nC_r = \frac{_nP_r}{r!} = \frac{\dfrac{n!}{(n-r)!}}{r!} = \frac{n!}{r!(n-r)!}$$

Combination Formula   The number of combinations of *n* objects taken *r* at a time is also denoted by $\binom{n}{r}$; that is,

$$\binom{n}{r} = {_nC_r} = \frac{n!}{r!(n-r)!}$$

EXAMPLE 5   **Auto Options**

An auto dealer is offering any 4 of 6 special options at one price on a specially equipped car being sold. How many different choices of specially equipped cars do you have?

Solution

The order in which you choose the options is not relevant, so we seek the number of combinations of 6 things taken 4 at a time. Thus the number of possible choices is

$$_6C_4 = \frac{6!}{4!(6-4)!} = \frac{6!}{4!2!} = 15$$

Calculator Note   See Appendix C, Section 7.5, for details on evaluating combinations with a calculator. ■

EXAMPLE 6   Project Teams

Six men and eight women are qualified to serve on a project team to revise operational procedures. How many different teams can be formed containing three men and three women?

Solution

Three men can be selected from six in $\binom{6}{3}$ ways, and three women can be selected from eight in $\binom{8}{3}$ ways. Therefore, three men *and* three women can be selected in

$$\binom{6}{3}\binom{8}{3} = \frac{6!}{3!3!} \cdot \frac{8!}{5!3!} = 1120 \text{ ways}$$ ■

CHECKPOINT   Determine whether each of the following counting problems can be solved by using permutations or combinations, and answer each question.

3. Find the number of ways in which 8 members of a space shuttle crew can be selected from 20 available astronauts.
4. The command structure on a space shuttle flight is determined by the order in which astronauts are selected for a flight. How many different command structures are possible if 8 astronauts are selected from the 20 that are available?
5. If 14 men and 6 women are available for a space shuttle flight, how many crews are possible that have 5 men and 3 women astronauts?

EXAMPLE 7   Little Caesar's Pizza Choices | APPLICATION PREVIEW |

During a national television advertising campaign, Little Caesar's Pizza stated that for $9.95, you could get 2 medium-sized pizzas, each with any of 0 to 5 toppings chosen from 11 that are available. The commercial asked the question, "How many different pairs of pizzas can you get?" Answer the question, if the first pizza has a thin crust and the second has a thick crust.

Solution

Because you can get 0, 1, 2, 3, 4, or 5 toppings from the 11 choices, the possible choices for the first pizza are

$$_{11}C_0 + _{11}C_1 + _{11}C_2 + _{11}C_3 + _{11}C_4 + _{11}C_5 = 1 + 11 + 55 + 165 + 330 + 462 = 1024$$

Because the same number of choices is available for the second pizza, the number of choices for two pizzas is

$$1024 \cdot 1024 = 1,048,576$$ ■

EXAMPLE **8** **Quality Control**

Suppose a batch of 35 CPUs has 10 defective CPUs. In how many ways can a sample of 15 from this batch have 5 defective CPUs?

Solution

We seek the number of ways to find 5 of 10 defective and 10 of 25 good CPUs. This is

$$_{10}C_5 \cdot {}_{25}C_{10} = 823,727,520$$

CHECKPOINT SOLUTIONS

1. Any of 35 items (26 letters and 9 digits) can be used in any of the six places, so the number of different license plates is

$$35 \cdot 35 \cdot 35 \cdot 35 \cdot 35 \cdot 35 = 35^6 = 1,838,265,625$$

2. (a) There are 5 answers available for the first question, 4 for the second, 3 for the third, 2 for the fourth, and 1 for the fifth, so there are

$$5 \cdot 4 \cdot 3 \cdot 2 \cdot 1 = 120$$

possible answer sheets. Note that this is the number of permutations of 5 things taken 5 at a time,

$$_5P_5 = 5! = 120$$

(b) There are 7 answers available for the first question, 6 for the second, 5 for the third, 4 for the fourth, and 3 for the fifth, so there are

$$7 \cdot 6 \cdot 5 \cdot 4 \cdot 3 = 2520$$

possible answer sheets. Note that this is the number of permutations of 7 things taken 5 at a time, $_7P_5 = 2520$.

3. The members of the crew are to be selected without regard to the order of selection, so we seek the number of combinations of 8 people selected from 20. There are

$$_{20}C_8 = 125,970$$

possible crews.

4. The order in which the members are selected is important, so we seek the number of permutations of 8 people selected from 20. There are

$$_{20}P_8 = 5,079,110,400$$

possible command structures.

5. As in Problem 3, the order of selection is not relevant, so we seek the number of combinations of 5 men out of 14 *and* 3 women out of 6. There are

$$_{14}C_5 \cdot {}_6C_3 = 2002 \cdot 20 = 40,040$$

possible crews that have 5 men and 3 women.

| EXERCISES | 7.5

Access end-of-section exercises online at **www.webassign.net**

- To use counting techniques to solve probability problems

Permutations, Combinations, and Probability

| APPLICATION PREVIEW |

Suppose that of the 20 prospective jurors for a trial, 12 favor the death penalty, whereas 8 do not, and that 12 are chosen at random from these 20. To find the probability that 7 of the 12 will favor the death penalty, we can use combinations to find the number of ways to choose 12 jurors from 20 prospective jurors and to find the number of ways to choose 7 from the 12 who favor the death penalty and 5 from the 8 who oppose it. (See Example 4.) In finding the probability that an event E occurs, we frequently use other counting principles to determine the number of ways in which the event can occur, $n(E)$, and the total number of outcomes, $n(S)$.

Recall that if an event E can happen in $n(E)$ ways out of a total of $n(S)$ equally likely possibilities, the probability of the occurrence of the event is

$$\Pr(E) = \frac{n(E)}{n(S)}$$

EXAMPLE **1** **Intelligence Tests**

A psychologist claims she can teach four-year-old children to spell three-letter words very quickly. To test her, one of her students was given cards with the letters A, C, D, K, and T on them and told to spell CAT. What is the probability the child will spell CAT by chance?

Solution

The number of different three-letter "words" that can be formed using these 5 cards is $n(S) = {}_5P_3 = 5!/(5-3)! = 60$. Because only one of those 60 arrangements will be CAT, $n(E) = 1$, and the probability that the child will be successful by guessing is $1/60$. ∎

EXAMPLE **2** **License Plates**

Suppose that a state issues license plates with three letters followed by three digits and that there are 12 three-letter words that are not permitted on license plates. If all possible plates are produced and a plate is selected at random, what is the probability that the plate is unacceptable and will not be issued?

Solution

The total number of license plates that could be produced with three letters followed by three digits is

$$N(S) = 26^3 \cdot 10^3 = 17{,}576{,}000$$

and the number that will be unacceptable is

$$N(E) = 12 \cdot 10^3 = 12{,}000$$

Thus the probability that a plate selected at random will be unacceptable is

$$\frac{12{,}000}{17{,}576{,}000} = \frac{12}{17{,}576} = \frac{3}{4394} \approx 0.0006827$$

∎

1. If three wires (red, black, and white) are randomly attached to a three-way switch (which has 3 poles to which wires can be attached), what is the probability that the wires will be attached at random in the one order that makes the switch work properly?

EXAMPLE 3   Quality Control

A manufacturing process for computer chips is such that 5 out of 100 chips are defective. If 10 chips are chosen at random from a box containing 100 newly manufactured chips, what is the probability

(a) that none of the chips will be defective?
(b) that 8 will be good and 2 defective?

Solution

(a) The number of ways to choose any 10 chips from 100 is $n(S) = {}_{100}C_{10}$, and 10 good chips can be chosen from the 95 good ones in $n(E) = {}_{95}C_{10}$ ways. Thus

$$\Pr(10 \text{ good}) = \Pr(E) = \frac{n(E)}{n(S)} = \frac{{}_{95}C_{10}}{{}_{100}C_{10}} = \frac{\binom{95}{10}}{\binom{100}{10}}$$

$$= \frac{\dfrac{95!}{10!85!}}{\dfrac{100!}{10!90!}} \approx 0.58375$$

(b) Eight good chips and two defective chips can be selected in

$$n(E) = {}_{95}C_8 \cdot {}_5C_2 = \binom{95}{8}\binom{5}{2} = \frac{95!}{8!87!} \cdot \frac{5!}{2!3!}$$

ways. Thus

$$\Pr(8 \text{ good and 2 defective}) = \frac{n(E)}{n(S)} = \frac{\dfrac{95!}{8!87!} \cdot \dfrac{5!}{2!3!}}{\dfrac{100!}{10!90!}} \approx 0.07022 \qquad \blacksquare$$

EXAMPLE 4   Jury Selection | APPLICATION PREVIEW |

Suppose that of the 20 prospective jurors for a trial, 12 favor the death penalty and 8 do not. If 12 jurors are chosen at random from these 20, what is the probability that 7 of the jurors will favor the death penalty?

Solution

The number of ways in which 7 jurors of the 12 will favor the death penalty is the number of ways in which 7 favor and 5 do not favor, or

$$n(E) = {}_{12}C_7 \cdot {}_8C_5 = \frac{12!}{7!5!} \cdot \frac{8!}{5!3!}$$

The total number of ways in which 12 jurors can be selected from 20 people is

$$n(S) = {}_{20}C_{12} = \frac{20!}{12!8!}$$

Thus the probability we seek is

$$\Pr(E) = \frac{n(E)}{n(S)} = \frac{\dfrac{12!}{7!5!} \cdot \dfrac{8!}{5!3!}}{\dfrac{20!}{12!8!}} \approx 0.3521$$

■

CHECKPOINT

2. If 5 people are chosen from a group that contains 4 men and 6 women, what is the probability that 3 men and 2 women will be chosen?

EXAMPLE 5 **Auto Keys**

Kletr/Shutterstock.com

Beginning in 1990, General Motors began to use a theft deterrent key on some of its cars. The key has six parts, with three patterns for each part, plus an electronic chip containing a code from 1 to 15. What is the probability that one of these keys selected at random will start a GM car requiring a key of this type?

Solution
Each of the six parts has three patterns, and the chip can have any of 15 codes, so there are

$$3 \cdot 3 \cdot 3 \cdot 3 \cdot 3 \cdot 3 \cdot 15 = 10{,}935$$

possible keys. Because there is only one key that will start the car, the probability that a key selected at random will start a given car is $1/10{,}935$.

■

CHECKPOINT SOLUTIONS

1. To find the total number of orders in which the wires can be attached, we note that the first pole can have any of the 3 wires attached to it, the second pole can have any of 2, and the third pole can have only the remaining wire, so there are $3 \cdot 2 \cdot 1 = 6$ possible different orders. Only one of these orders permits the switch to work properly, so the probability is $1/6$.
2. Five people can be chosen from 10 people in ${}_{10}C_5 = 252$ ways. Three men and 2 women can be chosen in ${}_4C_3 \cdot {}_6C_2 = 4 \cdot 15 = 60$ ways. Thus the probability is $60/252 = 5/21$.

| EXERCISES | 7.6

Access end-of-section exercises online at www.webassign.net **WebAssign**

OBJECTIVES 7.7

- To use transition matrices to find specific stages of Markov chains
- To write transition matrices for Markov chain problems
- To find steady-state vectors for certain Markov chain problems

Markov Chains

| APPLICATION PREVIEW |

Suppose a department store has determined that a woman who uses a credit card in the current month will use it again next month with probability 0.8. That is,

$$\Pr(\text{woman will use card next month} \mid \text{she used it this month}) = 0.8$$

and

$$\Pr(\text{woman will not use card next month} \mid \text{she used it this month}) = 0.2$$

Furthermore, suppose the store has determined that the probability that a woman who does not use a credit card this month will not use it next month is 0.7. That is,

Pr(woman will not use card next month | she did not use it this month) = 0.7

and

Pr(woman will use card next month | she did not use it this month) = 0.3

To find the probability that the woman will use the credit card in some later month, we use Markov chains and put the conditional probabilities in a transition matrix (see Example 1). Markov chains are named for the famous Russian mathematician A. A. Markov, who developed this theory in 1906. They are useful in the analysis of price movements, consumer behavior, laboratory animal behavior, and many other processes in business, the social sciences, and the life sciences.

Markov Chains   Suppose that a sequence of experiments with the same set of outcomes is performed, with the probabilities of the outcomes in one experiment depending on the outcomes of the previous experiment. The outcomes of the experiments are called **states,** and the sequence of the experiments forms a **Markov chain** if the next state solely depends on the present state and doesn't directly depend on the previous states. The changes of state are called **transitions,** the probability of moving from one state to another is called the **transition probability,** and these probabilities can often be organized into a **transition matrix.**

EXAMPLE 1   Credit Card Use | APPLICATION PREVIEW |

Create the transition matrix that contains the conditional probabilities from the problem described in the Application Preview.

Solution
If the card is used this month, the probabilities that the woman will or will not use the card next month can be represented by the probability tree

$$\text{Used current month} \begin{cases} 0.8 \quad \text{Used next month} \\ 0.2 \quad \text{Not used next month} \end{cases}$$

We can enter these numbers in the first row of the transition matrix:

$$\begin{array}{cc} & \textit{Next month} \\ & \begin{array}{cc} \text{Card} & \text{Card} \\ \text{used} & \text{not used} \end{array} \\ \textit{Current month} \begin{array}{c} \text{Card used} \\ \text{Card not used} \end{array} & \begin{bmatrix} 0.8 & 0.2 \\ & \end{bmatrix} \end{array}$$

If the card is not used this month, the probabilities that the woman will or will not use the card next month can be represented by the probability tree

$$\text{Not used current month} \begin{cases} 0.7 \quad \text{Not used next month} \\ 0.3 \quad \text{Used next month} \end{cases}$$

Putting these conditional probabilities in the second row completes the transition matrix:

$$\begin{array}{cc} & \textit{Next month} \\ & \begin{array}{cc} \text{Card} & \text{Card} \\ \text{used} & \text{not used} \end{array} \\ P = \textit{Current month} \begin{array}{c} \text{Card used} \\ \text{Card not used} \end{array} & \begin{bmatrix} 0.8 & 0.2 \\ 0.3 & 0.7 \end{bmatrix} \end{array}$$

In Example 1, there were two states for the Markov chain described. State 1 was "used next month" and state 2 was "not used next month." The transition matrix was a 2×2 matrix.

Transition Matrix Properties

A **transition matrix** for a Markov chain with n states is an $n \times n$ matrix characterized by the following properties.

1. The ij-entry in the matrix is Pr(moving from state i to state j) = Pr(state j | state i).
2. The entries are nonnegative.
3. The sum of the entries in each row equals 1.

Now suppose the department store data show that during the first month a credit card is received, the probabilities for each state are as follows:

$$Pr(\text{a woman used a credit card}) = 0.9$$
$$Pr(\text{a woman did not use a credit card}) = 0.1$$

These probabilities can be placed in the row matrix

$$A = \begin{bmatrix} 0.9 & 0.1 \end{bmatrix}$$

which is called the **initial probability vector** for the problem. This vector gives the probabilities for each state at the outset of the trials. In general, a probability vector is defined as follows.

Probability Vector

For a vector to be a **probability vector**, the sum of its entries must be 1 and each of its entries must be nonnegative.

Note that if a woman used her credit card in the initial month, the probability is 0.8 that she will use it the next month, and if she didn't use the card, the probability is 0.3 that she will use it the next month. Thus,

$$\begin{aligned}
Pr(\text{will use 2nd month}) &= Pr(\text{used 1st month}) \cdot Pr(\text{will use 2nd | used 1st}) \\
&\quad + Pr(\text{not used 1st month}) \cdot Pr(\text{will use 2nd | not used 1st}) \\
&= (0.9)(0.8) + (0.1)(0.3) \\
&= 0.72 + 0.03 = 0.75
\end{aligned}$$

Note that this is one element of the product matrix AP. Similarly, the second element of the product matrix is the probability that a woman will not use the card in the second month.

$$AP = \underset{\substack{\text{Initial} \\ \text{month}}}{\begin{bmatrix} 0.9 & 0.1 \end{bmatrix}} \underset{\substack{\text{Transition} \\ \text{matrix}}}{\begin{bmatrix} 0.8 & 0.2 \\ 0.3 & 0.7 \end{bmatrix}} = \underset{\substack{\text{Second} \\ \text{month}}}{\begin{bmatrix} 0.75 & 0.25 \end{bmatrix}}$$

Thus

$$Pr(\text{a woman will use a credit card in 2nd month}) = 0.75$$

and

$$Pr(\text{a woman will not use a credit card in 2nd month}) = 0.25$$

To find the probabilities for each state for the third month, we multiply the second-month probability vector times the transition matrix.

$$\underset{\substack{\text{Second} \\ \text{month}}}{\begin{bmatrix} 0.75 & 0.25 \end{bmatrix}} \underset{\substack{\text{Transition} \\ \text{matrix}}}{\begin{bmatrix} 0.8 & 0.2 \\ 0.3 & 0.7 \end{bmatrix}} = \underset{\substack{\text{Third} \\ \text{month}}}{\begin{bmatrix} 0.675 & 0.325 \end{bmatrix}}$$

Continuing in this manner, we can find the probabilities for later months.

EXAMPLE 2   **Politics**

Suppose that in a certain city the Democratic, Republican, and Consumer parties have members of their parties on the city council. The probability of a member of one of these parties winning in any election depends on the proportional membership of his or her party at the time of the election. The probabilities for all these parties are given by the following transition matrix P.

$$
\begin{array}{c}
\textit{Party in office} \\[2pt]
\begin{array}{r}
\text{Democrat} \\
P = \text{Republican} \\
\text{Consumer}
\end{array}
\end{array}
\quad
\begin{array}{c}
\textit{Party in office next term} \\[2pt]
\begin{array}{ccc}
\text{D} & \text{R} & \text{C}
\end{array} \\
\begin{bmatrix}
0.5 & 0.4 & 0.1 \\
0.4 & 0.5 & 0.1 \\
0.3 & 0.3 & 0.4
\end{bmatrix}
\end{array}
$$

Suppose that the current distribution of parties on the council is given in the initial probability vector $A_0 = [0.4 \ \ 0.4 \ \ 0.2]$. The probable distribution of parties after the third election from now is given by the probability vector for the third election. Find this probability vector.

Solution

We multiply the initial probability vector by the transition matrix to get the probability vector for the first election.

$$
A_1 = A_0 P = \begin{bmatrix} 0.4 & 0.4 & 0.2 \end{bmatrix}
\begin{bmatrix}
0.5 & 0.4 & 0.1 \\
0.4 & 0.5 & 0.1 \\
0.3 & 0.3 & 0.4
\end{bmatrix}
= \begin{bmatrix} 0.42 & 0.42 & 0.16 \end{bmatrix}
$$

We multiply the probability vector for the first election by the transition matrix to get the probability vector for the second election.

$$
A_2 = A_1 P = \begin{bmatrix} 0.42 & 0.42 & 0.16 \end{bmatrix}
\begin{bmatrix}
0.5 & 0.4 & 0.1 \\
0.4 & 0.5 & 0.1 \\
0.3 & 0.3 & 0.4
\end{bmatrix}
= \begin{bmatrix} 0.426 & 0.426 & 0.148 \end{bmatrix}
$$

We multiply the probability vector for the second election by the transition matrix to get the probability vector for the third election. This probability vector gives the probable distribution after the third election.

$$
A_3 = A_2 P = \begin{bmatrix} 0.426 & 0.426 & 0.148 \end{bmatrix}
\begin{bmatrix}
0.5 & 0.4 & 0.1 \\
0.4 & 0.5 & 0.1 \\
0.3 & 0.3 & 0.4
\end{bmatrix}
= \begin{bmatrix} 0.4278 & 0.4278 & 0.1444 \end{bmatrix}
$$

Thus the probable distribution of parties on the council after the third election is approximately 43% Democrat, 43% Republican, and 14% Consumer.   ■

Steady-State Vectors

The multiplications done in Example 2 can be written in a second way, which gives us a general method to find the probability vectors for later elections.

$$
\begin{aligned}
A_1 &= A_0 P \\
A_2 &= A_1 P = A_0 P P = A_0 P^2 \\
A_3 &= A_2 P = A_0 P^2 P = A_0 P^3
\end{aligned}
$$

Thus the probability vector for the third election can be found by multiplying the initial probability vector A_0 by the transition matrix P three times—that is, multiplying A_0 by P^3.

$$
A_3 = A_0 P^3
$$

In general, we have

> If A_0 represents the $1 \times n$ initial probability vector and P is the $n \times n$ transition matrix for a Markov chain, then mth probability vector A_m is
>
> $$A_m = A_0 P^m$$

We can use this formula to compute the probability vectors for large numbers of elections. In Example 2 the probability vector for the 10th election (to five decimal places) is

$$A_{10} = A_0 P^{10} = \begin{bmatrix} 0.4 & 0.4 & 0.2 \end{bmatrix} P^{10} = \begin{bmatrix} 0.42857 & 0.42857 & 0.14286 \end{bmatrix}$$

Calculator Note

Figure 7.13 shows this same calculation on a graphing calculator. See Appendix C, Section 7.7, for details. ∎

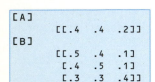
```
[A]
      [[.4  .4  .2]]
[B]
      [[.5  .4  .1]
       [.4  .5  .1]
       [.3  .3  .4]]
```

```
[A][B]^10
.42857 .42857 .14286
```

Figure 7.13

Note also that the probability vector for the 21st election (to five decimal places) is

$$A_{21} = A_0 P^{21} = \begin{bmatrix} 0.4 & 0.4 & 0.2 \end{bmatrix} P^{21} = \begin{bmatrix} 0.42857 & 0.42857 & 0.14286 \end{bmatrix}$$

Repeated multiplications will not change the probability vectors (for at least five decimal places), so this system appears to be stabilizing. The vector that is being approached has the form $\begin{bmatrix} V_1 & V_2 & V_3 \end{bmatrix}$ such that

$$\begin{bmatrix} V_1 & V_2 & V_3 \end{bmatrix} \cdot P = \begin{bmatrix} V_1 & V_2 & V_3 \end{bmatrix}$$

The exact probability vector for which this is true in Example 2 is $\begin{bmatrix} \dfrac{3}{7} & \dfrac{3}{7} & \dfrac{1}{7} \end{bmatrix}$, because

$$\begin{bmatrix} \dfrac{3}{7} & \dfrac{3}{7} & \dfrac{1}{7} \end{bmatrix} \begin{bmatrix} 0.5 & 0.4 & 0.1 \\ 0.4 & 0.5 & 0.1 \\ 0.3 & 0.3 & 0.4 \end{bmatrix} = \begin{bmatrix} \dfrac{3}{7} & \dfrac{3}{7} & \dfrac{1}{7} \end{bmatrix}$$

This vector is called the **fixed-probability vector** for the given transition matrix. Because this vector is where the process eventually stabilizes, it is also called the **steady-state vector.**

Rather than trying to guess this steady-state vector by repeated multiplication, we can solve the following equation for V_1, V_2, and V_3:

$$\begin{bmatrix} V_1 & V_2 & V_3 \end{bmatrix} \begin{bmatrix} 0.5 & 0.4 & 0.1 \\ 0.4 & 0.5 & 0.1 \\ 0.3 & 0.3 & 0.4 \end{bmatrix} = \begin{bmatrix} V_1 & V_2 & V_3 \end{bmatrix}$$

$$
\begin{array}{l}
0.5V_1 + 0.4V_2 + 0.3V_3 = V_1 \\
0.4V_1 + 0.5V_2 + 0.3V_3 = V_2 \\
0.1V_1 + 0.1V_2 + 0.4V_3 = V_3
\end{array}
\quad \text{or} \quad
\begin{array}{l}
-0.5V_1 + 0.4V_2 + 0.3V_3 = 0 \\
0.4V_1 - 0.5V_2 + 0.3V_3 = 0 \\
0.1V_1 + 0.1V_2 - 0.6V_3 = 0
\end{array}
$$

Solving by Gauss-Jordan elimination gives

$$
\begin{bmatrix} -0.5 & 0.4 & 0.3 & | & 0 \\ 0.4 & -0.5 & 0.3 & | & 0 \\ 0.1 & 0.1 & -0.6 & | & 0 \end{bmatrix}
\xrightarrow[\text{Switch } R_1 \ \& \ R_3]{10 \ (\text{each row})}
\begin{bmatrix} 1 & 1 & -6 & | & 0 \\ 4 & -5 & 3 & | & 0 \\ -5 & 4 & 3 & | & 0 \end{bmatrix}
$$

$$
\xrightarrow[5R_1 + R_3 \to R_3]{-4R_1 + R_2 \to R_2}
\begin{bmatrix} 1 & 1 & -6 & | & 0 \\ 0 & -9 & 27 & | & 0 \\ 0 & 9 & -27 & | & 0 \end{bmatrix}
\xrightarrow[\substack{\text{then} \\ -R_2 + R_1 \to R_1 \\ -9R_2 + R_3 \to R_3}]{-\frac{1}{9} R_2 \to R_2}
\begin{bmatrix} 1 & 0 & -3 & | & 0 \\ 0 & 1 & -3 & | & 0 \\ 0 & 0 & 0 & | & 0 \end{bmatrix}
$$

Thus $V_1 = 3V_3$ $V_2 = 3V_3$ $V_3 = $ any real number

Any vector that satisfies these conditions is a solution of the system. Because $[V_1 \ V_2 \ V_3]$ is a probability vector, we have $V_1 + V_2 + V_3 = 1$. Substituting gives $3V_3 + 3V_3 + V_3 = 1$, so $V_3 = 1/7$ and the steady-state vector is

$$\begin{bmatrix} \dfrac{3}{7} & \dfrac{3}{7} & \dfrac{1}{7} \end{bmatrix}$$

Calculator Note A graphing calculator can be used to find the steady-state vector found above. See Appendix C, Section 7.7, for details.

As we have seen, the Markov chain in Example 2 has a steady-state vector, but not every Markov chain has one. If the transition matrix P for a Markov chain is a special matrix called a **regular transition matrix,** there is a unique steady-state vector for the Markov chain. A transition matrix P is a regular transition matrix if some power of P contains only positive entries. Note that a regular transition matrix may contain zeros; the only requirement is that some power of P contain all positive entries. In this text we limit ourselves to regular transition matrices. These matrices are related to **regular Markov chains** as follows.

Regular Markov Chain

- If P is a regular transition matrix, the sequence $P, P^2, P^3, \ldots$ approaches a matrix in which all of its rows are equal.
- If the transition matrix for a Markov chain is regular, the chain is called a **regular Markov chain,** and this chain has a steady-state vector.

Note also that if a steady-state matrix exists for a Markov chain, then the effect of the initial probabilities diminishes as the number of steps in the process increases. This means that the probability vectors for the elections in Example 2 would approach the steady-state vector regardless of the initial probabilities. For example, even if all members of the council are Democrats at the initial stage (that is, the initial probability vector is $[1 \ 0 \ 0]$), the probability vector after the fourth election is $[0.4292 \ 0.4291 \ 0.1417]$, which approximates the steady-state vector $[3/7 \ 3/7 \ 1/7]$ found previously.

CHECKPOINT

1. Use the transition matrix $P = \begin{bmatrix} 0.1 & 0.3 & 0.6 \\ 0.3 & 0.5 & 0.2 \\ 0.1 & 0.1 & 0.8 \end{bmatrix}$ and the initial probability vector

$A_0 = [0.2 \ 0.3 \ 0.5]$ to
(a) find the second probability vector (two steps after the initial probability vector).
(b) find the steady-state vector associated with the matrix.

EXAMPLE 3　Educational Attainment

Suppose that in a certain city the probabilities that a woman with less than a high school education has a daughter with less than a high school education, with a high school degree, and with education beyond high school are 0.2, 0.6, and 0.2, respectively. The probabilities that a woman with a high school degree has a daughter with less than a high school education, with a high school degree, and with education beyond high school are 0.1, 0.5, and 0.4, respectively. The probabilities that a woman with education beyond high school has a daughter with less than a high school education, with a high school degree, and with education beyond high school are 0.1, 0.1, and 0.8, respectively. The population of women in the city is now 60% with less than a high school education, 30% with a high school degree, and 10% with education beyond high school.

(a) What is the transition matrix for this information?
(b) What is the probable distribution of women according to educational level two generations from now?
(c) What is the steady-state vector for this information?

Solution

(a) The transition matrix is

	Daughter's educational level		
Mother's educational level	Less than H.S.	H.S. degree	More than H.S.
Less than H.S.	0.2	0.6	0.2
H.S. degree	0.1	0.5	0.4
More than H.S.	0.1	0.1	0.8

(b) The probable distribution after one generation is

$$[0.6 \quad 0.3 \quad 0.1] \begin{bmatrix} 0.2 & 0.6 & 0.2 \\ 0.1 & 0.5 & 0.4 \\ 0.1 & 0.1 & 0.8 \end{bmatrix} = [0.16 \quad 0.52 \quad 0.32]$$

The probable distribution after two generations is

$$[0.16 \quad 0.52 \quad 0.32] \begin{bmatrix} 0.2 & 0.6 & 0.2 \\ 0.1 & 0.5 & 0.4 \\ 0.1 & 0.1 & 0.8 \end{bmatrix} = [0.116 \quad 0.388 \quad 0.496]$$

(c) The steady-state vector is found by solving

$$[V_1 \quad V_2 \quad V_3] \begin{bmatrix} 0.2 & 0.6 & 0.2 \\ 0.1 & 0.5 & 0.4 \\ 0.1 & 0.1 & 0.8 \end{bmatrix} = [V_1 \quad V_2 \quad V_3]$$

$$\begin{cases} 0.2V_1 + 0.1V_2 + 0.1V_3 = V_1 \\ 0.6V_1 + 0.5V_2 + 0.1V_3 = V_2 \\ 0.2V_1 + 0.4V_2 + 0.8V_3 = V_3 \end{cases} \text{ or } \begin{cases} -0.8V_1 + 0.1V_2 + 0.1V_3 = 0 \\ 0.6V_1 - 0.5V_2 + 0.1V_3 = 0 \\ 0.2V_1 + 0.4V_2 - 0.2V_3 = 0 \end{cases}$$

Using Gauss-Jordan elimination or technology gives

$$\begin{bmatrix} -0.8 & 0.1 & 0.1 & | & 0 \\ 0.6 & -0.5 & 0.1 & | & 0 \\ 0.2 & 0.4 & -0.2 & | & 0 \end{bmatrix} \rightarrow \begin{bmatrix} 1 & 0 & -\frac{3}{17} & | & 0 \\ 0 & 1 & -\frac{7}{17} & | & 0 \\ 0 & 0 & 0 & | & 0 \end{bmatrix}$$

Thus $V_1 = \frac{3}{17}V_3$ and $V_2 = \frac{7}{17}V_3$. The probability vector that satisfies these conditions and $V_1 + V_2 + V_3 = 1$ must satisfy

$$\frac{3}{17}V_3 + \frac{7}{17}V_3 + V_3 = 1 \quad \text{or} \quad \frac{27}{17}V_3 = 1 \quad \text{so} \quad V_3 = \frac{17}{27}$$

Thus the steady-state vector is

$$\begin{bmatrix} \dfrac{3}{27} & \dfrac{7}{27} & \dfrac{17}{27} \end{bmatrix}$$

■

CHECKPOINT SOLUTIONS

1. (a) The first-stage probability vector is

$$A_1 = A_0 P = [0.2 \quad 0.3 \quad 0.5] \begin{bmatrix} 0.1 & 0.3 & 0.6 \\ 0.3 & 0.5 & 0.2 \\ 0.1 & 0.1 & 0.8 \end{bmatrix} = [0.16 \quad 0.26 \quad 0.58]$$

and that for the second stage is

$$A_2 = A_1 P = [0.16 \quad 0.26 \quad 0.58] \begin{bmatrix} 0.1 & 0.3 & 0.6 \\ 0.3 & 0.5 & 0.2 \\ 0.1 & 0.1 & 0.8 \end{bmatrix} = [0.152 \quad 0.236 \quad 0.612]$$

(b) To find the steady-state vector, we solve

$$[V_1 \quad V_2 \quad V_3] \begin{bmatrix} 0.1 & 0.3 & 0.6 \\ 0.3 & 0.5 & 0.2 \\ 0.1 & 0.1 & 0.8 \end{bmatrix} = [V_1 \quad V_2 \quad V_3]$$

Multiplying and combining like terms give the following system of equations.

$$\begin{cases} -0.9V_1 + 0.3V_2 + 0.1V_3 = 0 \\ 0.3V_1 - 0.5V_2 + 0.1V_3 = 0 \\ 0.6V_1 + 0.2V_2 - 0.2V_3 = 0 \end{cases}$$

We solve this by using the augmented matrix

$$\begin{bmatrix} -0.9 & 0.3 & 0.1 & | & 0 \\ 0.3 & -0.5 & 0.1 & | & 0 \\ 0.6 & 0.2 & -0.2 & | & 0 \end{bmatrix}$$

The solution gives the steady-state vector $\begin{bmatrix} \dfrac{1}{7} & \dfrac{3}{14} & \dfrac{9}{14} \end{bmatrix}$.

| EXERCISES | 7.7

Access end-of-section exercises online at **www.webassign.net**

KEY TERMS AND FORMULAS

Section	Key Terms	Formulas
7.1	Equally likely events	$\Pr(E) = k/n$
	Sample space	
	Certain event	$\Pr(E) = 1$
	Impossible event	$\Pr(E) = 0$
	Odds	
	Empirical probability	
7.2	Intersection of events	$\Pr(E \text{ and } F) = \Pr(E \cap F)$
	Union of events	$\Pr(E \text{ or } F) = \Pr(E \cup F)$
		$\qquad = \Pr(E) + \Pr(F) - \Pr(E \cap F)$
	Complement	$\Pr(\text{not } E) = \Pr(E') = 1 - \Pr(E)$
	Mutually exclusive events	$\Pr(E \text{ or } F) = \Pr(E) + \Pr(F)$
7.3	Conditional probability	$\Pr(E \mid F) = \dfrac{\Pr(E \cap F)}{\Pr(F)}$
	Product Rule	$\Pr(E \cap F) = \Pr(F) \cdot \Pr(E \mid F)$
	Independent events	$\Pr(A \mid B) = \Pr(A)$ and
		$\qquad \Pr(A \cap B) = \Pr(A) \cdot \Pr(B)$
7.4	Probability tree	
	Bayes' formula and trees	$\Pr(E_1 \mid F_1) = \dfrac{\text{Branch product on path leading through } E_1 \text{ to } F_1}{\text{Sum of all branch products on paths leading to } F_1}$

Section	Key Terms	Formulas
7.5	Fundamental Counting Principle	$n(A) \cdot n(B)$
	n-factorial	$n! = n(n-1)\cdots(3)(2)(1)$ and $0! = 1$
	Permutations	$_nP_r = \dfrac{n!}{(n-r)!}$
	Combinations	$_nC_r = \dfrac{n!}{r!(n-r)!}$
7.6	Probability using permutations and combinations	
7.7	Markov chains	
	Vectors	
	Initial probability vector	
	Fixed-probability or steady-state vector	

REVIEW EXERCISES

1. If one ball is drawn from a bag containing nine balls numbered 1 through 9, what is the probability that the ball's number is
   (a) odd?
   (b) divisible by 3?
   (c) odd and divisible by 3?

2. Suppose one ball is drawn from a bag containing nine red balls numbered 1 through 9 and three white balls numbered 10, 11, 12. What is the probability that
   (a) the ball is red?
   (b) the ball is odd-numbered?
   (c) the ball is white and even-numbered?
   (d) the ball is white or odd-numbered?

3. If the probability that an event E occurs is 3/7, what are the odds that
   (a) E will occur?   (b) E will not occur?

4. Suppose that a fair coin is tossed two times. Construct an equiprobable sample space for the experiment and determine each of the following probabilities.
   (a) Pr(0 heads)   (b) Pr(1 head)   (c) Pr(2 heads)

5. Suppose that a fair coin is tossed three times. Construct an equiprobable sample space for the experiment and determine each of the following probabilities.
   (a) Pr(2 heads)   (b) Pr(3 heads)   (c) Pr(1 head)

6. A card is drawn at random from an ordinary deck of 52 playing cards. What is the probability that it is a queen or a jack?

7. A deck of 52 cards is shuffled. A card is drawn, it is replaced, the pack is again shuffled, and a second card is drawn. What is the probability that each card drawn is an ace, king, queen, or jack?

8. If the probability of winning a game is 1/4 and there can be no ties, what is the probability of losing the game?

9. A card is drawn from a deck of 52 playing cards. What is the probability that it is an ace or a 10?

10. A card is drawn at random from a deck of playing cards. What is the probability that it is a king or a red card?

11. A bag contains 4 red balls numbered 1, 2, 3, 4 and 5 white balls numbered 5, 6, 7, 8, 9. A ball is drawn. What is the probability that the ball
    (a) is red and even-numbered?
    (b) is red or even-numbered?
    (c) is white or odd-numbered?

12. A bag contains 4 red balls and 3 black balls. Two balls are drawn at random from the bag without replacement. Find the probability that both balls are red.

13. A box contains 2 red balls and 3 black balls. Two balls are drawn from the box without replacement. Find the probability that the second ball is red, given that the first ball is black.

14. A bag contains 8 white balls, 5 red balls, and 7 black balls. If three balls are drawn at random from the bag, with replacement, what is the probability that the first two balls are red and the third is black?

15. A bag contains 8 white balls, 5 red balls, and 7 black balls. If three balls are drawn at random from the bag, without replacement, what is the probability that the first two balls are red and the third is black?

16. A bag contains 3 red, 5 black, and 8 white balls. Three balls are drawn, without replacement. What is the probability that one of each color is drawn?

17. An urn contains 4 red and 6 white balls. One ball is drawn, it is not replaced, and a second ball is drawn. What is the probability that one ball is white and one is red?

18. Urn I contains 3 red and 4 white balls and urn II contains 5 red and 2 white balls. An urn is selected and a ball is drawn.
    (a) What is the probability that urn I is selected and a red ball is drawn?
    (b) What is the probability that a red ball is selected?
    (c) If a red ball is selected, what is the probability that urn I was selected?

19. Bag A contains 3 red, 6 black, and 5 white balls, and bag B contains 4 red, 5 black, and 7 white balls. A bag is selected at random and a ball is drawn. If the ball is white, what is the probability that bag B was selected?

20. Compute $_6P_2$.   21. Compute $_7C_3$.

22. How many 3-letter sets of initials are possible?

23. How many combinations of 8 things taken 5 at a time are possible?

24. Explain why each given matrix is not a transition matrix for a Markov chain.

    (a) $\begin{bmatrix} 0.2 & 0.7 & 0.1 \\ 0.3 & 0.2 & 0.5 \end{bmatrix}$ (b) $\begin{bmatrix} 0.4 & 0.7 \\ 0.6 & 0.3 \end{bmatrix}$

25. For the initial probability vector $[0.1 \quad 0.9]$ and the transition matrix A, find the next two probability vectors.

$$A = \begin{bmatrix} 0.4 & 0.6 \\ 0.8 & 0.2 \end{bmatrix}$$

26. Find the steady-state vector associated with the transition matrix.

$$\begin{bmatrix} 0.6 & 0.4 \\ 0.1 & 0.9 \end{bmatrix}$$

APPLICATIONS

27. *Senior citizens* In a certain city, 30,000 citizens out of 80,000 are over 50 years of age. What is the probability that a citizen selected at random will be 50 years old or younger?

28. *Newspapers* According to funny2.com, the odds that a person aged 18–29 reads a newspaper regularly are 1 to 3. What is the probability that such a person selected at random reads a newspaper regularly?

29. *United Nations* Of 100 job applicants to the United Nations, 30 speak French, 40 speak German, and 12 speak both French and German. If an applicant is chosen at random, what is the probability that the applicant speaks French or German?

Productivity **Suppose that in a study of leadership style versus industrial productivity, the following data were obtained. Use these empirical data to answer Problems 30–32.**

| | Leadership Style | | | |
Productivity	Demo-cratic	Authori-tarian	Laissez-faire	Total
Low	40	15	40	95
Medium	25	75	10	110
High	25	30	20	75
Total	90	120	70	280

30. Find the probability that a person chosen at random has a democratic style and medium productivity.

31. Find the probability that an individual chosen at random has high productivity or an authoritarian style.

32. Find the probability that an individual chosen at random has an authoritarian style, given that he or she has medium productivity.

33. *Quality control* A product must pass an initial inspection, where the probability that it will be rejected is 0.2. If it passes this inspection, it must also pass a second inspection where the probability that it will be rejected is 0.1. What is the probability that it will pass both inspections?

34. *Color blindness* Sixty men out of 1000 and 3 women out of 1000 are color blind. A person is picked at random from a group containing 10 men and 10 women.
    (a) What is the probability that the person is color blind?
    (b) What is the probability that the person is a man if the person is color blind?

35. *Purchasing* A regional survey found that 70% of all families who indicated an intention to buy a new car bought a new car within 3 months, that 10% of families who did not indicate an intention to buy a new car bought one within 3 months, and that 22% indicated an intention to buy a new car. If a family chosen at random bought a car, find the probability that the family had not previously indicated an intention to buy a car.

36. *Management* A personnel director ranks 4 applicants for a job. How many rankings are possible?

37. *Management* An organization wants to select a president, vice president, secretary, and treasurer. If 8 people are willing to serve and each of them is eligible for any of the offices, in how many different ways can the offices be filled?

38. *Utilities* A utility company sends teams of 4 people each to perform repairs. If it has 12 qualified people, how many different ways can people be assigned to a team?

39. *Committees* An organization wants to select a committee of 4 members from a group of 8 eligible members. How many different committees are possible?

40. *Juries* A jury can be deadlocked if one person disagrees with the rest. There are 12 ways in which a jury can be deadlocked if one person disagrees, because any

one of the 12 jurors could disagree. In how many ways can a jury be deadlocked if
(a) 2 people disagree?
(b) 3 people disagree?

41. *License plates* Because Atlanta was the site of the 1996 Olympics, the state of Georgia released vanity license plates containing the Olympic symbol, and a large number of these plates were sold. Because of the symbol, each plate could contain at most 5 digits and/or letters. How many possible plates could have been produced using from 1 to 5 digits or letters?

42. *Blood types* In a book describing the mass murder of 18 people in northern California, a policewoman was quoted as stating that there are blood groups O, A, B, and AB, positive and negative, blood types M, N, MN, and P, and 8 Rh blood types, so there are 288 unique groups (*Source:* Joseph Harrington and Robert Berger, *Eye of Evil,* St. Martin's Press, 1993). Comment on this conclusion.

43. *Scheduling* A college registrar adds 4 new courses to the list of offerings for the spring semester. If he added the course names in random order at the end of the list, what is the probability that these 4 courses are listed in alphabetical order?

44. *Lottery* Pennsylvania's Daily Number pays 500 to 1 to people who play the winning 3-digit number exactly as it is drawn. However, players can "box" a number so they can win $80 from a $1 bet if the 3 digits they pick come out in any order. What is the probability that a "boxed" number with 3 different digits will be a winner?

45. *Lottery* If a person "boxes" (see Problem 44) a 4-digit number with 4 different digits in Pennsylvania's Pick Four Lottery, what is the probability that the person will win?

46. *Quality control* A supplier has 200 compact discs, of which 10% are known to be defective. If a music store purchases 10 of these discs, what is the probability that
(a)   none of the discs is defective?
(b) 2 of the discs are defective?

47. *Stocks* Mr. Way must sell stocks from 3 of the 5 companies whose stocks he owns so that he can send his children to college. If he chooses the companies at random, what is the probability that the 3 companies will be the 3 with the best future earnings?

48. *Quality control* A sample of 6 fuses is drawn from a lot containing 10 good fuses and 2 defective fuses. Find the probability that the number of defective fuses is
(a) exactly 1.   (b) at least 1.

Income levels **Use the following information for Problems 49 and 50. Suppose people in a certain community are classified as being low-income (L), middle-income (M), or high-income (H). Suppose further that the probabilities for children being in a given state depend on which state their parents were in, according to the following Markov chain transition matrix.**

Income level of parents	Income level of children		
	L	M	H
L	0.15	0.6	0.25
M	0.15	0.35	0.5
H	0	0.2	0.8

49. If the initial distribution probabilities of families in a certain population are [0.7 0.2 0.1], find the distribution for the next three generations.

50. Find the expected long-term probability distribution of families by income level.

---

7   CHAPTER TEST

A bag contains three white balls numbered 1, 2, 3 and four black balls numbered 4, 5, 6, 7. If one ball is drawn at random, find the probability of each event described in Problems 1–3.

1. (a) The ball is odd-numbered.
   (b) The ball is white.
2. (a) The ball is black and even-numbered.
   (b) The ball is white or even-numbered.
3. (a) The ball is red.
   (b) The ball is numbered less than 8.

A bag contains three white balls numbered 1, 2, 3 and four black balls numbered 4, 5, 6, 7. Two balls are drawn without replacement. Find the probability of each event described in Problems 4–9.

4. The sum of the numbers is 7.
5. Both balls are white.

6. (a) The first ball is white, and the second is black.
   (b) A white ball and a black ball are drawn.
7. Both balls are odd-numbered.
8. The second ball is white. (Use a tree.)
9. The first ball is black, given that the second ball is white.
10. A cat hits the letters on a computer keyboard three times. What is the probability that the word RAT appears?
11. In a certain region of the country with a large number of cars, 45% of the cars are from Asian manufacturers, 35% are American, and 20% are European. If three cars are chosen at random, what is the probability that two American cars are chosen?
12. In a group of 20 people, 15 are right-handed, 4 are left-handed, and 1 is ambidextrous. What is the probability that a person selected at random is
(a) left-handed?   (b) ambidextrous?

13. In a group of 20 people, 15 are right-handed, 4 are left-handed, and 1 is ambidextrous. If 2 people are selected at random, without replacement, use a tree to find the following probabilities.
    (a) Both are left-handed.
    (b) One is right-handed, and one is left-handed.
    (c) Two are right-handed.
    (d) Two are ambidextrous.

14. The state of Arizona has a "6 of 42" game in which 6 balls are drawn without replacement from 42 balls numbered 1–42. The order in which the balls are drawn does not matter.
    (a) How many different drawing results are possible?
    (b) What is the probability that a person holding one ticket will win (match all the numbers drawn)?

15. The multistate "5 of 50" game draws 5 balls without replacement from 50 balls numbered 1–50.
    (a) How many different drawing results are possible if order does not matter?
    (b) What is the probability that a person holding one ticket will win (match all the numbers drawn)?

16. Computer chips come from two suppliers, with 80% coming from supplier 1 and 20% coming from supplier 2. Six percent of the chips from supplier 1 are defective, and 8% of the chips from supplier 2 are defective. If a chip is chosen at random, what is the probability that it is defective?

17. A placement test is given by a university to predict student success in a calculus course. On average, 70% of students who take the test pass it, and 87% of those who pass the test also pass the course, whereas 8% of those who fail the test pass the course.
    (a) What is the probability that a student taking the placement test and the calculus course will pass the course?
    (b) If a student passed the course, what is the probability that he or she passed the test?

18. Lactose intolerance affects about 20% of non-Hispanic white Americans and 50% of white Hispanic Americans. Nine percent of Americans are Hispanic whites and 75.6% are non-Hispanic whites (*Source:* Jean Carper, "Eat Smart," *USA Weekend*). If a white American is chosen at random and is lactose-intolerant, what is the probability that he or she is Hispanic?

19. A car rental firm has 350 cars. Seventy of the cars have only defective windshield wipers and 25 have only defective tail lights. Two hundred of the cars have no defects; the remainder have other defects. What is the probability that a car chosen at random
    (a) has defective windshield wipers?
    (b) has defective tail lights?
    (c) does not have defective tail lights?

20. NACO Body Shops has found that 14% of the cars on the road need to be painted and that 3% need major body work and painting. What is the probability that a car selected at random needs major body work if it is known that it needs to be painted?

21. Garage door openers have 10 on-off switches on the opener in the garage and on the remote control used to open the door. The door opens when the remote control is pressed and all the switches on both the opener and the remote agree.
    (a) How many different sequences of the on-off switches are possible on the door opener?
    (b) If the switches on the remote control are set at random, what is the probability that it will open a given garage door?
    (c) Remote controls and the openers in the garage for a given brand of door opener have the controls and openers matched in one of three sequences. If an owner does not change the sequence of switches on her or his remote control and opener after purchase, what is the probability that a different new remote control purchased from this company will open that owner's door?
    (d) What should the owner of a new garage door opener do to protect the contents of his or her garage (and home)?

22. A publishing company has determined that a new edition of an existing mathematics textbook will be readopted by 80% of its current users and will be adopted by 7% of the users of other texts if the text is not changed radically. To determine whether it should change the book radically to attract more sales, it uses Markov chains. Assume that the text in question currently has 25% of its possible market.
    (a) Create the transition matrix for this chain.
    (b) Find the probability vector for the text three editions later and, from that, determine the percent of the market for that future edition.
    (c) Find the steady-state vector for this text to determine what percent of its market this text will have if this policy is continued.

I. Phone Numbers

Primarily because of the rapid growth in the use of cellular phones, the number of available phone numbers had to be increased. This was accomplished by increasing the number of available area codes in the United States. Investigate how this was done by completing the following, remembering that an area code cannot begin with a 0 or a 1.

1. Prior to 1995, all area codes had a 0 or a 1 as the middle digit. How many area codes were available then?
2. Consult nanpa.com to see the one number that is still not used as the middle of an area code.
3. Assuming that all the middle-digit restrictions except the one mentioned in Question 2 were removed in 1995, how many additional area codes became available?
4. What further changes could be made to the area code (other than adding a fourth digit) that would increase the available phone numbers? How many times more numbers would be available?
5. After the area code, every number mentioned on television dramas has 555 as its first three digits. How many phone numbers are available in each area code of TV Land?
6. Find how many area codes Los Angeles has, then determine how many different phone numbers are available in TV Land's Los Angeles.

II. Competition in the Telecommunications Industry

Of the 100,000 persons in a particular city who have cellular telephones, 10,000 use AT&T, 85,000 use Verizon, and 5000 use Sprint. A research firm has surveyed cell users to determine the level of customer satisfaction and to learn what company each expects to be using in a year's time. The results are depicted below.

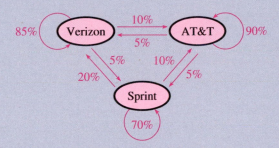

Although Verizon clearly dominates this city's cellular service now, the situation could change in the future.

1. Calculate the percent of the cell users who will be using each company in a year's time.
2. What will the long-term end result be if this pattern of change continues year after year?
3. You should have found, in your answer to Question 1, that the number of Sprint customers increases in a year's time. Explain how this can happen when Sprint loses 30% of its customers and gains only 5% of AT&T's and Verizon's customers.
4. AT&T managers have been presented with two proposals for increasing their market share, and each proposal has the same cost. It is estimated that plan A would increase their customer retention rate from 90% to 95% through an enhanced customer services department. Plan B, an advertising campaign targeting Verizon's customers, is intended to increase the percent of people switching from Verizon to AT&T from 10% to 11%. If the company has sufficient funds to implement one plan only, which one would you advise it to approve and why?
5. Suppose that because of budget constraints, AT&T implements neither plan. A year later it faces the same choice. What should it do then?

Further Topics in Probability; Data Description

Proehl Studios/Corbis

The probability that the birth of three children will result in two boys and a girl can be found efficiently by treating it as a binomial experiment. In this chapter, we discuss how a set of data can be described with descriptive statistics, including mode, median, mean, and standard deviation. Descriptive statistics are used by businesses to summarize data about advertising effectiveness, production costs, and profit. We continue our discussion of probability by considering binomial probability distributions, discrete probability distributions, and normal probability distributions. Social and behavioral scientists collect data about carefully selected samples and use probability distributions to reach conclusions about the populations from which the samples were drawn.

The topics and applications discussed in this chapter include the following.

Prerequisite Problem Type	For Section	Answer	Section for Review
If a fair coin is tossed 3 times, find $\Pr(2$ heads and 1 tail$)$.	**8.1**	$\frac{3}{8}$	7.4 Probability trees
Evaluate: (a) $\binom{4}{3}$ (b) $\binom{5}{2}$ (c) $\binom{24}{0}$	**8.1** **8.3**	 (a) 4 (b) 10 (c) 1	7.5 Counting: Permutations and Combinations
Expand: (a) $(p + q)^2$ (b) $(p + q)^3$	**8.3**	(a) $p^2 + 2pq + q^2$ (b) $p^3 + 3p^2q + 3pq^2 + q^3$	0.5 Special products
Graph all real numbers satisfying: (a) $100 \leq x \leq 115$ (b) $-1 \leq z \leq 2$	**8.4**	 (a) graph on number line: 100, 110, 120 (x-axis) (b) graph on number line: $-3\ -2\ -1\ 0\ 1\ 2\ 3$ (z-axis)	1.1 Linear inequalities
If 68% of the IQ scores of adults lie between 85 and 115, what is the probability that an adult chosen at random will have an IQ score between 85 and 115?	**8.4**	0.68	7.1 Probability

- To solve probability problems related to binomial experiments

Binomial Probability Experiments

| APPLICATION PREVIEW |

It has been determined with probability 0.01 that any child whose mother lives in a specific area near a chemical dump will be born with a birth defect. To find the probability that 2 of 10 children whose mothers live in the area will be born with birth defects, we could use a probability tree that lists all the possibilities and then select the paths that give 2 children with birth defects. However, it is easier to find the probability by using a formula developed in this section. (See Example 3.) This formula can be used when the probability experiment is a binomial probability experiment.

Suppose you have a coin that is biased so that when the coin is tossed the probability of getting a head is 2/3 and the probability of getting a tail is 1/3. What is the probability of getting 2 heads in 3 tosses of this coin?

Suppose the question were "What is the probability that the first two tosses will be heads and the third will be tails?" In that case, the answer would be

$$\Pr(HHT) = \frac{2}{3} \cdot \frac{2}{3} \cdot \frac{1}{3} = \frac{4}{27}$$

However, the original question did not specify the order, so we must consider other orders that will give us 2 heads and 1 tail. We can use a tree to find all the possibilities. See Figure 8.1. We can see that there are 3 paths through the tree that correspond to 2 heads and 1 tail in 3 tosses. In this problem we can find the probability of 2 heads in 3 tosses by finding the probability for any one path of the tree (such as HHT) and then multiplying that probability by the number of paths that result in 2 heads. Notice that the 3 paths in the tree that result in 2 heads correspond to the different orders in which 2 heads and 1 tail can occur.

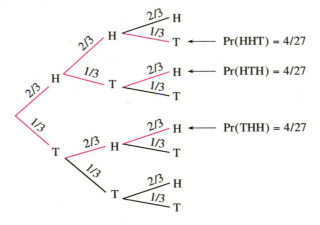

Because the probability for each successful path is 4/27,

$$\Pr(2H \text{ and } 1T) = 3 \cdot \frac{4}{27} = \frac{4}{9}$$

Figure 8.1

This is the number of ways that we can pick 2 positions out of 3 for the heads to occur, which is $_3C_2 = \binom{3}{2}$. Thus the probability of 2 heads resulting from 3 tosses is

$$\Pr(2H \text{ and } 1T) = \binom{3}{2}\frac{2}{3} \cdot \frac{2}{3} \cdot \frac{1}{3} = \binom{3}{2}\left(\frac{2}{3}\right)^2\left(\frac{1}{3}\right)^1 = \frac{4}{9}$$

The experiment just discussed (tossing a coin 3 times) is an example of a general class of experiments called **binomial probability experiments.** These experiments are also called **Bernoulli experiments,** or **Bernoulli trials,** after the 18th-century mathematician Jacob Bernoulli.

Binomial Probability Experiment

A **binomial probability experiment** satisfies the following properties:
1. There are n repeated trials of the experiment.
2. Each trial results in one of two outcomes, with one denoted success (S) and the other failure (F).
3. The trials are independent, with the probability of success, p, the same for every trial. The probability of failure is $q = 1 - p$ for every trial.
4. The probability of x successes in n trials is

$$Pr(x) = \binom{n}{x} p^x q^{n-x}$$

$Pr(x)$ is called a **binomial probability**.

EXAMPLE 1   Binomial Experiment

A die is rolled 4 times and the number of times a 6 results is recorded.

(a) Is the experiment a binomial experiment?
(b) What is the probability that three 6's will result?

Solution
(a) Yes. There are 4 trials, resulting in success (a 6) or failure (not a 6). The result of each roll is independent of previous rolls; the probability of success on each roll is $= 1/6$.
(b) There are $n = 4$ trials with $Pr(\text{rolling a } 6) = p = 1/6$ and $Pr(\text{not rolling a } 6) = q = 5/6$, and we seek the probability of $x = 3$ successes (because we want three 6's). Thus

$$Pr(\text{three 6's}) = \binom{4}{3}\left(\frac{1}{6}\right)^3\left(\frac{5}{6}\right)^1 = \frac{4!}{1!3!}\left(\frac{1}{216}\right)\left(\frac{5}{6}\right) = \frac{5}{324}$$ ∎

CHECKPOINT

Determine whether the following experiments satisfy the conditions for a binomial probability experiment.
1. A coin is tossed 8 times and the number of times a tail occurs is recorded.
2. A die is rolled twice and the sum that occurs is recorded.
2. A ball is drawn from a bag containing 4 white and 7 red balls and then replaced. This is done a total of 6 times, and the number of times a red ball is drawn is recorded.

EXAMPLE 2   Quality Control

A manufacturer of motorcycle parts guarantees that a box of 24 parts will contain at most 1 defective part. If the records show that the manufacturer's machines produce 1% defective parts, what is the probability that a box of parts will satisfy the guarantee?

Solution
The problem is a binomial experiment problem with $n = 24$ and (considering getting a defective part as a success) $p = 0.01$. Then the probability that the manufacturer will have no more than 1 defective part in a box is

$$Pr(x \leq 1) = Pr(0) + Pr(1) = \binom{24}{0}(0.01)^0(0.99)^{24} + \binom{24}{1}(0.01)^1(0.99)^{23}$$

$$= 1(1)(0.7857) + 24(0.01)(0.7936) = 0.9762$$

We interpret this to mean that there is a 97.62% likelihood that there will be no more than 1 defective part in a box of 24 parts. ∎

EXAMPLE **3** **Birth Defects** | **APPLICATION PREVIEW** |

It has been found that the probability is 0.01 that any child whose mother lives in an area near a chemical dump will be born with a birth defect. Suppose that 10 children whose mothers live in the area are born in a given month.

(a) What is the probability that 2 of the 10 will be born with birth defects?
(b) What is the probability that at least 2 of them will have birth defects?

Solution

(a) Note that we may consider a child having a birth defect as a success for this problem even though it is certainly not true in reality. Each of the 10 births is independent, with probability of success $p = 0.01$ and probability of failure $q = 0.99$. Therefore, the experiment is a binomial experiment, and

$$\Pr(2) = \binom{10}{2}(0.01)^2(0.99)^8 = 0.00415$$

(b) The probability of at least 2 of the children having birth defects is

$$1 - [\Pr(0) + \Pr(1)] = 1 - \left[\binom{10}{0}(0.01)^0(0.99)^{10} + \binom{10}{1}(0.01)^1(0.99)^9\right]$$
$$= 1 - [0.90438 + 0.09135] = 0.00427$$

We interpret this to mean that there is a 0.427% likelihood that at least 2 of these children will have birth defects. ■

Technology Note

For details of finding probabilities in a binomial experiment with a calculator or Excel, see Appendices C and D, Section 8.1, and the Online Excel Guide. ■

CHECKPOINT

4. A bag contains 5 red balls and 3 white balls. If 3 balls are drawn, with replacement, from the bag, what is the probability that
(a) two balls will be red?   (b) at least two balls will be red?

CHECKPOINT SOLUTIONS

1. A tail is a success on each trial, and all of the conditions of a binomial probability distribution are satisfied.
2. Each trial in this experiment does not result in a success or failure, so this is not a binomial probability experiment.
3. A red ball is a success on each trial, and each condition of a binomial distribution is satisfied.
4. (a) $\Pr(2 \text{ red}) = \binom{3}{2}\left(\frac{5}{8}\right)^2\left(\frac{3}{8}\right)$

$$= \frac{225}{512}$$

(b) $\Pr(\text{at least 2 red}) = \binom{3}{2}\left(\frac{5}{8}\right)^2\left(\frac{3}{8}\right) + \binom{3}{3}\left(\frac{5}{8}\right)^3\left(\frac{3}{8}\right)^0$

$$= \frac{225}{512} + \frac{125}{512} = \frac{175}{256}$$

| **EXERCISES** | **8.1**

Access end-of-section exercises online at **www.webassign.net**

OBJECTIVES

8.2

- To set up frequency tables and construct frequency histograms for sets of data
- To find the mode of a set of scores (numbers)
- To find the median of a set of scores
- To find the mean of a set of scores
- To find the range of a set of data
- To find the variance and standard deviation of a set of data

Data Description

| APPLICATION PREVIEW |

In modern business, a vast amount of data is collected for use in making decisions about the production, distribution, and sale of merchandise. Businesses also collect and summarize data about advertising effectiveness, production costs, sales, wages, and profits. Behavioral scientists attempt to reach conclusions about general behavioral characteristics by studying the characteristics of small samples of people. For example, election predictions are based on careful sampling of the votes; correct predictions are frequently announced on television even though only 5% of the vote has been counted. Life scientists can use statistical methods with laboratory animals to detect substances that may be dangerous to humans.

The following table gives the numbers of days per year that San Diego, California, failed to meet acceptable air quality standards for the years 1990–2005. We will see in this section how to display these data and how to summarize them by finding the mean and standard deviation of the data. (See Example 8.)

Year	Unacceptable Days	Year	Unacceptable Days	Year	Unacceptable Days
1990	96	1996	31	2002	20
1991	67	1997	14	2003	20
1992	66	1998	33	2004	16
1993	58	1999	16	2005	7
1994	46	2000	31		
1995	48	2001	30		

Source: U.S. Environmental Protection Agency, Office of Air Quality Planning and Standards

Statistical data that have been collected can be visualized with a graph that shows the relationships among various quantities. Statistical graphs include **pie charts** (see Figure 8.2), **bar graphs,** and **frequency histograms.**

How Adults and Adolescents Became Infected with HIV

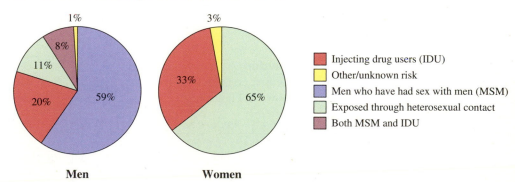

Figure 8.2

Men

Women

- Injecting drug users (IDU)
- Other/unknown risk
- Men who have had sex with men (MSM)
- Exposed through heterosexual contact
- Both MSM and IDU

Source: www.avert.org/statsum.htm (Percents do not add to 100% due to rounding.)

Bar graphs are useful in showing qualitative data as well as quantitative data. Bar graphs are often used to represent frequencies of data in different periods of time. Figure 8.3 is a bar graph that shows the industrywide net income in the property-casualty industry for the years 2002–2007.

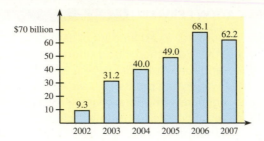

Figure 8.3

Spreadsheet Note Details for constructing bar graphs with Excel can be found in Appendix D, Section 8.2, and the Online Excel Guide.

Data can also be organized in a **frequency table,** which lists data values (individually or grouped in intervals called classes) along with their corresponding frequencies. A **frequency histogram** is a special bar graph that gives a graphical view of a frequency table. It is constructed by putting the data values or data classes along the horizontal axis and putting the frequencies with which they occur along the vertical axis.

EXAMPLE 1 **Frequency Histogram**

Construct a histogram from the frequency table.

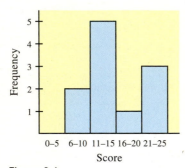

Interval	Frequency
0–5	0
6–10	2
11–15	5
16–20	1
21–25	3

Solution
The histogram is shown in Figure 8.4.

Figure 8.4

EXAMPLE 2 **Frequency Histogram**

Construct a frequency histogram for the following scores: 38, 37, 36, 40, 35, 40, 38, 37, 36, 37, 39, 38.

Solution
We first construct the frequency table and then use the table to construct the histogram shown in Figure 8.5.

Score	Frequency
35	1
36	2
37	3
38	3
39	1
40	2

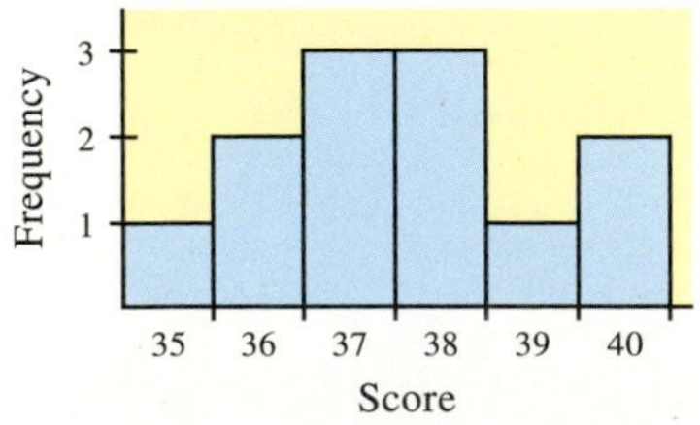

Figure 8.5

Calculator Note Graphing calculators can be used to plot the histogram of a set of data. The width of each bar of the histogram can be determined by the calculator or adjusted by the user when the histogram is plotted. See Appendix C, Section 8.2, for details. ■

Measures of Central Tendency   A set of data can be described by listing all the scores or by drawing a frequency histogram. But we often wish to describe a set of scores by giving the *average score* or the typical score in the set. There are three types of measures that are called *averages,* or measures of central tendency; **mode, median,** and **mean.**

To determine what value represents the most "typical" score in a set of scores, we use the **mode** of the scores.

Mode	The **mode** of a set of scores is the score that occurs most frequently. The mode of a set of scores may not be unique.

That is, the mode is the most popular score—the one that is most likely to occur. The mode can be readily determined from a frequency table or frequency histogram because it is determined according to the frequencies of the scores. The score associated with the highest bar on a histogram is the mode.

EXAMPLE 3   Mode

(a) Find the mode of the following scores: 10, 4, 3, 6, 4, 2, 3, 4, 5, 6, 8, 10, 2, 1, 4, 3.
(b) Determine the mode of the scores shown in the histogram in Figure 8.6.
(c) Find the mode of the following scores: 4, 3, 2, 4, 6, 5, 5, 7, 6, 5, 7, 3, 1, 7, 2.

Solution

(a) Arranging the scores in order gives: 1, 2, 2, 3, 3, 3, 4, 4, 4, 4, 5, 6, 6, 8, 10, 10. The mode is 4 because it occurs more frequently than any other score.
(b) The most frequent score, as the histogram reveals, is 2.

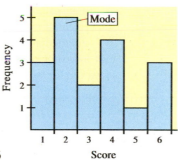

Figure 8.6

(c) Arranging the scores in order gives: 1, 2, 2, 3, 3, 4, 4, 5, 5, 5, 6, 6, 7, 7, 7. Both 5 and 7 occur three times. Thus they are both modes. This set of data is said to be *bimodal* because it has two modes. ■

Although the mode of a set of scores tells us the most frequent score, it does not always represent the typical performance, or central tendency, of the set of scores. For example, the scores 2, 3, 3, 3, 5, 8, 9, 10, 13 have a mode of 3. But 3 is not a good measure of central tendency for this set of scores, for it is nowhere near the middle of the distribution.

One value that gives a better measure of central tendency is the **median.**

Median

The **median** is the score or point above and below which an equal number of the scores lie when the scores are arranged in ascending or descending order.

If there is an odd number of scores, we find the median by ranking the scores from smallest to largest and picking the middle score.

If the number of scores is even, there will be no middle score when the scores are ranked, so the *median* is the point that is the average *of the two middle scores.*

EXAMPLE 4 **Median**

Find the median of each of the following sets of scores.

(a) 2, 3, 16, 5, 15, 38, 18, 17, 12
(b) 3, 2, 6, 8, 12, 4, 3, 2, 1, 6

Solution

(a) We rank the scores from smallest to largest: 2, 3, 5, 12, 15, 16, 17, 18, 38. The median is the middle score, which is 15. Note that there are 4 scores above and 4 scores below 15.
(b) We rank the scores from smallest to largest: 1, 2, 2, 3, 3, 4, 6, 6, 8, 12. The two scores that lie in the middle of the distribution are 3 and 4. Thus the median is a point that is midway between 3 and 4; that is, it is the arithmetic average of 3 and 4. The median is

$$\frac{3+4}{2} = \frac{7}{2} = 3.5$$

In this case we say the median is a point because there is no *score* that is 3.5. ∎

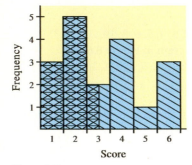

Figure 8.7

The median is a measure of central tendency that is easy to find and that is not influenced by extreme scores. On a histogram, the area to the left of the median is equal to the area to the right of the median. The scores in the histogram shown in Figure 8.7 are 1, 1, 1, 2, 2, 2, 2, 2, 3, 3, 4, 4, 4, 4, 5, 6, 6, 6, and the median is 3. Note that the area of the histogram to the left of 3 equals the area to the right of 3.

Some calculators have keys or functions for finding the median of a set of data. Use of such calculators is valuable when the set of data is large.

The median is the most easily interpreted measure of central tendency, and it is the best indicator of central tendency when the set of scores contains a few extreme values. However, the most frequently used measure of central tendency is the **mean.**

Mean

Suppose x is a variable representing the individual scores in a sample of n scores and Σx represents the sum of the scores. Then the symbol $\bar{x}$ is used to represent the **mean** of a set of scores in a sample and the formula for the mean is

$$\bar{x} = \frac{\text{sum of scores}}{\text{number of scores}} = \frac{\Sigma x}{n}$$

The mean is used most often as a measure of central tendency because it is more useful than the median in the general applications of statistics.

EXAMPLE 5 **Mean**

Find the mean of the following sample of numbers: 12, 8, 7, 10, 6, 14, 7, 6, 12, 9.

Solution

$$\bar{x} = \frac{\Sigma x}{10} = \frac{12 + 8 + 7 + 10 + 6 + 14 + 7 + 6 + 12 + 9}{10} = \frac{91}{10}$$

so $\bar{x} = 9.1$.

Note that the mean need not be one of the numbers (scores) given.   ■

If the data are given in a frequency table or a frequency histogram, we can use a more efficient formula for finding the mean.

Mean of Grouped Data

If a set of n scores is grouped into k classes with x representing the score in a given class and $f(x)$ representing the number of scores in that class, then $n = \Sigma f(x)$ and the mean of the data is given by

$$\bar{x} = \frac{\Sigma(x \cdot f(x))}{\Sigma f(x)}$$

with each value of x used once.

The following table gives a set of data values, x, and their frequencies.

x	1	2	3	4
$f(x)$	3	1	4	2

The mean of the data in this frequency table is

$$\bar{x} = \frac{\Sigma(x \cdot f(x))}{\Sigma f(x)} = \frac{1(3) + 2(1) + 3(4) + 4(2)}{3 + 1 + 4 + 2} = 2.5$$

EXAMPLE 6 **Test Scores**

Find the mean of the following sample of test scores for a math class.

Scores	Class Marks	Frequencies
40–49	44.5	2
50–59	54.5	0
60–69	64.5	6
70–79	74.5	12
80–89	84.5	8
90–99	94.5	2

Solution

For the purpose of finding the mean of interval data, we assume that all scores within an interval are represented by the class mark (midpoint of the interval). Thus

$$\bar{x} = \frac{(44.5)(2) + (54.5)(0) + (64.5)(6) + (74.5)(12) + (84.5)(8) + (94.5)(2)}{30} = 74.5 \ \blacksquare$$

If a set of data is represented by a histogram, the mean of the data is at the point where the histogram could be balanced. The mean of the data in Figure 8.8 is

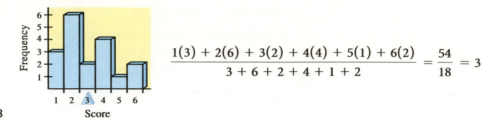

$$\frac{1(3) + 2(6) + 3(2) + 4(4) + 5(1) + 6(2)}{3 + 6 + 2 + 4 + 1 + 2} = \frac{54}{18} = 3$$

Figure 8.8

Any of the three measures of central tendency (mode, median, mean) may be referred to as an average. The measure of central tendency that should be used with data depends on the purpose for which the data were collected. The mode is the number that is used if we want the most popular value. For example, the mode salary for a company is the salary that is received by most of the employees. The median salary for the company is the salary that falls in the middle of the distribution. The mean salary is the arithmetic average of the salaries. It is quite possible that the mode, median, and mean salaries for a company will be different. Figure 8.9 illustrates how the mode, median, and mean salaries might look for a company.

Payroll of Ace Cap Company

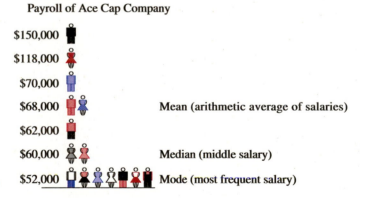

Figure 8.9

Which of the three measures represents the "average" salary? To say the average salary is $68,000 (the mean) helps conceal two things: the bosses' (or owners') large salaries ($150,000 and $118,000), and the laborers' comparatively small salaries ($52,000). The mode tells us more: The most common salary in this company is $52,000. The median gives us more information about these salaries than any other single figure. It tells us that if the salaries are ranked from lowest to highest, then the salary that lies in the middle is $60,000.

Because any of the three measures may be referred to as the average of a set of data, we must be careful to avoid being misled. The owner of the Ace Cap Company will probably state that the average salary for his company is $68,000 because it makes it appear that he pays high salaries. The local union president will likely claim that the mode ($52,000) is the "average" salary. Advertising agencies may also make use of an "average" that will make their product appear in the best light. We run the risk of being misled if we do not know which average is being cited.

One reason we may be misled concerning these three measures is that they frequently fall very close to each other. Also, since the mean is the most frequently used, we tend to associate it with the word *average*.

CHECKPOINT

Consider the data 10, 12, 12, 8, 30, 12, 8, 10, 5, 23.
1. Find the mode.
2. Find the mean.
3. Find the median.

Measures of Dispersion

Although the mean of a set of data is useful in locating the center of the distribution of the data, it doesn't tell us as much about the distribution as we might think at first. For example, a basketball team with five 6-foot players is quite different from a team with one 6-foot, two 5-foot, and two 7-foot players. The distributions of the heights of these teams differ not in the mean height but in how the heights *vary* from the mean.

One measure of how a distribution varies is the **range** of the distribution.

Range

> The **range** of a set of numbers is the difference between the largest and smallest numbers in the set.

Consider the following set of heights of players on a basketball team, in inches: 69, 70, 75, 69, 73, 78, 74, 73, 78, 71. The range of this set of heights is $78 - 69 = 9$ inches. Note that this range is determined by only two numbers and does not give any information about how the other heights vary.

If we calculated the deviation of each score from the mean, $(x - \bar{x})$, and then attempted to average the deviations, we would see that the deviations add to 0. In an effort to find a meaningful measure of dispersion about the mean, statisticians developed the **variance,** which squares the deviations (to make them all positive) and then averages the squared deviations.

Frequently we do not have all the data for a population and must use a sample of these data. To estimate the variance of a population from a sample, statisticians compensate for the fact that there is usually less variability in a sample than in the population itself by summing the squared deviations and dividing them by $n - 1$ rather than averaging them.

To get a measure comparable to the original deviations before they were squared, the **standard deviation,** which is the square root of the variance, was introduced. The formulas for the variance and standard deviation of sample data follow.

Variance and Standard Deviation

$$\text{Sample Variance} \quad s^2 = \frac{\Sigma(x - \bar{x})^2}{n - 1}$$

$$\text{Sample Standard Deviation} \quad s = \sqrt{\frac{\Sigma(x - \bar{x})^2}{n - 1}}$$

The standard deviation of a sample is a measure of the concentration of the scores about their mean. The smaller the standard deviation, the closer the scores lie to the mean. For example, the histograms in Figure 8.10 on the next page describe two sets of data with the same means and the same ranges. From looking at the histograms, we see that more of the data are concentrated about the mean $\bar{x} = 4$ in Figure 8.10(b) than in Figure 8.10(a). In the following example we will see that the standard deviation is smaller for the data in Figure 8.10(b) than in Figure 8.10(a).

EXAMPLE 7   **Standard Deviation**

Figure 8.10(a) is the histogram for the sample data 1, 1, 1, 3, 3, 4, 4, 5, 6, 6, 7, 7. Figure 8.10(b) is the histogram for the data 1, 2, 3, 3, 4, 4, 4, 4, 5, 5, 6, 7. Both sets of data have a range of 6 and a mean of 4. Find the standard deviations of these samples.

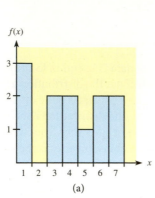

Figure 8.10   (a)   (b)

Solution

For the sample data in Figure 8.10(a), we have

$$\Sigma(x - \bar{x})^2 = (1 - 4)^2 + (1 - 4)^2 + (1 - 4)^2 + (3 - 4)^2 + (3 - 4)^2 + (4 - 4)^2$$
$$+ (4 - 4)^2 + (5 - 4)^2 + (6 - 4)^2 + (6 - 4)^2 + (7 - 4)^2 + (7 - 4)^2$$
$$= 56$$

$$s^2 = \frac{\Sigma(x - \bar{x})^2}{n - 1} = \frac{56}{11} \approx 5.0909 \qquad \text{and} \qquad s = \sqrt{s^2} = \sqrt{5.0909} \approx 2.26$$

For the sample data in Figure 8.10(b), we have

$$\Sigma(x - \bar{x})^2 = (1 - 4)^2 + (2 - 4)^2 + (3 - 4)^2 + (3 - 4)^2 + (4 - 4)^2 + (4 - 4)^2$$
$$+ (4 - 4)^2 + (4 - 4)^2 + (5 - 4)^2 + (5 - 4)^2 + (6 - 4)^2 + (7 - 4)^2$$
$$= (-3)^2 + (-2)^2 + (-1)^2 + (-1)^2 + 0^2 + 0^2 + 0^2 + 0^2 + 1^2 + 1^2$$
$$+ 2^2 + 3^2$$
$$= 30$$

$$s^2 = \frac{30}{11} \approx 2.7273 \qquad \text{and} \qquad s = \sqrt{\frac{30}{11}} \approx 1.65 \qquad \blacksquare$$

Technology Note Most calculators have keys or built-in functions that compute the mean and standard deviation of a set of data. The ease of using calculators to find the standard deviation makes it the preferred method. Excel also has commands for computing means and standard deviations.

For example, by using a calculator, we find that the mean of the heights of the basketball team members mentioned earlier (69, 70, 75, 69, 73, 78, 74, 73, 78, and 71 inches) is $\bar{x} = 73$ inches, and the sample standard deviation of these heights is $s = 3.333$ inches. See Appendices C and D, Section 8.2 and the Online Excel Guide. $\blacksquare$

 EXAMPLE 8   **Air Quality | APPLICATION PREVIEW |**

The table on the next page gives the number of days per year that San Diego, California, failed to meet acceptable air quality standards for the years 1990–2005.

(a) Construct a bar graph to represent the number of unacceptable days in San Diego per year.

(b) Find the mean number of unacceptable days per year for the 16-year period.

(c) What is the standard deviation of the data?

Year	Unacceptable Days	Year	Unacceptable Days	Year	Unacceptable Days
1990	96	1996	31	2002	20
1991	67	1997	14	2003	20
1992	66	1998	33	2004	16
1993	58	1999	16	2005	7
1994	46	2000	31		
1995	48	2001	30		

Source: U.S. Environmental Protection Agency, Office of Air Quality Planning and Standards

Solution

(a) The bar graph displaying the number of unacceptable days per year is shown in Figure 8.11.

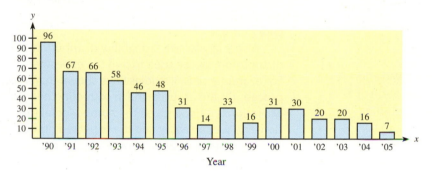

Figure 8.11

(b) The mean number of unacceptable days per year over this 16-year period is

$$\bar{x} = \frac{\sum x}{n} = \frac{599}{16} = 37.4375 \approx 37.4$$

(c) The standard deviation of the number of unacceptable days over this period is

$$s = 24.341922 \approx 24.3$$

This shows that the air quality fluctuated considerably over this period of time. ■

CHECKPOINT

4. Find the standard deviation of the data:

$$10, 12, 12, 8, 30, 12, 8, 10, 5, 23$$

CHECKPOINT SOLUTIONS

1. The mode (most frequent) score is 12.
2. The mean is

$$\frac{10 + 12 + 12 + 8 + 30 + 12 + 8 + 10 + 5 + 23}{10} = 13$$

3. Arranging the 10 scores from smallest to largest gives 5, 8, 8, 10, 10, 12, 12, 12, 23, 30. Thus the median is the average of the fifth and sixth scores.

$$\frac{10 + 12}{2} = 11$$

4. Using technology, the standard deviation is found to be

$$s = 7.63$$

| EXERCISES | 8.2

Access end-of-section exercises online at **www.webassign.net**   **ENHANCED WebAssign**

OBJECTIVES

8.3

- To identify random variables
- To verify that a table or formula describes a discrete probability distribution
- To compute the mean and expected value of a discrete probability distribution
- To make decisions by using expected value
- To find the variance and standard deviation of a discrete probability distribution
- To find the mean and standard deviation of a binomial distribution
- To graph a binomial distribution
- To expand a binomial to a power, using the binomial formula

Discrete Probability Distributions; The Binomial Distribution

| APPLICATION PREVIEW |

The T. J. Cooper Insurance Company insures 100,000 automobiles. Company records over a 5-year period give the following payouts and their probabilities during each 6-month period.

Payment	$500,000	$250,000	$125,000	$25,000	$5000
Probability	0.0001	0.001	0.002	0.008	0.02

The company can determine how much to charge each driver by finding the expected value of the discrete probability distribution determined by this table. (See Example 4.)

A special discrete probability distribution, the binomial probability distribution, can be used to find the expected value for binomial experiments. For example, clinical studies have shown that 5% of photorefractive keratectomy procedures are unsuccessful in correcting nearsightedness. Because each eye operation is independent, the number of failures follows the binomial distribution, and we can easily find the expected number of unsuccessful operations per 1000 operations. (See Example 7.) In this section, we will determine if a table or formula describes a probability distribution, compute the mean and standard deviation of a binomial distributions, and make decisions using expected value.

Discrete Probability Distributions

If a player rolls a die and receives $1 for each dot on the face she rolls, the amount of money won on one roll can be represented by the variable x. If the die is rolled once, the following table gives the possible outcomes of the experiment and their probabilities.

x	1	2	3	4	5	6
Pr(x)	1/6	1/6	1/6	1/6	1/6	1/6

Note that we have used the variable x to denote the possible numerical outcomes of this experiment. Because x results from a probability experiment, we call x a **random variable,** and because there are a finite number of possible values of x, we say that x is a discrete random variable. Whenever all possible values of a random variable can be listed (or counted), the random variable is a **discrete random variable.**

If x represents the possible number of heads that can occur in the toss of four coins, we can use the sample space {0, 1, 2, 3, 4} to list the possible numerical values of x. Table 8.1 shows the values that the random variable of the coin-tossing experiment can assume and how the probability is distributed over these values.

| TABLE 8.1 |

x	Pr(x)
0	1/16
1	1/4
2	3/8
3	1/4
4	1/16

Discrete Probability Distribution

A table, graph, or formula that assigns to each value of a **discrete random variable** x a probability Pr(x) describes a **discrete probability distribution** if the following two conditions hold.

(i) $0 \leq \text{Pr}(x) \leq 1$, for any value of x.
(ii) The sum of all the probabilities is 1. We use Σ to denote "the sum of" and write $\Sigma \text{Pr}(x) = 1$ where the sum is taken over all values of x.

EXAMPLE   1   ## Discrete Probability Distribution

An experiment consists of selecting a ball from a bag containing 15 balls, each with a number 1 through 5. If the probability of selecting a ball with the number x on it is $\Pr(x) = x/15$, where x is an integer and $1 \le x \le 5$, verify that $f(x) = \Pr(x) = x/15$ describes a discrete probability distribution for the random variable x.

Solution
For each integer $x(1 \le x \le 5)$, $\Pr(x)$ satisfies $0 \le \Pr(x) \le 1$.

$$\sum \Pr(x) = \Pr(1) + \Pr(2) + \Pr(3) + \Pr(4) + \Pr(5)$$

$$= \frac{1}{15} + \frac{2}{15} + \frac{3}{15} + \frac{4}{15} + \frac{5}{15} = 1$$

Hence $\Pr(x) = x/15$ describes a discrete probability distribution.   ■

Probability Density Histograms
We can visualize the possible values of a discrete random variable and their associated probabilities by constructing a graph. This graph, called a **probability density histogram,** is designed so that centered over each value of the discrete random variable x along the horizontal axis is a bar having width equal to 1 unit and height (and thus, area) equal to $\Pr(x)$. Figure 8.12 gives the probability density histogram for the experiment described in Example 1.

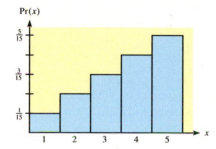

Figure 8.12

Mean and Expected Value
We found the **mean** of a set of n numbers grouped into k classes by using the formula

$$\frac{\sum [x \cdot f(x)]}{n}, \text{ or with } \sum \left[x \cdot \frac{f(x)}{n} \right]$$

We can also find the **theoretical mean** of a probability distribution by using the values x of the random variable, and the probabilities that those values will occur. This mean is also called the **expected value** of x because it gives the average outcome of a probability experiment.

Mean and Expected Value

If x is a discrete random variable with values $x_1, x_2, \ldots, x_n$, then the **mean, μ,** of the distribution is the **expected value of x,** denoted by $E(x)$, and is given by

$$\mu = E(x) = \sum [x \Pr(x)] = x_1 \Pr(x_1) + x_2 \Pr(x_2) + \cdots + x_n \Pr(x_n)$$

EXAMPLE	2	**Testing**

A five-question multiple choice test has 5 possible answers to each question. If a student guesses the answer to each question, the possible numbers of answers she could get correct and the probability of getting each number are given in the table. What is the expected number of answers she will get correct?

x	$Pr(x)$
0	1024/3125
1	256/625
2	128/625
3	32/625
4	4/625
5	1/3125

Solution

The expected value of this distribution is

$$E(x) = 0 \cdot \frac{1024}{3125} + 1 \cdot \frac{256}{625} + 2 \cdot \frac{128}{625} + 3 \cdot \frac{32}{625} + 4 \cdot \frac{4}{625} + 5 \cdot \frac{1}{3125} = 1$$

This means that a student can expect to get 1 answer correct if she guesses at every answer. ■

EXAMPLE	3	**Raffle**

A fire company sells chances to win a new car, with each ticket costing $10. If the car is worth $36,000 and the company sells 6000 tickets, what is the expected value for a person buying one ticket?

Solution

For a person buying one ticket, there are two possible outcomes from the drawing: winning or losing. The probability of winning is 1/6000, and the amount won would be $36,000 − $10 (the cost of the ticket). The probability of losing is 5999/6000, and the amount lost is written as a winning of −$10. Thus the expected value is

$$35,990\left(\frac{1}{6000}\right) + (-10)\left(\frac{5999}{6000}\right) = \frac{35,990}{6000} - \frac{59,990}{6000} = -4$$

Thus, on the average, a person can expect to lose $4 on every ticket. It is not possible to lose exactly $4 on a ticket—the person will either win $36,000 or lose $10; the expected value means that the fire company will make $4 on each ticket it sells if it sells all 6000. ■

Many decisions in business and science are made on the basis of what the outcomes of specific decisions will be, so expected value is very important in decision making and planning.

EXAMPLE	4	**Insurance**	APPLICATION PREVIEW	

The T. J. Cooper Insurance Company insures 100,000 cars. Company records over a 5-year period indicate that during each 6-month period it will pay out the following amounts for accidents.

$500,000 with probability 0.0001	$25,000 with probability 0.008
$250,000 with probability 0.001	$5000 with probability 0.02
$125,000 with probability 0.002	$0 with probability 0.9689

How could the company use the expected payout per car for each 6-month period to help determine its rates?

Solution

The expected value of the company's payments is

$$\$500,000(0.0001) + \$250,000(0.001) + \$125,000(0.002) + \$25,000(0.008)$$
$$+ \$5000(0.02) = \$50 + \$250 + \$250 + \$200 + \$100 = \$850$$

Thus the average premium the company would charge each driver per car for a 6-month period would be

$$\$850 + \text{operating costs} + \text{profit} + \text{reserve for bad years}$$ ■

CHECKPOINT

1. Consider a game in which you roll a die and receive $1 for each spot that occurs.
   (a) What are the expected winnings from this game?
   (b) If you paid $4 to play this game, how much would you lose, on average, each time you played the game?

2. An experiment has the following possible outcomes for the random variable x: 1, 2, 3, 4, 6, 7, 8, 9, 10. The probability that the value x occurs is $x/50$. Find the expected value of x for the experiment.

Measures of Dispersion   Figure 8.13 shows two probability histograms that have the same mean (expected value). However, these histograms look very different. This is because the histograms differ in the **dispersion,** or **variation,** of the values of the random variables.

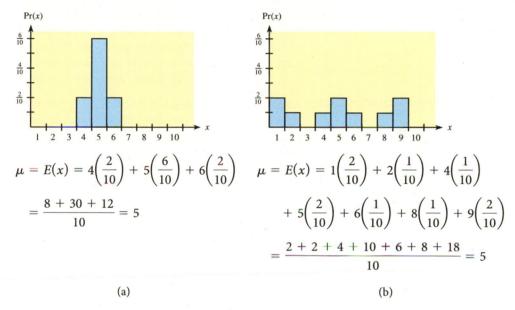

$$\mu = E(x) = 4\left(\frac{2}{10}\right) + 5\left(\frac{6}{10}\right) + 6\left(\frac{2}{10}\right)$$

$$= \frac{8 + 30 + 12}{10} = 5$$

$$\mu = E(x) = 1\left(\frac{2}{10}\right) + 2\left(\frac{1}{10}\right) + 4\left(\frac{1}{10}\right)$$

$$+ 5\left(\frac{2}{10}\right) + 6\left(\frac{1}{10}\right) + 8\left(\frac{1}{10}\right) + 9\left(\frac{2}{10}\right)$$

$$= \frac{2 + 2 + 4 + 10 + 6 + 8 + 18}{10} = 5$$

Figure 8.13   (a)   (b)

Thus we need a way to indicate how distributions with the same mean, such as the ones above, differ. As with sample data, we can find the **variance** of a probability distribution by finding the mean of the *squares* of these deviations. The square root of this measure, called the **standard deviation,** is very useful in describing how the values of x are concentrated about the mean of the probability distribution.

Variance and Standard Deviation

If x is a discrete random variable with values $x_1, x_2, \ldots, x_n$ and mean μ, then the **variance** of the distribution of x is

$$\sigma^2 = \sum (x - \mu)^2 \text{Pr}(x)$$

The **standard deviation** is

$$\sigma = \sqrt{\sigma^2} = \sqrt{\sum (x - \mu)^2 \text{Pr}(x)}$$

EXAMPLE 5 **Standard Deviations of Distributions**

Find the standard deviations of the distributions described in (a) Figure 8.13(a) and (b) Figure 8.13(b).

Solution

(a) $\sigma^2 = \Sigma(x - \mu)^2 \Pr(x)$

$$= (4 - 5)^2 \cdot \frac{2}{10} + (5 - 5)^2 \cdot \frac{6}{10} + (6 - 5)^2 \cdot \frac{2}{10} = 0.4$$

$\sigma = \sqrt{0.4} \approx 0.632$

(b) $\sigma^2 = (1 - 5)^2 \cdot \frac{2}{10} + (2 - 5)^2 \cdot \frac{1}{10} + (4 - 5)^2 \cdot \frac{1}{10} + (5 - 5)^2 \cdot \frac{2}{10}$

$$+ (6 - 5)^2 \cdot \frac{1}{10} + (8 - 5)^2 \cdot \frac{1}{10} + (9 - 5)^2 \cdot \frac{2}{10} = 8.4$$

$\sigma = \sqrt{8.4} \approx 2.898$ ∎

As we can see by referring to Figure 8.13 and to the standard deviations calculated in Example 5, smaller values of σ indicate that the values of x are clustered nearer μ (as is true with sample mean and standard deviation). When σ is large, the values of x are more widely dispersed from μ. As we saw for sample data in Section 8.2, a large standard deviation s indicates that the data are widely dispersed from the sample mean $\bar{x}$.

The Binomial Distribution

A special discrete probability distribution is the **binomial probability distribution**, which describes all possible outcomes of a binomial experiment with their probabilities. Table 8.2 gives the binomial probability distribution for the experiment of tossing a fair coin 6 times, with x equal to the number of heads.

■ | **TABLE 8.2** |

x	$\Pr(x)$	x	$\Pr(x)$
0	$\binom{6}{0}\left(\frac{1}{2}\right)^0\left(\frac{1}{2}\right)^6 = \frac{1}{64}$	4	$\binom{6}{4}\left(\frac{1}{2}\right)^4\left(\frac{1}{2}\right)^2 = \frac{15}{64}$
1	$\binom{6}{1}\left(\frac{1}{2}\right)^1\left(\frac{1}{2}\right)^5 = \frac{6}{64}$	5	$\binom{6}{5}\left(\frac{1}{2}\right)^5\left(\frac{1}{2}\right)^1 = \frac{6}{64}$
2	$\binom{6}{2}\left(\frac{1}{2}\right)^2\left(\frac{1}{2}\right)^4 = \frac{15}{64}$	6	$\binom{6}{6}\left(\frac{1}{2}\right)^6\left(\frac{1}{2}\right)^0 = \frac{1}{64}$
3	$\binom{6}{3}\left(\frac{1}{2}\right)^3\left(\frac{1}{2}\right)^3 = \frac{20}{64}$		

Binomial Probability Distribution

If x is a variable that assumes the values $0, 1, 2, \ldots, r, \ldots, n$ with probabilities

$$\binom{n}{0}p^0q^n, \quad \binom{n}{1}p^1q^{n-1}, \quad \binom{n}{2}p^2q^{n-2}, \quad \ldots, \quad \binom{n}{r}p^rq^{n-r}, \quad \ldots, \quad \binom{n}{n}p^nq^0,$$

respectively, where p is the probability of success and $q = 1 - p$ is the probability of failure, then x is called a **binomial variable**.

The values of x and their corresponding probabilities described above form the **binomial probability distribution**.

Recall that $\binom{n}{r} = {_nC_r}$ can be evaluated easily on a calculator.

CHECKPOINT

3. Show that the values of x and the associated probabilities in Table 8.2 satisfy the conditions for a discrete probability distribution.

Mean and Standard Deviation

The expected value of the number of successes (heads) for the coin toss experiment in Table 8.2 is given by

$$E(x) = 0 \cdot \frac{1}{64} + 1 \cdot \frac{6}{64} + 2 \cdot \frac{15}{64} + 3 \cdot \frac{20}{64} + 4 \cdot \frac{15}{64} + 5 \cdot \frac{6}{64} + 6 \cdot \frac{1}{64} = \frac{192}{64} = 3$$

This expected number of successes seems reasonable. If the probability of success is $1/2$, we would expect to succeed on half of the 6 trials. We could also use the formula for expected value to see that the expected number of heads in 100 tosses is $100(1/2) = 50$. For any binomial distribution the expected number of successes is given by np, where n is the number of trials and p is the probability of success.

Mean of a Binomial Distribution

The theoretical **mean** (or **expected value**) of any binomial distribution is

$$\mu = np$$

where n is the number of trials in the corresponding binomial experiment and p is the probability of success on each trial.

A simple formula can also be developed for the standard deviation of a binomial distribution.

Standard Deviation of a Binomial Distribution

The **standard deviation** of a binomial distribution is

$$\sigma = \sqrt{npq}$$

where n is the number of trials, p is the probability of success on each trial, and $q = 1 - p$.

The standard deviation of the binomial distribution corresponding to the number of heads resulting when a coin is tossed 16 times is

$$\sigma = \sqrt{16 \cdot \frac{1}{2} \cdot \frac{1}{2}} = \sqrt{4} = 2$$

EXAMPLE 6   Nursing Homes

According to the *New England Journal of Medicine,* there is a 43% chance that a person 65 years or older will enter a nursing home sometime in his or her lifetime. Of 100 people in this age group, what is the expected number of people who will eventually enter a nursing home, and what is the standard deviation?

Solution

The number of people who will eventually enter a nursing home follows a binomial distribution, with the probability that any one person in this group will enter a home equal to $p = 0.43$ and the probability that he or she will not equal to $q = 0.57$. Thus the expected number out of 100 that will eventually enter a nursing home is

$$E(x) = np = 100(0.43) = 43$$

The standard deviation of this distribution is

$$\sigma = \sqrt{npq} = \sqrt{(100)(0.43)(0.57)} \approx 4.95$$ ■

EXAMPLE   7   Eye Surgery | APPLICATION PREVIEW |

Clinical studies have shown that 5% of patients operated on for nearsightedness with photorefractive keratectomy (PRK) still need to wear glasses. Find the expected number of unsuccessful operations (patients still wearing glasses) per 1000 operations.

Solution

Each operation is independent, so the number of failures follows the binomial distribution. The probability of an unsuccessful operation is $p = 0.05$ and $n = 1000$. Thus the expected number of unsuccessful operations in 1000 operations is

$$E(x) = \mu = 1000(0.05) = 50$$ ■

CHECKPOINT

4. A fair cube is colored so that 4 faces are green and 2 are red. The cube is tossed 9 times. If success is defined as a red face being up, find
   (a) the mean number of red faces and
   (b) the standard deviation of the number of red faces that occur.

Binomial Formula

The binomial probability distribution is closely related to the powers of a binomial. For example, if a binomial experiment has 3 trials with probability of success p on each trial, then the binomial probability distribution is given in Table 8.3, with the values of x written from 3 to 0. Compare this with the expansion of $(p + q)^3$, which we introduced in Chapter 0, "Algebraic Concepts."

$$(p + q)^3 = p^3 + 3p^2q + 3pq^2 + q^3$$

■ | TABLE 8.3 | ■

x	$Pr(x)$
3	$\binom{3}{3}p^3q^0 = p^3$
2	$\binom{3}{2}p^2q = 3p^2q$
1	$\binom{3}{1}pq^2 = 3pq^2$
0	$\binom{3}{0}p^0q^3 = q^3$

The formula we can use to expand a binomial $(a + b)$ to any positive integer power n is as follows.

Binomial Formula

$$(a + b)^n = \binom{n}{n}a^n + \binom{n}{n-1}a^{n-1}b + \binom{n}{n-2}a^{n-2}b^2 + \cdots$$

$$+ \binom{n}{2}a^2b^{n-2} + \binom{n}{1}ab^{n-1} + \binom{n}{0}b^n$$

EXAMPLE 8   Binomial Formula

Expand $(x + y)^4$.

Solution

$$(x + y)^4 = \binom{4}{4}x^4 + \binom{4}{3}x^3y + \binom{4}{2}x^2y^2 + \binom{4}{1}xy^3 + \binom{4}{0}y^4$$

$$= x^4 + 4x^3y + 6x^2y^2 + 4xy^3 + y^4$$

CHECKPOINT SOLUTIONS

1. (a) $E(x) = \$1\left(\dfrac{1}{6}\right) + \$2\left(\dfrac{1}{6}\right) + \$3\left(\dfrac{1}{6}\right) + \$4\left(\dfrac{1}{6}\right) + \$5\left(\dfrac{1}{6}\right) + \$6\left(\dfrac{1}{6}\right)$

   $= \$3.50$

   (b) $\$4 - \$3.50 = \$.50$. You would lose $\$0.50$, on average, from each play.

2. The expected value of x for this experiment is

$$E(x) = 1 \cdot \frac{1}{50} + 2 \cdot \frac{2}{50} + 3 \cdot \frac{3}{50} + 4 \cdot \frac{4}{50} + 6 \cdot \frac{6}{50} + 7 \cdot \frac{7}{50} + 8 \cdot \frac{8}{50}$$

$$+ 9 \cdot \frac{9}{50} + 10 \cdot \frac{10}{50} = 7.2$$

3. Each value of the random variable x has a probability, with $0 \leq \Pr(x) \leq 1$, and with $\Sigma \Pr(x) = 1$. Thus the conditions of a probability distribution are satisfied.

4. $\Pr(\text{red face}) = \dfrac{1}{3} = p$, so $q = \dfrac{2}{3}$ and $n = 9$.

   (a) $\mu = np = 9\left(\dfrac{1}{3}\right) = 3$ (b) $\sigma = \sqrt{npq} = \sqrt{9\left(\dfrac{1}{3}\right)\left(\dfrac{2}{3}\right)} = \sqrt{2}$

| EXERCISES | 8.3

Access end-of-section exercises online at **www.webassign.net**

OBJECTIVES **8.4**

- To calculate the probability that a random variable following the standard normal distribution has values in a certain interval
- To convert normal distribution values to standard normal values (z-scores)
- To find the probability that normally distributed values lie in a certain interval

Normal Probability Distribution

■ | **APPLICATION PREVIEW** | ■

Discrete probability distributions are important, but they do not apply to many kinds of measurements. For example, the weights of people, the heights of trees, and the IQ scores of college students cannot be measured with whole numbers because each of them can assume any one of an infinite number of values on a measuring scale, and the values cannot be listed or counted, as is possible for a discrete random variable. These measurements follow a special continuous probability distribution called the normal distribution.

IQ scores follow the normal distribution with a mean of 100 and a standard deviation of 15. The graph of this normal distribution is a bell-shaped curve that is symmetric about the mean of 100. We can use this information to find the probability that a person picked at random will have an IQ between 85 and 115. (See Example 1.)

The **normal distribution** is perhaps the most important probability distribution, because so many measurements that occur in nature follow this particular distribution.

Normal Distribution

The **normal distribution** has the following properties.

1. Its graph is a bell-shaped curve like that in Figure 8.14. The graph is called the **normal curve.** It approaches but never touches the horizontal axis as it extends in both directions. The figure shows the percent of scores that lie under the curve and within given intervals.*

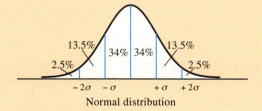

Figure 8.14   Normal distribution

2. The curve is symmetric about a vertical center line that passes through the value that is the mean, the median, *and* the mode of the distribution. That is, the mean, median, and mode are the same for a normal distribution. (That is why they are all called average.)
3. A normal distribution is completely determined when its mean μ and its standard deviation σ are known.
4. Approximately 68% of all scores lie within 1 standard deviation of the mean. Approximately 95% of all scores lie within 2 standard deviations of the mean. More than 99.5% of all scores will lie within 3 standard deviations of the mean.

*Percents shown are approximate.

| EXAMPLE | 1 | IQ Scores | APPLICATION PREVIEW | |

IQ scores follow a normal distribution with mean 100 and standard deviation 15.
(a) What percent of the scores will be
   (i) between 100 and 115?
   (ii) between 85 and 115?
   (iii) between 70 and 130?
   (iv) greater than 130?
(b) Find the probability that a person picked at random has an IQ score
   (i) between 85 and 115.
   (ii) greater than 130.

Solution

(a) Because the mean is 100 and the standard deviation is 15, IQs of 85 and 115 are each 1 standard deviation from 100, and IQs of 70 and 130 are each 2 standard deviations from 100. The approximate percents associated with these values are shown on the graph of a normal distribution in Figure 8.15. We can use this graph to answer the questions.
   (i) 34%   (ii) 68%   (iii) 95%   (iv) 2.5%

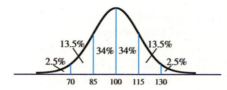

Figure 8.15

(b) The total area under the normal curve is 1. The area under the curve from value x_1 to value x_2 represents the percent of the scores that lie between x_1 and x_2. Thus the percent of the scores between x_1 and x_2 equals the *area* under the curve from x_1 to x_2 and both *represent the probability* that a score chosen at random will lie between x_1 and x_2. Thus,
   (i) the probability that a person chosen at random has an IQ between 85 and 115 is 0.68, and
   (ii) the probability that a person chosen at random has an IQ greater than 130 is 0.025. ■

| EXAMPLE | 2 | **Stock Market Price Swings** |

Big stock market price swings have been common in the 2000s, and the most common measure of volatility is the standard deviation. From 1926 through 2010, stocks have an average gain of 9.7% per year, with a standard deviation of 21.4 percentage points. If the prices are normally distributed and this continues over your investing life, the annual gains on your stocks would range from what low to what high 68% of the time? (*Source: Money,* August 2010)

Solution

Sixty-eight percent of the time the annual return on stocks landed 21.4 percentage points below or above 9.7%. If this continues, your results would range from an 11.7% loss to a 31.1% gain. Thirty-two percent of the time, the gains or losses would be more extreme. ■

CHECKPOINT

1. The mean of a normal distribution is 25 and the standard deviation is 3.
   (a) What percent of the scores will be between 22 and 28?
   (b) What is the probability that a score chosen at random will be between 22 and 28?

We have seen that 34% of the scores lie between 100 and 115 in the normal distribution graph in Figure 8.15. We can write this as

$$\Pr(100 \le x \le 115) = 0.34$$

The mean and standard deviation of the IQ scores are given to be 100 and 15, respectively, so a score of 115 is 15 points above the mean, which is 1 standard deviation above the mean.

z-Scores   Because the probability of obtaining a score from a normal distribution is always related to how many standard deviations the score is away from the mean, it is desirable to convert all scores from a normal distribution to **standard scores,** or **z-scores.** The z-score for any score x is found by determining how many standard deviations x is from the mean μ. The formula for converting scores to z-scores follows.

z-Scores

If σ is the standard deviation of the population data, then the number of standard deviations that x is from the mean μ is given by the **z-score**

$$z = \frac{x - \mu}{\sigma}$$

This formula enables us to convert scores from any normal distribution to a distribution of z-scores with no units of measurement.

The distribution of z-scores will always be a normal distribution with mean 0 and standard deviation 1. This distribution is called the **standard normal distribution.** Figure 8.16 shows the graph of the standard normal distribution, with approximate percents shown. The total area under the curve is 1, with 0.5 on either side of the mean 0.

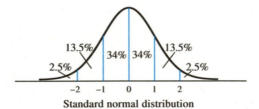

Figure 8.16   Standard normal distribution

By comparing Figure 8.16 with Figure 8.14, we see that each unit from 0 in the standard normal distribution corresponds to 1 standard deviation from the mean of the normal distribution.

We can use a table to determine more accurately the area under the standard normal curve between two z-scores. Appendix B gives the area under the standard normal curve from $z = 0$ to $z = z_0$, for values of z_0 from 0 to 3. As with the normal curve, the area under the curve from $z = 0$ to $z = z_0$ is the probability that a z-score lies between 0 and z_0.

EXAMPLE 3   Standard Normal Distribution

(a) Find the area under the standard normal curve from $z = 0$ to $z = 1.50$.
(b) Find $\Pr(0 \le z \le 1.50)$.

Solution

(a) Looking in the column headed by z in Appendix B, we see 1.50. Across from 1.50 in the column headed by A is 0.4332. Thus the area under the standard normal curve between $z = 0$ and $z = 1.50$ is $A_{1.50} = 0.4432$. (See Figure 8.17.)

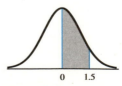

Figure 8.17

(b) The area under the curve from 0 to 1.50 equals the probability that z lies between 0 and 1.50. Thus

$$\Pr(0 \leq z \leq 1.50) = A_{1.50} = 0.4332$$ ■

The following facts, which are direct results of the symmetry of the normal curve about μ, are useful in calculating probabilities using Appendix B.

1. $\Pr(-z_0 \leq z \leq 0) = \Pr(0 \leq z \leq z_0) = A_{z_0}$
2. $\Pr(z \geq 0) = \Pr(z \leq 0) = 0.5$

EXAMPLE 4   Probabilities with z-Scores

Find the following probabilities for the random variable z with standard normal distribution.

(a) $\Pr(-1 \leq z \leq 0)$
(b) $\Pr(-1 \leq z \leq 1.5)$
(c) $\Pr(1 \leq z \leq 1.5)$
(d) $\Pr(z > 2)$
(e) $\Pr(z < 1.35)$

Solution

(a) $\Pr(-1 \leq z \leq 0) = \Pr(0 \leq z \leq 1) = A_1 = 0.3413$
(b) We find $\Pr(-1 \leq z \leq 1.5)$ by using A_1 and $A_{1.5}$ follows:

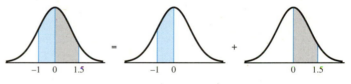

$$
\begin{aligned}
\Pr(-1 \leq z \leq 1.5) &= \Pr(-1 \leq z \leq 0) + \Pr(0 \leq z \leq 1.5) \\
&= \qquad A_1 \qquad + \qquad A_{1.5} \\
&= \qquad 0.3413 \qquad + \qquad 0.4332 \\
&= 0.7745
\end{aligned}
$$

(c) We find $\Pr(1 \leq z \leq 1.5)$ by using A_1 and $A_{1.5}$ follows:

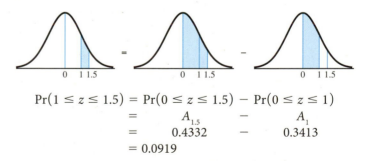

$$
\begin{aligned}
\Pr(1 \leq z \leq 1.5) &= \Pr(0 \leq z \leq 1.5) - \Pr(0 \leq z \leq 1) \\
&= \qquad A_{1.5} \qquad - \qquad A_1 \\
&= \qquad 0.4332 \qquad - \qquad 0.3413 \\
&= 0.0919
\end{aligned}
$$

(d) We find $\Pr(z > 2)$ by using 0.5 and A_2 as follows:

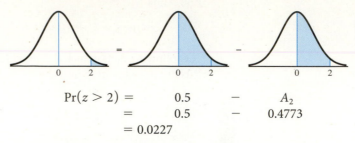

$$\begin{aligned}\Pr(z > 2) &= &0.5 &- &A_2\\ &= &0.5 &- &0.4773\\ &= 0.0227\end{aligned}$$

(e) We find $\Pr(z < 1.35)$ by using 0.5 and $A_{1.35}$ as follows:

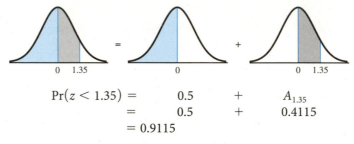

$$\begin{aligned}\Pr(z < 1.35) &= &0.5 &+ &A_{1.35}\\ &= &0.5 &+ &0.4115\\ &= 0.9115\end{aligned}$$

CHECKPOINT

2. For the standard normal distribution, find the following probabilities.
    (a) $\Pr(0 \le z \le 2.5)$
    (b) $\Pr(z > 2.5)$
    (c) $\Pr(z \le 2.5)$

Recall that if a population follows a normal distribution, but with mean and/or standard deviation different from 0 and 1, respectively, we can convert the scores to standard scores, or z-scores, by using the formula

$$z = \frac{x - \mu}{\sigma}$$

Thus

$$\Pr(a \le x \le b) = \Pr(z_a \le z \le z_b)$$

where

$$z_a = \frac{a - \mu}{\sigma} \quad \text{and} \quad z_b = \frac{b - \mu}{\sigma}$$

and we can use Appendix B to find these probabilities. For example, the normal distribution of IQ scores has mean $\mu = 100$ and standard deviation $\sigma = 15$. Thus the z-score for $x = 115$ is

$$z = \frac{115 - 100}{15} = 1$$

A z-score of 1 indicates that 115 is 1 standard deviation above the mean (see Figure 8.18), and

$$\Pr(100 \le x \le 115) = \Pr(0 \le z \le 1) = 0.3413$$

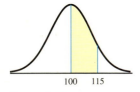

Figure 8.18

EXAMPLE **5**   **Height**

If the mean height of a population of students is $\mu = 68$ inches with standard deviation $\sigma = 3$ inches, what is the probability that a person chosen at random from the population will be between

(a) 68 and 74 inches tall?   (b) 65 and 74 inches tall?

Solution

(a) To find $\Pr(68 \le x \le 74)$, we convert 68 and 74 to z-scores.

$$\text{For 68: } z = \frac{68 - 68}{3} = 0 \quad \text{For 74: } z = \frac{74 - 68}{3} = 2$$

Thus

$$\Pr(68 \le x \le 74) = \Pr(0 \le z \le 2) = A_2 = 0.4773$$

(b) The z-score for 65 is

$$z = \frac{65 - 68}{3} = -1$$

and the z-score for 74 is 2. Thus

$$\Pr(65 \le x \le 74) = \Pr(-1 \le z \le 2)$$
$$= A_1 + A_2$$
$$= 0.3413 + 0.4773 = 0.8186$$

CHECKPOINT

3. For the normal distribution with mean 70 and standard deviation 5, find the following probabilities.
    (a) $\Pr(70 \le x \le 77)$
    (b) $\Pr(68 \le x \le 73)$
    (c) $\Pr(x > 75)$

EXAMPLE **6**   **Demand**

Recessionary pressures have forced an electrical supplies distributor to consider reducing its inventory. It has determined that, over the past year, the daily demand for 48-in. fluorescent light fixtures is normally distributed with a mean of 432 and a standard deviation of 86. If the distributor decides to reduce its daily inventory to 500 units, what percent of the time will its inventory be insufficient to meet the demand? Will this cost the distributor business?

Solution

The distribution is normal, with $\mu = 432$ and $\sigma = 86$. The probability that more than 500 units will be demanded is

$$\Pr(x > 500) = \Pr\left(z > \frac{500 - 432}{86}\right) = \Pr(z > 0.79)$$
$$= 0.5 - A_{0.79} = 0.5 - 0.2852 = 0.2148$$

The distributor will have an insufficient inventory 21.48% of the time. This will probably result in a loss of customers, so the distributor should not decrease the inventory to 500 units per day.

Technology Note   The steps for calculating normal probabilities with a graphing calculator and with Excel, and the steps for graphing a normal distribution with a calculator, are shown in Appendices C and D, Section 8.4, and the Online Excel Guide.

EXAMPLE 7 **Graphs of Normal Distributions**

The function

$$f(x) = \frac{1}{\sigma\sqrt{2\pi}} e^{-(x-\mu)^2/(2\sigma^2)}$$

gives the equation for a normal curve with mean μ and standard deviation σ.

(a) Graph the following normal curves to see how different means yield different distributions.

$$\mu = 0, \sigma = 1, y = \frac{1}{\sqrt{2\pi}} e^{-x^2/2} \quad \text{and}$$

$$\mu = 3, \sigma = 1, y = \frac{1}{\sqrt{2\pi}} e^{-(x-3)^2/2}$$

If the standard deviations are the same, how does increasing the size of the mean affect the graph of the distribution?

(b) To see how normal distributions differ for different standard deviations, graph the normal distributions with

$$\mu = 0, \sigma = 3 \quad \text{and}$$
$$\mu = 0, \sigma = 2$$

What happens to the height of the curve as the standard deviation decreases? Do the data appear to be clustered nearer the mean when the standard deviation is smaller?

Solution

(a) The graphs are shown in Figure 8.19. The graphs are the same size but are centered over different means. A different mean causes the graph to shift to the right or left.

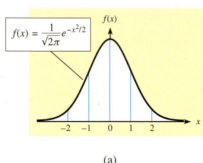

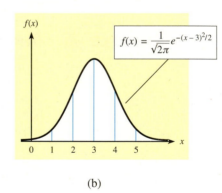

Figure 8.19   (a)   (b)

(b) The graphs are shown in Figure 8.20 on the next page. The height of the graph will be larger if the standard deviation decreases. Yes, the data will be clustered closer to the mean when the standard deviation is smaller.

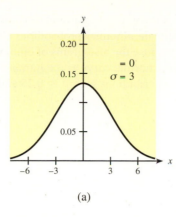

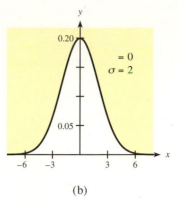

(a)   (b)

Figure 8.20

1. (a) These scores lie within 1 standard deviation of the mean, 25, so 68% lie in this range.
    (b) The percent of the scores in this range is 68%, so the probability that a score chosen at random from this normal distribution will be in this range is 0.68.
2. (a) $\Pr(0 \le z \le 2.5) = A_{2.5} = 0.4938$
    (b) $\Pr(z > 2.5) = 0.5 - A_{2.5} = 0.5 - 0.4938 = 0.0062$
    (c) $\Pr(z \le 2.5) = 0.5 + A_{2.5} = 0.5 + 0.4938 = 0.9938$
3. (a) $\Pr(70 \le x \le 77) = \Pr(0 \le z \le 1.4) = A_{1.4} = 0.4192$
    (b) $\Pr(68 \le x \le 73) = \Pr(-0.4 \le z \le 0.6) = A_{0.4} + A_{0.6}$
        $$= 0.1554 + 0.2258 = 0.3812$$
    (c) $\Pr(x > 75) = \Pr(z > 1) = 0.5 - A_1 = 0.5 - 0.3413 = 0.1587$

| EXERCISES | 8.4

Access end-of-section exercises online at **www.webassign.net**

OBJECTIVE   8.5

- To use a normal approximation to compute a binomial probability

The Normal Curve Approximation to the Binomial Distribution

| APPLICATION PREVIEW |

An insurance company has 20,000 houses insured for fire damage in a certain city. If the probability that any one of these insured houses will have fire damage is 0.003, what is the probability that more than 50 of them will have fire damage? (See Example 2.) In this section we see how to apply normal approximation to solve a problem such as this.

Binomial probability calculations can be tedious when a large number of trials are involved (especially if we do not use technology). For example, to compute the probability requested in the Application Preview using the binomial probability distribution, we would have to compute the sum of 19,950 terms that use $i = 51$ through $i = 20,000$, which we represent by

$$\Pr(x > 50) = \sum_{i=51}^{20,000} \left(_{20,000}C_i\right)(0.003)^i(0.997)^{20,000-i}$$

Even with technology, this computation would be cumbersome. Fortunately, normal curves can be used to approximate binomial probabilities closely if the number of trials is sufficiently large. This is because the histograms of binomial distributions more closely fit the shape of the area under a normal curve as the number of trials increases. To illustrate this, consider the binomial distributions with $p = 0.25$ and $n = 10$, $n = 25$, and $n = 50$. (See Figure 8.21.)

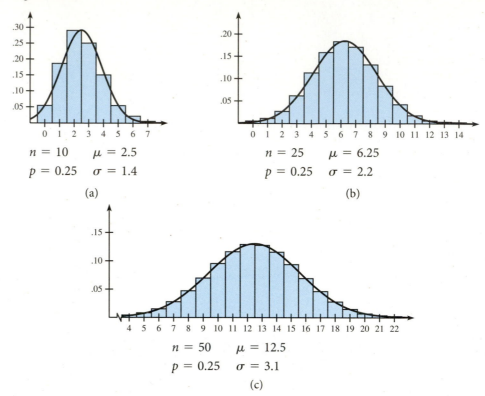

$$n = 10 \qquad \mu = 2.5$$
$$p = 0.25 \qquad \sigma = 1.4$$

(a)

$$n = 25 \qquad \mu = 6.25$$
$$p = 0.25 \qquad \sigma = 2.2$$

(b)

$$n = 50 \qquad \mu = 12.5$$
$$p = 0.25 \qquad \sigma = 3.1$$

(c)

Figure 8.21

Normal distributions can be used to approximate binomial probabilities under the following conditions.

Normal Approximation to the Binomial

For a binomial probability experiment with n trials and with the probability of success on each trial equal to p and probability of failure equal to $q = 1 - p$:

1. If $np \geq 5$ and $nq \geq 5$, then probabilities from a binomial probability distribution can be approximated by using a normal distribution with mean $\mu = np$ and standard deviation $\sigma = \sqrt{npq}$.
2. Because a binomial distribution uses only nonnegative integers and the normal distribution is continuous, the binomial probability that the integer x occurs is approximated by finding the normal probability over the interval $[x - 0.5, x + 0.5]$.

Under these conditions, we can approximate a binomial probability as follows.

Normal Approximation of a Binomial Probability

Procedure	Example
To find a binomial probability with a normal approximation:	Sam can win $5000 if he can roll a die 18 times and get a number divisible by 3 on exactly 12 of the rolls. What is the probability that he will win the $5000?

(continued)

Normal Approximation of a Binomial Probability *(continued)*

Procedure	**Example**
1. Confirm that $np \geq 5$ and $nq \geq 5$, so that probabilities from a binomial probability distribution can be approximated by using a normal distribution.	1. Because $n = 18$ and the probability of rolling a number divisible by 3 is $p = 2/6 = 1/3$, we have $$np = 18 \cdot \frac{1}{3} = 6 \geq 5 \text{ and } nq = 18 \cdot \frac{2}{3} = 12 \geq 5$$
2. Compute the mean $\mu = np$ and the standard deviation $\sigma = \sqrt{npq}$ for use in the normal approximation.	2. $\mu = np = 6$ and $\sigma = \sqrt{npq} = \sqrt{4} = 2$
3. Use the continuity correction to convert the binomial range of values to the corresponding interval of normal distribution values.	3. We find the probability that x is in the continuous interval from $12 - 0.5 = 11.5$ to $12 + 0.05 = 12.5$. We approximate $\Pr(x = 12)$ by finding the normal probability approximation $\Pr(11.5 \leq x \leq 12.5)$.
4. Convert the x-values of the normal distribution to standard distribution z-scores and find the probabilities.	4. Computing the z-scores gives $$z_{11.5} = \frac{x - \mu}{\sigma} = \frac{11.5 - 6}{2} = \frac{5.5}{2} = 2.75 \text{ and}$$ $$z_{12.5} = \frac{12.5 - 6}{2} = \frac{6.5}{2} = 3.25$$ So $$\Pr(2.75 \leq z \leq 3.25) = A_{3.25} - A_{2.75}$$ $$= 0.4994 - 0.4970 = 0.0024$$ Thus the probability is approximately 0.0024.

The probability found above approximates the binomial probability

$$\Pr(x = 12) = {}_{18}C_{12}\left(\frac{1}{3}\right)^{12}\left(\frac{2}{3}\right)^{6} \approx 0.0031$$

The normal approximation would be closer if the number of trials (rolls) were larger. This binomial probability is found directly with relative ease, so the normal approximation is not particularly useful here. But it is very useful and gives an excellent approximation in solving the following problem.

EXAMPLE **1**   ## Centenarians

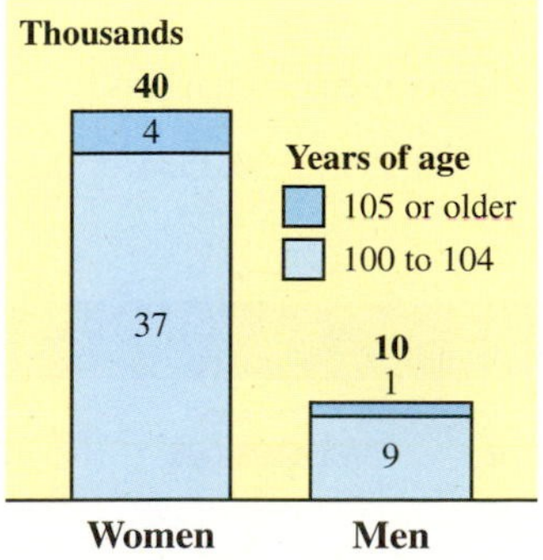

Thousands

Women   **Men**

Note: Components do not add to total because of rounding.

Figure 8.22

Figure 8.22 shows that eighty percent of U.S. centenarians are women (*Source*: U.S. Bureau of the Census, "PCT3 Sex by Age," http://factfinder.census.gov/). In a group of 6000 centenarians belonging to AARP, what is the probability that no more than 4750 will be women?

Solution

The problem is a binomial experiment with $n = 6000$ and $p = 0.80$. Because $np = 6000(0.80) = 4800$ and $nq = 6000(0.20) = 1200$, we can apply the normal approximation using

$$\mu = np = 4800 \quad \text{and} \quad \sigma = \sqrt{npq} = \sqrt{6000(0.80)(0.20)} = 30.984$$

We find the binomial probability $\Pr(x \leq 4750)$ by using the normal approximation $\Pr(x \leq 4750.5)$. The z-score for 4750.5 is

$$z = \frac{4750.5 - 4800}{30.984} \approx -1.60$$

Thus the probability that no more than 4750 of the centenarians will be women is

$$\Pr(x \le 4750.5) = \Pr(z \le -1.60) = 0.5 - A_{1.60} = 0.5 - 0.4452 = 0.0548 \qquad \blacksquare$$

EXAMPLE **2**   **Fire Insurance** | **APPLICATION PREVIEW** |

In a certain city, an insurance company has 20,000 houses insured for fire damage. If the probability that any one of these insured houses will have fire damage is 0.003, what is the probability that more than 50 of them will have fire damage?

Solution

We seek the binomial probability $\Pr(x > 50)$, where $n = 20{,}000$ and $p = 0.003$. First note that $\mu = np = 20{,}000(0.003) = 60$, $nq = 20{,}000(0.997) = 19{,}940$, and $\sigma = \sqrt{20{,}000(0.003)(0.997)} \approx 7.73$. Because np and nq are both significantly greater than 5, we are confident that the normal approximation will be very close to the binomial probability.

In the normal approximation we find $\Pr(x \ge 50.5)$ because this gives an interval with $x > 50$. The z-score for 50.5 is

$$z = \frac{50.5 - 60}{7.73} \approx -1.23$$

and the probability is

$$\Pr(x \ge 50.5) = \Pr(z \ge -1.23) = 0.5 + A_{1.23} = 0.5 + 0.3907 = 0.8907$$

Thus the approximate probability that more than 50 of these houses will have fire damage is 0.89.

$\blacksquare$

EXAMPLE **3**   **Centenarians**

If 0.15% of the U.S. population lives to be 100 years of age (*Source:* U.S. Bureau of the Census), what is the probability that between 8 and 12 (inclusive) of the 3400 citizens of Springfield will live to be centenarians?

Solution

The problem is a binomial experiment with $n = 3400$ and $p = 0.0015$. Because $np = 3400(0.0015) = 5.1$ and $nq = 3400(0.9985) = 3394.9$, we can apply the normal approximation, using

$$\mu = np = 5.1 \qquad \text{and} \qquad \sigma = \sqrt{npq} = \sqrt{3400(0.0015)(0.9985)} \approx 2.257$$

We seek the binomial probability $\Pr(8 \le x \le 12)$ and approximate it with the normal approximation by finding $\Pr(7.5 \le x \le 12.5)$. The z-scores for 7.5 and for 12.5 are

$$z_{7.5} = \frac{7.5 - 5.1}{2.257} \approx 1.06 \qquad \text{and} \qquad z_{12.5} = \frac{12.5 - 5.1}{2.257} \approx 3.28$$

Thus the probability that between 8 and 12 of the citizens will become centenarians is

$$\Pr(7.5 \le x \le 12.5) = \Pr(1.06 \le z \le 3.28) = A_{3.28} - A_{1.06} = 0.4995 - 0.3554 = 0.1441$$

$\blacksquare$

| **EXERCISES** | **8.5**

Access end-of-section exercises online at **www.webassign.net**

KEY TERMS AND FORMULAS

Section	Key Terms	Formulas
8.1	Binomial probability experiment	$\Pr(x) = \binom{n}{x} p^x q^{n-x}$
8.2	Descriptive statistics Frequency histogram Mode Median	
	Mean Range	$\bar{x} = \dfrac{\Sigma x}{n}$
	Sample variance	$s^2 = \dfrac{\Sigma(x - \bar{x})^2}{n - 1}$
	Sample standard deviation	$s = \sqrt{\dfrac{\Sigma(x - \bar{x})^2}{n - 1}}$
8.3	Discrete probability distribution Random variable Probability density histogram Expected value	$E(x) = \Sigma[x\Pr(x)]$
	Mean	$\mu = E(x) = \Sigma[x\Pr(x)]$
	Variance	$\sigma^2 = \Sigma(x - \mu)^2 \Pr(x)$
	Standard deviation Binomial probability distribution	$\sigma = \sqrt{\sigma^2}$
	Mean	$\mu = np$
	Standard deviation	$\sigma = \sqrt{npq}$
	Binomial formula	$(a + b)^n = \binom{n}{n} a^n + \binom{n}{n-1} a^{n-1} b$ $+ \cdots + \binom{n}{1} ab^{n-1} + \binom{n}{0} b^n$
8.4	Normal distribution Standard normal distribution	
	z-scores	$z = \dfrac{x - \mu}{\sigma}$
8.5	Normal approximation to the binomial distribution	$np \geq 5$ and $nq \geq 5$ $\Pr(x)$ uses interval $[x - 0.5, x + 0.5]$

REVIEW EXERCISES

1. If the probability of success on each trial of an experiment is 0.4, what is the probability of 5 successes in 7 trials?
2. A bag contains 5 black balls and 7 white balls. If we draw 4 balls, with each one replaced before the next is drawn, what is the probability that
   (a) 2 balls are black?
   (b) at least 2 balls are black?
3. If a die is rolled 4 times, what is the probability that a number greater than 4 is rolled at least 2 times?

Consider the data in the frequency table, and use these data in Problems 4–7.

Score	Frequency
1	4
2	6
3	8
4	3
5	5

4. Construct a frequency histogram for the data.
5. What is the mode of the data?
6. What is the mean of the data?
7. What is the median of the data?
8. Construct a frequency histogram for the following set of test scores:
   14, 16, 15, 14, 17, 16, 12, 12, 13, 14.
9. What is the median of the scores in Problem 8?
10. What is the mode of the scores in Problem 8?
11. What is the mean of the scores in Problem 8?
12. Find the mean, variance, and standard deviation of the following data: 4, 3, 4, 6, 8, 0, 2.
13. Find the mean, variance, and standard deviation of the following data: 3, 2, 1, 5, 1, 4, 0, 2, 1, 1.
14. For the following probability distribution, find the expected value $E(x)$.

x	Pr(x)
1	0.2
2	0.3
3	0.4
4	0.1

Determine whether each function or table in Problems 15–18 represents a discrete probability distribution.

15. $\Pr(x) = x/15, x = 1, 2, 3, 4, 5$
16. $\Pr(x) = x/55, x = 1, 2, 3, 4, 5, 6, 7, 8, 9$

17.
x	Pr(x)
3	1/6
4	1/3
5	1/3
6	1/6

18.
x	Pr(x)
2	1/2
3	1/4
4	−1/4
5	1/2

19. For the probability distribution shown, find the expected value $E(x)$.

x	1	2	3	4
Pr(x)	0.4	0.3	0.2	0.1

20. For the discrete probability distribution described by $\Pr(x) = x/16, x = 1, 2, 3, 4, 6$, find the
   (a) mean.
   (b) variance.
   (c) standard deviation.

21. For the probability distribution shown, find the
   (a) mean.
   (b) variance.
   (c) standard deviation.

x	Pr(x)
1	1/12
2	1/6
3	1/3
4	5/12

22. A coin has been altered so that the probability that a head will occur is 2/3. If the coin is tossed 6 times, give the mean and standard deviation of the distribution of the number of heads.
23. Suppose a pair of dice is thrown 18 times. How many times would we expect a sum of 7 to occur?
24. Expand $(x + y)^5$.
25. What is the area under the standard normal curve between $z = -1.6$ and $z = 1.9$?
26. If z is a standard normal score, find $\Pr(-1 \le z \le -0.5)$.
27. Find $\Pr(1.23 \le z \le 2.55)$.
28. If a variable x is normally distributed, with $\mu = 25$ and $\sigma = 5$, find $\Pr(25 \le x \le 30)$.
29. If a variable x is normally distributed, with $\mu = 25$ and $\sigma = 5$, find $\Pr(20 \le x \le 30)$.
30. If a variable x is normally distributed, with $\mu = 25$ and $\sigma = 5$, find $\Pr(30 \le x \le 35)$.

For the binomial distributions in Problems 31 and 32, will the normal distribution be a good approximation if

31. $n = 25, p = 0.15$?
32. $n = 15, p = 0.4$?

For the binomial experiments in Problems 33–36, find the probability of

33. 50 successes in 200 trials when $p = 0.2$.
34. at least 43 successes in 60 trials when $p = 0.8$.
35. between 19 and 36 successes (inclusive) in 80 trials when $p = 0.4$.
36. no more than 132 successes in 250 trials when $p = 0.6$.

APPLICATIONS

37. *Genetics* Suppose the probability that a certain couple will have a blond child is 1/4. If they have 6 children, what is the probability that 2 of them will be blond?
38. *Sampling* Suppose 70% of a population opposes a proposal and a sample of size 5 is drawn from the population. What is the probability that the majority of the sample will favor the proposal?
39. *Crime* Suppose 20% of the population are victims of crime. Out of 5 people selected at random, find the probability that 2 are crime victims.

40. *Disease* One person in 100,000 develops a certain disease. Calculate
    (a) Pr(exactly 1 person in 100,000 has the disease).
    (b) Pr(at least 1 person in 100,000 has the disease).

Farm families **The distribution of farm families (as a percent of the population) in a 50-county survey is given in the table. Use these data in Problems 41–43.**

Percent	Number of Counties
10–19	5
20–29	16
30–39	25
40–49	3
50–59	1

41. Make a frequency histogram for these data.
42. Find the mean percent of farm families in these 50 counties.
43. Find the standard deviation of the percent of farm families in these 50 counties.
44. *Cancer testing* The probability that a man with prostate cancer will test positive with the prostate specific antigen (PSA) test is 0.91. If 500 men with prostate cancer are tested, how many would we expect to test positive?
45. *Fraud* A company selling substandard drugs to developing countries sold 2,000,000 capsules with 60,000 of them empty (*Source:* "60 Minutes"). If a person gets 100 randomly chosen capsules from this company, what is the expected number of empty capsules that this person will get?
46. *Racing* Suppose you paid $2 for a bet on a horse and as a result you will win $100 if your horse wins a given race. Find the expected value of the bet if the odds that the horse will win are 3 to 12.
47. *Lotteries* A state lottery pays $500 to anyone who selects the correct 3-digit number from a random drawing. If it costs $1 to play, then the probability distribution for the amount won is given in the table. What are the expected winnings of a person who plays?

x	Pr(x)
−1	0.999
499	0.001

48. *Crime* Refer to Problem 39.
    (a) What is the expected number of people in this sample that are crime victims?
    (b) Find the probability that exactly the expected number of people in this sample are victims of crime.
49. *Testing* Suppose the mean SAT score for students admitted to a university is 1000, with a standard deviation of 200. Suppose that a student is selected at random. If the scores are normally distributed, find the probability that the student's SAT score is
    (a) between 1000 and 1400.
    (b) between 1200 and 1400.
    (c) greater than 1400.
50. *Net worth* Suppose the mean net worth of the residents of Sun City, a retirement community, is $611,000 (*Source: The Island Packet*). If their net worth is normally distributed with a standard deviation of $96,000, what percent of the residents have net worths between $700,000 and $800,000?
51. *Genetics* The ratio of females to males in the U.S. population is 51 to 49. If 100 people are selected at random from the population, find the probability that at least 53 will be female.
52. *Quality control* Records over 10 years indicate that 5% of Acer televisions will have some defective components. If the latest models have the same number of defects, what is the probability that 6 or more of 100 sets will have defective components.
53. *Management* Suppose that a particular Walgreens estimates that it will earn 10% of the printer ink business by offering an ink cartridge refill program. If the estimate is correct, what is the probability that more than 800 of the 7500 persons who need ink will use this Walgreens' program?
54. *Revenue* The Islands Realty Company estimates that 5% of the families who inquire about time-share villas will purchase one. If this is true, what is the probability that no more than 20 of 500 families who inquire about time-share villas will make a purchase?

---

8   CHAPTER TEST

1. If a coin is "loaded" so that the probability that a head will occur on each toss is 1/3, what is the probability that
   (a) 3 heads will result from 5 tosses of the coin?
   (b) at least 3 heads will occur in 5 tosses of the coin?
2. Suppose the coin in Problem 1 is tossed 12 times,
   (a) What is the expected number of heads?
   (b) Find the mean, variance, and standard deviation of the distribution of heads.

3. If x is a random variable, what properties must Pr(x) satisfy in order to be a discrete probability distribution function?
4. Find the expected value of the variable x for the probability distribution given in the table.

x	3	4	5	6	7	8
Pr(x)	0.2	0.3	0.1	0.1	0.2	0.1

5. For the probability distribution shown, find the mean, variance, and standard deviation.

x	10	12	15	18	20	25
Pr(x)	0.1	0.3	0.1	0.2	0.1	0.2

6. For the sample data in the table, find the mean, median, and mode.

Score	Frequency
20	5
21	7
22	3
23	4
24	2

7. If the variable x is normally distributed, with a mean of 16 and a standard deviation of 6, find the following probabilities.
   (a) $\Pr(14 \leq x \leq 22)$
   (b) $\Pr(x \leq 22)$
   (c) $\Pr(22 \leq x \leq 24)$

8. For the normal distribution with mean 70 and standard deviation 12, find:
   (a) $\Pr(73 \leq x \leq 97)$
   (b) $\Pr(65 \leq x \leq 84)$
   (c) $\Pr(x > 84)$

9. Use a normal approximation to a binomial distribution to find the probability that $x \geq 10$ if $n = 50$ and $p = 0.3$.

10. Use a normal approximation to a binomial distribution to find the probability that $x \leq 25$ if $n = 120$ and $p = 0.2$.

The table gives the 2009 ages of householders in primary families. Use this table in Problems 11 and 12.

Age of Householder (years)	Number (thousands)
15–24	3395
25–34	13,347
35–44	17,547
45–54	18,119
55–64	13,462
65+	12,978

Source: U.S. Bureau of the Census

11. Construct a histogram of the data.

12. Use class marks and the given number (frequency) in each class to find the mean and standard deviation of the ages of the householders. Use 74 as the class mark for the 65 + class.

13. The table gives the projected 2020 population of the United States.
    (a) Use the class marks and the given number (frequency) in each class to find the mean age.
    (b) In third world countries where the birth rate is higher and the life span is shorter, which group,

under 30, 30–59, or 60 and over, would be most likely to be the largest? How would this affect a third world country's mean age relative to that of the United States?

Age	Class Mark	Number (millions)
Under 15	7	68.1
15–29	22	67.3
30–44	37	66.8
45–59	52	63.3
60–74	67	53.3
Over 75	82	22.5

Source: U.S. Bureau of the Census

14. The table gives selected 2008 year-end data for cellular phone use in 28 countries worldwide.
    (a) Use class marks and the number of countries (frequency) to find the mean number of cellular telephone subscribers per 100 of population.
    (b) If these data were reexamined in 2012, how would you expect the mean found in (a) might have changed? Explain.

Number of Subscribers (per 100 pop.)	Number of Countries
90–99	8
80–89	5
70–79	4
60–69	3
50–59	4
40–49	4

Source: International Telecommunications Union

15. Suppose that 2% of the computer chips shipped to an assembly plant are defective.
    (a) What is the probability that 10 of 100 chips chosen from the shipment will be defective?
    (b) What is the expected number of defective chips in the shipment if it contains 1500 chips?

The table gives the percents of women who become pregnant during 1 year of use with different types of contraceptives. The "Typical Use" column includes failures from incorrect or inconsistent use. Use this table in Problems 16–18.

Failure Rates Over 1 Year

Contraceptive Type	Correct and Consistent Use	Typical Use
Spermicides	6%	21%
Diaphragm	6	18
Male latex condom	3	12
Unprotected	N/A	85

Source: R. A. Hatcher, J. Trussel, F. Stewart, et al., *Contraceptive Technology*, Irvington Pub.

16. If a group of 30 women use diaphragms correctly and consistently, how many of these 30 women could be expected to become pregnant in a year?

17. If a group of 30 women use diaphragms typically, how many of these 30 women could be expected to become pregnant in a year?

18. If spermicides and male latex condoms are used correctly by a group of 30 couples, how many of the 30 women could be expected to become pregnant in a year? Assume that these uses are independent events.

19. A new light bulb has an expected life of 18,620 hours, with a standard deviation of 4012 hours. If the hours of life of these bulbs are normally distributed, find the probability that one of these bulbs chosen at random will last
    (a) less than 10,000 hours.
    (b) more than 24,000 hours.
    (c) between 15,000 and 21,000 hours.

20. Hargray Communications offered a bundled package of television cable, broadband Internet, and basic telephone service for $79.99 for the first 3 months. If the company expects 1% of the viewers of the ad will purchase the service, what is the probability that at least 120 of 11,000 viewers of the ad will make the purchase?

I. Lotteries

One of the reasons why the Powerball Lottery (offered by 42 states, the U.S. Virgin Islands, and the District of Columbia) is so large is that the odds of winning are so small and the pot continues to grow until someone wins. To win the Powerball Jackpot, a player's 5 game balls and 1 red powerball must match those chosen. Since Powerball began on April 19, 1992, the game has been changed. Each change has lowered the probability of winning by placing more balls in play. See the table, which compares the original and the 2011 powerball games.

Type of Game	Original Powerball	2011 Powerball
Game	5-of-45	5-of-59
Powerballs	1-of-45	1-of-39

Source: powerball.com

1. Use the information in the table to find the probability of winning the original 5-of-45 game and that of winning the 2011 5-of-59 game. Note that the order in which the first five numbers occur does not matter, and the Powerball is not used in this game.

2. Use the information in the table to find the probability of winning the Powerball Jackpot in the original game and in the 2011 game. Note that the order in which the first five numbers occur does not matter. Approximate odds for 2011 Powerball can be found at http://www.powerball.com.

   The minimum payoff for the Jackpot has increased from $5 million to $20 million, and the table gives the payoffs for each game.

Payoffs

Match	Original Game Payout	2011 Game Payout
5 + Powerball	Jackpot	Jackpot
5	$100,000	$200,000
4 + Powerball	5,000	10,000
4	100	100
3 + Powerball	100	100
3	5	7
2 + Powerball	5	7
1 + Powerball	2	4
Powerball	1	3

Source: powerball.com

3. Assume that a share of the Jackpot is $20 million, and find the expected winnings from one ticket in the original game and in the 2011 game. The sponsors of the Powerball game claim that making the probability lower improves the game. How do they justify that claim?

II. Statistics in Medical Research; Hypothesis Testing

Bering Research Laboratories is involved in a national study organized by the National Centers for Disease Control to determine whether a new birth control pill affects the blood pressures of women who are using it. A random sample of 100 women using the pill was selected, and the participants' blood pressures were measured. The mean blood pressure was $\bar{x} = 112.5$. If the blood pressures of all women are normally distributed, with mean $\mu = 110$ and standard deviation $\sigma = 12$, is the mean of this sample sufficiently far from the mean of the population of all women to indicate that the birth control pill affects blood pressure?

To answer this question, researchers at Bering Labs first need to know how the means of samples drawn from a normal population are distributed. As a result of the **Central Limit Theorem,** if all possible samples of size n are drawn from a normal population that has mean μ and standard deviation σ, then the distribution of the means of these samples will also be normally distributed with mean μ but with standard deviation given by

$$\sigma_{\bar{x}} = \frac{\sigma}{\sqrt{n}}$$

To conclude whether women using this birth control pill have blood pressures that are significantly different from those of all women in the population, researchers must decide whether the mean $\bar{x}$ of their sample is so far from the population mean μ that it is unlikely that the sample was chosen from the population.

In this statistical test, the researchers will assume that the sample has been drawn from the population of women with mean $\mu = 110$ unless the blood pressures collected in the sample are so different from those of the population that it is not reasonable to make this assumption. In statistics, we can say that the assumption is unreasonable if the probability that the sample mean, $\bar{x} = 112.5$, could be drawn from this population, with mean 110, is less than 0.05. If this probability is less than 0.05, we say that the sample is drawn from a population with a mean different from 110, which strongly suggests that the women who have taken the birth control pill have, on average, blood pressures different from those of the women who are not using the pill. That is, there is statistically significant evidence that the pill affects the blood pressures of women.

Because 95% of all normally distributed data points lie within 2 standard deviations of the mean, researchers at Bering Labs need to determine whether the sample mean $\bar{x} = 112.5$ is more than 2 standard deviations from $\mu = 110$. To determine how many standard deviations $\bar{x} = 112.5$ is from $\mu = 110$, they compute the z-score for $\bar{x}$ with the formula

$$z_{\bar{x}} = \frac{\bar{x} - \mu}{\sigma/\sqrt{n}}$$

Use this to determine whether there is sufficient evidence that the new birth control pill affects blood pressure.

Derivatives

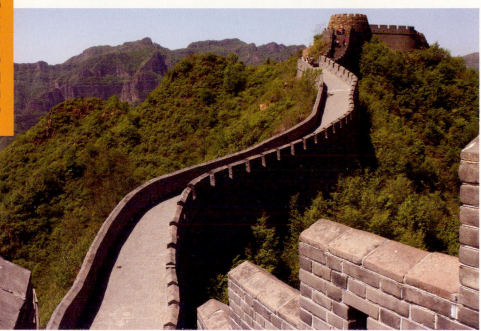

Contax6b/Shutterstock.com

If a firm receives $30,000 in revenue during a 30-day month, its average revenue per day is $30,000/30 = $1000. This does not necessarily mean the actual revenue was $1000 on any one day, just that the average was $1000 per day. Similarly, if a person drove 50 miles in one hour, the average velocity was 50 miles per hour, but the driver could still have received a speeding ticket for traveling 70 miles per hour.

The smaller the time interval, the nearer the average velocity will be to the instantaneous velocity (the speedometer reading). Similarly, changes in revenue over a smaller number of units can give information about the instantaneous rate of change of revenue. The mathematical bridge from average rates of change to instantaneous rates of change is the limit.

This chapter is concerned with *limits* and *rates of change*. We will see that the *derivative* of a function can be used to determine instantaneous rates of change.

The topics and applications discussed in this chapter include the following.

Prerequisite Problem Type	For Section	Answer	Section for Review
If $f(x) = \dfrac{x^2 - x - 6}{x + 2}$, then find (a) $f(-3)$ (b) $f(-2.5)$ (c) $f(-2.1)$ (d) $f(-2)$	9.1–9.9	(a) -6 (b) -5.5 (c) -5.1 (d) undefined	1.2 Function notation
Factor: (a) $x^2 - x - 6$ (b) $x^2 - 4$ (c) $x^2 + 3x + 2$	9.1 9.7	(a) $(x + 2)(x - 3)$ (b) $(x - 2)(x + 2)$ (c) $(x + 1)(x + 2)$	0.6 Factoring
Write as a power: (a) $\sqrt{t}$ (b) $\dfrac{1}{x}$ (c) $\dfrac{1}{\sqrt[3]{x^2 + 1}}$	9.4–9.8	(a) $t^{1/2}$ (b) x^{-1} (c) $(x^2 + 1)^{-1/3}$	0.3, 0.4 Exponents and radicals
Simplify: (a) $\dfrac{4(x + h)^2 - 4x^2}{h}$, if $h \neq 0$ (b) $(2x^3 + 3x + 1)(2x) + (x^2 + 4)(6x^2 + 3)$ (c) $\dfrac{x(3x^2) - x^3(1)}{x^2}$, if $x \neq 0$	9.3 9.5 9.7	(a) $8x + 4h$ (b) $10x^4 + 33x^2 + 2x + 12$ (c) $2x$	0.5 Simplifying algebraic expressions
Simplify: (a) $\dfrac{x^2 - x - 6}{x + 2}$ if $x \neq -2$ (b) $\dfrac{x^2 - 4}{x - 2}$ if $x \neq 2$	9.1 9.7	(a) $x - 3$ (b) $x + 2$	0.7 Simplifying fractions
If $f(x) = 3x^2 + 2x$, find $\dfrac{f(x + h) - f(x)}{h}$, if $h \neq 0$	9.3	$6x + 3h + 2$	1.2 Function notation
Find the slope of the line passing through $(1, 2)$ and $(2, 4)$.	9.3	2	1.3 Slopes
Write the equation of the line passing through $(1, 5)$ with slope 8.	9.3 9.4 9.6	$y = 8x - 3$	1.3 Point-slope equation of a line

OBJECTIVES   **9.1**

- To use graphs and numerical tables to find limits of functions, when they exist
- To find limits of polynomial functions
- To find limits of rational functions

Limits

■ | **APPLICATION PREVIEW** |

Although everyone recognizes the value of eliminating any and all particulate pollution from smokestack emissions of factories, company owners are concerned about the cost of removing this pollution. Suppose that USA Steel has shown that the cost C of removing p percent of the particulate pollution from the emissions at one of its plants is

$$C = C(p) = \frac{7300p}{100 - p}$$

To investigate the cost of removing as much of the pollution as possible, we can evaluate the limit as p (the percent) approaches 100 from values less than 100. (See Example 6.) Using a limit is important in this case, because this function is undefined at $p = 100$ (it is impossible to remove 100% of the pollution).

In various applications we have seen the importance of the slope of a line as a rate of change. In particular, the slope of a linear total cost, total revenue, or profit function for a product tells us the marginals or rates of change of these functions. When these functions are not linear, how do we define marginals (and slope)?

We can get an idea about how to extend the concept of slope (and rate of change) to functions that are not linear. Observe that for many curves, if we take a very close (or "zoom-in") view near a point, the curve appears straight. See Figure 9.1. We can think of the slope of the "straight" line as the slope of the curve. The mathematical process used to obtain this "zoom-in" view is the process of taking limits. The concept of limit is essential to the study of calculus.

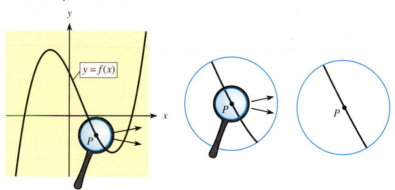

Figure 9.1   When we zoom in near point P, the curve appears straight.

Concept of a Limit   We have used the notation $f(c)$ to indicate the value of a function $f(x)$ at $x = c$. If we need to discuss a value that $f(x)$ approaches as x approaches c, we use the idea of a *limit*. For example, if

$$f(x) = \frac{x^2 - x - 6}{x + 2}$$

then we know that $x = -2$ is not in the domain of $f(x)$, so $f(-2)$ does not exist even though $f(x)$ exists for every value of $x \neq -2$. Figure 9.2 on the next page shows the graph of $y = f(x)$ with an open circle where $x = -2$. The open circle indicates that $f(-2)$ does not exist but shows that points near $x = -2$ have functional values that lie on the line on either side of the open circle. Even though $f(-2)$ is not defined, the figure shows that as x approaches -2 from either side of -2, the graph approaches the open circle at $(-2, -5)$ and the values of $f(x)$ approach -5. Thus -5 is the limit of $f(x)$ as x approaches -2, and we write

$$\lim_{x \to -2} f(x) = -5, \quad \text{or} \quad f(x) \to -5 \quad \text{as} \quad x \to -2$$

| TABLE 9.1 |

Left of –2	
x	$f(x) = \dfrac{x^2 - x - 6}{x + 2}$
–3.000	–6.000
–2.500	–5.500
–2.100	–5.100
–2.010	–5.010
–2.001	–5.001

Right of –2	
x	$f(x) = \dfrac{x^2 - x - 6}{x + 2}$
–1.000	–4.000
–1.500	–4.500
–1.900	–4.900
–1.990	–4.990
–1.999	–4.999

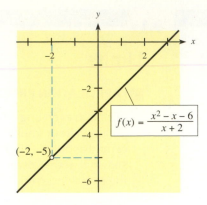

$$f(x) = \frac{x^2 - x - 6}{x + 2}$$

$(-2, -5)$

Figure 9.2

This conclusion is fairly obvious from the graph, but it is not so obvious from the equation for $f(x)$.

We can use the values near $x = -2$ in Table 9.1 to help verify that $f(x) \to -5$ as $x \to -2$. Note that to the left of -2, the values of $f(x)$ get very close to -5 as x gets very close to -2, and to the right of -2, the values of $f(x)$ get very close to -5 as x gets very close to -2. Hence, Table 9.1 and Figure 9.2 indicate that the value of $f(x)$ approaches -5 as x approaches -2 from both sides of $x = -2$.

From our discussion of the graph in Figure 9.2 and Table 9.1, we see that as x approaches -2 from either side of -2, the limit of the function is the value L that the function approaches. This limit L is not necessarily the value of the function at $x = -2$. This leads to our intuitive definition of **limit.**

Limit

Let $f(x)$ be a function defined on an open interval containing c, except perhaps at c. Then

$$\lim_{x \to c} f(x) = L$$

is read "the **limit** of $f(x)$ as x approaches c equals L." The number L exists if we can make values of $f(x)$ as close to L as we desire by choosing values of x sufficiently close to c. When the values of $f(x)$ do not approach a single finite value L as x approaches c, we say the limit does not exist.

As the definition states, a limit as $x \to c$ can exist only if the function approaches a single finite value as x approaches c from both the left and right of c.

EXAMPLE 1 Limits

Figure 9.3 shows three functions for which the limit exists as x approaches 2. Use this figure to find the following.

(a) $\lim\limits_{x \to 2} f(x)$ and $f(2)$ (if it exists)

(b) $\lim\limits_{x \to 2} g(x)$ and $g(2)$ (if it exists)

(c) $\lim\limits_{x \to 2} h(x)$ and $h(2)$ (if it exists)

Figure 9.3

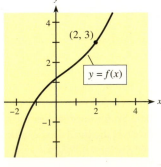

(a)

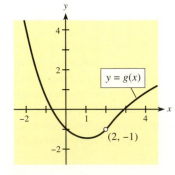

(b)

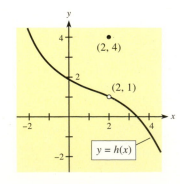

(c)

Solution

(a) From the graph in Figure 9.3(a), we see that as x approaches 2 from both the left and the right, the graph approaches the point $(2, 3)$. Thus $f(x)$ approaches the single value 3. That is,

$$\lim_{x \to 2} f(x) = 3$$

The value of $f(2)$ is the y-coordinate of the point on the graph at $x = 2$. Thus $f(2) = 3$.

(b) Figure 9.3(b) shows that as x approaches 2 from both the left and the right, the graph approaches the open circle at $(2, -1)$. Thus

$$\lim_{x \to 2} g(x) = -1$$

The figure also shows that at $x = 2$ there is no point on the graph. Thus $g(2)$ is undefined.

(c) Figure 9.3(c) shows that

$$\lim_{x \to 2} h(x) = 1$$

The figure also shows that at $x = 2$ there is a point on the graph at $(2, 4)$. Thus $h(2) = 4$, and we see that $\lim_{x \to 2} h(x) \neq h(2)$. ■

As Example 1 shows, the limit of the function as x approaches c may or may not be the same as the value of the function at $x = c$.

In Example 1 we saw that the limit as x approaches 2 meant the limit as x approaches 2 from both the left and the right. We can also consider limits only from the left or only from the right; these are called **one-sided limits**.

One-Sided Limits

Limit from the Right: $\lim_{x \to c^+} f(x) = L$

means the values of $f(x)$ approach the value L as $x \to c$ but $x > c$.

Limit from the Left: $\lim_{x \to c^-} f(x) = M$

means the values of $f(x)$ approach the value M as $x \to c$ but $x < c$.

Note that when one or both one-sided limits fail to exist, then the limit does not exist. Also, when the one-sided limits differ, such as if $L \neq M$ above, then the values of $f(x)$ do not approach a *single* value as x approaches c, and $\lim_{x \to c} f(x)$ does not exist.

EXAMPLE 2   **One-Sided Limits**

Using the functions graphed in Figure 9.4, determine why the limit as $x \to 2$ does not exist for

(a) $f(x)$.

(b) $g(x)$.

(c) $h(x)$.

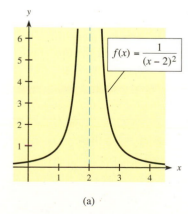

$f(x) = \dfrac{1}{(x-2)^2}$

(a)

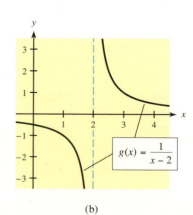

$g(x) = \dfrac{1}{x-2}$

(b)

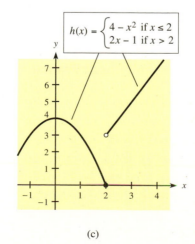

$h(x) = \begin{cases} 4 - x^2 & \text{if } x \leq 2 \\ 2x - 1 & \text{if } x > 2 \end{cases}$

(c)

Figure 9.4

Solution

(a) As $x \to 2$ from the left side and the right side of $x = 2$, $f(x)$ increases without bound, which we denote by saying that $f(x)$ approaches $+\infty$ as $x \to 2$. In this case, $\lim_{x \to 2} f(x)$ does not exist [denoted by $\lim_{x \to 2} f(x)$ DNE] because $f(x)$ does not approach a finite value as $x \to 2$. In this case, we write

$$f(x) \to +\infty \text{ as } x \to 2$$

The graph has a vertical asymptote at $x = 2$.

(b) As $x \to 2$ from the left, $g(x)$ approaches $-\infty$, and as $x \to 2$ from the right, $g(x)$ approaches $+\infty$, so $g(x)$ does not approach a finite value as $x \to 2$. Therefore, the limit does not exist. The graph of $y = g(x)$ has a vertical asymptote at $x = 2$.
In this case we summarize by writing

$$\lim_{x \to 2^-} g(x) \text{ DNE} \quad \text{or} \quad g(x) \to -\infty \text{ as } x \to 2^-$$
$$\lim_{x \to 2^+} g(x) \text{ DNE} \quad \text{or} \quad g(x) \to +\infty \text{ as } x \to 2^+$$

and

$$\lim_{x \to 2} g(x) \text{ DNE}$$

(c) As $x \to 2$ from the left, the graph approaches the point at $(2, 0)$, so $\lim_{x \to 2^-} h(x) = 0$. As $x \to 2$ from the right, the graph approaches the open circle at $(2, 3)$, so $\lim_{x \to 2^+} h(x) = 3$. Because these one-sided limits differ, $\lim_{x \to 2} h(x)$ does not exist. ■

Examples 1 and 2 illustrate the following two important facts regarding limits.

The Limit

1. The limit is said to exist only if the following conditions are satisfied:
   (a) The limit L is a finite value (real number).
   (b) The limit as x approaches c from the left equals the limit as x approaches c from the right. That is, we must have

   $$\lim_{x \to c^-} f(x) = \lim_{x \to c^+} f(x)$$

   Figure 9.4 and Example 2 illustrate cases where $\lim_{x \to c} f(x)$ does not exist.

2. The limit of a function as x approaches c is independent of the value of the function at c. When $\lim_{x \to c} f(x)$ exists, the value of the function at c may be (i) the same as the limit, (ii) undefined, or (iii) defined but different from the limit (see Figure 9.3 and Example 1).

CHECKPOINT

1. Can $\lim_{x \to c^-} f(x)$ exist if $f(c)$ is undefined?
2. Does $\lim_{x \to c} f(x)$ exist if $f(c) = 0$?
3. Does $f(c) = 1$ if $\lim_{x \to c} f(x) = 1$?
4. If $\lim_{x \to c^-} f(x) = 0$, does $\lim_{x \to c^-} f(x)$ exist?

Properties of Limits; Algebraic Evaluation

We have seen that the value of the limit of a function as $x \to c$ will not always be the same as the value of the function at $x = c$. However, there are many functions for which the limit and the functional value agree [see Figure 9.3(a)], and for these functions we can easily evaluate limits.

Properties of Limits

If k is a constant, $\lim_{x \to c} f(x) = L$, and $\lim_{x \to c} g(x) = M$, then the following are true.

I. $\lim_{x \to c} k = k$

II. $\lim_{x \to c} x = c$

III. $\lim_{x \to c} [f(x) \pm g(x)] = L \pm M$

IV. $\lim_{x \to c} [f(x) \cdot g(x)] = LM$

V. $\lim_{x \to c} \dfrac{f(x)}{g(x)} = \dfrac{L}{M}$ if $M \neq 0$

VI. $\lim_{x \to c} \sqrt[n]{f(x)} = \sqrt[n]{\lim_{x \to c} f(x)} = \sqrt[n]{L}$, provided that $L > 0$ when n is even.

If f is a polynomial function, then Properties I–IV imply that $\lim\limits_{x \to c} f(x)$ can be found by evaluating $f(c)$. Moreover, if h is a rational function whose denominator is not zero at $x = c$, then Property V implies that $\lim\limits_{x \to c} h(x)$ can be found by evaluating $h(c)$. The following summarizes these observations and recalls the definitions of polynomial and rational functions.

Function	Definition	Limit
Polynomial function	The function $$f(x) = a_n x^n + a_{n-1} x^{n-1} + \cdots + a_1 x + a_0,$$ where $a_n \neq 0$ and n is a positive integer, is called a **polynomial function** of degree n.	$\lim\limits_{x \to c} f(x) = f(c)$ for all values c (by Properties I–IV)
Rational function	The function $$h(x) = \frac{f(x)}{g(x)}$$ where both $f(x)$ and $g(x)$ are polynomial functions, is called a **rational function**.	$\lim\limits_{x \to c} h(x) = \lim\limits_{x \to c} \frac{f(x)}{g(x)} = \frac{f(c)}{g(c)}$ when $g(c) \neq 0$ (by Property V)

EXAMPLE 3   Limits

Find the following limits, if they exist.

(a) $\lim\limits_{x \to -1} (x^3 - 2x)$ (b) $\lim\limits_{x \to 4} \dfrac{x^2 - 4x}{x - 2}$

Solution

(a) Note that $f(x) = x^3 - 2x$ is a polynomial, so

$$\lim_{x \to -1} f(x) = f(-1) = (-1)^3 - 2(-1) = 1$$

Figure 9.5(a) shows the graph of $f(x) = x^3 - 2x$.

(b) Note that this limit has the form

$$\lim_{x \to c} \frac{f(x)}{g(x)}$$

where $f(x)$ and $g(x)$ are polynomials and $g(c) \neq 0$. Therefore, we have

$$\lim_{x \to 4} \frac{x^2 - 4x}{x - 2} = \frac{4^2 - 4(4)}{4 - 2} = \frac{0}{2} = 0$$

Figure 9.5(b) shows the graph of $g(x) = \dfrac{x^2 - 4x}{x - 2}$.

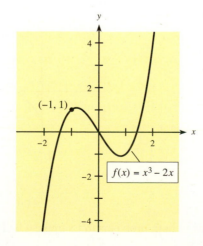

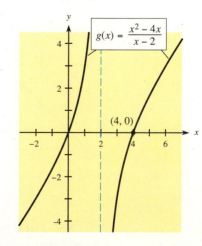

Figure 9.5   (a)   (b)

We have seen that we can use Property V to find the limit of a rational function $f(x)/g(x)$ as long as the denominator is *not* zero. If the limit of the denominator of $f(x)/g(x)$ *is* zero, then there are two possible cases.

Rational Functions: Evaluating Limits of the Form $\lim\limits_{x \to c} \dfrac{f(x)}{g(x)}$ **where** $\lim\limits_{x \to c} g(x) = 0$

Type I. If $\lim\limits_{x \to c} f(x) = 0$ and $\lim\limits_{x \to c} g(x) = 0$, then $\lim\limits_{x \to c} \dfrac{f(x)}{g(x)}$ has the **0/0 indeterminate form** at $x = c$. We can factor $x - c$ from $f(x)$ and $g(x)$, reduce the fraction, and then find the limit of the resulting expression, if it exists.

Type II. If $\lim\limits_{x \to c} f(x) \neq 0$ and $\lim\limits_{x \to c} g(x) = 0$, then $\lim\limits_{x \to c} \dfrac{f(x)}{g(x)}$ does not exist. In this case, the values of $f(x)/g(x)$ become unbounded as x approaches c; the line $x = c$ is a vertical asymptote.

EXAMPLE 4   **0/0 Indeterminate Form**

Evaluate the following limits, if they exist.

(a) $\lim\limits_{x \to 2} \dfrac{x^2 - 4}{x - 2}$

(b) $\lim\limits_{x \to 1} \dfrac{x^2 - 3x + 2}{x^2 - 1}$

Solution

(a) This limit has the 0/0 indeterminate form at $x = 2$ because both the numerator and denominator equal zero when $x = 2$. Thus we can factor $x - 2$ from both the numerator and the denominator and reduce the fraction. (We can divide by $x - 2$ because $x - 2 \neq 0$ while $x \to 2$.)

$$\lim_{x \to 2} \frac{x^2 - 4}{x - 2} = \lim_{x \to 2} \frac{(x - 2)(x + 2)}{x - 2} = \lim_{x \to 2} (x + 2) = 4$$

Figure 9.6(a) shows the graph of $f(x) = (x^2 - 4)/(x - 2)$. Note the open circle at $(2, 4)$.

(b) By substituting 1 for x in $(x^2 - 3x + 2)/(x^2 - 1)$, we see that the expression has the 0/0 indeterminate form at $x = 1$, so $x - 1$ is a factor of both the numerator and the denominator. (We can then reduce the fraction because $x - 1 \neq 0$ while $x \to 1$.)

$$\lim_{x \to 1} \frac{x^2 - 3x + 2}{x^2 - 1} = \lim_{x \to 1} \frac{(x - 1)(x - 2)}{(x - 1)(x + 1)} = \lim_{x \to 1} \frac{x - 2}{x + 1}$$

$$= \frac{1 - 2}{1 + 1} = \frac{-1}{2} \quad \text{(by Property V)}$$

Figure 9.6(b) shows the graph of $g(x) = (x^2 - 3x + 2)/(x^2 - 1)$. Note the open circle at $\left(1, -\tfrac{1}{2}\right)$.   ∎

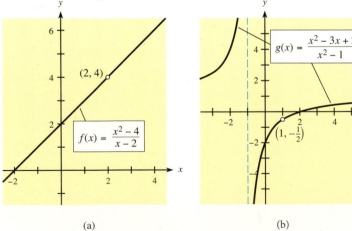

Figure 9.6   (a)   (b)

Note that although both problems in Example 4 had the 0/0 indeterminate form, they had different answers.

EXAMPLE **5**   **Limit with *a*/0 Form**

Find $\lim\limits_{x \to 1} \dfrac{x^2 + 3x + 2}{x - 1}$, if it exists.

Solution

Substituting 1 for *x* in the function results in 6/0, so this limit has the form *a*/0, with $a \neq 0$, and is like the Type II form discussed previously. Hence the limit does not exist. Because the numerator is not zero when $x = 1$, we know that $x - 1$ is *not* a factor of the numerator, and we cannot divide numerator and denominator as we did in Example 4. Table 9.2 confirms that this limit does not exist, because the values of the expression become unbounded as *x* approaches 1.

TABLE 9.2			
Left of *x* = 1		**Right of *x* = 1**	
x	$\dfrac{x^2 + 3x + 2}{x - 1}$	*x*	$\dfrac{x^2 + 3x + 2}{x - 1}$
0	−2	2	12
0.5	−7.5	1.5	17.5
0.7	−15.3	1.2	35.2
0.9	−55.1	1.1	65.1
0.99	−595.01	1.01	605.01
0.999	−5995.001	1.001	6005.001
0.9999	−59,995.0001	1.0001	60,005.0001
$\lim\limits_{x \to 1^-} \dfrac{x^2 + 3x + 2}{x - 1}$ DNE		$\lim\limits_{x \to 1^+} \dfrac{x^2 + 3x + 2}{x - 1}$ DNE	
$(f(x) \to -\infty$ as $x \to 1^-)$		$(f(x) \to +\infty$ as $x \to 1^+)$	

The left-hand and right-hand limits do not exist. Thus $\lim\limits_{x \to 1} \dfrac{x^2 + 3x + 2}{x - 1}$ does not exist. ∎

In Example 5, even though the left-hand and right-hand limits do not exist (see Table 9.2), knowledge that the functional values are unbounded (that is, that they become infinite) is helpful in graphing. The graph is shown in Figure 9.7. We see that $x = 1$ is a vertical asymptote.

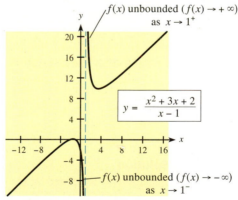

Figure 9.7

EXAMPLE **6**   **Cost-Benefit** | **APPLICATION PREVIEW** |

USA Steel has shown that the cost *C* of removing *p* percent of the particulate pollution from the smokestack emissions at one of its plants is

$$C = C(p) = \frac{7300p}{100 - p}$$

Investigate the cost of removing as much of the pollution as possible.

© iStockphoto.com/Stefan Baum

Solution
First note that the costs of removing 90% and 99% of the pollution are found as follows:

Removing 90%: $C(90) = \dfrac{7300(90)}{100 - 90} = \dfrac{657,000}{10} = 65,700$ (dollars)

Removing 99%: $C(99) = \dfrac{7300(99)}{100 - 99} = \dfrac{722,700}{1} = 722,700$ (dollars)

The cost of removing 100% of the pollution is undefined because the denominator of the function is 0 when $p = 100$. To see what the cost approaches as p approaches 100 from values smaller than 100, we evaluate $\lim\limits_{x \to 100^-} \dfrac{7300p}{100 - p}$. This limit has the Type II form for rational functions. Thus $\dfrac{7300\,p}{100 - p} \to +\infty$ as $x \to 100^-$, which means that as the amount of pollution that is removed approaches 100%, the cost increases without bound. (That is, it is impossible to remove 100% of the pollution.)   ■

CHECKPOINT

5. Evaluate the following limits (if they exist).

(a) $\lim\limits_{x \to -3} \dfrac{2x^2 + 5x - 3}{x^2 - 9}$ (b) $\lim\limits_{x \to 5} \dfrac{x^2 - 3x - 3}{x^2 - 8x + 1}$ (c) $\lim\limits_{x \to -3/4} \dfrac{4x}{4x + 3}$

In Problems 6–9, assume that f, g, and h are polynomials.

6. Does $\lim\limits_{x \to c} f(x) = f(c)$?

7. Does $\lim\limits_{x \to c} \dfrac{g(x)}{h(x)} = \dfrac{g(c)}{h(c)}$?

8. If $g(c) = 0$ and $h(c) = 0$, can we be certain that

(a) $\lim\limits_{x \to c} \dfrac{g(x)}{h(x)} = 0$?   (b) $\lim\limits_{x \to c} \dfrac{g(x)}{h(x)}$ exists?

9. If $g(c) \neq 0$ and $h(c) = 0$, what can be said about $\lim\limits_{x \to c} \dfrac{g(x)}{h(x)}$ and $\lim\limits_{x \to c} \dfrac{h(x)}{g(x)}$?

Limits of Piecewise Defined Functions

As we noted in Section 2.4, "Special Functions and Their Graphs," many applications are modeled by piecewise defined functions. To see how we evaluate a limit involving a piecewise defined function, consider the following example.

EXAMPLE 7   Limits of a Piecewise Defined Function

Find $\lim\limits_{x \to 1^-} f(x)$, $\lim\limits_{x \to 1^+} f(x)$, and $\lim\limits_{x \to 1} f(x)$, if they exist, for

$$f(x) = \begin{cases} x^2 + 1 & \text{if } x \leq 1 \\ x + 2 & \text{if } x > 1 \end{cases}$$

Solution
Because $f(x)$ is defined by $x^2 + 1$ when $x < 1$,

$$\lim\limits_{x \to 1^-} f(x) = \lim\limits_{x \to 1^-} (x^2 + 1) = 2$$

Because $f(x)$ is defined by $x + 2$ when $x > 1$,

$$\lim\limits_{x \to 1^+} f(x) = \lim\limits_{x \to 1^+} (x + 2) = 3$$

And because

$$2 = \lim\limits_{x \to 1^-} f(x) \neq \lim\limits_{x \to 1^+} f(x) = 3$$

$\lim\limits_{x \to 1} f(x)$ does not exist.   ■

| TABLE 9.3 |

Left of 1	
x	$f(x) = x^2 + 1$
0.1	1.01
0.9	1.81
0.99	1.98
0.999	1.998
0.9999	1.9998

Right of 1	
x	$f(x) = x + 2$
1.2	3.2
1.01	3.01
1.001	3.001
1.0001	3.0001
1.00001	3.00001

Table 9.3 and Figure 9.8 show these results numerically and graphically.

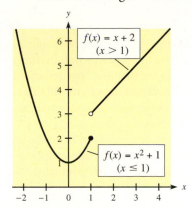

Figure 9.8

Calculator Note We have used graphical, numerical, and algebraic methods to understand and evaluate limits. Graphing calculators can be especially effective when exploring limits graphically or numerically. See Appendix C, Section 9.1, for details. ■

EXAMPLE 8   Limits: Graphically, Numerically, and Algebraically

Consider the following limits.

(a) $\displaystyle\lim_{x \to 5} \frac{x^2 + 2x - 35}{x^2 - 6x + 5}$ (b) $\displaystyle\lim_{x \to -1} \frac{2x}{x + 1}$

Investigate each limit by using the following methods.

(i) Graphically: Graph the function with a graphing calculator and trace near the limiting x-value.

(ii) Numerically: Use the table feature of a graphing calculator to evaluate the function very close to the limiting x-value.

(iii) Algebraically: Use properties of limits and algebraic techniques.

Solution

(a) $\displaystyle\lim_{x \to 5} \frac{x^2 + 2x - 35}{x^2 - 6x + 5}$

(i) Figure 9.9(a) shows the graph of $y = (x^2 + 2x - 35)/(x^2 - 6x + 5)$. Tracing near $x = 5$ shows y-values getting close to 3.

(ii) Figure 9.9(b) shows a table for $y_1 = (x^2 + 2x - 35)/(x^2 - 6x + 5)$ with x-values approaching 5 from both sides (note that the function is undefined at $x = 5$). Again, the y-values approach 3 as x approaches 5 from both sides.

Both (i) and (ii) strongly suggest $\displaystyle\lim_{x \to 5} \frac{x^2 + 2x - 35}{x^2 - 6x + 5} = 3.$

(iii) Algebraic evaluation of this limit confirms what the graph and the table suggest.

$$\lim_{x \to 5} \frac{x^2 + 2x - 35}{x^2 - 6x + 5} = \lim_{x \to 5} \frac{(x + 7)(x - 5)}{(x - 1)(x - 5)} = \lim_{x \to 5} \frac{x + 7}{x - 1} = \frac{12}{4} = 3$$

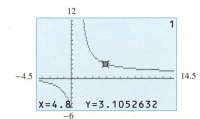

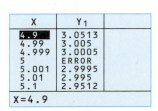

X	Y₁	
4.9	3.0513	
4.99	3.005	
4.999	3.0005	
5	ERROR	
5.001	2.9995	
5.01	2.995	
5.1	2.9512	

X=4.9

Figure 9.9   (a)   (b)

(b) $\displaystyle\lim_{x\to-1}\frac{2x}{x+1}$

(i) Figure 9.10(a) shows the graph of $y = 2x/(x + 1)$; it indicates a break in the graph near $x = -1$. Evaluation confirms that the break occurs at $x = -1$ and also suggests that the function becomes unbounded near $x = -1$. In addition, we can see that as x approaches -1 from opposite sides, the function is headed in different directions. All this suggests that the limit does not exist.

(ii) Figure 9.10(b) shows a graphing calculator table of values for $y_1 = 2x/(x + 1)$ and with x-values approaching $x = -1$ from both sides. The table reinforces our preliminary conclusion from the graph that the limit does not exist, because the function is unbounded near $x = -1$.

(iii) Algebraically we see that this limit has the form $-2/0$. Thus $\displaystyle\lim_{x\to-1}\frac{2x}{x+1}$ DNE.

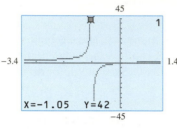

(a)

X	Y₁	
-1.1	22	
-1.01	202	
-1.001	2002	
-1	ERROR	
-.999	-1998	
-.99	-198	
-.9	-18	
X=-.9		

Figure 9.10   (b)   ■

We could also use the graphing and table features of spreadsheets to explore limits.

CHECKPOINT SOLUTIONS

1. Yes. For example, Figure 9.2 and Table 9.1 show that this is possible for $f(x) = \dfrac{x^2 - x - 6}{x + 2}$. Remember that $\displaystyle\lim_{x\to c} f(x)$ does not depend on $f(c)$.

2. Not necessarily. Figure 9.4(c) shows the graph of $y = h(x)$ with $h(2) = 0$, but $\displaystyle\lim_{x\to2} h(x)$ does not exist.

3. Not necessarily. Figure 9.3(c) shows the graph of $y = h(x)$ with $\displaystyle\lim_{x\to2} h(x) = 1$ but $h(2) = 4$.

4. Not necessarily. For example, Figure 9.4(c) shows the graph of $y = h(x)$ with $\displaystyle\lim_{x\to2^-} h(x) = 0$, but with $\displaystyle\lim_{x\to2^+} h(x) = 2$, so the limit does not exist. Recall that if $\displaystyle\lim_{x\to c^-} f(x) = \lim_{x\to c^+} f(x) = L$, then $\displaystyle\lim_{x\to c} f(x) = L$.

5. (a) $\displaystyle\lim_{x\to-3}\frac{2x^2 + 5x - 3}{x^2 - 9} = \lim_{x\to-3}\frac{(2x - 1)(x + 3)}{(x + 3)(x - 3)} = \lim_{x\to-3}\frac{2x - 1}{x - 3} = \frac{-7}{-6} = \frac{7}{6}$

   (b) $\displaystyle\lim_{x\to5}\frac{x^2 - 3x - 3}{x^2 - 8x + 1} = \frac{7}{-14} = -\frac{1}{2}$

   (c) Substituting $x = -3/4$ gives $-3/0$, so $\displaystyle\lim_{x\to-3/4}\frac{4x}{4x + 3}$ does not exist.

6. Yes, Properties I–IV yield this result.

7. Not necessarily. If $h(c) \neq 0$, then this is true. Otherwise, it is not true.

8. For both (a) and (b), $g(x)/h(x)$ has the 0/0 indeterminate form at $x = c$. In this case we can make no general conclusion about the limit. It is possible for the limit to exist (and be zero or nonzero) or not to exist. Consider the following 0/0 indeterminate forms.

   (i) $\lim\limits_{x \to 0} \dfrac{x^2}{x} = \lim\limits_{x \to 0} x = 0$ (ii) $\lim\limits_{x \to 0} \dfrac{x(x + 1)}{x} = \lim\limits_{x \to 0} (x + 1) = 1$

   (iii) $\lim\limits_{x \to 0} \dfrac{x}{x^2} = \lim\limits_{x \to 0} \dfrac{1}{x}$, which does not exist

9. $\lim\limits_{x \to c} \dfrac{g(x)}{h(x)}$ does not exist and $\lim\limits_{x \to c} \dfrac{h(x)}{g(x)} = 0$

| EXERCISES | 9.1

Access end-of-section exercises online at **www.webassign.net**

OBJECTIVES   9.2

- To determine whether a function is continuous or discontinuous at a point
- To determine where a function is discontinuous
- To find limits at infinity and horizontal asymptotes

Continuous Functions; Limits at Infinity

■ | APPLICATION PREVIEW |

Suppose that a friend of yours and her husband have a taxable income of $137,300 and she tells you that she doesn't want to make any more money because that would put them in a higher tax bracket. She makes this statement because the tax rate schedule for married taxpayers filing a joint return (shown in the table) appears to have a jump in taxes for taxable income at $137,300.

Schedule Y-1—If your filing status is Married filing jointly or Qualifying widow(er)

If taxable income is over—	But not over—	The tax is:
$0	$16,750	10% of the amount over $0
$16,750	$68,000	$1,675.00 plus 15% of the amount over $16,750
$68,000	$137,300	$9,362.50 plus 25% of the amount over $68,000
$137,300	$209,250	$26,687.50 plus 28% of the amount over $137,300
$209,250	$373,650	$46,833.50 plus 33% of the amount over $209,250
$373,650	no limit	$101,085.50 plus 35% of the amount over $373,650

Source: Internal Revenue Service

To see whether the couple's taxes would jump to some higher level, we will write the function that gives income tax for married taxpayers as a function of taxable income and show that the function is continuous (see Example 3). That is, we will see that the tax paid does not jump at $137,300 even though the tax on income above $137,300 is collected at a higher rate. In this section, we will show how to determine whether a function is continuous, and we will investigate some different types of discontinuous functions.

Continuous Functions

We have found that $f(c)$ is the same as the limit as $x \to c$ for any polynomial function $f(x)$ and any real number c. Any function for which this special property holds is called a **continuous function**. The graphs of such functions can be drawn without lifting the pencil from the paper, and graphs of others may have holes, vertical asymptotes, or jumps that make it impossible to draw them without lifting the pencil. In general, we define continuity of a function at the value $x = c$ as follows.

Continuity at a Point

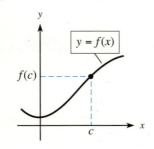

The function f is **continuous at $x = c$** if *all* of the following conditions are satisfied.

1. $f(c)$ exists   2. $\lim_{x \to c} f(x)$ exists   3. $\lim_{x \to c} f(x) = f(c)$

The figure at the left illustrates these three conditions.
If one or more of the conditions above do not hold, we say the function is **discontinuous at $x = c$.**

If a function is discontinuous at one or more points, it is called a **discontinuous function.** Figure 9.11 shows graphs of some functions that are discontinuous at $x = 2$.

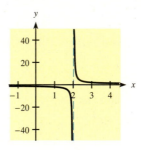

(a) $f(x) = \dfrac{1}{x - 2}$

$\lim_{x \to 2} f(x)$ and $f(2)$ do not exist.

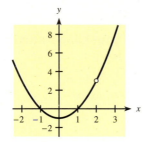

(b) $f(x) = \dfrac{x^3 - 2x^2 - x + 2}{x - 2}$

$f(2)$ does not exist.

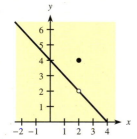

(c) $f(x) = \begin{cases} 4 - x & \text{if } x \neq 2 \\ 4 & \text{if } x = 2 \end{cases}$

$\lim_{x \to 2} f(x) = 2 \neq 4 = f(2)$

Figure 9.11

In the previous section, we saw that if f is a polynomial function, then $\lim_{x \to c} f(x) = f(c)$ for every real number c, and also that $\lim_{x \to c} h(x) = h(c)$ if $h(x) = \dfrac{f(x)}{g(x)}$ is a rational function and $g(c) \neq 0$. Thus, by definition, we have the following.

Polynomial and Rational Functions

Every polynomial function is continuous for all real numbers.
Every rational function is continuous at all values of x except those that make the denominator 0.

EXAMPLE **1**   **Discontinuous Functions**

For what values of x, if any, are the following functions continuous?

(a) $h(x) = \dfrac{3x + 2}{4x - 6}$ (b) $f(x) = \dfrac{x^2 - x - 2}{x^2 - 4}$

Solution

(a) This is a rational function, so it is continuous for all values of x except for those that make the denominator, $4x - 6$, equal to 0. Because $4x - 6 = 0$ at $x = 3/2$, $h(x)$ is continuous for all real numbers except $x = 3/2$. Figure 9.12(a) shows a vertical asymptote at $x = 3/2$.

(b) This is a rational function, so it is continuous everywhere except where the denominator is 0. To find the zeros of the denominator, we factor $x^2 - 4$.

$$f(x) = \frac{x^2 - x - 2}{x^2 - 4} = \frac{x^2 - x - 2}{(x - 2)(x + 2)}$$

Because the denominator is 0 for $x = 2$ and for $x = -2$, $f(2)$ and $f(-2)$ do not exist (recall that division by 0 is undefined). Thus the function is continuous except at $x = 2$ and $x = -2$. The graph of this function (see Figure 9.12(b)) shows a hole at $x = 2$ and a vertical asymptote at $x = -2$. ■

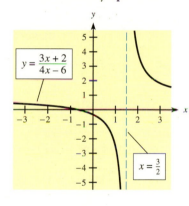
(a)

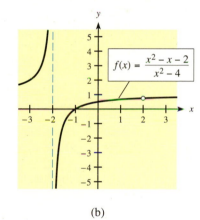
(b)

Figure 9.12

CHECKPOINT

1. Find any x-values where the following functions are discontinuous.

(a) $f(x) = x^3 - 3x + 1$ (b) $g(x) = \dfrac{x^3 - 1}{(x - 1)(x + 2)}$

If the pieces of a piecewise defined function are polynomials, the only values of x where the function might be discontinuous are those at which the definition of the function changes.

EXAMPLE **2**   **Piecewise Defined Functions**

Determine the values of x, if any, for which the following functions are discontinuous.

(a) $g(x) = \begin{cases} (x + 2)^3 + 1 & \text{if } x \le -1 \\ 3 & \text{if } x > -1 \end{cases}$ (b) $f(x) = \begin{cases} 4 - x^2 & \text{if } x < 2 \\ x - 2 & \text{if } x \ge 2 \end{cases}$

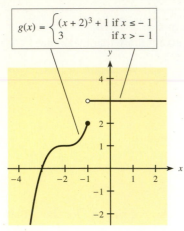

$$g(x) = \begin{cases} (x+2)^3 + 1 & \text{if } x \le -1 \\ 3 & \text{if } x > -1 \end{cases}$$

Figure 9.13

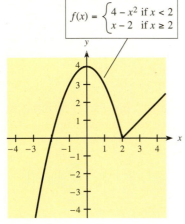

$$f(x) = \begin{cases} 4 - x^2 & \text{if } x < 2 \\ x - 2 & \text{if } x \ge 2 \end{cases}$$

Figure 9.14

Solution

(a) $g(x)$ is a piecewise defined function in which each part is a polynomial. Thus, to see whether a discontinuity exists, we need only check the value of x for which the definition of the function changes—that is, at $x = -1$. Note that $x = -1$ satisfies $x \le -1$, so $g(-1) = (-1 + 2)^3 + 1 = 2$. Because $g(x)$ is defined differently for $x < -1$ and $x > -1$, we use left- and right-hand limits.

For $x \to -1^-$, we know that $x < -1$, so $g(x) = (x + 2)^3 + 1$:

$$\lim_{x \to -1^-} g(x) = \lim_{x \to -1^-} [(x + 2)^3 + 1] = (-1 + 2)^3 + 1 = 2$$

Similarly, for $x \to -1^+$, we know that $x > -1$, so $g(x) = 3$:

$$\lim_{x \to -1^+} g(x) = \lim_{x \to -1^+} 3 = 3$$

Because the left- and right-hand limits differ, $\lim_{x \to -1} g(x)$ does not exist, so $g(x)$ is discontinuous at $x = -1$. This result is confirmed by examining the graph of g, shown in Figure 9.13.

(b) As with $g(x)$, $f(x)$ is continuous everywhere except perhaps at $x = 2$, where the definition of $f(x)$ changes. Because $x = 2$ satisfies $x \ge 2$, $f(2) = 2 - 2 = 0$. The left- and right-hand limits are

Left: $\lim_{x \to 2^-} f(x) = \lim_{x \to 2^-} (4 - x^2) = 4 - 2^2 = 0$

Right: $\lim_{x \to 2^+} f(x) = \lim_{x \to 2^+} (x - 2) = 2 - 2 = 0$

Because the right- and left-hand limits are equal, we conclude that $\lim_{x \to 2} f(x) = 0$. The limit is equal to the functional value, or:

$$\lim_{x \to 2} f(x) = f(2)$$

so we conclude that f is continuous at $x = 2$ and thus f is continuous for all values of x. This result is confirmed by the graph of f, shown in Figure 9.14. ∎

EXAMPLE 3   Taxes | APPLICATION PREVIEW |

The tax rate schedule for married taxpayers filing a joint return (shown in the table) appears to have a jump in taxes for taxable income at $137,300.

Schedule Y-1—If your filing status is Married filing jointly or Qualifying widow(er)

If taxable income is over—	But not over—	The tax is:
$0	$16,750	10% of the amount over $0
$16,750	$68,000	$1,675.00 plus 15% of the amount over $16,750
$68,000	$137,300	$9,362.50 plus 25% of the amount over $68,000
$137,300	$209,250	$26,687.50 plus 28% of the amount over $137,300
$209,250	$373,650	$46,833.50 plus 33% of the amount over $209,250
$373,650	no limit	$101,085.50 plus 35% of the amount over $373,650

Source: Internal Revenue Service

(a) Use the table and write the function that gives income tax for married taxpayers as a function of taxable income, x.

(b) Is the function in part (a) continuous at $x = 137{,}300$?

(c) A married friend of yours and her husband have a taxable income of $137,300, and she tells you that she doesn't want to make any more money because doing so would put her in a higher tax bracket. What would you tell her to do if she is offered a raise?

Solution

(a) The function that gives the tax due for married taxpayers is

$$T(x) = \begin{cases} 0.10x & \text{if } 0 \le x \le 16{,}750 \\ \$1675.00 + 0.15(x - 16{,}750) & \text{if } 16{,}750 < x \le 68{,}000 \\ \$9362.50 + 0.25(x - 68{,}000) & \text{if } 68{,}000 < x \le 137{,}300 \\ \$26{,}687.50 + 0.28(x - 137{,}300) & \text{if } 137{,}300 < x \le 209{,}250 \\ \$46{,}833.50 + 0.33(x - 209{,}250) & \text{if } 209{,}250 < x \le 373{,}650 \\ \$101{,}085.50 + 0.35(x - 373{,}650) & \text{if } x > 373{,}650 \end{cases}$$

(b) If this function is continous at $x = 137{,}300$, there is no jump in taxes at $137,300. We examine the three conditions for continuity at $x = 137{,}300$

(i) $T(137{,}300) = 26{,}687.50$, so $T(137{,}300)$ exists.

(ii) Because the function is piecewise defined near 137,300, we evaluate $\lim\limits_{x \to 137{,}300} T(x)$ by evaluating one-sided limits:

From the left, we evaluate $\lim\limits_{x \to 137{,}300^-} T(x)$:

$$\lim\limits_{x \to 137{,}300^-} [9362.50 + 0.25(x - 68{,}000)] = 26{,}687.50$$

From the right, we evaluate $\lim\limits_{x \to 137{,}300^+} T(x)$:

$$\lim\limits_{x \to 137{,}300^+} [26{,}687.50 + 0.28(x - 137{,}300)] = 26{,}687.50$$

Because these one-sided limits agree, the limit exists and is

$$\lim\limits_{x \to 137{,}300} T(x) = 26{,}687.50.$$

(iii) Because $\lim\limits_{x \to 137{,}300} T(x) = T(137{,}300) = 26{,}687.50$, the function is continuous at 137,300.

(c) If your friend earned more than $137,300, she and her husband would pay taxes at a higher rate on the money earned *above* the $137,300, but it would not increase the tax rate on any income *up to* $137,300. Thus she should take any raise that is offered. ■

2. If $f(x)$ and g(x) are polynomials, $h(x) = \begin{cases} f(x) & \text{if } x \le a \\ g(x) & \text{if } x > a \end{cases}$ is continuous everywhere except perhaps at _____.

Limits at Infinity We have seen that the graph of $y = 1/x$ has a vertical asymptote at $x = 0$ (shown in Figure 9.15(a) on the next page). By graphing $y = 1/x$ and evaluating the function for very large x-values, we can see that $y = 1/x$ never becomes negative for positive x-values regardless of how large the x-value is. Although no value of x makes $1/x$ equal to 0, it is easy to see that $1/x$ approaches 0 as x gets very large. This is denoted by

$$\lim\limits_{x \to +\infty} \frac{1}{x} = 0$$

and means that the line $y = 0$ (the x-axis) is a horizontal asymptote for $y = 1/x$. We also see that $y = 1/x$ approaches 0 as x decreases without bound, and we denote this by

$$\lim\limits_{x \to -\infty} \frac{1}{x} = 0$$

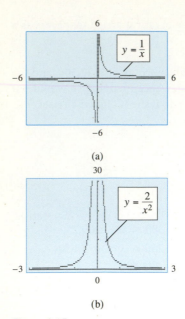

Figure 9.15

These limits for $f(x) = 1/x$ can also be established with numerical tables.

x	$f(x) = 1/x$	x	$f(x) = 1/x$
100	0.01	-100	-0.01
100,000	0.00001	$-100,000$	-0.00001
100,000,000	0.00000001	$-100,000,000$	-0.00000001
$\downarrow$	$\downarrow$	$\downarrow$	$\downarrow$
$+\infty$	0	$-\infty$	0
$\displaystyle\lim_{x \to +\infty} \frac{1}{x} = 0$		$\displaystyle\lim_{x \to -\infty} \frac{1}{x} = 0$	

We can use the graph of $y = 2/x^2$ in Figure 9.15(b) to see that the x-axis $(y = 0)$ is a horizontal asymptote and that

$$\lim_{x \to +\infty} \frac{2}{x^2} = 0 \quad \text{and} \quad \lim_{x \to -\infty} \frac{2}{x^2} = 0$$

By using graphs and/or tables of values, we can generalize the results for the functions shown in Figure 9.15 and conclude the following.

Limits at Infinity

If c is any constant, then

1. $\displaystyle\lim_{x \to +\infty} c = c \quad$ and $\quad \displaystyle\lim_{x \to -\infty} c = c.$

2. $\displaystyle\lim_{x \to +\infty} \frac{c}{x^p} = 0$, where $p > 0$.

3. $\displaystyle\lim_{x \to -\infty} \frac{c}{x^n} = 0$, where $n > 0$ is any integer.

In order to use these properties for finding the limits of rational functions as x approaches $+\infty$ or $-\infty$, we first divide each term of the numerator and denominator by the highest power of x present and then determine the limit of the resulting expression.

EXAMPLE 4   Limits at Infinity

Find each of the following limits, if they exist.

(a) $\displaystyle\lim_{x \to +\infty} \frac{2x - 1}{x + 2}$ (b) $\displaystyle\lim_{x \to -\infty} \frac{x^2 + 3}{1 - x}$

Solution

(a) The highest power of x present is x^1, so we divide each term in the numerator and denominator by x and then use the properties for limits at infinity.

$$\lim_{x \to +\infty} \frac{2x - 1}{x + 2} = \lim_{x \to +\infty} \frac{\dfrac{2x}{x} - \dfrac{1}{x}}{\dfrac{x}{x} + \dfrac{2}{x}} = \lim_{x \to +\infty} \frac{2 - \dfrac{1}{x}}{1 + \dfrac{2}{x}}$$

$$= \frac{2 - 0}{1 + 0} = 2 \quad \text{(by Properties 1 and 2)}$$

Figure 9.16(a) shows the graph of this function with the y-coordinates of the graph approaching 2 as x approaches $+\infty$ and as x approaches $-\infty$. That is, $y = 2$ is a horizontal asymptote. Note also that there is a discontinuity (vertical asymptote) where $x = -2$.

(b) We divide each term in the numerator and denominator by x^2 and then use the properties.

$$\lim_{x \to -\infty} \frac{x^2 + 3}{1 - x} = \lim_{x \to -\infty} \frac{\dfrac{x^2}{x^2} + \dfrac{3}{x^2}}{\dfrac{1}{x^2} - \dfrac{x}{x^2}} = \lim_{x \to -\infty} \frac{1 + \dfrac{3}{x^2}}{\dfrac{1}{x^2} - \dfrac{1}{x}}$$

This limit does not exist because the numerator approaches 1 and the denominator approaches 0 through positive values. Thus

$$\frac{x^2 + 3}{1 - x} \to +\infty \text{ as } x \to -\infty$$

The graph of this function, shown in Figure 9.16(b), has y-coordinates that increase without bound as x approaches $-\infty$ and that decrease without bound as x approaches $+\infty$. (There is no horizontal asymptote.) Note also that there is a vertical asymptote at $x = 1$. ■

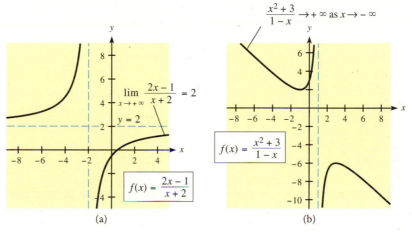

Figure 9.16   (a)   (b)

In our work with limits at infinity, we have mentioned horizontal asymptotes several times. The connection between these concepts follows.

Limits at Infinity and Horizontal Asymptotes

If $\lim_{x \to \infty} f(x) = b$ or $\lim_{x \to -\infty} f(x) = b$, where b is a constant, then the line $y = b$ is a horizontal asymptote for the graph of $y = f(x)$. Otherwise, $y = f(x)$ has no horizontal asymptotes.

CHECKPOINT

3. (a) Evaluate $\displaystyle\lim_{x \to +\infty} \frac{x^2 - 4}{2x^2 - 7}$.

   (b) What does part (a) say about horizontal asymptotes for $f(x) = (x^2 - 4)/(2x^2 - 7)$?

Calculator Note We can use the graphing and table features of a graphing calculator to help locate and investigate discontinuities and limits at infinity (horizontal asymptotes). A graphing calculator can be used to focus our attention on a possible discontinuity and to support or suggest appropriate algebraic calculations. See Appendix C, Section 9.2, for details. ■

 EXAMPLE 5   **Limits with Technology**

Use a graphing utility to investigate the continuity of the following functions.

(a) $f(x) = \dfrac{x^2 + 1}{x + 1}$ (b) $g(x) = \dfrac{x^2 - 2x - 3}{x^2 - 1}$

(c) $h(x) = \dfrac{|x + 1|}{x + 1}$ (d) $k(x) = \begin{cases} \dfrac{-x^2}{2} - 2x & \text{if } x \le -1 \\[2mm] \dfrac{x}{2} + 2 & \text{if } x > -1 \end{cases}$

Solution

(a) Figure 9.17(a) shows that $f(x)$ has a discontinuity (vertical asymptote) near $x = -1$. Because $f(-1)$ DNE, we know that $f(x)$ is not continuous at $x = -1$.

(b) Figure 9.17(b) shows that $g(x)$ is discontinuous (vertical asymptote) near $x = 1$, and this looks like the only discontinuity. However, the denominator of $g(x)$ is zero at $x = 1$ and $x = -1$, so $g(x)$ must have discontinuities at both of these x-values. Evaluating or using the table feature confirms that $x = -1$ is a discontinuity (a hole, or missing point). The figure also shows a horizontal asymptote; evaluation of $\lim_{x \to \infty} g(x)$ confirms this is the line $y = 1$.

(c) Figure 9.17(c) shows a discontinuity (jump) at $x = -1$. We also see that $h(-1)$ DNE, which confirms the observations from the graph.

(d) The graph in Figure 9.17(d) appears to be continuous. The only "suspicious" x-value is $x = -1$, where the formula for $k(x)$ changes. Evaluating $k(-1)$ and examining a table near $x = -1$ indicates that $k(x)$ is continuous there. Algebraic evaluations of the two one-sided limits confirm this. ◼

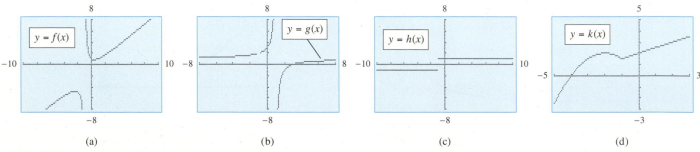

(a)　　　　　(b)　　　　　(c)　　　　　(d)

Figure 9.17

Continous Functions and Limits at Infinity

- The following information is useful in discussing continuity of functions.

  A. A polynomial function is continuous everywhere.

  B. A rational function is a function of the form $\dfrac{f(x)}{g(x)}$, where $f(x)$ and $g(x)$ are polynomials.

  1. If $g(x) \neq 0$ at any value of x, the function is continuous everywhere.
  2. If $g(c) = 0$, the function is discontinuous at $x = c$.
     (a) If $g(c) = 0$ and $f(c) \neq 0$, then there is a vertical asymptote at $x = c$.
     (b) If $g(c) = 0$ and $\lim\limits_{x \to c} \dfrac{f(x)}{g(x)} = L$, then the graph has a missing point at (c, L).

  C. A piecewise defined function may have a discontinuity at any x-value where the function changes its formula. One-sided limits must be used to see whether the limit exists.

- The following steps are useful when we are evaluating limits at infinity for a rational function $f(x) = p(x)/q(x)$.

  1. Divide both $p(x)$ and $q(x)$ by the highest power of x found in either polynomial.
  2. Use the properties of limits at infinity to complete the evaluation.

CHECKPOINT SOLUTIONS

1. (a) This is a polynomial function, so it is continuous at all values of x (discontinuous at none).

   (b) This is a rational function. It is discontinuous at $x = 1$ and $x = -2$ because these values make its denominator 0.

2. $x = a$

3. (a) $\lim\limits_{x \to +\infty} \dfrac{x^2 - 4}{2x^2 - 7} = \lim\limits_{x \to +\infty} \dfrac{1 - \dfrac{4}{x^2}}{2 - \dfrac{7}{x^2}} = \dfrac{1 - 0}{2 - 0} = \dfrac{1}{2}$

   (b) The line $y = 1/2$ is a horizontal asymptote.

| EXERCISES | 9.2

Access end-of-section exercises online at www.webassign.net WebAssign

OBJECTIVES

- To define and find average rates of change
- To define the derivative as a rate of change
- To use the definition of derivative to find derivatives of functions
- To use derivatives to find slopes of tangents to curves

9.3

Rates of Change and Derivatives

| APPLICATION PREVIEW |

In Chapter 1, "Linear Equations and Functions," we studied linear revenue functions and defined the marginal revenue for a product as the rate of change of the revenue function. For linear revenue functions, this rate is also the slope of the line that is the graph of the revenue function. In this section, we will define marginal revenue as the rate of change of the revenue function, even when the revenue function is not linear.

Thus, if an oil company's revenue (in thousands of dollars) is given by

$$R = 100x - x^2, \quad x \geq 0$$

where x is the number of thousands of barrels of oil sold per day, we can find and interpret the marginal revenue when 20,000 barrels are sold (see Example 4).

We will discuss the relationship between the *instantaneous rate of change* of a function at a given point and the slope of the line tangent to the graph of the function at that point. We will see how the derivative of a revenue function can be used to find both the slope of its tangent line and its marginal revenue.

Average Rates of Change

For linear functions, we have seen that the slope of the line measures the average rate of change of the function and can be found from any two points on the line. However, for a function that is not linear, the slope between different pairs of points no longer always gives the same number, but it can be interpreted as an **average rate of change.**

Average Rate of Change

The **average rate of change** of a function $y = f(x)$ from $x = a$ to $x = b$ is defined by

$$\text{Average rate of change} = \frac{f(b) - f(a)}{b - a}$$

The figure shows that this average rate is the same as the slope of the segment (or secant line) joining the points $(a, f(a))$ and $(b, f(b))$.

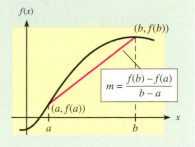

EXAMPLE 1 Total Cost

Suppose a company's total cost in dollars to produce x units of its product is given by

$$C(x) = 0.01x^2 + 25x + 1500$$

Find the average rate of change of total cost for the second 100 units produced (from $x = 100$ to $x = 200$).

Solution

The average rate of change of total cost from $x = 100$ to $x = 200$ units is

$$\frac{C(200) - C(100)}{200 - 100} = \frac{[0.01(200)^2 + 25(200) + 1500] - [0.01(100)^2 + 25(100) + 1500]}{100}$$

$$= \frac{6900 - 4100}{100} = \frac{2800}{100} = 28 \text{ dollars per unit}$$

EXAMPLE 2 **Elderly in the Work Force**

Figure 9.18 shows the percents of elderly men and of elderly women in the work force in selected years from 1950 to 2008. For these years, find and interpret the average rate of change of the percent of (a) elderly men in the work force and (b) elderly women in the work force.

Elderly in the Labor Force, 1950–2008

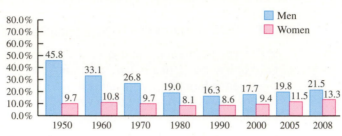

Figure 9.18   *Source:* Bureau of the Census, U.S. Department of Commerce

Solution

(a) From 1950 to 2008, the annual average rate of change in the percent of elderly men in the work force is

$$\frac{\text{Change in men's percent}}{\text{Change in years}} = \frac{21.5 - 45.8}{2008 - 1950} = \frac{-24.3}{58} \approx -0.419 \text{ percentage points per year}$$

This means that from 1950 to 2008, *on average*, the percent of elderly men in the work force dropped by 0.419 percentage points per year.

(b) Similarly, the average rate of change for women is

$$\frac{\text{Change in women's percent}}{\text{Change in years}} = \frac{13.3 - 9.7}{2008 - 1950} = \frac{3.6}{58} \approx 0.062 \text{ percentage points per year}$$

In like manner, this means that from 1950 to 2008, *on average*, the percent of elderly women in the work force increased by 0.062 percentage points each year. ◼

Instantaneous Rates of Change: Velocity

Another common rate of change is velocity. For instance, if we travel 200 miles in our car over a 4-hour period, we know that we averaged 50 mph. However, during that trip there may have been times when we were traveling on an Interstate at faster than 50 mph and times when we were stopped at a traffic light. Thus, for the trip we have not only an average velocity but also instantaneous velocities (or instantaneous speeds as displayed on the speedometer). Let's see how average velocity can lead us to instantaneous velocity.

Suppose a ball is thrown straight upward at 64 feet per second from a spot 96 feet above ground level. The equation that describes the height y of the ball after x seconds is

$$y = f(x) = 96 + 64x - 16x^2$$

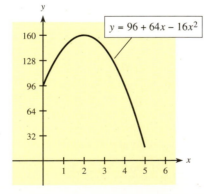

Figure 9.19 shows the graph of this function for $0 \le x \le 5$. The average velocity of the ball over a given time interval is the change in the height divided by the length of time that has passed. Table 9.4 on the next page shows some average velocities over time intervals beginning at $x = 1$.

Figure 9.19

■ | TABLE 9.4 |

AVERAGE VELOCITIES

| Time (seconds) | | | Height (feet) | | | Average Velocity (ft/sec) |
Beginning	Ending	Change (Δx)	Beginning	Ending	Change (Δy)	($\Delta y/\Delta x$)
1	2	1	144	160	16	$16/1 = 16$
1	1.5	0.5	144	156	12	$12/0.5 = 24$
1	1.1	0.1	144	147.04	3.04	$3.04/0.1 = 30.4$
1	1.01	0.01	144	144.3184	0.3184	$0.3184/0.01 = 31.84$

In Table 9.4, the smaller the time interval, the more closely the average velocity approximates the instantaneous velocity at $x = 1$. Thus the instantaneous velocity at $x = 1$ is closer to 31.84 feet per second than to 30.4 feet per second.

If we represent the change in time by h, then the average velocity from $x = 1$ to $x = 1 + h$ approaches the instantaneous velocity at $x = 1$ as h approaches 0. (Note that h can be positive or negative.) This is illustrated in the following example.

EXAMPLE   3   Velocity

Suppose a ball is thrown straight upward so that its height $f(x)$ (in feet) is given by the equation

$$f(x) = 96 + 64x - 16x^2$$

where x is time (in seconds).
(a) Find the average velocity from $x = 1$ to $x = 1 + h$.
(b) Find the instantaneous velocity at $x = 1$.

Solution
(a) Let h represent the change in x (time) from 1 to $1 + h$. Then the corresponding change in $f(x)$ (height) is

$$f(1 + h) - f(1) = [96 + 64(1 + h) - 16(1 + h)^2] - [96 + 64 - 16]$$
$$= 96 + 64 + 64h - 16(1 + 2h + h^2) - 144$$
$$= 16 + 64h - 16 - 32h - 16h^2 = 32h - 16h^2$$

The average velocity V_{av} is the change in height divided by the change in time.

$$V_{av} = \frac{f(1 + h) - f(1)}{1 + h - 1} = \frac{32h - 16h^2}{h} = 32 - 16h$$

(b) The instantaneous velocity V is the limit of the average velocity as h approaches 0.

$$V = \lim_{h \to 0} V_{av} = \lim_{h \to 0} (32 - 16h) = 32 \text{ feet per second}$$ ■

Note that average velocity is found over a time interval. Instantaneous velocity is usually called **velocity**, and it can be found at any time x, as follows.

Velocity

Suppose that an object moving in a straight line has its position y at time x given by $y = f(x)$. Then the **velocity** function for the object at time x is

$$V = \lim_{h \to 0} \frac{f(x + h) - f(x)}{h}$$

provided that this limit exists.

The instantaneous rate of change of any function (commonly called *rate of change*) can be found in the same way we find velocity. The function that gives this instantaneous rate of change of a function f is called the **derivative** of f.

Derivative

If f is a function defined by $y = f(x)$, then the **derivative** of $f(x)$ at any value x, denoted $f'(x)$, is

$$f'(x) = \lim_{h \to 0} \frac{f(x + h) - f(x)}{h}$$

if this limit exists. If $f'(c)$ exists, we say that f is **differentiable** at c.

The following procedure illustrates how to find the derivative of a function $y = f(x)$ at any value x.

Derivative Using the Definition

Procedure	**Example**
To find the derivative of $y = f(x)$ at any value x:	Find the derivative of $f(x) = 4x^2$.
1. Let h represent the change in x from x to $x + h$.	1. The change in x from x to $x + h$ is h.
2. The corresponding change in $y = f(x)$ is $$f(x + h) - f(x)$$	2. The change in $f(x)$ is $$\begin{aligned} f(x + h) - f(x) &= 4(x + h)^2 - 4x^2 \\ &= 4(x^2 + 2xh + h^2) - 4x^2 \\ &= 4x^2 + 8xh + 4h^2 - 4x^2 \\ &= 8xh + 4h^2 \end{aligned}$$
3. Form the difference quotient $\dfrac{f(x + h) - f(x)}{h}$ and simplify.	3. $$\begin{aligned} \frac{f(x + h) - f(x)}{h} &= \frac{8xh + 4h^2}{h} \\ &= 8x + 4h \end{aligned}$$
4. Find $\displaystyle\lim_{h \to 0} \frac{f(x + h) - f(x)}{h}$ to determine $f'(x)$, the derivative of $f(x)$.	4. $$f'(x) = \lim_{h \to 0} \frac{f(x + h) - f(x)}{h}$$ $$f'(x) = \lim_{h \to 0} (8x + 4h) = 8x$$

Note that in the example above, we could have found the derivative of the function $f(x) = 4x^2$ at a particular value of x, say $x = 3$, by evaluating the derivative formula at that value:

$$f'(x) = 8x \quad \text{so} \quad f'(3) = 8(3) = 24$$

In addition to $f'(x)$, the derivative at any point x may be denoted by

$$\frac{dy}{dx}, \quad y', \quad \frac{d}{dx} f(x), \quad D_x y, \quad \text{or} \quad D_x f(x)$$

We can, of course, use variables other than x and y to represent functions and their derivatives. For example, we can represent the derivative of the function $p = 2q^2 - 1$ by dp/dq.

CHECKPOINT

1. Find the average rate of change of $f(x) = 30 - x - x^2$ over $[1, 4]$.
2. For the function $y = f(x) = x^2 - x + 1$, find
   (a) $f(x + h) - f(x)$.
   (b) $\dfrac{f(x + h) - f(x)}{h}$.
   (c) $f'(x) = \displaystyle\lim_{h \to 0} \dfrac{f(x + h) - f(x)}{h}$.
   (d) $f'(2)$.

For linear functions, we defined the **marginal revenue** for a product as the rate of change of the total revenue function for the product. If the total revenue function for a product is not linear, we define the marginal revenue for the product as the instantaneous rate of change, or the derivative, of the revenue function.

Marginal Revenue

Suppose that the total revenue function for a product is given by $R = R(x)$, where x is the number of units sold. Then the **marginal revenue** at x units is

$$\overline{MR} = R'(x) = \lim_{h \to 0} \frac{R(x + h) - R(x)}{h}$$

provided that the limit exists.

Note that the marginal revenue (derivative of the revenue function) can be found by using the steps in the Procedure/Example box on the preceding page. These steps can also be combined, as they are in Example 4.

EXAMPLE 4   Revenue | APPLICATION PREVIEW |

Suppose that an oil company's revenue (in thousands of dollars) is given by the equation

$$R = R(x) = 100x - x^2, \ x \geq 0$$

where x is the number of thousands of barrels of oil sold each day.

(a) Find the function that gives the marginal revenue at any value of x.
(b) Find the marginal revenue when 20,000 barrels are sold (that is, at $x = 20$).

Solution
(a) The marginal revenue function is the derivative of $R(x)$.

$$R'(x) = \lim_{h \to 0} \frac{R(x + h) - R(x)}{h} = \lim_{h \to 0} \frac{[100(x + h) - (x + h)^2] - (100x - x^2)}{h}$$

$$= \lim_{h \to 0} \frac{100x + 100h - (x^2 + 2xh + h^2) - 100x + x^2}{h}$$

$$= \lim_{h \to 0} \frac{100h - 2xh - h^2}{h} = \lim_{h \to 0} (100 - 2x - h) = 100 - 2x$$

Thus, the marginal revenue function is $\overline{MR} = R'(x) = 100 - 2x$.
(b) The function found in part (a) gives the marginal revenue at *any* value of x. To find the marginal revenue when 20 units are sold, we evaluate $R'(20)$.

$$R'(20) = 100 - 2(20) = 60$$

Hence the marginal revenue at $x = 20$ is 60 thousand dollars per thousand barrels of oil. Because the marginal revenue is used to approximate the revenue from the sale of one additional unit, we interpret $R'(20) = 60$ to mean that the expected revenue from the sale of the next thousand barrels (after 20,000) will be approximately $60,000. [*Note:* The actual revenue from this sale is $R(21) - R(20) = 1659 - 1600 = 59$ thousand dollars.] ■

Tangent to a Curve   Just as average rates of change are connected with slopes, so are instantaneous rates (derivatives). In fact, the slope of the graph of a function at any point is the same as the derivative at that point. To show this, we define the slope of a curve at a point on the curve as the slope of the line tangent to the curve at the point.

In geometry, a **tangent** to a circle is defined as a line that has one point in common with the circle. [See Figure 9.20(a) on the next page.] This definition does not apply to all curves, as Figure 9.20(b) shows. Many lines can be drawn through the point A that touch the curve only at A. One of the lines, line l, looks like it is tangent to the curve.

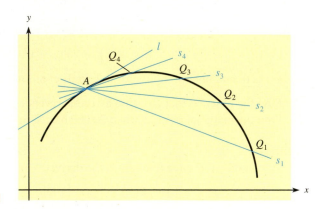

Figure 9.20   (a)   (b)

In Figure 9.21, the line l represents the tangent line to the curve at point A and shows that secant lines (s_1, s_2, etc.) through A approach line l as the second points (Q_1, Q_2, etc.) approach A. (For points and secant lines to the left of A, there would be a similar figure and conclusion.) This means that as we choose points on the curve closer and closer to A (from both sides of A), the limiting position of the secant lines through A is the **tangent line** to the curve at A.

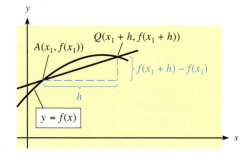

Figure 9.21

Figure 9.22

From Figure 9.22, we see that the slope of an arbitrary secant line through $A(x_1, f(x_1))$ and $Q(x_1 + h, f(x_1 + h))$ is given by

$$m_{AQ} = \frac{f(x_1 + h) - f(x_1)}{h}$$

Thus, as Q approaches A, the slope of the secant line AQ approaches the **slope of the tangent line at A**, and we have the following.

Slope of the Tangent

The **slope of the tangent** to the graph of $y = f(x)$ at point $A(x_1, f(x_1))$ is

$$m = \lim_{h \to 0} \frac{f(x_1 + h) - f(x_1)}{h}$$

if this limit exists. That is, $m = f'(x_1)$, the derivative at $x = x_1$.

EXAMPLE 5   **Slope of the Tangent**

Find the slope of $y = f(x) = x^2$ at the point $A(2, 4)$.

Solution

The formula for the slope of the tangent to $y = f(x)$ at $(2, 4)$ is

$$m = f'(2) = \lim_{h \to 0} \frac{f(2 + h) - f(2)}{h}$$

Thus for $f(x) = x^2$, we have

$$m = f'(2) = \lim_{h \to 0} \frac{(2 + h)^2 - 2^2}{h}$$

Taking the limit immediately would result in both the numerator and the denominator approaching 0. To avoid this, we simplify the fraction before taking the limit.

$$m = \lim_{h \to 0} \frac{4 + 4h + h^2 - 4}{h} = \lim_{h \to 0} \frac{4h + h^2}{h} = \lim_{h \to 0} (4 + h) = 4$$

Thus the slope of the tangent to $y = x^2$ at $(2, 4)$ is 4 (see Figure 9.23). ■

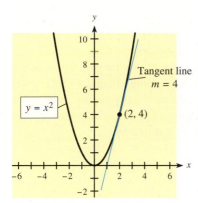

Figure 9.23

The statement "the slope of the tangent to the curve at $(2, 4)$ is 4" is frequently simplified to the statement "the slope of the curve at $(2, 4)$ is 4." Knowledge that the slope is a positive number on an interval tells us that the function is increasing on that interval, which means that a point moving along the graph of the function rises as it moves to the right on that interval. If the derivative (and thus the slope) is negative on an interval, the curve is decreasing on the interval; that is, a point moving along the graph falls as it moves to the right on that interval.

EXAMPLE 6   **Tangent Line**

Given $y = f(x) = 3x^2 + 2x + 11$, find

(a)   the derivative of $f(x)$ at any point $(x, f(x))$.
(b)   the slope of the tangent to the curve at $(1, 16)$.
(c)   the equation of the line tangent to $y = 3x^2 + 2x + 11$ at $(1, 16)$.

Solution

(a)   The derivative of $f(x)$ at any value x is denoted by $f'(x)$ and is

$$\begin{aligned}
y' = f'(x) &= \lim_{h \to 0} \frac{f(x + h) - f(x)}{h} \\
&= \lim_{h \to 0} \frac{[3(x + h)^2 + 2(x + h) + 11] - (3x^2 + 2x + 11)}{h} \\
&= \lim_{h \to 0} \frac{3(x^2 + 2xh + h^2) + 2x + 2h + 11 - 3x^2 - 2x - 11}{h} \\
&= \lim_{h \to 0} \frac{6xh + 3h^2 + 2h}{h} = \lim_{h \to 0} (6x + 3h + 2) = 6x + 2
\end{aligned}$$

(b)   The derivative is $f'(x) = 6x + 2$, so the slope of the tangent to the curve at $(1, 16)$ is $f'(1) = 6(1) + 2 = 8$.

(c)   The equation of the tangent line uses the given point $(1, 16)$ and the slope $m = 8$. Using $y - y_1 = m(x - x_1)$ gives $y - 16 = 8(x - 1)$, or $y = 8x + 8$. ■

The discussion in this section indicates that the derivative of a function has several interpretations.

Interpretations of the Derivative

For a given function, each of the following means "find the **derivative**."

1. Find the **velocity** of an object moving in a straight line.
2. Find the **instantaneous rate of change** of a function.
3. Find the **marginal revenue** for a given revenue function.
4. Find the **slope of the tangent** to the graph of a function.

That is, all the terms printed in boldface are mathematically the same, and the answers to questions about any one of them give information about the others.

Note in Figure 9.23 that near the point of tangency at (2, 4), the tangent line and the function look coincident. In fact, if we graphed both with a graphing calculator and repeatedly zoomed in near the point (2, 4), the two graphs would eventually appear as one. Try this for yourself. Thus the derivative of $f(x)$ at the point where $x = a$ can be approximated by finding the slope between $(a, f(a))$ and a second point that is nearby.

In addition, we know that the slope of the tangent to $f(x)$ at $x = a$ is defined by

$$f'(a) = \lim_{h \to 0} \frac{f(a + h) - f(a)}{h}$$

Hence we could also estimate $f'(a)$—that is, the slope of the tangent at $x = a$—by evaluating

$$\frac{f(a + h) - f(a)}{h} \quad \text{when } h \approx 0 \text{ and } h \neq 0$$

EXAMPLE 7   **Approximating the Slope of the Tangent Line**

(a) Let $f(x) = 2x^3 - 6x^2 + 2x - 5$. Use $\dfrac{f(a + h) - f(a)}{h}$ and two values of h to make estimates of the slope of the tangent to $f(x)$ at $x = 3$ on opposite sides of $x = 3$.

(b) Use the following table of values of x and $g(x)$ to estimate $g'(3)$.

x	1	1.9	2.7	2.9	2.999	3	3.002	3.1	4	5
$g(x)$	1.6	4.3	11.4	10.8	10.513	10.5	10.474	10.18	6	−5

Solution

The table feature of a graphing calculator can facilitate the following calculations.

(a) We can use $h = 0.0001$ and $h = -0.0001$ as follows:

With $h = 0.0001$: $f'(3) \approx \dfrac{f(3 + 0.0001) - f(3)}{0.0001} = \dfrac{f(3.0001) - f(3)}{0.0001} \approx 20$

With $h = -0.0001$: $f'(3) \approx \dfrac{f(3 + (-0.0001)) - f(3)}{-0.0001} = \dfrac{f(2.9999) - f(3)}{-0.0001} \approx 20$

(b) We use the given table and measure the slope between (3, 10.5) and another point that is nearby (the closer, the better). Using (2.999, 10.513), we obtain

$$g'(3) \approx \frac{y_2 - y_1}{x_2 - x_1} = \frac{10.5 - 10.513}{3 - 2.999} = \frac{-0.013}{0.001} = -13 \qquad \blacksquare$$

Calculator Note   Most graphing calculators have a feature called the **numerical derivative** (usually denoted by nDer or nDeriv) that can approximate the derivative of a function at a point. See the steps used in Appendix C, Section 9.3. On most calculators this feature uses a calculation similar to our method in part (a) of Example 7 and produces the same estimate. The numerical derivative of $f(x) = 2x^3 - 6x^2 + 2x - 5$ with respect to x at $x = 3$ can be found as follows on many graphing calculators:

$$\text{nDeriv}(2x^3 - 6x^2 + 2x - 5, x, 3) = 20 \qquad \blacksquare$$

Differentiability and Continuity   So far we have talked about how the derivative is defined, what it represents, and how to find it. However, there are functions for which derivatives do not exist at every value of x. Figure 9.24 shows some common cases where $f'(c)$ does not exist but where $f'(x)$ exists for all other values of x. These cases occur where there is a discontinuity, a corner, or a vertical tangent line.

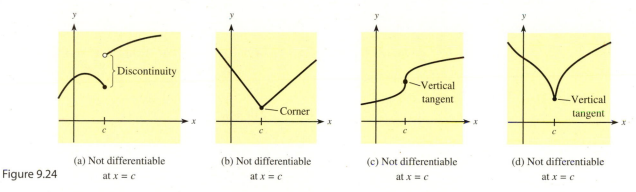

Figure 9.24

(a) Not differentiable at $x = c$ (b) Not differentiable at $x = c$ (c) Not differentiable at $x = c$ (d) Not differentiable at $x = c$

From Figure 9.24 we see that a function may be continuous at $x = c$ even though $f'(c)$ does not exist. Thus continuity does not imply differentiability at a point. However, differentiability does imply continuity.

Differentiability Implies Continuity

If a function f is differentiable at $x = c$, then f is continuous at $x = c$.

CHECKPOINT

3. Which of the following are given by $f'(c)$?
   (a) The slope of the tangent when $x = c$
   (b) The y-coordinate of the point where $x = c$
   (c) The instantaneous rate of change of $f(x)$ at $x = c$
   (d) The marginal revenue at $x = c$, if $f(x)$ is the revenue function
4. Must a graph that has no discontinuity, corner, or cusp at $x = c$ be differentiable at $x = c$?

Calculator Note   We can use a graphing calculator to explore the relationship between secant lines and tangent lines. For example, if the point (a, b) lies on the graph of $y = x^2$, then the equation of the secant line to $y = x^2$ from $(1, 1)$ to (a, b) has the equation

$$y - 1 = \frac{b - 1}{a - 1}(x - 1), \quad \text{or} \quad y = \frac{b - 1}{a - 1}(x - 1) + 1$$

Figure 9.25 illustrates the secant lines for three different choices for the point (a, b).

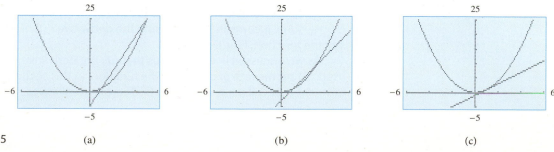

Figure 9.25   (a)   (b)   (c)

We see that as the point (a, b) moves closer to $(1, 1)$, the secant line looks more like the tangent line to $y = x^2$ at $(1, 1)$. Furthermore, (a, b) approaches $(1, 1)$ as $a \to 1$, and the slope of the secant approaches the following limit.

$$\lim_{a \to 1} \frac{b - 1}{a - 1} = \lim_{a \to 1} \frac{a^2 - 1}{a - 1} = \lim_{a \to 1} (a + 1) = 2$$

This limit, 2, is the slope of the tangent line at $(1, 1)$. That is, the derivative of $y = x^2$ at $(1, 1)$ is 2. [Note that a graphing calculator's calculation of the numerical derivative of $f(x) = x^2$ with respect to x at $x = 1$ gives $f'(1) = 2$.] ■

CHECKPOINT SOLUTIONS

1. $\dfrac{f(4) - f(1)}{4 - 1} = \dfrac{10 - 28}{3} = \dfrac{-18}{3} = -6$

2. (a) $f(x + h) - f(x) = [(x + h)^2 - (x + h) + 1] - (x^2 - x + 1)$
$$= x^2 + 2xh + h^2 - x - h + 1 - x^2 + x - 1$$
$$= 2xh + h^2 - h$$

   (b) $\dfrac{f(x + h) - f(x)}{h} = \dfrac{2xh + h^2 - h}{h}$
$$= 2x + h - 1$$

   (c) $f'(x) = \lim_{h \to 0} \dfrac{f(x + h) - f(x)}{h} = \lim_{h \to 0} (2x + h - 1)$
$$= 2x - 1$$

   (d) $f'(x) = 2x - 1$, so $f'(2) = 3$.

3. Parts (a), (c), and (d) are given by $f'(c)$. The y-coordinate where $x = c$ is given by $f(c)$.

4. No. Figure 9.24(c) shows such an example.

| EXERCISES | 9.3

Access end-of-section exercises online at **www.webassign.net**

OBJECTIVES 9.4

- To find derivatives of powers of x
- To find derivatives of constant functions
- To find derivatives of functions involving constant coefficients
- To find derivatives of sums and differences of functions

Derivative Formulas

| APPLICATION PREVIEW |

For more than 50 years, U.S. total personal income has experienced steady growth. With Bureau of Economic Analysis, U.S. Department of Commerce data for selected years from 1960 and projected to 2018, U.S. total personal income I, in billions of current dollars, can be modeled by

$$I = I(t) = 6.29t^2 - 51.7t + 601$$

where t is the number of years past 1960. We can find the rate of growth of total U.S. personal income in 2015 by using the derivative $I'(t)$ of the total personal income function. (See Example 8.)

As we discussed in the previous section, the derivative of a function can be used to find the rate of change of the function. In this section we will develop formulas that will make it easier to find certain derivatives.

Derivative of $f(x) = x^n$ We can use the definition of derivative to show the following:

$$\text{If } f(x) = x^2, \text{ then } f'(x) = 2x.$$
$$\text{If } f(x) = x^3, \text{ then } f'(x) = 3x^2.$$
$$\text{If } f(x) = x^4, \text{ then } f'(x) = 4x^3.$$
$$\text{If } f(x) = x^5, \text{ then } f'(x) = 5x^4.$$

Do you recognize a pattern here? We can use the definition of derivative and the binomial formula to verify the pattern and find a formula for the derivative of $f(x) = x^n$. If n is a positive integer, then

$$f'(x) = \lim_{h \to 0} \frac{f(x + h) - f(x)}{h} = \lim_{h \to 0} \frac{(x + h)^n - x^n}{h}$$

Because we are assuming that n is a positive integer, we can use the binomial formula, developed in Section 8.3, to expand $(x + h)^n$. This formula can be stated as follows:

$$(x + h)^n = x^n + nx^{n-1}h + \frac{n(n - 1)}{1 \cdot 2}x^{n-2}h^2 + \cdots + h^n$$

Thus the derivative formula yields

$$f'(x) = \lim_{h \to 0} \frac{\left[x^n + nx^{n-1}h + \dfrac{n(n - 1)}{1 \cdot 2}x^{n-2}h^2 + \cdots + h^n\right] - x^n}{h}$$

$$= \lim_{h \to 0} \left[nx^{n-1} + \frac{n(n - 1)}{1 \cdot 2}x^{n-2}h + \cdots + h^{n-1}\right]$$

Note that each term after nx^{n-1} contains h as a factor, so all terms except nx^{n-1} will approach 0 as $h \to 0$. Thus

$$f'(x) = nx^{n-1}$$

Even though we proved this derivative rule only for the case when n is a positive integer, the rule applies for any real number n.

Powers of x Rule   If $f(x) = x^n$, where n is a real number, then $f'(x) = nx^{n-1}$.

EXAMPLE 1   Powers of x Rule

Find the derivatives of the following functions.
(a) $y = x^{14}$ (b) $f(x) = x^{-2}$ (c) $y = x$ (d) $g(x) = x^{1/3}$

Solution
(a) If $y = x^{14}$, then $dy/dx = 14x^{14-1} = 14x^{13}$.
(b) The Powers of x Rule applies for all real values. Thus for $f(x) = x^{-2}$, we have

$$f'(x) = -2x^{-2-1} = -2x^{-3} = \frac{-2}{x^3}$$

(c) If $y = x$, then $dy/dx = 1x^{1-1} = x^0 = 1$. (Note that $y = x$ is a line with slope 1.)

(d) The Powers of x Rule applies to $y = x^{1/3}$.

$$g'(x) = \frac{1}{3}x^{1/3-1} = \frac{1}{3}x^{-2/3} = \frac{1}{3x^{2/3}}$$ ■

In Example 1 we took the derivative with respect to x of *both sides* of each equation. We denote the operation "take the derivative with respect to x" by $\dfrac{d}{dx}$. Thus for $y = x^{14}$, in part (a), we can use this notation on both sides of the equation.

$$\frac{d}{dx}(y) = \frac{d}{dx}(x^{14}) \quad \text{gives} \quad \frac{dy}{dx} = 14x^{13}$$

Similarly, we can use this notation to indicate the derivative of an expression.

$$\frac{d}{dx}(x^{-2}) = -2x^{-3}$$

The differentiation rules are stated and proved for the independent variable x, but they also apply to other independent variables. The following examples illustrate differentiation involving variables other than x and y.

EXAMPLE 2 Derivatives

Find the derivatives of the following functions.

(a) $u(s) = s^8$ (b) $p = q^{2/3}$ (c) $C(t) = \sqrt{t}$ (d) $s = \dfrac{1}{\sqrt{t}}$

Solution
(a) If $u(s) = s^8$, then $u'(s) = 8s^{8-1} = 8s^7$.
(b) If $p = q^{2/3}$, then

$$\frac{dp}{dq} = \frac{2}{3}q^{2/3-1} = \frac{2}{3}q^{-1/3} = \frac{2}{3q^{1/3}}$$

(c) Writing $\sqrt{t}$ in its equivalent form, $t^{1/2}$, using the derivative formula, and writing the derivative in radical form gives $C'(t) = \dfrac{1}{2}t^{1/2-1} = \dfrac{1}{2}t^{-1/2} = \dfrac{1}{2} \cdot \dfrac{1}{t^{1/2}} = \dfrac{1}{2\sqrt{t}}$

(d) Writing $1/\sqrt{t}$ as $\dfrac{1}{t^{1/2}} = t^{-1/2}$, taking the derivative, and writing the derivative in a form similar to that of the original function gives

$$\frac{ds}{dt} = -\frac{1}{2}t^{-1/2-1} = -\frac{1}{2}t^{-3/2} = -\frac{1}{2} \cdot \frac{1}{t^{3/2}} = -\frac{1}{2\sqrt{t^3}}$$ ■

EXAMPLE 3 Tangent Line

Write the equation of the tangent line to the graph of $y = x^3$ at $x = 1$.

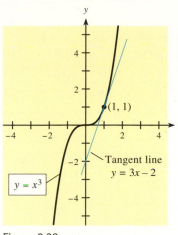

Figure 9.28

Solution

Writing the equation of the tangent line to $y = x^3$ at $x = 1$ involves three steps.

1. Evaluate the function to find the point of tangency.
   At $x = 1$: $y = (1)^3 = 1$, so the point is $(1, 1)$

2. Evaluate the derivative to find the slope of the tangent.
   At any point: $m_{\tan} = y' = 3x^2$
   At $x = 1$: $m_{\tan} = y'|_{x=1} = 3(1^2) = 3$

3. Use $y - y_1 = m(x - x_1)$ with the point $(1, 1)$ and slope $m = 3$.
   $$y - 1 = 3(x - 1) \Rightarrow y = 3x - 3 + 1 \Rightarrow y = 3x - 2$$

   Figure 9.28 shows the graph of $y = x^3$ and the tangent line at $x = 1$. ■

Derivative of a Constant

A function of the form $y = f(x) = c$, where c is a constant, is called a **constant function**. We can show that the derivative of a constant function is 0, as follows.

$$f'(x) = \lim_{h \to 0} \frac{f(x + h) - f(x)}{h} = \lim_{h \to 0} \frac{c - c}{h} = \lim_{h \to 0} 0 = 0$$

We can state this rule formally.

Constant Function Rule

If $f(x) = c$, where c is a constant, then $f'(x) = 0$.

EXAMPLE 4 Derivative of a Constant

Find the derivative of the function defined by $y = 4$.

Solution

Because 4 is a constant, $\dfrac{dy}{dx} = 0$.

Recall that the function defined by $y = 4$ has a horizontal line as its graph. Thus the slope of the line (and the derivative of the function) is 0. ■

Derivative of $y = c \cdot u(x)$

We now can take derivatives of constant functions and powers of x. But we do not yet have a rule for taking derivatives of functions of the form $f(x) = 4x^5$ or $g(t) = \frac{1}{2}t^2$. The following rule provides a method for handling functions of this type.

Coefficient Rule

If $f(x) = c \cdot u(x)$, where c is a constant and $u(x)$ is a differentiable function of x, then $f'(x) = c \cdot u'(x)$.

The Coefficient Rule says that the derivative of a constant times a function is the constant times the derivative of the function.

Using Properties of Limits II and IV (from Section 9.1, "Limits"), we can show

$$\lim_{h \to 0} c \cdot g(h) = c \cdot \lim_{h \to 0} g(h)$$

We can use this result to verify the Coefficient Rule. If $f(x) = c \cdot u(x)$, then

$$f'(x) = \lim_{h \to 0} \frac{f(x + h) - f(x)}{h} = \lim_{h \to 0} \frac{c \cdot u(x + h) - c \cdot u(x)}{h}$$

$$= \lim_{h \to 0} c \cdot \left[\frac{u(x + h) - u(x)}{h} \right] = c \cdot \lim_{h \to 0} \frac{u(x + h) - u(x)}{h}$$

so $f'(x) = c \cdot u'(x)$.

EXAMPLE 5   Coefficient Rule for Derivatives

Find the derivatives of the following functions.

(a) $f(x) = 4x^5$ (b) $g(t) = \dfrac{1}{2}t^2$ (c) $p = \dfrac{5}{\sqrt{q}}$

Solution

(a) $f'(x) = 4(5x^4) = 20x^4$

(b) $g'(t) = \frac{1}{2}(2t) = t$

(c) $p = \dfrac{5}{\sqrt{q}} = 5q^{-1/2}$,   so $\dfrac{dp}{dq} = 5\left(-\dfrac{1}{2}q^{-3/2}\right) = -\dfrac{5}{2\sqrt{q^3}}$

EXAMPLE 6   World Tourism

Over the past 20 years, world tourism has grown into one of the world's major industries. Since 1990 the receipts from world tourism y, in billions of dollars, can be modeled by the function

$$y = 113.54x^{0.52584}$$

where x is the number of years past 1985 (*Source:* World Tourism Organization).

(a) Name the function that models the rate of change of the receipts from world tourism.
(b) Find the function from part (a).
(c) Find the rate of change in world tourism in 2012.

Solution

(a) The rate of change of world tourism receipts is modeled by the derivative.

(b) $\dfrac{dy}{dx} = 113.54(0.52584x^{0.52584 - 1}) \approx 59.704x^{-0.47416}$

(c) $\dfrac{dy}{dx}\bigg|_{x=27} = 59.704(27^{-0.47416}) \approx 12.511$

Thus, the model estimates that world tourism changed by about $12.511 billion per year in 2012.

Derivatives of Sums and Differences

In Example 6 of Section 9.3, "Rates of Change and Derivatives," we found the derivative of $f(x) = 3x^2 + 2x + 11$ to be $f'(x) = 6x + 2$. This result, along with the results of several of the derivatives calculated in the exercises for that section, suggest that we can find the derivative of a function by finding the derivatives of its terms and combining them. The following rules state this formally.

Sum Rule

If $f(x) = u(x) + v(x)$, where u and v are differentiable functions of x, then $f'(x) = u'(x) + v'(x)$.

We can prove this rule as follows. If $f(x) = u(x) + v(x)$, then

$$f'(x) = \lim_{h \to 0} \frac{f(x + h) - f(x)}{h} = \lim_{h \to 0} \frac{[u(x + h) + v(x + h)] - [u(x) + v(x)]}{h}$$

$$= \lim_{h \to 0} \left[\frac{u(x + h) - u(x)}{h} + \frac{v(x + h) - v(x)}{h} \right]$$

$$= \lim_{h \to 0} \frac{u(x + h) - u(x)}{h} + \lim_{h \to 0} \frac{v(x + h) - v(x)}{h}$$

$$= u'(x) + v'(x)$$

A similar rule applies to the difference of functions.

Difference Rule	If $f(x) = u(x) - v(x)$, where u and v are differentiable functions of x, then $f'(x) = u'(x) - v'(x)$.

EXAMPLE 7 Sum and Difference Rules

Find the derivatives of the following functions.

(a) $y = 3x + 5$ 　　　　　(b) $f(x) = 4x^3 - 2x^2 + 5x - 3$

(c) $p = \frac{1}{3}q^3 + 2q^2 - 3$ 　　(d) $u(x) = 5x^4 + x^{1/3}$

(e) $y = 4x^3 + \sqrt{x}$ 　　　　(f) $s = 5t^6 - \dfrac{1}{t^2}$

Solution

(a) $y' = 3 \cdot 1 + 0 = 3$

(b) The rules regarding the derivatives of sums and differences of two functions also apply if more than two functions are involved. We may think of the functions that are added and subtracted as terms of the function f. Then it would be correct to say that we may take the derivative of a function term by term. Thus,

$$f'(x) = 4(3x^2) - 2(2x) + 5(1) - 0 = 12x^2 - 4x + 5$$

(c) $\dfrac{dp}{dq} = \frac{1}{3}(3q^2) + 2(2q) - 0 = q^2 + 4q$

(d) $u'(x) = 5(4x^3) + \frac{1}{3}x^{-2/3} = 20x^3 + \dfrac{1}{3x^{2/3}}$

(e) We may write the function as $y = 4x^3 + x^{1/2}$, so

$$y' = 4(3x^2) + \frac{1}{2}x^{-1/2} = 12x^2 + \frac{1}{2x^{1/2}} = 12x^2 + \frac{1}{2\sqrt{x}}$$

(f) We may write $s = 5t^6 - 1/t^2$ as $s = 5t^6 - t^{-2}$, so

$$\frac{ds}{dt} = 5(6t^5) - (-2t^{-3}) = 30t^5 + 2t^{-3} = 30t^5 + \frac{2}{t^3}$$
■

Each derivative in Example 7 has been *simplified*. This means that the final form of the derivative contains no negative exponents and the use of radicals or fractional exponents matches the original problem.

Also, in part (a) of Example 7, we saw that the derivative of $y = 3x + 5$ is 3. Because the slope of a line is the same at all points on the line, it is reasonable that the derivative of a linear equation is a constant. In particular, the slope of the graph of the equation $y = mx + b$ is m at all points on its graph because the derivative of $y = mx + b$ is $y' = f'(x) = m$.

EXAMPLE 8 Personal Income | APPLICATION PREVIEW |

Suppose that t is the number of years past 1960 and that

$$I = I(t) = 6.29t^2 - 51.7t + 601$$

models the U.S. total personal income in billions of current dollars. For 2015, find the model's prediction for

(a) the U.S. total personal income.
(b) the rate of change of U.S. total personal income.

Solution

(a) For 2015, $t = 55$ and

$$I(55) = 6.29(55^2) - 51.7(55) + 601 \approx 16{,}785 \text{ billion (current dollars)}$$

(b) The rate of change of U.S. total personal income is given by

$$I'(t) = 12.58t - 51.7$$

The predicted rate for 2015 is

$$I'(55) = \$640.2 \text{ billion (current dollars) per year}$$

This predicts that U.S. total personal income will change by about $640.2 billion from 2015 to 2016. ∎

CHECKPOINT

1. True or false: The derivative of a constant times a function is equal to the constant times the derivative of the function.
2. True or false: The derivative of the sum of two functions is equal to the sum of the derivatives of the two functions.
3. True or false: The derivative of the difference of two functions is equal to the difference of the derivatives of the two functions.
4. Does the Coefficient Rule apply to $f(x) = x^n/c$, where c is a constant? Explain.
5. Find the derivative of each of the following functions.

   (a) $f(x) = x^{10} - 10x + 5$ (b) $s = \dfrac{1}{t^5} - 10^7 + 1$

6. Find the slope of the line tangent to $f(x) = x^3 - 4x^2 + 1$ at $x = -1$.

EXAMPLE 9   Horizontal Tangents

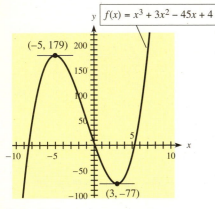

Figure 9.29

Find all points on the graph of $f(x) = x^3 + 3x^2 - 45x + 4$ where the tangent line is horizontal.

Solution

A horizontal line has slope equal to 0. Thus, to find the desired points, we solve $f'(x) = 0$.

$$f'(x) = 3x^2 + 6x - 45$$

We solve $3x^2 + 6x - 45 = 0$ as follows:

$$3x^2 + 6x - 45 = 0 \Rightarrow 3(x^2 + 2x - 15) = 0 \Rightarrow 3(x + 5)(x - 3) = 0$$

Solving $3(x + 5)(x - 3) = 0$ gives $x = -5$ and $x = 3$. The y-coordinates for these x-values come from $f(x)$. The desired points are $(-5, f(-5)) = (-5, 179)$ and $(3, f(3)) = (3, -77)$. Figure 9.29 shows the graph of $y = f(x)$ with these points and the tangent lines at them indicated. ∎

Marginal Revenue   The marginal revenue $R'(x)$ is used to estimate the change in revenue caused by the sale of one additional unit.

EXAMPLE 10   Revenue

Suppose that a manufacturer of a product knows that because of the demand for this product, his revenue is given by

$$R(x) = 1500x - 0.02x^2, \quad 0 \le x \le 1000$$

where x is the number of units sold and $R(x)$ is in dollars.

(a) Find the marginal revenue at $x = 500$.

(b) Find the change in revenue caused by the increase in sales from 500 to 501 units.

Solution

(a) The marginal revenue for any value of x is

$$R'(x) = 1500 - 0.04x$$

The marginal revenue at $x = 500$ is

$$R'(500) = 1500 - 20 = 1480 \text{ (dollars per unit)}$$

We can interpret this to mean that the approximate revenue from the sale of the 501st unit will be $1480.

(b) The revenue at $x = 500$ is $R(500) = 745{,}000$, and the revenue at $x = 501$ is $R(501) = 746{,}479.98$, so the change in revenue is

$$R(501) - R(500) = 746{,}479.98 - 745{,}000 = 1479.98 \text{ (dollars)}$$

Notice that the marginal revenue at $x = 500$ is a good estimate of the revenue from the 501st unit. ■

Calculator Note   We have mentioned that graphing calculators have a numerical derivative feature that can be used to estimate the derivative of a function at a specific value of x. This feature can also be used to check the derivative of a function that has been computed with a formula. See Appendix C, Section 9.4, for details. We graph both the derivative calculated with a formula and the numerical derivative. If the two graphs lie on top of one another, the computed derivative agrees with the numerical derivative. Figure 9.30 illustrates this idea for the derivative of $f(x) = \frac{1}{3}x^3 - 2x^2 + 4$. Figure 9.30(a) shows $f'(x) = x^2 - 4x$ as y_1 and the calculator's numerical derivative of $f(x)$ as y_2. Figure 9.30(b) shows the graphs of both y_1 and y_2 (the graphs are coincident).

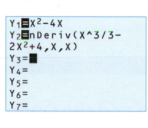

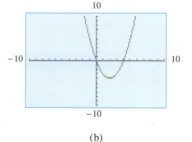

Figure 9.30   (a)   (b)   ■

 EXAMPLE 11   Comparing $f(x)$ and $f'(x)$

(a) Graph $f(x) = x^3 - 3x + 3$ and its derivative $f'(x)$ on the same set of axes so that all values of x that make $f'(x) = 0$ are in the x-range.

(b) Investigate the graph of $y = f(x)$ near values of x where $f'(x) = 0$. Does the graph of $y = f(x)$ appear to turn at values where $f'(x) = 0$?

(c) Compare the interval of x values where $f'(x) < 0$ with the interval where the graph of $y = f(x)$ is decreasing from left to right.

(d) What is the relationship between the intervals where $f'(x) > 0$ and where the graph of $y = f(x)$ is increasing from left to right?

Solution

(a) The graphs of $f(x) = x^3 - 3x + 3$ and $f'(x) = 3x^2 - 3$ are shown in Figure 9.31 on the next page.

(b) The values where $f'(x) = 0$ are the x-intercepts, $x = -1$ and $x = 1$. The graph of $y = x^3 - 3x + 3$ appears to turn at both these values.

(c) $f'(x) < 0$ where the graph of $y = f'(x)$ is below the x-axis, for $-1 < x < 1$. The graph of $y = f(x)$ appears to be decreasing on this interval.

(d) They appear to be the same intervals.

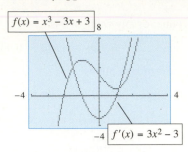

Figure 9.31

CHECKPOINT SOLUTIONS

1. True, by the Coefficient Rule.
2. True, by the Sum Rule.
3. True, by the Difference Rule.
4. Yes, $f(x) = x^n/c = (1/c)x^n$, so the coefficient is $(1/c)$.
5. (a) $f'(x) = 10x^9 - 10$
   (b) Note that $s = t^{-5} - 10^7 + 1$ and that 10^7 and 1 are constants.

$$\frac{ds}{dt} = -5t^{-6} = \frac{-5}{t^6}$$

6. The slope of the tangent at $x = -1$ is $f'(-1)$.

$$f'(x) = 3x^2 - 8x, \quad f'(-1) = 3(-1)^2 - 8(-1) = 11$$

| **EXERCISES** | **9.4**

Access end-of-section exercises online at **www.webassign.net**   **WebAssign**

OBJECTIVES   **9.5**

- To use the Product Rule to find the derivatives of certain functions
- To use the Quotient Rule to find the derivatives of certain functions

The Product Rule and the Quotient Rule

■ **| APPLICATION PREVIEW |**

When medicine is administered, reaction (measured in change of blood pressure or temperature) can be modeled by

$$R = m^2\left(\frac{c}{2} - \frac{m}{3}\right)$$

where c is a positive constant and m is the amount of medicine absorbed into the blood.* The rate of change of R with respect to m is the sensitivity of the body to medicine. To find an expression for sensitivity as a function of m, we calculate dR/dm. We can find this derivative with the Product Rule for derivatives. See Example 5.

*Source: R. M. Thrall et al., *Some Mathematical Models in Biology*, U.S. Department of Commerce, 1967.

Product Rule

We have simple formulas for finding the derivatives of the sums and differences of functions. But we are not so lucky with products. The derivative of a product is *not* the product of the derivatives. To see this, we consider the function $f(x) = x \cdot x$. Because this function is $f(x) = x^2$, its derivative is $f'(x) = 2x$. But the product of the derivatives of x and x would give $1 \cdot 1 = 1 \neq 2x$. Thus we need a different formula to find the derivative of a product. This formula is given by the **Product Rule.**

Product Rule

> If $f(x) = u(x) \cdot v(x)$, where u and v are differentiable functions of x, then
>
> $$f'(x) = u(x) \cdot v'(x) + v(x) \cdot u'(x)$$

Thus the derivative of a product of two functions is the first function times the derivative of the second plus the second function times the derivative of the first.

We can prove the Product Rule as follows. If $f(x) = u(x) \cdot v(x)$, then

$$\lim_{h \to 0} \frac{f(x+h) - f(x)}{h} = \lim_{h \to 0} \frac{u(x+h) \cdot v(x+h) - u(x) \cdot v(x)}{h}$$

Subtracting and adding $u(x+h) \cdot v(x)$ in the numerator gives

$$f'(x) = \lim_{h \to 0} \frac{u(x+h) \cdot v(x+h) - u(x+h) \cdot v(x) + u(x+h) \cdot v(x) - u(x) \cdot v(x)}{h}$$

$$= \lim_{h \to 0} \left\{ u(x+h) \left[\frac{v(x+h) - v(x)}{h} \right] + v(x) \left[\frac{u(x+h) - u(x)}{h} \right] \right\}$$

Properties III and IV of limits (from Section 9.1, "Limits") give

$$f'(x) = \lim_{h \to 0} u(x+h) \cdot \lim_{h \to 0} \frac{v(x+h) - v(x)}{h} + \lim_{h \to 0} v(x) \cdot \lim_{h \to 0} \frac{u(x+h) - u(x)}{h}$$

Because u is differentiable and hence continuous, it follows that $\lim_{h \to 0} u(x+h) = u(x)$, so we have the formula we seek:

$$f'(x) = u(x) \cdot v'(x) + v(x) \cdot u'(x)$$

EXAMPLE 1 Product Rule

(a) Find dy/dx if $y = (2x^3 + 3x + 1)(x^2 + 4)$.
(b) Find the slope of the tangent to the graph of
$y = f(x) = (4x^3 + 5x^2 - 6x + 5)(x^3 - 4x^2 + 1)$ at $x = 1$.

Solution

(a) Using the Product Rule with $u(x) = 2x^3 + 3x + 1$ and $v(x) = x^2 + 4$, we have

$$\frac{dy}{dx} = (2x^3 + 3x + 1)(2x) + (x^2 + 4)(6x^2 + 3)$$

$$= 4x^4 + 6x^2 + 2x + 6x^4 + 3x^2 + 24x^2 + 12$$

$$= 10x^4 + 33x^2 + 2x + 12$$

(b) $f'(x) = (4x^3 + 5x^2 - 6x + 5)(3x^2 - 8x) + (x^3 - 4x^2 + 1)(12x^2 + 10x - 6)$
If we substitute $x = 1$ into $f'(x)$, we find that the slope of the curve at $x = 1$ is $f'(1) = 8(-5) + (-2)(16) = -72$. ∎

We could, of course, avoid using the Product Rule by multiplying the two factors before taking the derivative. But multiplying the factors first might involve more work than using the Product Rule.

Quotient Rule   For a quotient of two functions, we might be tempted to take the derivative of the numerator divided by the derivative of the denominator; but this is incorrect. With the example $f(x) = x^3/x$ (which equals x^2 if $x \neq 0$), this approach would give $3x^2/1 = 3x^2$ as the derivative, rather than $2x$. Thus, finding the derivative of a function that is the quotient of two functions requires the **Quotient Rule.**

Quotient Rule

If $f(x) = \dfrac{u(x)}{v(x)}$, where u and v are differentiable functions of x, with $v(x) \neq 0$, then

$$f'(x) = \frac{v(x) \cdot u'(x) - u(x) \cdot v'(x)}{[v(x)]^2}$$

The preceding formula says that the derivative of a quotient is the denominator times the derivative of the numerator minus the numerator times the derivative of the denominator, all divided by the square of the denominator.

To see that this rule is reasonable, again consider the function $f(x) = x^3/x$, $x \neq 0$. Using the Quotient Rule, with $u(x) = x^3$ and $v(x) = x$, we get

$$f'(x) = \frac{x(3x^2) - x^3(1)}{x^2} = \frac{3x^3 - x^3}{x^2} = \frac{2x^3}{x^2} = 2x$$

We see that $f'(x) = 2x$ is the correct derivative. The proof of the Quotient Rule is left for the student in Problem 37 of the exercises in this section.

EXAMPLE 2   Quotient Rule

(a) If $f(x) = \dfrac{x^2 - 4x}{x + 5}$, find $f'(x)$.

(b) If $f(x) = \dfrac{x^3 - 3x^2 + 2}{x^2 - 4}$, find the instantaneous rate of change of $f(x)$ at $x = 3$.

Solution

(a) Using the Quotient Rule with $u(x) = x^2 - 4x$ and $v(x) = x + 5$, we get

$$f'(x) = \frac{(x + 5)(2x - 4) - (x^2 - 4x)(1)}{(x + 5)^2}$$

$$= \frac{2x^2 + 6x - 20 - x^2 + 4x}{(x + 5)^2} = \frac{x^2 + 10x - 20}{(x + 5)^2}$$

(b) We evaluate $f'(x)$ at $x = 3$ to find the desired rate of change. Using the Quotient Rule with $u(x) = x^3 - 3x^2 + 2$ and $v(x) = x^2 - 4$, we get

$$f'(x) = \frac{(x^2 - 4)(3x^2 - 6x) - (x^3 - 3x^2 + 2)(2x)}{(x^2 - 4)^2}$$

$$= \frac{(3x^4 - 6x^3 - 12x^2 + 24x) - (2x^4 - 6x^3 + 4x)}{(x^2 - 4)^2} = \frac{x^4 - 12x^2 + 20x}{(x^2 - 4)^2}$$

Thus, the instantaneous rate of change at $x = 3$ is $f'(3) = 33/25 = 1.32$ ∎

EXAMPLE 3   Quotient Rule

Use the Quotient Rule to find the derivative of $y = 1/x^3$.

Solution

Letting $u(x) = 1$ and $v(x) = x^3$, we get

$$y' = \frac{x^3(0) - 1(3x^2)}{(x^3)^2} = -\frac{3x^2}{x^6} = -\frac{3}{x^4}$$

Note that we could have found the derivative more easily by rewriting y.

$$y = 1/x^3 = x^{-3} \quad \text{gives} \quad y' = -3x^{-4} = -\frac{3}{x^4}$$ ■

Recall that we proved the Powers of x Rule for positive integer powers and assumed that it was true for all real number powers. In Problem 38 of the exercises in this section, you will be asked to use the Quotient Rule to show that the Powers of x Rule applies to negative integers.

It is not necessary to use the Quotient Rule when the denominator of the function in question contains only a constant. For example, the function $y = (x^3 - 3x)/3$ can be written $y = \frac{1}{3}(x^3 - 3x)$, so the derivative is $y' = \frac{1}{3}(3x^2 - 3) = x^2 - 1$.

CHECKPOINT

1. True or false: The derivative of the product of two functions is equal to the product of the derivatives of the two functions.
2. True or false: The derivative of the quotient of two functions is equal to the quotient of the derivatives of the two functions.
3. Find $f'(x)$ for each of the following.
   (a) $f(x) = (x^{12} + 8x^5 - 7)(10x^7 - 4x + 19)$ Do not simplify.
   (b) $f(x) = \dfrac{2x^4 + 3}{3x^4 + 2}$ Simplify.
4. If $y = \frac{4}{3}(x^2 + 3x - 4)$, does finding y' require the Product Rule? Explain.
5. If $y = f(x)/c$, where c is a constant, does finding y' require the Quotient Rule? Explain.

EXAMPLE 4   Marginal Revenue

Suppose that the revenue function for a flash drive is given by

$$R(x) = 10x + \frac{100x}{3x + 5}$$

where x is the number of flash drives sold and R is in dollars.

(a) Find the marginal revenue function.
(b) Find the marginal revenue when $x = 15$.

Solution

(a) We must use the Quotient Rule to find the marginal revenue (the derivative).

$$\overline{MR} = R'(x) = 10 + \frac{(3x + 5)(100) - 100x(3)}{(3x + 5)^2}$$

$$= 10 + \frac{300x + 500 - 300x}{(3x + 5)^2} = 10 + \frac{500}{(3x + 5)^2}$$

$$= 10 + \frac{500}{(3x + 5)^2}$$

(b) The marginal revenue when $x = 15$ is $R'(15)$.

$$R'(15) = 10 + \frac{500}{[(3)(15) + 5]^2} = 10 + \frac{500}{(50)^2} = 10.20 \text{ (dollars per unit)}$$

Recall that $R'(15)$ estimates the revenue from the sale of the 16th flash drive. ∎

$R'(15) = \$10.20$

EXAMPLE 5 Sensitivity to a Drug | APPLICATION PREVIEW |

When medicine is administered, reaction (measured in change of blood pressure or temperature) can be modeled by

$$R = m^2\left(\frac{c}{2} - \frac{m}{3}\right)$$

where c is a positive constant and m is the amount of medicine absorbed into the blood. The rate of change of R with respect to m is the sensitivity of the body to medicine. Find an expression for sensitivity s as a function of m.

Solution

The sensitivity is the rate of change of R with respect to m, or the derivative. Thus

$$s = \frac{dR}{dm} = m^2\left(0 - \frac{1}{3}\right) + \left(\frac{c}{2} - \frac{1}{3}m\right)(2m)$$

$$= -\frac{1}{3}m^2 + mc - \frac{2}{3}m^2 = mc - m^2 \qquad ∎$$

CHECKPOINT SOLUTIONS

1. False. The derivative of a product is equal to the first function times the derivative of the second plus the second function times the derivative of the first. That is,

$$\frac{d}{dx}(fg) = f \cdot \frac{dg}{dx} + g \cdot \frac{df}{dx}$$

2. False. The derivative of a quotient is equal to the denominator times the derivative of the numerator minus the numerator times the derivative of the denominator, all divided by the square of the denominator. That is,

$$\frac{d}{dx}\left(\frac{f}{g}\right) = \frac{g \cdot f' - f \cdot g'}{g^2}$$

3. (a) $f'(x) = (x^{12} + 8x^5 - 7)(70x^6 - 4) + (10x^7 - 4x + 19)(12x^{11} + 40x^4)$

   (b) $f'(x) = \dfrac{(3x^4 + 2)(8x^3) - (2x^4 + 3)(12x^3)}{(3x^4 + 2)^2}$

   $= \dfrac{24x^7 + 16x^3 - 24x^7 - 36x^3}{(3x^4 + 2)^2} = \dfrac{-20x^3}{(3x^4 + 2)^2}$

4. No; y' can be found with the Coefficient Rule:

$$y' = \frac{4}{3}(2x + 3)$$

5. No; y' can be found with the Coefficient Rule:

$$y' = \left(\frac{1}{c}\right)f'(x)$$

| EXERCISES | 9.5

Access end-of-section exercises online at www.webassign.net

OBJECTIVES 9.6

- To use the Chain Rule to differentiate functions
- To use the Power Rule to differentiate functions

The Chain Rule and the Power Rule

| APPLICATION PREVIEW |

The demand x for a product is given by

$$x = \frac{98}{\sqrt{2p + 1}} - 1$$

where p is the price per unit. To find how fast demand is changing when price is $24, we take the derivative of x with respect to p. If we write this function with a power rather than a radical, it has the form

$$x = 98(2p + 1)^{-1/2} - 1$$

The formulas learned so far cannot be used to find this derivative. We use a new formula, the Power Rule, to find this derivative. (See Example 6.) In this section we will discuss the Chain Rule and the Power Rule, which is one of the results of the Chain Rule, and we will use these formulas to solve applied problems.

Composite Functions Recall from Section 1.2, "Functions," that if f and g are functions, then the composite functions g of f (denoted $g \circ f$) and f of g (denoted $f \circ g$) are defined as follows:

$$(g \circ f)(x) = g(f(x)) \quad \text{and} \quad (f \circ g)(x) = f(g(x))$$

EXAMPLE 1   Composite Function

If $f(x) = 3x^2$ and $g(x) = 2x - 1$, find $F(x) = f(g(x))$.

Solution
Substituting $g(x) = 2x - 1$ for x in $f(x)$ gives

$$f(g(x)) = f(2x - 1) = 3(2x - 1)^2 \qquad \text{or} \qquad F(x) = 3(2x - 1)^2 \qquad ■$$

Chain Rule We could find the derivative of the function $F(x) = 3(2x - 1)^2$ by expanding the expression $3(2x - 1)^2$. Then

$$F(x) = 3(4x^2 - 4x + 1) = 12x^2 - 12x + 3$$

so $F'(x) = 24x - 12$. But we can also use a very powerful rule, called the **Chain Rule**, to find derivatives of composite functions. If we write the composite function $y = f(g(x))$ in the form $y = f(u)$, where $u = g(x)$, we state the Chain Rule as follows.

Chain Rule

If f and g are differentiable functions with $y = f(u)$, and $u = g(x)$, then y is a differentiable function of x, and

$$\frac{dy}{dx} = f'(u) \cdot g'(x)$$

or, equivalently,

$$\frac{dy}{dx} = \frac{dy}{du} \cdot \frac{du}{dx}$$

Note that dy/du represents the derivative of $y = f(u)$ *with respect to u* and du/dx represents the derivative of $u = g(x)$ *with respect to x*. For example, if $y = 3(2x - 1)^2$, then the outside function, f, is the squaring function, and the inside function, g, is $2x - 1$, so we may write $y = f(u) = 3u^2$, where $u = g(x) = 2x - 1$. Then the derivative is

$$\frac{dy}{dx} = \frac{dy}{du} \cdot \frac{du}{dx} = 6u \cdot 2 = 12u$$

To write this derivative in terms of x, we substitute $2x - 1$ for u. Thus

$$\frac{dy}{dx} = 12(2x - 1) = 24x - 12$$

Note that we get the same result by using the Chain Rule as we did by expanding $f(x) = 3(2x - 1)^2$. The Chain Rule is important because it is not always possible to rewrite the function as a polynomial. Consider the following example.

EXAMPLE 2 Chain Rule

If $y = \sqrt{x^2 - 1}$, find $\dfrac{dy}{dx}$.

Solution

If we write this function as $y = f(u) = \sqrt{u} = u^{1/2}$, with $u = x^2 - 1$, we can find the derivative.

$$\frac{dy}{dx} = \frac{dy}{du} \cdot \frac{du}{dx} = \frac{1}{2} \cdot u^{-1/2} \cdot 2x = u^{-1/2} \cdot x = \frac{1}{\sqrt{u}} \cdot x = \frac{x}{\sqrt{u}}$$

To write this derivative in terms of x alone, we substitute $x^2 - 1$ for u. Then

$$\frac{dy}{dx} = \frac{x}{\sqrt{x^2 - 1}}$$

Note that we could not find the derivative of a function like that of Example 2 by the methods learned previously. However, if

$$y = \frac{1}{x^2 + 3x + 1}$$

then we can find the derivative with the Chain Rule (using $u = x^2 + 3x + 1$ and $y = u^{-1}$) or with the Quotient Rule.

EXAMPLE 3   Allometric Relationships

The relationship between the length L (in meters) and weight W (in kilograms) of a species of fish in the Pacific Ocean is given by $W = 10.375L^3$. The rate of growth in length is given by $\dfrac{dL}{dt} = 0.36 - 0.18L$, where t is measured in years.

(a) Determine a formula for the rate of growth in weight $\dfrac{dW}{dt}$ in terms of L.

(b) If a fish weighs 30 kilograms, approximate its rate of growth in weight using the formula found in part (a).

Solution

(a) The rate of change uses the Chain Rule, as follows:

$$\frac{dW}{dt} = \frac{dW}{dL} \cdot \frac{dL}{dt} = 31.125L^2 (0.36 - 0.18L) = 11.205L^2 - 5.6025L^3$$

(b) From $W = 10.375L^3$ and $W = 30$ kg, we can find L by solving

$$30 = 10.375L^3$$

$$\frac{30}{10.375} = L^3 \quad \text{so} \quad L = \sqrt[3]{\frac{30}{10.375}} \approx 1.4247 \text{ m}$$

Hence, the rate of growth in weight is

$$\frac{dW}{dt} = 11.205(1.4247)^2 - 5.6025(1.4247)^3 \approx 6.542 \text{ kilograms/year} \quad\blacksquare$$

Power Rule   The Chain Rule is very useful and will be extremely important with functions that we will study later. A special case of the Chain Rule, called the **Power Rule,** is useful for the algebraic functions we have studied so far, composite functions where the outside function is a power.

Power Rule

If $y = u^n$, where u is a differentiable function of x, then

$$\frac{dy}{dx} = nu^{n-1} \cdot \frac{du}{dx}$$

EXAMPLE 4   Power Rule

(a) If $y = (x^2 - 4x)^6$, find $\dfrac{dy}{dx}$.   (b) If $p = \dfrac{4}{3q^2 + 1}$, find $\dfrac{dp}{dq}$.

Solution

(a) This has the form $y = u^n = u^6$, with $u = x^2 - 4x$. Thus, by the Power Rule,

$$\frac{dy}{dx} = nu^{n-1} \cdot \frac{du}{dx} = 6u^5 (2x - 4)$$

Substituting $x^2 - 4x$ for u gives

$$\frac{dy}{dx} = 6(x^2 - 4x)^5 (2x - 4) = (12x - 24)(x^2 - 4x)^5$$

(b) We can use the Power Rule to find dp/dq if we write the equation in the form

$$p = 4(3q^2 + 1)^{-1}$$

Then

$$\frac{dp}{dq} = 4\left[-1(3q^2 + 1)^{-2}(6q)\right] = \frac{-24q}{(3q^2 + 1)^2}$$ ∎

The derivative of the function in Example 4(b) can also be found by using the Quotient Rule, but the Power Rule provides a more efficient method.

EXAMPLE 5 Power Rule with Radicals

Find the derivatives of (a) $y = 3\sqrt[3]{x^2 - 3x + 1}$ and (b) $g(x) = \dfrac{1}{\sqrt{(x^2 + 1)^3}}$.

Solution

(a) Because $y = 3(x^2 - 3x + 1)^{1/3}$, we can make use of the Power Rule with $u = x^2 - 3x + 1$.

$$y' = 3\left(nu^{n-1}\frac{du}{dx}\right) = 3\left[\frac{1}{3}u^{-2/3}(2x - 3)\right]$$

$$= (x^2 - 3x + 1)^{-2/3}(2x - 3) = \frac{2x - 3}{(x^2 - 3x + 1)^{2/3}}$$

(b) Writing $g(x)$ as a power gives $g(x) = (x^2 + 1)^{-3/2}$. Then

$$g'(x) = -\frac{3}{2}(x^2 + 1)^{-5/2}(2x) = -3x \cdot \frac{1}{(x^2 + 1)^{5/2}} = \frac{-3x}{\sqrt{(x^2 + 1)^5}}$$ ∎

CHECKPOINT

1. (a) If $f(x) = (3x^4 + 1)^{10}$, does $f'(x) = 10(3x^4 + 1)^9$?
   (b) If $f(x) = (2x + 1)^5$, does $f'(x) = 10(2x + 1)^4$?
   (c) If $f(x) = \dfrac{[u(x)]^n}{c}$, where c is a constant, does $f'(x) = \dfrac{n[u(x)]^{n-1} \cdot u'(x)}{c}$?

2. (a) If $f(x) = \dfrac{12}{2x^2 - 1}$, find $f'(x)$ by using the Power Rule (not the Quotient Rule).

   (b) If $f(x) = \dfrac{\sqrt{x^3 - 1}}{3}$, find $f'(x)$ by using the Power Rule (not the Quotient Rule).

EXAMPLE 6 Demand | APPLICATION PREVIEW |

The demand for x hundred units of a product is given by

$$x = 98(2p + 1)^{-1/2} - 1$$

where p is the price per unit in dollars. Find the rate of change of the demand with respect to price when $p = 24$.

Solution

The rate of change of demand with respect to price is

$$\frac{dx}{dp} = 98\left[-\frac{1}{2}(2p + 1)^{-3/2}(2)\right] = -98(2p + 1)^{-3/2}$$

When $p = 24$, the rate of change is

$$\frac{dx}{dp}\bigg|_{p=24} = -98(48 + 1)^{-3/2} = -98 \cdot \frac{1}{49^{3/2}} = -\frac{2}{7}$$

This means that when the price is $24, demand is changing at the rate of $-2/7$ hundred units per dollar, or if the price changes by $1, demand will change by about $-200/7$ units. ∎

CHECKPOINT SOLUTIONS

1. (a) No, $f'(x) = 10(3x^4 + 1)^9 (12x^3)$.   (b) Yes   (c) Yes
2. (a) $f(x) = 12(2x^2 - 1)^{-1}$

$$f'(x) = -12(2x^2 - 1)^{-2}(4x) = \frac{-48x}{(2x^2 - 1)^2}$$

(b) $f(x) = \dfrac{1}{3}(x^3 - 1)^{1/2}$

$$f'(x) = \frac{1}{6}(x^3 - 1)^{-1/2}(3x^2) = \frac{x^2}{2\sqrt{x^3 - 1}}$$

| EXERCISES | 9.6

Access end-of-section exercises online at **www.webassign.net**

OBJECTIVE   9.7

- To use derivative formulas separately and in combination with each other

Using Derivative Formulas

| APPLICATION PREVIEW |

Suppose the weekly revenue function for a product is given by

$$R(x) = \frac{36,000,000x}{(2x + 500)^2}$$

where $R(x)$ is the dollars of revenue from the sale of x units. We can find marginal revenue by finding the derivative of the revenue function. This revenue function contains both a quotient and a power, so its derivative is found by using both the Quotient Rule and the Power Rule. But first we must decide the order in which to apply these formulas. (See Example 4.)

We have used the Power Rule to find the derivatives of functions like

$$y = (x^3 - 3x^2 + x + 1)^5$$

but we have not found the derivatives of functions like

$$y = [(x^2 + 1)(x^3 + x + 1)]^5$$

This function is different because the function u (which is raised to the fifth power) is the product of two functions, $(x^2 + 1)$ and $(x^3 + x + 1)$. The equation is of the form $y = u^5$, where $u = (x^2 + 1)(x^3 + x + 1)$. This means that the Product Rule should be used to find du/dx. Then

$$\frac{dy}{dx} = 5u^4 \cdot \frac{du}{dx}$$
$$= 5[(x^2 + 1)(x^3 + x + 1)]^4 [(x^2 + 1)(3x^2 + 1) + (x^3 + x + 1)(2x)]$$
$$= 5[(x^2 + 1)(x^3 + x + 1)]^4 (5x^4 + 6x^2 + 2x + 1)$$
$$= (25x^4 + 30x^2 + 10x + 5)[(x^2 + 1)(x^3 + x + 1)]^4$$

A different type of problem involving the Power Rule and the Product Rule is finding the derivative of $y = (x^2 + 1)^5 (x^3 + x + 1)$. We may think of y as the *product* of two functions, one of which is a power. Thus the fundamental formula we should use is the Product Rule. The two functions are $u(x) = (x^2 + 1)^5$ and $v(x) = x^3 + x + 1$. The Product Rule gives

$$\frac{dy}{dx} = u(x) \cdot v'(x) + v(x) \cdot u'(x)$$
$$= (x^2 + 1)^5 (3x^2 + 1) + (x^3 + x + 1)[5(x^2 + 1)^4(2x)]$$

Note that the Power Rule was used to find $u'(x)$, since $u(x) = (x^2 + 1)^5$.
We can simplify dy/dx by factoring $(x^2 + 1)^4$ from both terms:

$$\frac{dy}{dx} = (x^2 + 1)^4[(x^2 + 1)(3x^2 + 1) + (x^3 + x + 1) \cdot 5 \cdot 2x]$$

$$= (x^2 + 1)^4(13x^4 + 14x^2 + 10x + 1)$$

EXAMPLE 1 Power of a Quotient

If $y = \left(\dfrac{x^2}{x - 1}\right)^5$, find y'.

Solution
We again have an equation of the form $y = u^n$, but this time u is a quotient. Thus we will need the Quotient Rule to find du/dx.

$$y' = nu^{n-1} \cdot \frac{du}{dx} = 5u^4 \frac{(x - 1) \cdot 2x - x^2 \cdot 1}{(x - 1)^2}$$

Substituting for u and simplifying give

$$y' = 5\left(\frac{x^2}{x - 1}\right)^4 \cdot \frac{2x^2 - 2x - x^2}{(x - 1)^2} = \frac{5x^8(x^2 - 2x)}{(x - 1)^6} = \frac{5x^{10} - 10x^9}{(x - 1)^6}$$ ∎

EXAMPLE 2 Quotient of Two Powers

Find $f'(x)$ if $f(x) = \dfrac{(x - 1)^2}{(x^4 + 3)^3}$.

Solution
This function is the quotient of two functions, $(x - 1)^2$ and $(x^4 + 3)^3$, so we must use the Quotient Rule to find the derivative of $f(x)$, but taking the derivatives of $(x - 1)^2$ and $(x^4 + 3)^3$ will require the Power Rule.

$$f'(x) = \frac{[v(x) \cdot u'(x) - u(x) \cdot v'(x)]}{[v(x)]^2}$$

$$= \frac{(x^4 + 3)^3[2(x - 1)(1)] - (x - 1)^2[3(x^4 + 3)^2 \, 4x^3]}{[(x^4 + 3)^3]^2}$$

$$= \frac{2(x^4 + 3)^3(x - 1) - 12x^3(x - 1)^2(x^4 + 3)^2}{(x^4 + 3)^6}$$

We see that 2, $(x^4 + 3)^2$, and $(x - 1)$ are all factors in both terms of the numerator, so we can factor them from both terms and reduce the fraction.

$$f'(x) = \frac{2(x^4 + 3)^2(x - 1)[(x^4 + 3) - 6x^3(x - 1)]}{(x^4 + 3)^6}$$

$$= \frac{2(x - 1)(-5x^4 + 6x^3 + 3)}{(x^4 + 3)^4}$$ ∎

EXAMPLE 3 Product with a Power

Find $f'(x)$ if $f(x) = (x^2 - 1)\sqrt{3 - x^2}$.

Solution

The function is the product of two functions, $x^2 - 1$ and $\sqrt{3 - x^2}$. Therefore, we will use the Product Rule to find the derivative of $f(x)$, but the derivative of $\sqrt{3 - x^2} = (3 - x^2)^{1/2}$ will require the Power Rule.

$$f'(x) = u(x) \cdot v'(x) + v(x) \cdot u'(x)$$

$$= (x^2 - 1)\left[\frac{1}{2}(3 - x^2)^{-1/2}(-2x)\right] + (3 - x^2)^{1/2}(2x)$$

$$= (x^2 - 1)[-x(3 - x^2)^{-1/2}] + (3 - x^2)^{1/2}(2x)$$

$$= \frac{-x^3 + x}{(3 - x^2)^{1/2}} + 2x(3 - x^2)^{1/2}$$

We can combine these terms over the common denominator $(3 - x^2)^{1/2}$ as follows:

$$f'(x) = \frac{-x^3 + x}{(3 - x^2)^{1/2}} + \frac{2x(3 - x^2)^1}{(3 - x^2)^{1/2}} = \frac{-x^3 + x + 6x - 2x^3}{(3 - x^2)^{1/2}} = \frac{-3x^3 + 7x}{(3 - x^2)^{1/2}}$$ ■

We should note that in Example 3 we could have written $f'(x)$ in the form

$$f'(x) = (-x^3 + x)(3 - x^2)^{-1/2} + 2x(3 - x^2)^{1/2}$$

Now the factor $(3 - x^2)$, to different powers, is contained in both terms of the expression. Thus we can factor $(3 - x^2)^{-1/2}$ from both terms. (We choose the $-1/2$ power because it is the smaller of the two powers.) Dividing $(3 - x^2)^{-1/2}$ into the first term gives $(-x^3 + x)$, and dividing it into the second term gives $2x(3 - x^2)^1$. (Why?) Thus we have

$$f'(x) = (3 - x^2)^{-1/2}[(-x^3 + x) + 2x(3 - x^2)] = \frac{-3x^3 + 7x}{(3 - x^2)^{1/2}}$$

which agrees with our previous answer.

CHECKPOINT

1. If a function has the form $y = [u(x)]^n \cdot v(x)$, where n is a constant, we begin to find the derivative by using the _____ Rule and then use the _____ Rule to find the derivative of $[u(x)]^n$.
2. If a function has the form $y = [u(x)/v(x)]^n$, where n is a constant, we begin to find the derivative by using the _____ Rule and then use the _____ Rule.
3. Find the derivative of each of the following and simplify.

    (a) $f(x) = 3x^4(2x^4 + 7)^5$ (b) $g(x) = \dfrac{(4x + 3)^7}{2x - 9}$

EXAMPLE 4   **Revenue | APPLICATION PREVIEW |**

Suppose that the weekly revenue function for a product is given by

$$R(x) = \frac{36{,}000{,}000x}{(2x + 500)^2}$$

where $R(x)$ is the dollars of revenue from the sale of x units.

(a) Find the marginal revenue function.
(b) Find the marginal revenue when 50 units are sold.

Solution

(a) $\overline{MR} = R'(x)$

$$= \frac{(2x + 500)^2(36{,}000{,}000) - 36{,}000{,}000x[2(2x + 500)^1(2)]}{(2x + 500)^4}$$

$$= \frac{36{,}000{,}000(2x + 500)(2x + 500 - 4x)}{(2x + 500)^4} = \frac{36{,}000{,}000(500 - 2x)}{(2x + 500)^3}$$

(b) $\overline{MR}(50) = R'(50) = \dfrac{36{,}000{,}000(500 - 100)}{(100 + 500)^3} = \dfrac{36{,}000{,}000(400)}{(600)^3} = \dfrac{200}{3} \approx 66.67$

The marginal revenue is \$66.67 when 50 units are sold. That is, the predicted revenue from the sale of the 51st unit is approximately \$66.67. ■

It may be helpful to review the formulas needed to find the derivatives of various types of functions. Table 9.5 presents examples of different types of functions and the formulas needed to find their derivatives.

| TABLE 9.5 |

SUMMARY OF DERIVATIVE FORMULAS

Examples	Formulas
$f(x) = 14$	If $f(x) = c$, then $f'(x) = 0$.
$y = x^4$	If $f(x) = x^n$, then $f'(x) = nx^{n-1}$.
$g(x) = 5x^3$	If $g(x) = cf(x)$, then $g'(x) = cf'(x)$.
$y = 3x^2 + 4x$	If $f(x) = u(x) + v(x)$, then $f'(x) = u'(x) + v'(x)$.
$y = (x^2 - 2)(x + 4)$	If $f(x) = u(x) \cdot v(x)$, then $f'(x) = u(x) \cdot v'(x) + v(x) \cdot u'(x)$.
$f(x) = \dfrac{x^3}{x^2 + 1}$	If $f(x) = \dfrac{u(x)}{v(x)}$ then $f'(x) = \dfrac{v(x) \cdot u'(x) - u(x) \cdot v'(x)}{[v(x)]^2}$.
$y = (x^3 - 4x)^{10}$	If $y = u^n$ and $u = g(x)$, then $\dfrac{dy}{dx} = nu^{n-1} \cdot \dfrac{du}{dx}$.
$y = \left(\dfrac{x - 1}{x^2 + 3}\right)^3$	Power Rule, then Quotient Rule to find $\dfrac{du}{dx}$, where $u = \dfrac{x - 1}{x^2 + 3}$.
$y = (x + 1)\sqrt{x^3 + 1}$	Product Rule, then Power Rule to find $v'(x)$, where $v(x) = \sqrt{x^3 + 1}$.
$y = \dfrac{(x^2 - 3)^4}{x + 1}$	Quotient Rule, then Power Rule to find the derivative of the numerator.

CHECKPOINT SOLUTIONS

1. Product, Power
2. Power, Quotient
3. (a) $f'(x) = 3x^4[5(2x^4 + 7)^4(8x^3)] + (2x^4 + 7)^5(12x^3)$

   $= 120x^7(2x^4 + 7)^4 + 12x^3(2x^4 + 7)^5$

   $= 12x^3(2x^4 + 7)^4[10x^4 + (2x^4 + 7)] = 12x^3(12x^4 + 7)(2x^4 + 7)^4$

   (b) $g'(x) = \dfrac{(2x - 9)[7(4x + 3)^6(4)] - (4x + 3)^7(2)}{(2x - 9)^2}$

   $= \dfrac{2(4x + 3)^6[14(2x - 9) - (4x + 3)]}{(2x - 9)^2} = \dfrac{2(24x - 129)(4x + 3)^6}{(2x - 9)^2}$

| **EXERCISES** | **9.7** |

Access end-of-section exercises online at **www.webassign.net**

| **OBJECTIVE** | **9.8** |

- To find second derivatives and higher-order derivatives of certain functions

Higher-Order Derivatives

| APPLICATION PREVIEW |

Since cell phones were introduced, their popularity has increased enormously. Figure 9.32(a) shows a graph of the millions of worldwide cellular subscribers (actual and projected) as a function of the number of years past 1995 (*Source:* International Telecommunications Union). Note that the number of subscribers is always increasing and that the rate of change of that number (as seen from tangent lines to the graph) is always positive. However, the tangent lines shown in Figure 9.32(b) indicate that the rate of change of the number of subscribers is greater at *B* than at either *A* or *C*.

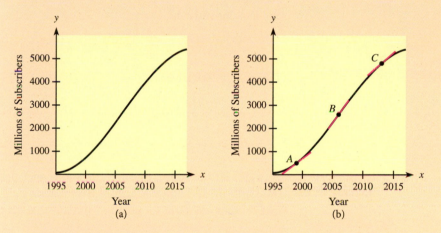

Figure 9.32

Furthermore, the rate of change of the number of subscribers (the slopes of tangents) increases from *A* to *B* and then decreases from *B* to *C*. To learn how the rate of change of the number of subscribers is changing, we are interested in finding the derivative of the rate of change of the number of subscribers—that is, the derivative of the derivative of the number of subscribers. (See Example 4.) This is called the second derivative. In this section we will discuss second and higher-order derivatives.

Second Derivatives
Because the derivative of a function is itself a function, we can take a derivative of the derivative. The derivative of a first derivative is called a **second derivative.** We can find the second derivative of a function f by differentiating it twice. If f' represents the first derivative of a function, then f'' represents the second derivative of that function.

EXAMPLE 1   Second Derivative

(a) Find the second derivative of $y = x^4 - 3x^2 + x^{-2}$.
(b) If $f(x) = 3x^3 - 4x^2 + 5$, find $f''(x)$.

Solution

(a) The first derivative is $y' = 4x^3 - 6x - 2x^{-3}$.
The second derivative, which we may denote by y'', is

$$y'' = 12x^2 - 6 + 6x^{-4}$$

(b) The first derivative is $f'(x) = 9x^2 - 8x$.
The second derivative is $f''(x) = 18x - 8$.   ■

It is also common to use $\dfrac{d^2y}{dx^2}$ and $\dfrac{d^2f(x)}{dx^2}$ to denote the second derivative of a function.

EXAMPLE 2   Second Derivative

If $y = \sqrt{2x - 1}$, find d^2y/dx^2.

Solution
The first derivative is

$$\frac{dy}{dx} = \frac{1}{2}(2x - 1)^{-1/2}(2) = (2x - 1)^{-1/2}$$

The second derivative is

$$\frac{d^2y}{dx^2} = -\frac{1}{2}(2x - 1)^{-3/2}(2) = -(2x - 1)^{-3/2}$$

$$= \frac{-1}{(2x - 1)^{3/2}} = \frac{-1}{\sqrt{(2x - 1)^3}}$$ ■

Higher-Order Derivatives   We can also find third, fourth, fifth, and higher derivatives, continuing indefinitely. The third, fourth, and fifth derivatives of a function f are denoted by f''', $f^{(4)}$, and $f^{(5)}$, respectively. Other notations for the third and fourth derivatives include

$$y''' = \frac{d^3y}{dx^3} = \frac{d^3f(x)}{dx^3}, \qquad y^{(4)} = \frac{d^4y}{dx^4} = \frac{d^4f(x)}{dx^4}$$

EXAMPLE 3   Higher-Order Derivatives

Find the first four derivatives of $f(x) = 4x^3 + 5x^2 + 3$.

Solution

$$f'(x) = 12x^2 + 10x, \quad f''(x) = 24x + 10, \quad f'''(x) = 24, \quad f^{(4)}(x) = 0   ■$$

Just as the first derivative, $f'(x)$, can be used to determine the rate of change of a function $f(x)$, the second derivative, $f''(x)$, can be used to determine the rate of change of $f'(x)$.

EXAMPLE 4   Worldwide Cellular Subscriberships | APPLICATION PREVIEW |

By using the International Telecommunications Union data, the millions of worldwide cellular subscribers (both actual and projected) can be modeled by

$$C(t) = -0.895t^3 + 30.6t^2 + 2.99t + 55.6$$

where t is the number of years past 1995.

(a) Find the instantaneous rate of change of the worldwide cellular subscribers function.
(b) Find the instantaneous rate of change of the function found in part (a).
(c) Find where the function in part (b) equals zero. Then explain how the rate of change of the number of cellular subscribers is changing for one t-value before and one after this value.

Solution

(a) The instantaneous rate of change of $C(t)$ is

$$C'(t) = -2.685t^2 + 61.2t + 2.99$$

(b) The instantaneous rate of change of $C'(t)$ is

$$C''(t) = -5.37t + 61.2$$

(c) $0 = -5.37t + 61.2 \Rightarrow 5.37t = 61.2 \quad$ so $\quad t = \dfrac{61.2}{5.37} \approx 11.4$

For $t = 11$, which is less than 11.4,

$$C''(11) = -5.37(11) + 61.2 = 2.13$$

This means that for $t = 11$, the rate of change of the number of cellular subscribers is changing at the rate of 2.13 million subscribers per year, per year. Thus the number of cellular subscribers is increasing at an increasing rate. See Figure 9.32(a) earlier in this section. For $t = 12$,

$$C''(12) = -5.37(12) + 61.2 = -3.24$$

This means that the rate of change of the number of cellular subscribers is changing at the rate of -3.24 million subscribers per year, per year. Thus the number of cellular subscribers is (still) increasing but at a decreasing (slower) rate. See Figure 9.32(a). ∎

EXAMPLE 5 Rate of Change of a Derivative

Let $f(x) = 3x^4 + 6x^3 - 3x^2 + 4$.

(a) How fast is $f(x)$ changing at $(1, 10)$?
(b) How fast is $f'(x)$ changing at $(1, 10)$?
(c) Is $f'(x)$ increasing or decreasing at $(1, 10)$?

Solution

(a) Because $f'(x) = 12x^3 + 18x^2 - 6x$, we have

$$f'(1) = 12 + 18 - 6 = 24$$

Thus the rate of change of $f(x)$ at $(1, 10)$ is 24 (y units per x unit).
(b) Because $f''(x) = 36x^2 + 36x - 6$, we have

$$f''(1) = 66$$

Thus the rate of change of $f'(x)$ at $(1, 10)$ is 66 (y units per x unit per x unit).
(c) Because $f''(1) = 66 > 0$, $f'(x)$ is increasing at $(1, 10)$. ∎

EXAMPLE 6 Acceleration

Suppose that a particle travels according to the equation

$$s = 100t - 16t^2 + 200$$

where s is the distance in feet and t is the time in seconds. Then ds/dt is the velocity, and $d^2s/dt^2 = dv/dt$ is the acceleration of the particle. Find the acceleration.

Solution

The velocity is $v = ds/dt = 100 - 32t$ feet per second, and the acceleration is

$$\frac{dv}{dt} = \frac{d^2s}{dt^2} = -32 \text{ (feet/second)/second} = -32 \text{ feet/second}^2 \qquad \blacksquare$$

CHECKPOINT

Suppose that the distance a particle travels is given by

$$s = 4x^3 - 12x^2 + 6$$

where s is in feet and x is in seconds.

1. Find the function that describes the velocity of this particle.
2. Find the function that describes the acceleration of this particle.
3. Is the acceleration always positive?
4. When does the *velocity* of this particle increase?

Calculator Note

We can use the numerical derivative feature of a graphing calculator to find the second derivative of a function at a point. Figure 9.33 shows how the numerical derivative feature of a graphing calculator can be used to find $f''(2)$ if $f(x) = \sqrt{x^3 - 1}$. See Appendix C, Section 9.8, for details.

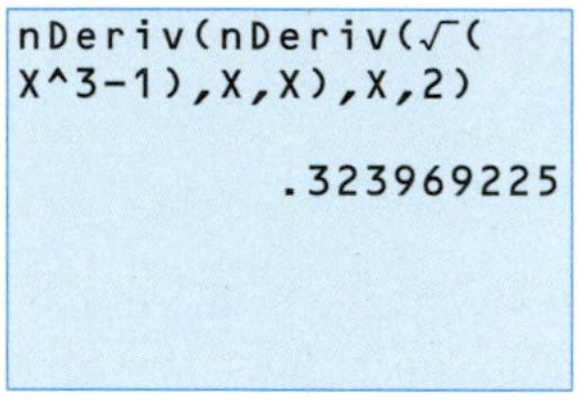

```
nDeriv(nDeriv(√(
X^3-1),X,X),X,2)

             .323969225
```

Figure 9.33

The figure shows that $f''(2) = 0.323969225 \approx 0.32397$. We can check this result by calculating $f''(x)$ with formulas.

$$f'(x) = \frac{1}{2}(x^3 - 1)^{-\frac{1}{2}}(3x^2)$$

$$f''(x) = \frac{1}{2}(x^3 - 1)^{-\frac{1}{2}}(6x) + (3x^2)\left[-\frac{1}{4}(x^3 - 1)^{-\frac{3}{2}}(3x^2)\right]$$

$$f''(2) = 0.3239695483 \approx 0.32397$$

Thus we see that the numerical derivative approximation is quite accurate. $\qquad \blacksquare$

CHECKPOINT SOLUTIONS

1. The velocity is described by $s'(x) = 12x^2 - 24x$.
2. The acceleration is described by $s''(x) = 24x - 24$.
3. No; the acceleration is positive when $s''(x) > 0$—that is, when $24x - 24 > 0$. It is zero when $24x - 24 = 0$ and negative when $24x - 24 < 0$. Thus acceleration is negative when $x < 1$ second, zero when $x = 1$ second, and positive when $x > 1$ second.
4. The velocity increases when the acceleration is positive. Thus the velocity is increasing after 1 second.

| EXERCISES | 9.8

Access end-of-section exercises online at **www.webassign.net**

Applications: Marginals and Derivatives

| APPLICATION PREVIEW |

If the total cost in dollars to produce x kitchen blenders is given by

$$C(x) = 0.001x^3 - 0.3x^2 + 32x + 2500$$

what is the marginal cost function and what does it tell us about how costs are changing when $x = 80$ and $x = 200$ blenders are produced? See Example 3.

In Section 1.6, "Applications of Functions in Business and Economics," we defined the marginals for linear total cost, total revenue, and profit functions as the rate of change or slope of the respective function. In Section 9.3, we extended the notion of marginal revenue to nonlinear total revenue functions by defining marginal revenue as the derivative of total revenue. In this section, we extend the notion of marginal to nonlinear functions for any total cost or profit function.

Marginal Revenue   As we saw earlier, the instantaneous rate of change (the derivative) of the revenue function gives the **marginal revenue function.**

Marginal Revenue   If $R = R(x)$ is the total revenue function for a commodity, then the **marginal revenue function** is $\overline{MR} = R'(x)$.

Recall that if the demand function for a product in a monopoly market is $p = f(x)$, then the total revenue from the sale of x units is

$$R(x) = px = f(x) \cdot x$$

EXAMPLE   1   Revenue and Marginal Revenue

If the demand for a product in a monopoly market is given by

$$p = 16 - 0.02x$$

where x is the number of units and p is the price per unit, (a) find the total revenue function, and (b) find the marginal revenue for this product at $x = 40$.

Solution

(a) The total revenue function is

$$R(x) = px = (16 - 0.02x)x = 16x - 0.02x^2$$

(b) The marginal revenue function is

$$\overline{MR} = R'(x) = 16 - 0.04x$$

At $x = 40$, $R'(40) = 16 - 1.6 = 14.40$ dollars per unit. Thus the 41st item sold will increase the total revenue by approximately $14.40.   ■

The marginal revenue is an approximation or estimate of the revenue gained from the sale of 1 additional unit. We have used marginal revenue in Example 1 to find that the revenue from the sale of the 41st item will be approximately $14.40. The actual increase in revenue from the sale of the 41st item is

$$R(41) - R(40) = 622.38 - 608 = \$14.38$$

Marginal revenue (and other marginals) can be used to predict for more than one additional unit. For instance, in Example 1, $\overline{MR}(40) = \$14.40$ per unit means that the expected or approximate revenue for the 41st through the 45th items sold would be $5(\$14.40) = \$72.00$. The actual revenue for these 5 items is $R(45) - R(40) = \$679.50 - \$608 = \$71.50$.

EXAMPLE 2   Maximum Revenue

Use the graphs in Figure 9.34 to determine the x-value where the revenue function has its maximum. What is happening to the marginal revenue at and near this x-value?

Solution
Figure 9.34(a) shows that the total revenue function has a maximum value at $x = 400$. After that, the total revenue function decreases. This means that the total revenue will be reduced each time a unit is sold if more than 400 are produced and sold. The graph of the marginal revenue function in Figure 9.34(b) shows that the marginal revenue is positive to the left of 400. This indicates that the rate at which the total revenue is changing is positive until 400 units are sold; thus the total revenue is increasing. Then, at 400 units, the rate of change is 0. After 400 units are sold, the marginal revenue is negative, which indicates that the total revenue is now decreasing. It is clear from looking at either graph that 400 units should be produced and sold to maximize the total revenue function $R(x)$. That is, the *total revenue function has its maximum at $x = 400$.* ∎

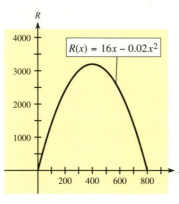

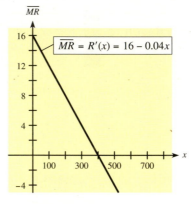

Figure 9.34   (a) Total revenue function   (b) Marginal revenue function

Marginal Cost   As with marginal revenue, the derivative of a total cost function gives the **marginal cost function.**

Marginal Cost

If $C = C(x)$ is a total cost function for a commodity, then its derivative, $\overline{MC} = C'(x)$, is the **marginal cost function.**

Notice that the linear total cost function with equation

$$C(x) = 300 + 6x \qquad \text{(in dollars)}$$

has marginal cost $6 per unit because its slope is 6. Taking the derivative of $C(x)$ also gives

$$\overline{MC} = C'(x) = 6$$

which verifies that the marginal cost is $6 per unit at all levels of production.

EXAMPLE 3   Marginal Cost | APPLICATION PREVIEW |

Suppose the daily total cost in dollars for a certain factory to produce x kitchen blenders is

$$C(x) = 0.001x^3 - 0.3x^2 + 32x + 2500$$

(a) Find the marginal cost function for these blenders.
(b) Find and interpret the marginal cost when $x = 80$ and $x = 200$.

Solution

(a) The marginal cost function is the derivative of $C(x)$.

$$\overline{MC} = C'(x) = 0.003x^2 - 0.6x + 32$$

(b) $C'(80) = 0.003(80)^2 - 0.6(80) + 32 = 3.2$ dollars per unit
$C'(200) = 0.003(200)^2 - 0.6(200) + 32 = 32$ dollars per unit

These values for the marginal cost can be used to *estimate* the amount that total cost would change if production were increased by one blender. Thus,

$$C'(80) = 3.2$$

means total cost would increase by *about* $3.20 if an 81st blender were produced. Note that

$$C(81) - C(80) = 3.141 \approx \$3.14$$

is the actual increase in total cost for an 81st blender.

Also, $C'(200) = 32$ means that total cost would increase by *about* $32 if a 201st blender were produced.

Whenever $C'(x)$ is positive it means that an additional unit produced adds to or increases the total cost. In addition, the value of the derivative (or marginal cost) measures how fast $C(x)$ is increasing. Thus, our calculations above indicate that $C(x)$ is increasing faster at $x = 200$ than at $x = 80$. The graphs of the total cost function and the marginal cost function in Figure 9.35(a) and (b) also illustrate these facts. ∎

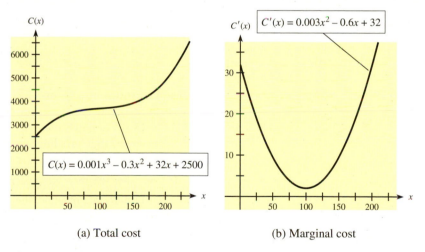

Figure 9.35   (a) Total cost   (b) Marginal cost

The graphs of many marginal cost functions tend to be U-shaped; they eventually will rise, even though there may be an initial interval where they decrease. Looking at the marginal cost graph in Figure 9.35(b), we see that marginal cost reaches its minimum near $x = 100$. We can also see this in Figure 9.35(a) by noting that tangent lines drawn to the total cost graph would have slopes that decrease until about $x = 100$ and then would increase.

Because producing more units can never reduce the total cost of production, the following properties are valid for total cost functions.

1. The total cost can never be negative. If there are fixed costs, the cost of producing 0 units is positive; otherwise, the cost of producing 0 units is 0.
2. The total cost function is always increasing; the more units produced, the higher the total cost. Thus the marginal cost is always positive.
3. There may be limitations on the units produced, such as those imposed by plant space.

C(x)

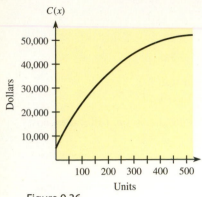

Figure 9.36

C(x)

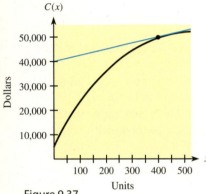

Figure 9.37

EXAMPLE 4   Total and Marginal Cost

Suppose the graph in Figure 9.36 shows the monthly total cost for producing a product.

(a) Estimate the total cost of producing 400 items.
(b) Estimate the cost of producing the 401st item.
(c) Will producing the 151st item cost more or less than producing the 401st item?

Solution

(a) The total cost of producing 400 items is the height of the total cost graph when $x = 400$, or about $50,000.

(b) The approximate cost of the 401st item is the marginal cost when $x = 400$, or the slope of the tangent line drawn to the graph at $x = 400$. Figure 9.37 shows the total cost graph with a tangent line at $x = 400$.

Note that the tangent line passes through the point $(0, 40,000)$, so we can find the slope by using the points $(0, 40,000)$ and $(400, 50,000)$.

$$m = \frac{y_2 - y_1}{x_2 - x_1} = \frac{50,000 - 40,000}{400 - 0} = \frac{10,000}{400} = 25$$

Thus the marginal cost at $x = 400$ is 25 dollars per item, so the approximate cost of the 401st item is $25.

(c) From Figure 9.37 we can see that if a tangent line to the graph were drawn where $x = 150$, it would be steeper than the one at $x = 400$. Because the slope of the tangent is the marginal cost and the marginal cost predicts the cost of the next item, this means that it would cost more to produce the 151st item than to produce the 401st. ∎

CHECKPOINT

Suppose the total cost function for a commodity is $C(x) = 0.01x^3 - 0.9x^2 + 33x + 3000$.

1. Find the marginal cost function.
2. What is the marginal cost if $x = 50$ units are produced?
3. Use marginal cost to estimate the cost of producing the 51st unit.
4. Calculate $C(51) - C(50)$ to find the actual cost of producing the 51st unit.
5. True or false: For products that have linear cost functions, the actual cost of producing the $(x + 1)$st unit is equal to the marginal cost at x.

Marginal Profit

As with marginal cost and marginal revenue, the derivative of a profit function for a commodity will give us the **marginal profit function** for the commodity.

Marginal Profit

If $P = P(x)$ is the profit function for a commodity, then the **marginal profit function** is $\overline{MP} = P'(x)$.

EXAMPLE 5   Marginal Profit

If the total profit, in thousands of dollars, for a product is given by $P(x) = 20\sqrt{x + 1} - 2x - 22$, what is the marginal profit at a production level of 15 units?

Solution

The marginal profit function is

$$\overline{MP} = P'(x) = 20 \cdot \frac{1}{2}(x + 1)^{-1/2} - 2 = \frac{10}{\sqrt{x + 1}} - 2$$

If 15 units are produced, the marginal profit is

$$P'(15) = \frac{10}{\sqrt{15 + 1}} - 2 = \frac{1}{2}$$

This means that the profit from the sale of the 16th unit is approximately $\frac{1}{2}$ (thousand dollars), or $500. ■

In a **competitive market,** each firm is so small that its actions in the market cannot affect the price of the product. The price of the product is determined in the market by the intersection of the market demand curve (from all consumers) and the market supply curve (from all firms that supply this product). The firm can sell as little or as much as it desires at the given market price, which it cannot change.

Therefore, a firm in a competitive market has a total revenue function given by $R(x) = px$, where p is the market equilibrium price for the product and x is the quantity sold.

EXAMPLE 6   Profit in a Competitive Market

A firm in a competitive market must sell its product for $200 per unit. The cost per unit (per month) is $80 + x$, where x represents the number of units sold per month. Find the marginal profit function.

Solution

If the cost per unit is $80 + x$, then the total cost of x units is given by the equation $C(x) = (80 + x)x = 80x + x^2$. The revenue per unit is $200, so the total revenue is given by $R(x) = 200x$. Thus the profit function is

$$P(x) = R(x) - C(x) = 200x - (80x + x^2), \quad \text{or} \quad P(x) = 120x - x^2$$

The marginal profit is $P'(x) = 120 - 2x$. ■

The marginal profit in Example 6 is not always positive, so producing and selling a certain number of items will maximize profit. Note that the marginal profit will be negative (that is, profit will decrease) if more than 60 items per month are produced. We will discuss methods of maximizing total revenue and profit, and of minimizing average cost, in the next chapter.

CHECKPOINT   If the total profit function for a product is $P(x) = 20\sqrt{x + 1} - 2x - 22$, then the marginal profit is

$$P'(x) = \frac{10}{\sqrt{x + 1}} - 2 \quad \text{and} \quad P''(x) = \frac{-5}{\sqrt{(x + 1)^3}}$$

6. Is $P''(x) < 0$ for all values of $x \geq 0$?
7. Is the marginal profit decreasing for all $x \geq 0$?

EXAMPLE 7   Marginal Profit

Figure 9.38 on the next page shows graphs of a company's total revenue and total cost functions.

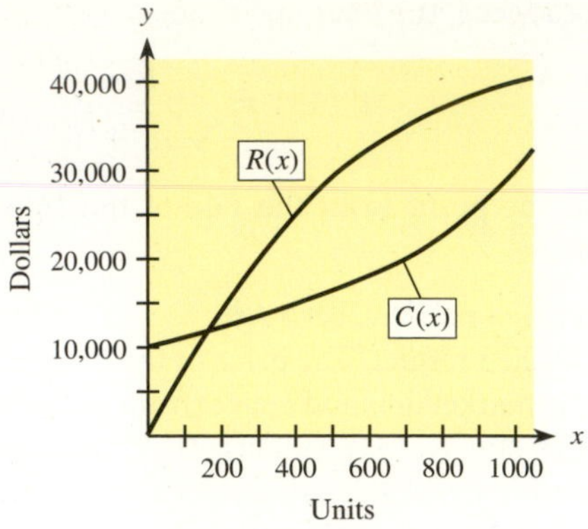

y

40,000

30,000 $\boxed{R(x)}$

Dollars

20,000

$\boxed{C(x)}$

10,000

200   400   600   800   1000   x

Units

Figure 9.38

(a) If 100 units are being produced and sold, will producing the 101st item increase or decrease the profit?
(b) If 300 units are being produced and sold, will producing the 301st item increase or decrease the profit?
(c) If 1000 units are being produced and sold, will producing the 1001st item increase or decrease the profit?

Solution

At a given x-value, the slope of the tangent line to each function gives the marginal at that x-value.

(a) At $x = 100$, we can see that the graph of $R(x)$ is steeper than the graph of $C(x)$. Thus, the tangent line to $R(x)$ will be steeper than the tangent line to $C(x)$. Hence $\overline{MR}(100) > \overline{MC}(100)$, which means that the revenue from the 101st item will exceed the cost. Therefore, profit will increase when the 101st item is produced. Note that at $x = 100$, total costs are greater than total revenue, so the company is losing money but should still sell the 101st item because it will reduce the amount of loss.
(b) At $x = 300$, we can see that the tangent line to $R(x)$ again will be steeper than the tangent line to $C(x)$. Hence $\overline{MR}(300) > \overline{MC}(300)$, which means that the revenue from the 301st item will exceed the cost. Therefore, profit will increase when the 301st item is produced.
(c) At $x = 1000$, we can see that the tangent line to $C(x)$ will be steeper than the tangent line to $R(x)$. Hence $\overline{MC}(1000) > \overline{MR}(1000)$, which means that the cost of the 1001st item will exceed the revenue. Therefore, profit will decrease when the 1001st item is produced. ∎

EXAMPLE 8 Profit and Marginal Profit

In Example 5, we found that the profit (in thousands of dollars) for a company's products is given by $P(x) = 20\sqrt{x + 1} - 2x - 22$ and its marginal profit is given by $P'(x) = \dfrac{10}{\sqrt{x + 1}} - 2$.

(a) Use the graphs of $P(x)$ and $P'(x)$ to determine the relationship between the two functions.
(b) When is the marginal profit 0? What is happening to profit at this level of production?

Solution

(a) By comparing the graphs of the two functions (shown in Figure 9.39 on the next page), we see that for $x > 0$, profit $P(x)$ is increasing over the interval where the marginal profit $P'(x)$ is positive, and profit is decreasing over the interval where the marginal profit $P'(x)$ is negative.

(b) By using ZERO or SOLVER, or by using algebra, we see that $P'(x) = 0$ when $x = 24$. This level of production ($x = 24$) is where profit is maximized, at 30 (thousand dollars). ■

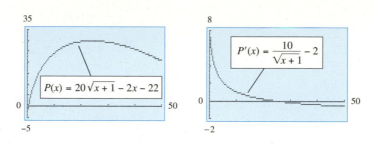

Figure 9.39

CHECKPOINT SOLUTIONS

1. $\overline{MC} = C'(x) = 0.03x^2 - 1.8x + 33$
2. $C'(50) = 18$
3. $C'(50) = 18$, so it will cost approximately $18 to produce the 51st unit.
4. $C(51) - C(50) = 3668.61 - 3650 = 18.61$
5. True   6. Yes   7. Yes, because $P''(x) < 0$ for $x \geq 0$.

| EXERCISES | 9.9

Access end-of-section exercises online at www.webassign.net

KEY TERMS AND FORMULAS

Section	Key Terms	Formulas
9.1	Limit	
	One-sided limits	
	Properties of limits	
	0/0 indeterminate form	
9.2	Continuous function	
	Vertical asymptote	
	Horizontal asymptote	
	Limit at infinity	
9.3	Average rate of change of f over $[a, b]$	$\dfrac{f(b) - f(a)}{b - a}$
	Average velocity	
	Velocity	
	Instantaneous rate of change	
	Derivative	$f'(x) = \lim\limits_{h \to 0} \dfrac{f(x + h) - f(x)}{h}$
	Marginal revenue	$\overline{MR} = R'(x)$
	Tangent line	
	Secant line	
	Slope of a curve	
	Differentiability and continuity	

Section	Key Terms	Formulas
9.4	Powers of x Rule	$\dfrac{d(x^n)}{dx} = nx^{n-1}$
	Constant Function Rule	$\dfrac{d(c)}{dx} = 0$ for constant c
	Coefficient Rule	$\dfrac{d}{dx}[c \cdot f(x)] = c \cdot f'(x)$
	Sum Rule	$\dfrac{d}{dx}(u + v) = \dfrac{du}{dx} + \dfrac{dv}{dx}$
	Difference Rule	$\dfrac{d}{dx}(u - v) = \dfrac{du}{dx} - \dfrac{dv}{dx}$
9.5	Product Rule	$\dfrac{d}{dx}(uv) = uv' + vu'$
	Quotient Rule	$\dfrac{d}{dx}\left(\dfrac{u}{v}\right) = \dfrac{vu' - uv'}{v^2}$
9.6	Chain Rule	$\dfrac{dy}{dx} = \dfrac{dy}{du} \cdot \dfrac{du}{dx}$
	Power Rule	$\dfrac{d}{dx}(u^n) = nu^{n-1}\dfrac{du}{dx}$
9.8	Second derivative; third derivative; higher-order derivatives	
9.9	Marginal revenue function	$\overline{MR} = R'(x)$
	Marginal cost function	$\overline{MC} = C'(x)$
	Marginal profit function	$\overline{MP} = P'(x)$

REVIEW EXERCISES

In Problems 1–6, use the graph of $y = f(x)$ in Figure 9.40 to find the functional values and limits, if they exist.

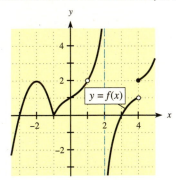

Figure 9.40

1. (a) $f(-2)$ (b) $\lim\limits_{x \to -2} f(x)$
2. (a) $f(-1)$ (b) $\lim\limits_{x \to -1} f(x)$
3. (a) $f(4)$ (b) $\lim\limits_{x \to 4^-} f(x)$
4. (a) $\lim\limits_{x \to 4^+} f(x)$ (b) $\lim\limits_{x \to 4} f(x)$
5. (a) $f(1)$ (b) $\lim\limits_{x \to 1} f(x)$
6. (a) $f(2)$ (b) $\lim\limits_{x \to 2} f(x)$

In Problems 7–20, find each limit, if it exists.

7. $\lim\limits_{x \to 4} (3x^2 + x + 3)$
8. $\lim\limits_{x \to 4} \dfrac{x^2 - 16}{x + 4}$
9. $\lim\limits_{x \to -1} \dfrac{x^2 - 1}{x + 1}$
10. $\lim\limits_{x \to 5} \dfrac{x^2 - 6x + 5}{x^2 - 5x}$
11. $\lim\limits_{x \to 2} \dfrac{4x^3 - 8x^2}{4x^3 - 16x}$
12. $\lim\limits_{x \to -\frac{1}{2}} \dfrac{x^2 - \frac{1}{4}}{6x^2 + x - 1}$
13. $\lim\limits_{x \to 3} \dfrac{x^2 - 16}{x - 3}$
14. $\lim\limits_{x \to -3} \dfrac{x^2 - x - 12}{2x - 6}$
15. $\lim\limits_{x \to 1} \dfrac{x^2 - 9}{x - 3}$
16. $\lim\limits_{x \to 2} \dfrac{x^2 - 8}{x - 2}$

17. $\lim\limits_{x \to 1} f(x)$ where $f(x) = \begin{cases} 4 - x^2 & \text{if } x < 1 \\ 4 & \text{if } x = 1 \\ 2x + 1 & \text{if } x > 1 \end{cases}$

18. $\lim\limits_{x \to -2} f(x)$ where $f(x) = \begin{cases} x^3 - x & \text{if } x < -2 \\ 2 - x^2 & \text{if } x \geq -2 \end{cases}$

19. $\lim\limits_{h \to 0} \dfrac{3(x + h)^2 - 3x^2}{h}$

20. $\lim\limits_{h \to 0} \dfrac{[(x + h) - 2(x + h)^2] - (x - 2x^2)}{h}$

In Problems 21 and 22, use tables to investigate each limit. Check your result analytically or graphically.

21. $\lim\limits_{x \to 2} \dfrac{x^2 + 10x - 24}{x^2 - 5x + 6}$ 22. $\lim\limits_{x \to -\frac{1}{2}} \dfrac{x^2 + \frac{1}{6}x - \frac{1}{6}}{x^2 + \frac{5}{6}x + \frac{1}{6}}$

Use the graph of $y = f(x)$ in Figure 9.40 to answer the questions in Problems 23 and 24.

23. Is $f(x)$ continuous at
    (a) $x = -1$?   (b) $x = 1$?
24. Is $f(x)$ continuous at
    (a) $x = -2$?   (b) $x = 2$?

In Problems 25–30, suppose that

$$f(x) = \begin{cases} x^2 + 1 & \text{if } x \leq 0 \\ x & \text{if } 0 < x < 1 \\ 2x^2 - 1 & \text{if } x \geq 1 \end{cases}$$

25. What is $\lim\limits_{x \to -1} f(x)$?
26. What is $\lim\limits_{x \to 0} f(x)$, if it exists?
27. What is $\lim\limits_{x \to 1} f(x)$, if it exists?
28. Is $f(x)$ continuous at $x = 0$?
29. Is $f(x)$ continuous at $x = 1$?
30. Is $f(x)$ continuous at $x = -1$?

For the functions in Problems 31–34, determine which are continuous. Identify discontinuities for those that are not continuous.

31. $y = \dfrac{x^2 + 25}{x - 5}$ 32. $y = \dfrac{x^2 - 3x + 2}{x - 2}$

33. $f(x) = \begin{cases} x + 2 & \text{if } x \leq 2 \\ 5x - 6 & \text{if } x > 2 \end{cases}$

34. $y = \begin{cases} x^4 - 3 & \text{if } x \leq 1 \\ 2x - 3 & \text{if } x > 1 \end{cases}$

In Problems 35 and 36, use the graphs to find (a) the points of discontinuity, (b) $\lim\limits_{x \to +\infty} f(x)$, and (c) $\lim\limits_{x \to -\infty} f(x)$.

35.

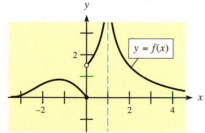

36.

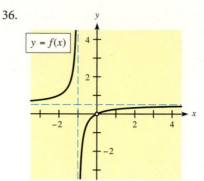

In Problems 37 and 38, evaluate the limits, if they exist. Then state what each limit tells about any horizontal asymptotes.

37. $\lim\limits_{x \to -\infty} \dfrac{2x^2}{1 - x^2}$ 38. $\lim\limits_{x \to +\infty} \dfrac{3x^{2/3}}{x + 1}$

39. Find the average rate of change of

$$f(x) = 2x^4 - 3x + 7 \text{ over } [-1, 2]$$

In Problems 40 and 41, decide whether the statements are true or false.

40. $\lim\limits_{h \to 0} \dfrac{f(x + h) - f(x)}{h}$ gives the formula for the slope of the tangent and the instantaneous rate of change of $f(x)$ at any value of x.

41. $\lim\limits_{h \to 0} \dfrac{f(c + h) - f(c)}{h}$ gives the equation of the tangent line to $f(x)$ at $x = c$.

42. Use the definition of derivative to find $f'(x)$ for $f(x) = 3x^2 + 2x - 1$.

43. Use the definition of derivative to find $f'(x)$ if $f(x) = x - x^2$.

Use the graph of $y = f(x)$ in Figure 9.40 to answer the questions in Problems 44–46.

44. Explain which is greater: the average rate of change of f over $[-3, 0]$ or over $[-1, 0]$.

45. Is $f(x)$ differentiable at
    (a) $x = -1$?   (b) $x = 1$?

46. Is $f(x)$ differentiable at
    (a) $x = -2$?   (b) $x = 2$?

47. Let $f(x) = \dfrac{\sqrt[3]{4x}}{(3x^2 - 10)^2}$. Approximate $f'(2)$
    (a) by using the numerical derivative feature of a graphing calculator, and
    (b) by evaluating $\dfrac{f(2 + h) - f(2)}{h}$ with $h = 0.0001$.

48. Use the given table of values for $g(x)$ to
    (a) find the average rate of change of $g(x)$ over $[2, 5]$.
    (b) approximate $g'(4)$ as accurately as possible.

x	2	2.3	3.1	4	4.3	5
$g(x)$	13.2	12.1	9.7	12.2	14.3	18.1

Use the following graph of $f(x)$ to complete Problems 49 and 50.

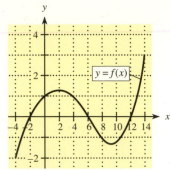

49. Estimate $f'(4)$.
50. Rank the following from smallest to largest and explain.
    A: $f'(2)$ B: $f'(6)$
    C: the average rate of change of $f(x)$ over $[2, 10]$
51. If $c = 4x^5 - 6x^3$, find c'.
52. If $f(x) = 10x^9 - 5x^6 + 4x - 2^7 + 19$, find $f'(x)$.
53. If $p = q + \sqrt{7}$, find dp/dq.
54. If $y = \sqrt{x}$, find y'.
55. If $f(z) = \sqrt[3]{z^4}$, find $f'(z)$.
56. If $v(x) = 4/\sqrt[3]{x}$, find $v'(x)$.
57. If $y = \dfrac{1}{x} - \dfrac{1}{\sqrt{x}}$, find y'.
58. If $f(x) = \dfrac{3}{2x^2} - \sqrt[3]{x} + 4^5$, find $f'(x)$.
59. Write the equation of the line tangent to the graph of $y = 3x^5 - 6$ at $x = 1$.
60. Write the equation of the line tangent to the curve $y = 3x^3 - 2x$ at the point where $x = 2$.

In Problems 61 and 62, (a) find all x-values where the slope of the tangent equals zero, (b) find points (x, y) where the slope of the tangent equals zero, and (c) use a graphing utility to graph the function and label the points found in part (b).
61. $f(x) = x^3 - 3x^2 + 1$ 62. $f(x) = x^6 - 6x^4 + 8$
63. If $f(x) = (3x - 1)(x^2 - 4x)$, find $f'(x)$.
64. Find y' if $y = (x^4 + 3)(3x^3 + 1)$.
65. If $p = \dfrac{5q^3}{2q^3 + 1}$, find $\dfrac{dp}{dq}$.
66. Find $\dfrac{ds}{dt}$ if $s = \dfrac{\sqrt{t}}{(3t + 1)}$.
67. Find $\dfrac{dy}{dx}$ for $y = \sqrt{x}\,(3x + 2)$.
68. Find $\dfrac{dC}{dx}$ for $C = \dfrac{5x^4 - 2x^2 + 1}{x^3 + 1}$.
69. If $y = (x^3 - 4x^2)^3$, find y'.
70. If $y = (5x^6 + 6x^4 + 5)^6$, find y'.
71. If $y = (2x^4 - 9)^9$, find $\dfrac{dy}{dx}$.

72. Find $g'(x)$ if $g(x) = \dfrac{1}{\sqrt{x^3 - 4x}}$.
73. Find $f'(x)$ if $f(x) = x^2(2x^4 + 5)^8$.
74. Find S' if $S = \dfrac{(3x + 1)^2}{x^2 - 4}$.
75. Find $\dfrac{dy}{dx}$ if $y = [(3x + 1)(2x^3 - 1)]^{12}$.
76. Find y' if $y = \left(\dfrac{x + 1}{1 - x^2}\right)^3$.
77. Find y' if $y = x\sqrt{x^2 - 4}$.
78. Find $\dfrac{dy}{dx}$ if $y = \dfrac{x}{\sqrt[3]{3x - 1}}$.

In Problems 79 and 80, find the second derivatives.
79. $y = \sqrt{x} - x^2$ 80. $y = x^4 - \dfrac{1}{x}$

In Problems 81 and 82, find the fifth derivatives.
81. $y = (2x + 1)^4$
82. $y = \dfrac{(1 - x)^6}{24}$
83. If $\dfrac{dy}{dx} = \sqrt{x^2 - 4}$, find $\dfrac{d^3y}{dx^3}$.
84. If $\dfrac{d^2y}{dx^2} = \dfrac{x}{x^2 + 1}$, find $\dfrac{d^4y}{dx^4}$.

APPLICATIONS

Cost, revenue, and profit **In Problems 85–88, assume that a company's monthly total revenue and total cost (both in dollars) are given by**

$$R(x) = 140x - 0.01x^2 \quad \text{and} \quad C(x) = 60x + 70{,}000$$

where x is the number of units. (Let $P(x)$ denote the profit function.)
85. Find (a) $\lim\limits_{x \to 4000} R(x)$, (b) $\lim\limits_{x \to 4000} C(x)$, and (c) $\lim\limits_{x \to 4000} P(x)$.
86. Find and interpret (a) $\lim\limits_{x \to 0^+} C(x)$ and (b) $\lim\limits_{x \to 1000} P(x)$.
87. If $\overline{R}(x) = \dfrac{R(x)}{x}$ and $\overline{C}(x) = \dfrac{C(x)}{x}$ are, respectively, the company's average revenue per unit and average cost per unit, find
    (a) $\lim\limits_{x \to 0^+} \overline{R}(x)$.   (b) $\lim\limits_{x \to 0^+} \overline{C}(x)$.
88. Evaluate and explain the meanings of
    (a) $\lim\limits_{x \to \infty} C(x)$.   (b) $\lim\limits_{x \to \infty} \overline{C}(x) = \lim\limits_{x \to \infty} \dfrac{C(x)}{x}$.

Elderly in the work force **The graph shows the percent of elderly men and women in the work force for selected years from 1950 to 2008. Use this graph in Problems 89 and 90.**
89. For the period from 1990 to 2008, find and interpret the annual average rate of change of
    (a) elderly men in the work force and
    (b) elderly women in the work force.
90. (a) Find the annual average rate of change of the percent of elderly men in the work force from 1950 to 1960 and from 2000 to 2008.

(b) Find the annual average rate of change of the percent of elderly women in the work force from 1950 to 1960 and from 2000 to 2008.

Elderly in the Labor Force, 1950–2008

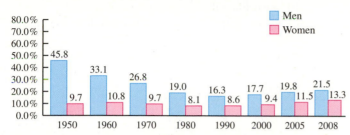

Source: Bureau of the Census, U.S. Department of Commerce

91. **Demand** Suppose that the demand for x units of a product is given by $x = (100/p) - 1$, where p is the price per unit of the product. Find and interpret the rate of change of demand with respect to price if the price is
(a) $10.   (b) $20.

92. **Severe weather ice makers** Thunderstorms severe enough to produce hail develop when an upper-level low (a pool of cold air high in the atmosphere) moves through a region where there is warm, moist air at the surface. These storms create an updraft that draws the moist air into subfreezing air above 10,000 feet. Data from the National Weather Service indicates that the strength of the updraft, as measured by its speed s in mph, affects the size of the hail according to

$$h = 0.000595s^{1.922}$$

where h is the diameter of the hail (in inches). Find and interpret $h(100)$ and $h'(100)$.

93. **Revenue** The graph shows the revenue function for a commodity. Will the $(A + 1)$st item sold or the $(B + 1)$st item sold produce more revenue? Explain.

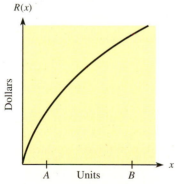

94. **Revenue** In a 100-unit apartment building, when the price charged per apartment rental is $(830 + 30x)$ dollars, then the number of apartments rented is $100 - x$ and the total revenue for the building is

$$R(x) = (830 + 30x)(100 - x)$$

where x is the number of $30 rent increases (and also the resulting number of unrented apartments). Find the marginal revenue when $x = 10$. Does this tell you that the rent should be raised (causing more vacancies) or lowered? Explain.

95. **Productivity** Suppose the productivity of a worker (in units per hour) after x hours of training and time on the job is given by

$$P(x) = 3 + \frac{70x^2}{x^2 + 1000}$$

(a) Find and interpret $P(20)$.
(b) Find and interpret $P'(20)$.

96. **Demand** The demand q for a product at price p is given by

$$q = 10,000 - 50\sqrt{0.02p^2 + 500}$$

Find the rate of change of demand with respect to price.

97. **Supply** The number of units x of a product that is supplied at price p is given by

$$x = \sqrt{p - 1}, \quad p \geq 1$$

If the price p is $10, what is the rate of change of the supply with respect to the price, and what does it tell us?

98. **Acceleration** Suppose an object moves so that its distance to a sensor, in feet, is given by

$$s(t) = 16 + 140t + 8\sqrt{t}$$

where t is the time in seconds. Find the acceleration at time $t = 4$ seconds.

99. **Profit** Suppose a company's profit (in dollars) is given by

$$P(x) = 70x - 0.1x^2 - 5500$$

where x is the number of units. Find and interpret $P'(300)$ and $P''(300)$.

In Problems 100–107, cost, revenue, and profit are in dollars and x is the number of units.

100. **Cost** If the cost function for a particular good is $C(x) = 3x^2 + 6x + 600$, what is the
(a) marginal cost function?
(b) marginal cost if 30 units are produced?
(c) interpretation of your answer in part (b)?

101. **Cost** If the total cost function for a commodity is $C(x) = 400 + 5x + x^3$, what is the marginal cost when 4 units are produced, and what does it mean?

102. **Revenue** The total revenue function for a commodity is $R = 40x - 0.02x^2$, with x representing the number of units.
(a) Find the marginal revenue function.
(b) At what level of production will marginal revenue be 0?

103. **Profit** If the total revenue function for a product is given by $R(x) = 60x$ and the total cost function is given by $C = 200 + 10x + 0.1x^2$, what is the marginal profit at $x = 10$? What does the marginal profit at $x = 10$ predict?

104. **Revenue** The total revenue function for a commodity is given by $R = 80x - 0.04x^2$.
    (a) Find the marginal revenue function.
    (b) What is the marginal revenue at $x = 100$?
    (c) Interpret your answer in part (b).

105. **Revenue** If the revenue function for a product is

$$R(x) = \frac{60x^2}{2x + 1}$$

find the marginal revenue.

106. **Profit** A firm has monthly costs given by

$$C = 45,000 + 100x + x^3$$

where x is the number of units produced per month. The firm can sell its product in a competitive market for \$4600 per unit. Find the marginal profit.

107. **Profit** A small business has weekly costs of

$$C = 100 + 30x + \frac{x^2}{10}$$

where x is the number of units produced each week. The competitive market price for this business's product is \$46 per unit. Find the marginal profit.

108. **Cost, revenue, and profit** The graph shows the total revenue and total cost functions for a company. Use the graph to decide (and justify) at which of points A, B, and C
    (a) the revenue from the next item will be least.
    (b) the profit will be greatest.
    (c) the profit from the sale of the next item will be greatest.
    (d) the next item sold will reduce the profit.

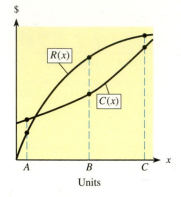

Units

---

9 CHAPTER TEST

1. Evaluate the following limits, if they exist. Use algebraic methods.
    (a) $\displaystyle\lim_{x \to -2} \frac{4x - x^2}{4x - 8}$
    (b) $\displaystyle\lim_{x \to \infty} \frac{8x^2 - 4x + 1}{2 + x - 5x^2}$
    (c) $\displaystyle\lim_{x \to 7} \frac{x^2 - 5x - 14}{x^2 - 6x - 7}$
    (d) $\displaystyle\lim_{x \to -5} \frac{5x - 25}{x + 5}$

2. (a) Write the limit definition for $f'(x)$.
    (b) Use the definition from (a) to find $f'(x)$ for $f(x) = 3x^2 - x + 9$.

3. Let $f(x) = \dfrac{4x}{x^2 - 8x}$. Identify all x-values where $f(x)$ is *not* continuous.

4. Use derivative formulas to find the derivative of each of the following. Simplify, except for part (d).
    (a) $B = 0.523W - 5176$
    (b) $p = 9t^{10} - 6t^7 - 17t + 23$

    (c) $y = \dfrac{3x^3}{2x^7 + 11}$
    (d) $f(x) = (3x^5 - 2x + 3)(4x^{10} + 10x^4 - 17)$
    (e) $g(x) = \frac{3}{4}(2x^5 + 7x^3 - 5)^{12}$
    (f) $y = (x^2 + 3)(2x + 5)^6$
    (g) $f(x) = 12\sqrt{x} - \dfrac{10}{x^2} + 17$

5. Find $\dfrac{d^3y}{dx^3}$ for $y = x^3 - x^{-3}$.

6. Let $f(x) = x^3 - 3x^2 - 24x - 10$.
    (a) Write the equation of the line tangent to the graph of $y = f(x)$ at $x = -1$.
    (b) Find all points (both x- and y-coordinates) where $f'(x) = 0$.

7. Find the average rate of change of $f(x) = 4 - x - 2x^2$ over $[1, 6]$.

8. Use the given tables to evaluate the following limits, if they exist.
    (a) $\displaystyle\lim_{x \to 5} f(x)$ (b) $\displaystyle\lim_{x \to 5} g(x)$ (c) $\displaystyle\lim_{x \to 5} g(x)$

x	4.99	4.999	→5←	5.001	5.01
$f(x)$	2.01	2.001	→?←	1.999	1.99

x	4.99	4.999	→5←	5.001	5.01
$g(x)$	−3.99	−3.999	→?←	6.999	6.99

9. Use the definition of continuity to investigate whether $g(x)$ is continuous at $x = -2$. Show your work.

$$g(x) = \begin{cases} 6 - x & \text{if } x \le -2 \\ x^3 & \text{if } x > -2 \end{cases}$$

In Problems 10 and 11, suppose a company has its total cost for a product given by $C(x) = 200x + 10{,}000$ dollars and its total revenue given by $R(x) = 250x - 0.01x^2$ dollars, where x is the number of units produced and sold.

10. (a) Find the marginal revenue function.
    (b) Find $R(72)$ and $R'(72)$ and tell what each represents or predicts.
11. (a) Form the profit function for this product.
    (b) Find the marginal profit function.
    (c) Find the marginal profit when $x = 1000$, and then write a sentence that interprets this result.
12. Suppose that $f(x)$ is a differentiable function. Use the table of values to approximate $f'(3)$ as accurately as possible.

x	2	2.5	2.999	3	3.01	3.1
$f(x)$	0	18.4	44.896	45	46.05	56.18

13. Use the graph to perform the evaluations (a)–(f) and to answer (g)–(i). If no value exists, so indicate.
    (a) $f(1)$
    (b) $\lim_{x \to 6} f(x)$
    (c) $\lim_{x \to 3^-} f(x)$
    (d) $\lim_{x \to -4} f(x)$
    (e) $\lim_{x \to -\infty} f(x)$
    (f) Estimate $f'(4)$.
    (g) Find all x-values where $f'(x)$ does not exist.

(h) Find all x-values where $f(x)$ is not continuous.
(i) Rank from smallest to largest: $f'(-2), f'(2)$, and the average rate of change of $f(x)$ over $[-2, 2]$.

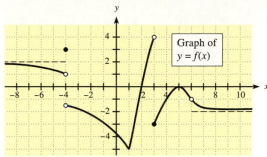

14. Given that the line $y = \frac{2}{3}x - 8$ is tangent to the graph of $y = f(x)$ at $x = 6$, find
    (a) $f'(6)$.
    (b) $f(6)$.
    (c) the instantaneous rate of change of $f(x)$ with respect to x at $x = 6$.

15. The graph shows the total revenue and total cost functions for a company. Use the graph to decide (and justify) at which of points A, B, and C
    (a) profit is the greatest.
    (b) there is a loss.
    (c) producing and selling another item will increase profit.
    (d) the next item sold will decrease profit.

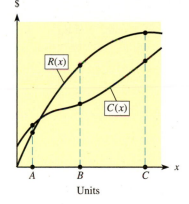

I. Marginal Return to Sales

A tire manufacturer studying the effectiveness of television advertising and other promotions on sales of its GRIPPER-brand tires attempted to fit data it had gathered to the equation

$$S = a_0 + a_1 x + a_2 x^2 + b_1 y$$

where S is sales revenue in millions of dollars, x is millions of dollars spent on television advertising, y is millions of dollars spent on other promotions, and a_0, a_1, a_2, and b_1 are constants. The data, gathered in two different regions of the country where expenditures for other promotions were kept constant (at B_1 and B_2), resulted in the following quadratic equations relating TV advertising and sales.

$$\text{Region 1: } S_1 = 30 + 20x - 0.4x^2 + B_1$$
$$\text{Region 2: } S_2 = 20 + 36x - 1.3x^2 + B_2$$

The company wants to know how to make the best use of its advertising dollars in the regions and whether the current allocation could be improved. Advise management about current advertising effectiveness, allocation of additional expenditures, and reallocation of current advertising expenditures by answering the following questions.

1. In the analysis of sales and advertising, **marginal return to sales** is usually used, and it is given by dS_1/dx for Region 1 and dS_2/dx for Region 2.
   (a) Find $\dfrac{dS_1}{dx}$ and $\dfrac{dS_2}{dx}$.
   (b) If $10 million is being spent on TV advertising in each region, what is the marginal return to sales in each region?
2. Which region would benefit more from additional advertising expenditure, if $10 million is currently being spent in each region?
3. If any additional money is made available for advertising, in which region should it be spent?
4. How could money already being spent be reallocated to produce more sales revenue?

II. Tangent Lines and Optimization in Business and Economics

In business and economics, common questions of interest include how to maximize profit or revenue, how to minimize average costs, and how to maximize average productivity. For questions such as these, the description of how to obtain the desired maximum or minimum represents an optimal (or best possible) solution to the problem. And the process of finding an optimal solution is called optimization. Answering these questions often involves tangent lines. In this project, we examine how tangent lines can be used to minimize average costs.

Suppose that Wittage, Inc. manufactures paper shredders for home and office use and that its weekly total costs (in dollars) for x shredders are given by

$$C(x) = 0.03x^2 + 12.75x + 6075$$

The average cost per unit for x units (denoted by $\overline{C}(x)$) is the total cost divided by the number of units. Thus, the average cost function for Wittage, Inc. is

$$\overline{C}(x) = \frac{C(x)}{x} = \frac{0.03x^2 + 12.75x + 6075}{x} = \frac{0.03x^2}{x} + \frac{12.75x}{x} + \frac{6075}{x}$$

$$\overline{C}(x) = 0.03x + 12.75 + \frac{6075}{x}$$

Wittage, Inc. would like to know how the company can use marginal costs to gain information about minimizing average costs. To investigate this relationship, answer the following questions.

1. Find Wittage's marginal cost function, then complete the following table.

x	100	200	300	400	500	600
$\overline{C}(x)$						
$\overline{MC}(x)$						

2. (a) For what x-values in the table is $\overline{C}$ getting smaller?
   (b) For these x-values, is $\overline{MC} < \overline{C}, \overline{MC} > \overline{C}$, or $\overline{MC} = \overline{C}$?
3. (a) For what x-values in the table is $\overline{C}$ getting larger?
   (b) For these x-values, is $\overline{MC} < \overline{C}, \overline{MC} > \overline{C}$, or $\overline{MC} = \overline{C}$?
4. Between what pair of consecutive x-values in the table will $\overline{C}$ reach its minimum?
5. Interpret (a) $\overline{C}(200)$ and $\overline{MC}(200)$ and (b) $\overline{C}(400)$ and $\overline{MC}(400)$.
6. On the basis of what $\overline{C}$ and $\overline{MC}$ tell us, and the work so far, complete the following statements by filling the blank with *increase, decrease,* or *stay the same.* Explain your reasoning.
   (a) $\overline{MC} < \overline{C}$ means that if Wittage makes one more unit, then $\overline{C}$ will _____ .
   (b) $\overline{MC} > \overline{C}$ means that if Wittage makes one more unit, then $\overline{C}$ will _____ .
7. (a) On the basis of your answers to Question 6, how will $\overline{C}$ and $\overline{MC}$ be related when $\overline{C}$ is minimized? Write a careful statement.
   (b) Use your idea from part (a) to find x where $\overline{C}$ is minimized. Then check your calculations by graphing $\overline{C}$ and locating the minimum from the graph.

Next, let's examine graphically the relationship that emerged in Question 7 to see whether it is true in general. Figure 9.41 shows a typical total cost function (Wittage's total cost function has a similar shape) and an arbitrary point on the graph.

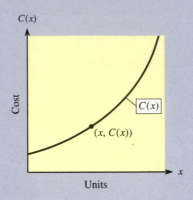

Figure 9.41

8. (a) Measure the slope of the line joining the point $(x, C(x))$ and the origin.
    (b) How is this slope related to the function $\overline{C}$ at $(x, C(x))$?
9. Figure 9.42(a) shows a typical total cost function with several points labeled.
    (a) Explain how you can tell at a glance that $\overline{C}(x)$ is decreasing as the level of production moves from A to B to C to D.
    (b) Explain how you can tell at a glance that $\overline{C}(x)$ is increasing as the level of production moves from Q to R to S to T.

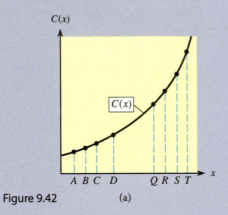
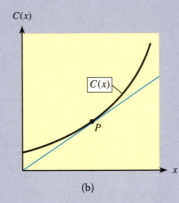

Figure 9.42   (a)   (b)

10. In Figure 9.42(b), a line from $(0, 0)$ is drawn tangent to $C(x)$ and the point of tangency is labeled P. At point P in Figure 9.42(b), what represents
     (a) the average cost, $\overline{C}(x)$?
     (b) the marginal cost, $\overline{MC}(x)$?
     (c) Explain how Figure 9.42 confirms that $\overline{C}(x)$ is minimized when $\overline{MC}(x) = \overline{C}(x)$.

On the basis of the numerical approach for Wittage's total cost function and the general graphical approach, what recommendation would you give Wittage about determining the number of units that will minimize average costs (even if the total cost function changes)?

10

Applications of Derivatives

Tischenko Irina/Shutterstock.com

The derivative can be used to determine where a function has a "turning point" on its graph, so that we can determine where the graph reaches its highest or lowest point within a particular interval. These points are called the relative maxima and relative minima, respectively, and are useful in sketching the graph of the function. The techniques for finding these points are also useful in solving applied problems, such as finding the maximum profit, the minimum average cost, and the maximum productivity. The second derivative can be used to find points of inflection of the graph of a function and to find the point of diminishing returns in certain applications.

The topics and applications discussed in this chapter include the following.

SECTIONS	APPLICATIONS
10.1 Relative Maxima and Minima: Curve Sketching	Advertising, marijuana use
10.2 Concavity: Points of Inflection **Second derivative test**	Water purity, profit
10.3 Optimization in Business and Economics **Maximizing revenue** **Minimizing average cost** **Maximizing profit**	Maximizing revenue, minimizing average cost, maximizing profit
10.4 Applications of Maxima and Minima	Company growth, minimizing cost, postal restriction, production costs, property development
10.5 Rational Functions: More Curve Sketching **Asymptotes** **More curve sketching**	Production costs

Prerequisite Problem Type	For Section	Answer	Section for Review
Write $\frac{1}{3}(x^2 - 1)^{-2/3}(2x)$ with positive exponents.	10.2	$\dfrac{2x}{3(x^2 - 1)^{2/3}}$	0.3, 0.4 Exponents and radicals
Factor: (a) $x^3 - x^2 - 6x$ (b) $8000 - 80x - 3x^2$	10.1 10.2 10.3	(a) $x(x - 3)(x + 2)$ (b) $(40 - x)(200 + 3x)$	0.6 Factoring
(a) For what values of x is $\dfrac{2}{3\sqrt[3]{x + 2}}$ undefined? (b) For what values of x is $\frac{1}{3}(x^2 - 1)^{-2/3}(2x)$ undefined?	10.1 10.2 10.5	(a) $x = -2$ (b) $x = -1, x = 1$	1.2 Domains of functions
If $f(x) = \frac{1}{3}x^3 - x^2 - 3x + 2$, and $f'(x) = x^2 - 2x - 3$, (a) find $f(-1)$. (b) find $f'(-2)$.	10.1	(a) $\dfrac{11}{3}$ (b) 5	1.2 Function notation
(a) Solve $0 = x^2 - 2x - 3$. (b) If $f'(x) = 3x^2 - 3$, what values of x make $f'(x) = 0$?	10.1–10.5	(a) $x = -1, x = 3$ (b) $x = -1, x = 1$	2.1 Solving quadratic equations
Does $\displaystyle\lim_{x \to -2} \dfrac{2x - 4}{3x + 6}$ exist?	10.5	No; unbounded	9.1 Limits
(a) Find $f''(x)$ if $f(x) = x^3 - 4x^2 + 3$. (b) Find $P''(x)$ if $P(x) = 48x - 1.2x^2$.	10.2	(a) $f''(x) = 6x - 8$ (b) $P''(x) = -2.4$	9.8 Higher-order derivatives
Find the derivatives: (a) $y = \dfrac{1}{3}x^3 - x^2 - 3x + 2$ (b) $f = x + 2\left(\dfrac{80{,}000}{x}\right)$ (c) $p(t) = 1 + \dfrac{4t}{t^2 + 16}$ (d) $y = (x + 2)^{2/3}$ (e) $y = \sqrt[3]{x^2 - 1}$	10.1 10.2 10.3 10.4	(a) $y' = x^2 - 2x - 3$ (b) $f' = 1 - \dfrac{160{,}000}{x^2}$ (c) $p'(t) = \dfrac{64 - 4t^2}{(t^2 + 16)^2}$ (d) $y' = \dfrac{2}{3(x + 2)^{1/3}}$ (e) $y' = \dfrac{2x}{3(x^2 - 1)^{2/3}}$	9.4, 9.5, 9.6 Derivatives

OBJECTIVES

10.1

- To find relative maxima and minima and horizontal points of inflection of functions
- To sketch graphs of functions by using information about maxima, minima, and horizontal points of inflection

Relative Maxima and Minima: Curve Sketching

| APPLICATION PREVIEW |

When a company initiates an advertising campaign, there is typically a surge in weekly sales. As the effect of the campaign lessens, sales attributable to it usually decrease. For example, suppose a company models its weekly sales revenue during an advertising campaign by

$$S = \frac{100t}{t^2 + 100}$$

where t is the number of weeks since the beginning of the campaign. The company would like to determine accurately when the revenue function is increasing, when it is decreasing, and when sales revenue is maximized. (See Example 4.)

In this section we will use the derivative of a function to decide whether the function is increasing or decreasing on an interval and to find where the function has relative maximum points and relative minimum points. We will use the information about derivatives of functions to graph the functions and to solve applied problems.

Recall that we can find the maximum value or the minimum value of a quadratic function by finding the vertex of its graph. But for functions of higher degree, the special features of their graphs may be harder to find accurately, even when a graphing utility is used. In addition to intercepts and asymptotes, we can use the first derivative as an aid in graphing. The first derivative identifies the "turning points" of the graph, which help us determine the general shape of the graph and choose a viewing window that includes the interesting points of the graph. Note that for a quadratic function $f(x) = ax^2 + bx + c$, the solution to $f'(x) = 2ax + b = 0$ is $x = -b/2a$. This solution is the x-coordinate of the vertex (turning point) of $f(x)$, found in Section 2.2.

In Figure 10.1(a) we see that the graph of $y = \frac{1}{3}x^3 - x^2 - 3x + 2$ has two "turning points," at $\left(-1, \frac{11}{3}\right)$ and $(3, -7)$. The curve has a relative maximum at $\left(-1, \frac{11}{3}\right)$ because this point is higher than any other point "near" it on the curve; the curve has a relative minimum at $(3, -7)$ because this point is lower than any other point "near" it on the curve. A formal definition follows.

Relative Maxima and Minima

The point $(x_1, f(x_1))$ is a **relative maximum point** of the function f if there is an interval around x_1 on which $f(x_1) \geq f(x)$ for all x in the interval. In this case, we say the relative maximum *occurs* at $x = x_1$ and the relative maximum is $f(x_1)$.

The point $(x_2, f(x_2))$ is a **relative minimum point** of the function f if there is an interval around x_2 on which $f(x_2) \leq f(x)$ for all x in the interval. In this case, we say the relative minimum *occurs* at $x = x_2$ and the relative minimum is $f(x_2)$.

In order to determine whether a turning point of a function is a maximum point or a minimum point, it is frequently helpful to know what the graph of the function does in intervals on either side of the turning point. We say a function is **increasing** on an interval if the values of the function increase as the x-values increase (that is, if the graph rises as we move from left to right on the interval). Similarly, a function is **decreasing** on an interval if the values of the function decrease as the x-values increase (that is, if the graph falls as we move from left to right on the interval).

We have seen that if the slope of a line is positive, then the linear function is increasing and its graph is rising. Similarly, if $f(x)$ is differentiable over an interval and if each tangent line to the curve over that interval has positive slope, then the curve is rising over the interval and the function is increasing. Because the derivative of the function gives the slope of the tangent to the curve, we see that if $f'(x) > 0$ on an interval, then $f(x)$ is increasing on that interval. A similar conclusion can be reached when the derivative is negative on the interval.

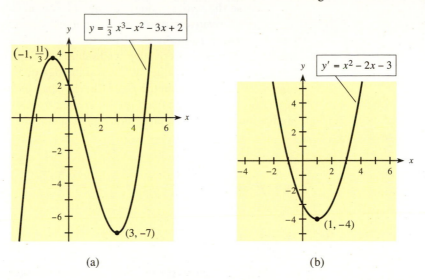

Figure 10.1   (a)   (b)

Increasing and Decreasing Functions

If f is a function that is differentiable on an interval (a, b), then

if $f'(x) > 0$ for all x in (a, b), f is increasing on (a, b).
if $f'(x) < 0$ for all x in (a, b), f is decreasing on (a, b).

Figure 10.1(a) shows the graph of a function, and Figure 10.1(b) shows the graph of its derivative. The figures show that the graph of $y = f(x)$ is increasing for the same x-values for which the graph of $y' = f'(x)$ is above the x-axis (when $f'(x) > 0$). Similarly, the graph of $y = f(x)$ is decreasing for the same x-values ($-1 < x < 3$) for which the graph of $y' = f'(x)$ is below the x-axis (when $f'(x) < 0$).

The derivative $f'(x)$ can change signs only at values of x at which $f'(x) = 0$ or $f'(x)$ is undefined. We call these values of x **critical values.** The point corresponding to a critical value for x is a **critical point.*** Because a curve changes from increasing to decreasing at a relative maximum and from decreasing to increasing at a relative minimum (see Figure 10.1(a)), we have the following.

Relative Maximum and Minimum

If f has a relative maximum or a relative minimum at $x = x_0$, then $f'(x_0) = 0$ or $f'(x_0)$ is undefined.

Figure 10.2 shows a function with two relative maxima, one at $x = x_1$ and the second at $x = x_3$, and one relative minimum at $x = x_2$. At $x = x_1$ and $x = x_2$, we see that $f'(x) = 0$, and at $x = x_3$ the derivative does not exist.

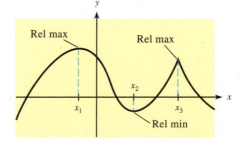

Figure 10.2

Thus we can find relative maxima and minima for a curve by finding values of x for which the function has critical points. The behavior of the derivative to the left and right

---

*There may be some critical values at which both $f'(x)$ and $f(x)$ are undefined. Critical points do not occur at these values, but studying the derivative on either side of such values may be of interest.

of (and near) these points will tell us whether they are relative maxima, relative minima, or neither.

Because the critical values are the only values at which the graph can have turning points, the derivative cannot change sign anywhere except at a critical value. Thus, in an interval between two critical values, the sign of the derivative at any value in the interval will be the sign of the derivative at all values in the interval.

Using the critical values of $f(x)$ and the sign of $f'(x)$ between those critical values, we can create a **sign diagram for $f'(x)$**. The sign diagram for the graph in Figure 10.2 is shown in Figure 10.3. This sign diagram was created from the graph of f, but it is also possible to predict the shape of a graph from a sign diagram.

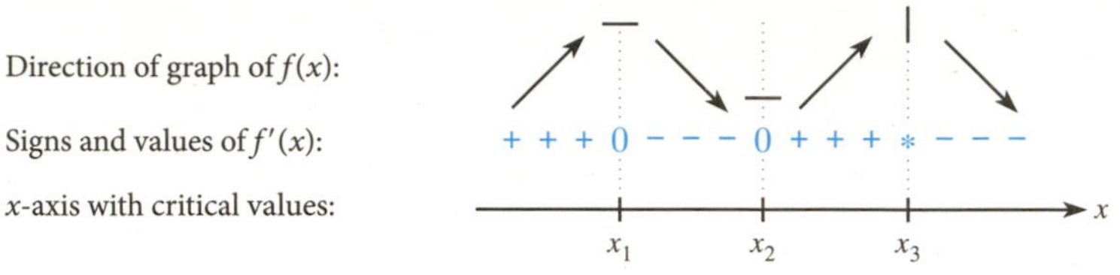

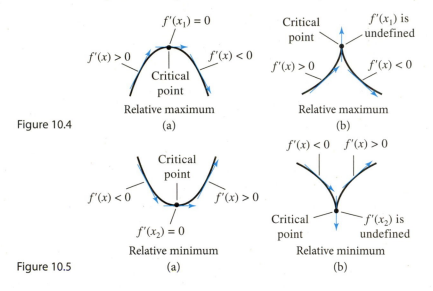

Direction of graph of $f(x)$:

Signs and values of $f'(x)$: $+ + + 0 - - - 0 + + + * - - -$

x-axis with critical values:

                       x_1 x_2 x_3

Figure 10.3   *means $f'(x_3)$ is undefined

Figure 10.4 shows the two ways that a function can have a relative maximum at a critical point, and Figure 10.5 shows the two ways for a relative minimum.

$f'(x_1) = 0$

$f'(x) > 0$ $f'(x) < 0$

Critical point

Relative maximum
(a)

Critical point $f'(x_1)$ is undefined

$f'(x) > 0$ $f'(x) < 0$

Relative maximum
(b)

Figure 10.4

Critical point

$f'(x) < 0$ $f'(x) > 0$

$f'(x_2) = 0$

Relative minimum
(a)

$f'(x) < 0$ $f'(x) > 0$

Critical point $f'(x_2)$ is undefined

Relative minimum
(b)

Figure 10.5

The preceding discussion suggests the following procedure for finding relative maxima and minima of a function.

First-Derivative Test

Procedure

To find relative maxima and minima of a function:

1. Find the first derivative of the function.
2. Set the derivative equal to 0, and solve for values of x that satisfy $f'(x) = 0$. These are called **critical values.** Values that make $f'(x)$ undefined are also critical values.
3. Substitute the critical values into the *original function* to find the **critical points.**

Example

Find the relative maxima and minima of $f(x) = \frac{1}{3}x^3 - x^2 - 3x + 2$.

1. $f'(x) = x^2 - 2x - 3$
2. $0 = x^2 - 2x - 3 = (x + 1)(x - 3)$ has solutions $x = -1, x = 3$. No values of x make $f'(x) = x^2 - 2x - 3$ undefined. Critical values are -1 and 3.
3. $f(-1) = \frac{11}{3}$ $f(3) = -7$
   The critical points are $\left(-1, \frac{11}{3}\right)$ and $(3, -7)$.

First-Derivative Test *(continued)*

Procedure	Example

4. Evaluate $f'(x)$ at a value of x to the left and one to the right of each critical point to develop a sign diagram.
   (a) If $f'(x) > 0$ to the left and $f'(x) < 0$ to the right of the critical value, the critical point is a relative maximum point.
   (b) If $f'(x) < 0$ to the left and $f'(x) > 0$ to the right of the critical value, the critical point is a relative minimum point.

4. $f'(-2) = 5 > 0$ and $f'(0) = -3 < 0$
   Thus $(-1, 11/3)$ is a relative maximum point.
   $f'(2) = -3 < 0$ and $f'(4) = 5 > 0$
   Thus $(3, -7)$ is a relative minimum point. The sign diagram for $f'(x)$ is

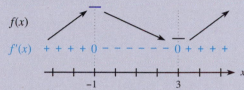

5. Use the information from the sign diagram and selected points to sketch the graph.

5. The information from this sign diagram is shown in Figure 10.6(a). Plotting additional points gives the graph of the function, which is shown in Figure 10.6(b).

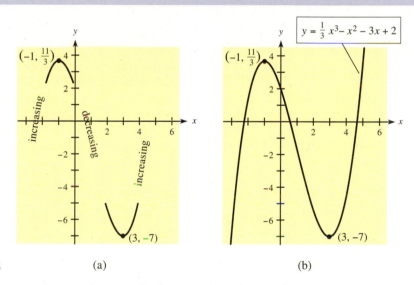

Figure 10.6   (a)   (b)

Because the critical values are the only x-values at which the graph can have turning points, we can test to the left and right of each critical value by testing to the left of the smallest critical value, then testing a value *between* each two successive critical values, and then testing to the right of the largest critical value. The following example illustrates this procedure.

EXAMPLE 1   Maxima and Minima

Find the relative maxima and minima of $f(x) = \frac{1}{4}x^4 - \frac{1}{3}x^3 - 3x^2 + 8$, and sketch its graph.

Solution

1. $f'(x) = x^3 - x^2 - 6x$
2. Setting $f'(x) = 0$ gives $0 = x^3 - x^2 - 6x$. Solving for x gives

$$0 = x(x - 3)(x + 2)$$

$x = 0$	$x - 3 = 0$	$x + 2 = 0$
	$x = 3$	$x = -2$

Thus the critical values are $x = 0$, $x = 3$, and $x = -2$.

3. Substituting the critical values into the original function gives the critical points:

$$f(-2) = \tfrac{8}{3}, \text{ so } \left(-2, \tfrac{8}{3}\right) \text{ is a critical point.}$$
$$f(0) = 8, \text{ so } (0, 8) \text{ is a critical point.}$$
$$f(3) = -\tfrac{31}{4}, \text{ so } \left(3, -\tfrac{31}{4}\right) \text{ is a critical point.}$$

4. Testing $f'(x)$ to the left of the smallest critical value, then between the critical values, and finally to the right of the largest critical value will give the sign diagram. Evaluating $f'(x)$ at the test values $x = -3, x = -1, x = 1,$ and $x = 4$ gives the signs to determine relative maxima and minima.

The sign diagram for $f'(x)$ is

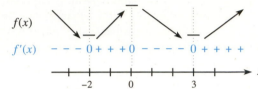

Thus we have

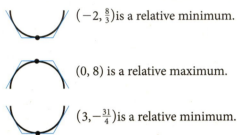

$\left(-2, \tfrac{8}{3}\right)$ is a relative minimum.

$(0, 8)$ is a relative maximum.

$\left(3, -\tfrac{31}{4}\right)$ is a relative minimum.

5. Figure 10.7(a) shows the graph of the function near the critical points, and Figure 10.7(b) shows the graph of the function. ■

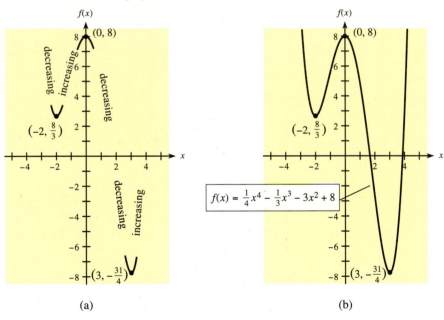

Figure 10.7   (a)   (b)

Note that we substitute the critical values into the *original function* $f(x)$ to find the y-values of the critical points, but we test for relative maxima and minima by substituting values near the critical values into the *derivative of the function*, $f'(x)$.

Only four values were needed to test three critical points in Example 1. This method will work *only if* the critical values are tested in order from smallest to largest.

If the first derivative of f is 0 at x_0 but does not change from positive to negative or from negative to positive as x passes through x_0, then the critical point at x_0 is neither a relative maximum nor a relative minimum. In this case we say that f has a **horizontal point of inflection** (abbreviated HPI) at x_0.

EXAMPLE 2

Maxima, Minima, and Horizontal Points of Inflection

Find the relative maxima, relative minima, and horizontal points of inflection of $h(x) = \frac{1}{4}x^4 - \frac{2}{3}x^3 - 2x^2 + 8x + 4$, and sketch its graph.

Solution

1. $h'(x) = x^3 - 2x^2 - 4x + 8$
2. $0 = x^3 - 2x^2 - 4x + 8$ or $0 = x^2(x - 2) - 4(x - 2)$. Therefore, we have $0 = (x - 2)(x^2 - 4)$. Thus $x = -2$ and $x = 2$ are solutions.
3. The critical points are $\left(-2, -\frac{32}{3}\right)$ and $\left(2, \frac{32}{3}\right)$.
4. Using test values, such as $x = -3$, $x = 0$, and $x = 3$, gives the sign diagram for $h'(x)$.

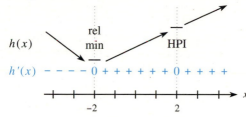

5. Figure 10.8(a) shows the graph of the function near the critical points, and Figure 10.8(b) shows the graph of the function. ◼

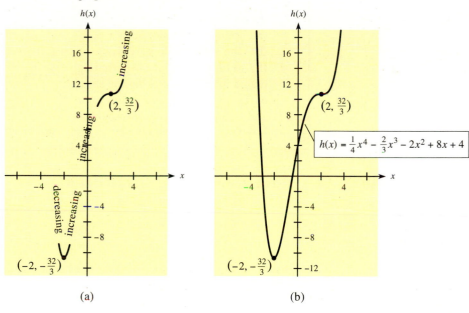

Figure 10.8   (a)   (b)

Technology Note   See Appendices C and D, Section 10.1, for details on using a graphing calculator and Excel to find critical values and to determine maxima and minima. See also the Online Excel Guide. ◼

EXAMPLE 3

Undefined Derivatives

Find the relative maxima and minima (if any) of the graph of $y = (x + 2)^{2/3}$.

Solution

1. $y' = f'(x) = \frac{2}{3}(x + 2)^{-1/3} = \dfrac{2}{3\sqrt[3]{x + 2}}$
2. $0 = \dfrac{2}{3\sqrt[3]{x + 2}}$ has no solutions; $f'(x)$ is undefined at $x = -2$.
3. $f(-2) = 0$, so the critical point is $(-2, 0)$.
4. The sign diagram for $f'(x)$ is

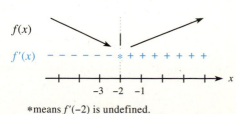

Thus a relative minimum occurs at $(-2, 0)$.

5. Figure 10.9(a) shows the graph of the function near the critical point, and Figure 10.9(b) shows the graph. ■

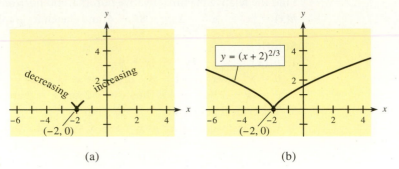

Figure 10.9   (a)   (b)

CHECKPOINT

1. The x-values of critical points are found where $f'(x)$ is _____ or _____.
2. Decide whether the following are true or false.
   (a) If $f'(1) = 7$, then $f(x)$ is increasing at $x = 1$.
   (b) If $f'(-2) = 0$, then a relative maximum or a relative minimum occurs at $x = -2$.
   (c) If $f'(-3) = 0$ and $f'(x)$ changes from positive on the left to negative on the right of $x = -3$, then a relative minimum occurs at $x = -3$.
3. If $f(x) = 7 + 3x - x^3$, then $f'(x) = 3 - 3x^2$. Use these functions to decide whether the following statements are true or false.
   (a) The only critical value is $x = 1$.
   (b) The critical points are $(1, 0)$ and $(-1, 0)$.
4. If $f'(x)$ has the following partial sign diagram, make a "stick-figure" sketch of $f(x)$ and label where any maxima and minima occur. Assume that $f(x)$ is defined for all real numbers.

   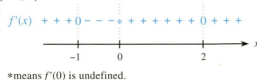

   *means $f'(0)$ is undefined.

EXAMPLE 4   Advertising | APPLICATION PREVIEW |

The weekly sales S of a product during an advertising campaign are given by

$$S = \frac{100t}{t^2 + 100}, \quad 0 \le t \le 20$$

where t is the number of weeks since the beginning of the campaign and S is in thousands of dollars.

(a) Over what interval are sales increasing? Decreasing?
(b) What are the maximum weekly sales?
(c) Sketch the graph for $0 \le t \le 20$.

Solution

(a) To find where S is increasing, we first find $S'(t)$

$$S'(t) = \frac{(t^2 + 100)100 - (100t)2t}{(t^2 + 100)^2}$$

$$= \frac{10,000 - 100t^2}{(t^2 + 100)^2}$$

We see that $S'(t) = 0$ when $10,000 - 100t^2 = 0$, or

$$100(100 - t^2) = 0$$
$$100(10 + t)(10 - t) = 0$$
$$t = -10 \quad \text{or} \quad t = 10$$

Because $S'(t)$ is never undefined ($t^2 + 100 \ne 0$ for any real t) and because $0 \le t \le 20$, our only critical value is $t = 10$. Testing $S'(t)$ to the left and right of $t = 10$ gives the sign diagram.

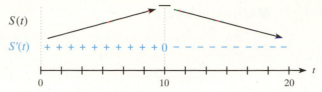

Hence, S is increasing on the interval $[0, 10)$ and decreasing on the interval $(10, 20]$.

(b) Because S is increasing to the left of $t = 10$ and S is decreasing to the right of $t = 10$, the maximum value of S occurs at $t = 10$ and is

$$S = S(10) = \frac{100(10)}{10^2 + 100} = \frac{1000}{200} = 5 \text{ (thousand dollars)}$$

(c) Plotting some additional points gives the graph; see Figure 10.10.

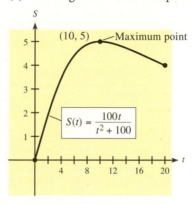

Figure 10.10

Calculator Note   With a graphing calculator, choosing an appropriate viewing window is the key to understanding the graph of a function. See Appendix C, Section 10.1, for details.

 EXAMPLE 5   **Critical Points and Viewing Windows**

Find the critical values of $f(x) = 0.0001x^3 + 0.003x^2 - 3.6x + 5$. Use them to determine an appropriate viewing window. Then sketch the graph.

Solution

The critical points are helpful in graphing the function. We begin by finding $f'(x)$.

$$f'(x) = 0.0003x^2 + 0.006x - 3.6$$

Now we solve $f'(x) = 0$ to find the critical values.

$$0 = 0.0003x^2 + 0.006x - 3.6$$
$$0 = 0.0003(x^2 + 20x - 12,000)$$
$$0 = 0.0003(x + 120)(x - 100)$$
$$x = -120 \quad \text{or} \quad x = 100$$

We choose a window that includes $x = -120$ and $x = 100$, and use VALUE to find $f(-120) = 307.4$ and $f(100) = -225$. We then graph $y = f(x)$ in a window that includes $y = -225$ and $y = 307.4$ (see Figure 10.11). On this graph we can see that $(-120, 307.4)$ is a relative maximum and that $(100, -225)$ is a relative minimum.

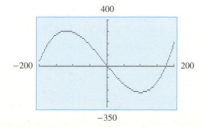

Figure 10.11

MODELING EXAMPLE 6   **Marijuana Use**

The table gives the percents of high school seniors who claimed to have tried marijuana for selected years from 1990 to 2006.

(a) With x as the number of years past 1990, find a function that models these data.

(b) During what years does the model indicate that the maximum and minimum use occurred?

Year	Percent	Year	Percent
1990	40.7	2000	48.8
1991	36.7	2001	49.8
1993	35.3	2002	47.8
1995	41.7	2003	46.7
1997	49.6	2004	45.7
1998	49.1	2005	44.8
1999	47.7	2006	42.3

Source: National Institute on Drug Abuse

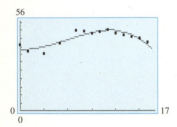

Figure 10.12

Solution

(a) A model for the data is $y = f(x) = -0.0187x^3 + 0.3329x^2 - 0.3667x + 38.10$. (See Figure 10.12.)

(b) We use $f'(x) = -0.0561x^2 + 0.6658x - 0.3667$ and solve $f'(x) = 0$ to find the critical values. We can solve

$$-0.0561x^2 + 0.6658x - 0.3667 = 0$$

with the quadratic formula or with a graphing calculator. The two solutions are approximately

$$x = 0.58 \quad \text{and} \quad x = 11.29.$$

The sign diagram shows that $x = 0.58$ gives a relative minimum (during 1991) and that $x = 11.29$ gives a relative maximum (during 2002).

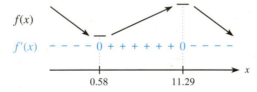

 Spreadsheet Note   The Solver feature of Excel can be used to solve optimization problems from calculus. Details are given in Appendix D, Section 10.1, and the Online Excel Guide.

CHECKPOINT SOLUTIONS

1. $f'(x) = 0$ or $f'(x)$ is undefined.

2. (a) True. $f(x)$ is increasing when $f'(x) > 0$.
   (b) False. There may be a horizontal point of inflection at $x = -2$. (See Figure 10.8 earlier in this section.)
   (c) False. A relative maximum occurs at $x = -3$.

3. (a) False. Critical values are solutions to $3 - 3x^2 = 0$, or $x = 1$ and $x = -1$.
   (b) False. y-coordinates of critical points come from $f(x) = 7 + 3x - x^3$. Thus critical points are $(1, 9)$ and $(-1, 5)$.

4.

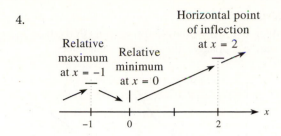

| EXERCISES | **10.1**

Access end-of-section exercises online at **www.webassign.net**

OBJECTIVES   10.2

- To find points of inflection of graphs of functions
- To use the second-derivative test to graph functions

Concavity; Points of Inflection

| APPLICATION PREVIEW |

Suppose that a retailer wishes to sell his store and uses the graph in Figure 10.13 to show how profits have increased since he opened the store and the potential for profit in the future. Can we conclude that profits will continue to grow, or should we be concerned about future earnings?

Note that although profits are still increasing in 2013, they seem to be increasing more slowly than in previous years. Indeed, they appear to have been growing at a decreasing rate since about 2011. We can analyze this situation more carefully by identifying the point at which profit begins to grow more slowly, that is, the *point of diminishing returns*. (See Example 3.)

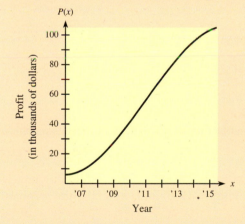

Figure 10.13

Just as we used the first derivative to determine whether a curve was increasing or decreasing on a given interval, we can use the second derivative to determine whether the curve is concave up or concave down on an interval.

Concavity   A curve is said to be **concave up** on an interval (a, b) if at each point on the interval the curve is above its tangent at the point (see Figure 10.14(a) on the next page). If the curve is below all its tangents on a given interval, it is **concave down** on the interval (Figure 10.14(b)).

Looking at Figure 10.14(a), we see that the *slopes* of the tangent lines increase over the interval where the graph is concave up. Because $f'(x)$ gives the slopes of those tangents, it follows that $f'(x)$ is increasing over the interval where $f(x)$ is concave up. However, we know that $f'(x)$ is increasing when its derivative, $f''(x)$, is positive. That is, if the second derivative is positive, the curve is concave up.

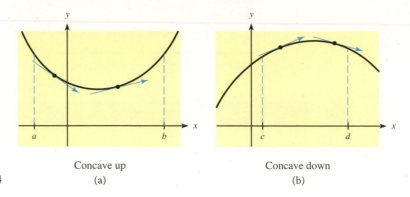

Figure 10.14

Concave up	Concave down
(a)	(b)

Similarly, if the second derivative of a function is negative over an interval, the slopes of the tangents to the graph decrease over that interval. This happens when the tangent lines are above the graph, as in Figure 10.14(b), so the graph must be concave down on this interval.

Thus we see that the second derivative can be used to determine the concavity of a curve.

Concave Up and Concave Down

Assume that the first and second derivatives of function f exist.
The function f is **concave up** on an interval I, if $f''(x) > 0$ on I, and **concave up** at the point $(a, f(a))$, if $f''(a) > 0$.
The function f is **concave down** on an interval I, if $f''(x) < 0$ on I, and **concave down** at the point $(a, f(a))$, if $f''(a) < 0$.

EXAMPLE 1 Concavity at a Point

Is the graph of $f(x) = x^3 - 4x^2 + 3$ concave up or down at the point
(a) $(1, 0)$?   (b) $(2, -5)$?

Solution
(a) We must find $f''(x)$ before we can answer this question.

$$f'(x) = 3x^2 - 8x \quad f''(x) = 6x - 8$$

Then $f''(1) = 6(1) - 8 = -2$, so the graph is concave down at $(1, 0)$.
(b) Because $f''(2) = 6(2) - 8 = 4$, the graph is concave up at $(2, -5)$. The graph of $f(x) = x^3 - 4x^2 + 3$ is shown in Figure 10.15(a). ∎

Points of Inflection   Looking at the graph of $y = x^3 - 4x^2 + 3$ [Figure 10.15(a)], we see that the curve is concave down on the left and concave up on the right. Thus it has changed from concave down to concave up, and from Example 1 we would expect the concavity to change somewhere between $x = 1$ and $x = 2$. Figure 10.15(b) shows the graph of $y'' = f''(x) = 6x - 8$, and we can see that $y'' = 0$ when $x = \frac{4}{3}$ and that $y'' < 0$ for $x < \frac{4}{3}$ and $y'' > 0$ for $x > \frac{4}{3}$. Thus the second derivative changes sign at $x = \frac{4}{3}$, so the concavity of the graph of $y = f(x)$ changes at $x = \frac{4}{3}, y = -\frac{47}{27}$. The point where concavity changes is called a **point of inflection.**

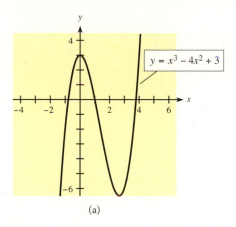

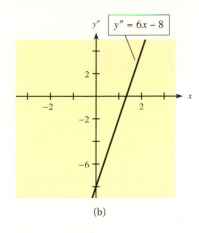

Figure 10.15 (a) (b)

Point of Inflection

A point (x_0, y_0) on the graph of a function f is called a **point of inflection** if the curve is concave up on one side of the point and concave down on the other side. The second derivative at this point, $f''(x_0)$, will be 0 or undefined.

In general, we can find points of inflection and information about concavity as follows.

Finding Points of Inflection and Concavity

Procedure	Example

To find the point(s) of inflection of a curve and intervals where it is concave up and where it is concave down:

Find the points of inflection and concavity of the graph of
$$y = \frac{x^4}{2} - x^3 + 5.$$

1. Find the second derivative of the function.

2. Set the second derivative equal to 0, and solve for x. Potential points of inflection occur at these values of x or at values of x where $f(x)$ is defined and $f''(x)$ is undefined.

3. Find the potential points of inflection.

4. If the second derivative has opposite signs on the two sides of one of these values of x, a point of inflection occurs.
   The curve is concave up where $f''(x) > 0$ and concave down where $f''(x) < 0$.
   The changes in the sign of $f''(x)$ correspond to changes in concavity and occur at points of inflection.

1. $y' = f'(x) = 2x^3 - 3x^2$
   $y'' = f''(x) = 6x^2 - 6x$

2. $0 = 6x^2 - 6x = 6x(x - 1)$ has solutions
   $x = 0, x = 1$.
   $f''(x)$ is defined everywhere.

3. $(0, 5)$ and $(1, \frac{9}{2})$ are potential points of inflection.

4. A **sign diagram for $f''(x)$** is

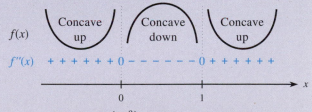

$(0, 5)$ and $(1, \frac{9}{2})$ are points of inflection.

See the graph in Figure 10.16 on the next page.

The graph of $y = \frac{1}{2}x^4 - x^3 + 5$ is shown in Figure 10.16. Note the points of inflection at $(0, 5)$ and $(1, \frac{9}{2})$. The point of inflection at $(0, 5)$ is a horizontal point of inflection because $f'(x)$ is also 0 at $x = 0$.

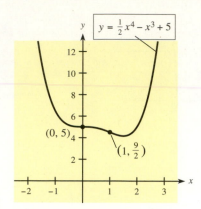

Figure 10.16

EXAMPLE 2 Water Purity

Suppose that a real estate developer wishes to remove pollution from a small lake so that she can sell lakefront homes on a "crystal clear" lake. The graph in Figure 10.17 shows the relation between dollars spent on cleaning the lake and the purity of the water. The point of inflection on the graph is called the **point of diminishing returns** on her investment because it is where the *rate* of return on her investment changes from increasing to decreasing. Show that the rate of change in the purity of the lake, $f'(x)$, is maximized at this point, $x = c$. Assume that $f(c)$, $f'(c)$, and $f''(c)$ are defined.

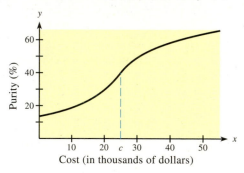

Figure 10.17

Solution

Because $x = c$ is a point of inflection for $f(x)$, we know that the concavity must change at $x = c$. From the figure we see the following.

$$x < c: \quad f(x) \text{ is concave up, so } f''(x) > 0.$$
$$f''(x) > 0 \text{ means that } f'(x) \text{ is increasing.}$$
$$x > c: \quad f(x) \text{ is concave down, so } f''(x) < 0.$$
$$f''(x) < 0 \text{ means that } f'(x) \text{ is decreasing.}$$

Thus $f'(x)$ has $f'(c)$ as its relative maximum.

EXAMPLE 3 Diminishing Returns | APPLICATION PREVIEW |

Suppose the annual profit for a store (in thousands of dollars) is given by

$$P(x) = -0.2x^3 + 3x^2 + 6$$

where x is the number of years past 2006. If this model is accurate, find the point of diminishing returns for the profit.

Solution

The point of diminishing returns occurs at the point of inflection. Thus we seek the point where the graph of this function changes from concave up to concave down, if such a point exists.

$$P'(x) = -0.6x^2 + 6x$$
$$P''(x) = -1.2x + 6$$
$$P''(x) = 0 \quad \text{when} \quad 0 = -1.2x + 6 \quad \text{or} \quad \text{when } x = 5$$

Thus $x = 5$ is a possible point of inflection. We test $P''(x)$ to the left and the right of $x = 5$.

$$P''(4) = 1.2 > 0 \Rightarrow \text{concave up to the left of } x = 5$$
$$P''(6) = -1.2 < 0 \Rightarrow \text{concave down to the right of } x = 5$$

Thus $(5, 56)$ is the point of inflection for the graph, and the point of diminishing returns for the profit is when $x = 5$ (in the year 2011) and is $P(5) = 56$ thousand dollars. Figure 10.18 shows the graphs of $P(x)$, $P'(x)$, and $P''(x)$. At $x = 5$, we see that the point of diminishing returns on the graph of $P(x)$ corresponds to the maximum point of the graph of $P'(x)$ and the zero (or x-intercept) of the graph of $P''(x)$. ■

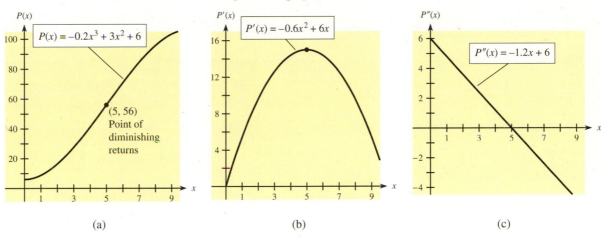

Figure 10.18   (a)   (b)   (c)

CHECKPOINT

1. If $f''(x) > 0$, then $f(x)$ is concave _____.
2. At what value of x does the graph $y = \frac{1}{3}x^3 - 2x^2 + 2x$ have a point of inflection?
3. On the graph below, locate any points of inflection (approximately) and label where the curve satisfies $f''(x) > 0$ and $f''(x) < 0$.

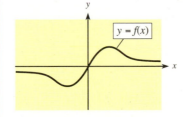

4. Determine whether the following is true or false. If $f''(0) = 0$, then $f(x)$ has a point of inflection at $x = 0$.

Second-Derivative Test

We can use information about points of inflection and concavity to help sketch graphs. For example, if we know that the curve is concave up at a critical point where $f'(x) = 0$, then the point must be a relative minimum because the tangent to the curve is horizontal at the critical point, and only a point at the bottom of a "concave up" curve could have a horizontal tangent (see Figure 10.19(a) on the next page).

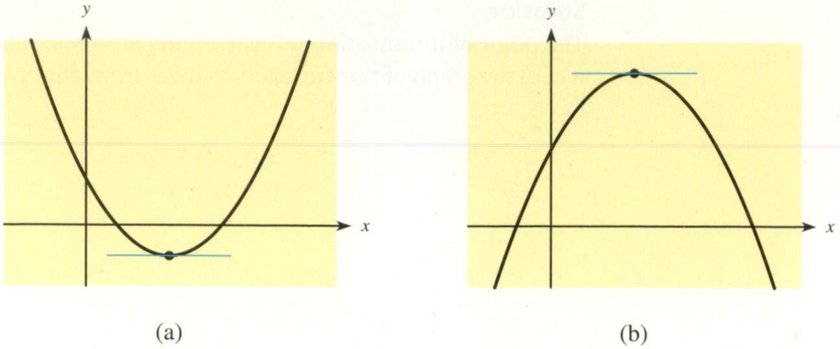

y

(a)   (b)

Figure 10.19

On the other hand, if the curve is concave down at a critical point where $f'(x) = 0$, then the point is a relative maximum (see Figure 10.19(b)).

Thus we can use the **second-derivative test** to determine whether a critical point where $f'(x) = 0$ is a relative maximum or minimum.

Second-Derivative Test

Procedure	Example
To find relative maxima and minima of a function:	Find the relative maxima and minima of $y = f(x) = \frac{1}{3}x^3 - x^2 - 3x + 2$.
1. Find the critical values of the function.	1. $f'(x) = x^2 - 2x - 3$ $0 = (x - 3)(x + 1)$ has solutions $x = -1$ and $x = 3$. No values of x make $f'(x)$ undefined.
2. Substitute the critical values into $f(x)$ to find the critical points.	2. $f(-1) = \frac{11}{3}$ $f(3) = -7$ The critical points are $(-1, \frac{11}{3})$ and $(3, -7)$.
3. Evaluate $f''(x)$ at each critical value for which $f'(x) = 0$. (a) If $f''(x_0) < 0$, a relative maximum occurs at x_0. (b) If $f''(x_0) > 0$, a relative minimum occurs at x_0. (c) If $f''(x_0) = 0$, or $f''(x_0)$ is undefined, the second-derivative test fails; use the first-derivative test.	3. $f''(x) = 2x - 2$ $f''(-1) = 2(-1) - 2 = -4 < 0$, so $(-1, \frac{11}{3})$ is a relative maximum point. $f''(3) = 2(3) - 2 = 4 > 0$, so $(3, -7)$ is a relative minimum point. (The graph is shown in Figure 10.20.)

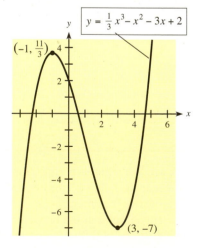

Figure 10.20

Because the second-derivative test (just shown) and the first-derivative test (in Section 10.1) are both methods for classifying critical values, let's compare the advantages and disadvantages of the second derivative test.

Advantage of the Second-Derivative Test	Disadvantages of the Second-Derivative Test
It is quick and easy for many functions.	The second derivative is difficult to find for some functions. The second-derivative test sometimes fails to give results.

EXAMPLE 4 Maxima, Minima, and Points of Inflection

Find the relative maxima and minima and points of inflection of $y = 3x^4 - 4x^3 - 2$.

Solution

$$y' = f'(x) = 12x^3 - 12x^2$$

Solving $0 = 12x^3 - 12x^2 = 12x^2(x - 1)$ gives $x = 1$ and $x = 0$. Thus the critical points are $(1, -3)$ and $(0, -2)$.

$$y'' = f''(x) = 36x^2 - 24x$$
$$f''(1) = 12 > 0 \Rightarrow (1, -3) \text{ is a relative minimum point.}$$
$$f''(0) = 0 \Rightarrow \text{the second-derivative test fails.}$$

Because the second-derivative test fails, we must use the first-derivative test at the critical point $(0, -2)$. A sign diagram for $f'(x)$ shows that $(0, -2)$ is a horizontal point of inflection.

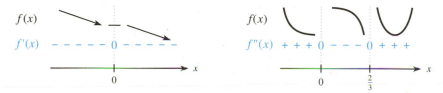

We look for points of inflection by setting $f''(x) = 0$ and solving for x. We find that $0 = 36x^2 - 24x$ has solutions $x = 0$ and $x = \frac{2}{3}$. The sign diagram for $f''(x)$ shows points of inflection at both of these x-values, that is at the points $(0, -2)$ and $(\frac{2}{3}, -\frac{70}{27})$. The point $(0, -2)$ is a special point, where the curve changes concavity *and* has a horizontal tangent (see Figure 10.21). ■

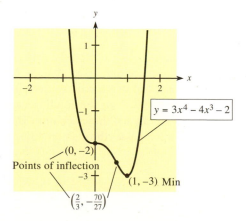

Figure 10.21

Calculator Note

We can use a graphing calculator to explore the relationships among f, f', and f'', as we did in the previous section for f and f'. See Appendix C, Section 10.2, for details. ∎

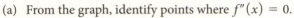

EXAMPLE 5 | **Concavity from the Graph of $y = f(x)$**

Figure 10.22 shows the graph of $f(x) = \frac{1}{6}(2x^3 - 3x^2 - 12x + 12)$.

(a) From the graph, identify points where $f''(x) = 0$.
(b) From the graph, observe intervals where $f''(x) > 0$ and where $f''(x) < 0$.
(c) Check the conclusions from parts (a) and (b) by calculating $f''(x)$ and graphing it.

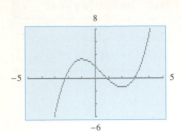

Figure 10.22

Solution

(a) From Figure 10.22, the point of inflection appears to be near $x = \frac{1}{2}$, so we expect $f''(x) = 0$ at (or very near) $x = \frac{1}{2}$.

(b) We see that the graph is concave down (so $f''(x) < 0$) to the left of the point of inflection. That is, $f''(x) < 0$ when $x < \frac{1}{2}$. Similarly, $f''(x) > 0$ when $x = \frac{1}{2}$.

(c) $f(x) = \frac{1}{6}(2x^3 - 3x^2 - 12x + 12)$
$f'(x) = \frac{1}{6}(6x^2 - 6x - 12) = x^2 - x - 2$
$f''(x) = 2x - 1$
Thus $f''(x) = 0$ when $x = \frac{1}{2}$.
Figure 10.23 shows the graph of $f''(x) = 2x - 1$. We see that the graph crosses the x-axis ($f''(x) = 0$) when $x = \frac{1}{2}$, is below the x-axis ($f''(x) < 0$) when $x < \frac{1}{2}$, and is above the x-axis ($f''(x) > 0$) when $x > \frac{1}{2}$. This verifies our conclusions from parts (a) and (b). ∎

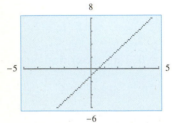

Figure 10.23

Spreadsheet Note

We can investigate the function in Example 5 by using Excel to find the outputs of $f(x) = \frac{1}{6}(2x^3 - 3x^2 - 12x + 12)$, $f'(x)$, and $f''(x)$ for values of x at and near the critical points. See Appendix D, Section 10.2, and the Online Excel Guide for details.

	A	B	C	D
1	x	$f(x)$	$f'(x)$	$f''(x)$
2	−2	1.3333	4	−5
3	−1	3.1667	0	−3
4	0	2	−2	−1
5	1/2	0.9167	−2.25	0
6	1	−0.1667	−2	1
7	2	−1.3333	0	3
8	3	0.5	4	5

By looking at column C, we see that $f'(x)$ changes from positive to negative as x passes through the value $x = -1$ and that the derivative changes from negative to positive as x passes through the value $x = 2$. Thus the graph has a relative maximum at $x = -1$ and a relative minimum at $x = 2$. By looking at column D, we can also see that $f''(x)$ changes from negative to positive as x passes through the value $x = 1/2$, so the graph has a point of inflection at $x = 1/2$. ∎

EXAMPLE 6

Concavity from the Graph of $y = f'(x)$

Figure 10.24 shows the graph of $f'(x) = -x^2 - 2x$. Use the graph of $f'(x)$ to do the following.

(a) Find intervals on which $f(x)$ is concave down and concave up.
(b) Find x-values at which $f(x)$ has a point of inflection.
(c) Check the conclusions from parts (a) and (b) by finding $f''(x)$ and graphing it.
(d) For $f(x) = \frac{1}{3}(9 - x^3 - 3x^2)$, calculate $f'(x)$ to verify that this could be $f(x)$.

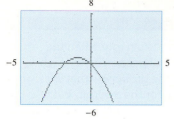

Figure 10.24

Solution

(a) Concavity for $f(x)$ can be found from the sign of $f''(x)$. Because $f''(x)$ is the first derivative of $f'(x)$, wherever the graph of $f'(x)$ is increasing, it follows that $f''(x) > 0$. Thus $f''(x) > 0$ and $f(x)$ is concave up when $x < -1$. Similarly, $f''(x) < 0$, and $f(x)$ is concave down, when $f'(x)$ is decreasing—that is, when $x > -1$.

(b) From (a) we know that $f''(x)$ changes sign at $x = -1$, so $f(x)$ has a point of inflection at $x = -1$. Note that $f'(x)$ has its maximum at the x-value where $f(x)$ has a point of inflection. In fact, points of inflection for $f(x)$ will correspond to relative extrema for $f'(x)$.

(c) For $f'(x) = -x^2 - 2x$, we have $f''(x) = -2x - 2$. Figure 10.25 shows the graph of $y = f''(x)$ and verifies our conclusions from (a) and (b).

(d) If $f(x) = \frac{1}{3}(9 - x^3 - 3x^2)$, then $f'(x) = \frac{1}{3}(-3x^2 - 6x) = -x^2 - 2x$. Figure 10.26 shows the graph of $f(x) = \frac{1}{3}(9 - x^3 - 3x^2)$. Note that the point of inflection and the concavity correspond to what we discovered in parts (a) and (b). ■

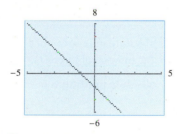

Figure 10.25

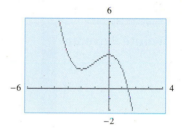

Figure 10.26

The relationships among $f(x)$, $f'(x)$, and $f''(x)$ that we explored in Example 6 can be summarized as follows.

$f(x)$	Concave Up	Concave Down	Point of Inflection	
$f'(x)$	increasing	decreasing	maximum	minimum
$f''(x)$	positive $(+)$	negative $(-)$	$(+)$ to $(-)$	$(-)$ to $(+)$

CHECKPOINT SOLUTIONS

1. Up
2. The second derivative of $y = \frac{1}{3}x^3 - 2x^2 + 2x$ is $y'' = 2x - 4$, so $y'' = 0$ at $x = 2$. Because $y'' < 0$ for $x < 2$ and $y'' > 0$ for $x > 2$, a point of inflection occurs at $x = 2$.
3. Points of inflection at A, B, and C
   $f''(x) < 0$ to the left of A and between B and C
   $f''(x) > 0$ between A and B and to the right of C
4. (a) False. For example, if $f(x) = x^4$, then $f'(x) = 4x^3$ and $f''(x) = 12x^2$. Note that $f''(0) = 0$, but $f''(x)$ does not change sign from $x < 0$ to $x > 0$.

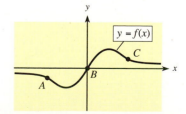

| EXERCISES | **10.2**

Access end-of-section exercises online at **www.webassign.net** **ENHANCED WebAssign**

10.3

Optimization in Business and Economics

| APPLICATION PREVIEW |

Suppose a travel agency charges $300 per person for a certain tour when the tour group has 25 people, but will reduce the price by $10 per person for each additional person above the 25. In order to determine the group size that will maximize the revenue, the agency must create a model for the revenue. With such a model, the techniques of calculus can be useful in determining the maximum revenue. (See Example 2.)

Most companies (such as the travel agency) are interested in obtaining the greatest possible revenue or profit. Similarly, manufacturers of products are concerned about producing their products for the lowest possible average cost per unit. Therefore, rather than just finding the relative maxima or relative minima of a function, we will consider where the absolute maximum or absolute minimum of a function occurs in a given interval.

In this section we will discuss how to find the absolute extrema of a function and then use these techniques to solve applications involving revenue, cost, and profit.

As their name implies, **absolute extrema** are the functional values that are the largest or smallest values over the entire domain of the function (or over the interval of interest).

Absolute Extrema

The value $f(a)$ is the **absolute maximum** of f if $f(a) \geq f(x)$ for all x in the domain of f (or over the interval of interest).

The value $f(b)$ is the **absolute minimum** of f if $f(b) \leq f(x)$ for all x in the domain of f (or over the interval of interest).

Let us begin by considering the graph of $y = (x - 1)^2$, shown in Figure 10.28(a). This graph has a relative minimum at $(1, 0)$. Note that the relative minimum is the lowest point on the graph. In this case, the point $(1, 0)$ is an **absolute minimum point**, and 0 is the absolute minimum for the function. Similarly, when there is a point that is the highest point on the graph over the domain of the function, we call the point an **absolute maximum point** of the graph of the function.

In Figure 10.28(a), we see that there is no relative maximum. However, if the domain of the function is restricted to the interval $\left[\frac{1}{2}, 2\right]$, then we get the graph shown in Figure 10.28(b). In this case, there is an absolute maximum of 1 at the point $(2, 1)$ and the absolute minimum of 0 is still at $(1, 0)$.

If the domain of $y = (x - 1)^2$ is restricted to the interval $[2, 3]$, the resulting graph is that shown in Figure 10.28(c). In this case, the absolute minimum is 1 and occurs at the point $(2, 1)$, and the absolute maximum is 4 and occurs at $(3, 4)$.

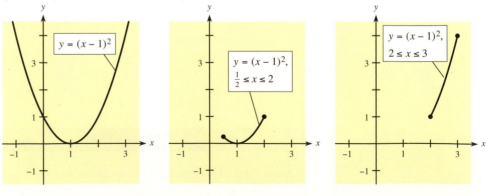

Figure 10.28   (a)   (b)   (c)

As the preceding discussion indicates, if the domain of a continuous function is limited to a closed interval, the absolute maximum or minimum may occur at an endpoint of the domain. In testing functions with limited domains for absolute maxima and minima, we must compare the function values at the endpoints of the domain with the function values at the critical values found by taking derivatives. In the management, life, and social sciences, a limited domain occurs very often, because many quantities are required to be positive, or at least nonnegative.

Maximizing Revenue

Because the marginal revenue is the first derivative of the total revenue, it should be obvious that the total revenue function will have a critical point at the point where the marginal revenue equals 0. With the total revenue function $R(x) = 16x - 0.02x^2$, the point where $R'(x) = 0$ is clearly a maximum because $R(x)$ is a parabola that opens downward. But the domain may be limited, the revenue function may not always be a parabola, or the critical point may not always be a maximum, so it is important to verify where the maximum value occurs.

EXAMPLE 1   Revenue

The total revenue in dollars for a firm is given by

$$R(x) = 8000x - 40x^2 - x^3$$

where x is the number of units sold per day. If only 50 units can be sold per day, find the number of units that must be sold to maximize revenue. Find the maximum revenue.

Solution

This revenue function is limited to x in the interval $[0, 50]$. Thus, the maximum revenue will occur at a critical value in this interval or at an endpoint. $R'(x) = 8000 - 80x - 3x^2$, so we must solve $8000 - 80x - 3x^2 = 0$ for x:

$$(40 - x)(200 + 3x) = 0 \quad \text{means} \quad 40 - x = 0 \quad \text{or} \quad 200 + 3x = 0$$

Thus
$$x = 40 \quad \text{or} \quad x = -\frac{200}{3}$$

We reject the negative value of x and then use either the second-derivative test or the first-derivative test with a sign diagram.

Second-Derivative Test	First-Derivative Test
$R''(x) = -80 - 6x$, $R''(40) = -320 < 0$, so 40 is a relative maximum.	

These tests show that a relative maximum occurs at $x = 40$, giving revenue $R(40) = \$192,000$. Checking the endpoints $x = 0$ and $x = 50$ gives $R(0) = \$0$ and $R(50) = \$175,000$. Thus $R = \$192,000$ at $x = 40$ is the (absolute) maximum revenue. ■

CHECKPOINT

1. True or false: If $R(x)$ is the revenue function, we find all possible points where $R(x)$ could be maximized by solving $\overline{MR} = 0$ for x.

EXAMPLE 2   Revenue | APPLICATION PREVIEW |

A travel agency will plan tours for groups of 25 or larger. If the group contains exactly 25 people, the price is $300 per person. However, the price per person is reduced by $10 for each additional person above the 25. What size group will produce the largest revenue for the agency?

Solution

The total revenue is

$$R = (\text{number of people})(\text{price per person})$$

The table shows how the revenue is changed by increases in the size of the group.

Number in Group	Price per Person	Revenue
25	300	7500
25 + 1	300 − 10	7540
25 + 2	300 − 20	7560
⋮	⋮	⋮
25 + x	300 − 10x	(25 + x)(300 − 10x)

Thus when x is the number of people added to the 25, the total revenue will be

$$R = R(x) = (25 + x)(300 - 10x) \quad \text{or} \quad R(x) = 7500 + 50x - 10x^2$$

This is a quadratic function, so its graph is a parabola that is concave down. A maximum will occur at its vertex, where $R'(x) = 0$.

$R'(x) = 50 - 20x$, and the solution to $0 = 50 - 20x$ is $x = 2.5$. Thus adding 2.5 people to the group should maximize the total revenue. But we cannot add half a person, so we will test the total revenue function for 27 people and 28 people. This will determine the optimal group size because $R(x)$ is concave down for all x.

For $x = 2$ (giving 27 people) we get $R(2) = 7500 + 50(2) - 10(2)^2 = 7560$. For $x = 3$ (giving 28 people) we get $R(3) = 7500 + 50(3) - 10(3)^2 = 7560$. Note that both 27 and 28 people give the same total revenue and that this revenue is greater than the revenue for 25 people. Thus the revenue is maximized at either 27 or 28 people in the group. (See Figure 10.29.) ■

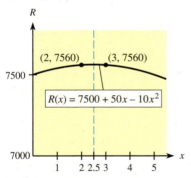

Figure 10.29

Minimizing Average Cost

Because the total cost function is always increasing for $x \geq 0$, the number of units that will make the total cost a minimum is always $x = 0$ units, which gives an absolute minimum. However, it is more useful to find the number of units that will make the **average cost** per unit a minimum.

Average Cost

If the total cost function is $C = C(x)$, then the per unit **average cost function** is

$$\overline{C} = \frac{C(x)}{x}$$

Note that the average cost per unit is undefined if no units are produced.

We can use derivatives to find the minimum of the average cost function, as the next example shows.

EXAMPLE 3 **Average Cost**

If the total cost function for a commodity is given by $C = \frac{1}{4}x^2 + 4x + 100$ dollars, where x represents the number of units produced, producing how many units will result in a minimum *average cost* per unit? Find the minimum average cost.

Solution

Begin by finding the average cost function and its derivative:

$$\overline{C} = \frac{\frac{1}{4}x^2 + 4x + 100}{x} = \frac{1}{4}x + 4 + \frac{100}{x}$$

$$\overline{C}' = \overline{C}'(x) = \frac{1}{4} - \frac{100}{x^2}$$

Setting $\overline{C}' = 0$ gives

$$0 = \frac{1}{4} - \frac{100}{x^2} \quad \text{so} \quad 0 = x^2 - 400 \text{ and } x = \pm 20$$

Because the quantity produced must be positive, 20 units should minimize the average cost per unit. We show that it is an absolute minimum by using the second derivative.

$$\overline{C}''(x) = \frac{200}{x^3} \quad \text{so} \quad \overline{C}''(x) > 0 \text{ when } x > 0$$

Thus the minimum average cost per unit occurs if 20 units are produced. The graph of the average cost per unit is shown in Figure 10.30. The minimum average cost per unit is $\overline{C}(20) = \$14$. ∎

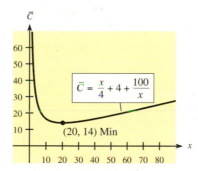

$$\overline{C} = \frac{x}{4} + 4 + \frac{100}{x}$$

(20, 14) Min

Figure 10.30

CHECKPOINT

2. If $C(x) = 0.01x^2 + 20x + 2500$, form $\overline{C}(x)$, the average cost function, and find the minimum average cost.

The graph of the cost function for the commodity in Example 3 is shown in Figure 10.31, along with several lines that join $(0, 0)$ to a point of the form $(x, C(x))$ on the total cost graph. Note that the slope of each of these lines has the form

$$\frac{C(x) - 0}{x - 0} = \frac{C(x)}{x}$$

so the slope of each line is the *average cost* for the given number of units at the point on $C(x)$. Note that the line from the origin to the point on the curve where $x = 20$ is tangent to the total cost curve. Hence, when $x = 20$, the slope of this line represents both the derivative of the cost function (the *marginal cost*) and the *average cost*. All lines from the origin to points with x-values larger than 20 or smaller than 20 are steeper (and therefore have greater slopes) than the line to the point where $x = 20$. Thus the minimum average cost occurs where the average cost equals the marginal cost. You will be asked to show this analytically in Problems 21 and 22 of the 10.3 Exercises.

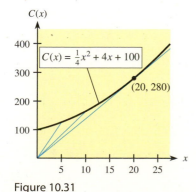

$C(x) = \frac{1}{4}x^2 + 4x + 100$

(20, 280)

Figure 10.31

Maximizing Profit   We have defined the marginal profit function as the derivative of the profit function. That is,

$$\overline{MP} = P'(x)$$

In this chapter we have seen how to use the derivative to find maxima and minima for various functions. Now we can apply those same techniques in the context of **profit maximization.** We can use marginal profit to maximize profit functions.

If there is a physical limitation on the number of units that can be produced in a given period of time, then the endpoints of the interval caused by these limitations should also be checked.

EXAMPLE 4   Profit

Suppose that the production capacity for a certain commodity cannot exceed 30. If the total profit for this commodity is

$$P(x) = 4x^3 - 210x^2 + 3600x - 200 \text{ dollars}$$

where x is the number of units sold, find the number of items that will maximize profit.

Solution

The restrictions on capacity mean that $P(x)$ is restricted by $0 \le x \le 30$. The marginal profit function is

$$P'(x) = 12x^2 - 420x + 3600$$

Setting $P'(x)$ equal to 0, we get

$$0 = 12(x - 15)(x - 20)$$

so $P'(x) = 0$ at $x = 15$ and $x = 20$. A sign diagram for $P'(x)$ tests these critical values.

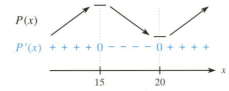

Thus, at (15, 20,050) the total profit function has a *relative* maximum, but we must check the endpoints (0 and 30) before deciding whether it is the absolute maximum.

$$P(0) = -200 \text{ and } P(30) = 26,800$$

Thus the absolute maximum profit is $26,800, and it occurs at the endpoint, $x = 30$. Figure 10.32 shows the graph of the profit function.   ■

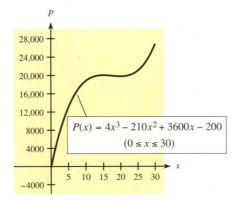

Figure 10.32

In a **monopolistic market,** the seller who has a monopoly controls the supply of a product and can force the price higher by limiting supply.

If the demand function for the product is $p = f(x)$, total revenue for the sale of x units is $R(x) = px = f(x) \cdot x$. Note that the price p is fixed by the market in a competitive market but varies with output for the monopolist.

If $\overline{C} = \overline{C}(x)$ represents the average cost per unit sold, then the total cost for the x units sold is $C = \overline{C} \cdot x = \overline{C}x$. Because we have both total cost and total revenue as a function of the quantity, x, we can maximize the profit function, $P(x) = px - \overline{C}x$, where p represents the demand function $p = f(x)$ and $\overline{C}$ represents the average cost function $\overline{C} = \overline{C}(x)$.

| EXAMPLE | 5 | **Profit in a Monopoly Market** |

MONOPOLY

One big player controls supply and dictates pricing.

The price of a product in dollars is related to the number of units x demanded daily by

$$p = 168 - 0.2x$$

A monopolist finds that the daily average cost for this product is

$$\overline{C} = 120 + x \ \text{ dollars}$$

(a)   How many units must be sold daily to maximize profit?
(b)   What is the selling price at this "optimal" level of production?
(c)   What is the maximum possible daily profit?

Solution

(a)   The total revenue function for the product is

$$R(x) = px = (168 - 0.2x)x = 168x - 0.2x^2$$

and the total cost function is

$$C(x) = \overline{C} \cdot x = (120 + x)x = 120x + x^2$$

Thus the profit function is

$$P(x) = R(x) - C(x) = 168x - 0.2x^2 - (120x + x^2) = 48x - 1.2x^2$$

Then $P'(x) = 48 - 2.4x$, so $P'(x) = 0$ when $x = 20$. We see that $P''(20) = -2.4$, so by the second-derivative test, $P(x)$ has a maximum at $x = 20$. That is, selling 20 units will maximize profit.

(b)   The selling price is determined by $p = 168 - 0.2x$, so the price that will result from supplying 20 units per day is $p = 168 - 0.2(20) = 164$. That is, the "optimal" selling price is $164 per unit.

(c)   The profit at $x = 20$ is $P(20) = 48(20) - 1.2(20)^2 = 960 - 480 = 480$. Thus the maximum possible profit is $480 per day.

In a **competitive market,** each firm is so small that its actions in the market cannot affect the price of the product. The price of the product is determined in the market by the intersection of the market demand curve (from all consumers) and market supply curve (from all firms that supply this product). The firm can sell as little or as much as it desires at the market equilibrium price.

Therefore, a firm in a competitive market has a total revenue function given by $R(x) = px$, where p is the market equilibrium price for the product and x is the quantity sold.

COMPETITION

Several players are so small that individual actions cannot affect product

| EXAMPLE | 6 | **Profit in a Competitive Market** |

A firm in a competitive market must sell its product for $200 per unit. The average cost per unit (per month) is $\overline{C} = 80 + x$, where x is the number of units sold per month. How many units should be sold to maximize profit?

Solution

If the average cost per unit is $\overline{C} = 80 + x$, then the total cost of x units is given by $C(x) = (80 + x)x = 80x + x^2$. The revenue per unit is \$200, so the total revenue is given by $R(x) = 200x$. Thus the profit function is

$$P(x) = R(x) - C(x) = 200x - (80x + x^2), \quad \text{or} \quad P(x) = 120x - x^2$$

Then $P'(x) = 120 - 2x$. Setting $P'(x) = 0$ and solving for x gives $x = 60$. Because $P''(60) = -2$, the profit is maximized when the firm sells 60 units per month.   ■

CHECKPOINT

3. (a) If $p = 5000 - x$ gives the demand function in a monopoly market, find $R(x)$, if it is possible with this information.
    (b) If $p = 5000 - x$ gives the demand function in a competitive market, find $R(x)$, if it is possible with this information.
4. If $R(x) = 400x - 0.25x^2$ and $C(x) = 150x + 0.25x^2 + 8500$ are the total revenue and total cost (both in dollars) for x units of a product, find the number of units that gives maximum profit and find the maximum profit.

Calculator Note

As we have seen, graphing calculators can be used to locate maximum values. In addition, if it is difficult to determine critical values algebraically, we may be able to approximate them graphically. For example, if

$$P(x) = 2500 - \frac{3000}{x + 1} - 12x - x^2 \quad \text{then} \quad P'(x) = \frac{3000}{(x + 1)^2} - 12 - 2x$$

Finding the critical values by solving $P'(x) = 0$ is difficult unless we use a graphing approach. Figure 10.33(a) shows the graph of $P(x)$, and Figure 10.33(b) shows the graph of $P'(x)$. These figures indicate that the maximum occurs near $x = 10$.

By adjusting the viewing window for $P'(x)$, we obtain the graph in Figure 10.34. This shows that $P'(x) = 0$ when $x = 9$. The maximum is $P(9) = 2500 - 300 - 108 - 81 = 2011$. See Appendix C, Section 10.3, for additional details.   ■

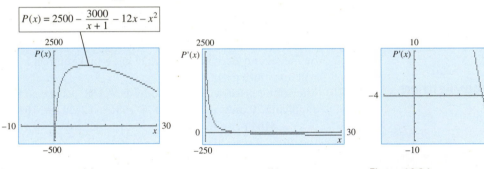

Figure 10.33   (a)   (b)   Figure 10.34

CHECKPOINT SOLUTIONS

1. False. $\overline{MR} = R'(x)$, but there may also be critical points where $R'(x)$ is undefined, or $R(x)$ may be maximized at endpoints of a restricted domain.

2. $$\overline{C}(x) = \frac{C(x)}{x} = \frac{0.01x^2 + 20x + 2500}{x} = 0.01x + 20 + \frac{2500}{x}$$

$$\overline{C}'(x) = 0.01 - \frac{2500}{x^2}$$

Solving $\overline{C}'(x) = 0$ gives

$$0.01 - \frac{2500}{x^2} = 0 \quad \text{so} \quad x^2 = 250,000 \quad \text{and} \quad x = \pm 500$$

Using $\overline{C}''(x) = 5000/x^3$ we see that $\overline{C}''(500) > 0$ so $x = 500$ gives the minimum average cost per unit of $\overline{C}(500) = 5 + 20 + 5 = 30$ dollars per unit.

3. (a) $R(x) = p \cdot x = (5000 - x)x = 5000x - x^2$
    (b) In a competitive market, $R(x) = p \cdot x$, where p is the constant equilibrium price. Thus we need to know the supply function and find the equilibrium price before we can form $R(x)$.

4. $P(x) = R(x) - C(x) = 250x - 0.5x^2 - 8500$ and $P'(x) = 250 - x$. $P'(x) = 0$ when $x = 250$ and $P''(x) = -1$ for all x. Hence the second-derivative test shows that the (absolute) maximum profit occurs when $x = 250$ and is $P(250) = \$22{,}750$.

| EXERCISES | 10.3

Access end-of-section exercises online at **www.webassign.net**

OBJECTIVE   10.4

- To apply the procedures for finding maxima and minima to solve problems from the management, life, and social sciences

Applications of Maxima and Minima

| APPLICATION PREVIEW |

Suppose that a company needs 1,000,000 items during a year and that preparation costs are $800 for each production run. Suppose further that it costs the company $6 to produce each item and $1 to store each item for up to a year. Find the number of units that should be produced in each production run so that the total costs of production and storage are minimized. This question is answered using an inventory cost model (see Example 4).

This inventory-cost determination is a typical example of the kinds of questions and important business applications that require the use of the derivative for finding maxima and minima. As managers, workers, or consumers, we may be interested in such things as maximum revenue, maximum profit, minimum cost, maximum medical dosage, maximum utilization of resources, and so on.

If we have functions that model cost, revenue, or population growth, we can apply the methods of this chapter to find the maxima and minima of those functions.

EXAMPLE   1   Company Growth

Suppose that a new company begins production in 2010 with eight employees and the growth of the company over the next 10 years is predicted by

$$N = N(t) = 8\left(1 + \frac{160t}{t^2 + 16}\right), \quad 0 \le t \le 10$$

where N is the number of employees t years after 2010.

Determine in what year the number of employees will be maximized and the maximum number of employees.

Solution

This function will have a relative maximum when $N'(t) = 0$.

$$N'(t) = 8\left[\frac{(t^2 + 16)(160) - (160t)(2t)}{(t^2 + 16)^2}\right]$$

$$= 8\left[\frac{160t^2 + 2560 - 320t^2}{(t^2 + 16)^2}\right]$$

$$= 8\left[\frac{2560 - 160t^2}{(t^2 + 16)^2}\right]$$

Because $N'(t) = 0$ when its numerator is 0 (note that this denominator is never 0), we must solve

$$2560 - 160t^2 = 0$$
$$160(4 + t)(4 - t) = 0$$
so
$$t = -4 \quad \text{or} \quad t = 4$$

We are interested only in positive t-values, so we test $t = 4$.

$$\left.\begin{array}{l} N'(0) = 8\left[\dfrac{2560}{256}\right] > 0 \\[3mm] N'(10) = 8\left[\dfrac{-13,440}{(116)^2}\right] < 0 \end{array}\right\} \Rightarrow \textit{relative maximum}$$

The relative maximum is

$$N(4) = 8\left(1 + \frac{640}{32}\right) = 168$$

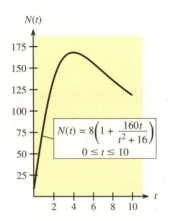

At $t = 0$, the number of employees is $N(0) = 8$, and it increases to $N(4) = 168$. After $t = 4$ (in 2014), $N(t)$ decreases to $N(10) = 118$ (approximately), so $N(4) = 168$ is the maximum number of employees. Figure 10.35 verifies these conclusions. ■

Figure 10.35

Sometimes we must develop the function we need from the statement of the problem. In this case, it is important to understand what is to be maximized or minimized and to express that quantity as a function of *one* variable.

EXAMPLE 2 | Minimizing Cost

A farmer needs to enclose a rectangular pasture containing 1,600,000 square feet. Suppose that along the road adjoining his property he wants to use a more expensive fence and that he needs no fence on one side perpendicular to the road because a river bounds his property on that side. If the fence costs $15 per foot along the road and $10 per foot along the two remaining sides that must be fenced, what dimensions of his rectangular field will minimize his cost?

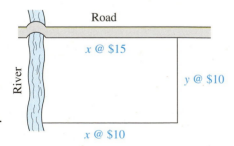

Figure 10.36

Solution

In Figure 10.36, x represents the length of the pasture along the road (and parallel to the road) and y represents the width. The cost function for the fence used is

$$C = 15x + 10y + 10x = 25x + 10y$$

We cannot use a derivative to find where C is minimized unless we write C as a function of x or y only. Because the area of the rectangular field must be 1,600,000 square feet, we have

$$A = xy = 1,600,000$$

Solving for y in terms of x and substituting give C as a function of x.

$$y = \frac{1{,}600{,}000}{x}$$

$$C = 25x + 10\left(\frac{1{,}600{,}000}{x}\right) = 25x + \frac{16{,}000{,}000}{x}$$

The derivative of C with respect to x is

$$C'(x) = 25 - \frac{16{,}000{,}000}{x^2}$$

and we find the relative minimum of C as follows:

$$0 = 25 - \frac{16{,}000{,}000}{x^2}$$
$$0 = 25x^2 - 16{,}000{,}000$$
$$25x^2 = 16{,}000{,}000$$
$$x^2 = 640{,}000 \Rightarrow x = \pm800$$

We use $x = 800$ feet

Testing to see whether $x = 800$ gives the minimum cost, we find

$$C''(x) = \frac{32{,}000{,}000}{x^3}$$

$C''(x) > 0$ for $x > 0$, so $C(x)$ is concave up for all positive x. Thus $x = 800$ gives the absolute minimum, and $C(800) = 40{,}000$ is the minimum cost. The other dimension of the rectangular field is $y = 1{,}600{,}000/800 = 2000$ feet. Figure 10.37 verifies that $C(x)$ reaches its minimum (of 40,000) at $x = 800$. ■

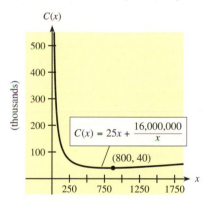

Figure 10.37

EXAMPLE 3   Postal Restrictions

Postal restrictions limit the size of packages sent through the mail. If the restrictions are that the length plus the girth may not exceed 108 inches, find the volume of the largest box with square cross section that can be mailed.

Solution
Let l equal the length of the box, and let s equal a side of the square end. See Figure 10.38. The volume we seek to maximize is given by

$$V = s^2 l$$

We can use the restriction that girth plus length equals 108,

$$4s + l = 108$$

to express V as a function of s or l. Because $l = 108 - 4s$, the equation for V becomes

$$V = s^2(108 - 4s) = 108s^2 - 4s^3$$

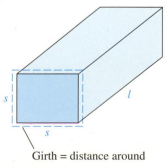

Girth = distance around

Figure 10.38

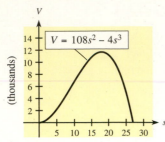

Figure 10.39

Thus we can use dV/ds to find the critical values.

$$\frac{dV}{ds} = 216s - 12s^2$$

$$0 = s(216 - 12s)$$

The critical values are $s = 0$, $s = \frac{216}{12} = 18$. The critical value $s = 0$ will not maximize the volume, for in this case, $V = 0$. Testing to the left and right of $s = 18$ gives

$$V'(17) > 0 \quad \text{and} \quad V'(19) < 0$$

Thus $s = 18$ inches and $l = 108 - 4(18) = 36$ inches yield a maximum volume of 11,664 cubic inches. Once again we can verify our results graphically. Figure 10.39 shows that $V = 108s^2 - 4s^3$ achieves its maximum when $s = 18$. ■

CHECKPOINT

Suppose we want to find the minimum value of $C = 5x + 2y$ and we know that x and y must be positive and that $xy = 1000$.
1. What equation do we differentiate to solve this problem?
2. Find the critical values.
3. Find the minimum value of C.

We next consider **inventory cost models**, in which x items are produced in each production run and items are removed from inventory at a fixed constant rate. Because items are removed at a constant rate, the average number stored at any time is $x/2$. Also, when $x = 0$, new items must be added to inventory from a production run. Thus the number of units in storage changes with time and is illustrated in Figure 10.40. In these models there are costs associated with both production and storage, but lowering one of these costs means increasing the other. To see how inventory cost models work, consider the following example.

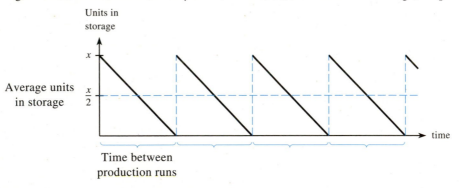

Figure 10.40

EXAMPLE 4   **Inventory Cost Model** | APPLICATION PREVIEW |

Suppose that a company needs 1,000,000 items during a year and that preparation costs are $800 for each production run. Suppose further that it costs the company $6 to produce each item and $1 to store an item for up to a year. If each production run consists of x items, find x so that the total costs of production and storage are minimized.

Solution
The total production costs are given by

$$\left(\begin{array}{c}\text{No. of}\\\text{runs}\end{array}\right)\left(\begin{array}{c}\text{cost}\\\text{per run}\end{array}\right) + \left(\begin{array}{c}\text{no. of}\\\text{items}\end{array}\right)\left(\begin{array}{c}\text{cost}\\\text{per item}\end{array}\right) = \left(\frac{1{,}000{,}000}{x}\right)(\$800) + (1{,}000{,}000)(\$6)$$

The total storage costs are

$$\left(\begin{array}{c}\text{Average}\\\text{no. stored}\end{array}\right)\left(\begin{array}{c}\text{storage cost}\\\text{per item}\end{array}\right) = \left(\frac{x}{2}\right)(\$1)$$

Thus the total costs of production and storage are

$$C = \left(\frac{1,000,000}{x}\right)(800) + 6,000,000 + \frac{x}{2} = \frac{800,000,000}{x} + 6,000,000 + \frac{x}{2}$$

We wish to find x so that C is minimized.

$$C' = \frac{-800,000,000}{x^2} + \frac{1}{2}$$

If $x > 0$, critical values occur when $C' = 0$.

$$0 = \frac{-800,000,000}{x^2} + \frac{1}{2}$$

$$\frac{800,000,000}{x^2} = \frac{1}{2}$$

$$1,600,000,000 = x^2$$

$$x = \pm 40,000$$

Because x must be positive, we test $x = 40,000$ with the second derivative.

$$C''(x) = \frac{1,600,000,000}{x^3}, \text{ so } C''(40,000) > 0$$

Note that $x = 40,000$ yields an absolute minimum value of C, because $C'' > 0$ for all $x > 0$. That is, production runs of 40,000 items yield minimum total costs for production and storage.   ■

Technology Note   Problems of the types we've studied in this section could also be solved (at least approximately) with a graphing calculator or Excel. With this approach, our first goal is still to express the quantity to be maximized or minimized as a function of one variable. Then that function can be graphed, and the (at least approximate) optimal value can be obtained from the graph. See Appendices C and D, Section 10.4, and the Online Excel Guide for details.   ■

 EXAMPLE 5   Property Development

A developer of a campground has to pay for utility line installation to the community center from a transformer on the street at the corner of her property. Because of local restrictions, the lines must be underground on her property. Suppose that the costs are $50 per meter along the street and $100 per meter underground. How far from the transformer should the line enter the property to minimize installation costs? See Figure 10.41.

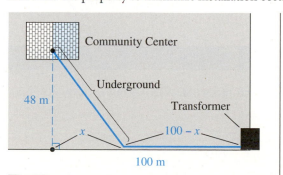

Figure 10.41

Solution
If the developer had the cable placed underground from the community center perpendicular to the street and then to the transformer, then $x = 0$ in Figure 10.41 and the cost would be

$$\$100(48) + \$50(100) = \$9800$$

It may be possible to save some money by placing the cable on a diagonal to the street, but then only $x \leq 100$ makes sense. By using the Pythagorean Theorem, we find that the length of the underground cable that meets the street x meters closer to the transformer is

$$\sqrt{48^2 + x^2} = \sqrt{2304 + x^2} \text{ meters}$$

Thus the cost C of installation is given by

$$C = 100\sqrt{2304 + x^2} + 50(100 - x) \text{ dollars}$$

Figure 10.42 shows the graph of this function over an interval for x that contains $0 \leq x \leq 100$. Because any extrema must occur in this interval, we can find the minimum by using the MIN command on a calculator or Excel. See Appendices C and D, Section 10.4, and the Online Excel Guide for the steps. The minimum cost is \$9156.92, when $x = 27.7$ meters (that is, when the cable meets the street 72.3 meters from the transformer). ■

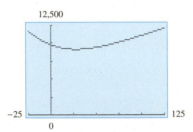

Figure 10.42

CHECKPOINT SOLUTIONS

1. We must differentiate C, but first C must be expressed as a function of one variable: $xy = 1000$ means that $y = 1000/x$. If we substitute $1000/x$ for y in $C = 5x + 2y$, we get

$$C(x) = 5x + 2\left(\frac{1000}{x}\right) = 5x + \frac{2000}{x}$$

   Find $C'(x)$ to solve the problem.

2. $C'(x) = 5 - 2000/x^2$, so $C'(x) = 0$ when

$$5 - \frac{2000}{x^2} = 0, \text{ or } x = \pm 20$$

   Because x must be positive, the only critical value is $x = 20$.

3. $C''(x) = 4000/x^3$, so $C''(x) > 0$ for all $x > 0$. Thus $x = 20$ yields the minimum value. Also, when $x = 20$, we have $y = 50$, so the minimum value of C is $C = 5(20) + 2(50) = 200$.

| EXERCISES | 10.4

Access end-of-section exercises online at **www.webassign.net**

- To locate horizontal asymptotes
- To locate vertical asymptotes
- To sketch graphs of functions that have vertical and/or horizontal asymptotes

Rational Functions: More Curve Sketching

| APPLICATION PREVIEW |

Suppose that the total cost of producing a shipment of a product is

$$C(x) = 5000x + \frac{125{,}000}{x}, \quad x > 0$$

where x is the number of machines used in the production process. To find the number of machines that will minimize the total cost, we find the minimum value of this rational function. (See Example 3.) The graph of this function contains a vertical asymptote at $x = 0$. We will discuss graphs and applications involving asymptotes in this section.

The procedures for using the first-derivative test and the second-derivative test are given in previous sections, but none of the graphs discussed in those sections contains vertical asymptotes or horizontal asymptotes. In this section, we consider how to use information about asymptotes along with the first and second derivatives, and we present a unified approach to curve sketching.

Asymptotes   In Section 2.4, "Special Functions and Their Graphs," we first discussed asymptotes and saw that they are important features of the graphs that have them. Then, in our discussion of limits in Sections 9.1 and 9.2, we discovered the relationship between certain limits and asymptotes. The formal definition of **vertical asymptotes** uses limits.

Vertical Asymptote

The line $x = x_0$ is a **vertical asymptote** of the graph of $y = f(x)$ if the values of $f(x)$ approach $+\infty$ or $-\infty$ as x approaches x_0 (from the left or the right).

From our work with limits, recall that a vertical asymptote will occur on the graph of a function at an x-value at which the denominator (but not the numerator) of the function is equal to zero. These observations allow us to determine where vertical asymptotes occur.

Vertical Asymptote of a Rational Function

The graph of the rational function

$$h(x) = \frac{f(x)}{g(x)}$$

has a vertical asymptote at $x = c$ if $g(c) = 0$ and $f(c) \neq 0$.

Because a **horizontal asymptote** tells us the behavior of the values of the function (y-coordinates) when x increases or decreases without bound, we use limits at infinity to determine the existence of horizontal asymptotes.

Horizontal Asymptote

The graph of a rational function $y = f(x)$ will have a **horizontal asymptote** at $y = b$, for a constant b, if

$$\lim_{x \to +\infty} f(x) = b \quad \text{or} \quad \lim_{x \to -\infty} f(x) = b$$

Otherwise, the graph has no horizontal asymptote.

For a rational function f, $\lim_{x \to +\infty} f(x) = b$ if and only if $\lim_{x \to -\infty} f(x) = b$, so we only need to find one of these limits to locate a horizontal asymptote. In Problems 37 and 38 in the 9.2 Exercises, the following statements regarding horizontal asymptotes of the graphs of rational functions were proved.

Horizontal Asymptotes of Rational Functions

Consider the rational function $y = \dfrac{f(x)}{g(x)} = \dfrac{a_n x^n + \cdots + a_1 x + a_0}{b_m x^m + \cdots + b_1 x + b_0}$.

1. If $n < m$ (that is, if the degree of the numerator is less than that of the denominator), a horizontal asymptote occurs at $y = 0$ (the x-axis).

2. If $n = m$ (that is, if the degree of the numerator equals that of the denominator), a horizontal asymptote occurs at $y = \dfrac{a_n}{b_m}$ (the ratio of the leading coefficients).

3. If $n > m$ (that is, if the degree of the numerator is greater than that of the denominator), there is no horizontal asymptote.

EXAMPLE **1**   **Vertical and Horizontal Asymptotes**

Find any vertical and horizontal asymptotes for

(a) $f(x) = \dfrac{2x - 1}{x + 2}$ (b) $f(x) = \dfrac{x^2 + 3}{1 - x}$

Solution

(a) The denominator of this function is 0 at $x = -2$, and because this value does not make the numerator 0, there is a vertical asymptote at $x = -2$. Because the function is rational, with the degree of the numerator equal to that of the denominator and with the ratio of the leading coefficients equal to 2, the graph of the function has a horizontal asymptote at $y = 2$. The graph is shown in Figure 10.43(a).

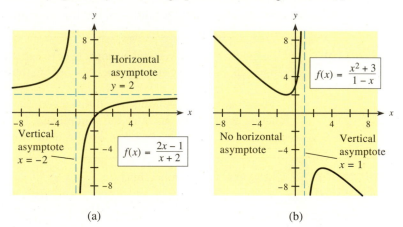

Figure 10.43   (a)   (b)

(b) At $x = 1$, the denominator of $f(x)$ is 0 and the numerator is not, so a vertical asymptote occurs at $x = 1$. The function is rational with the degree of the numerator greater than that of the denominator, so there is no horizontal asymptote. The graph is shown in Figure 10.43(b). ■

More Curve Sketching   We now extend our first- and second-derivative techniques of curve sketching to include functions that have asymptotes.

In general, the following steps are helpful when we sketch the graph of a function.

Graphing Guidelines

1. Determine the domain of the function. The domain may be restricted by the nature of the problem or by the equation.
2. Look for vertical asymptotes, especially if the function is a rational function.
3. Look for horizontal asymptotes, especially if the function is a rational function.
4. Find the relative maxima and minima by using the first-derivative test or the second-derivative test.
5. Use the second derivative to find the points of inflection if this derivative is easily found.
6. Use other information (intercepts, for example) and plot additional points to complete the sketch of the graph.

EXAMPLE 2

Graphing with Asymptotes

Sketch the graph of the function $f(x) = \dfrac{x^2}{(x + 1)^2}$.

Solution

1. The domain is the set of all real numbers except $x = -1$.
2. Because $x = -1$ makes the denominator 0 and does not make the numerator 0, there is a vertical asymptote at $x = -1$.
3. Because $\dfrac{x^2}{(x + 1)^2} = \dfrac{x^2}{x^2 + 2x + 1}$

   the function is rational with the degree of the numerator equal to that of the denominator and with the ratio of the leading coefficients equal to 1. Hence, the graph of the function has a horizontal asymptote at $y = 1$.
4. To find any maxima and minima, we first find $f'(x)$.

$$f'(x) = \frac{(x + 1)^2\,(2x) - x^2\,[2(x + 1)]}{(x + 1)^4} = \frac{2x(x + 1)[(x + 1) - x]}{(x + 1)^4} = \frac{2x}{(x + 1)^3}$$

Thus $f'(x) = 0$ when $x = 0$ (and $y = 0$), and $f'(x)$ is undefined at $x = -1$ (where the vertical asymptote occurs). Using $x = 0$ and $x = -1$ gives the following sign diagram for $f'(x)$. The sign diagram shows that the critical point $(0, 0)$ is a relative minimum and shows how the graph approaches the vertical asymptote at $x = -1$.

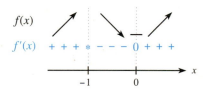

*$x = -1$ is a vertical asymptote.

5. The second derivative is

$$f''(x) = \frac{(x + 1)^3\,(2) - 2x[3(x + 1)^2]}{(x + 1)^6}$$

Factoring $(x + 1)^2$ from the numerator and simplifying give

$$f''(x) = \frac{2 - 4x}{(x + 1)^4}$$

We can see that $f''(0) = 2 > 0$, so the second-derivative test also shows that $(0, 0)$ is a relative minimum. We see that $f''(x) = 0$ when $x = \frac{1}{2}$. Checking $f''(x)$ between $x = -1$ (where it is undefined) and $x = \frac{1}{2}$ shows that the graph is concave up on this

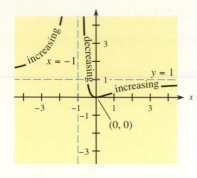

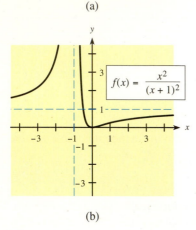

Figure 10.44

interval. Note also that $f''(x) < 0$ for $x > \frac{1}{2}$, so the point $(\frac{1}{2}, \frac{1}{9})$ is a point of inflection. Also see the sign diagram for $f''(x)$.

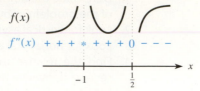

*$x = -1$ is a vertical asymptote.

6. To see how the graph approaches the horizontal asymptote, we check $f(x)$ for large values of $|x|$.

$$f(-100) = \frac{(-100)^2}{(-99)^2} = \frac{10,000}{9,801} > 1, \quad f(100) = \frac{100^2}{101^2} = \frac{10,000}{10,201} < 1$$

Thus the graph has the characteristics shown in Figure 10.44(a). The graph is shown in Figure 10.44(b). ■

When we wish to learn about a function $f(x)$ or sketch its graph, it is important to understand what information we obtain from $f(x)$, from $f'(x)$, and from $f''(x)$. The following summary may be helpful.

■ | DERIVATIVES AND GRAPHS |

Source	Information Provided
$f(x)$	y-coordinates; horizontal asymptotes, vertical asymptotes; domain restrictions
$f'(x)$	Increasing $[f'(x) > 0]$; decreasing $[f'(x) < 0]$; critical points $[f'(x) = 0 \text{ or } f'(x) \text{ undefined}]$; sign-diagram tests for maxima and minima
$f''(x)$	Concave up $[f''(x) > 0]$; concave down $[f''(x) < 0]$; possible points of inflection $[f''(x) = 0 \text{ or } f''(x) \text{ undefined}]$; sign-diagram tests for points of inflection; second-derivative test for maxima and minima

CHECKPOINT

1. Let $f(x) = \dfrac{2x + 10}{x - 1}$ and decide whether the following are true or false.
   (a) $f(x)$ has a vertical asymptote at $x = 1$.
   (b) $f(x)$ has $y = 2$ as its horizontal asymptote.

2. Let $f(x) = \dfrac{x^3 - 16}{x} + 1$; then $f'(x) = \dfrac{2x^3 + 16}{x^2}$ and $f''(x) = \dfrac{2x^3 - 32}{x^3}$.
   Use these to determine whether the following are true or false.
   (a) There are no asymptotes.
   (b) $f'(x) = 0$ when $x = -2$.
   (c) A partial sign diagram for $f'(x)$ is

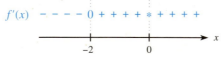

*means $f'(0)$ is undefined.

   (d) There is a relative minimum at $x = -2$.
   (e) A partial sign diagram for $f''(x)$ is

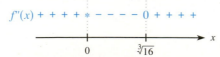

*means $f''(0)$ is undefined.

   (f) There are points of inflection at $x = 0$ and $x = \sqrt[3]{16}$.

EXAMPLE 3   Production Costs | APPLICATION PREVIEW |

Suppose that the total cost of producing a shipment of a certain product is

$$C(x) = 5000x + \frac{125,000}{x}, \quad x > 0$$

where x is the number of machines used in the production process.

(a) Determine any asymptotes for $C(x)$.
(b) How many machines should be used to minimize the total cost?
(c) Graph this total cost function.

Solution

(a) Writing this function with all terms over a common denominator gives

$$C(x) = 5000x + \frac{125,000}{x} = \frac{5000x^2 + 125,000}{x}$$

The domain of $C(x)$ does not include 0, and $C \to +\infty$ as $x \to 0^+$, so there is a vertical asymptote at $x = 0$. Thus the cost increases without bound as the number of machines used in the process approaches zero. Because the numerator has a higher degree than the denominator, there is no horizontal asymptote.

(b) Finding the derivative of $C(x)$ gives

$$C'(x) = 5000 - \frac{125,000}{x^2} = \frac{5000x^2 - 125,000}{x^2}$$

Setting $C'(x) = 0$ and solving for x gives the critical values of x.

$$0 = \frac{5000(x + 5)(x - 5)}{x^2}$$

$$x = 5 \quad \text{or} \quad x = -5$$

Because $C''(x) = 250,000x^{-3} = \dfrac{250,000}{x^3}$ is positive for all $x > 0$, using 5 machines minimizes the cost at $C(5) = 50,000$.

(c) The graph is shown in Figure 10.45.   ■

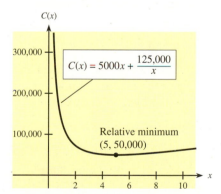

Figure 10.45

Calculator Note   The procedures previously outlined in this section are necessary to generate a complete and accurate graph. With a graphing calculator, the graph of a function is easily generated as long as the viewing window dimensions are appropriate. We frequently need information provided by derivatives to obtain a window that shows all features of a graph. See Appendix C, Section 10.5, for details.   ■

15

-10 ⊢⊣⊣⊣⊣⊣ 10

-15

Figure 10.46

EXAMPLE 4 Horizontal and Vertical Asymptotes

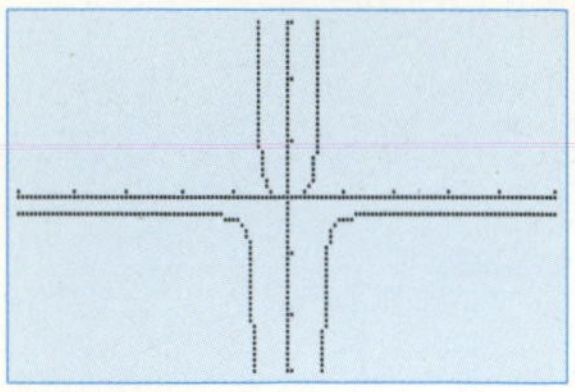

Figure 10.46 shows the graph of $f(x) = \dfrac{71x^2}{28(3 - 2x^2)}$.

(a) Determine whether the function has horizontal or vertical asymptotes, and estimate where they occur.
(b) Check your conclusions to part (a) analytically.

Solution

(a) The graph appears to have a horizontal asymptote somewhere between $y = -1$ and $y = -2$, perhaps near $y = -1.5$. Also, there are two vertical asymptotes located approximately at $x = 1.25$ and $x = -1.25$.

(b) The function

$$f(x) = \frac{71x^2}{28(3 - 2x^2)} = \frac{71x^2}{84 - 56x^2}$$

is a rational function with the degree of the numerator equal to that of the denominator and with the ratio of the leading coefficients equal to $-71/56$. Thus the graph of the function has a horizontal asymptote at $y = -71/56 \approx -1.27$.

Vertical asymptotes occur at x-values where $28(3 - 2x^2) = 0$. Solving gives

$$3 - 2x^2 = 0 \quad \text{or} \quad 3 = 2x^2 \quad \text{so} \quad \frac{3}{2} = x^2$$

$$\text{Thus} \quad \pm\sqrt{\frac{3}{2}} = x \quad \text{or} \quad x \approx \pm 1.225 \qquad ■$$

An accurate graph shows all features, but not necessarily the details of any feature. Even with a graphing calculator, sometimes analytic methods are needed to determine an appropriate viewing window.

EXAMPLE 5 Graphing with Technology

The standard viewing window of the graph of $f(x) = \dfrac{x + 10}{x^2 + 300}$ appears blank (check and see). Find any asymptotes, maxima, and minima, and determine an appropriate viewing window. Sketch the graph.

Solution

Because $x^2 + 300 = 0$ has no real solution, there are no vertical asymptotes. The function is rational with the degree of the numerator less than that of the denominator, so the horizontal asymptote is $y = 0$, which is the x-axis.

We then find an appropriate viewing window by locating the critical points.

$$f'(x) = \frac{(x^2 + 300)(1) - (x + 10)(2x)}{(x^2 + 300)^2}$$

$$= \frac{x^2 + 300 - 2x^2 - 20x}{(x^2 + 300)^2} = \frac{300 - 20x - x^2}{(x^2 + 300)^2}$$

$f'(x) = 0$ when the numerator is zero. Thus

$$300 - 20x - x^2 = 0$$
$$0 = x^2 + 20x - 300$$
$$0 = (x + 30)(x - 10)$$
$$x + 30 = 0 \qquad x - 10 = 0$$

$$x = -30 \qquad x = 10$$

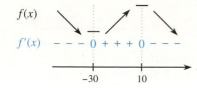

$f(x)$

$f'(x)$ $-\,-\,-\;0\;+\,+\,+\;0\;-\,-\,-$

$-30 \qquad 10$

x

The critical points are $x = -30$, $y = -\frac{1}{60} \approx -0.01666667$ and $x = 10$, $y = \frac{1}{20} = 0.05$. A sign diagram for $f'(x)$ is shown at the left.

Without using the information above, a graphing calculator may not give a useful graph. An x-range that includes -30 and 10 is needed. Because $y = 0$ is a horizontal asymptote, these relative extrema are absolute, and the y-range must be quite small for the shape of the graph to be seen clearly. Figure 10.47 shows the graph. ∎

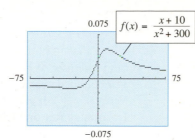

0.075

$f(x) = \dfrac{x + 10}{x^2 + 300}$

-75 75

-0.075

Figure 10.47

CHECKPOINT SOLUTIONS

1. (a) True, $x = 1$ makes the denominator of $f(x)$ equal to zero, whereas the numerator is nonzero.
   (b) True, the function is rational with the degree of the numerator equal to that of the denominator and with the ratio of the leading coefficients equal to 2.
2. (a) False. There are no horizontal asymptotes, but $x = 0$ is a vertical asymptote.
   (b) True
   (c) True
   (d) True. The relative minimum point is $(-2, f(-2)) = (-2, 13)$.
   (e) True
   (f) False. There is a point of inflection only at $(\sqrt[3]{16}, 1)$. At $x = 0$ the vertical asymptote occurs, so there is no point on the graph and hence no point of inflection.

| EXERCISES | 10.5

Access end-of-section exercises online at **www.webassign.net** ENHANCED **Web**Assign

KEY TERMS AND FORMULAS

Section	Key Terms	Formulas
10.1	Relative maxima and minima	
	Increasing	$f'(x) > 0$
	Decreasing	$f'(x) < 0$
	Critical points	$f'(x) = 0$ or $f'(x)$ undefined
	Sign diagram for $f'(x)$	
	First-derivative test	
	Horizontal point of inflection	
10.2	Concave up	$f''(x) > 0$
	Concave down	$f''(x) < 0$
	Point of inflection	May occur where $f''(x) = 0$ or $f''(x)$ undefined
	Sign diagram for $f''(x)$	
	Second-derivative test	
10.3	Absolute extrema	
	Average cost	$\overline{C}(x) = C(x)/x$

Section	Key Terms	Formulas
	Profit maximization	
	Competitive market	$R(x) = p \cdot x$ where p = equilibrium price
	Monopolistic market	$R(x) = p \cdot x$ where $p = f(x)$ is the demand function
10.4	Inventory cost models	
10.5	Asymptotes	
	Horizontal: $y = b$	$f(x) \to b$ as $x \to \infty$ or $x \to -\infty$
	Vertical: $x = c$	y unbounded near $x = c$
	For rational function	Vertical: $x = c$ if $g(c) = 0$ and $f(c) \neq 0$
	$y = f(x)/g(x)$	Horizontal: Key on highest-power terms of $f(x)$ and $g(x)$

REVIEW EXERCISES

In Problems 1–4, find all critical points and determine whether they are relative maxima, relative minima, or horizontal points of inflection.

1. $y = -x^2$
2. $p = q^2 - 4q - 5$
3. $f(x) = 1 - 3x + 3x^2 - x^3$
4. $f(x) = \dfrac{3x}{x^2 + 1}$

In Problems 5–10:

(a) Find all critical values, including those at which $f'(x)$ is undefined.

(b) Find the relative maxima and minima, if any exist.

(c) Find the horizontal points of inflection, if any exist.

(d) Sketch the graph.

5. $y = x^3 + x^2 - x - 1$
6. $f(x) = 4x^3 - x^4$
7. $f(x) = x^3 - \dfrac{15}{2}x^2 - 18x + \dfrac{3}{2}$
8. $y = 5x^7 - 7x^5 - 1$
9. $y = x^{2/3} - 1$
10. $y = x^{2/3}(x - 4)^2$

11. Is the graph of $y = x^4 - 3x^3 + 2x - 1$ concave up or concave down at $x = 2$?

12. Find intervals on which the graph of $y = x^4 - 2x^3 - 12x^2 + 6$ is concave up and intervals on which it is concave down, and find points of inflection.

13. Find the relative maxima, relative minima, and points of inflection of the graph of $y = x^3 - 3x^2 - 9x + 10$.

In Problems 14 and 15, find any relative maxima, relative minima, and points of inflection, and sketch each graph.

14. $y = x^3 - 12x$
15. $y = 2 + 5x^3 - 3x^5$
16. Given $R = 280x - x^2$, find the absolute maximum and minimum for R when (a) $0 \leq x \leq 200$ and (b) $0 \leq x \leq 100$.

17. Given $y = 6400x - 18x^2 - \dfrac{x^3}{3}$, find the absolute maximum and minimum for y when (a) $0 \leq x \leq 50$ and (b) $0 \leq x \leq 100$.

In Problems 18 and 19, use the graphs to find the following items.

(a) vertical asymptotes

(b) horizontal asymptotes

(c) $\lim\limits_{x \to +\infty} f(x)$

(d) $\lim\limits_{x \to -\infty} f(x)$

18.

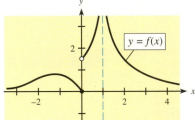

19.

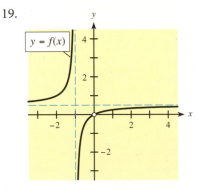

In Problems 20 and 21, find any horizontal asymptotes and any vertical asymptotes.

20. $y = \dfrac{3x + 2}{2x - 4}$

21. $y = \dfrac{x^2}{1 - x^2}$

In Problems 22–24:
(a) Find any horizontal and vertical asymptotes.
(b) Find any relative maxima and minima.
(c) Sketch each graph.

22. $y = \dfrac{3x}{x+2}$ 23. $y = \dfrac{8(x-2)}{x^2}$

24. $y = \dfrac{x^2}{x-1}$

 In Problems 25 and 26, a function and its graph are given.
(a) Use the graph to determine (estimate) x-values where $f'(x) > 0$, where $f'(x) < 0$, and where $f'(x) = 0$.
(b) Use the graph to determine x-values where $f''(x) > 0$, where $f''(x) < 0$, and where $f''(x) = 0$.
(c) Check your conclusions to (a) by finding $f'(x)$ and graphing it with a graphing calculator.
(d) Check your conclusions to (b) by finding $f''(x)$ and graphing it with a graphing calculator.

25. $f(x) = x^3 - 4x^2 + 4x$

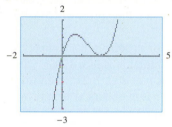

26. $f(x) = 0.0025x^4 + 0.02x^3 - 0.48x^2 + 0.08x + 4$

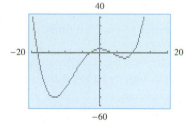

 In Problems 27 and 28, $f'(x)$ and its graph are given.
(a) Use the graph of $f'(x)$ to determine (estimate) where the graph of $f(x)$ is increasing, where it is decreasing, and where it has relative extrema.
(b) Use the graph of $f'(x)$ to determine where $f''(x) > 0$, where $f''(x) < 0$, and where $f''(x) = 0$.
(c) Verify that the given $f(x)$ has $f'(x)$ as its derivative, and graph $f(x)$ to check your conclusions in part (a).
(d) Calculate $f''(x)$ and graph it to check your conclusions in part (b).

27. $f'(x) = x^2 + 4x - 5$ $\left(\text{for } f(x) = \dfrac{x^3}{3} + 2x^2 - 5x\right)$

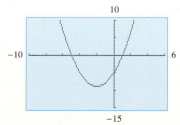

28. $f'(x) = 6x^2 - x^3$ $\left(\text{for } f(x) = 2x^3 - \dfrac{x^4}{4}\right)$

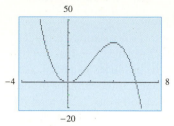

 In Problems 29 and 30, $f''(x)$ and its graph are given.
(a) Use the graph to determine (estimate) where the graph of $f(x)$ is concave up, where it is concave down, and where it has points of inflection.
(b) Verify that the given $f(x)$ has $f''(x)$ as its second derivative, and graph $f(x)$ to check your conclusions in part (a).

29. $f''(x) = 4 - x$ $\left(\text{for } f(x) = 2x^2 - \dfrac{x^3}{6}\right)$

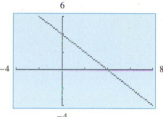

30. $f''(x) = 6 - x - x^2$ $\left(\text{for } f(x) = 3x^2 - \dfrac{x^3}{6} - \dfrac{x^4}{12}\right)$

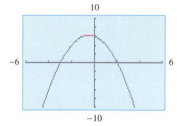

APPLICATIONS

In Problems 31–36, cost, revenue, and profit are in dollars and x is the number of units.

31. *Cost* Suppose the total cost function for a product is

$$C(x) = 3x^2 + 15x + 75$$

How many units will minimize the average cost? Find the minimum average cost.

32. *Revenue* Suppose the total revenue function for a product is given by

$$R(x) = 32x - 0.01x^2$$

(a) How many units will maximize the total revenue? Find the maximum revenue.
(b) If production is limited to 1500 units, how many units will maximize the total revenue? Find the maximum revenue.

33. **Profit** Suppose the profit function for a product is

$$P(x) = 1080x + 9.6x^2 - 0.1x^3 - 50,000$$

Find the maximum profit.

34. **Profit** How many units (x) will maximize profit if $R(x) = 46x - 0.01x^2$ and $C(x) = 0.05x^2 + 10x + 1100$?

35. **Profit** A product can be produced at a total cost of $C(x) = 800 + 4x$, where x is the number produced and is limited to at most 150 units. If the total revenue is given by $R(x) = 80x - \frac{1}{4}x^2$, determine the level of production that will maximize the profit.

36. **Average cost** The total cost function for a product is $C = 2x^2 + 54x + 98$. How many units must be produced to minimize average cost?

37. **Marginal profit** The figure shows the graph of a marginal profit function for a company. At what level of sales will profit be maximized? Explain.

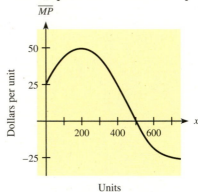

38. **Productivity—diminishing returns** Suppose the productivity P of an individual worker (in number of items produced per hour) is a function of the number of hours of training t according to

$$P(t) = 5 + \frac{95t^2}{t^2 + 2700}$$

Find the number of hours of training at which the rate of change of productivity is maximized. (That is, find the point of diminishing returns.)

39. **Output** The figure shows a typical graph of output y (in thousands of dollars) as a function of capital investment I (also in thousands of dollars).

    (a) Is the point of diminishing returns closest to the point at which $I = 20$, $I = 60$, or $I = 120$? Explain.

    (b) The average output per dollar of capital investment is defined as the total output divided by the amount of capital investment; that is,

    $$\text{Average output} = \frac{f(I)}{I}$$

    Calculate the slope of a line from $(0, 0)$ to an arbitrary point $(I, f(I))$ on the output graph. How is this slope related to the average output?

(c) Is the maximum average output attained when the capital investment is closest to $I = 40$, to $I = 70$, or to $I = 140$? Explain.

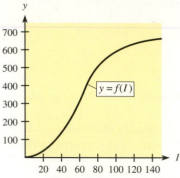

40. **Revenue** MMR II Extreme Bike Shop sells 54 basic-style mountain bikes per month at a price of $385 each. Market research indicates that MMR II could sell 10 more of these bikes if the price were $25 lower. At what selling price will MMR II maximize the revenue from these bikes?

41. **Profit** If in Problem 40 the mountain bikes cost the shop $200 each, at what selling price will MMR II's profit be a maximum?

42. **Profit** Suppose that for a product in a competitive market, the demand function is $p = 1200 - 2x$ and the supply function is $p = 200 + 2x$, where x is the number of units and p is in dollars. A firm's average cost function for this product is

$$\overline{C}(x) = \frac{12,000}{x} + 50 + x$$

Find the maximum profit. (*Hint:* First find the equilibrium price.)

43. **Profit** The monthly demand function for x units of a product sold at $\$p$ per unit by a monopoly is $p = 800 - x$, and its average cost is $\overline{C} = 200 + x$.
    (a) Determine the quantity that will maximize profit.
    (b) Find the selling price at the optimal quantity.

44. **Profit** Suppose that in a monopolistic market, the demand function for a commodity is

$$p = 7000 - 10x - \frac{x^2}{3}$$

where x is the number of units and p is in dollars. If a company's average cost function for this commodity is

$$\overline{C}(x) = \frac{40,000}{x} + 600 + 8x$$

find the maximum profit.

45. **Reaction to a drug** The reaction R to an injection of a drug is related to the dose x (in milligrams) according to

$$R(x) = x^2\left(500 - \frac{x}{3}\right)$$

Find the dose that yields the maximum reaction.

46. *Productivity* The number of parts produced per hour by a worker is given by

$$N = 4 + 3t^2 - t^3$$

where t is the number of hours on the job without a break. If the worker starts at 8 A.M., when will she be at maximum productivity during the morning?

47. *Population* Population estimates show that the equation $P = 300 + 10t - t^2$ represents the size of the graduating class of a high school, where t represents the number of years after 2010, $0 \leq t \leq 10$. What will be the largest graduating class in the next 10 years?

48. *Night brightness* Suppose that an observatory is to be built between cities A and B, which are 30 miles apart. For the best viewing, the observatory should be located where the night brightness from these cities is minimum. If the night brightness of city A is 8 times that of city B, then the night brightness b between the two cities and x miles from A is given by

$$b = \frac{8k}{x^2} + \frac{k}{(30 - x)^2}$$

where k is a constant. Find the best location for the observatory; that is, find x that minimizes b.

49. *Product design* A playpen manufacturer wants to make a rectangular enclosure with maximum play area. To remain competitive, he wants the perimeter of the base to be only 16 feet. What dimensions should the playpen have?

50. *Printing design* A printed page is to contain 56 square inches and have a $\frac{3}{4}$-inch margin at the bottom and 1-inch margins at the top and on both sides. Find the dimensions that minimize the size of the page (and hence the costs for paper).

51. *Drug sensitivity* The reaction R to an injection of a drug is related to the dose x, in milligrams, according to

$$R(x) = x^2\left(500 - \frac{x}{3}\right)$$

The sensitivity to the drug is defined by dR/dx. Find the dose that maximizes sensitivity.

52. *Per capita health care costs* For the years from 2000 and projected to 2018, the U.S. per capita out-of-pocket cost for health care C (in dollars) can be modeled by the function

$$C(t) = 0.118t^3 - 2.51t^2 + 40.2t + 677$$

where t is the number of years past 2000 (*Source:* U.S. Centers for Medicare and Medicaid Services).
   (a) When does the rate of change of health care costs per capita reach its minimum?
   (b) On a graph of $C(t)$, what feature occurs at the t-value found in part (a)?

53. *Inventory cost model* A company needs to produce 288,000 items per year. Production costs are $1500 to prepare for a production run and $30 for each item produced. Inventory costs are $1.50 per year for each item stored. Find the number of items that should be produced in each run so that the total costs of production and storage are minimum.

54. *Average cost* Suppose the total cost of producing x units of a product is given by

$$C(x) = 4500 + 120x + 0.05x^2 \quad \text{dollars}$$

   (a) Find any asymptotes of the average cost function $\overline{C}(x) = C(x)/x$.
   (b) Graph the average cost function.

55. *Market share* Suppose a company's percent share of the market (actual and projected) for a new product t quarters after its introduction is given by

$$M(t) = \frac{3.8t^2 + 3}{0.1t^2 + 1}$$

   (a) Find the company's market share when the product is introduced.
   (b) Find any horizontal asymptote of the graph of $M(t)$, and write a sentence that explains the meaning of this asymptote.

10	CHAPTER TEST

Find the local maxima, local minima, points of inflection, and asymptotes, if they exist, for each of the functions in Problems 1–3. Graph each function.

1. $f(x) = x^3 + 6x^2 + 9x + 3$
2. $y = 4x^3 - x^4 - 10$
3. $y = \dfrac{x^2 - 3x + 6}{x - 2}$

In Problems 4–6, use the function $y = 3x^5 - 5x^3 + 2$.

4. Over what intervals is the graph of this function concave up?
5. Find the points of inflection of this function.

6. Find the relative maxima and minima of this function.
7. Find the absolute maximum and minimum for $f(x) = 2x^3 - 15x^2 + 3$ on the interval $[-2, 8]$.
8. Find all horizontal and vertical asymptotes of the function

$$f(x) = \frac{200x - 500}{x + 300}.$$

9. Use the graph of $y = f(x)$ and the indicated points to complete the chart. Enter $+$, $-$, or 0, according to whether f, f', and f'' are positive, negative, or zero at each point.

Point	f	f'	f"
A			
B			
C			

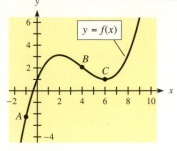

10. Use the figure to complete the following.
    (a) $\lim\limits_{x \to -\infty} f(x) = ?$
    (b) What is the vertical asymptote?
    (c) Find the horizontal asymptote.

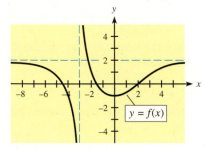

11. If $f(6) = 10$, $f'(6) = 0$, and $f''(6) = -3$, what can we conclude about the point on the graph of $y = f(x)$ where $x = 6$? Explain.

12. Using U.S. Census Bureau data since 1870, the average number of acres per U.S. farm can be modeled with the function

    $$A(t) = -0.000377t^3 + 0.1021t^2 - 4.731t + 174.3$$

    where t is the number of years past 1870.
    (a) Find the years when the model has a relative maximum and a relative minimum.
    (b) What changes in our society and in the economy during the last half of the 20th century have caused $A(t)$ to increase then?

13. The revenue function for a product is $R(x) = 164x$ dollars and the cost function for the product is

    $$C(x) = 0.01x^2 + 20x + 300 \quad \text{dollars}$$

    where x is the number of units produced and sold.
    (a) How many units of the product should be sold to obtain maximum profit?
    (b) What is the maximum possible profit?

14. The cost of producing x units of a product is given by

    $$C(x) = 100 + 20x + 0.01x^2 \quad \text{dollars}$$

    How many units should be produced to minimize average cost?

15. A firm sells 100 TV sets per month at $300 each, but market research indicates that it can sell 1 more set per month for each $2 reduction of the price. At what price will the revenue be maximized?

16. An open-top box is made by cutting squares from the corners of a piece of tin and folding up the sides. If the piece of tin was originally 20 centimeters on a side, how long should the sides of the removed squares be to maximize the resulting volume?

17. A company estimates that it will need 784,000 items during the coming year. It costs $420 to manufacture each item, $2500 to prepare for each production run, and $5 per year for each item stored. How many units should be in each production run so that the total costs of production and storage are minimized?

18. **Modeling** The table gives the number of women age 16 years and older (in millions) in the U.S. civilian work force for selected years from 1950 and projected to 2050.
    (a) Use x as the number of years past 1950 to create a cubic model using these data.
    (b) During what year does the model indicate that the rate of change of the number of women in the work force reached its maximum?
    (c) What feature of the graph of the function found in part (a) is the result found in part (b)?

Year	Women in the Work Force (in millions)
1950	18.4
1960	23.2
1970	31.5
1980	45.5
1990	56.8
2000	65.6
2010	75.5
2015	78.6
2020	79.2
2030	81.6
2040	86.5
2050	91.5

Source: U.S. Bureau of Labor Statistics

I. Production Management

Metal Containers, Inc. is reviewing the way it submits bids on U.S. Army contracts. The army often requests open-top boxes, with square bases and of specified volumes. The army also specifies the materials for the boxes, and the base is usually made of a different material than the sides. The box is assembled by riveting a bracket at each of the eight corners. For Metal Containers, the total cost of producing a box is the sum of the cost of the materials for the box and the labor costs associated with affixing each bracket.

Instead of estimating each job separately, the company wants to develop an overall approach that will allow it to cost out proposals more easily. To accomplish this, company managers need you to devise a formula for the total cost of producing each box and determine the dimensions that allow a box of specified volume to be produced at minimum cost. Use the following notation to help you solve this problem.

Cost of the material for the base $= A$ per square unit

Cost of the material for the sides $= B$ per square unit

Cost of each bracket $= C$

Cost to affix each bracket $= D$

Length of the sides of the base $= x$

Height of the box $= h$

Volume specified by the army $= V$

1. Write an expression for the company's total cost in terms of these quantities.
2. At the time an order is received for boxes of a specified volume, the costs of the materials and labor will be fixed and only the dimensions will vary. Find a formula for each dimension of the box so that the total cost is a minimum.
3. The army requests bids on boxes of 48 cubic feet with base material costing the container company $12 per square foot and side material costing $8 per square foot. Each bracket costs $5, and the associated labor cost is $1 per bracket. Use your formulas to find the dimensions of the box that meet the army's requirements at a minimum cost. What is this cost?

Metal Containers asks you to determine how best to order the brackets it uses on its boxes. You are able to obtain the following information: The company uses approximately 100,000 brackets a year, and the purchase price of each is $5. It buys the same number of brackets (say, n) each time it places an order with the supplier, and it costs $60 to process each order. Metal Containers also has additional costs associated with storing, insuring, and financing its inventory of brackets. These carrying costs amount to 15% of the average value of inventory annually. The brackets are used steadily and deliveries are made just as inventory reaches zero, so that inventory fluctuates between zero and n brackets.

4. If the total annual cost associated with the bracket supply is the sum of the annual purchasing cost and the annual carrying costs, what order size n would minimize the total cost?
5. In the general case of the bracket-ordering problem, the order size n that minimizes the total cost of the bracket supply is called the economic order quantity, or EOQ. Use the following notations to determine a general formula for the EOQ.

Fixed cost per order $= F$

Unit cost $= C$

Quantity purchased per year $= P$

Carrying cost (as a decimal rate) $= r$

II. Room Pricing in the Off-Season (Modeling)

The data in the table, from a survey of resort hotels with comparable rates on Hilton Head Island, show that room occupancy during the off-season (November through February) is related to the price charged for a basic room.

Price per Day	Occupancy Rate, %
$104	53
134	47
143	46
149	45
164	40
194	32

The goal is to use these data to help answer the following questions.

A. What price per day will maximize the daily off-season revenue for a typical hotel in this group if it has 200 rooms available?
B. Suppose that for this typical hotel the daily cost is $5510 plus $30 per occupied room. What price will maximize the profit for this hotel in the off-season?

The price per day that will maximize the off-season profit for this typical hotel applies to this group of hotels. To find the room price per day that will maximize the daily revenue and the room price per day that will maximize the profit for this hotel (and thus the group of hotels) in the off-season, complete the following.

1. Multiply each occupancy rate by 200 to get the hypothetical room occupancy. Create the revenue data points that compare the price with the revenue, R, which is equal to price times the room occupancy.
2. Use technology to create an equation that models the revenue, R, as a function of the price per day, x.
3. Use maximization techniques to find the price that these hotels should charge to maximize the daily revenue.
4. Use technology to get the occupancy as a function of the price, and use the occupancy function to create a daily cost function.
5. Form the profit function.
6. Use maximization techniques to find the price that will maximize the profit.

Kenneth Sponsler/Shutterstock.com

11

Derivatives Continued

In this chapter we will develop derivative formulas for logarithmic and exponential functions, focusing primarily on base *e* exponentials and logarithms. We will apply logarithmic and exponential functions and use their derivatives to solve maximization and minimization problems in the management and life sciences.

We will also develop methods for finding the derivative of one variable with respect to another even when the first variable is not a function of the other. This method is called implicit differentiation. We will use implicit differentiation with respect to time to solve problems involving rates of change of two or more variables. These problems are called related-rates problems.

The topics and applications discussed in this chapter include the following.

Prerequisite Problem Type	For Section	Answer	Section for Review
(a) Simplify: $\dfrac{1}{(3x)^{1/2}} \cdot \dfrac{1}{2}(3x)^{-1/2} \cdot 3$ (b) Write with a positive exponent: $\sqrt{x^2 - 1}$	**11.1**	(a) $\dfrac{1}{2x}$ (b) $(x^2 - 1)^{1/2}$	0.4 Rational exponents
(a) Vertical lines have _____ slopes. (b) Horizontal lines have _____ slopes.	**11.3**	(a) Undefined (b) 0	1.3 Slopes
Write the equation of the line passing through $(-2, -2)$ with slope 5.	**11.3**	$y = 5x + 8$	1.3 Equations of lines
Solve: $x^2 + y^2 - 9 = 0$, for y.	**11.3**	$y = \pm\sqrt{9 - x^2}$	2.1 Quadratic equations
(a) Write $\log_a(x + h) - \log_a x$ as an expression involving one logarithm. (b) Does $\ln x^4 = 4 \ln x$? (c) Does $\dfrac{x}{h} \log_a\left(\dfrac{x + h}{x}\right) = \log_a\left(1 + \dfrac{h}{x}\right)^{x/h}$? (d) Expand $\ln(xy)$ to separate x and y. (e) If $y = a^x$, then $x = $ _____. (f) Simplify $\ln e^x$.	**11.1** **11.2**	(a) $\log_a\left(\dfrac{x + h}{x}\right)$ (b) Yes (c) Yes (d) $\ln x + \ln y$ (e) $\log_a y$ (f) x	5.2 Logarithms
Find the derivative of (a) $y = x^2 - 2x - 2$ (b) $T(q) = 400q - \dfrac{4}{3}q^2$ (c) $y = \sqrt{9 - x^2}$	**11.1–11.5**	(a) $y' = 2x - 2$ (b) $T'(q) = 400 - \dfrac{8}{3}q$ (c) $y' = -x(9 - x^2)^{-1/2}$	9.4–9.6 Derivatives
If $\dfrac{dy}{dx} = \dfrac{-1}{2\sqrt{x}}$, find the slope of the tangent to $y = f(x)$ at $x = 4$.	**11.3**	Slope $= -\dfrac{1}{4}$	9.3–9.7 Derivatives

OBJECTIVE

11.1

- To find derivatives of logarithmic functions

Derivatives of Logarithmic Functions

▮ | APPLICATION PREVIEW | ▮

The table shows the expected life spans at birth for people born in certain years in the United States, with projections to 2020. These data can be modeled by the function $l(x) = 11.249 + 14.244 \ln x$, where x is the number of years past 1900. The graph of this function is shown in Figure 11.1. If we wanted to use this model to find the rate of change of life span with respect to the number of years past 1900, we would need the derivative of this function and hence the derivative of the logarithmic function $\ln x$. (See Example 4.)

Year	Life Span (years)	Year	Life Span (years)
1920	54.1	1985	74.7
1930	59.7	1990	75.4
1940	62.9	1995	75.8
1950	68.2	2000	77.0
1960	69.7	2005	77.9
1970	70.8	2010	78.1
1975	72.6	2015	78.9
1980	73.7	2020	79.5

Source: National Center for Health Statistics

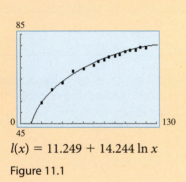

$l(x) = 11.249 + 14.244 \ln x$

Figure 11.1

Logarithmic Functions   Recall that we define the **logarithmic function** $y = \log_a x$ as follows.

Logarithmic Function

For $a > 0$ and $a \neq 1$, the **logarithmic function**

$$y = \log_a x \quad \text{(logarithmic form)}$$

has domain $x > 0$, base a, and is defined by

$$a^y = x \quad \text{(exponential form)}$$

The a is called the **base** in both $\log_a x = y$ and $a^y = x$, and y is the *logarithm* in $\log_a x = y$ and the *exponent* in $a^y = x$. Thus **a logarithm is an exponent.** Although logarithmic functions can have any base a, where $a > 0$ and $a \neq 1$, most problems in calculus and many of the applications to the management, life, and social sciences involve logarithms with base e, called **natural logarithms.** In this section we'll see why this base is so important. Recall that the natural logarithmic function $(y = \log_e (x))$ is written $y = \ln x$; see Figure 11.2 for the graph.

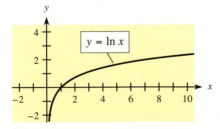

Figure 11.2

Derivative of $y = \ln x$ From Figure 11.2 we see that for $x > 0$, the graph of $y = \ln x$ is always increasing, so the slope of the tangent line to any point must be positive. This means that the derivative of $y = \ln x$ is always positive. Figure 11.3 shows the graph of $y = \ln x$ with tangent lines drawn at several points and with their slopes indicated.

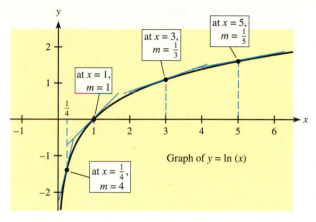

Figure 11.3

Note in Figure 11.3 that at each point where a tangent line is drawn, the slope of the tangent line is the reciprocal of the x-coordinate. In fact, this is true for every point on $y = \ln x$, so the slope of the tangent at any point is given by $1/x$. Thus we have the following:

Derivative of $y = \ln x$	If $y = \ln x$, then $\dfrac{dy}{dx} = \dfrac{1}{x}$.

This formula for the derivative of $y = \ln x$ is proved at the end of this section.

EXAMPLE   1   Derivatives Involving $\ln x$

(a) If $y = x^3 + 3 \ln x$, find dy/dx.   (b) If $y = x^2 \ln x$, find y'.

Solution

(a) $\dfrac{dy}{dx} = 3x^2 + 3\left(\dfrac{1}{x}\right) = 3x^2 + \dfrac{3}{x}$

(b) By the Product Rule,

$$y' = x^2 \cdot \dfrac{1}{x} + (\ln x)(2x) = x + 2x \ln x$$

We can use the Chain Rule to find the formula for the derivative of $y = \ln u$, where $u = f(x)$.

Derivatives of Natural Logarithmic Functions	If $y = \ln u$, where u is a differentiable function of x, then $$\dfrac{dy}{dx} = \dfrac{1}{u} \cdot \dfrac{du}{dx}$$

EXAMPLE   2   Derivatives of $y = \ln (u)$

Find the derivative for each of the following.

(a) $f(x) = \ln (x^4 - 3x + 7)$ (b) $f(x) = \frac{1}{3} \ln (2x^6 - 3x + 2)$

(c) $g(x) = \dfrac{\ln (2x + 1)}{2x + 1}$

Solution

(a) $f'(x) = \dfrac{1}{x^4 - 3x + 7}(4x^3 - 3) = \dfrac{4x^3 - 3}{x^4 - 3x + 7}$

(b) $f'(x)$ is $\frac{1}{3}$ of the derivative of $\ln(2x^6 - 3x + 2)$.

$$f'(x) = \frac{1}{3} \cdot \frac{1}{2x^6 - 3x + 2}(12x^5 - 3)$$

$$= \frac{4x^5 - 1}{2x^6 - 3x + 2}$$

(c) We begin with the Quotient Rule.

$$g'(x) = \frac{(2x + 1)\dfrac{1}{2x + 1}(2) - [\ln(2x + 1)]2}{(2x + 1)^2}$$

$$= \frac{2 - 2\ln(2x + 1)}{(2x + 1)^2}$$

CHECKPOINT

1. If $y = \ln(3x^2 + 2)$, find y'.
2. If $y = \ln x^6$, find y'.

Using Properties of Logarithms   A logarithmic function of products, quotients, or powers, such as $y = \ln[x(x^5 - 2)^{10}]$, can be rewritten with properties of logarithms so that finding the derivative is much easier. The properties of logarithms, which were introduced in Section 5.2, are stated here for logarithms with an arbitrary base a (with $a > 0$ and $a \neq 1$) and for natural logarithms.

Properties of Logarithms

Let M, N, p, and a be real numbers with $M > 0$, $N > 0$, $a > 0$, and $a \neq 1$.

Base a Logarithms

I. $\log_a(a^x) = x$ (for any real x)

II. $a^{\log_a(x)} = x$ (for $x > 0$)

III. $\log_a(MN) = \log_a M + \log_a N$

IV. $\log_a\left(\dfrac{M}{N}\right) = \log_a M - \log_a N$

V. $\log_a(M^p) = p\log_a M$

Natural Logarithms

I. $\ln(e^x) = x$

II. $e^{\ln x} = x$

III. $\ln(MN) = \ln M + \ln N$

IV. $\ln\left(\dfrac{M}{N}\right) = \ln M - \ln N$

V. $\ln(M^p) = p\ln M$

For example, to find the derivative of $f(x) = \ln\sqrt[3]{2x^6 - 3x + 2}$, it is easier to rewrite this as follows:

$$f(x) = \ln[(2x^6 - 3x + 2)^{1/3}] = \tfrac{1}{3}\ln(2x^6 - 3x + 2)$$

Then take the derivative, as in Example 2(b).

EXAMPLE 3   **Logarithm Properties and Derivatives**

Use logarithm properties to find the derivatives for

(a) $y = \ln[x(x^5 - 2)^{10}]$.   (b) $f(x) = \ln\left(\dfrac{\sqrt[3]{3x + 5}}{x^2 + 11}\right)^4$.

Solution

(a) We use logarithm Properties III and V to rewrite the function.

$$y = \ln x + \ln (x^5 - 2)^{10} \qquad \text{Property III}$$
$$y = \ln x + 10 \ln (x^5 - 2) \qquad \text{Property V}$$

We now take the derivative.

$$\frac{dy}{dx} = \frac{1}{x} + 10 \cdot \frac{1}{x^5 - 2} \cdot 5x^4$$
$$= \frac{1}{x} + \frac{50x^4}{x^5 - 2}$$

(b) Again we begin by using logarithm properties.

$$f(x) = \ln \left(\frac{\sqrt[3]{3x + 5}}{x^2 + 11} \right)^4 = 4 \ln \left(\frac{\sqrt[3]{3x + 5}}{x^2 + 11} \right) \qquad \text{Property V}$$
$$f(x) = 4[\tfrac{1}{3} \ln (3x + 5) - \ln (x^2 + 11)] \qquad \text{Properties IV and V}$$

We now take the derivative.

$$f'(x) = 4 \left(\frac{1}{3} \cdot \frac{1}{3x + 5} \cdot 3 - \frac{1}{x^2 + 11} \cdot 2x \right)$$

$$= 4 \left(\frac{1}{3x + 5} - \frac{2x}{x^2 + 11} \right) = \frac{4}{3x + 5} - \frac{8x}{x^2 + 11} \qquad ■$$

| **CHECKPOINT** | 3. If $y = \ln \sqrt[3]{x^2 + 1}$, find y'. |
| | 4. Find $f'(x)$ for $f(x) = \ln \left[\dfrac{2x^4}{(5x + 7)^5} \right]$. |

EXAMPLE 4   **Life Span**   **| APPLICATION PREVIEW |**

Assume that the average life span (in years) for people born from 1920 and projected to 2020 can be modeled by

$$l(x) = 11.249 + 14.244 \ln x$$

where x is the number of years past 1900.
(a) Find the function that models the rate of change of life span.
(b) Does $l(x)$ have a maximum value for $x > 0$?
(c) Evaluate $\lim_{x \to \infty} l'(x)$.
(d) What do the results of parts (b) and (c) tell us about the average life span?

Solution

(a) The rate of change of life span is given by the derivative.

$$l'(x) = 0 + 14.244 \left(\frac{1}{x} \right) = \frac{14.244}{x}$$

(b) For $x > 0$, we see that $l'(x) > 0$. Hence $l(x)$ is increasing for all values of $x > 0$, so $l(x)$ never achieves a maximum value. That is, there is no maximum life span.

(c) $\lim_{x \to \infty} l'(x) = \lim_{x \to \infty} \dfrac{14.244}{x} = 0$

(d) If this model is accurate, life span will continue to increase, but at an ever slower rate.   ■

EXAMPLE 5   **Cost**

Suppose the cost function for x skateboards is given by

$$C(x) = 18{,}250 + 615 \ln (4x + 10)$$

where $C(x)$ is in dollars. Find the marginal cost when 100 units are produced, and explain what it means.

Solution

Marginal cost is given by $C'(x)$.

$$\overline{MC} = C'(x) = 615\left(\frac{1}{4x + 10}\right)(4) = \frac{2460}{4x + 10}$$

$$\overline{MC}(100) = \frac{2460}{4(100) + 10} = \frac{2460}{410} = 6$$

When 100 units are produced, the marginal cost is 6. This means that the approximate cost of producing the 101st skateboard is $6. ∎

Derivative of
$y = \log_a(x)$

So far we have found derivatives of natural logarithmic functions. If we have a logarithmic function with a base other than e, then we can use the **change-of-base formula.**

Change-of-Base Formula

To express a logarithm base a as a natural logarithm, use

$$\log_a x = \frac{\ln x}{\ln a}$$

We can apply this change-of-base formula to find the derivative of a logarithm with any base, as the following example illustrates.

EXAMPLE 6

Derivative of $y = \log_a(u)$

If $y = \log_4(x^3 + 1)$, find dy/dx.

Solution

By using the change-of-base formula, we have

$$y = \log_4(x^3 + 1) = \frac{\ln(x^3 + 1)}{\ln 4} = \frac{1}{\ln 4} \cdot \ln(x^3 + 1)$$

Thus

$$\frac{dy}{dx} = \frac{1}{\ln 4} \cdot \frac{1}{x^3 + 1} \cdot 3x^2 = \frac{3x^2}{(x^3 + 1)\ln 4}$$ ∎

Note that this formula means that logarithms with bases other than e will have the *constant* $1/\ln a$ as a factor in their derivatives (as Example 6 had $1/\ln 4$ as a factor). This means that derivatives involving natural logarithms have a simpler form, and we see why base e logarithms are used more frequently in calculus.

EXAMPLE 7

Critical Values

Let $f(x) = x \ln x - x$. Use the graph of the derivative of $f(x)$ for $x > 0$ to answer the following questions.

(a) At what value a does the graph of $f'(x)$ cross the x-axis (that is, where is $f'(x) = 0$)?
(b) What value a is a critical value for $y = f(x)$?
(c) Does $f(x)$ have a relative maximum or a relative minimum at $x = a$?

Solution

$f'(x) = x \cdot \dfrac{1}{x} + \ln x - 1 = \ln x$. The graph of $f'(x) = \ln x$ is shown in Figure 11.4.

(a) The graph crosses the *x*-axis at $x = 1$, so $f'(a) = 0$ if $a = 1$.

(b) $a = 1$ is a critical value of $f(x)$.

(c) Because $f'(x)$ is negative for $x < 1$, $f(x)$ is decreasing for $x < 1$.

Because $f'(x)$ is positive for $x > 1$, $f(x)$ is increasing for $x > 1$.

Therefore, $f(x)$ has a relative minimum at $x = 1$.

The graph of $f(x) = x \ln x - x$ is shown in Figure 11.5. It has a relative minimum at the point $(1, -1)$. ■

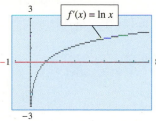

Figure 11.4

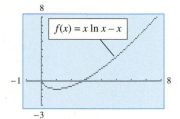

Figure 11.5

Calculator Note   Critical values can also be found by using the SOLVER feature of a graphing calculator. See Appendix C, Section 11.1, for details. ■

Proof That $\dfrac{d}{dx}(\ln x) = \dfrac{1}{x}$ For completeness, we now include the formal proof that if $y = \ln x$, then $dy/dx = 1/x$.

$$\frac{dy}{dx} = \lim_{h \to 0} \frac{f(x + h) - f(x)}{h}$$

$$= \lim_{h \to 0} \frac{\ln (x + h) - \ln x}{h}$$

$$= \lim_{h \to 0} \frac{\ln \left(\dfrac{x + h}{x} \right)}{h} \qquad \text{Property IV}$$

$$= \lim_{h \to 0} \frac{x}{x} \cdot \frac{1}{h} \ln \left(\frac{x + h}{x} \right) \qquad \text{Introduce } \frac{x}{x}$$

$$= \lim_{h \to 0} \frac{1}{x} \cdot \frac{x}{h} \ln \left(1 + \frac{h}{x} \right)$$

$$= \lim_{h \to 0} \frac{1}{x} \ln \left(1 + \frac{h}{x} \right)^{x/h} \qquad \text{Property V}$$

$$= \frac{1}{x} \lim_{h \to 0} \left[\ln \left(1 + \frac{h}{x} \right)^{x/h} \right]^{*}$$

$$= \frac{1}{x} \ln \left[\lim_{h \to 0} \left(1 + \frac{h}{x} \right)^{x/h} \right]$$

If we let $a = \dfrac{h}{x}$, then $h \to 0$ means $a \to 0$, and we have

$$\frac{dy}{dx} = \frac{1}{x} \ln \left[\lim_{a \to 0} (1 + a)^{1/a} \right]$$

In Problem 47 in the 9.1 Exercises, we saw that $\lim\limits_{a \to 0} (1 + a)^{1/a} = e$. Hence,

$$\frac{dy}{dx} = \frac{1}{x} \ln e = \frac{1}{x}$$

---

*The next step uses a new limit property for continuous composite functions. In particular, $\lim\limits_{x \to c} \ln (f(x)) = \ln \left[\lim\limits_{x \to c} f(x) \right]$ when $\lim\limits_{x \to c} f(x)$ exists and is positive.

CHECKPOINT SOLUTIONS

1. $y' = \dfrac{6x}{3x^2 + 2}$ 2. $y' = \dfrac{1}{x^6} \cdot 6x^5 = \dfrac{6}{x}$

3. By first using logarithm Property V, we get $y = \ln (x^2 + 1)^{1/3} = \frac{1}{3} \ln (x^2 + 1)$. Then

$$y' = \frac{1}{3} \cdot \frac{2x}{x^2 + 1} = \frac{2x}{3(x^2 + 1)}$$

4. By using logarithm Properties IV and V, we get

$$f(x) = \ln (2x^4) - \ln [(5x + 7)^5] = \ln (2x^4) - 5 \ln (5x + 7)$$

Thus

$$f'(x) = \frac{1}{2x^4} \cdot 8x^3 - 5 \cdot \frac{1}{5x + 7} \cdot 5 = \frac{4}{x} - \frac{25}{5x + 7}$$

| EXERCISES | 11.1

Access end-of-section exercises online at **www.webassign.net**   **ENHANCED** **WebAssign**

OBJECTIVE   11.2

- To find derivatives of exponential functions

Derivatives of Exponential Functions

| APPLICATION PREVIEW |

We saw in Chapter 6, "Mathematics of Finance," that the amount that accrues when $100 is invested at 8%, compounded continuously, is

$$S(t) = 100e^{0.08t}$$

where t is the number of years. If we want to find the rate at which the money in this account is growing at the end of 1 year, then we need to find the derivative of this exponential function. (See Example 4.)

Derivative of $y = e^x$ Just as base e logarithms are most convenient, we begin by focusing on base e exponentials. Figure 11.6(a) shows the graph of $f(x) = e^x$, and Figure 11.6(b) shows the same graph with tangent lines drawn to several points.

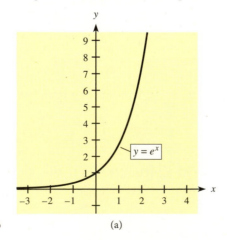

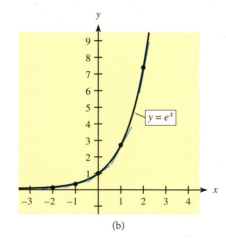

Figure 11.6   (a)   (b)

Note in Figure 11.6(b) that when $x < 0$, tangent lines have slopes near 0, much like the y-coordinates of $f(x) = e^x$. Furthermore, as x increases to $x = 0$ and for $x > 0$, the slopes of the tangents increase just as the values of the function do. This suggests that the function that gives the slope of the tangent to the graph of $f(x) = e^x$ (that is, the derivative) is similar to the function itself. In fact, the derivative of $f(x) = e^x$ is exactly the function itself, which we prove as follows:

From logarithms Property I, we know that

$$\ln e^x = x$$

Taking the derivative, with respect to x, of both sides of this equation, we have

$$\frac{d}{dx}(\ln e^x) = \frac{d}{dx}(x)$$

Using the Chain Rule for logarithms gives

$$\frac{1}{e^x} \cdot \frac{d}{dx}(e^x) = 1$$

and solving for $\dfrac{d}{dx}(e^x)$ yields

$$\frac{d}{dx}(e^x) = e^x$$

Thus we can conclude the following.

Derivative of $y = e^x$ If $y = e^x$, then $\dfrac{dy}{dx} = e^x$.

EXAMPLE 1   Derivative of an Exponential Function

If $y = 3e^x + 4x - 11$, find $\dfrac{dy}{dx}$.

Solution

$$\frac{dy}{dx} = 3e^x + 4$$ ∎

Derivative of $y = e^u$ As with logarithmic functions, the Chain Rule permits us to expand our derivative formulas.

Derivatives of Exponential Functions   If $y = e^u$, where u is a differentiable function of x, then

$$\frac{dy}{dx} = e^u \cdot \frac{du}{dx}$$

EXAMPLE 2   Derivatives of $y = e^u$

(a) If $f(x) = e^{4x^3}$, find $f'(x)$.

(b) If $s = 3te^{3t^2 + 5t}$, find ds/dt.

(c) If $u = w/e^{3w}$, find u'.

Solution

(a) $f'(x) = e^{4x^3} \cdot (12x^2) = 12x^2 \, e^{4x^3}$

(b) We begin with the Product Rule.

$$\frac{ds}{dt} = 3t \cdot e^{3t^2 + 5t}(6t + 5) + e^{3t^2 + 5t} \cdot 3$$

$$= (18t^2 + 15t)e^{3t^2 + 5t} + 3e^{3t^2 + 5t}$$

$$= 3e^{3t^2 + 5t}(6t^2 + 5t + 1)$$

(c) The function is a quotient. Using the Quotient Rule gives

$$u' = \frac{(e^{3w})(1) - (w)(e^{3w} \cdot 3)}{(e^{3w})^2} = \frac{e^{3w} - 3we^{3w}}{e^{6w}} = \frac{1 - 3w}{e^{3w}} \qquad \blacksquare$$

EXAMPLE | **3** | **Derivatives and Logarithmic Properties**

If $y = e^{\ln x^2}$, find y'.

Solution

$$y' = e^{\ln x^2} \cdot \frac{1}{x^2} \cdot 2x = \frac{2}{x} e^{\ln x^2}$$

By logarithm Property II (see the previous section), $e^{\ln u} = u$, and we can simplify the derivative to

$$y' = \frac{2}{x} \cdot x^2 = 2x$$

Note that if we had used this property *before* taking the derivative, we would have had

$$y = e^{\ln x^2} = x^2$$

Then the derivative is $y' = 2x$. $\qquad \blacksquare$

CHECKPOINT | 1. If $y = 2e^{4x}$, find y'.   2. If $y = e^{x^2 + 6x}$, find y'.   3. If $s = te^{t^2}$, find ds/dt.

EXAMPLE | **4** | **Future Value** | **APPLICATION PREVIEW** |

When \$100 is invested at 8% compounded continuously, the amount that accrues after t years, which is called the future value, is $S(t) = 100e^{0.08t}$. At what rate is the money in this account growing

(a) at the end of 1 year?   (b) at the end of 10 years?

Solution

The rate of growth of the money is given by

$$S'(t) = 100e^{0.08t}(0.08) = 8e^{0.08t}$$

(a) The rate of growth of the money at the end of 1 year is

$$S'(1) = 8e^{0.08} \approx 8.666$$

Thus the future value will change by about \$8.67 during the next year.

(b) The rate of growth of the money at the end of 10 years is

$$S'(10) = 8e^{0.08(10)} \approx 17.804$$

Thus the future value will change by about \$17.80 during the next year.   ■

EXAMPLE 5   Revenue

North Forty, Inc. is a manufacturer of wilderness camping equipment. The revenue function for its best-selling tent, the Sierra, can be modeled by the function

$$R(x) = 250xe^{(1-0.01x)}$$

where $R(x)$ is the revenue in thousands of dollars from the sale of x thousand Sierra tents. Find the marginal revenue when 75,000 tents are sold, and explain what it means.

Solution

The marginal revenue function is given by $R'(x)$, and to find this derivative we use the Product Rule.

$$R'(x) = \overline{MR} = 250x[e^{(1-0.01x)} \cdot (-0.01)] + e^{(1-0.01x)}(250)$$
$$\overline{MR} = 250e^{(1-0.01x)}(1 - 0.01x)$$

To find the marginal revenue when 75,000 tents are sold, we use $x = 75$.

$$\overline{MR}(75) = 250e^{(1-0.75)}(1 - 0.75) \approx 80.25$$

This means that the sale of one (thousand) more Sierra tents will yield approximately \$80.25 (thousand) in additional revenue.   ■

Lilyana Vynogradova/Shutterstock.com

CHECKPOINT

4. If the sales of a product are given by $S = 1000e^{-0.2x}$, where x is the number of days after the end of an advertising campaign, what is the rate of decline in sales 20 days after the end of the campaign?

Derivative of $y = a^u$ In a manner similar to that used to find the derivative of $y = e^x$, we can develop a formula for the derivative of $y = a^x$ for any base $a > 0$ and $a \neq 1$.

Derivative of $y = a^u$	If $y = a^x$, with $a > 0$, $a \neq 1$, then
	$$\frac{dy}{dx} = a^x \ln a$$
	If $y = a^u$, with $a > 0$, $a \neq 1$, where u is a differentiable function of x, then
	$$\frac{dy}{dx} = a^u \frac{du}{dx} \ln a$$

EXAMPLE 6   Derivatives of $y = a^u$

(a) If $y = 4^x$, find dy/dx.   (b) If $y = 5^{x^2+x}$, find y'.

Solution

(a) $\dfrac{dy}{dx} = 4^x \ln 4$

(b) $y' = 5^{x^2+x}(2x + 1) \ln 5$

 Calculator Note   We can make use of a graphing calculator to find maxima and minima of functions involving enponentials. See Appendix C, Section 11.2, for details.

 EXAMPLE 7   Technology and Critical Values

For the function $y = e^x - 3x^2$, complete the following.

(a) Approximate the critical values of the function to four decimal places.

(b) Determine whether relative maxima or relative minima occur at the critical values.

Solution

(a) The derivative is $y' = e^x - 6x$. Using the ZERO feature of a graphing calculator, we find that $y' = 0$ at $x \approx 0.2045$ (see Figure 11.7(a)) and at $x \approx 2.8331$.

(b) From the graph of $y' = e^x - 6x$ in Figure 11.7(a), we can observe where $y' > 0$ and where $y' < 0$. From this we can make a sign diagram to determine relative maxima and relative minima.

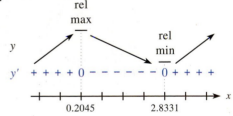

The graph of $y = e^x - 3x^2$ in Figure 11.7(b) shows that the relative maximum point is $(0.2045, 1.1015)$ and that the relative minimum point is $(2.8331, -7.0813)$.

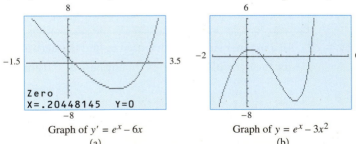

Graph of $y' = e^x - 6x$ Graph of $y = e^x - 3x^2$
(a)   (b)

Figure 11.7

 **Calculator Note**   The SOLVER and TABLE features of a graphing calculator can also be used to find maxima and minima of the function in Example 7. See Appendix C, Section 11.2, for details.

CHECKPOINT SOLUTIONS

1. $y' = 2e^{4x}(4) = 8e^{4x}$

2. $y' = (2x + 6)e^{x^2 + 6x}$

3. By the Product Rule, $\dfrac{ds}{dt} = e^{t^2}(1) + t[e^{t^2}(2t)] = e^{t^2} + 2t^2 e^{t^2}$.

4. The rate of decline is given by dS/dx.

$$\frac{dS}{dx} = 1000e^{-0.2x}(-0.2) = -200e^{-0.2x}$$

$$\left.\frac{dS}{dx}\right|_{x=20} = -200e^{(-0.2)(20)}$$

$$= -200e^{-4}$$

$$\approx -3.663 \text{ sales/day}$$

| **EXERCISES** | | **11.2** |

Access end-of-section exercises online at **www.webassign.net**

| **OBJECTIVES** | | **11.3** |

- To find derivatives by using implicit differentiation
- To find slopes of tangents by using implicit differentiation

Implicit Differentiation

█ | APPLICATION PREVIEW |

In the retail electronics industry, suppose the monthly demand for Precision, Inc. headphones is given by

$$p = \frac{10{,}000}{(x + 1)^2}$$

where p is the price in dollars per set of headphones and x is demand in hundreds of sets of headphones. If we want to find the rate of change of the quantity demanded with respect to the price, then we need to find dx/dp. (See Example 8.) Although we can solve this equation for x so that dx/dp can be found, the resulting equation does not define x as a function of p. In this case, and in other cases where we cannot solve equations for the variables we need, we can find derivatives with a technique called implicit differentiation.

Up to this point, functions involving x and y have been written in the form $y = f(x)$, defining y as an *explicit function* of x. However, not all equations involving x and y can be written in the form $y = f(x)$, and we need a new technique for taking their derivatives. For example, solving $y^2 = x$ for y gives $y = \pm\sqrt{x}$ so that y is not a function of x. We can write $y = \sqrt{x}$ and $y = -\sqrt{x}$, but then finding the derivative $\dfrac{dy}{dx}$ at a point on the graph of $y^2 = x$ would require determining which of these functions applies before taking the derivative. Alternatively, we can *imply* that y is a function of x and use a technique called **implicit differentiation.** When y is an implied function of x, we find $\dfrac{dy}{dx}$ by differentiating both sides of the equation with respect to x and then algebraically solving for $\dfrac{dy}{dx}$.

EXAMPLE   1   **Implicit Differentiation**

(a) Use implicit differentiation to find $\dfrac{dy}{dx}$ for $y^2 = x$.

(b) Find the slopes of the tangents to the graph of $y^2 = x$ at the points $(4, 2)$ and $(4, -2)$.

Solution

(a) First take the derivative of both sides of the equation with respect to x.

$$\frac{d}{dx}(y^2) = \frac{d}{dx}(x)$$

Because y is an implied, or implicit, function of x, we can think of $y^2 = x$ as meaning $[u(x)]^2 = x$ for some function u. We use the Chain Rule to take the derivative of y^2 in the same way we would for $[u(x)]^2$. Thus

$$\frac{d}{dx}(y^2) = \frac{d}{dx}(x) \quad \text{gives} \quad 2y\frac{dy}{dx} = 1$$

Solving for $\dfrac{dy}{dx}$ gives $\dfrac{dy}{dx} = \dfrac{1}{2y}$.

(b) To find the slopes of the tangents at the points $(4, 2)$ and $(4, -2)$, we use the coordinates of the points to evaluate $\dfrac{dy}{dx}$ at those points.

$$\left.\frac{dy}{dx}\right|_{(4, 2)} = \frac{1}{2(2)} = \frac{1}{4} \quad \text{and} \quad \left.\frac{dy}{dx}\right|_{(4, -2)} = \frac{1}{2(-2)} = \frac{-1}{4}$$

Figure 11.8 shows the graph of $y^2 = x$ with tangent lines drawn at $(4, 2)$ and $(4, -2)$. ∎

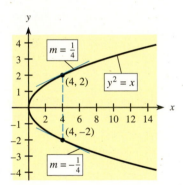

Figure 11.8

As noted previously, solving $y^2 = x$ for y gives the two functions $y = \sqrt{x} = x^{1/2}$ and $y = -\sqrt{x} = -x^{1/2}$; see Figures 11.9(a) and (b).

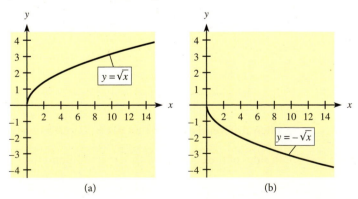

Figure 11.9   (a)   (b)

Let us now compare the results obtained in Example 1 with the results from the derivatives of the two functions $y = \sqrt{x}$ and $y = -\sqrt{x}$. The derivatives of these two functions are as follows:

$$\text{For } y = x^{1/2}, \text{ then } \frac{dy}{dx} = \frac{1}{2}x^{-1/2} = \frac{1}{2\sqrt{x}}$$

$$\text{For } y = -x^{1/2}, \text{ then } \frac{dy}{dx} = -\frac{1}{2}x^{-1/2} = \frac{-1}{2\sqrt{x}}$$

We cannot find the slope of a tangent line at $x = 4$ unless we also know the y-coordinate and hence which function and which derivative to use.

$$\text{At } (4, 2) \text{ use } y = \sqrt{x} \text{ so } \frac{dy}{dx}\bigg|_{(4,\,2)} = \frac{1}{2\sqrt{4}} = \frac{1}{4}$$

$$\text{At } (4, -2) \text{ use } y = -\sqrt{x} \text{ so } \frac{dy}{dx}\bigg|_{(4,\,-2)} = \frac{-1}{2\sqrt{4}} = \frac{-1}{4}$$

These are the same results that we obtained directly after implicit differentiation. In this example, we could easily solve for y in terms of x and work the problem two different ways. However, sometimes solving explicitly for y is difficult or impossible. For these functions, implicit differentiation is necessary and extends our ability to find derivatives.

EXAMPLE 2 Slope of a Tangent

Find the slope of the tangent to the graph of $x^2 + y^2 - 9 = 0$ at $(\sqrt{5}, 2)$.

Solution
We find the derivative dy/dx from $x^2 + y^2 - 9 = 0$ by taking the derivative term by term on both sides of the equation.

$$\frac{d}{dx}(x^2) + \frac{d}{dx}(y^2) + \frac{d}{dx}(-9) = \frac{d}{dx}(0)$$

$$2x + 2y \cdot \frac{dy}{dx} + 0 = 0$$

Solving for dy/dx gives

$$\frac{dy}{dx} = -\frac{2x}{2y} = -\frac{x}{y}$$

The slope of the tangent to the curve at $(\sqrt{5}, 2)$ is the value of the derivative at this point. Evaluating $dy/dx = -x/y$ at $(\sqrt{5}, 2)$ gives the slope of the tangent as $-\sqrt{5}/2$. ■

As we have seen in Examples 1 and 2, the Chain Rule yields $\frac{d}{dx}(y^n) = ny^{n-1}\frac{dy}{dx}$. Just as we needed this result to find derivatives of powers of y, the next example discusses how to find the derivative of xy.

EXAMPLE 3 Equation of a Tangent Line

Write the equation of the tangent to the graph of $x^3 + xy + 4 = 0$ at the point $(2, -6)$.

Solution
Note that the second term is the *product* of x and y. Because we are assuming that y is a function of x, and because x is a function of x, we must use the Product Rule to find $\frac{d}{dx}(xy)$.

$$\frac{d}{dx}(xy) = x \cdot 1\frac{dy}{dx} + y \cdot 1 = x\frac{dy}{dx} + y$$

Thus we have

$$\frac{d}{dx}(x^3) + \frac{d}{dx}(xy) + \frac{d}{dx}(4) = \frac{d}{dx}(0)$$

$$3x^2 + \left(x\frac{dy}{dx} + y\right) + 0 = 0$$

Solving for dy/dx gives $\quad \dfrac{dy}{dx} = \dfrac{-3x^2 - y}{x}$

The slope of the tangent to the curve at $x = 2$, $y = -6$ is

$$m = \frac{-3(2)^2 - (-6)}{2} = -3$$

The equation of the tangent line is

$$y - (-6) = -3[x - (2)], \quad \text{or} \quad y = -3x \qquad \blacksquare$$

Technology Note A graphing utility can be used to graph the function of Example 3 and the line that is tangent to the curve at $(2, -6)$. To graph the equation, we solve the equation for y, getting

$$y = \frac{-x^3 - 4}{x}$$

The graph of the equation and the line that is tangent to the curve at $(2, -6)$ are shown in Figure 11.10. $\qquad \blacksquare$

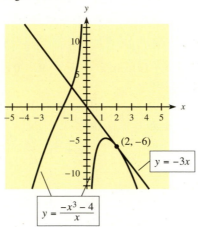

Figure 11.10

EXAMPLE 4 **Implicit Differentiation**

Find dy/dx if $x^4 + 5xy^4 = 2y^2 + x - 1$.

Solution
By viewing $5xy^4$ as the product $(5x)(y^4)$ and differentiating implicitly, we get

$$\frac{d}{dx}(x^4) + \frac{d}{dx}(5xy^4) = \frac{d}{dx}(2y^2) + \frac{d}{dx}(x) - \frac{d}{dx}(1)$$

$$4x^3 + 5x\left(4y^3\frac{dy}{dx}\right) + y^4(5) = 4y\frac{dy}{dx} + 1$$

To solve for $\dfrac{dy}{dx}$, we first rewrite the equation with terms containing $\dfrac{dy}{dx}$ on one side and the other terms on the other side.

$$20xy^3\frac{dy}{dx} - 4y\frac{dy}{dx} = 1 - 4x^3 - 5y^4$$

Now $\dfrac{dy}{dx}$ is a factor of one side. To complete the solution, we factor out $\dfrac{dy}{dx}$ and divide both sides by its coefficient.

$$(20xy^3 - 4y)\frac{dy}{dx} = 1 - 4x^3 - 5y^4$$

$$\frac{dy}{dx} = \frac{1 - 4x^3 - 5y^4}{20xy^3 - 4y}$$

∎

CHECKPOINT

1. Find the following:

   (a) $\dfrac{d}{dx}(x^3)$ (b) $\dfrac{d}{dx}(y^4)$ (c) $\dfrac{d}{dx}(x^2y^5)$

2. Find $\dfrac{dy}{dx}$ for $x^3 + y^4 = x^2y^5$.

EXAMPLE 5

Production

Suppose that a company's weekly production output is $384,000 and that this output is related to hours of labor x and dollars of capital investment y by

$$384{,}000 = 30x^{1/3}\,y^{2/3}$$

(This relationship is an example of a Cobb-Douglas production function, studied in more detail in Chapter 14.) Find and interpret the rate of change of capital investment with respect to labor hours when labor hours are 512 and capital investment is $64,000.

Solution

The desired rate of change is given by the value of dy/dx when $x = 512$ and $y = 64{,}000$. Taking the derivative implicitly gives

$$\frac{d}{dx}(384{,}000) = \frac{d}{dx}(30x^{1/3}\,y^{2/3})$$

$$0 = 30x^{1/3}\left(\frac{2}{3}y^{-1/3}\frac{dy}{dx}\right) + y^{2/3}(10x^{-2/3})$$

$$0 = \frac{20x^{1/3}}{y^{1/3}}\cdot\frac{dy}{dx} + \frac{10y^{2/3}}{x^{2/3}}$$

$$\frac{-20x^{1/3}}{y^{1/3}}\cdot\frac{dy}{dx} = \frac{10y^{2/3}}{x^{2/3}}$$

Multiplying both sides by $\dfrac{-y^{1/3}}{20x^{1/3}}$ gives

$$\frac{dy}{dx} = \left(\frac{10y^{2/3}}{x^{2/3}}\right)\left(\frac{-y^{1/3}}{20x^{1/3}}\right) = \frac{-y}{2x}$$

When $x = 512$ and $y = 64{,}000$, we obtain

$$\frac{dy}{dx} = \frac{-64{,}000}{2(512)} = -62.5$$

This means that when labor hours are 512 and capital investment is $64,000, if labor hours change by 1 hour, then capital investment could decrease by about $62.50. ∎

EXAMPLE 6 **Horizontal and Vertical Tangents**

(a) At what point(s) does $x^2 + 4y^2 - 2x + 4y - 2 = 0$ have a horizontal tangent?

(b) At what point(s) does it have a vertical tangent?

Solution

(a) First we find the derivative implicitly, with y' representing $\dfrac{dy}{dx}$.

$$2x + 8y \cdot y' - 2 + 4y' - 0 = 0$$

We isolate the y' terms, factor out y', and solve for y'.

$$8yy' + 4y' = 2 - 2x$$

$$(8y + 4)y' = 2 - 2x$$

$$y' = \frac{2 - 2x}{8y + 4} = \frac{1 - x}{4y + 2}$$

Horizontal tangents will occur where $y' = 0$; that is, where $1 - x = 0$, or $x = 1$. We can now find the corresponding y-value(s) by substituting 1 for x in the original equation and solving.

$$1 + 4y^2 - 2 + 4y - 2 = 0$$
$$4y^2 + 4y - 3 = 0$$
$$(2y - 1)(2y + 3) = 0$$
$$y = \frac{1}{2} \quad \text{or} \quad y = -\frac{3}{2}$$

Thus horizontal tangents occur at $\left(1, \frac{1}{2}\right)$ and $\left(1, -\frac{3}{2}\right)$; see Figure 11.11.

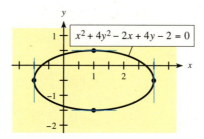

Figure 11.11

$x^2 + 4y^2 - 2x + 4y - 2 = 0$

(b) Vertical tangents will occur where the derivative is undefined—that is, where $4y + 2 = 0$, or $y = -\frac{1}{2}$. To find the corresponding x-value(s), we substitute $-\frac{1}{2}$ in the equation for y and solve for x.

$$x^2 + 4\left(-\frac{1}{2}\right)^2 - 2x + 4\left(-\frac{1}{2}\right) - 2 = 0$$

$$x^2 - 2x - 3 = 0$$

$$(x - 3)(x + 1) = 0$$

$$x = 3 \quad \text{or} \quad x = -1$$

Thus vertical tangents occur at $\left(3, -\frac{1}{2}\right)$ and $\left(-1, -\frac{1}{2}\right)$; see Figure 11.11.

EXAMPLE 7   **Implicit Derivatives with Logarithms and Exponentials**

Find dy/dx for each of the following.
(a) $\ln xy = 6$ (b) $4x^2 + e^{xy} = 6y$

Solution
(a) Using the properties of logarithms, we have

$$\ln x + \ln y = 6$$

which leads to the implicit derivative

$$\frac{1}{x} + \frac{1}{y}\frac{dy}{dx} = 0$$

Solving gives $\dfrac{1}{y}\dfrac{dy}{dx} = -\dfrac{1}{x}$ so $\dfrac{dy}{dx} = -\dfrac{y}{x}$

(b) We take the derivative of both sides of $4x^2 + e^{xy} = 6y$ and obtain

$$8x + e^{xy}\left(x\frac{dy}{dx} + y\right) = 6\frac{dy}{dx}$$

$$8x + xe^{xy}\frac{dy}{dx} + ye^{xy} = 6\frac{dy}{dx}$$

$$8x + ye^{xy} = 6\frac{dy}{dx} - xe^{xy}\frac{dy}{dx}$$

$$8x + ye^{xy} = (6 - xe^{xy})\frac{dy}{dx}$$ so $\dfrac{8x + ye^{xy}}{6 - xe^{xy}} = \dfrac{dy}{dx}$ ∎

EXAMPLE 8   **Demand** | **APPLICATION PREVIEW** |

Suppose the demand for Precision, Inc. headphones is given by

$$p = \frac{10,000}{(x + 1)^2}$$

where p is the price per set in dollars and x is in hundreds of headphone sets demanded. Find the rate of change of demand with respect to price when 19 (hundred) sets are demanded.

Solution
The rate of change of demand with respect to price is dx/dp. Using implicit differentiation, we get

$$\frac{d}{dp}(p) = \frac{d}{dp}\left[\frac{10,000}{(x + 1)^2}\right] = \frac{d}{dp}[10,000(x + 1)^{-2}]$$

$$1 = 10,000\left[-2(x + 1)^{-3}\frac{dx}{dp}\right]$$

$$1 = \frac{-20,000}{(x + 1)^3} \cdot \frac{dx}{dp}$$

$$\frac{(x + 1)^3}{-20,000} = \frac{dx}{dp}$$

When 19 (hundred) headphone sets are demanded we use $x = 19$, and the rate of change of demand with respect to price is

$$\frac{dx}{dp}\bigg|_{x=19} = \frac{(19+1)^3}{-20,000} = \frac{8000}{-20,000} = -0.4$$

This result means that when 19 (hundred) headphone sets are demanded, if the price per set is increased by \$1, then the expected change in demand is a decrease of 0.4 hundred, or 40, headphone sets. ▪

Calculator Note

To graph a function with a graphing calculator, we need to write y as an *explicit* function of x (such as $y = \sqrt{4 - x^2}$). If an equation defines y as an *implicit* function of x, we have to solve for y in terms of x before we can use the graphing calculator. Sometimes we cannot solve for y, and other times, such as for

$$x^{2/3} + y^{2/3} = 8^{2/3}$$

y cannot be written as a single function of x. If this equation is solved for y as a single function and a graphing calculator is used to graph that function, the resulting graph usually shows only the portion of the graph that lies in Quadrants I and II. The complete graph is shown in Figure 11.12. In general, for the graph of an implicitly defined function, a graphing calculator must be used carefully (and sometimes cannot be used at all). ▪

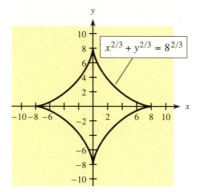

Figure 11.12

CHECKPOINT SOLUTIONS

1. (a) $\dfrac{d}{dx}(x^3) = 3x^2$ (b) $\dfrac{d}{dx}(y^4) = 4y^3\dfrac{dy}{dx}$

    (c) $\dfrac{d}{dx}(x^2y^5) = x^2\left(5y^4\dfrac{dy}{dx}\right) + y^5(2x)$ (by the Product Rule)

2. For $x^3 + y^4 = x^2y^5$, we can use the answers to Question 1 to obtain

$$3x^2 + 4y^3\frac{dy}{dx} = x^2\left(5y^4\frac{dy}{dx}\right) + y^5(2x)$$

$$3x^2 + 4y^3\frac{dy}{dx} = 5x^2y^4\frac{dy}{dx} + 2xy^5$$

$$3x^2 - 2xy^5 = (5x^2y^4 - 4y^3)\frac{dy}{dx}$$

$$\frac{3x^2 - 2xy^5}{5x^2y^4 - 4y^3} = \frac{dy}{dx}$$

| EXERCISES | 11.3

Access end-of-section exercises online at **www.webassign.net**

OBJECTIVE

11.4

- To use implicit differentiation to solve problems that involve related rates

Related Rates

| APPLICATION PREVIEW |

According to Poiseuille's law, the flow of blood F is related to the radius r of the vessel according to

$$F = kr^4$$

where k is a constant. When the radius of a blood vessel is reduced, such as by cholesterol deposits, the flow of blood is also restricted. Drugs can be administered that increase the radius of the blood vessel and, hence, the flow of blood. The rate of change of the blood flow and the rate of change of the radius of the blood vessel are time rates of change that are related to each other, so they are called related rates. We can use these related rates to find the percent rate of change in the blood flow that corresponds to the percent rate of change in the radius of the blood vessel caused by the drug. (See Example 2.)

Related Rates   We have seen that the derivative represents the instantaneous rate of change of one variable with respect to another. When the derivative is taken with respect to time, it represents the rate at which that variable is changing with respect to time (or the velocity). For example, if distance x is measured in miles and time t in hours, then dx/dt is measured in miles per hour and indicates how fast x is changing. Similarly, if V represents the volume (in cubic feet) of water in a swimming pool and t is time (in minutes), then dV/dt is measured in cubic feet per minute (ft^3/min) and might measure the rate at which the pool is being filled with water or being emptied.

Sometimes, two (or more) quantities that depend on time are also related to each other. For example, the height of a tree h (in feet) is related to the radius r (in inches) of its trunk, and this relationship can be modeled by

$$h = kr^{2/3}$$

where k is a constant.* Of course, both h and r are also related to time, so the rates of change dh/dt and dr/dt are related to each other. Thus they are called **related rates.**

The specific relationship between dh/dt and dr/dt can be found by differentiating $h = kr^{2/3}$ implicitly with respect to time t.

EXAMPLE 1   Tree Height and Trunk Radius

Suppose that for a certain type of tree, the height of the tree (in feet) is related to the radius of its trunk (in inches) by

$$h = 15r^{2/3}$$

Suppose that the rate of change of r is $\frac{3}{4}$ inch per year. Find how fast the height is changing when the radius is 8 inches.

*T. McMahon, "Size and Shape in Biology," *Science* 179 (1979): 1201.

Solution

To find how the rates dh/dt and dr/dt are related, we differentiate $h = 15r^{2/3}$ implicitly with respect to time t.

$$\frac{dh}{dt} = 10r^{-1/3}\frac{dr}{dt}$$

Using $r = 8$ inches and $dr/dt = \frac{3}{4}$ inch per year gives

$$\frac{dh}{dt} = 10(8)^{-1/3}(3/4) = \frac{15}{4} = 3\tfrac{3}{4} \text{ feet per year} \qquad \blacksquare$$

Percent Rates of Change   The work in Example 1 shows how to obtain related rates, but the different units (feet per year and inches per year) may be somewhat difficult to interpret. For this reason, many applications in the life sciences deal with **percent rates of change.** The percent rate of change of a quantity is the rate of change of the quantity divided by the quantity.

EXAMPLE 2   **Blood Flow | APPLICATION PREVIEW |**

Poiseuille's law expresses the flow of blood F as a function of the radius r of the vessel according to

$$F = kr^4$$

where k is a constant. When the radius of a blood vessel is restricted, such as by cholesterol deposits, drugs can be administered that will increase the radius of the blood vessel (and hence the blood flow). Find the percent rate of change of the flow of blood that corresponds to the percent rate of change of the radius of a blood vessel caused by the drug.

Solution

We seek the percent rate of change of flow, $(dF/dt)/F$, that results from a given percent rate of change of the radius $(dr/dt)/r$. We first find the related rates of change by differentiating

$$F = kr^4$$

implicitly with respect to time.

$$\frac{dF}{dt} = k\left(4r^3\frac{dr}{dt}\right)$$

Then the percent rate of change of flow can be found by dividing both sides of the equation by F.

$$\frac{\dfrac{dF}{dt}}{F} = \frac{4kr^3\dfrac{dr}{dt}}{F}$$

If we replace F on the right side of the equation with kr^4 and reduce, we get

$$\frac{\dfrac{dF}{dt}}{F} = \frac{4kr^3\dfrac{dr}{dt}}{kr^4} = 4\left(\frac{\dfrac{dr}{dt}}{r}\right)$$

Thus we see that the percent rate of change of the flow of blood is 4 times the corresponding percent rate of change of the radius of the blood vessel. This means that a drug that

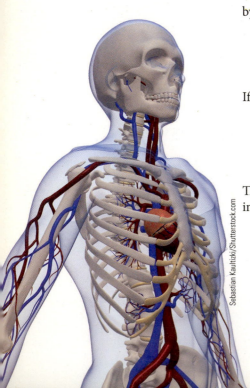

would cause a 12% increase in the radius of a blood vessel at a certain time would produce a corresponding 48% increase in blood flow through that vessel at that time. ■

Solving Related-Rates Problems

In the examples above, the equation relating the time-dependent variables has been given. For some problems, the original equation relating the variables must first be developed from the statement of the problem. These problems can be solved with the aid of the following procedure.

Solving a Related-Rates Problem

Procedure	**Example**
To solve a related-rates problem:	Sand falls at a rate of 5 ft³/min on a conical pile, with the diameter always equal to the height of the pile. At what rate is the height increasing when the pile is 10 ft high?
1. Use geometric and/or physical conditions to write an equation that relates the time-dependent variables.	1. The conical pile has its volume given by $$V = \frac{1}{3}\pi r^2 h$$
2. Substitute into the equation values or relationships that are true at *all times*.	2. The radius $r = \frac{1}{2}h$ at all times, so $$V = \frac{1}{3}\pi\left(\frac{1}{4}h^2\right)h = \frac{\pi}{12}h^3$$
3. Differentiate both sides of the equation implicitly with respect to time. This equation is valid for all times.	3. $\dfrac{dV}{dt} = \dfrac{\pi}{12}\left(3h^2 \dfrac{dh}{dt}\right) = \dfrac{\pi}{4}h^2\dfrac{dh}{dt}$
4. Substitute the values that are known at the instant specified, and solve the equation.	4. $\dfrac{dV}{dt} = 5$ at all times, so when $h = 10$, $$5 = \frac{\pi}{4}(10^2)\frac{dh}{dt}$$
5. Solve for the specified quantity at the given time.	5. $\dfrac{dh}{dt} = \dfrac{20}{100\pi} = \dfrac{1}{5\pi}\,(\text{feet/minute})$

Note that you should *not* substitute numerical values for any quantity that varies with time until after the derivative is taken. If values are substituted before the derivative is taken, that quantity will have the constant value resulting from the substitution and hence will have a derivative equal to zero.

EXAMPLE 3 Hot Air Balloon

A hot air balloon has a velocity of 50 feet per minute and is flying at a constant height of 500 feet. An observer on the ground is watching the balloon approach. How fast is the distance between the balloon and the observer changing when the balloon is 1000 feet from the observer?

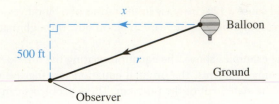

Figure 11.13

Solution

If we let r be the distance between the balloon and the observer and let x be the horizontal distance from the balloon to a point directly above the observer, then we see that these quantities are related by the equation

$$x^2 + 500^2 = r^2 \quad (\text{See Figure 11.13})$$

Because the distance x is decreasing, we know that dx/dt must be negative. Thus we are given that $dx/dt = -50$ at all times, and we need to find dr/dt when $r = 1000$. Taking the derivative with respect to t of both sides of the equation $x^2 + 500^2 = r^2$ gives

$$2x\frac{dx}{dt} + 0 = 2r\frac{dr}{dt}$$

Using $dx/dt = -50$ and $r = 1000$, we get

$$2x(-50) = 2000\frac{dr}{dt}$$

$$\frac{dr}{dt} = \frac{-100x}{2000} = \frac{-x}{20}$$

Using $r = 1000$ in $x^2 + 500^2 = r^2$ gives $x^2 = 750{,}000$. Thus $x = 500\sqrt{3}$, and

$$\frac{dr}{dt} = \frac{-500\sqrt{3}}{20} = -25\sqrt{3} \approx -43.3$$

The distance is decreasing at 43.3 feet per minute. ■

CHECKPOINT

1. If V represents volume, write a mathematical symbol that represents "the rate of change of volume with respect to time."
2. (a) Differentiate $x^2 + 64 = y^2$ implicitly with respect to time.
   (b) Suppose that we know that y is increasing at 2 units per minute. Use part (a) to find the rate of change of x at the instant when $x = 6$ and $y = 10$.
3. True or false: In solving a related-rates problem, we substitute all numerical values into the equation before we take derivatives.

EXAMPLE 4 Spread of an Oil Slick

Suppose that oil is spreading in a circular pattern from a leak at an offshore rig. If the rate at which the radius of the oil slick is growing is 1 foot per minute at what rate is the area of the oil slick growing when the radius is 600 feet?

Solution

The area of the circular oil slick is given by

$$A = \pi r^2$$

where r is the radius. The rate at which the area is changing is

$$\frac{dA}{dt} = 2\pi r\frac{dr}{dt}$$

Using $r = 600$ feet and $dr/dt = 1$ foot per minute gives

$$\frac{dA}{dt} = 2\pi \,(600 \text{ ft})\,(1 \text{ ft/min}) = 1200\pi \text{ ft}^2/\text{min}$$

Thus when the radius of the oil slick is 600 feet, the area is growing at the rate of 1200π square feet per minute, or approximately 3770 square feet per minute. ■

CHECKPOINT SOLUTIONS

1. dV/dt

2. (a) $2x\dfrac{dx}{dt} + 0 = 2y\dfrac{dy}{dt}$

   $x\dfrac{dx}{dt} = y\dfrac{dy}{dt}$

   (b) Use $\dfrac{dy}{dt} = 2$, $x = 6$, and $y = 10$ in part (a) to obtain

   $6\dfrac{dx}{dt} = 10(2)$ or $\dfrac{dx}{dt} = \dfrac{20}{6} = \dfrac{10}{3}$

3. False. The numerical values for any quantity that is varying with time should not be substituted until after the derivative is taken.

| EXERCISES | **11.4**

Access end-of-section exercises online at **www.webassign.net**

OBJECTIVES   **11.5**

- To find the elasticity of demand
- To find the tax per unit that will maximize tax revenue

Applications in Business and Economics

| APPLICATION PREVIEW |

Suppose that the demand for a product is given by

$$p = \frac{1000}{(q + 1)^2}$$

We can measure how sensitive the demand for this product is to price changes by finding the elasticity of demand. (See Example 2.) This elasticity can be used to measure the effects that price changes have on total revenue. We also consider taxation in a competitive market, which examines how a tax levied on goods shifts market equilibrium. We will find the tax per unit that, despite changes in market equilibrium, maximizes tax revenue.

Elasticity of Demand   We know from the law of demand that consumers will respond to changes in prices; if prices increase, the quantities demanded will decrease. But the degree of responsiveness of the consumers to price changes will vary widely for different products. For example, a price increase in insulin will not greatly decrease the demand for it by diabetics, but a price increase in clothes may cause consumers to buy less and wear their old clothes longer. When the response to price changes is considerable, we say the demand is *elastic*. When price changes cause relatively small changes in demand for a product, the demand is said to be *inelastic* for that product.

Economists measure the **elasticity of demand** as follows.

Elasticity

The **elasticity of demand** at the point (q_A, p_A) is

$$\eta = -\frac{p}{q} \cdot \frac{dq}{dp}\bigg|_{(q_A,\, p_A)}$$

EXAMPLE 1 **Elasticity**

Find the elasticity of the demand function $p + 5q = 100$ when

(a) the price is $40. (b) the price is $60. (c) the price is $50.

Solution

Solving the demand function for q gives $q = 20 - \frac{1}{5}p$. Then $dq/dp = -\frac{1}{5}$ and

$$\eta = -\frac{p}{q}\left(-\frac{1}{5}\right)$$

(a) When $p = 40$, $q = 12$ and $\eta = -\frac{p}{q}\left(-\frac{1}{5}\right)\Big|_{(12,\,40)} = -\frac{40}{12}\left(-\frac{1}{5}\right) = \frac{2}{3}$.

(b) When $p = 60$, $q = 8$ and $\eta = -\frac{p}{q}\left(-\frac{1}{5}\right)\Big|_{(8,\,60)} = -\frac{60}{8}\left(-\frac{1}{5}\right) = \frac{3}{2}$.

(c) When $p = 50$, $q = 10$ and $\eta = -\frac{p}{q}\left(-\frac{1}{5}\right)\Big|_{(10,\,50)} = -\frac{50}{10}\left(-\frac{1}{5}\right) = 1$. ∎

Note that in Example 1 the demand equation was $p + 5q = 100$, so the demand "curve" is a straight line, with slope $m = -5$. But the elasticity was $\eta = \frac{2}{3}$ at $(12, 40)$, $\eta = \frac{3}{2}$ at $(8, 60)$, and $\eta = 1$ at $(10, 50)$. This illustrates that the elasticity of demand may be different at different points on the demand curve, even though the slope of the demand "curve" is constant. (See Figure 11.14.)

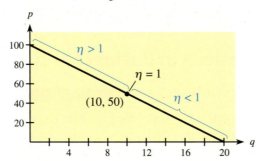

Figure 11.14

This example shows that the elasticity of demand is more than just the slope of the demand curve, which is the rate at which the demand is changing. Recall that the elasticity measures the consumers' degree of responsiveness to a price change.

Economists use η to measure how responsive demand is to price at different points on the demand curve for a product.

Elasticity of Demand

- If $\eta > 1$, the demand is **elastic,** and the percent decrease in demand is greater than the corresponding percent increase in price.
- If $\eta < 1$, the demand is **inelastic,** and the percent decrease in demand is less than the corresponding percent increase in price.
- If $\eta = 1$, the demand is **unitary elastic,** and the percent decrease in demand is approximately equal to the corresponding percent increase in price.

We can also use implicit differentiation to find dq/dp in evaluating the point elasticity of demand.

EXAMPLE 2 **Elasticity | APPLICATION PREVIEW |**

The demand for a certain product is given by

$$p = \frac{1000}{(q + 1)^2}$$

where p is the price per unit in dollars and q is demand in units of the product. Find the elasticity of demand with respect to price when $q = 19$.

Solution

To find the elasticity, we need to find dq/dp. Using implicit differentiation, we get the following:

$$\frac{d}{dp}(p) = \frac{d}{dp}\left[1000\,(q + 1)^{-2}\right]$$

$$1 = 1000\left[-2\,(q + 1)^{-3}\frac{dq}{dp}\right]$$

$$1 = \frac{-2000}{(q + 1)^3}\frac{dq}{dp}$$

$$\frac{(q + 1)^3}{-2000} = \frac{dq}{dp}$$

When $q = 19$, we have $p = 1000/(19 + 1)^2 = 1000/400 = 5/2$ and

$$\left.\frac{dq}{dp}\right|_{(q=19)} = \frac{(19 + 1)^3}{-2000} = \frac{8000}{-2000} = -4$$

The elasticity of demand when $q = 19$ is

$$\eta = \frac{-p}{q}\cdot\frac{dq}{dp} = -\frac{(5/2)}{19}\cdot(-4) = \frac{10}{19} < 1$$

Thus the demand for this product is inelastic. ◼

Elasticity and Revenue

Elasticity is related to revenue in a special way. We can see how by computing the derivative with respect to p of the revenue function

$$R = pq$$

$$\frac{dR}{dp} = p\cdot\frac{dq}{dp} + q\cdot 1$$

$$= \frac{q}{q}\cdot p\cdot\frac{dq}{dp} + q = q\cdot\frac{p}{q}\cdot\frac{dq}{dp} + q$$

$$= q(-\eta) + q$$

$$= q(1 - \eta)$$

From this we can summarize the relationship of elasticity and revenue.

Elasticity and Revenue

The rate of change of revenue R with respect to price p is related to elasticity in the following way.

- Elastic $(\eta > 1)$ means $\dfrac{dR}{dp} < 0.$ $\begin{cases}\text{Hence if price increases, revenue decreases,}\\ \text{and if price decreases, revenue increases.}\end{cases}$

- Inelastic $(\eta < 1)$ means $\dfrac{dR}{dp} > 0.$ $\begin{cases}\text{Hence if price increases, revenue increases,}\\ \text{and if price decreases, revenue decreases.}\end{cases}$

- Unitary elastic $(\eta = 1)$ means $\dfrac{dR}{dp} = 0.$ Hence an increase or decrease in price will not change revenue. Revenue is optimized at this point.

EXAMPLE **3**   **Elasticity and Revenue**

The demand for a product is given by

$$p = 10\sqrt{100 - q}, \quad 0 \le q \le 100$$

(a) Find the point at which demand is of unitary elasticity, and find intervals in which the demand is inelastic and in which it is elastic.
(b) Find q where revenue is increasing, where it is decreasing, and where it is maximized.
(c) Use a graphing utility to show the graph of the revenue function $R = pq$, with $0 \le q \le 100$, and confirm the results from part (b).

Solution
The elasticity is

$$\eta = \frac{-p}{q} \cdot \frac{dq}{dp} = -\frac{10\sqrt{100 - q}}{q} \cdot \frac{dq}{dp}$$

Finding dq/dp implicitly, we have

$$1 = 10\left[\frac{1}{2}(100 - q)^{-1/2}\left(-\frac{dq}{dp}\right)\right] = \frac{-5}{\sqrt{100 - q}} \cdot \frac{dq}{dp}$$

so

$$\frac{dq}{dp} = \frac{-\sqrt{100 - q}}{5}$$

Thus

$$\eta = \left(-\frac{10\sqrt{100 - q}}{q}\right)\left(\frac{-\sqrt{100 - q}}{5}\right) = \frac{200 - 2q}{q}$$

(a) Unitary elasticity occurs where $\eta = 1$.

$$1 = \frac{200 - 2q}{q}$$
$$q = 200 - 2q$$
$$3q = 200$$
$$q = 66\tfrac{2}{3}$$

so unitary elasticity occurs when $66\tfrac{2}{3}$ units are sold, at a price of $57.74. For values of q between 0 and $66\tfrac{2}{3}$, $\eta > 1$ and demand is elastic. For values of q between $66\tfrac{2}{3}$ and 100, $\eta < 1$ and demand is inelastic.

(b) When q increases over $0 < q < 66\tfrac{2}{3}$, p decreases, so $\eta > 1$ means R increases. Similarly, when q increases over $66\tfrac{2}{3} < q < 100$, p decreases, so $\eta < 1$ means R decreases. Revenue is maximized where $\eta = 1$, at $q = 66\tfrac{2}{3}$ and $p = 57.74$.

(c) The graph of this revenue function,

$$R = 10q\sqrt{100 - q}$$

is shown in Figure 11.15. The maximum revenue is $3849 at $q = 66\tfrac{2}{3}$.

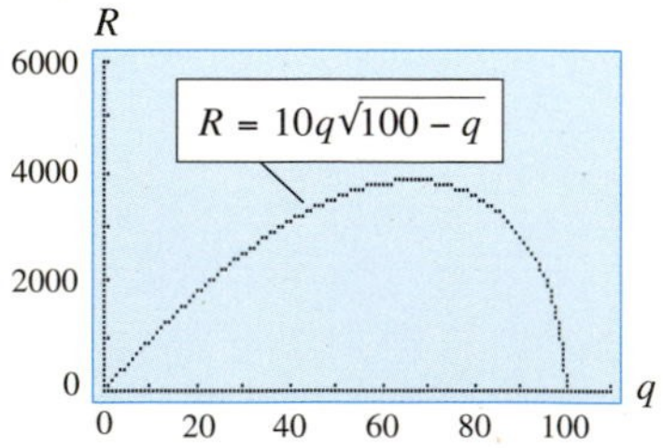

Figure 11.15

1. Write the formula for point elasticity, η.
2. (a) If $\eta > 1$, the demand is called _____.
   (b) If $\eta < 1$, the demand is called _____.
   (c) If $\eta = 1$, the demand is called _____.
3. Find the elasticity of demand for $q = \dfrac{100}{p} - 1$ when $p = 10$ and $q = 9$.

Taxation in a Competitive Market

Many taxes imposed by governments are "hidden." That is, the tax is levied on goods produced, and the producers must pay the tax. Of course, the tax becomes a cost to the producers, and they pass that cost on to the consumer in the form of higher prices for goods.

Suppose the government imposes a tax of t dollars on each unit produced and sold by producers. If we are in pure competition in which the consumers' demand depends only on price, the *demand function* will not change. The tax will change the supply function, of course, because at each level of output q, the firm will want to charge a price $p + t$ per unit, where p is the original price per unit and t is the tax per unit.

The graphs of the market demand function, the original market supply function, and the market supply function after taxes are shown in Figure 11.16. Because the tax added to each item is constant, the graph of the new supply function is t units above the original supply function. If $p = f(q)$ defines the original supply function, then $p = f(q) + t$ defines the supply function after taxation.

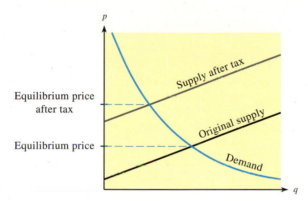

Figure 11.16

Note in this case that after the taxes are imposed, *no* items are supplied at the price that was the equilibrium price before taxation. After the taxes are imposed, the consumers simply have to pay more for the product. Because taxation does not change the demand curve, the quantity purchased at market equilibrium will be less than it was before taxation. Thus governments planning taxes should recognize that they will not collect taxes on the original equilibrium quantity. They will collect on the *new* equilibrium quantity, a quantity reduced by their taxation. Thus a large tax on each item may reduce the quantity demanded at the new market equilibrium so much that very little revenue results from the tax!

If the tax revenue is represented by $T = tq$, where t is the tax per unit and q is the equilibrium quantity of the supply and demand functions after taxation, we can use the following procedure for maximizing the total tax revenue in a competitive market.

Maximizing Total Tax Revenue

Procedure	Example
To find the tax per item (under pure competition) that will maximize total tax revenue:	Suppose the demand and supply functions are given by $p = 600 - q$ and $p = 200 + \frac{1}{3}q$, respectively, where $\$p$ is the price per unit and q is the number of units. Find the tax t that will maximize the total tax revenue T.
1. Write the supply function after taxation.	1. $p = 200 + \frac{1}{3}q + t$
2. Set the demand function and the new supply function equal, and solve for t in terms of q.	2. $600 - q = 200 + \frac{1}{3}q + t$ $400 - \frac{4}{3}q = t$
3. Form the total tax revenue function, $T = tq$, by multiplying the expression for t by q, and then take its derivative with respect to q.	3. $T = tq = 400q - \frac{4}{3}q^2$ $T'(q) = \frac{dT}{dq} = 400 - \frac{8}{3}q$
4. Set $T' = 0$, and solve for q. This is the q that should maximize T. Use the second-derivative test to verify it.	4. $0 = 400 - \frac{8}{3}q$ $q = 150$ $T''(q) = -\frac{8}{3}$. Thus T is maximized at $q = 150$.
5. Substitute the value of q in the equation for t (in Step 2). This is the value of t that will maximize T.	5. $t = 400 - \frac{4}{3}(150) = 200$ A tax of $200 per item will maximize the total tax revenue. The total tax revenue for the period would be $200 \cdot (150) = \$30,000$.

Note that in the example just given, if a tax of $300 were imposed, market equilibrium would occur at $q = 75$, and the total tax revenue the government would receive would be

$$(\$300)(75) = \$22,500$$

Thus, with a tax of $300 per unit rather than $200, the government's tax revenue would be $22,500 rather than $30,000. In addition, with a $300 tax, suppliers would sell only 75 units rather than 150, and consumers would pay $p = 200 + 25 + 300 = \$525$ per item, which is $75 more than the price with a $200 tax. Thus everyone would suffer if the tax rate were raised to $300.

CHECKPOINT

4. For problems involving taxation in a competitive market, if supply is $p = f(q)$ and demand is $p = g(q)$, is the tax t added to $f(q)$ or to $g(q)$?

EXAMPLE 4 Maximizing Tax Revenue

The demand and supply functions for a product are $p = 900 - 20q - \frac{1}{3}q^2$ and $p = 200 + 10q$, respectively, where p is in dollars and q is the number of units. Find the tax per unit that will maximize the tax revenue T.

Solution

After taxation, the supply function is $p = 200 + 10q + t$, where t is the tax per unit. The demand function will meet the new supply function where

$$900 - 20q - \frac{1}{3}q^2 = 200 + 10q + t$$

so

$$t = 700 - 30q - \frac{1}{3}q^2$$

Then the total tax T is $T = tq = 700q - 30q^2 - \frac{1}{3}q^3$, and we maximize T as follows:

$$T'(q) = 700 - 60q - q^2$$

$$0 = -(q + 70)(q - 10)$$

$$q = 10 \quad \text{or} \quad q = -70$$

$$T''(q) = -60 - 2q, \text{ so } T''(q) < 0 \text{ for } q \geq 0$$

Thus the curve is concave down for $q \geq 0$, and $q = 10$ gives an absolute maximum for the tax revenue T. The maximum possible tax revenue is

$$T(10) \approx \$3666.67$$

The tax per unit that maximizes T is

$$t = 700 - 30(10) - \frac{1}{3}(10)^2 \approx \$366.67$$

An infamous example of a tax increase that resulted in decreased tax revenue and economic disaster is the "luxury tax" enacted in 1991. This was a 10% excise tax on the sale of expensive jewelry, furs, airplanes, certain expensive boats, and luxury automobiles. The Congressional Joint Tax Committee had estimated that the luxury tax would raise $6 million from airplanes alone, but it raised only $53,000 while it destroyed the small-airplane market (one company lost $130 million and 480 jobs in a single year). It also capsized the boat market. The luxury tax was repealed at the end of 1993 (except for automobiles).*

CHECKPOINT SOLUTIONS

1. $\eta = \dfrac{-p}{q} \cdot \dfrac{dq}{dp}$

2. (a) elastic   (b) inelastic   (c) unitary elastic

3. $\dfrac{dq}{dp} = \dfrac{-100}{p^2}$ and $\dfrac{dq}{dp} = -1$ when $p = 10, q = 9$

   $\eta = \dfrac{-10}{9}(-1) = \dfrac{10}{9}(\text{elastic})$

4. Tax t is added to supply: $p = f(q) + t$.

*Fortune, Sept. 6, 1993; Motor Trend, December 1993.

| EXERCISES | 11.5

Access end-of-section exercises online at **www.webassign.net**

KEY TERMS AND FORMULAS

Section	Key Terms	Formulas
11.1	Logarithmic function	$y = \log_a x$, defined by $x = a^y$
	Natural logarithm	$\ln x = \log_e x$; $y = \ln(x)$ means $e^y = x$
	Logarithmic Properties I–V	$\ln e^x = x$; $e^{\ln x} = x$
	for natural logarithms	$\ln(MN) = \ln M + \ln N$
		$\ln(M/N) = \ln M - \ln N$
		$\ln(M^p) = p(\ln M)$
	Change-of-base formula	$\log_a x = \dfrac{\ln x}{\ln a}$
	Derivatives of logarithmic functions	$\dfrac{d}{dx}(\ln x) = \dfrac{1}{x}$
		$\dfrac{d}{dx}(\ln u) = \dfrac{1}{u} \cdot \dfrac{du}{dx}$
11.2	Derivatives of exponential functions	$\dfrac{d}{dx}(e^x) = e^x$
		$\dfrac{d}{dx}e^u = e^u \dfrac{du}{dx}$
		$\dfrac{d}{dx}a^u = a^u \dfrac{du}{dx}\ln a$
11.3	Implicit differentiation	$\dfrac{d}{dx}(y^n) = ny^{n-1}\dfrac{dy}{dx}$
11.4	Related rates	Differentiate implicitly with respect to time, t
	Percent rates of change	
11.5	Elasticity of demand	$\eta = \dfrac{-p}{q} \cdot \dfrac{dq}{dp}$
	Elastic	$\eta > 1$
	Inelastic	$\eta < 1$
	Unitary elastic	$\eta = 1$
	Taxation in competitive market	
	Supply function after taxation	$p = f(q) + t$

In Problems 1–12, find the indicated derivative.

1. If $y = e^{3x^2 - x}$, find dy/dx.

2. If $y = \ln e^{x^2}$, find y'.

3. If $p = \ln\left(\dfrac{q}{q^2 - 1}\right)$, find $\dfrac{dp}{dq}$.

4. If $y = xe^{x^2}$, find dy/dx.

5. If $f(x) = 5e^{2x} - 40e^{-0.1x} + 11$, find $f'(x)$.

6. If $g(x) = (2e^{3x+1} - 5)^3$, find $g'(x)$.

7. If $y = \ln(3x^4 + 7x^2 - 12)$, find dy/dx.

8. If $s = \dfrac{3}{4}\ln(x^{12} - 2x^4 + 5)$, find ds/dx.

9. If $y = 3^{3x-4}$, find dy/dx.

10. If $y = 1 + \log_8(x^{10})$, find dy/dx.

11. If $y = \dfrac{\ln x}{x}$, find $\dfrac{dy}{dx}$.

12. If $y = \dfrac{1 + e^{-x}}{1 - e^{-x}}$, find $\dfrac{dy}{dx}$.

13. Write the equation of the line tangent to $y = 4e^{x^3}$ at $x = 1$.

14. Write the equation of the line tangent to $y = x \ln x$ at $x = 1$.

In Problems 15–20, find the indicated derivative.

15. If $y \ln x = 5y^2 + 11$, find dy/dx.

16. Find dy/dx for $e^{xy} = y$.

17. Find dy/dx for $y^2 = 4x - 1$.

18. Find dy/dx for $x^2 + 3y^2 + 2x - 3y + 2 = 0$.

19. Find dy/dx for $3x^2 + 2x^3y^2 - y^5 = 7$.

20. Find the second derivative y'' if $x^2 + y^2 = 1$.

21. Find the slope of the tangent to the curve $x^2 + 4x - 3y^2 + 6 = 0$ at $(3, 3)$.

22. Find the points where tangents to the graph of the equation in Problem 21 are horizontal.

23. Suppose $3x^2 - 2y^3 = 10y$, where x and y are differentiable functions of t. If $dx/dt = 2$, find dy/dt when $x = 10$ and $y = 5$.

24. A right triangle with legs of lengths x and y has its area given by
$$A = \frac{1}{2}xy$$
If the rate of change of x is 2 units per minute and the rate of change of y is 5 units per minute, find the rate of change of the area when $x = 4$ and $y = 1$.

APPLICATIONS

25. *Deforestation* One of the major causes of rain forest deforestation is agricultural and residential development. The number of hectares destroyed in a particular year t can be modeled by
$$y = -3.91435 + 2.62196 \ln t$$
where $t = 0$ represents 1950.

(a) Find $y'(t)$.

(b) Find and interpret $y(50)$ and $y'(50)$.

26. *Nonmarital childbearing* The percent of live births to unmarried mothers for the years 1970–2007 can be modeled by the function
$$y = -33.410 + 18.035 \ln x$$
where x is the number of years past 1960.

(a) What does the model predict the rate of change of this percent to be in 2015?

(b) Is the percent of live births to unmarried mothers predicted to increase or decrease after 2015?

27. *Compound interest* If the future value of $1000 invested for n years at 12%, compounded continuously, is given by
$$S = 1000e^{0.12n} \text{ dollars}$$
find the rate at which the future value is growing after 1 year.

28. *Compound interest*

(a) In Problem 27, find the rate of growth of the future value after 2 years.

(b) How much faster is the future value growing at the end of 2 years than after 1 year?

29. *Radioactive decay* A breeder reactor converts stable uranium-238 into the isotope plutonium-239. The decay of this isotope is given by
$$A(t) = A_0 e^{-0.00002876t}$$
where $A(t)$ is the amount of isotope at time t, in years, and A_0 is the original amount. This isotope has a half-life of 24,101 years (that is, half of it will decay away in 24,101 years).

(a) At what rate is $A(t)$ decaying at this point in time?

(b) At what rate is $A(t)$ decaying after 1 year?

(c) Is the rate of decay at its half-life greater or less than after 1 year?

30. *Marginal cost* The average cost of producing x units of a product is $\bar{C} = 600e^{x/600}$ dollars per unit. What is the marginal cost when 600 units are produced?

31. *Inflation* The impact of inflation on a $20,000 pension can be measured by the purchasing power P of $20,000 after t years. For an inflation rate of 5% per year, compounded annually, P is given by
$$P = 20{,}000e^{-0.0495t}$$
At what rate is purchasing power changing when $t = 10$? (*Source: Viewpoints*, VALIC)

32. *Evaporation* A spherical droplet of water evaporates at a rate of 1 mm^3/min. Find the rate of change of the radius when the droplet has a radius of 2.5 mm.

33. *Worker safety* A sign is being lowered over the side of a building at the rate of 2 ft/min. A worker handling

a guide line is 7 ft away from a spot directly below the sign. How fast is the worker taking in the guide line at the instant the sign is 25 ft from the worker's hands? See the figure.

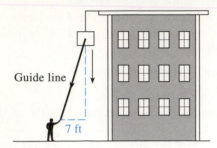

Guide line

7 ft

34. **Environment** Suppose that in a study of water birds, the relationship between the area A of wetlands (in square miles) and the number of different species S of birds found in the area was determined to be

$$S = kA^{1/3}$$

where k is constant. Find the percent rate of change of the number of species in terms of the percent rate of change of the area.

35. **Taxes** Can increasing the tax per unit sold actually lead to a decrease in tax revenues?

36. **Taxes** Suppose the demand and supply functions for a product are

$$p = 2800 - 8q - \frac{q^2}{3} \quad \text{and} \quad p = 400 + 2q$$

respectively, where p is in dollars and q is the number of units. Find the tax per unit t that will maximize the tax revenue T, and find the maximum tax revenue.

37. **Taxes** Suppose the supply and demand functions for a product are

$$p = 40 + 20q \quad \text{and} \quad p = \frac{5000}{q + 1}$$

respectively, where p is in dollars and q is the number of units. Find the tax t that maximizes the tax revenue T, and find the maximum tax revenue.

38. **Elasticity** A demand function is given by

$$pq = 27$$

where p is in dollars and q is the number of units.
(a) Find the elasticity of demand at $(9, 3)$.
(b) How will a price increase affect total revenue?

39. **Elasticity** Suppose the demand for a product is given by

$$p^2(2q + 1) = 10,000$$

where p is in dollars and q is the number of units.
(a) Find the elasticity of demand when $p = \$20$.
(b) How will a price increase affect total revenue?

40. **Elasticity** Suppose the weekly demand function for a product is given by

$$p = 100e^{-0.1q}$$

where p is the price in dollars and q is the number of tons demanded.
(a) What is the elasticity of demand when the price is $\$36.79$ and the quantity demanded is 10?
(b) How will a price increase affect total revenue?

41. **Revenue** A product has the demand function

$$p = 100 - 0.5q$$

where p is in dollars and q is the number of units.
(a) Find the elasticity $\eta(q)$ as a function of q, and graph the function

$$f(q) = \eta(q)$$

(b) Find where $f(q) = 1$, which gives the quantity for which the product has unitary elasticity.
(c) The revenue function for this product is

$$R(q) = pq = (100 - 0.5q)q$$

Graph $R(q)$ and find the q-value for which the maximum revenue occurs.
(d) What is the relationship between elasticity and maximum revenue?

---

11   CHAPTER TEST

In Problems 1–8, find the derivative of each function.

1. $y = 5e^{x^3} + x^2$

2. $y = 4\ln(x^3 + 1)$

3. $y = \ln(x^4 + 1)^3$

4. $f(x) = 10(3^{2x})$

5. $S = te^{t^4}$

6. $y = \dfrac{e^{x^3 + 1}}{x}$

7. $y = \dfrac{\ln x}{x}$

8. $g(x) = 2\log_5(4x + 7)$

9. Find y' if $3x^4 + 2y^2 + 10 = 0$.

10. Let $x^2 + y^2 = 100$.
If $\dfrac{dx}{dt} = 2$, find $\dfrac{dy}{dt}$ when $x = 6$ and $y = 8$.

11. Find y' if $xe^y = 10y$.

12. Suppose the weekly revenue and weekly cost (both in dollars) for a product are given by
$R(x) = 300x - 0.001x^2$ and $C(x) = 4000 + 30x$,
respectively, where x is the number of units produced

and sold. Find the rate at which profit is changing with respect to time when the number of units produced and sold is 50 and is increasing at a rate of 5 units per week.

13. Suppose the demand for a product is $p^2 + 3p + q = 1500$, where p is in dollars and q is the number of units. Find the elasticity of demand at $p = 30$. If the price is raised to $31, does revenue increase or decrease?

14. Suppose the demand function for a product is given by $(p + 1)q^2 = 10,000$, where p is the price and q is the quantity. Find the rate of change of quantity with respect to price when $p = \$99$.

15. The sales of a product are given by $S = 80,000e^{-0.4t}$, where S is the daily sales and t is the number of days after the end of an advertising campaign. Find the rate of sales decay 10 days after the end of the ad campaign.

16. The table gives the total U.S. expenditures (in billions of dollars) for health services and supplies for selected years from 2000 and projected to 2018. Given that the model for these data is

$$y = 1319e^{0.062t}$$

where $t = 0$ in 2000.
(a) Find the function that models the rate of change.
(b) Find the model's value for the rate of change of U.S. expenditures for health services and supplies in 2005 and in 2020.

Year	Health Expenditures	Year	Health Expenditures
2000	1264	2010	2458
2002	1498	2012	2746
2004	1733	2014	3107
2006	1976	2016	3556
2008	2227	2018	4086

Source: U.S. Centers for Medicare and Medicaid Services

17. Suppose the demand and supply functions for a product are $p = 1100 - 5q$ and $p = 20 + 0.4q$, respectively, where p is in dollars and q is the number of units. Find the tax per unit t that will maximize the tax revenue $T = tq$.

18. *Modeling* Projections indicate that the percent of U.S. adults with diabetes could dramatically increase, and already in 2007 this disease had cost the country almost $175 billion. The table gives the percent of U.S. adults with diabetes for selected years from 2010 and projected to 2050.
(a) Find a logarithmic model, $y = f(x)$, for these data. Use $x = 0$ to represent 2000.
(b) Find the function that models the rate of change of the percent of U.S. adults with diabetes.
(c) Find and interpret $f(25)$ and $f'(25)$.

Year	Percent	Year	Percent
2010	15.7	2035	29.0
2015	18.9	2040	31.4
2020	21.1	2045	32.1
2025	24.2	2050	34.3
2030	27.2		

Source: Centers for Disease Control and Prevention

19. By using data from the U.S. Bureau of Labor Statistics for the years 1968–2008, the purchasing power P of a 1983 dollar can be modeled by the function

$$P(t) = 3.443(1.046)^{-t}$$

where t is the number of years past 1960.

　　Use the model to find the function that models the rate of change of the purchasing power of the 1983 dollar and use that function to predict the rate of change in the year 2015.

I. Inflation

Hollingsworth Pharmaceuticals specializes in manufacturing generic medicines. Recently it developed an antibiotic with outstanding profit potential. The new antibiotic's total costs, sales, and sales growth, as well as projected inflation, are described as follows.

Total monthly costs, in dollars, to produce x units (1 unit is 100 capsules):

$$C(x) = \begin{cases} 15{,}000 + 10x & 0 \le x \le 11{,}000 \\ 15{,}000 + 10x + 0.001(x - 11{,}000)^2 & x \ge 11{,}000 \end{cases}$$

Sales: 10,000 units per month and growing at 1.25% per month, compounded continuously
Selling price: $34 per unit
Inflation: Approximately 0.25% per month, compounded continuously, affecting both total costs and selling price

Company owners are pleased with the sales growth but are concerned about the projected increase in variable costs when production levels exceed 11,000 units per month. The consensus is that improvements eventually can be made that will reduce costs at higher production levels, thus altering the current cost function model. To plan properly for these changes, Hollingsworth Pharmaceuticals would like you to determine when the company's profits will begin to decrease. To help you determine this, answer the following.

1. If inflation is assumed to be compounded continuously, the selling price and total costs must be multiplied by the factor $e^{0.0025t}$. In addition, if sales growth is assumed to be compounded continuously, then sales must be multiplied by a factor of the form e^{rt}, where r is the monthly sales growth rate (expressed as a decimal) and t is time in months. Use these factors to write each of the following as a function of time t:
   (a) selling price p per unit (including inflation).
   (b) number of units x sold per month (including sales growth).
   (c) total revenue. (Recall that $R = px$.)
2. Determine how many months it will be before monthly sales exceed 11,000 units.
3. If you restrict your attention to total costs when $x \ge 11{,}000$, then, after expanding and collecting like terms, $C(x)$ can be written as follows:

$$C(x) = 136{,}000 - 12x + 0.001x^2 \text{ for } x \ge 11{,}000$$

   Use this form for $C(x)$ with your result from Question 1(b) and with the inflationary factor $e^{0.0025t}$ to express these total costs as a function of time.
4. Form the profit function that would be used when monthly sales exceed 11,000 units by using the total revenue function from Question 1(c) and the total cost function from Question 3. This profit function should be a function of time t.
5. Find how long it will be before the profit is maximized. You may have to solve $P'(t) = 0$ by using a graphing calculator or computer to find the t-intercept of the graph of $P'(t)$. In addition, because $P'(t)$ has large numerical coefficients, you may want to divide both sides of $P'(t) = 0$ by 1000 before solving or graphing.

II. Knowledge Workers (Modeling)

The representation of women in leadership positions in our economy has changed dramatically in the past 20 years. In this application we'll investigate whether some predictions made in 1997 are still valid.

In January 1997, *Working Woman* made the following points about the U.S. economy and the place of women in the economy.

- The telecommunications industry employs more people than the auto and auto parts industries combined.
- More Americans make semiconductors than make construction equipment.
- Almost twice as many Americans make surgical and medical instruments as make plumbing and heating products.
- The ratio of male to female knowledge workers (engineers, scientists, technicians, professionals, and senior managers) was 3 to 2 in 1983. The table, which gives the numbers (in millions) of male and female knowledge workers from 1983 to 1997, shows how that ratio is changing.

Year	Female Knowledge Workers (millions)	Male Knowledge Workers (millions)	Year	Female Knowledge Workers (millions)	Male Knowledge Workers (millions)
1983	11.0	15.4	1991	16.1	18.4
1984	11.6	15.9	1992	16.7	18.66
1985	12.3	16.3	1993	17.3	18.7
1986	12.9	16.7	1994	18.0	19.0
1987	13.6	16.8	1995	18.5	19.8
1988	14.3	17.6	1996	19.0	19.6
1989	15.3	18.1	1997	19.5	19.8
1990	15.9	18.6			

Source: Working Woman, January 1997

To compare how the growth in the number of female knowledge workers compares with that of male knowledge workers, do the following:

1. Find a logarithmic equation (with $x = 0$ representing 1980) that models the number of females, and find a logarithmic equation (with $x = 0$ representing 1980) that models the number of males.
2. Use the models found in (1) to find the rate of growth with respect to time of (a) the number of female knowledge workers and (b) the number of male knowledge workers.
3. Compare the two rates of growth in the year 2000 and determine which rate is larger.
4. If these models indicate that it is possible for the number of females to equal the number of males, during what year do they indicate that this will occur?
5. Research this topic on the Internet to determine whether the trends of your models are supported by current data. Be sure to cite your sources and include recent data.

12

Indefinite Integrals

Vasily Smirnov/Shutterstock.com

When we know the derivative of a function, it is often useful to determine the function itself. For example, accountants can use linear regression to translate information about marginal cost into a linear equation defining (approximately) the marginal cost function and then use the process of antidifferentiation (or integration) as part of finding the (approximate) total cost function. We can also use integration to find total revenue functions from marginal revenue functions, to optimize profit from information about marginal cost and marginal revenue, and to find national consumption from information about marginal propensity to consume.

Integration can also be used in the social and life sciences to predict growth or decay from expressions giving rates of change. For example, we can find equations for population size from the rate of change of growth, we can write equations for the number of radioactive atoms remaining in a substance if we know the rate of the decay of the substance, and we can determine the volume of blood flow from information about rate of flow.

The topics and applications discussed in this chapter include the following.

SECTIONS	APPLICATIONS
12.1 Indefinite Integrals	Revenue, population growth
12.2 The Power Rule	Revenue, productivity
12.3 Integrals Involving Exponential and Logarithmic Functions	Real estate inflation, population growth
12.4 Applications of the Indefinite Integral in Business and Economics Total Cost and Profit National Consumption and Savings	Cost, revenue, maximum profit, national consumption and savings
12.5 Differential Equations Solution of Differential Equations Separable Differential Equations Applications of Differential Equations	Carbon-14 dating, drug in an organ

Prerequisite Problem Type	For Section	Answer	Section for Review
Write as a power: (a) $\sqrt{x}$ (b) $\sqrt{x^2 - 9}$	**12.1.–12.4**	(a) $x^{1/2}$ (b) $(x^2 - 9)^{1/2}$	0.4 Radicals
Expand $(x^2 + 4)^2$.	**12.2**	$x^4 + 8x^2 + 16$	0.5 Special powers
Divide $x^4 - 2x^3 + 4x^2 - 7x - 1$ by $x^2 - 2x$.	**12.3**	$x^2 + 4 + \dfrac{x - 1}{x^2 - 2x}$	0.5 Division
Find the derivative of (a) $f(x) = 2x^{1/2}$ (b) $u = x^3 - 3x$	**12.1** **12.2** **12.3**	(a) $f'(x) = x^{-1/2}$ (b) $u' = 3x^2 - 3$	9.4 Derivatives
If $y = \dfrac{(x^2 + 4)^6}{6}$, what is y'?	**12.2**	$y' = (x^2 + 4)^5 2x$	9.6 Derivatives
(a) If $y = \ln u$, what is y'? (b) If $y = e^u$, what is y'?	**12.3**	(a) $y' = \dfrac{1}{u} \cdot u'$ (b) $y' = e^u \cdot u'$	11.1, 11.2 Derivatives
Solve for y: $\ln y = kt + C$	**12.5**	$y = e^{kt + C}$	5.2 Logarithmic functions
Solve for k: $0.5 = e^{5730k}$	**12.5**	$k \approx -0.00012097$	5.3 Exponential equations

OBJECTIVE 12.1

- To find certain indefinite integrals

Indefinite Integrals

| APPLICATION PREVIEW |

In our study of the theory of the firm, we have worked with total cost, total revenue, and profit functions and have found their marginal functions. In practice, it is often easier for a company to measure marginal cost, revenue, and profit and use these data to form marginal functions from which it can find total cost, revenue, and profit functions. For example, Jarus Technologies manufactures computer motherboards, and the company's sales records show that the marginal revenue (in dollars per unit) for its motherboards is given by

$$\overline{MR} = 300 - 0.2x$$

where x is the number of units sold. If we want to use this function to find the total revenue function for Jarus Technologies' motherboards, we need to find $R(x)$ from $\overline{MR} = R'(x)$. (See Example 7.) In this situation, we need to be able to reverse the process of differentiation. This reverse process is called antidifferentiation, or integration.

Indefinite Integrals

We have studied procedures for and applications of finding derivatives of a given function. We now turn our attention to reversing this process of differentiation. When we know the derivative of a function, the process of finding the function itself is called **antidifferentiation.** For example, if the derivative of a function is $2x$, we know that the function could be $f(x) = x^2$ because $\dfrac{d}{dx}(x^2) = 2x$. But the function could also be $f(x) = x^2 + 4$ because $\dfrac{d}{dx}(x^2 + 4) = 2x$. It is clear that any function of the form $f(x) = x^2 + C$, where C is an arbitrary constant, will have $f'(x) = 2x$ as its derivative. Thus we say that the **general antiderivative** of $f'(x) = 2x$ is $f(x) = x^2 + C$, where C is an arbitrary constant.

The process of finding an antiderivative is also called **integration.** The function that results when integration takes place is called an **indefinite integral** or, more simply, an **integral.** We can denote the indefinite integral (that is, the general antiderivative) of a function $f(x)$ by $\int f(x)\, dx$. Thus we can write $\int 2x\, dx$ to indicate the general antiderivative of the function $f(x) = 2x$. The expression is read as "the integral of $2x$ with respect to x." In this case, $2x$ is called the **integrand.** The **integral sign,** $\int$, indicates the process of integration, and the dx indicates that the integral is to be taken with respect to x. Because the antiderivative of $2x$ is $x^2 + C$, we can write

$$\int 2x\, dx = x^2 + C$$

EXAMPLE 1 Antidifferentiation

If $f'(x) = 3x^2$, what is $f(x)$?

Solution

The derivative of the function $f(x) = x^3$ is $f'(x) = 3x^2$. But other functions also have this derivative. They will all be of the form $f(x) = x^3 + C$, where C is a constant. Thus we write

$$\int 3x^2\, dx = x^3 + C$$

EXAMPLE 2   **Integration**

If $f'(x) = x^3$, what is $f(x)$?

Solution

We know that $\dfrac{d}{dx}(x^4) = 4x^3$, so the derivative of $f(x) = \frac{1}{4}x^4$ is $f'(x) = x^3$. Thus

$$f(x) = \int f'(x)\,dx = \int x^3\,dx = \tfrac{1}{4}x^4 + C$$

Powers of x Formula   It is easily seen that

$$\int x^4\,dx = \frac{x^5}{5} + C \quad \text{because} \quad \frac{d}{dx}\left(\frac{x^5}{5} + C\right) = x^4$$

$$\int x^5\,dx = \frac{x^6}{6} + C \quad \text{because} \quad \frac{d}{dx}\left(\frac{x^6}{6} + C\right) = x^5$$

In general, we have the following.

Powers of x Formula

$$\int x^n\,dx = \frac{x^{n+1}}{n+1} + C \quad (\text{for } n \neq -1)$$

In the Powers of x Formula, we see that $n \neq -1$ is essential, because if $n = -1$, then the denominator $n + 1 = 0$. We will discuss the case when $n = -1$ later. (Can you think what function has $1/x$ as its derivative?)

In addition, we can see that this Powers of x Formula applies for any $n \neq -1$ by noting that

$$\frac{d}{dx}\left(\frac{x^{n+1}}{n+1} + C\right) = \frac{d}{dx}\left(\frac{1}{n+1}x^{n+1} + C\right) = \frac{n+1}{n+1}x^n = x^n$$

EXAMPLE 3   **Powers of x Formula**

Evaluate $\int x^{-1/2}\,dx$.

Solution

Using the formula, we get

$$\int x^{-1/2}\,dx = \frac{x^{-1/2+1}}{-1/2+1} + C = \frac{x^{1/2}}{1/2} + C = 2x^{1/2} + C$$

We can check by noting that the derivative of $2x^{1/2} + C$ is $x^{-1/2}$. ✔

Note that the indefinite integral in Example 3 is a function (actually a number of functions, one for each value of C). Graphs of several members of this family of functions are shown in Figure 12.1. Note that at any given x-value, the tangent line to each curve has the same slope, indicating that all family members have the same derivative.

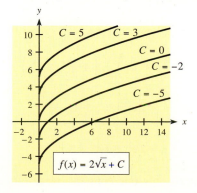

Figure 12.1

EXAMPLE 4 **Powers of *x* Formula**

Find (a) $\int \sqrt[3]{x}\, dx$ and (b) $\int \frac{1}{x^2}\, dx$.

Solution

(a) $\int \sqrt[3]{x}\, dx = \int x^{1/3}\, dx = \frac{x^{4/3}}{4/3} + C$

$\qquad = \frac{3}{4} x^{4/3} + C = \frac{3}{4} \sqrt[3]{x^4} + C$

(b) We write the power of *x* in the numerator so that the integral has the form in the formula above.

$$\int \frac{1}{x^2}\, dx = \int x^{-2}\, dx = \frac{x^{-2+1}}{-2+1} + C = \frac{x^{-1}}{-1} + C = \frac{-1}{x} + C \qquad \blacksquare$$

Other Formulas and Properties Other formulas will be useful in evaluating integrals. The following table shows how some new integration formulas result from differentiation formulas.

Integration Formulas	
Derivative	**Resulting Integral**
$\frac{d}{dx}(x) = 1$	$\int 1\, dx = \int dx = x + C$
$\frac{d}{dx}[c \cdot u(x)] = c \cdot \frac{d}{dx} u(x)$	$\int c\, u(x)\, dx = c \int u(x)\, dx$
$\frac{d}{dx}[u(x) \pm v(x)] = \frac{d}{dx} u(x) \pm \frac{d}{dx} v(x)$	$\int [u(x) \pm v(x)]\, dx = \int u(x)\, dx \pm \int v(x)\, dx$

These formulas indicate that we can integrate functions term by term just as we were able to take derivatives term by term.

EXAMPLE 5 **Using Integration Formulas**

Evaluate: (a) $\int 4\, dx$ (b) $\int 8x^5 dx$ (c) $\int (x^3 + 4x)\, dx$

Solution

(a) $\int 4\, dx = 4 \int dx = 4(x + C_1) = 4x + C$

(Because C_1 is an unknown constant, we can write $4C_1$ as the unknown constant C.)

(b) $\int 8x^5 dx = 8 \int x^5 dx = 8\left(\frac{x^6}{6} + C_1\right) = \frac{4x^6}{3} + C$

(c) $\int (x^3 + 4x)\, dx = \int x^3\, dx + \int 4x\, dx$

$\qquad = \left(\frac{x^4}{4} + C_1\right) + \left(4 \cdot \frac{x^2}{2} + C_2\right)$

$\qquad = \frac{x^4}{4} + 2x^2 + C_1 + C_2$

$\qquad = \frac{x^4}{4} + 2x^2 + C$

Note that we need only one constant because the sum of C_1 and C_2 is just a new constant. $\qquad \blacksquare$

EXAMPLE **6**   **Integral of a Polynomial**

Evaluate $\int (x^2 - 4)^2 \, dx$.

Solution

We expand $(x^2 - 4)^2$ so that the integrand is in a form that fits the basic integration formulas.

$$\int (x^2 - 4)^2 \, dx = \int (x^4 - 8x^2 + 16) \, dx = \frac{x^5}{5} - \frac{8x^3}{3} + 16x + C$$ ■

CHECKPOINT

1. True or false:

   (a) $\int (4x^3 - 2x) \, dx = \int 4x^3 \, dx - \int 2x \, dx$
   $= (x^4 + C) - (x^2 + C) = x^4 - x^2$

   (b) $\int \dfrac{1}{3x^2} \, dx = \dfrac{1}{3(x^3/3)} + C = \dfrac{1}{x^3} + C$

2. Evaluate $\int (2x^3 + x^{-1/2} - 4x^{-5}) \, dx$.

EXAMPLE **7**   **Revenue**   **| APPLICATION PREVIEW |**

Sales records at Jarus Technologies show that the rate of change of the revenue (that is, the marginal revenue) in dollars per unit for a motherboard is $\overline{MR} = 300 - 0.2x$, where x represents the quantity sold. Find the total revenue function for the product. Then find the total revenue from the sale of 1000 motherboards.

Solution

We know that the marginal revenue can be found by differentiating the total revenue function. That is,

$$R'(x) = 300 - 0.2x$$

Thus integrating the marginal revenue function gives the total revenue function.

$$R(x) = \int (300 - 0.2x) \, dx = 300x - 0.1x^2 + K^\star$$

We can use the fact that there is no revenue when no units are sold to evaluate K. Setting $x = 0$ and $R = 0$ gives $0 = 300(0) - 0.1(0)^2 + K$, so $K = 0$. Thus the total revenue function is

$$R(x) = 300x - 0.1x^2$$

The total revenue from the sale of 1000 motherboards is

$$R(1000) = 300(1000) - 0.1(1000^2) = \$200{,}000$$ ■

We can check that the $R(x)$ we found in Example 7 is correct by verifying that $R'(x) = 300 - 0.2x$ and $R(0) = 0$. Also, graphs can help us check the reasonableness of our result. Figure 12.2 on the next page shows the graphs of $\overline{MR} = 300 - 0.2x$ and of the $R(x)$ we found. Note that $R(x)$ passes through the origin, indicating $R(0) = 0$. Also, reading both graphs from left to right, we see that $R(x)$ increases when $\overline{MR} > 0$, attains its maximum when $\overline{MR} = 0$, and decreases when $\overline{MR} < 0$.

---

*Here we are using K rather than C to represent the constant of integration to avoid confusion between the constant C and the cost function $C = C(x)$.

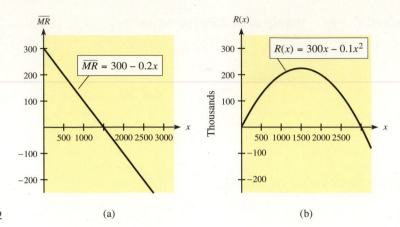

Figure 12.2   (a)   (b)

Calculator Note We mentioned in Chapter 9, "Derivatives," that graphing calculators have a numerical derivative feature that can be used to check graphically the derivative of a function that has been calculated with a formula. We can also use the numerical integration feature on graphing calculators to check our integration (if we assume temporarily that the constant of integration is 0). We do this by graphing the integral calculated with a formula and the numerical integral from the graphing calculator on the same set of axes. If the graphs lie on top of one another, the integrals agree. See Appendix C, Section 12.1, for details. Figure 12.3 illustrates this for the function $f(x) = 3x^2 - 2x + 1$. Its integral, with the constant of integration set equal to 0, is shown as $y_1 = x^3 - x^2 + x$ in Figure 12.3(a), and the graphs of the two integrals are shown in Figure 12.3(b). Of course, it is often easier to use the derivative to check integration. ∎

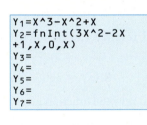

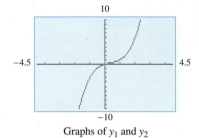

Graphs of y_1 and y_2

Figure 12.3   (a)   (b)

CHECKPOINT SOLUTIONS

1. (a) False, $\int (4x^3 - 2x)\,dx = \int 4x^3\,dx - \int 2x\,dx = (x^4 + C_1) - (x^2 + C_2)$
$$= x^4 - x^2 + C$$

(b) False, $\displaystyle\int \frac{1}{3x^2}\,dx = \int \frac{1}{3} \cdot \frac{1}{x^2}\,dx = \frac{1}{3}\int x^{-2}\,dx$
$$= \frac{1}{3} \cdot \frac{x^{-1}}{-1} + C = \frac{-1}{3x} + C$$

2. $\displaystyle\int (2x^3 + x^{-1/2} - 4x^{-5})\,dx = \frac{2x^4}{4} + \frac{x^{1/2}}{1/2} - \frac{4x^{-4}}{-4} + C$
$$= \frac{x^4}{2} + 2x^{1/2} + x^{-4} + C$$

| EXERCISES | **12.1**

Access end-of-section exercises online at **www.webassign.net**

OBJECTIVE **12.2**

- To evaluate integrals of the form $\int u^n \cdot u'\, dx = \int u^n du$ if $n \neq -1$

The Power Rule

| APPLICATION PREVIEW |

In the previous section, we saw that total revenue could be found by integrating marginal revenue. That is,

$$R(x) = \int \overline{MR}\, dx$$

For example, if the marginal revenue for a product is given by

$$\overline{MR} = \frac{600}{\sqrt{3x+1}} + 2$$

then

$$R(x) = \int \left[\frac{600}{\sqrt{3x+1}} + 2 \right] dx$$

To evaluate this integral, however, we need a more general formula than the Powers of x Formula. (See Example 8.)

In this section, we will extend the Powers of x Formula to a rule for powers of a function of x.

Differentials   Our goal in this section is to extend the Powers of x Formula,

$$\int x^n dx = \frac{x^{n+1}}{n+1} + C, \quad n \neq -1$$

to powers of a function of x. In order to do this, we must understand the symbol dx.

Recall from Section 9.3, "Rates of Change and Derivatives," that the derivative of $y = f(x)$ with respect to x can be denoted by dy/dx. As we will see, there are advantages to using dy and dx as separate quantities whose ratio dy/dx equals $f'(x)$.

Differentials   If $y = f(x)$ is a differentiable function with derivative $dy/dx = f'(x)$, then the **differential of x** is dx, and the **differential of y** is dy, where

$$dy = f'(x)\, dx$$

Although differentials are useful in certain approximation problems, we are interested in the differential notation at this time.

EXAMPLE **1**   **Differentials**

Find (a) dy if $y = x^3 - 4x^2 + 5$ and (b) du if $u = 5x^4 + 11$.

Solution

(a) $dy = f'(x)\, dx = (3x^2 - 8x)\, dx$

(b) If the dependent variable in a function is u, then $du = u'(x)\, dx$.

$$du = u'(x)\, dx = 20x^3\, dx$$

The Power Rule   In terms of our goal of extending the Powers of x Formula, we would suspect that if x is replaced by a function of x, then dx should be replaced by the differential of that function. Let's see whether this is true.

Recall that if $y = [u(x)]^n$, the derivative of y is

$$\frac{dy}{dx} = n[u(x)]^{n-1} \cdot u'(x)$$

Using this formula for derivatives, we can see that

$$\int n[u(x)]^{n-1} \cdot u'(x)\, dx = [u(x)]^n + C$$

It is easy to see that this formula is equivalent to the following formula, which is called the **Power Rule for Integration.**

Power Rule for Integration

$$\int [u(x)]^n \cdot u'(x)\, dx = \frac{[u(x)]^{n+1}}{n+1} + C, \quad n \neq -1$$

Using the fact that

$$du = u'(x)\, dx \quad \text{or} \quad du = u'\, dx$$

we can write the Power Rule in the following alternative form.

Power Rule (Alternative Form)

If $u = u(x)$, then

$$\int u^n\, du = \frac{u^{n+1}}{n+1} + C, \quad n \neq -1$$

Note that this formula has the same form as the formula

$$\int x^n\, dx = \frac{x^{n+1}}{n+1} + C, \quad n \neq -1$$

with the *function u substituted for x and du substituted for dx.*

EXAMPLE **2**   **Power Rule**

Evaluate $\int (3x^2 + 4)^5 \cdot 6x\, dx$.

Solution

To use the Power Rule, we must be sure that we have the function $u(x)$, its derivative $u'(x)$, and n.

$$u = 3x^2 + 4, \quad n = 5$$
$$u' = 6x$$

All required parts are present, so the integral is of the form

$$\int (3x^2 + 4)^5 \, 6x \, dx = \int u^5 \cdot u' \, dx = \int u^5 \, du$$

$$= \frac{u^6}{6} + C = \frac{(3x^2 + 4)^6}{6} + C$$

We can check the integration by noting that the derivative of

$$\frac{(3x^2 + 4)^6}{6} + C \quad \text{is} \quad (3x^2 + 4)^5 \cdot 6x$$

■

EXAMPLE 3   Power Rule

Evaluate $\int \sqrt{2x + 3} \cdot 2 \, dx$.

Solution

If we let $u = 2x + 3$, then $u' = 2$, and so we have

$$\int \sqrt{2x + 3} \cdot 2 \, dx = \int \sqrt{u} \, u' \, dx = \int \sqrt{u} \, du$$

$$= \int u^{1/2} \, du = \frac{u^{3/2}}{3/2} + C$$

Because $u = 2x + 3$, we have

$$\int \sqrt{2x + 3} \cdot 2 \, dx = \frac{2}{3}(2x + 3)^{3/2} + C$$

Check: The derivative of $\frac{2}{3}(2x + 3)^{3/2} + C$ is $(2x + 3)^{1/2} \cdot 2$. ✔

■

Some members of the family of functions given by

$$\int \sqrt{2x + 3} \cdot 2 \, dx = \frac{2}{3}(2x + 3)^{3/2} + C$$

are shown in Figure 12.4. Note from the graphs that the domain of each function is $x \geq -3/2$. This is because $2x + 3$ must be nonnegative so that $(2x + 3)^{3/2} = (\sqrt{2x + 3})^3$ is a real number.

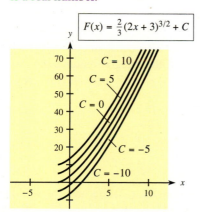

Figure 12.4

EXAMPLE 4 **Power Rule**

Evaluate $\int x^3(5x^4 + 11)^9 \, dx$.

Solution

If we let $u = 5x^4 + 11$, then $u' = 20x^3$. Thus we do not have an integral of the form $\int u^n \cdot u' \, dx$, as we had in Example 2 and Example 3; the factor 20 is not in the integrand. To get the integrand in the correct form, we can multiply by 20 and divide it out as follows:

$$\int x^3(5x^4 + 11)^9 \, dx = \int (5x^4 + 11)^9 \cdot x^3 \, dx = \int (5x^4 + 11)^9 \left(\frac{1}{20}\right)(20x^3) \, dx$$

Because $\frac{1}{20}$ is a constant factor, we can factor it outside the integral sign, getting

$$\int (5x^4 + 11)^9 \cdot x^3 \, dx = \frac{1}{20}\int (5x^4 + 11)^9(20x^3) \, dx$$

$$= \frac{1}{20}\int u^9 \cdot u' \, dx = \frac{1}{20} \cdot \frac{u^{10}}{10} + C$$

$$= \frac{1}{200}(5x^4 + 11)^{10} + C \qquad \blacksquare$$

EXAMPLE 5 **Power Rule**

Evaluate $\int 5x^2\sqrt{x^3 - 4} \, dx$.

Solution

If we let $u = x^3 - 4$, then $u' = 3x^2$. Thus we need the factor 3, rather than 5, in the integrand. If we first reorder the factors and then multiply by the constant factor 3 (and divide it out), we have

$$\int \sqrt{x^3 - 4} \cdot 5x^2 \, dx = \int \sqrt{x^3 - 4} \cdot \frac{5}{3}(3x^2) \, dx$$

$$= \frac{5}{3}\int (x^3 - 4)^{1/2} \cdot 3x^2 \, dx$$

This integral is of the form $\frac{5}{3}\int u^{1/2} \cdot u' \, dx$, resulting in

$$\frac{5}{3} \cdot \frac{u^{3/2}}{3/2} + C = \frac{5}{3} \cdot \frac{(x^3 - 4)^{3/2}}{3/2} + C = \frac{10}{9}(x^3 - 4)^{3/2} + C \qquad \blacksquare$$

Note that we can factor a constant outside the integral sign to obtain the integrand in the form we seek. But if the integral requires the introduction of a variable to obtain the form $u^n \cdot u' \, dx$, we *cannot* use this form and must try something else.

EXAMPLE 6 **Power Rule Fails**

Evaluate $\int (x^2 + 4)^2 \, dx$.

Solution

If we let $u = x^2 + 4$, then $u' = 2x$. Because we would have to introduce a variable to get u' in the integral, we cannot solve this problem by using the Power Rule. We must find another method. We can evaluate this integral by squaring and then integrating term by term.

$$\int (x^2 + 4)^2 \, dx = \int (x^4 + 8x^2 + 16) \, dx$$

$$= \frac{x^5}{5} + \frac{8x^3}{3} + 16x + C \qquad \blacksquare$$

Note that if we tried to introduce the factor $2x$ into the integral in Example 6, we would get

$$\int (x^2 + 4)^2 \, dx = \int (x^2 + 4)^2 \cdot \frac{1}{2x}(2x) \, dx$$

Although it is tempting to factor $1/2x$ outside the integral and use the Power Rule, this leads to an "answer" that does not check. That is, the derivative of the "answer" is not the integrand. (Try it and see.) To emphasize again, we can introduce only *a constant factor* to get an integral in the proper form.

EXAMPLE 7   Power Rule

Evaluate:

(a) $\displaystyle\int (2x^2 - 4x)^2 (x - 1) \, dx$ (b) $\displaystyle\int \frac{x^2 - 1}{(x^3 - 3x)^3} \, dx$

Solution

(a) If we want to treat this as an integral of the form $\int u^n u' \, dx$, we will have to let $u = 2x^2 - 4x$. Then $u' = 4x - 4$. Multiplying and dividing by 4 will give us this form.

$$\int (2x^2 - 4x)^2 (x - 1) \, dx = \int (2x^2 - 4x)^2 \cdot \frac{1}{4} \cdot 4(x - 1) \, dx$$

$$= \frac{1}{4}\int (2x^2 - 4x)^2 (4x - 4) \, dx$$

$$= \frac{1}{4}\int u^2 u' \, dx = \frac{1}{4} \cdot \frac{u^3}{3} + C$$

$$= \frac{1}{4} \frac{(2x^2 - 4x)^3}{3} + C$$

$$= \frac{1}{12}(2x^2 - 4x)^3 + C$$

(b) This integral can be treated as $\int u^{-3} u' \, dx$ if we let $u = x^3 - 3x$.

$$\int \frac{x^2 - 1}{(x^3 - 3x)^3} \, dx = \int (x^3 - 3x)^{-3}(x^2 - 1) \, dx$$

Then $u' = 3x^2 - 3$ and we can multiply and divide by 3 to get the form we need.

$$= \int (x^3 - 3x)^{-3} \cdot \frac{1}{3} \cdot 3(x^2 - 1) \, dx$$

$$= \frac{1}{3}\int (x^3 - 3x)^{-3}(3x^2 - 3) \, dx$$

$$= \frac{1}{3}\left[\frac{(x^3 - 3x)^{-2}}{-2} \right] + C$$

$$= \frac{-1}{6(x^3 - 3x)^2} + C$$

CHECKPOINT

1. Which of the following can be evaluated with the Power Rule?
    (a) $\int (4x^2 + 1)^{10}(8x \, dx)$
    (b) $\int (4x^2 + 1)^{10}(x \, dx)$
    (c) $\int (4x^2 + 1)^{10}(8 \, dx)$
    (d) $\int (4x^2 + 1)^{10} \, dx$

2. Which of the following is equal to $\int (2x^3 + 5)^{-2}(6x^2\,dx)$?

(a) $\dfrac{[(2x^4)/4 + 5x]^{-1}}{-1} \cdot \dfrac{6x^3}{3} + C$

(b) $\dfrac{(2x^3 + 5)^{-1}}{-1} \cdot \dfrac{6x^3}{3} + C$

(c) $\dfrac{(2x^3 + 5)^{-1}}{-1} + C$

3. True or false: Constants can be factored outside the integral sign.
4. Evaluate the following.

(a) $\int (x^3 + 9)^5(3x^2\,dx)$

(b) $\int (x^3 + 9)^{15}(x^2\,dx)$

(c) $\int (x^3 + 9)^2(x\,dx)$

EXAMPLE | **8** | **Revenue** | APPLICATION PREVIEW |

Suppose that the marginal revenue for a product is given by

$$\overline{MR} = \frac{600}{\sqrt{3x + 1}} + 2$$

Find the total revenue function.

Solution

$$R(x) = \int \overline{MR}\,dx = \int \left[\frac{600}{(3x + 1)^{1/2}} + 2 \right] dx$$

$$= \int 600(3x + 1)^{-1/2}\,dx + \int 2\,dx$$

$$= 600 \left(\frac{1}{3} \right) \int (3x + 1)^{-1/2}(3\,dx) + 2 \int dx$$

$$= 200 \frac{(3x + 1)^{1/2}}{1/2} + 2x + K$$

$$= 400\sqrt{3x + 1} + 2x + K$$

We know that $R(0) = 0$, so we have

$$0 = 400\sqrt{1} + 0 + K \quad \text{or} \quad K = -400$$

Thus the total revenue function is

$$R(x) = 400\sqrt{3x + 1} + 2x - 400 \qquad \blacksquare$$

Note in Example 8 that even though $R(0) = 0$, the constant of integration K was *not* 0. This is because $x = 0$ does not necessarily mean that $u(x)$ will also be 0.

The formulas we have stated and used in this and the previous section are all the result of "reversing" derivative formulas. We summarize in the following box.

Integration Formula		**Derivative Formula**
$\displaystyle\int dx = x + C$	because	$\dfrac{d}{dx}(x + C) = 1$
$\displaystyle\int x^n \, dx = \dfrac{x^{n+1}}{n+1} + C \ \ (n \neq -1)$	because	$\dfrac{d}{dx}\left(\dfrac{x^{n+1}}{n+1} + C\right) = x^n$
$\displaystyle\int [u(x) + v(x)] \, dx = \int u(x) \, dx + \int v(x) \, dx$	because	$\dfrac{d}{dx}[u(x) + v(x)] = \dfrac{du}{dx} + \dfrac{dv}{dx}$
$\displaystyle\int [u(x) - v(x)] \, dx = \int u(x) \, dx - \int v(x) \, dx$	because	$\dfrac{d}{dx}[u(x) - v(x)] = \dfrac{du}{dx} - \dfrac{dv}{dx}$
$\displaystyle\int cf(x) \, dx = c\int f(x) \, dx$	because	$\dfrac{d}{dx}[cf(x)] = c\left(\dfrac{df}{dx}\right)$
$\displaystyle\int u^n u' \, dx = \dfrac{u^{n+1}}{n+1} + C \ \ (n \neq -1)$	because	$\dfrac{d}{dx}\left(\dfrac{u^{n+1}}{n+1} + C\right) = u^n \cdot u'$

Note that there are no integration formulas that correspond to "reversing" the derivative formulas for a product or for a quotient. This means that functions that may be easy to differentiate can be quite difficult (or even impossible) to integrate. Hence, in general, integration is a more difficult process than differentiation. In fact, some functions whose derivatives can be readily found cannot be integrated, such as

$$f(x) = \sqrt{x^3 + 1} \quad \text{and} \quad g(x) = \frac{2x^2}{x^4 + 1}$$

In the next section, we will add to this list of basic integration formulas by "reversing" the derivative formulas for exponential and logarithmic functions.

CHECKPOINT SOLUTIONS

1. Expressions (a) and (b) can be evaluated with the Power Rule. For expression (a), if we let $u = 4x^2 + 1$, then $u' = 8x$ so that the integral becomes

$$\int u^{10} u' \, dx = \int u^{10} \, du$$

For expression (b), we let $u = 4x^2 + 1$ again. Then $u' = 8x$ and the integral becomes

$$\int (4x^2 + 1)^{10} \cdot x \, dx = \int (4x^2 + 1)^{10} \cdot \frac{1}{8}(8x) \, dx = \frac{1}{8}\int u^{10} u' \, dx = \frac{1}{8}\int u^{10} \, du$$

Expressions (c) and (d) do not fit the format of the Power Rule, because neither integral has an x with the dx outside the power (so they need to be multiplied out before integrating).

2. $u = 2x^3 + 5$, so $u' = 6x^2$

$$\int (2x^3 + 5)^{-2}(6x^2) \, dx = \int u^{-2}u' \, dx = \int u^{-2} \, du = u^{-1}/(-1) + C$$

$$= -u^{-1} + C = -(2x^3 + 5)^{-1} + C$$

Thus expression (c) is the correct choice.

3. True

4. (a) $\int (x^3 + 9)^5 (3x^2\, dx) = \int u^5 u'\, dx = \dfrac{u^6}{6} + C = \dfrac{(x^3 + 9)^6}{6} + C$

(b) $\int (x^3 + 9)^{15} (x^2\, dx) = \dfrac{1}{3} \int (x^3 + 9)^{15} (3x^2\, dx)$

$$= \dfrac{1}{3} \cdot \dfrac{(x^3 + 9)^{16}}{16} + C = \dfrac{(x^3 + 9)^{16}}{48} + C$$

(c) The Power Rule does not fit, so we expand the integrand.

$$\int (x^3 + 9)^2 (x\, dx) = \int (x^6 + 18x^3 + 81)x\, dx$$

$$= \int (x^7 + 18x^4 + 81x)\, dx$$

$$= \dfrac{x^8}{8} + \dfrac{18x^5}{5} + \dfrac{81x^2}{2} + C$$

| EXERCISES | 12.2

Access end-of-section exercises online at **www.webassign.net**   **WebAssign**

OBJECTIVES 12.3

- To evaluate integrals of the form $\int e^u u'\, dx$ or, equivalently, $\int e^u\, du$
- To evaluate integrals of the form $\int \dfrac{u'}{u}\, dx$ or, equivalently, $\int \dfrac{1}{u}\, du$

Integrals Involving Exponential and Logarithmic Functions

| APPLICATION PREVIEW |

The rate of growth of the market value of a home has typically exceeded the inflation rate. Suppose, for example, that the real estate market has an average annual inflation rate of 8%. Then the rate of change of the value of a house that cost $200,000 can be modeled by

$$\dfrac{dV}{dt} = 15.4e^{0.077t}$$

where V is the value of the home in thousands of dollars and t is the time in years since the home was purchased. To find the market value of such a home 10 years after it was purchased, we would first have to integrate dV/dt. That is, we must be able to integrate an exponential. (See Example 3.)

In this section, we consider integration formulas that result in natural logarithms and formulas for integrating exponentials.

Integrals Involving Exponential Functions

We know that

$$\dfrac{d}{dx}(e^x) = e^x \qquad \text{and} \qquad \dfrac{d}{dx}(e^u) = e^u \cdot u'$$

The corresponding integrals are given by the following.

Exponential Formula

If u is a function of x,

$$\int e^u \cdot u'\, dx = \int e^u\, du = e^u + C$$

In particular, $\int e^x\, dx = e^x + C$.

EXAMPLE 1 **Integral of an Exponential**

Evaluate $\int 5e^x \, dx$.

Solution

$\int 5e^x \, dx = 5\int e^x \, dx = 5e^x + C$

■

EXAMPLE 2 **Integral of $e^u \, du$**

Evaluate: (a) $\int 2xe^{x^2} \, dx$ (b) $\int \dfrac{x^2 \, dx}{e^{x^3}}$

Solution

(a) Letting $u = x^2$ implies that $u' = 2x$, and the integral is of the form $\int e^u \cdot u' \, dx$. Thus

$$\int 2xe^{x^2} \, dx = \int e^{x^2}(2x) \, dx = \int e^u \cdot u' \, dx = e^u + C = e^{x^2} + C$$

(b) In order to use $\int e^u \cdot u' \, dx$, we write the exponential in the numerator. Thus

$$\int \dfrac{x^2 \, dx}{e^{x^3}} = \int e^{-x^3}(x^2 \, dx)$$

This is *almost* of the form $\int e^u \cdot u' \, dx$. Letting $u = -x^3$ gives $u' = -3x^2$. Thus

$$\int e^{-x^3}(x^2 \, dx) = -\frac{1}{3}\int e^{-x^3}(-3x^2 \, dx) = -\frac{1}{3}e^{-x^3} + C = \dfrac{-1}{3e^{x^3}} + C$$

■

CHECKPOINT

1. True or false:

(a) $\int e^{x^2}(2x \, dx) = e^{x^2} \cdot x^2 + C$ (b) $\int e^{-3x} \, dx = -\frac{1}{3}e^{-3x} + C$

(c) $\int \dfrac{dx}{e^{3x}} = \frac{1}{3}\left(\dfrac{1}{e^{3x}}\right) + C$ (d) $\int e^{3x+1}(3 \, dx) = \dfrac{e^{3x+2}}{3x+2} + C$

EXAMPLE 3 **Real Estate Inflation** | **APPLICATION PREVIEW** |

As the housing market emerges from the collapse of 2008, suppose the rate of change of the value of a house that cost \$200,000 in 2011 can be modeled by

$$\frac{dV}{dt} = 15.4e^{0.077t}$$

where V is the market value of the home in thousands of dollars and t is the time in years since 2011.

(a) Find the function that expresses the value V in terms of t.
(b) Find the predicted value in 2021 (after 10 years).

Solution

(a) $V = \int \dfrac{dV}{dt} \, dt = \int 15.4e^{0.077t} \, dt$

$$V = 15.4\int e^{0.077t}\left(\dfrac{1}{0.077}\right)(0.077 \, dt)$$

$$V = 15.4\left(\dfrac{1}{0.077}\right)\int e^{0.077t}(0.077 \, dt)$$

$$V = 200e^{0.077t} + C$$

Using $V = 200$ (thousand) when $t = 0$, we have

$$200 = 200 + C$$
$$0 = C$$

Thus we have the value as a function of time given by

$$V = 200e^{0.077t}$$

(b) The value after 10 years is found by using $t = 10$.

$$V = 200e^{0.077(10)} = 200e^{0.77} \approx 431.95$$

Thus, in 2021, the predicted value of the home is $431,950. ∎

EXAMPLE 4   Graphs of Functions and Integrals

Figure 12.5 shows the graphs of $g(x) = 5e^{-x^2}$ and $h(x) = -10xe^{-x^2}$. One of these functions is $f(x)$ and the other is $\int f(x)\,dx$ with $C = 0$.

(a) Decide which of $g(x)$ and $h(x)$ is $f(x)$ and which is $\int f(x)\,dx$.
(b) How can the graph of $f(x)$ be used to locate and classify the extrema of $\int f(x)\,dx$?
(c) What feature of the graph of $f(x)$ occurs at the same x-values as the inflection points of the graph of $\int f(x)\,dx$?

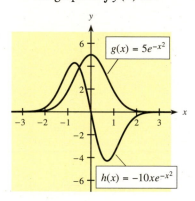

Figure 12.5

Solution

(a) The graph of $h(x)$ looks like the graph of $g'(x)$ because $h(x) > 0$ where $g(x)$ is increasing, $h(x) < 0$ where $g(x)$ is decreasing, and $h(x) = 0$ where $g(x)$ has its maximum. However, if $h(x) = g'(x)$, then, equivalently,

$$\int h(x)\,dx = \int g'(x)\,dx$$
$$= g(x) + C$$

so $h(x) = f(x)$ and $g(x) = \int f(x)\,dx$. We can verify this by noting that

$$\int -10xe^{-x^2}\,dx = 5\int e^{-x^2}(-2x\,dx)$$
$$= 5e^{-x^2} + C$$

(b) We know that $f(x)$ is the derivative of $\int f(x)\,dx$, so, as we saw in part (a), the x-intercepts of $f(x)$ locate the critical values and extrema of $\int f(x)\,dx$.

(c) The first derivative of $\int f(x)\,dx$ is $f(x)$, and its second derivative is $f'(x)$. Hence the inflection points of $\int f(x)\,dx$ occur where $f'(x) = 0$. But $f(x)$ has its extrema where $f'(x) = 0$. Thus the extrema of $f(x)$ occur at the same x-values as the inflection points of $\int f(x)\,dx$. ∎

Integrals Involving Logarithmic Functions

Recall that the Power Rule for integrals applies only if $n \neq -1$. That is,

$$\int u^n u' \, dx = \frac{u^{n+1}}{n+1} + C \quad \text{if } n \neq -1$$

The following formula applies when $n = -1$.

Logarithmic Formula

If u is a function of x, then

$$\int u^{-1} u' \, dx = \int \frac{u'}{u} \, dx = \int \frac{1}{u} \, du = \ln |u| + C$$

In particular, $\displaystyle\int \frac{1}{x} \, dx = \ln |x| + C.$

We use the absolute value of u in the integral because the logarithm is defined only when the quantity is positive. This logarithmic formula is a direct result of the fact that

$$\frac{d}{dx} (\ln |u|) = \frac{1}{u} \cdot u'$$

We can see this result by considering the following.

For $u > 0$: $\quad \dfrac{d}{dx} (\ln |u|) = \dfrac{d}{dx} (\ln u) = \dfrac{1}{u} \cdot u'$

For $u < 0$: $\quad \dfrac{d}{dx} (\ln |u|) = \dfrac{d}{dx} [\ln (-u)] = \dfrac{1}{(-u)} \cdot (-u') = \dfrac{1}{u} \cdot u'$

In addition to this verification, we can graphically illustrate the need for the absolute value sign. Figure 12.6(a) shows that $f(x) = 1/x$ is defined for $x \neq 0$, and from Figures 12.6(b) and 12.6(c), we see that $F(x) = \int 1/x \, dx = \ln |x|$ is also defined for $x \neq 0$, but $y = \ln x$ is defined only for $x > 0$.

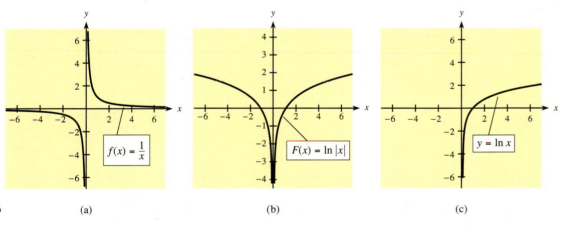

Figure 12.6   (a)   (b)   (c)

EXAMPLE 5   **Integral Resulting in a Logarithmic Function**

Evaluate $\displaystyle\int \frac{4}{4x+8} \, dx.$

Solution
This integral is of the form

$$\int \frac{u'}{u} \, dx = \ln |u| + C$$

with $u = 4x + 8$ and $u' = 4$. Thus

$$\int \frac{4}{4x + 8}\, dx = \ln |4x + 8| + C \qquad \blacksquare$$

Figure 12.7 shows several members of the family

$$F(x) = \int \frac{4\, dx}{4x + 8} = \ln |4x + 8| + C$$

We can choose different values for C and use a graphing utility to graph families of curves such as those in Figure 12.7.

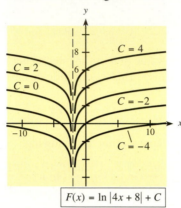

Figure 12.7

$F(x) = \ln |4x + 8| + C$

EXAMPLE **6** **Integral of** *du/u*

Evaluate $\displaystyle\int \frac{x - 3}{x^2 - 6x + 1}\, dx$.

Solution

This integral is of the form $\int (u'/u)\, dx$, *almost*. If we let $u = x^2 - 6x + 1$, then $u' = 2x - 6$. If we multiply (and divide) the numerator by 2, we get

$$\int \frac{x - 3}{x^2 - 6x + 1}\, dx = \frac{1}{2}\int \frac{2(x - 3)}{x^2 - 6x + 1}\, dx$$

$$= \frac{1}{2}\int \frac{2x - 6}{x^2 - 6x + 1}\, dx$$

$$= \frac{1}{2}\int \frac{u'}{u}\, dx = \frac{1}{2}\ln |u| + C$$

$$= \frac{1}{2}\ln |x^2 - 6x + 1| + C \qquad \blacksquare$$

EXAMPLE **7** **Population Growth**

Because the world contains only about 10 billion acres of arable land, world population is limited. Suppose that world population is limited to 40 billion people and that the rate of population growth per year is given by

$$\frac{dP}{dt} = k(40 - P)$$

where P is the population in billions at time t and k is a positive constant. Then the relationship between the year and the population during that year is given by the integral

$$t = \frac{1}{k}\int \frac{1}{40 - P}\, dP$$

where $40 - P > 0$ because 40 billion is the population's upper limit.

(a) Evaluate this integral to find the relationship.

(b) Use properties of logarithms and exponential functions to write P as a function of t.

Solution

(a) $t = \dfrac{1}{k}\displaystyle\int \dfrac{1}{40-P}\,dP = -\dfrac{1}{k}\displaystyle\int \dfrac{-dP}{40-P} = -\dfrac{1}{k}\ln|40-P| + C_1$

(b) $t = -\dfrac{1}{k}\ln(40-P) + C_1$ because $40 - P > 0$ means $|40 - P| = 40 - P$.

Solving this equation for P requires converting to exponential form.

$$-k(t - C_1) = \ln(40 - P)$$
$$e^{C_1 k - kt} = 40 - P$$
$$e^{C_1 k} \cdot e^{-kt} = 40 - P$$

Because $e^{C_1 k}$ is an unknown constant, we replace it with C and solve for P.

$$P = 40 - Ce^{-kt}$$ ■

If an integral contains a fraction in which the degree of the numerator is equal to or greater than that of the denominator, we should divide the denominator into the numerator as a first step.

EXAMPLE 8   **Integral Requiring Division**

Evaluate $\displaystyle\int \dfrac{x^4 - 2x^3 + 4x^2 - 7x - 1}{x^2 - 2x}\,dx$.

Solution

Because the numerator is of higher degree than the denominator, we begin by dividing $x^2 - 2x$ into the numerator.

$$
\begin{array}{r}
x^2 + 4 \\
x^2 - 2x \overline{)\,x^4 - 2x^3 + 4x^2 - 7x - 1} \\
\underline{x^4 - 2x^3 } \\
4x^2 - 7x - 1 \\
\underline{4x^2 - 8x } \\
x - 1
\end{array}
$$

Thus

$$\int \frac{x^4 - 2x^3 + 4x^2 - 7x - 1}{x^2 - 2x}\,dx = \int \left(x^2 + 4 + \frac{x-1}{x^2 - 2x}\right) dx$$

$$= \int (x^2 + 4)\,dx + \frac{1}{2}\int \frac{2(x-1)\,dx}{x^2 - 2x}$$

$$= \frac{x^3}{3} + 4x + \frac{1}{2}\ln|x^2 - 2x| + C$$ ■

CHECKPOINT

2. True or false:

(a) $\displaystyle\int \dfrac{3x^2\,dx}{x^3 + 4} = \ln|x^3 + 4| + C$ (b) $\displaystyle\int \dfrac{2x\,dx}{\sqrt{x^2 + 1}} = \ln|\sqrt{x^2 + 1}| + C$

(c) $\displaystyle\int \dfrac{2}{x}\,dx = 2\ln|x| + C$ (d) $\displaystyle\int \dfrac{x}{x+1}\,dx = x\displaystyle\int \dfrac{1}{x+1}\,dx = x\ln|x+1| + C$

(e) To evaluate $\displaystyle\int \dfrac{4x}{4x+1}\,dx$, our first step is to divide $4x + 1$ into $4x$.

3. (a) Divide $4x + 1$ into $4x$.   (b) Evaluate $\displaystyle\int \dfrac{4x}{4x+1}\,dx$.

CHECKPOINT SOLUTIONS

1. (a) False. The correct solution is $e^{x^2} + C$ (see Example 2a).
   (b) True
   (c) False; $\displaystyle\int \frac{dx}{e^{3x}} = \int e^{-3x}\, dx = -\frac{1}{3}e^{-3x} + C = \frac{-1}{3e^{3x}} + C$
   (d) False; $\displaystyle\int e^{3x+1}(3\, dx) = e^{3x+1} + C$

2. (a) True
   (b) False; $\displaystyle\int \frac{2x\, dx}{\sqrt{x^2 + 1}} = \int (x^2 + 1)^{-1/2}(2x\, dx)$

   $$= \frac{(x^2 + 1)^{1/2}}{1/2} + C = 2(x^2 + 1)^{1/2} + C$$

   (c) True
   (d) False. We cannot factor the variable x outside the integral sign.
   (e) True

3. (a) $4x + 1 \overline{)\,4x}$ with $\dfrac{}{4x + 1}$, $\dfrac{}{-1}$ so $\dfrac{4x}{4x + 1} = 1 - \dfrac{1}{4x + 1}$

   (b) $\displaystyle\int \frac{4x\, dx}{4x + 1} = \int \left(1 - \frac{1}{4x + 1}\right) dx = x - \frac{1}{4}\ln|4x + 1| + C$

| EXERCISES | 12.3

Access end-of-section exercises online at **www.webassign.net**   ENHANCED **WebAssign**

OBJECTIVES 12.4

- To use integration to find total cost functions from information involving marginal cost
- To optimize profit, given information regarding marginal cost and marginal revenue
- To use integration to find national consumption functions from information about marginal propensity to consume and marginal propensity to save

Applications of the Indefinite Integral in Business and Economics

| APPLICATION PREVIEW |

If we know that the consumption of a nation is $9 billion when income is $0 and the marginal propensity to save is 0.25, we can easily find the marginal propensity to consume and use integration to find the national consumption function. (See Example 6.)

In this section, we also use integration to derive total cost and profit functions from the marginal cost and marginal revenue functions. One of the reasons for the marginal approach in economics is that firms can observe marginal changes in real life. If they know the marginal cost and the total cost when a given quantity is sold, they can develop their total cost function.

Total Cost and Profit

We know that the marginal cost for a commodity is the derivative of the total cost function—that is, $\overline{MC} = C'(x)$, where $C(x)$ is the total cost function. Thus if we have the marginal cost function, we can integrate (or "reverse" the process of differentiation) to find the total cost. That is, $C(x) = \int \overline{MC}\, dx$.

If, for example, the marginal cost is $\overline{MC} = 4x + 3$, the total cost is given by

$$C(x) = \int \overline{MC}\, dx$$

$$= \int (4x + 3)\, dx$$

$$= 2x^2 + 3x + K$$

where K represents the constant of integration. We know that the total revenue is 0 if no items are produced, but the total cost may not be 0 if nothing is produced. The fixed costs accrue whether goods are produced or not. Thus the value of the constant of integration depends on the fixed costs FC of production.

Thus we cannot determine the total cost function from the marginal cost unless additional information is available to help us determine the fixed costs.

EXAMPLE **1**   **Total Cost**

Suppose the marginal cost function for a month for a certain product is $\overline{MC} = 3x + 50$, where x is the number of units and cost is in dollars. If the fixed costs related to the product amount to \$10,000 per month, find the total cost function for the month.

Solution

The total cost function is

$$C(x) = \int (3x + 50)\, dx$$

$$= \frac{3x^2}{2} + 50x + K$$

The constant of integration K is found by using the fact that $C(0) = FC = 10{,}000$. Thus

$$3(0)^2 + 50(0) + K = 10{,}000, \quad \text{so } K = 10{,}000$$

and the total cost for the month is given by

$$C(x) = \frac{3x^2}{2} + 50x + 10{,}000 \qquad \blacksquare$$

EXAMPLE **2**   **Cost**

Suppose monthly records show that the rate of change of the cost (that is, the marginal cost) for a product is $\overline{MC} = 3(2x + 25)^{1/2}$, where x is the number of units and cost is in dollars. If the fixed costs for the month are \$11,125, what would be the total cost of producing 300 items per month?

Solution

We can integrate the marginal cost to find the total cost function.

$$C(x) = \int \overline{MC}\, dx = \int 3(2x + 25)^{1/2}\, dx$$

$$= 3 \cdot \left(\frac{1}{2}\right) \int (2x + 25)^{1/2}(2\, dx)$$

$$= \left(\frac{3}{2}\right) \frac{(2x + 25)^{3/2}}{3/2} + K$$

$$= (2x + 25)^{3/2} + K$$

We can find K by using the fact that fixed costs are \$11,125.

$$C(0) = 11{,}125 = (25)^{3/2} + K$$
$$11{,}125 = 125 + K, \quad \text{or} \quad K = 11{,}000$$

Thus the total cost function is

$$C(x) = (2x + 25)^{3/2} + 11{,}000$$

and the cost of producing 300 items per month is

$$C(300) = (625)^{3/2} + 11{,}000$$
$$= 26{,}625 \quad \text{(dollars)} \qquad \blacksquare$$

It can be shown that the profit is usually maximized when $\overline{MR} = \overline{MC}$. To see that this does not always give us a maximum *positive* profit, consider the following facts concerning the manufacture of widgets over the period of a month.

1. The marginal revenue is $\overline{MR} = 400 - 30x$.
2. The marginal cost is $\overline{MC} = 20x + 50$.
3. When 5 widgets are produced and sold, the total cost is $1750. The profit *should* be maximized when $\overline{MR} = \overline{MC}$, or when $400 - 30x = 20x + 50$. Solving for x gives $x = 7$. To see whether our profit is maximized when 7 units are produced and sold, let us examine the profit function.

The profit function is given by $P(x) = R(x) - C(x)$, where

$$R(x) = \int \overline{MR}\, dx \quad \text{and} \quad C(x) = \int \overline{MC}\, dx$$

Integrating, we get

$$R(x) = \int (400 - 30x)\, dx = 400x - 15x^2 + K$$

but $R(0) = 0$ gives $K = 0$ for this total revenue function, so

$$R(x) = 400x - 15x^2$$

The total cost function is

$$C(x) = \int (20x + 50)\, dx = 10x^2 + 50x + K$$

The value of fixed cost can be determined by using the fact that 5 widgets cost $1750. This tells us that $C(5) = 1750 = 250 + 250 + K$, so $K = 1250$.

Thus the total cost is $C(x) = 10x^2 + 50x + 1250$. Thus, the profit is

$$P(x) = R(x) - C(x) = (400x - 15x^2) - (10x^2 + 50x + 1250)$$

Simplifying gives

$$P(x) = 350x - 25x^2 - 1250$$

We have found that $\overline{MR} = \overline{MC}$ if $x = 7$, and the graph of $P(x)$ is a parabola that opens downward, so profit is maximized at $x = 7$. But if $x = 7$, profit is

$$P(7) = 2450 - 1225 - 1250 = -25$$

That is, the production and sale of 7 items result in a loss of $25.

The preceding discussion indicates that even though setting $\overline{MR} = \overline{MC}$ may optimize profit, it does not indicate the level of profit or loss, as forming the profit function does.

If the widget firm is in a competitive market, and its optimal level of production results in a loss, it has two options. It can continue to produce at the optimal level in the short run until it can lower or eliminate its fixed costs, even though it is losing money; or it can take a larger loss (its fixed cost) by stopping production. Producing 7 units causes a loss of $25 per month, and ceasing production results in a loss of $1250 (the fixed cost) per month. If this firm and many others like it cease production, the supply will be reduced, causing an eventual increase in price. The firm can resume production when the price increase indicates that it can make a profit.

EXAMPLE 3   Maximum Profit

Given that $\overline{MR} = 200 - 4x$, $\overline{MC} = 50 + 2x$, and the total cost of producing 10 Wagbats is $700, at what level should the Wagbat firm hold production in order to maximize the profits?

Solution

Setting $\overline{MR} = \overline{MC}$, we can solve for the production level that maximizes profit.

$$200 - 4x = 50 + 2x$$
$$150 = 6x$$
$$25 = x$$

The level of production that should optimize profit is 25 units. To see whether 25 units maximizes profits or minimizes the losses (in the short run), we must find the total revenue and total cost functions.

$$R(x) = \int (200 - 4x)\, dx = 200x - 2x^2 + K$$
$$= 200x - 2x^2, \text{ because } K = 0$$
$$C(x) = \int (50 + 2x)\, dx = 50x + x^2 + K$$

We find K by noting that $C(x) = 700$ when $x = 10$.

$$700 = 50(10) + (10)^2 + K$$

so $K = 100$.

Thus the cost is given by $C = C(x) = 50x + x^2 + 100$. At $x = 25$, $R = R(25) = 200(25) - 2(25)^2 = \$3750$ and $C = C(25) = 50(25) + (25)^2 + 100 = \$1975$.

We see that the total revenue is greater than the total cost, so production should be held at 25 units, which results in a maximum profit. ■

Calculator Note If it is difficult to solve $\overline{MC} = \overline{MR}$ analytically, we can use a graphing calculator to solve this equation by finding the point of intersection of the graphs of $\overline{MC}$ and $\overline{MR}$. We may also be able to integrate $\overline{MC}$ and $\overline{MR}$ to find the functions $C(x)$ and $R(x)$ and then use a graphing calculator to graph them. From the graphs of $C(x)$ and $R(x)$ we can learn about these functions—and hence about profit. ■

 EXAMPLE **4**   **Cost, Revenue, and Profit**

Suppose that $\overline{MC} = 1.01(x + 190)^{0.01}$ and $\overline{MR} = (1/\sqrt{2x + 1}) + 2$, where x is the number of thousands of units and both revenue and cost are in thousands of dollars. Suppose further that fixed costs are $\$100,236$ and that production is limited to at most 180 thousand units.

(a) Determine $C(x)$ and $R(x)$ and graph them to determine whether a profit can be made.
(b) Estimate the level of production that yields maximum profit, and find the maximum profit.

Solution

(a) $C(x) = \int \overline{MC}\, dx = \int 1.01(x + 190)^{0.01}\, dx$

$$= 1.01 \frac{(x + 190)^{1.01}}{1.01} + K$$

When we say that fixed costs equal $\$100,236$, we mean $C(0) = 100.236$.

$$100.236 = C(0) = (190)^{1.01} + K$$
$$100.236 = 200.236 + K$$
$$-100 = K$$

Thus $C(x) = (x + 190)^{1.01} - 100$.

$$R(x) = \int \overline{MR}\, dx = \int [(2x + 1)^{-1/2} + 2]dx$$

$$= \frac{1}{2}\int (2x + 1)^{-1/2}(2\, dx) + \int 2\, dx$$

$$= \frac{\frac{1}{2}(2x + 1)^{1/2}}{1/2} + 2x + K$$

$R(0) = 0$ means

$$0 = R(0) = (1)^{1/2} + 0 + K, \text{ or } K = -1$$

Thus $R(x) = (2x + 1)^{1/2} + 2x - 1$.

The graphs of $C(x)$ and $R(x)$ are shown in Figure 12.8. (The x-range is chosen to include the production range from 0 to 180 (thousand) units. The y-range is chosen to extend beyond fixed costs of about 100 thousand dollars.)

From the figure we see that a profit can be made as long as the number of units sold exceeds about 95 (thousand). We could locate this break even value more precisely by using INTERSECT.

(b) From the graph we also see that $R(x) - C(x) = P(x)$ is at its maximum at the right edge of the graph. Because production is limited to at most 180 thousand units, profit will be maximized when $x = 180$ and the maximum profit is

$$P(180) = R(180) - C(180)$$
$$= [(361)^{1/2} + 360 - 1] - [(370)^{1.01} - 100]$$
$$\approx 85.46 \text{ (thousand dollars)} \qquad \blacksquare$$

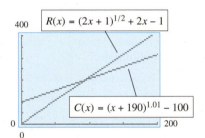

Figure 12.8

CHECKPOINT

1. True or false:
   (a) If $C(x) = \int \overline{MC}\, dx$, then the constant of integration equals the fixed costs.
   (b) If $R(x) = \int \overline{MR}\, dx$, then the constant of integration equals 0.
2. Find $C(x)$ if $\overline{MC} = \dfrac{100}{\sqrt{x + 1}}$ and fixed costs are $8000.

National Consumption and Savings

The consumption function is one of the basic ingredients in a larger discussion of how an economy can have persistent high unemployment or persistent high inflation. This study is often called **Keynesian analysis,** after its founder John Maynard Keynes (pronounced "canes").

If C represents national consumption (in billions of dollars), then a **national consumption function** has the form $C = f(y)$, where y is disposable national income (also in billions of dollars). The **marginal propensity to consume** is the derivative of the national consumption function with respect to y, or $dC/dy = f'(y)$. For example, suppose that

$$C = f(y) = 0.8y + 6$$

is a national consumption function; then the marginal propensity to consume is $f'(y) = 0.8$.

If we know the marginal propensity to consume, we can integrate with respect to y to find national consumption:

$$C = \int f'(y)\, dy = f(y) + K$$

We can find the unique national consumption function if we have additional information to help us determine the value of K, the constant of integration.

EXAMPLE 5   National Consumption

If consumption is \$6 billion when disposable income is \$0, and if the marginal propensity to consume is $dC/dy = 0.3 + 0.4/\sqrt{y}$ (in billions of dollars), find the national consumption function.

Solution

If

$$\frac{dC}{dy} = 0.3 + \frac{0.4}{\sqrt{y}}$$

then

$$C = \int \left(0.3 + \frac{0.4}{\sqrt{y}}\right) dy$$

$$= \int (0.3 + 0.4y^{-1/2})\, dy$$

$$= 0.3y + 0.4\frac{y^{1/2}}{1/2} + K = 0.3y + 0.8y^{1/2} + K$$

Now, if $C = 6$ when $y = 0$, then $6 = 0.3(0) + 0.8\sqrt{0} + K$. Thus the constant of integration is $K = 6$, and the consumption function is

$$C = 0.3y + 0.8\sqrt{y} + 6 \quad \text{(billions of dollars)} \qquad \blacksquare$$

If S represents national savings, we can assume that the disposable national income is given by $y = C + S$, or $S = y - C$. Then the **marginal propensity to save** is $dS/dy = 1 - dC/dy$.

EXAMPLE 6   Consumption and Savings | APPLICATION PREVIEW |

If the consumption is \$9 billion when income is \$0, and if the marginal propensity to save is 0.25, find the consumption function.

Solution

If $dS/dy = 0.25$, then $0.25 = 1 - dC/dy$, or $dC/dy = 0.75$. Thus

$$C = \int 0.75\, dy = 0.75y + K$$

If $C = 9$ when $y = 0$, then $9 = 0.75(0) + K$, or $K = 9$. Then the consumption function is $C = 0.75y + 9$ (billions of dollars). $\qquad \blacksquare$

CHECKPOINT

3. If the marginal propensity to save is

$$\frac{dS}{dy} = 0.7 - \frac{0.4}{\sqrt{y}}$$

find the marginal propensity to consume.

4. Find the national consumption function if the marginal propensity to consume is

$$\frac{dC}{dy} = \frac{1}{\sqrt{y+4}} + 0.2$$

and national consumption is $6.8 billion when disposable income is $0.

CHECKPOINT SOLUTIONS

1. (a) False. $C(0)$ equals the fixed costs. It may or may not be the constant of integration. (See Examples 1 and 2.)

   (b) False. We use $R(0) = 0$ to determine the constant of integration, but it may be nonzero. See Example 4.

2. $C(x) = 100\int (x+1)^{-1/2}\, dx = 100\left[\frac{(x+1)^{1/2}}{1/2}\right] + K$

   $C(x) = 200\sqrt{x+1} + K$

   When $x = 0$, $C(x) = 8000$, so

   $$8000 = 200\sqrt{1} + K$$
   $$7800 = K$$
   $$C(x) = 200\sqrt{x+1} + 7800$$

3. $\dfrac{dC}{dy} = 1 - \dfrac{dS}{dy} = 1 - \left(0.7 - \dfrac{0.4}{\sqrt{y}}\right) = 0.3 + \dfrac{0.4}{\sqrt{y}}$

4. $C(y) = \displaystyle\int\left(\frac{1}{\sqrt{y+4}} + 0.2\right) dy$

   $= \int \left[(y+4)^{-1/2} + 0.2\right] dy$

   $= \dfrac{(y+4)^{1/2}}{1/2} + 0.2y + K$

   $C(y) = 2\sqrt{y+4} + 0.2y + K$

   Using $C(0) = 6.8$ gives $6.8 = 2\sqrt{4} + 0 + K$, or $K = 2.8$.

   Thus $C(y) = 2\sqrt{y+4} + 0.2y + 2.8$.

| EXERCISES | 12.4

Access end-of-section exercises online at **www.webassign.net**   ENHANCED **WebAssign**

Differential Equations

■ **| APPLICATION PREVIEW |**

Carbon-14 dating, used to determine the age of fossils, is based on three facts. First, the half-life of carbon-14 is 5730 years. Second, the amount of carbon-14 in any living organism is essentially constant. Third, when an organism dies, the rate of change of carbon-14 in the organism is proportional to the amount present. If y represents the amount of carbon-14 present in the organism, then we can express the rate of change of carbon-14 by the differential equation

$$\frac{dy}{dt} = ky$$

where k is a constant and t is time in years. In this section, we study methods that allow us to find a function y that satisfies this differential equation, and then we use that function to date a fossil. (See Example 6.)

Recall that we introduced the derivative as an instantaneous rate of change and denoted the instantaneous rate of change of y with respect to time as dy/dt. For many growth or decay processes, such as carbon-14 decay, the rate of change of the amount of a substance with respect to time is proportional to the amount present. As we noted above, this can be represented by the equation

$$\frac{dy}{dt} = ky \quad (k = \text{constant})$$

An equation of this type, where y is an unknown function of x or t, is called a **differential equation** because it contains derivatives (or differentials). In this section, we restrict ourselves to differential equations where the highest derivative present in the equation is the first derivative. These differential equations are called **first-order differential equations.** Examples are

$$f'(x) = \frac{1}{x+1}, \quad \frac{dy}{dt} = 2t, \quad \text{and} \quad x\,dy = (y+1)\,dx$$

Solution of Differential Equations

The solution to a differential equation is a function [say $y = f(x)$] that, when used in the differential equation, results in an identity.

EXAMPLE 1 Differential Equation

Show that $y = 4e^{-5t}$ is a solution to $dy/dt + 5y = 0$.

Solution

We must show that substituting $y = 4e^{-5t}$ into the equation $dy/dt + 5y = 0$ results in an identity:

$$\frac{d}{dt}(4e^{-5t}) + 5(4e^{-5t}) = 0$$
$$-20e^{-5t} + 20e^{-5t} = 0$$
$$0 = 0$$

Thus $y = 4e^{-5t}$ is a solution.

Now that we know what it means for a function to be a solution to a differential equation, let us consider how to find solutions.

The most elementary differential equations are of the form

$$\frac{dy}{dx} = f(x)$$

where $f(x)$ is a continuous function. These equations are elementary to solve because the solutions are found by integration:

$$y = \int f(x)\, dx$$

EXAMPLE 2   **Solving a Differential Equation**

Find the solution of

$$f'(x) = \frac{1}{x + 1}$$

Solution
The solution is

$$f(x) = \int f'(x)\, dx = \int \frac{1}{x + 1}\, dx = \ln |x + 1| + C \qquad \blacksquare$$

The solution in Example 2, $f(x) = \ln |x + 1| + C$, is called the **general solution** because every solution to the equation has this form, and different values of C give different **particular solutions.** Figure 12.9 shows the graphs of several members of the family of solutions to this differential equation. (We cannot, of course, show all of them.)

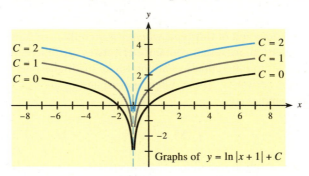

Figure 12.9

Graphs of $y = \ln |x + 1| + C$

We can find a particular solution to a differential equation when we know that the solution must satisfy additional conditions, such as **initial conditions** or **boundary conditions.** For instance, to find the particular solution to

$$f'(x) = \frac{1}{x + 1} \qquad \text{with the condition that} \qquad f(-2) = 2$$

we use $f(-2) = 2$ in the general solution, $f(x) = \ln |x + 1| + C.$

$$2 = f(-2) = \ln |-2 + 1| + C$$
$$2 = \ln |-1| + C \quad \text{so} \quad C = 2$$

Thus the particular solution is

$$f(x) = \ln |x + 1| + 2$$

and is shown in Figure 12.9 with $C = 2$.

We frequently denote the value of the solution function $y = f(t)$ at the initial time $t = 0$ as $y(0)$ instead of $f(0)$.

Calculator Note   We can use a graphing calculator to solve differential equations with boundary conditions. Details can be found in Appendix C, Section 12.5.   ∎

CHECKPOINT

1. Given $f'(x) = 2x - [1/(x + 1)], f(0) = 4$,
   (a) find the general solution to the differential equation.
   (b) find the particular solution that satisfies $f(0) = 4$.

Just as we can find the differential of both sides of an equation, we can find the solution to a differential equation of the form

$$G(y)\, dy = f(x)\, dx$$

by integrating both sides.

EXAMPLE 3   **Differential Equation with a Boundary Condition**

Solve $3y^2\, dy = 2x\, dx$, if $y(1) = 2$.

Solution
We find the general solution by integrating both sides.

$$\int 3y^2\, dy = \int 2x\, dx$$
$$y^3 + C_1 = x^2 + C_2$$
$$y^3 = x^2 + C, \qquad \text{where } C = C_2 - C_1$$

By using $y(1) = 2$, we can find C.

$$2^3 = 1^2 + C$$
$$7 = C$$

Thus the particular solution is given implicitly by

$$y^3 = x^2 + 7$$

Separable Differential Equations   It is frequently necessary to change the form of a differential equation before it can be solved by integrating both sides.
    For example, the equation

$$\frac{dy}{dx} = y^2$$

cannot be solved by simply integrating both sides of the equation with respect to x because we cannot evaluate $\int y^2\, dx$.
    However, we can multiply both sides of $dy/dx = y^2$ by dx/y^2 to obtain an equation that has all terms containing y on one side of the equation and all terms containing x on the other side. That is, we obtain

$$\frac{dy}{y^2} = dx$$

Separable Differential Equations

When a differential equation can be equivalently expressed in the form

$$g(y)\, dy = f(x)\, dx$$

we say that the equation is **separable**.
    The solution of a separable differential equation is obtained by integrating both sides of the equation after the variables have been separated.

EXAMPLE 4   **Separable Differential Equation**

Solve the differential equation

$$(x^2y + x^2)\, dy = x^3\, dx$$

Solution

To write the equation in separable form, we first factor x^2 from the left side and divide both sides by it.

$$x^2(y + 1)\, dy = x^3\, dx$$

$$(y + 1)\, dy = \frac{x^3}{x^2}\, dx$$

The equation is now separated, so we integrate both sides.

$$\int (y + 1)\, dy = \int x\, dx$$

$$\frac{y^2}{2} + y + C_1 = \frac{x^2}{2} + C_2$$

This equation, as well as the equation

$$y^2 + 2y - x^2 = C, \quad \text{where} \quad C = 2(C_2 - C_1)$$

gives the solution implicitly.

Note that we need not write both C_1 and C_2 when we integrate, because it is always possible to combine the two constants into one. ◼

EXAMPLE 5 **Separable Differential Equation**

Solve the differential equation

$$\frac{dy}{dt} = ky \quad (k = \text{constant})$$

Solution

To solve the equation, we write it in separated form and integrate both sides:

$$\frac{dy}{y} = k\, dt \implies \int \frac{dy}{y} = \int k\, dt \implies \ln |y| = kt + C_1$$

Assuming that $y > 0$ and writing this equation in exponential form gives

$$y = e^{kt + C_1}$$
$$y = e^{kt} \cdot e^{C_1} = Ce^{kt}, \quad \text{where } C = e^{C_1}$$

This solution,

$$y = Ce^{kt}$$

is the general solution to the differential equation $dy/dt = ky$ because all solutions have this form, with different values of C giving different particular solutions. The case of $y < 0$ is covered by values of $C < 0$. ◼

CHECKPOINT

2. True or false:
   (a) The general solution to $dy = (x/y)\, dx$ can be found from
   $$\int y\, dy = \int x\, dx$$
   (b) The first step in solving $dy/dx = -2xy^2$ is to separate it.
   (c) The equation $dy/dx = -2xy^2$ separates as $y^2\, dy = -2x\, dx$.
3. Suppose that $(xy + x)(dy/dx) = x^2 y + y$.
   (a) Separate this equation.   (b) Find the general solution.

In many applied problems that can be modeled with differential equations, we know conditions that allow us to obtain a particular solution.

Applications of Differential Equations

We now consider two applications that can be modeled by differential equations. These are radioactive decay and one-container mixture problems (as a model for drugs in an organ).

EXAMPLE **6**   **Carbon-14 Dating** | APPLICATION PREVIEW |

When an organism dies, the rate of change of the amount of carbon-14 present is proportional to the amount present and is represented by the differential equation

$$\frac{dy}{dt} = ky$$

where y is the amount present, k is a constant, and t is time in years. If we denote the initial amount of carbon-14 in an organism as y_0, then $y = y_0$ represents the amount present at time $t = 0$ (when the organism dies). Suppose that anthropologists discover a fossil that contains 1% of the initial amount of carbon-14. Find the age of the fossil. (Recall that the half-life of carbon-14 is 5730 years.)

Solution
We must find a particular solution to

$$\frac{dy}{dt} = ky$$

subject to the fact that when $t = 0$, $y = y_0$, and we must determine the value of k on the basis of the half-life of carbon-14 ($t = 5730$ years, $y = \frac{1}{2}y_0$ units.) From Example 5, we know that the general solution to the differential equation $dy/dt = ky$ is $y = Ce^{kt}$. Using $y = y_0$ when $t = 0$, we obtain $y_0 = C$, so the equation becomes $y = y_0e^{kt}$. Using $t = 5730$ and $y = \frac{1}{2}y_0$ in this equation gives

$$\frac{1}{2}y_0 = y_0e^{5730k} \quad \text{or} \quad 0.5 = e^{5730k}$$

Rewriting this equation in logarithmic form and then solving for k, we get

$$\ln(0.5) = 5730k$$
$$-0.69315 = 5730k$$
$$-0.00012097 = k$$

Thus the equation we seek is

$$y = y_0e^{-0.00012097t}$$

Using the fact that $y = 0.01y_0$ when the fossil was discovered, we can find its age t by solving

$$0.01y_0 = y_0e^{-0.00012097t} \quad \text{or} \quad 0.01 = e^{-0.00012097t}$$

Rewriting this in logarithmic form and then solving give

$$\ln(0.01) = -0.00012097t$$
$$-4.6051702 = -0.00012097t$$
$$38,069 \approx t$$

Thus the fossil is approximately 38,069 years old.   ■

The differential equation $dy/dt = ky$, which describes the decay of radioactive substances in Example 6, also models the rate of growth of an investment that is compounded continuously and the rate of decay of purchasing power due to inflation.

Another application of differential equations comes from a group of applications called *one-container mixture problems*. In problems of this type, there is a substance whose amount in a container is changing with time, and the goal is to determine the amount of the substance at any time t. The differential equations that model these problems are of the following form:

$$\begin{bmatrix} \text{Rate of change} \\ \text{of the amount} \\ \text{of the substance} \end{bmatrix} = \begin{bmatrix} \text{Rate at which} \\ \text{the substance} \\ \text{enters the container} \end{bmatrix} - \begin{bmatrix} \text{Rate at which} \\ \text{the substance} \\ \text{leaves the container} \end{bmatrix}$$

We consider this application as it applies to the amount of a drug in an organ.

EXAMPLE 7 **Drug in an Organ**

A liquid carries a drug into an organ of volume 300 cubic centimeters at a rate of 5 cubic centimeters per second, and the liquid leaves the organ at the same rate. If the concentration of the drug in the entering liquid is 0.1 grams per cubic centimeter, and if x represents the amount of drug in the organ at any time t, then using the fact that the rate of change of the amount of the drug in the organ, dx/dt, equals the rate at which the drug enters minus the rate at which it leaves, we have

$$\frac{dx}{dt} = \left(\frac{5 \text{ cc}}{\text{s}}\right)\left(\frac{0.1 \text{ g}}{\text{cc}}\right) - \left(\frac{5 \text{ cc}}{\text{s}}\right)\left(\frac{x \text{ g}}{300 \text{ cc}}\right)$$

or

$$\frac{dx}{dt} = 0.5 - \frac{x}{60} = \frac{30}{60} - \frac{x}{60} = \frac{30 - x}{60}, \quad \text{in grams per second}$$

Find the amount of the drug in the organ as a function of time t.

Solution

Multiplying both sides of the equation $\dfrac{dx}{dt} = \dfrac{30 - x}{60}$ by $\dfrac{dt}{(30 - x)}$ gives

$$\frac{dx}{30 - x} = \frac{1}{60} dt$$

The equation is now separated, so we can integrate both sides.

$$\int \frac{dx}{30 - x} = \int \frac{1}{60} dt$$

$$-\ln(30 - x) = \frac{1}{60}t + C_1 \quad (30 - x > 0)$$

$$\ln(30 - x) = -\frac{1}{60}t - C_1$$

Rewriting this in exponential form gives

$$30 - x = e^{-t/60 - C_1} = e^{-t/60} \cdot e^{-C_1}$$

Letting $C = e^{-C_1}$ yields

$$30 - x = Ce^{-t/60}$$

so

$$x = 30 - Ce^{-t/60}$$

and we have the desired function. ∎

CHECKPOINT SOLUTIONS

1. (a) $f(x) = \displaystyle\int\left(2x - \frac{1}{x + 1}\right)dx = x^2 - \ln|x + 1| + C$

   (b) If $f(0) = 4$, then $4 = 0^2 - \ln|1| + C$, so $C = 4$.
   Thus $f(x) = x^2 - \ln|x + 1| + 4$ is the particular solution.

2. (a) True, and the solution is

$$\int y \, dy = \int x \, dx \quad \Rightarrow \quad \frac{y^2}{2} = \frac{x^2}{2} + C$$

   (b) True

   (c) False. It separates as $dy/y^2 = -2x \, dx$, and the solution is

$$\int \frac{-dy}{y^2} = \int 2x \, dx \quad \Rightarrow \quad y^{-1} = x^2 + C$$

$$\frac{1}{y} = x^2 + C$$

$$y = \frac{1}{x^2 + C}$$

3. (a) $x(y + 1)\dfrac{dy}{dx} = (x^2 + 1)y$ (b) $\displaystyle\int\left(1 + \frac{1}{y}\right)dy = \int\left(x + \frac{1}{x}\right)dx$

$$\frac{y + 1}{y}\,dy = \frac{x^2 + 1}{x}\,dx \qquad\qquad y + \ln|y| = \frac{x^2}{2} + \ln|x| + C$$

$$\left(1 + \frac{1}{y}\right)dy = \left(x + \frac{1}{x}\right)dx \qquad \text{This is the general solution.}$$

| EXERCISES | 12.5

Access end-of-section exercises online at **www.webassign.net**

KEY TERMS AND FORMULAS

Section	Key Terms	Formulas
12.1	General antiderivative of $f'(x)$	$f(x) + C$
	Integral	$\displaystyle\int f(x)\,dx$
	Powers of x Formula	$\displaystyle\int x^n\,dx = \frac{x^{n+1}}{n + 1} + C \ \ (n \neq -1)$
	Integration Formulas	$\displaystyle\int dx = x + C$
		$\displaystyle\int cu(x)\,dx = c\int u(x)\,dx \ \ (c = \text{a constant})$
		$\displaystyle\int [u(x) \pm v(x)]\,dx = \int u(x)\,dx \pm \int v(x)\,dx$
12.2	Power Rule	$\displaystyle\int [u(x)]^n\,u'(x)\,dx = \frac{[u(x)]^{n+1}}{n + 1} + C \ \ (n \neq -1)$
12.3	Exponential Formula	$\displaystyle\int e^u\,u'\,dx = \int e^u\,du = e^u + C$
	Logarithmic Formula	$\displaystyle\int \frac{u'}{u}\,dx = \int \frac{1}{u}\,du = \ln|u| + C$
12.4	Total cost	$C(x) = \int \overline{MC}\,dx$
	Total revenue	$R(x) = \int \overline{MR}\,dx$
	Profit	
	Marginal propensity to consume	$\dfrac{dC}{dy}$
	Marginal propensity to save	$\dfrac{dS}{dy} = 1 - \dfrac{dC}{dy}$
	National consumption	$C = \displaystyle\int f'(y)\,dy = \int \frac{dC}{dy}\,dy$

Section	Key Terms	Formulas
12.5	Differential equations	
	Solutions	
	General	
	Particular	
	First order	$\dfrac{dy}{dx} = f(x) \Rightarrow y = \int f(x)\, dx$
	Separable	$g(y)\, dy = f(x)\, dx \Rightarrow \int g(y)\, dy = \int f(x)\, dx$
	Radioactive decay	$\dfrac{dy}{dt} = ky$
	Drug in an organ	Rate $=$ (rate in) $-$ (rate out)

REVIEW EXERCISES

Evaluate the integrals in Problems 1–26.

1. $\int x^6\, dx$

2. $\int x^{1/2}\, dx$

3. $\int (12x^3 - 3x^2 + 4x + 5)\, dx$

4. $\int 7(x^2 - 1)^2\, dx$

5. $\int 7x(x^2 - 1)^2\, dx$

6. $\int (x^3 - 3x^2)^5(x^2 - 2x)\, dx$

7. $\int (x^3 + 4)^2 3x\, dx$

8. $\int 5x^2(3x^3 + 7)^6\, dx$

9. $\int \dfrac{x^2}{x^3 + 1}\, dx$

10. $\int \dfrac{x^2}{(x^3 + 1)^2}\, dx$

11. $\int \dfrac{x^2\, dx}{\sqrt[3]{x^3 - 4}}$

12. $\int \dfrac{x^2\, dx}{x^3 - 4}$

13. $\int \dfrac{x^3 + 1}{x^2}\, dx$

14. $\int \dfrac{x^3 - 3x + 1}{x - 1}\, dx$

15. $\int y^2 e^{y^3}\, dy$

16. $\int (3x - 1)^{12}\, dx$

17. $\int \dfrac{3x^2}{2x^3 - 7}\, dx$

18. $\int \dfrac{5\, dx}{e^{4x}}$

19. $\int (x^3 - e^{3x})\, dx$

20. $\int xe^{1 + x^2}\, dx$

21. $\int \dfrac{6x^7}{(5x^8 + 7)^3}\, dx$

22. $\int \dfrac{7x^3}{\sqrt{1 - x^4}}\, dx$

23. $\int \left(\dfrac{e^{2x}}{2} + \dfrac{2}{e^{2x}} \right) dx$

24. $\int \left[x - \dfrac{1}{(x + 1)^2} \right] dx$

25. (a) $\int (x^2 - 1)^4 x\, dx$
   (b) $\int (x^2 - 1)^{10} x\, dx$
   (c) $\int (x^2 - 1)^7 3x\, dx$
   (d) $\int (x^2 - 1)^{-2/3} x\, dx$

26. (a) $\int \dfrac{2x\, dx}{x^2 - 1}$
   (b) $\int \dfrac{2x\, dx}{(x^2 - 1)^2}$
   (c) $\int \dfrac{3x\, dx}{\sqrt{x^2 - 1}}$
   (d) $\int \dfrac{3x\, dx}{x^2 - 1}$

In Problems 27–32, find the general solution to each differential equation.

27. $\dfrac{dy}{dt} = 4.6e^{-0.05t}$

28. $dy = (64 + 76x - 36x^2)\, dx$

29. $\dfrac{dy}{dx} = \dfrac{4x}{y - 3}$

30. $t\, dy = \dfrac{dt}{y + 1}$

31. $\dfrac{dy}{dx} = \dfrac{x}{e^y}$

32. $\dfrac{dy}{dt} = \dfrac{4y}{t}$

In Problems 33 and 34, find the particular solution to each differential equation.

33. $y' = \dfrac{x^2}{y + 1}, \quad y(0) = 4$

34. $y' = \dfrac{2x}{1 + 2y}, \quad y(2) = 0$

APPLICATIONS

35. *Revenue* If the marginal revenue for a month for a product is $\overline{MR} = 0.06x + 12$ dollars per unit, find the total revenue from the sale of $x = 800$ units of the product.

36. *Productivity* Suppose that the rate of change of production of the average worker at a factory is given by

$$\dfrac{dp}{dt} = 27 + 24t - 3t^2, \quad 0 \le t \le 8$$

where p is the number of units the worker produces in t hours. How many units will the average worker produce in an 8-hour shift? (Assume that $p = 0$ when $t = 0$.)

37. *Oxygen levels in water* The rate of change of the oxygen level (in mmol/l) per month in a body of water after an oil spill is given by

$$P'(t) = 400\left[\frac{5}{(t+5)^2} - \frac{50}{(t+5)^3}\right]$$

where t is the number of months after the spill. What function gives the oxygen level P at any time t if $P = 400$ mmol/l when $t = 0$?

38. *Bacterial growth* A population of bacteria grows at the rate

$$\frac{dp}{dt} = \frac{100,000}{(t+100)^2}$$

where p is the population and t is time. If the population is 1000 when $t = 1$, write the equation that gives the size of the population at any time t.

39. *Market share* The rate of change of the market share (as a percent) a firm expects for a new product is

$$\frac{dy}{dt} = 2.4e^{-0.04t}$$

where t is the number of months after the product is introduced.
    (a) Write the equation that gives the expected market share y at any time t. (Note that $y = 0$ when $t = 0$.)
    (b) What market share does the firm expect after 1 year?

40. *Revenue* If the marginal revenue for a product is $\overline{MR} = \dfrac{800}{x+2}$, find the total revenue function.

41. *Cost* The marginal cost for a product is $\overline{MC} = 6x + 4$ dollars per unit, and the cost of producing 100 items is $31,400.
    (a) Find the fixed costs.
    (b) Find the total cost function.

42. *Profit* Suppose a product has a daily marginal revenue $\overline{MR} = 46$ and a daily marginal cost $\overline{MC} = 30 + \frac{1}{5}x$, both in dollars per unit. If the daily fixed cost is $200, how many units will give maximum profit and what is the maximum profit?

43. *National consumption* If consumption is $8.5 billion when disposable income is $0, and if the marginal propensity to consume is

$$\frac{dC}{dy} = \frac{1}{\sqrt{2y+16}} + 0.6 \quad \text{(in billions of dollars)}$$

find the national consumption function.

44. *National consumption* Suppose that the marginal propensity to save is

$$\frac{dS}{dy} = 0.2 - 0.1e^{-2y} \quad \text{(in billions of dollars)}$$

and consumption is $7.8 billion when disposable income is $0. Find the national consumption function.

45. *Allometric growth* For many species of fish, the length L and weight W of a fish are related by

$$\frac{dW}{dL} = \frac{3W}{L}$$

The general solution to this differential equation expresses the allometric relationship between the length and weight of a fish. Find the general solution.

46. *Investment* When the interest on an investment is compounded continuously, the investment grows at a rate that is proportional to the amount in the account, so that if the amount present is P, then

$$\frac{dP}{dt} = kP$$

where P is in dollars, t is in years, and k is a constant.
    (a) Solve this differential equation to find the relationship.
    (b) Use properties of logarithms and exponential functions to write P as a function of t.
    (c) If $50,000 is invested (when $t = 0$) and the amount in the account after 10 years is $135,914, find the function that gives the value of the investment as a function of t.
    (d) In part (c), what does the value of k represent?

47. *Fossil dating* Radioactive beryllium is sometimes used to date fossils found in deep-sea sediment. The amount of radioactive material x satisfies

$$\frac{dx}{dt} = kx$$

Suppose that 10 units of beryllium are present in a living organism and that the half-life of beryllium is 4.6 million years. Find the age of a fossil if 20% of the original radioactivity is present when the fossil is discovered.

48. *Drug in an organ* Suppose that a liquid carries a drug into a 120-cc organ at a rate of 4 cc/s and leaves the organ at the same rate. If initially there is no drug in the organ and if the concentration of drug in the liquid is 3 g/cc, find the amount of drug in the organ as a function of time.

49. *Chemical mixture* A 300-gal tank initially contains a solution with 100 lb of a chemical. A mixture containing 2 lb/gal of the chemical enters the tank at 3 gal/min, and the well-stirred mixture leaves at the same rate. Find an equation that gives the amount of the chemical in the tank as a function of time. How long will it be before there is 500 lb of chemical in the tank?

12 CHAPTER TEST

Evaluate the integrals in Problems 1–8.

1. $\displaystyle\int (6x^2 + 8x - 7)\, dx$

2. $\displaystyle\int \left(4 + \sqrt{x} - \frac{1}{x^2}\right) dx$

3. $\displaystyle\int 5x^2(4x^3 - 7)^9\, dx$

4. $\displaystyle\int (3x^2 - 6x + 1)^{-3}(2x - 2)\, dx$

5. $\displaystyle\int \frac{s^3}{2s^4 - 5}\, ds$

6. $\displaystyle\int 100e^{-0.01x}\, dx$

7. $\displaystyle\int 5y^3 e^{2y^4 - 1}\, dy$

8. $\displaystyle\int \left(e^x + \frac{5}{x} - 1\right) dx$

9. Evaluate $\displaystyle\int \frac{x^2}{x + 1}\, dx$. Use long division.

10. If $\displaystyle\int f(x)\, dx = 2x^3 - x + 5e^x + C$, find $f(x)$.

In Problems 11 and 12, find the particular solution to each differential equation.

11. $y' = 4x^3 + 3x^2$, if $y(0) = 4$

12. $\dfrac{dy}{dx} = e^{4x}$, if $y(0) = 2$

13. Find the general solution of the separable differential equation $\dfrac{dy}{dx} = x^3 y^2$.

14. Suppose the rate of growth of the population of a city is predicted to be

$$\frac{dp}{dt} = 2000t^{1.04}$$

where p is the population and t is the number of years past 2010. If the population in the year 2010 is 50,000, what is the predicted population in the year 2020?

15. Suppose that the marginal cost for x units of a product is $\overline{MC} = 4x + 50$, the marginal revenue is $\overline{MR} = 500$, and the cost of the production and sale of 10 units is $1000. What is the profit function for this product?

16. Suppose the marginal propensity to save is given by

$$\frac{dS}{dy} = 0.22 - \frac{0.25}{\sqrt{0.5y + 1}} \quad \text{(in billions of dollars)}$$

and national consumption is $6.6 billion when disposable income is $0. Find the national consumption function.

17. A certain radioactive material has a half-life of 100 days. If the amount of material present, x, satisfies $\dfrac{dx}{dt} = kx$, where t is in days, how long will it take for 90% of the radioactivity to dissipate?

18. Suppose that a liquid carries a drug with concentration 0.1 g/cc into a 160-cc organ at the rate of 4 cc/sec, and leaves at the same rate. Find the amount of drug in the organ as a function of time t, if initially there is none in the organ.

I. Employee Production Rate

The manager of a plant has been instructed to hire and train additional employees to manufacture a new product. She must hire a sufficient number of new employees so that within 30 days they will be producing 2500 units of the product each day.

Because a new employee must learn an assigned task, production will increase with training. Suppose that research on similar projects indicates that production increases with training according to the learning curve, so that for the average employee, the rate of production per day is given by

$$\frac{dN}{dt} = be^{-at}$$

where N is the number of units produced per day after t days of training and a and b are constants that depend on the project. Because of experience with a similar project, the manager expects the rate for this project to be

$$\frac{dN}{dt} = 2.5e^{-0.05t}$$

The manager tested her training program with 5 employees and learned that the average employee could produce 11 units per day after 5 days of training. On the basis of this information, she must decide how many employees to hire and begin to train so that a month from now they will be producing 2500 units of the product per day. She estimates that it will take her 10 days to hire the employees, and thus she will have 15 days remaining to train them. She also expects a 10% attrition rate during this period.

How many employees would you advise the plant manager to hire? Check your advice by answering the following questions.

1. Use the expected rate of production and the results of the manager's test to find the function relating N and t—that is, $N = N(t)$.
2. Find the number of units the average employee can produce after 15 days of training. How many such employees would be needed to maintain a production rate of 2500 units per day?
3. Explain how you would revise this last result to account for the expected 10% attrition rate. How many new employees should the manager hire?

II. Supply and Demand

If p is the price in dollars of a given commodity at time t, then we can think of price as a function of time. Similarly, the number of units demanded by consumers q_d at any time, and the number of units supplied by producers q_s at any time, may also be considered as functions of time as well as functions of price.

Both the quantity demanded and the quantity supplied depend not only on the price at the time, but also on the direction and rate of change that consumers and producers ascribe to prices. For example, even when prices are high, if consumers feel that prices are rising, the demand may rise. Similarly, if prices are low but producers feel they may go lower, the supply may rise.

If we assume that prices are determined in the marketplace by supply and demand, then the equilibrium price is the one we seek.

Suppose the supply and demand functions for a certain commodity in a competitive market are given, in hundreds of units, by

$$q_s = 30 + p + 5\frac{dp}{dt}$$

$$q_d = 51 - 2p + 4\frac{dp}{dt}$$

where dp/dt denotes the rate of change of the price with respect to time. If, at $t = 0$, the market equilibrium price is \$12, we can express the market equilibrium price as a function of time.

Our goals are as follows.

A. To express the market equilibrium price as a function of time.
B. To determine whether there is price stability in the marketplace for this item (that is, to determine whether the equilibrium price approaches a constant over time).

To achieve these goals, do the following.

1. Set the expressions for q_s and q_d equal to each other.

2. Solve this equation for $\dfrac{dp}{dt}$.

3. Write this equation in the form $f(p)\, dp = g(t)\, dt$.

4. Integrate both sides of this separated differential equation.

5. Solve the resulting equation for p in terms of t.

6. Use the fact that $p = 12$ when $t = 0$ to find C, the constant of integration, and write the market equilibrium price p as a function of time t.

7. Find $\lim\limits_{t \to \infty} p$, which gives the price we can expect this product to approach. If this limit is finite, then for this item there is price stability in the marketplace. If $p \to +\infty$ as $t \to +\infty$, then price will continue to increase until economic conditions change.

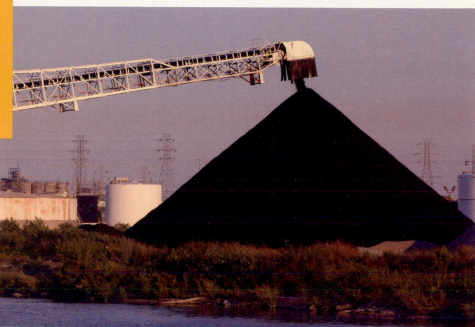
Gerald Bernard/Shutterstock.com

13 CHAPTER

Definite Integrals: Techniques of Integration

We saw some applications of the indefinite integral in Chapter 12. In this chapter we define the definite integral and discuss a theorem and techniques that are useful in evaluating or approximating it. We will also see how it can be used in many interesting applications, such as consumer's and producer's surplus and total value, present value, and future value of continuous income streams. Improper integrals can be used to find the capital value of a continuous income stream.

The topics and applications discussed in this chapter include the following.

Prerequisite Problem Type	For Section	Answer	Section for Review
Simplify: $\dfrac{1}{n^3}\left[\dfrac{n(n+1)(2n+1)}{6} - \dfrac{2n(n+1)}{2} + n\right]$	13.1	$\dfrac{2n^2 - 3n + 1}{6n^2}$	0.7 Fractions
(a) If $F(x) = \dfrac{x^4}{4} + 4x + C$, what is $F(4) - F(2)$? (b) If $F(x) = -\dfrac{1}{9}\ln\left(\dfrac{9 + \sqrt{81 - 9x^2}}{3x}\right)$ what is $F(3) - F(2)$?	13.2 13.5	(a) 68 (b) $\dfrac{1}{9}\ln\left(\dfrac{3 + \sqrt{5}}{2}\right)$	1.2 Function notation
Find the limit: (a) $\lim\limits_{n\to +\infty} \dfrac{n^2 + n}{2n^2}$ (b) $\lim\limits_{n\to +\infty} \dfrac{2n^2 - 3n + 1}{6n^2}$ (c) $\lim\limits_{b\to\infty}\left(1 - \dfrac{1}{b}\right)$ (d) $\lim\limits_{b\to\infty}\left(\dfrac{-100{,}000}{e^{0.10b}} + 100{,}000\right)$	13.1 13.7	(a) $\frac{1}{2}$ (b) $\frac{1}{3}$ (c) 1 (d) 100,000	9.2 Limits at infinity
Find the derivative of $y = \ln x$.	13.6	$\dfrac{1}{x}$	11.1 Derivatives of logarithmic functions
Integrate: (a) $\int (x^3 + 4)\,dx$ (b) $\int x\sqrt{x^2 - 9}\,dx$ (c) $\int e^{2x}\,dx$	13.2–13.7	(a) $\dfrac{x^4}{4} + 4x + C$ (b) $\frac{1}{3}(x^2 - 9)^{3/2} + C$ (c) $\frac{1}{2}e^{2x} + C$	12.1, 12.2, 12.3 Integration

Area Under a Curve

| APPLICATION PREVIEW |

One way to find the accumulated production (such as the production of ore from a mine) over a period of time is to graph the rate of production as a function of time and find the area under the resulting curve over a specified time interval. For example, if a coal mine produces at a rate of 30 tons per day, the production over 10 days ($30 \cdot 10 = 300$) could be represented by the area under the line $y = 30$ between $x = 0$ and $x = 10$ (see Figure 13.1).

Using area to determine the accumulated production is very useful when the rate-of-production function varies at different points in time. For example, if the rate of production (in tons per day) is represented by

$$y = 100e^{-0.1x}$$

where x represents the number of days, then the area under the curve (and above the x-axis) from $x = 0$ to $x = 10$ represents the total production over the 10-day period (see Figure 13.2(a) and Example 1). In order to determine the accumulated production and to solve other types of problems, we need a method for finding areas under curves. That is the goal of this section.

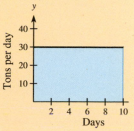

Figure 13.1

Area Under a Curve   To estimate the accumulated production for the example in the Application Preview, we approximate the area under the graph of the production rate function. We can find a rough approximation of the area under this curve by fitting two rectangles to the curve as shown in Figure 13.2(b). The area of the first rectangle is $5 \cdot 100 = 500$ square units, and the area of the second rectangle is $(10 - 5)\left[100e^{-0.1(5)}\right] \approx 5(60.65) = 303.25$ square units, so this rough approximation is 803.25 square units or 803.25 tons of ore. This approximation is clearly larger than the exact area under the curve. Why?

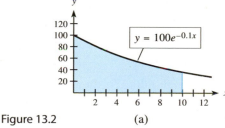   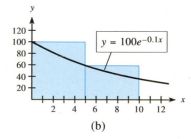   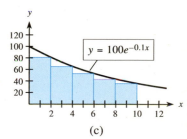

Figure 13.2   (a)   (b)   (c)

EXAMPLE 1   Ore Production | Application Preview |

Find another, more accurate approximation of the tons of ore produced by approximating the area under the curve in Figure 13.2(a). Fit five rectangles with equal bases inside the area under the curve $y = 100e^{-0.1x}$, and use them to approximate the area under the curve from $x = 0$ to $x = 10$ (see Figure 13.2(c)).

Solution
Each of the five rectangles has base 2, and the height of each rectangle is the value of the function at the right-hand endpoint of the interval forming its base. Thus the areas of the rectangles are as follows.

Rectangle	Base	Height	Area = Base × Height
1	2	$100e^{-0.1(2)} \approx 81.87$	$2(81.87) = 163.74$
2	2	$100e^{-0.1(4)} \approx 67.03$	$2(67.03) = 134.06$
3	2	$100e^{-0.1(6)} \approx 54.88$	$2(54.88) = 109.76$
4	2	$100e^{-0.1(8)} \approx 44.93$	$2(44.93) = 89.86$
5	2	$100e^{-0.1(10)} \approx 36.79$	$2(36.79) = 73.58$

The area under the curve is approximately equal to

$$163.74 + 134.06 + 109.76 + 89.86 + 73.58 = 571$$

so approximately 571 tons of ore are produced in the 10-day period. The area is actually 632.12, to 2 decimal places (or 632.12 tons of ore), so the approximation 571 is smaller than the actual area but is much better than the one we obtained with just two rectangles. In general, if we use bases of equal width, the approximation of the area under a curve improves when more rectangles are used. ■

Suppose that we wish to find the area between the curve $y = 2x$ and the x-axis from $x = 0$ to $x = 1$ (see Figure 13.3). As we saw in Example 1, one way to approximate this area is to use the areas of rectangles whose bases are on the x-axis and whose heights are the vertical distances from points on their bases to the curve. We can divide the interval $[0, 1]$ into n equal subintervals and use them as the bases of n rectangles whose heights are determined by the curve (see Figure 13.4). The width of each of these rectangles is $1/n$. Using the function value at the right-hand endpoint of each subinterval as the height of the rectangle, we get n rectangles as shown in Figure 13.4. Because part of each rectangle lies above the curve, the sum of the areas of the rectangles will overestimate the area.

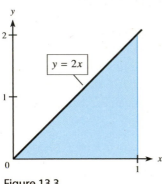

Figure 13.3

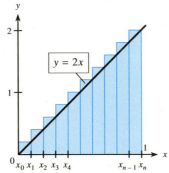

Figure 13.4

Then, with $y = f(x) = 2x$ and subinterval width $1/n$, the areas of the rectangles are as shown in the following table.

Rectangle	Base	Right Endpoint	Height	Area = Base × Height
1	$\dfrac{1}{n}$	$x_1 = \dfrac{1}{n}$	$f(x_1) = 2\left(\dfrac{1}{n}\right)$	$\dfrac{1}{n} \cdot \dfrac{2}{n} = \dfrac{2}{n^2}$
2	$\dfrac{1}{n}$	$x_2 = \dfrac{2}{n}$	$f(x_2) = 2\left(\dfrac{2}{n}\right)$	$\dfrac{1}{n} \cdot \dfrac{4}{n} = \dfrac{4}{n^2}$
3	$\dfrac{1}{n}$	$x_3 = \dfrac{3}{n}$	$f(x_3) = 2\left(\dfrac{3}{n}\right)$	$\dfrac{1}{n} \cdot \dfrac{6}{n} = \dfrac{6}{n^2}$
$\vdots$				
i	$\dfrac{1}{n}$	$x_i = \dfrac{i}{n}$	$f(x_i) = 2\left(\dfrac{i}{n}\right)$	$\dfrac{1}{n} \cdot \dfrac{2i}{n} = \dfrac{2i}{n^2}$
$\vdots$				
n	$\dfrac{1}{n}$	$x_n = \dfrac{n}{n}$	$f(x_n) = 2\left(\dfrac{n}{n}\right)$	$\dfrac{1}{n} \cdot \dfrac{2n}{n} = \dfrac{2n}{n^2}$

Note that $2i/n^2$ gives the area of the ith rectangle for *any* value of i. Thus for any value of n, this area can be approximated by the sum

$$A \approx \frac{2}{n^2} + \frac{4}{n^2} + \frac{6}{n^2} + \cdots + \frac{2i}{n^2} + \cdots + \frac{2n}{n^2}$$

In particular, we have the following approximations of this area for specific values of n (the number of rectangles).

$$n = 5: \qquad A \approx \frac{2}{25} + \frac{4}{25} + \frac{6}{25} + \frac{8}{25} + \frac{10}{25} = \frac{30}{25} = 1.20$$

$$n = 10: \qquad A \approx \frac{2}{100} + \frac{4}{100} + \frac{6}{100} + \cdots + \frac{20}{100} = \frac{110}{100} = 1.10$$

$$n = 100: \qquad A \approx \frac{2}{10,000} + \frac{4}{10,000} + \frac{6}{10,000} + \cdots + \frac{200}{10,000}$$
$$= \frac{10,100}{10,000} = 1.01$$

Figure 13.5 shows the rectangles associated with each of these approximations $(n = 5, 10,$ and $100)$ to the area under $y = 2x$ from $x = 0$ to $x = 1$. For larger n, the rectangles closely approximate the area under the curve.

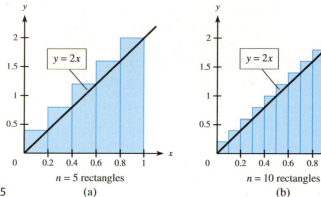

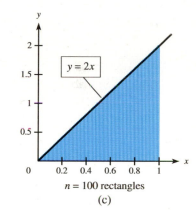

Figure 13.5   (a) $n = 5$ rectangles   (b) $n = 10$ rectangles   (c) $n = 100$ rectangles

We can find this sum for any n more easily if we observe that the common denominator is n^2 and that the numerator is twice the sum of the first n terms of an arithmetic sequence with first term 1 and last term n. As you may recall from Section 6.1, "Simple Interest; Sequences," the first n terms of this arithmetic sequence add to $n(n + 1)/2$. Thus the area is approximated by

$$A \approx \frac{2(1 + 2 + 3 + \cdots + n)}{n^2} = \frac{2[n(n + 1)/2]}{n^2} = \frac{n + 1}{n}$$

Using this formula, we see the following.

$$n = 5: \qquad A \approx \frac{5 + 1}{5} = \frac{6}{5} = 1.20$$

$$n = 10: \qquad A \approx \frac{10 + 1}{10} = \frac{11}{10} = 1.10$$

$$n = 100: \qquad A \approx \frac{100 + 1}{100} = \frac{101}{100} = 1.01$$

As Figure 13.5 indicated, as n gets larger, the number of rectangles increases, the area of each rectangle decreases, and the approximation becomes more accurate. If we let n increase without bound, the approximation approaches the exact area.

$$A = \lim_{n \to +\infty} \frac{n + 1}{n} = \lim_{n \to +\infty} \left(1 + \frac{1}{n}\right) = 1$$

We can see that this area is correct, for we are computing the area of a triangle with base 1 and height 2. The formula for the area of a triangle gives

$$A = \frac{1}{2}bh = \frac{1}{2} \cdot 1 \cdot 2 = 1$$

Summation Notation

A special notation exists that uses the Greek letter Σ (capital sigma) to express the sum of numbers or expressions. (We used sigma notation informally in Chapter 8, "Further Topics in Probability; Data Description.") We may indicate the sum of the n numbers $a_1, a_2, a_3, a_4, \ldots, a_n$ by

$$\sum_{i=1}^{n} a_i = a_1 + a_2 + a_3 + \cdots + a_n$$

This may be read as "The sum of a_i as i goes from 1 to n." The subscript i in a_i is replaced first by 1, then by 2, then by 3, . . . , until it reaches the value above the sigma. The i is called the **index of summation,** and it starts with the lower limit, 1, and ends with the upper limit, n. For example, if $x_1 = 2$, $x_2 = 3$, $x_3 = -1$, and $x_4 = -2$, then

$$\sum_{i=1}^{4} x_i = x_1 + x_2 + x_3 + x_4 = 2 + 3 + (-1) + (-2) = 2$$

The area of the triangle under $y = 2x$ that we discussed above was approximated by

$$A \approx \frac{2}{n^2} + \frac{4}{n^2} + \frac{6}{n^2} + \cdots + \frac{2i}{n^2} + \cdots + \frac{2n}{n^2}$$

Using **sigma notation,** we can write this sum as

$$A \approx \sum_{i=1}^{n} \left(\frac{2i}{n^2} \right)$$

Sigma notation allows us to represent the sums of the areas of the rectangles in an abbreviated fashion. Some formulas that simplify computations involving sums follow.

Sum Formulas

I. $\displaystyle\sum_{i=1}^{n} 1 = n$ IV. $\displaystyle\sum_{i=1}^{n} i = \frac{n(n+1)}{2}$

II. $\displaystyle\sum_{i=1}^{n} cx_i = c \sum_{i=1}^{n} x_i$ (c = constant)   V. $\displaystyle\sum_{i=1}^{n} i^2 = \frac{n(n+1)(2n+1)}{6}$

III. $\displaystyle\sum_{i=1}^{n} (x_i + y_i) = \sum_{i=1}^{n} x_i + \sum_{i=1}^{n} y_i$

We have found that the area of the triangle discussed above was approximated by

$$A \approx \sum_{i=1}^{n} \frac{2i}{n^2}$$

We can use formulas II and IV to simplify this sum as follows.

$$\sum_{i=1}^{n} \frac{2i}{n^2} = \frac{2}{n^2} \sum_{i=1}^{n} i = \frac{2}{n^2} \left[\frac{n(n+1)}{2} \right] = \frac{n+1}{n}$$

Note that this is the same formula we obtained previously using other methods.

We can also use these sum formulas to evaluate a particular sum. For example,

$$\sum_{i=1}^{100}(2i^2 - 3) = \sum_{i=1}^{100}2i^2 - \sum_{i=1}^{100}3(1) \qquad \text{Formula III}$$

$$= 2\sum_{i=1}^{100}i^2 - 3\sum_{i=1}^{100}1 \qquad \text{Formula II}$$

$$= 2\left[\frac{100(101)(201)}{6}\right] - 3(100) \qquad \text{Formulas I and V with } n = 100$$

$$= 676{,}400$$

Areas and Summation Notation   The following example shows that we can find the area by evaluating the function at the left-hand endpoints of the subintervals.

EXAMPLE 2   Area Under a Curve

Use rectangles to find the area under $y = x^2$ (and above the x-axis) from $x = 0$ to $x = 1$.

Solution

We again divide the interval $[0, 1]$ into n equal subintervals of length $1/n$. If we evaluate the function at the left-hand endpoints of these subintervals to determine the heights of the rectangles, the sum of the areas of the rectangles will underestimate the area (see Figure 13.6). Thus we have the information shown in the following table.

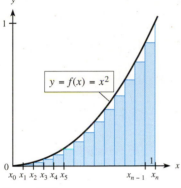

Figure 13.6

Rectangle	Base	Left Endpoint	Height	Area = Base × Height
1	$\dfrac{1}{n}$	$x_0 = 0$	$f(x_0) = 0$	$\dfrac{1}{n}\cdot 0 = 0$
2	$\dfrac{1}{n}$	$x_1 = \dfrac{1}{n}$	$f(x_1) = \dfrac{1}{n^2}$	$\dfrac{1}{n}\cdot\dfrac{1}{n^2} = \dfrac{1}{n^3}$
3	$\dfrac{1}{n}$	$x_2 = \dfrac{2}{n}$	$f(x_2) = \dfrac{4}{n^2}$	$\dfrac{1}{n}\cdot\dfrac{4}{n^2} = \dfrac{4}{n^3}$
4	$\dfrac{1}{n}$	$x_3 = \dfrac{3}{n}$	$f(x_3) = \dfrac{9}{n^2}$	$\dfrac{1}{n}\cdot\dfrac{9}{n^2} = \dfrac{9}{n^3}$
$\vdots$				
i	$\dfrac{1}{n}$	$x_{i-1} = \dfrac{i-1}{n}$	$\dfrac{(i-1)^2}{n^2}$	$\dfrac{(i-1)^2}{n^3}$
$\vdots$				
n	$\dfrac{1}{n}$	$x_{n-1} = \dfrac{n-1}{n}$	$\dfrac{(n-1)^2}{n^2}$	$\dfrac{(n-1)^2}{n^3}$

Thus: Area $= A \approx 0 + \dfrac{1}{n^3} + \dfrac{4}{n^3} + \dfrac{9}{n^3} + \cdots + \dfrac{(i-1)^2}{n^3} + \cdots + \dfrac{(n-1)^2}{n^3}.$

Note that $(i-1)^2/n^3 = (i^2 - 2i + 1)/n^3$ gives the area of the ith rectangle for *any* value of i. The sum of these areas may be written as

$$S = \sum_{i=1}^{n}\frac{i^2 - 2i + 1}{n^3} = \frac{1}{n^3}\left(\sum_{i=1}^{n}i^2 - 2\sum_{i=1}^{n}i + \sum_{i=1}^{n}1\right) \qquad \text{Formulas II and III}$$

$$= \frac{1}{n^3}\left[\frac{n(n+1)(2n+1)}{6} - \frac{2n(n+1)}{2} + n\right] \qquad \text{Formulas V, IV, and I}$$

$$= \frac{2n^3 + 3n^2 + n}{6n^3} - \frac{n^2 + n}{n^3} + \frac{n}{n^3} = \frac{2n^2 - 3n + 1}{6n^2}$$

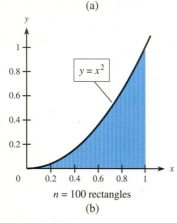

$n = 10$ rectangles
(a)

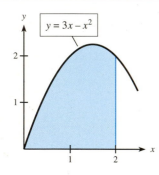

$n = 100$ rectangles
(b)

Figure 13.7

We can use this formula to find the approximate area (value of S) for different values of n.

If $n = 10$:   Area $\approx S(10) = \dfrac{200 - 30 + 1}{600} = 0.285$

If $n = 100$:   Area $\approx S(100) = \dfrac{20{,}000 - 300 + 1}{60{,}000} = 0.328$

Figure 13.7 shows the rectangles associated with each of these approximations.

As Figure 13.7 shows, the larger the value of n, the better the value of the sum approximates the exact area under the curve. If we let n increase without bound, we find the exact area.

$$A = \lim_{n \to \infty}\left(\frac{2n^2 - 3n + 1}{6n^2}\right) = \lim_{n \to \infty}\left(\frac{2 - \dfrac{3}{n} + \dfrac{1}{n^2}}{6}\right) = \frac{1}{3}$$

Note that the approximations with $n = 10$ and $n = 100$ are less than $\frac{1}{3}$. This is because all the rectangles are *under* the curve (see Figure 13.7). ■

Thus we see that we can determine the area under a curve $y = f(x)$ from $x = a$ to $x = b$ by dividing the interval $[a, b]$ into n equal subintervals of width $(b - a)/n$ and evaluating

$$A = \lim_{n \to \infty} S_R = \lim_{n \to \infty} \sum_{i=1}^{n} f(x_i)\left(\frac{b - a}{n}\right) \quad \text{(using right-hand endpoints)}$$

or

$$A = \lim_{n \to \infty} S_L = \lim_{n \to \infty} \sum_{i=1}^{n} f(x_{i-1})\left(\frac{b - a}{n}\right) \quad \text{(using left-hand endpoints)}$$

CHECKPOINT

1. For the interval $[0, 2]$, determine whether the following statements are true or false.

   (a) For 4 subintervals, each subinterval has width $\dfrac{1}{2}$.

   (b) For 200 subintervals, each subinterval has width $\dfrac{1}{100}$.

   (c) For n subintervals, each subinterval has width $\dfrac{2}{n}$.

   (d) For n subintervals, $x_0 = 0$, $x_1 = \dfrac{2}{n}$, $x_2 = 2\left(\dfrac{2}{n}\right), \ldots, x_i = i\left(\dfrac{2}{n}\right), \ldots, x_n = 2$.

2. Find the area under $y = f(x) = 3x - x^2$ from $x = 0$ to $x = 2$ using right-hand endpoints (see the figure). To accomplish this, use $\dfrac{b - a}{n} = \dfrac{2}{n}$, $x_i = \dfrac{2i}{n}$, and $f(x) = 3x - x^2$; find and simplify the following.

   (a) $f(x_i)$

   (b) $f(x_i)\left(\dfrac{b - a}{n}\right)$

   (c) $S_R = \displaystyle\sum_{i=1}^{n} f(x_i)\left(\frac{b - a}{n}\right)$

   (d) $A = \displaystyle\lim_{n \to \infty} S_R = \lim_{n \to \infty} \sum_{i=1}^{n} f(x_i)\left(\frac{b - a}{n}\right)$

Technology Note   Graphing calculators and spreadsheets can be used to approximate the area under a curve. These technologies are especially useful when summations approximating the area do not simplify easily. A calculator or Excel can be used to find the sums. Steps for using graphing calculators and Excel are shown in Appendices C and D, Section 13.1, and in the Online Excel Guide that accompanies this text. ■

EXAMPLE 3   **Estimating an Area with Technology**

To approximate the area under the graph of $y = \sqrt{x}$ on the interval $[0, 4]$, do the following.

(a) Use a calculator or computer to find S_L and S_R for $n = 8$ on the interval $[0, 4]$.
(b) Predict the area under the curve in the graph by evaluating S_L and S_R for larger values of n.

Solution

Figure 13.8(a) shows the graph of $y = \sqrt{x}$ with 8 rectangles whose heights are determined by evaluating the function at the left-hand endpoint of each interval (the first of these rectangles has height 0). Figure 13.8(b) shows the same graph with 8 rectangles whose heights are determined at the right-hand endpoints.

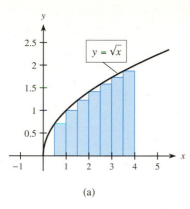

(a)

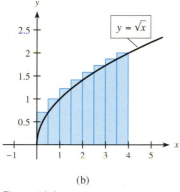

(b)

Figure 13.8

(a) The width of the ith rectangle is $\dfrac{b - a}{n} = \dfrac{4}{n}$, and its endpoints are $x_i = \dfrac{4i}{n}$ and $x_{i-1} = \dfrac{4(i - 1)}{n}$. The formula for the sum of the areas of n rectangles over the interval $[0, 4]$, using left-hand endpoints, is

$$S_L = \sum_{i=1}^{n} \sqrt{\frac{4(i - 1)}{n}} \frac{4}{n} = \frac{8}{n^{3/2}} \sum_{i=1}^{n} \sqrt{i - 1}$$

and the formula for the sum of the areas of n rectangles over the interval $[0, 4]$, using right-hand endpoints, is

$$S_R = \sum_{i=1}^{n} \sqrt{\frac{4(i)}{n}} \frac{4}{n} = \frac{8}{n^{3/2}} \sum_{i=1}^{n} \sqrt{i}$$

We have no formulas to write either of these summations in a simpler form, but we can find the sums by hand if n is small, and we can use calculators, computer programs, or spreadsheets to find the sums for large n. For $n = 8$, the sums are

$$S_L = \frac{1}{\sqrt{8}} \sum_{i=1}^{8} \sqrt{i - 1} \quad \text{and} \quad S_R = \frac{1}{\sqrt{8}} \sum_{i=1}^{8} \sqrt{i}$$

The following Excel output shows the sum for $n = 8$ rectangles.

	A	B	C	D
1	Area for	n=8		
2		i	SL	SR
3		1	0	.35355
4		2	.35355	.5
5		3	.5	.61237
6		4	.61237	.70711
7		5	.70711	.79057
8		6	.79057	.86603
9		7	.86603	.93541
10		8	.93541	1
11	Total		4.76504	5.76504

As line 11 shows, $S_L = 4.76504$ and $S_R = 5.76504$ for $n = 8$.

(b) Larger values of n give better approximations of the area under the curve (try some). For example, $n = 1000$ gives $S_L = 5.3293$ and $S_R = 5.3373$, and because the area under the curve is between these values, the area is approximately 5.33, to two decimal places. By using values of n larger than 1000, we can get even better approximations of the area.

CHECKPOINT SOLUTIONS

1. All parts are true.

2. (a) $f(x_i) = f\left(\dfrac{2i}{n}\right) = 3\left(\dfrac{2i}{n}\right) - \left(\dfrac{2i}{n}\right)^2 = \dfrac{6i}{n} - \dfrac{4i^2}{n^2}$

   (b) $f(x_i)\dfrac{b - a}{n} = \left(\dfrac{6i}{n} - \dfrac{4i^2}{n^2}\right)\left(\dfrac{2}{n}\right) = \dfrac{12i}{n^2} - \dfrac{8i^2}{n^3}$

   (c) $S_R = \displaystyle\sum_{i=1}^{n} f(x_i)\dfrac{b - a}{n} = \sum_{i=1}^{n}\left(\dfrac{12i}{n^2} - \dfrac{8i^2}{n^3}\right)$

   $\qquad = \displaystyle\sum_{i=1}^{n}\dfrac{12i}{n^2} - \sum_{i=1}^{n}\dfrac{8i^2}{n^3}$

   $\qquad = \dfrac{12}{n^2}\displaystyle\sum_{i=1}^{n}i - \dfrac{8}{n^3}\sum_{i=1}^{n}i^2$

   $\qquad = \dfrac{12}{n^2}\left[\dfrac{n(n + 1)}{2}\right] - \dfrac{8}{n^3}\left[\dfrac{n(n + 1)(2n + 1)}{6}\right]$

   $\qquad = \dfrac{6(n + 1)}{n} - \dfrac{4(n + 1)(2n + 1)}{3n^2}$

   (d) $A = \displaystyle\lim_{n\to\infty}\sum_{i=1}^{n} f(x_i)\dfrac{b - a}{n} = \lim_{n\to\infty}\left(\dfrac{6n + 6}{n} - \dfrac{8n^2 + 12n + 4}{3n^2}\right) = 6 - \dfrac{8}{3} = \dfrac{10}{3}$

| EXERCISES | 13.1

Access end-of-section exercises online at **www.webassign.net**

The Definite Integral: The Fundamental Theorem of Calculus

| APPLICATION PREVIEW |

Suppose that money flows continuously into a slot machine at a casino and grows at a rate given by $A'(t) = 100e^{0.1t}$, where t is the time in hours and $0 \leq t \leq 10$. Then the definite integral

$$\int_0^{10} 100\, e^{0.1t}\, dt$$

gives the total amount of money that accumulates over the 10-hour period, if no money is paid out. (See Example 6.)

In the previous section, we used the sums of areas of rectangles to approximate the areas under curves. In this section, we will see how such sums are related to the definite integral and how to evaluate definite integrals. In addition, we will see how definite integrals can be used to solve several types of applied problems.

Riemann Sums and the Definite Integral

In the previous section, we saw that we could determine the area under a curve $y = f(x)$ over a closed interval $[a, b]$ by using equal subintervals and the function values at either the left-hand endpoints or the right-hand endpoints of the subintervals. In fact, we can use subintervals that are not of equal length, and we can use any point within each subinterval, denoted by x_i^*, to determine the height of each rectangle. For the ith rectangle (for any i), if we denote the width as Δx_i then the height is $f(x_i^*)$ and the area is $f(x_i^*)\, \Delta x_i$. Then, if $[a, b]$ is divided into n subintervals, the sum of the areas of the n rectangles is

$$S = \sum_{i=1}^{n} f(x_i^*)\, \Delta x_i,$$

Such a sum is called a **Riemann sum** of f on $[a, b]$. Increasing the number of subintervals and making sure that every interval becomes smaller will in the long run improve the estimation. Thus for any subdivision of $[a, b]$ and any x_i^*, the exact area is given by

$$A = \lim_{\substack{\max \Delta x_i \to 0 \\ (n \to \infty)}} \sum_{i=1}^{n} f(x_i^*)\, \Delta x_i, \qquad \text{provided that this limit exists}$$

In addition to giving the exact area, this limit of the Riemann sum has other important applications and is called the **definite integral** of $f(x)$ over interval $[a, b]$.

Definite Integral

If f is a function on the interval $[a, b]$, then, for any subdivision of $[a, b]$ and any choice of x_i^* in the ith subinterval, the **definite integral** of f from a to b is

$$\int_a^b f(x)\, dx = \lim_{\substack{\max \Delta x_i \to 0 \\ (n \to \infty)}} \sum_{i=1}^{n} f(x_i^*)\, \Delta x_i$$

If f is a continuous function, and $\Delta x_i \to 0$ as $n \to \infty$, then the limit exists and we say that f is integrable on $[a, b]$.

Note that for some intervals, values of f may be negative. In this case, the product $f(x_i^*)\Delta x_i$ will be negative and can be thought of geometrically as a "signed area." (Remember that area is a positive number.) Thus a definite integral can be thought of geometrically as the sum of signed areas, just as a derivative can be thought of geometrically as the slope of a tangent line. In the case where $f(x)$ is positive for all x from a to b, the definite integral equals the area between the graph of $y = f(x)$ and the x-axis.

Fundamental Theorem of Calculus   The obvious question is how this definite integral is related to the indefinite integral (the antiderivative) discussed in Chapter 12. The connection between these two concepts is the most important result in calculus, because it connects derivatives, indefinite integrals, and definite integrals.

To help see the connection, consider the marginal revenue function

$$R'(x) = 300 - 0.2x$$

and the revenue function

$$R(x) = \int (300 - 0.2x)\, dx = 300x - 0.1x^2$$

that is the indefinite integral of the marginal revenue function. (See Figure 13.9(a).)

Using this revenue function, we can find the revenue from the sale of 1000 units to be

$$R(1000) = 300(1000) - 0.1(1000)^2 = 200{,}000 \text{ dollars}$$

and the revenue from the sale of 500 units to be

$$R(500) = 300(500) - 0.1(500)^2 = 125{,}000 \text{ dollars}$$

Thus the additional revenue received from the sale of 500 units to 1000 units is

$$200{,}000 - 125{,}000 = 75{,}000 \text{ dollars}$$

If we used the definition of the definite integral (or geometry) to find the area under the graph of the marginal revenue function from $x = 500$ to $x = 1000$, we would find that the area is 75,000 (see Figure 13.9(b)). Note in Figure 13.9(b) that the shaded area is given by

$$\text{Area} = \text{Area} (\Delta \text{ with base from 500 to 1500}) - \text{Area} (\Delta \text{ with base from 1000 to 1500})$$

Recall that the area of a triangle is $\frac{1}{2}$ (base)(height) so we have

$$\text{Area} = \tfrac{1}{2}(1000)(200) - \tfrac{1}{2}(500)(100) = 75{,}000$$

Thus we can also find this additional revenue when sales are increased from 500 to 1000 by evaluating the *definite integral*

$$\int_{500}^{1000} (300 - 0.2x)\, dx$$

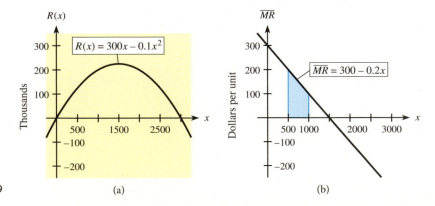

Figure 13.9   (a)   (b)

In general, the definite integral

$$\int_a^b f(x)\, dx$$

can be used to find the change in the function $F(x)$ when x changes from a to b, where $f(x)$ is the derivative of $F(x)$. This result is the **Fundamental Theorem of Calculus.**

Fundamental Theorem of Calculus	Let f be a continuous function on the closed interval $[a, b]$; then the definite integral of f exists on this interval, and

$$\int_a^b f(x)\, dx = F(b) - F(a)$$

where F is any function such that $F'(x) = f(x)$ for all x in $[a, b]$.

Stated differently, this theorem says that if the function F is an indefinite integral of a function f that is continuous on the interval $[a, b]$, then

$$\int_a^b f(x)\, dx = F(b) - F(a)$$

Thus, we apply the Fundamental Theorem of Calculus by using the following two steps.

1. Integration of $f(x)$: $\displaystyle\int_a^b f(x)\, dx = F(x)\Big|_a^b$

2. Evaluation of $F(x)$: $F(x)\Big|_a^b = F(b) - F(a)$

EXAMPLE   1   Definite Integral

Evaluate $\displaystyle\int_2^4 (x^3 + 4)\, dx$.

Solution

1. $\displaystyle\int_2^4 (x^3 + 4)\, dx = \frac{x^4}{4} + 4x + C\ \Big|_2^4$

2. $\displaystyle\qquad\qquad = \left[\frac{(4)^4}{4} + 4(4) + C \right] - \left[\frac{(2)^4}{4} + 4(2) + C \right]$

    $\displaystyle\qquad\qquad = (64 + 16 + C) - (4 + 8 + C)$

    $\displaystyle\qquad\qquad = 68 \quad \text{(Note that the C's subtract out.)}$ ■

Note that the Fundamental Theorem states that F can be *any* indefinite integral of f, so we need not add the constant of integration to the integral.

EXAMPLE   2   Fundamental Theorem

Evaluate $\displaystyle\int_1^3 (3x^2 + 6x)\, dx$.

Solution

$$\int_1^3 (3x^2 + 6x)\, dx = x^3 + 3x^2 \Big|_1^3$$

$$= (3^3 + 3 \cdot 3^2) - (1^3 + 3 \cdot 1^2)$$

$$= 54 - 4 = 50 \qquad \blacksquare$$

Properties   The properties of definite integrals given next follow from properties of summations.

1. $\displaystyle\int_a^b [f(x) \pm g(x)]\, dx = \int_a^b f(x)\, dx \pm \int_a^b g(x)\, dx$

2. $\displaystyle\int_a^b kf(x)\, dx = k \int_a^b f(x)\, dx,$ where k is a constant

The following example uses both of these properties.

EXAMPLE 3   Definite Integral

Evaluate $\displaystyle\int_3^5 (\sqrt{x^2 - 9} + 2)x\, dx$.

Solution

$$\int_3^5 (\sqrt{x^2 - 9} + 2)x\, dx = \int_3^5 \sqrt{x^2 - 9}(x\, dx) + \int_3^5 2x\, dx$$

$$= \frac{1}{2}\int_3^5 (x^2 - 9)^{1/2}(2x\, dx) + \int_3^5 2x\, dx$$

$$= \frac{1}{2}\left[\frac{2}{3}(x^2 - 9)^{3/2}\right]_3^5 + x^2\Big|_3^5$$

$$= \frac{1}{3}[(16)^{3/2} - (0)^{3/2}] + (25 - 9)$$

$$= \frac{64}{3} + 16 = \frac{64}{3} + \frac{48}{3} = \frac{112}{3} \qquad \blacksquare$$

In the integral $\int_a^b f(x)\, dx$, we call a the *lower limit* and b the *upper limit* of integration. Although we developed the definite integral with the assumption that the lower limit was less than the upper limit, the following properties permit us to evaluate the definite integral even when that is not the case.

3. $\displaystyle\int_a^a f(x)\, dx = 0$

4. If f is integrable on $[a, b]$, then

$$\int_b^a f(x)\, dx = -\int_a^b f(x)\, dx$$

The following examples illustrate these properties.

EXAMPLE 4   Properties of Definite Integrals

(a) Evaluate $\displaystyle\int_4^4 x^2\, dx$.   (b) Compare $\displaystyle\int_2^4 3x^2\, dx$ and $\displaystyle\int_4^2 3x^2\, dx$.

Solution

(a) $\displaystyle\int_4^4 x^2\,dx = \frac{x^3}{3}\bigg|_4^4 = \frac{4^3}{3} - \frac{4^3}{3} = 0$

This illustrates Property 3.

(b) $\displaystyle\int_2^4 3x^2\,dx = x^3\bigg|_2^4 = 4^3 - 2^3 = 56$ and $\displaystyle\int_4^2 3x^2\,dx = x^3\bigg|_4^2 = 2^3 - 4^3 = -56$

This illustrates Property 4. ■

Another property of definite integrals is called the additive property.

5. If f is continuous on some interval containing a, b, and c,* then

$$\int_a^c f(x)\,dx + \int_c^b f(x)\,dx = \int_a^b f(x)\,dx$$

EXAMPLE 5 **Properties of Definite Integrals**

Show that $\displaystyle\int_2^3 4x\,dx + \int_3^5 4x\,dx = \int_2^5 4x\,dx$.

Solution

$$\int_2^3 4x\,dx = 2x^2\bigg|_2^3 = 18 - 8 = 10 \quad\text{and}\quad \int_3^5 4x\,dx = 2x^2\bigg|_3^5 = 50 - 18 = 32$$

Thus $\displaystyle\int_2^5 4x\,dx = 2x^2\bigg|_2^5 = 50 - 8 = 42 = \int_2^3 4x\,dx + \int_3^5 4x\,dx$ ■

The Definite Integral and Areas

Let us now return to area problems, to see the relationship between the definite integral and the area under a curve. By the formula for the area of a triangle or by using rectangles and the limit definition of area, the area under the curve (line) $y = x$ from $x = 0$ to $x = 1$ can be shown to be $\frac{1}{2}$ (see Figure 13.10(a) on the next page). Using the definite integral to find the area gives

$$A = \int_0^1 x\,dx = \frac{x^2}{2}\bigg|_0^1 = \frac{1}{2} - 0 = \frac{1}{2}$$

In the previous section, we used rectangles to find that the area under $y = x^2$ from $x = 0$ to $x = 1$ was $\frac{1}{3}$ (see Figure 13.10(b)). Using the definite integral, we get

$$A = \int_0^1 x^2\,dx = \frac{x^3}{3}\bigg|_0^1 = \frac{1}{3} - 0 = \frac{1}{3}$$

which agrees with the answer obtained previously.

However, not every definite integral represents the area between the curve and the x-axis over an interval. For example,

$$\int_0^2 (x - 2)\,dx = \frac{x^2}{2} - 2x\bigg|_0^2 = (2 - 4) - (0) = -2$$

This would indicate that the area between the curve and the x-axis is negative, but area must be positive. A look at the graph of $y = x - 2$ (see Figure 13.10(c)) shows us what is happening.

*Note that c need not be between a and b.

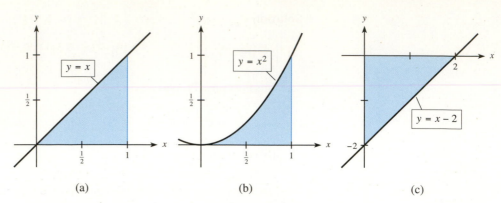

Figure 13.10   (a)   (b)   (c)

The region bounded by $y = x - 2$ and the x-axis between $x = 0$ and $x = 2$ is a triangle whose base is 2 and whose height is 2, so its area is $\frac{1}{2}bh = \frac{1}{2}(2)(2) = 2$. The integral has value -2 because $y = x - 2$ lies below the x-axis from $x = 0$ to $x = 2$, and the function values over the interval $[0, 2]$ are negative. Thus the value of the definite integral over *this* interval does not represent the area between the curve and the x-axis but, rather, represents the "signed" area as mentioned previously.

In general, the definite integral will give the **area under the curve** and above the x-axis only when $f(x) \geq 0$ for all x in $[a, b]$.

| **Area Under a Curve** | If f is a continuous function on $[a, b]$ and $f(x) \geq 0$ on $[a, b]$, then the exact area between $y = f(x)$ and the x-axis from $x = a$ to $x = b$ is given by $$\begin{matrix} \text{Area} \\ \text{(shaded)} \end{matrix} = \int_a^b f(x)\,dx$$ | 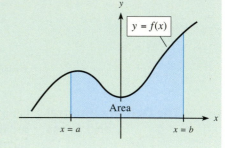 |

Note also that if $f(x) \leq 0$ for all x in $[a, b]$, then

$$\int_a^b f(x)\,dx = -\text{Area (between } f(x) \text{ and the } x\text{-axis)}$$

CHECKPOINT

1. True or false:

   (a) For any integral, we can omit the constant of integration (the $+C$).

   (b) $-\displaystyle\int_{-1}^3 f(x)\,dx = \int_3^{-1} f(x)\,dx$, if f is integrable on $[-1, 3]$.

   (c) The area between $f(x)$ and the x-axis on the interval $[a, b]$ is given by $\displaystyle\int_a^b f(x)\,dx$.

2. Evaluate:

   (a) $\displaystyle\int_0^3 (x^2 + 1)\,dx$ (b) $\displaystyle\int_0^3 (x^2 + 1)^4 x\,dx$

If the rate of growth of some function with respect to time t is $f'(t)$, then the total growth of the function during the period from $t = 0$ to $t = k$ can be found by evaluating the definite integral

$$\int_0^k f'(t)\,dt = f(t)\Big|_0^k = f(k) - f(0)$$

For nonnegative rates of growth, this definite integral (and thus growth) is the same as the area under the graph of $f'(t)$ from $t = 0$ to $t = k$.

EXAMPLE 6   **Income Stream** | **APPLICATION PREVIEW** |

Suppose that money flows continuously into a slot machine at a casino and grows at a rate given by

$$A'(t) = 100e^{0.1t}$$

where t is time in hours and $0 \leq t \leq 10$. Find the total amount that accumulates in the machine during the 10-hour period, if no money is paid out.

Solution
The total dollar amount is given by

$$A = \int_0^{10} 100e^{0.1t}\,dt = \frac{100}{0.1}\int_0^{10} e^{0.1t}(0.1)\,dt = 1000e^{0.1t}\Big|_0^{10} = 1000e - 1000 \approx 1718.28 \quad \blacksquare$$

In Section 8.4, "Normal Probability Distribution," we stated that the total area under the normal curve is 1 and that the area under the curve from value x_1 to value x_2 represents the probability that a score chosen at random will lie between x_1 and x_2.

The normal distribution is an example of a **continuous probability distribution** because the values of the random variable are considered over intervals rather than at discrete values. The statements above relating probability and area under the graph apply to other continuous probability distributions determined by **probability density functions.** In fact, if x is a continuous random variable with probability density function $f(x)$, then the probability that x is between a and b is

$$\Pr(a \leq x \leq b) = \int_a^b f(x)\,dx$$

EXAMPLE 7   **Product Life**

Suppose the probability density function for the life of a computer component is $f(x) = 0.10e^{-0.10x}$, where $x \geq 0$ is the number of years the component is in use. Find the probability that the component will last between 3 and 5 years.

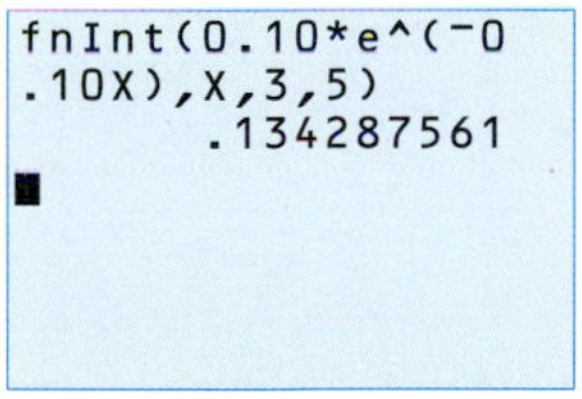

```
fnInt(0.10*e^(-0
.10X),X,3,5)
          .134287561
■
```

Solution
The probability that the component will last between 3 and 5 years is the area under the graph of the function between $x = 3$ and $x = 5$. The probability is given by

$$\Pr(3 \leq x \leq 5) = \int_3^5 0.10e^{-0.10x}\,dx = -e^{-0.10x}\Big|_3^5 = -e^{-0.5} + e^{-0.3}$$

Figure 13.11

$$\approx -0.6065 + 0.7408 = 0.1343 \quad \blacksquare$$

Calculator Note   Most graphing calculators have a numerical integration feature that can be used to get an accurate approximation of any definite integral. This feature can be used to evaluate definite integrals directly or to check those done with the Fundamental Theorem of Calculus. Figure 13.11 shows the numerical integration feature applied to the integral in Example 7. Note that when this answer is rounded to 4 decimal places, the results agree. Two methods of approximating definite integrals are shown in Appendix C, Section 13.1. ■

CHECKPOINT SOLUTIONS

1. (a) False. We can omit the constant of integration $(+C)$ only for definite integrals.
   (b) True   (c) False. Only if $f(x) \geq 0$ on $[a, b]$ is this true.

2. (a) $\displaystyle\int_0^3 (x^2 + 1)\,dx = \frac{x^3}{3} + x\Big|_0^3 = \left(\frac{27}{3} + 3\right) - (0 + 0) = 12$

   (b) $\displaystyle\int_0^3 (x^2 + 1)^4 x\,dx = \frac{1}{2}\int_0^3 (x^2 + 1)^4(2x\,dx)$

   $$= \frac{1}{2} \cdot \frac{(x^2 + 1)^5}{5}\Big|_0^3$$

   $$= \frac{1}{10}[(3^2 + 1)^5 - (1)^5] = \frac{1}{10}(10^5 - 1) = 9999.9$$

| **EXERCISES** | **13.2**

OBJECTIVES   **13.3**

- To find the area between two curves
- To find the average value of a function

Area Between Two Curves

| APPLICATION PREVIEW |

In economics, the **Lorenz curve** is used to represent the inequality of income distribution among different groups in the population of a country. The curve is constructed by plotting the cumulative percent of families at or below a given income level and the cumulative percent of total personal income received by these families. For example, the table shows the coordinates of some points on the Lorenz curve $y = L(x)$ that divide the income (for the United States in 2006) into 5 equal income levels (quintiles). The point (0.40, 0.120) is on the Lorenz curve because the families with incomes in the bottom 40% of the country received 12.0% of the total income in 2006. The graph of the Lorenz curve $y = L(x)$ is shown in Figure 13.12.

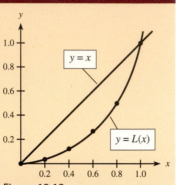

Figure 13.12

U.S. Income Distribution for 2006 (Points on the Lorenz Curve)

x, Cumulative Proportion of Families Below Income Level	y = L(x), Cumulative Proportion of Total Income
0	0
0.20	0.034
0.40	0.120
0.60	0.265
0.80	0.498
1	1

Source: U.S. Bureau of the Census

Equality of income would result if each family received an equal proportion of the total income, so that the bottom 20% would receive 20% of the total income, the bottom 40% would receive 40%, and so on. The Lorenz curve representing this would have the equation $y = x$.

The inequality of income distribution is measured by the Gini coefficient of income, which measures how far the Lorenz curve falls below $y = x$. It is defined as

$$\frac{\text{Area between } y = x \text{ and } y = L(x)}{\text{Area below } y = x}$$

Because the area of the triangle below $y = x$ and above the x-axis from $x = 0$ to $x = 1$ is 1/2, the Gini coefficient of income is

$$\frac{\text{Area between } y = x \text{ and } y = L(x)}{1/2} = 2 \cdot [\text{area between } y = x \text{ and } y = L(x)]$$

In this section we will use the definite integral to find the area between two curves. We will use the area between two curves to find the Gini coefficient of income (see Example 4) and to find average cost, average revenue, average profit, and average inventory.

Area Between Two Curves

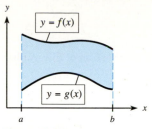

Figure 13.13

We have used the definite integral to find the area of the region between a curve and the x-axis over an interval where the curve lies above the x-axis. We can easily extend this technique to finding the area between two curves over an interval.

Suppose that the graphs of both $y = f(x)$ and $y = g(x)$ lie above the x-axis and that the graph of $y = f(x)$ lies above $y = g(x)$ throughout the interval from $x = a$ to $x = b$; that is, $f(x) \geq g(x)$ on $[a, b]$. (See Figure 13.13.)

Then Figures 13.14(a) and 13.14(b) show the areas under $y = f(x)$ and $y = g(x)$. Figure 13.14(c) shows how the difference of these two areas can be used to find the area of the region between the graphs of $y = f(x)$ and $y = g(x)$. That is,

$$\text{Area between the curves} = \int_a^b f(x)\, dx - \int_a^b g(x)\, dx$$

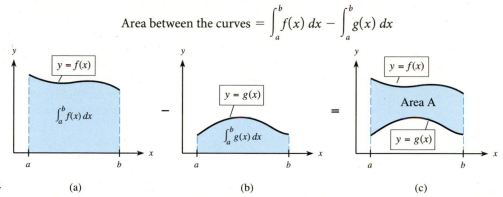

Figure 13.14   (a)   (b)   (c)

Although Figure 13.14(c) shows the graphs of both $y = f(x)$ and $y = g(x)$ lying above the x-axis, this difference of their integrals will always give the area between their graphs if both functions are continuous and if $f(x) \geq g(x)$ on the interval $[a, b]$. Using the fact that

$$\int_a^b f(x)\, dx - \int_a^b g(x)\, dx = \int_a^b [f(x) - g(x)]\, dx$$

we have the following result for the **area between two curves.**

Area Between Two Curves	If f and g are continuous functions on $[a, b]$ and if $f(x) \geq g(x)$ on $[a, b]$, then the area of the region bounded by $y = f(x), y = g(x), x = a$, and $x = b$ is $$A = \int_a^b [f(x) - g(x)]\, dx$$

EXAMPLE 1   Area Between Two Curves

Find the area of the region bounded by $y = x^2 + 4, y = x, x = 0$, and $x = 3$.

Solution

We first sketch the graphs of the functions on the same set of axes. The graph of the region is shown in Figure 13.15. Because $y = x^2 + 4$ lies above $y = x$ in the interval from $x = 0$ to $x = 3$, the area is

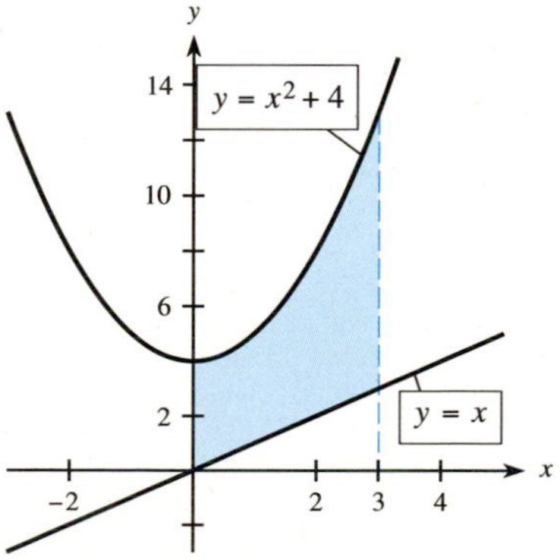

$$A = \int (\text{top curve} - \text{bottom curve})\, dx$$

$$A = \int_0^3 [(x^2 + 4) - x]\, dx = \left. \frac{x^3}{3} + 4x - \frac{x^2}{2} \right|_0^3$$

$$= \left(9 + 12 - \frac{9}{2} \right) - (0 + 0 - 0) = 16\tfrac{1}{2} \text{ square units} \quad ■$$

Figure 13.15

We are sometimes asked to find the area enclosed by two curves. In this case, we find the points of intersection of the curves to determine a and b.

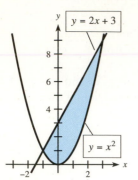

Figure 13.16

EXAMPLE 2   Area Enclosed by Two Curves

Find the area enclosed by $y = x^2$ and $y = 2x + 3$.

Solution
We first find a and b by finding the x-coordinates of the points of intersection of the graphs. Setting the y-values equal gives

$$x^2 = 2x + 3$$
$$x^2 - 2x - 3 = 0$$
$$(x - 3)(x + 1) = 0$$
$$x = 3, \quad x = -1$$

Thus with $a = -1$ and $b = 3$, we sketch the graphs of these functions on the same set of axes. Because the graphs do not intersect on the interval $(-1, 3)$, we can determine which function is larger on this interval by evaluating $2x + 3$ and x^2 at any value c where $-1 < c < 3$. Figure 13.16 shows the region between the graphs, with $2x + 3 \geq x^2$ from $x = -1$ to $x = 3$. The area of the enclosed region is

$$A = \int_{-1}^{3} [(2x + 3) - x^2]\, dx = x^2 + 3x - \frac{x^3}{3} \Big|_{-1}^{3}$$

$$= (9 + 9 - 9) - \left(1 - 3 + \frac{1}{3}\right) = 10\tfrac{2}{3} \text{ square units} \qquad \blacksquare$$

Some graphs enclose two or more regions because they have more than two points of intersection.

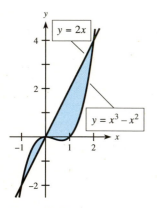

Figure 13.17

EXAMPLE 3   A Region with Two Sections

Find the area of the region enclosed by the graphs of
$$y = f(x) = x^3 - x^2 \qquad \text{and} \qquad y = g(x) = 2x.$$

Solution
To find the points of intersection of the graphs, we solve $f(x) = g(x)$, or $x^3 - x^2 = 2x$.

$$x^3 - x^2 - 2x = 0$$
$$x(x - 2)(x + 1) = 0$$
$$x = 0, \quad x = 2, \quad x = -1$$

Graphing these functions between $x = -1$ and $x = 2$, we see that for any x-value in the interval $(-1, 0)$, $f(x) \geq g(x)$, so $f(x) \geq g(x)$ for the region enclosed by the curves from $x = -1$ to $x = 0$. But evaluating the functions for any x-value in the interval $(0, 2)$ shows that $f(x) \leq g(x)$ for the region enclosed by the curves from $x = 0$ to $x = 2$. See Figure 13.17.

Thus we need one integral to find the area of the region from $x = -1$ to $x = 0$ and a second integral to find the area from $x = 0$ to $x = 2$. The area is found by summing these two integrals.

$$A = \int_{-1}^{0} [(x^3 - x^2) - (2x)]\, dx + \int_{0}^{2} [(2x) - (x^3 - x^2)]\, dx$$

$$= \int_{-1}^{0} (x^3 - x^2 - 2x)\, dx + \int_{0}^{2} (2x - x^3 + x^2)\, dx$$

$$= \left(\frac{x^4}{4} - \frac{x^3}{3} - x^2\right)\Big|_{-1}^{0} + \left(x^2 - \frac{x^4}{4} + \frac{x^3}{3}\right)\Big|_{0}^{2}$$

$$= \left[(0) - \left(\frac{1}{4} - \frac{-1}{3} - 1\right)\right] + \left[\left(4 - \frac{16}{4} + \frac{8}{3}\right) - (0)\right] = \frac{37}{12}$$

The area between the curves is $\frac{37}{12}$ square units. $\qquad \blacksquare$

Calculator Note Most graphing calculators have a numerical integration feature, and some can perform both symbolic and numerical integration. Appendix C, Section 13.2, shows two methods of approximating the area between two curves with a graphing calculator. ■

CHECKPOINT

1. True or false:
   (a) Over the interval $[a, b]$, the area between the continuous functions $f(x)$ and $g(x)$ is

   $$\int_a^b [f(x) - g(x)]\, dx$$

   (b) If $f(x) \geq g(x)$ and the area between $f(x)$ and $g(x)$ is given by

   $$\int_a^b [f(x) - g(x)]\, dx$$

   then $x = a$ and $x = b$ represent the left and right boundaries, respectively, of the region.
   (c) To find points of intersection of $f(x)$ and $g(x)$, solve $f(x) = g(x)$.

2. Consider the functions $f(x) = x^2 + 3x - 9$ and $g(x) = \dfrac{1}{4}x^2$.

   (a) Find the points of intersection of $f(x)$ and $g(x)$.
   (b) Determine which function is greater than the other between the points found in part (a).
   (c) Set up the integral used to find the area between the curves in the interval between the points found in part (a).
   (d) Find the area.

EXAMPLE **4** **Income Distribution** | APPLICATION PREVIEW |

The inequality of income distribution is measured by the **Gini coefficient** of income, which is defined as

$$\frac{\text{Area between } y = x \text{ and } y = L(x)}{\text{Area below } y = x} = \frac{\displaystyle\int_0^1 [x - L(x)]\, dx}{1/2}$$

$$= 2\int_0^1 [x - L(x)]\, dx$$

The function $y = L(x) = 2.266x^4 - 3.175x^3 + 2.056x^2 - 0.148x$ models the 2006 income distribution data in the Application Preview.

(a) Use this $L(x)$ to find the Gini coefficient of income for 2006.
(b) If the Census Bureau Gini coefficient of income for 1991 is 0.428, during which year is the distribution of income more nearly equal?

Solution
(a) The Gini coefficient of income for 2006 is

$$2\int_0^1 [x - L(x)]\, dx = 2\int_0^1 [x - (2.266x^4 - 3.175x^3 + 2.056x^2 - 0.148x)]\, dx$$

$$= 2[-0.453x^5 + 0.794x^4 - 0.685x^3 + 0.574x^2]_0^1 \approx 0.460$$

(b) Absolute equality of income would occur if the Gini coefficient of income were 0; and smaller coefficients indicate more nearly equal incomes. Thus the distribution of income was more nearly equal in 1991 than in 2006. ■

Average Value   If the graph of $y = f(x)$ lies on or above the x-axis from $x = a$ to $x = b$, then the area between the graph and the x-axis is

$$A = \int_a^b f(x)\, dx \quad \text{(See Figure 13.18(a))}$$

The area A is also the area of a rectangle with base equal to $b - a$ and height equal to the **average value** (or average height) of the function $y = f(x)$ (see Figure 13.18(b)).

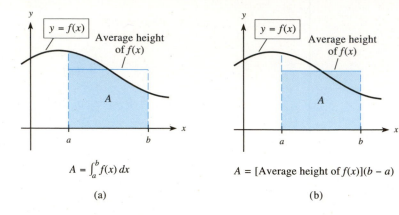

$$A = \int_a^b f(x)\, dx \qquad\qquad A = [\text{Average height of } f(x)](b - a)$$

Figure 13.18   (a)   (b)

Thus the average value of the function in Figure 13.18 is

$$\frac{A}{b - a} = \frac{1}{b - a}\int_a^b f(x)\, dx$$

Even if $f(x) \le 0$ on all or part of the interval $[a, b]$, we can find the average value by using the integral. Thus we have the following.

Average Value   The **average value** of a continuous function $y = f(x)$ over the interval $[a, b]$ is

$$\text{Average value} = \frac{1}{b - a}\int_a^b f(x)\, dx$$

EXAMPLE 5   **Average Cost**

Suppose that the cost in dollars for x table lamps is given by $C(x) = 400 + x + 0.3x^2$.

(a) What is the average value of $C(x)$ for 10 to 20 units?
(b) Find the average cost per unit if 40 lamps are produced.

Solution
(a) The average value of $C(x)$ from $x = 10$ to $x = 20$ is

$$\frac{1}{20 - 10}\int_{10}^{20}(400 + x + 0.3x^2)\, dx = \frac{1}{10}\left(400x + \frac{x^2}{2} + 0.1x^3\right)\Big|_{10}^{20}$$

$$= \frac{1}{10}[(8000 + 200 + 800) - (4000 + 50 + 100)]$$

$$= 485 \quad \text{(dollars)}$$

Thus for any number of lamps between 10 and 20 the *average total cost* for that number of lamps is $485.

(b) The average cost per unit if 40 units are produced is the average cost function evaluated at $x = 40$. The average cost function is

$$\overline{C}(x) = \frac{C(x)}{x} = \frac{400}{x} + 1 + 0.3x$$

Thus the *average cost per unit* if 40 units are produced is

$$\overline{C}(40) = \frac{400}{40} + 1 + 0.3(40) = 23 \quad [\text{dollars per unit (i.e., lamp)}]$$ ■

CHECKPOINT | 3. Find the average value of $f(x) = x^2 - 4$ over $[-1, 3]$.

EXAMPLE 6   **Average Value of a Function**

Consider the functions $f(x) = x^2 - 4$ and $g(x) = x^3 - 4x$. For each function, do the following.

(a) Graph the function on the interval $[-3, 3]$.
(b) On the graph, "eyeball" the average value (height) of each function on $[-2, 2]$.
(c) Compute the average value of the function over the interval $[-2, 2]$.

Solution
For $f(x) = x^2 - 4$:

(a) The graph of $f(x) = x^2 - 4$ is shown in Figure 13.19(a).
(b) The average height of $f(x)$ over $[-2, 2]$ appears to be near -2.
(c) The average value of $f(x)$ over the interval is given by

$$\frac{1}{2 - (-2)} \int_{-2}^{2} (x^2 - 4)\, dx = \frac{1}{4}\left(\frac{x^3}{3} - 4x\right)\Big|_{-2}^{2}$$

$$= \left(\frac{8}{12} - 2\right) - \left(-\frac{8}{12} + 2\right) = \frac{4}{3} - 4 = -\frac{8}{3} = -2\tfrac{2}{3}$$

For $g(x) = x^3 - 4x$:
(a) The graph of $g(x) = x^3 - 4x$ is shown in Figure 13.19(b).
(b) The average height of $g(x)$ over $[-2, 2]$ appears to be approximately 0.
(c) The average value of $g(x)$ is given by

$$\frac{1}{2 - (-2)} \int_{-2}^{2} (x^3 - 4x)\, dx = \frac{1}{4}\left(\frac{x^4}{4} - \frac{4x^2}{2}\right)\Big|_{-2}^{2}$$

$$= \left(\frac{16}{16} - \frac{16}{8}\right) - \left(\frac{16}{16} - \frac{16}{8}\right) = 0$$ ■

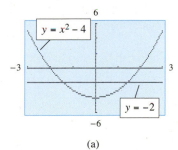

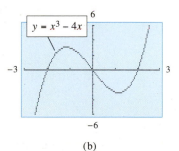

Figure 13.19   (a)   (b)

CHECKPOINT SOLUTIONS

1. (a) False. This is true only if $f(x) \geq g(x)$ over $[a, b]$.
   (b) True   (c) True

2. (a) Solve $f(x) = g(x)$, or $x^2 + 3x - 9 = \frac{1}{4}x^2$.

$$\frac{3}{4}x^2 + 3x - 9 = 0$$
$$3x^2 + 12x - 36 = 0$$
$$x^2 + 4x - 12 = 0$$
$$(x + 6)(x - 2) = 0$$
$$x = -6, x = 2$$

The points of intersection are $(-6, 9)$ and $(2, 1)$.

(b) Evaluating $g(x)$ and $f(x)$ at any point in the interval $(-6, 2)$ shows that $g(x) > f(x)$, so $g(x) \geq f(x)$ on $[-6, 2]$.

(c) $A = \int_{-6}^{2} \left[\frac{1}{4}x^2 - (x^2 + 3x - 9) \right] dx$

(d) $A = \frac{x^3}{12} - \frac{x^3}{3} - \frac{3x^2}{2} + 9x \Big|_{-6}^{2} = 64$ square units

3. $\dfrac{1}{3 - (-1)} \displaystyle\int_{-1}^{3} (x^2 - 4)\, dx = \frac{1}{4}\left(\frac{x^3}{3} - 4x \right)\Big|_{-1}^{3}$

$\qquad\qquad\qquad\qquad = \frac{1}{4}\left[(9 - 12) - \left(-\frac{1}{3} + 4 \right) \right]$

$\qquad\qquad\qquad\qquad = \frac{1}{4}\left(\frac{-20}{3} \right) = -\frac{5}{3}$

| EXERCISES | 13.3

Access end-of-section exercises online at **www.webassign.net**   WebAssign

OBJECTIVES 13.4

- To use definite integrals to find total income, present value, and future value of continuous income streams
- To use definite integrals to find the consumer's surplus
- To use definite integrals to find the producer's surplus

Applications of Definite Integrals in Business and Economics

| APPLICATION PREVIEW |

Suppose the oil pumped from a well is considered as a continuous income stream with annual rate of flow (in thousands of dollars per year) at time t years given by

$$f(t) = 600e^{-0.2(t+5)}$$

The company can use a definite integral involving $f(t)$ to estimate the well's present value over the next 10 years (see Example 2).

The definite integral can be used in a number of applications in business and economics. In addition to the present value, the definite integral can be used to find the total income over a fixed number of years from a continuous income stream.

Definite integrals also can be used to determine the savings realized in the marketplace by some consumers (called consumer's surplus) and some producers (called producer's surplus).

Continuous Income Streams

An oil company's profits depend on the amount of oil that can be pumped from a well. Thus we can consider a pump at an oil field as producing a **continuous stream of income** for the owner. Because both the pump and the oil field "wear out" with time, the continuous stream of income is a function of time. Suppose $f(t)$ is the (annual) *rate* of flow of income from this pump; then we can find the total income from the rate of income by using integration. In particular, the total income for k years is given by

$$\text{Total income} = \int_{0}^{k} f(t)\, dt$$

EXAMPLE 1   Oil Revenue

A small oil company considers the continuous pumping of oil from a well as a continuous income stream with its annual rate of flow at time t given by

$$f(t) = 600e^{-0.2t}$$

in thousands of dollars per year. Find an estimate of the total income from this well over the next 10 years.

Solution

$$\text{Total income} = \int_0^{10} f(t)\, dt = \int_0^{10} 600e^{-0.2t}\, dt$$

$$= \frac{600}{-0.2} e^{-0.2t} \Big|_0^{10} \approx 2594 \quad (\text{to the nearest integer})$$

Thus the total income is approximately $2,594,000. ∎

In addition to the total income from a continuous income stream, the **present value** of the stream is also important. The present value is the value today of a continuous income stream that will be providing income in the future. The present value is useful in deciding when to replace machinery or what new equipment to select.

To find the present value of a continuous stream of income with rate of flow $f(t)$, we first graph the function $f(t)$ and divide the time interval from 0 to k into n subintervals of width Δt_i, $i = 1$ to n.

The total amount of income is the area under this curve between $t = 0$ and $t = k$. We can approximate the amount of income in each subinterval by finding the area of the rectangle in that subinterval. (See Figure 13.20.)

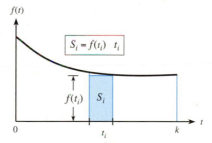

Figure 13.20

We have shown that the future value S that accrues if P is invested for t years at an annual rate r, compounded continuously, is $S = Pe^{rt}$. Thus the present value of the investment that yields the single payment of $S after t years is

$$P = \frac{S}{e^{rt}} = Se^{-rt}$$

The contribution to S in the ith subinterval is $S_i = f(t_i)\, \Delta t_i$, and the present value of this amount is $P_i = f(t_i)\, \Delta t_i e^{-rt_i}$. Thus the total present value of S can be approximated by

$$\sum_{i=1}^{n} f(t_i)\, \Delta t_i e^{-rt_i}$$

This approximation improves as $\Delta t_i \to 0$ with the present value given by

$$\lim_{\Delta t_i \to 0} \sum_{i=1}^{n} f(t_i) \, \Delta t_i e^{-rt_i}$$

This limit gives the **present value** as a definite integral.

Present Value of a Continuous Income Stream	If $f(t)$ is the rate of continuous income flow earning interest at rate r, compounded continuously, then the **present value of the continuous income stream** is $$\text{Present value} = \int_0^k f(t)e^{-rt}\, dt$$ where $t = 0$ to $t = k$ is the time interval.

EXAMPLE 2 Present Value | APPLICATION PREVIEW |

Suppose that the oil company in Example 1 is planning to sell one of its wells because of its remote location. Suppose further that the company wants to use the present value of this well over the next 10 years to help establish its selling price. If the company determines that the annual rate of flow is

$$f(t) = 600e^{-0.2(t+5)}$$

in thousands of dollars per year, and if money is worth 10%, compounded continuously, find this present value.

Solution

$$\text{Present value} = \int_0^{10} f(t)e^{-rt}\, dt$$

$$= \int_0^{10} 600e^{-0.2(t+5)}e^{-0.1t}\, dt = \int_0^{10} 600e^{-0.3t-1}\, dt$$

If $u = -0.3t - 1$, then $u' = -0.3$ and we get

$$\frac{1}{-0.3} \int 600e^{-0.3t-1}(-0.3\, dt) = \frac{600}{-0.3}e^{-0.3t-1}\bigg|_0^{10}$$

$$= -2000(e^{-4} - e^{-1}) \approx 699 \quad (\text{to the nearest integer})$$

Thus the present value is $699,000. ∎

Recall that the future value of a continuously compounded investment at rate r after k years is Pe^{rk}, where P is the amount invested (or the present value). Thus, for a continuous income stream, the **future value** is found as follows.

Future Value of a Continuous Income Stream	If $f(t)$ is the rate of continuous income flow for k years earning interest at rate r, compounded continuously, then the **future value of the continuous income stream** is $$FV = e^{rk}\int_0^k f(t)e^{-rt}\, dt$$

EXAMPLE 3 Future Value

If the rate of flow of income from an asset is $1000e^{0.02t}$, in millions of dollars per year, and if the income is invested at 6% compounded continuously, find the future value of the asset 4 years from now.

Solution

The future value is given by

$$FV = e^{rk} \int_0^k f(t)e^{-rt}\, dt$$

$$= e^{(0.06)4} \int_0^4 1000e^{0.02t}e^{-0.06t}\, dt = e^{0.24} \int_0^4 1000e^{-0.04t}\, dt$$

$$= e^{0.24}(-25{,}000e^{-0.04t})\big|_0^4 = -25{,}000e^{0.24}(e^{-0.16} - 1)$$

$$\approx 4699.05 \quad \text{(millions of dollars)}$$ ■

CHECKPOINT

1. Suppose that a continuous income stream has an annual rate of flow given by $f(t) = 5000e^{-0.01t}$, and suppose that money is worth 7% compounded continuously. Create the integral used to find
   (a) the total income for the next 5 years.   (b) the present value for the next 5 years.
   (c) the future value 5 years from now.

Consumer's Surplus

Suppose that the demand for a product is given by $p = f(x)$ and that the supply of the product is described by $p = g(x)$. The price p_1 where the graphs of these functions intersect is the **equilibrium price** (see Figure 13.21(a)). As the demand curve shows, some consumers (but not all) would be willing to pay more than $\$p_1$ for the product.

For example, some consumers would be willing to buy x_3 units if the price were $\$p_3$. Those consumers willing to pay more than $\$p_1$ are benefiting from the lower price. The total gain for all those consumers willing to pay more than $\$p_1$ is called the **consumer's surplus**, and under proper assumptions the area of the shaded region in Figure 13.21(a) represents this consumer's surplus.

Looking at Figure 13.21(b), we see that if the demand curve has equation $p = f(x)$, the consumer's surplus is given by the area between $f(x)$ and the x-axis from 0 to x_1, *minus* the area of the rectangle denoted *TR*:

$$CS = \int_0^{x_1} f(x)\, dx - p_1 x_1$$

Note that with equilibrium price p_1 and equilibrium quantity x_1, the product $p_1 x_1$ is the area of the rectangle that represents the total dollars spent by consumers and received as revenue by producers (see Figure 13.21(b)).

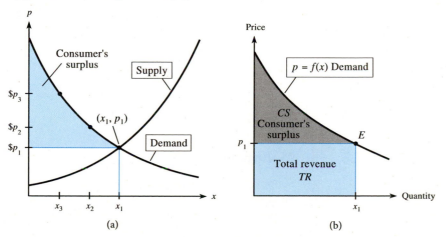

Figure 13.21   (a)   (b)

EXAMPLE 4   Consumer's Surplus

The demand function for x units of a product is $p = 1020/(x + 1)$ dollars. If the equilibrium price is $20, what is the consumer's surplus?

Solution

We must first find the quantity that will be purchased at this price. Letting $p = 20$ and solving for x, we get

$$20 = \frac{1020}{x + 1} \qquad \text{so} \qquad 20(x + 1) = 1020$$

Thus $\quad 20x + 20 = 1020 \quad$ or $\quad 20x = 1000 \quad$ so $\quad x = 50$

Thus the equilibrium point is (50, 20). The consumer's surplus is given by

$$CS = \int_0^{x_1} f(x)\, dx - p_1 x_1 = \int_0^{50} \frac{1020}{x + 1}\, dx - 20 \cdot 50$$

$$= 1020 \ln |x + 1| \Big|_0^{50} - 1000$$

$$= 1020(\ln 51 - \ln 1) - 1000$$

$$\approx 4010.46 - 1000 = 3010.46$$

The consumer's surplus is $3010.46. ∎

EXAMPLE 5 Consumer's Surplus

A product's demand function is $p = \sqrt{49 - 6x}$ and its supply function is $p = x + 1$, where p is the price per unit in dollars and x is the number of units. Find the equilibrium point and the consumer's surplus there.

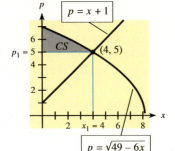

Figure 13.22

Solution

We can determine the equilibrium point by solving the two equations simultaneously.

$$\sqrt{49 - 6x} = x + 1$$
$$49 - 6x = (x + 1)^2$$
$$0 = x^2 + 8x - 48$$
$$0 = (x + 12)(x - 4)$$
$$x = 4 \quad \text{or} \quad x = -12$$

Thus the equilibrium quantity is 4 and the equilibrium price is $5 ($x = -12$ is not a solution). The graphs of the supply and demand functions are shown in Figure 13.22.

The consumer's surplus is given by

$$CS = \int_0^4 f(x)\, dx - p_1 x_1 = \int_0^4 \sqrt{49 - 6x}\, dx - 5 \cdot 4$$

$$= -\frac{1}{6}\int_0^4 \sqrt{49 - 6x}\, (-6\, dx) - 20 = -\frac{1}{9}(49 - 6x)^{3/2}\Big|_0^4 - 20$$

$$= -\frac{1}{9}[(25)^{3/2} - (49)^{3/2}] - 20 = -\frac{1}{9}(125 - 343) - 20 \approx 4.22$$

The consumer's surplus is $4.22. ∎

EXAMPLE 6 Monopoly Market

Suppose a monopoly has its total cost (in dollars) for a product given by $C(x) = 60 + 2x^2$. Suppose also that demand is given by $p = 30 - x$, where p is in dollars and x is the number of units. Find the consumer's surplus at the point where the monopoly has maximum profit.

Solution

We must first find the point where the profit function is maximized. Because the demand for x units is $p = 30 - x$, the total revenue is

$$R(x) = (30 - x)x = 30x - x^2$$

Thus the profit function is

$$P(x) = R(x) - C(x)$$
$$P(x) = 30x - x^2 - (60 + 2x^2)$$
$$P(x) = 30x - 60 - 3x^2$$

Then $P'(x) = 30 - 6x$. So $0 = 30 - 6x$ has the solution $x = 5$.

Because $P''(5) = -6 < 0$, the profit for the monopolist is maximized when $x = 5$ units are sold at price $p = 30 - x = 25$ dollars per unit.

If $f(x)$ is the demand function, the consumer's surplus at $x = 5$, $p = 25$ is given by

$$CS = \int_0^5 f(x)\, dx - 5 \cdot 25 = \int_0^5 (30 - x)\, dx - 125$$

$$= 30x - \frac{x^2}{2}\bigg|_0^5 - 125 = \left(150 - \frac{25}{2}\right) - 125 = 12.50$$

The consumer's surplus is $12.50. ∎

MONEY ON THE TABLE

Another way to understand consumer surplus is through the popular concept of "money on the table." Imagine you have $300 to spend on an iPod. When you get to the Apple store, the model you want costs only $250. You're delighted to spend only $250, but you **would** have spent all your money if the iPod had been priced at $300. In this example, Apple left money—$50 to be exact—on the table (or in your pocket). That's a consumer's surplus.

GRAVY

Now think of the iPod example in another way. Suppose Apple meets all its profit targets at a price of $200 per iPod. The fact that you pay $250 means that there is a producer surplus. Apple would sell that iPod at the $200 price point, but because you were willing to part with $250, the company receives an extra $50—or gravy—for its bottom line. That is, a producer's surplus.

Producer's Surplus

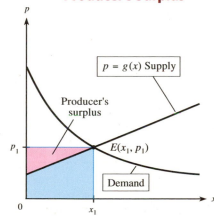

$p = g(x)$ Supply

Producer's surplus

p_1 $E(x_1, p_1)$

Demand

0 x_1

Figure 13.23

When a product is sold at the equilibrium price, some producers will also benefit, for they would have sold the product at a lower price. The area between the line $p = p_1$ and the supply curve (from $x = 0$ to $x = x_1$) gives the producer's surplus (see Figure 13.23).

If the supply function is $p = g(x)$, the **producer's surplus** is given by the area between the graph of $p = g(x)$ and the x-axis from 0 to x_1 *subtracted from* $p_1 x_1$, the area of the rectangle shown in Figure 13.23.

$$PS = p_1 x_1 - \int_0^{x_1} g(x)\, dx$$

Note that $p_1 x_1$ represents the total revenue at the equilibrium point.

EXAMPLE 7   Producer's Surplus

Suppose that the supply function for x million units of a product is $p = x^2 + x$ dollars per unit. If the equilibrium price is $20 per unit, what is the producer's surplus?

Solution

Because $p = 20$, we can find the equilibrium quantity x as follows:

$$20 = x^2 + x$$
$$0 = x^2 + x - 20$$
$$0 = (x + 5)(x - 4)$$
$$x = -5, \quad x = 4$$

The equilibrium point is $x = 4$ million units, $p = \$20$. The producer's surplus is given by

$$PS = 20 \cdot 4 - \int_0^4 (x^2 + x) \, dx$$

$$= 80 - \left(\frac{x^3}{3} + \frac{x^2}{2} \right) \Big|_0^4$$

$$= 80 - \left(\frac{64}{3} + 8 \right)$$

$$\approx 50.67$$

The producer's surplus is $\$50.67$ million. See Figure 13.24.

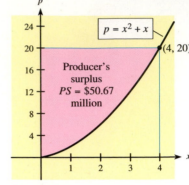

Figure 13.24

EXAMPLE 8   **Producer's Surplus**

The demand function for a product is $p = \sqrt{49 - 6x}$ and the supply function is $p = x + 1$. Find the producer's surplus.

Solution

We found the equilibrium point for these functions to be $(4, 5)$ in Example 5 (see Figure 13.22 earlier in this section). The producer's surplus is

$$PS = 5 \cdot 4 - \int_0^4 (x + 1) \, dx = 20 - \left(\frac{x^2}{2} + x \right) \Big|_0^4$$

$$= 20 - (8 + 4) = 8$$

The producer's surplus is $\$8$.

CHECKPOINT

2. Suppose that for a certain product, the supply function is $p = f(x)$, the demand function is $p = g(x)$, and the equilibrium point is (x_1, p_1). Decide whether the following are true or false.

   (a) $CS = \int_0^{x_1} f(x) \, dx - p_1 x_1$ (b) $PS = \int_0^{x_1} f(x) \, dx - p_1 x_1$

3. If demand is $p = \dfrac{100}{x + 1}$, supply is $p = x + 1$, and the market equilibrium is $(9, 10)$, create the integral used to find the
   (a) consumer's surplus.   (b) producer's surplus.

EXAMPLE 9   **Consumer's Surplus and Producer's Surplus**

Suppose that for x units of a certain product, the demand function is $p = 200e^{-0.01x}$ dollars and the supply function is $p = \sqrt{200x + 49}$ dollars.

(a) Use a graphing calculator to find the market equilibrium point.
(b) Find the consumer's surplus.
(c) Find the producer's surplus.

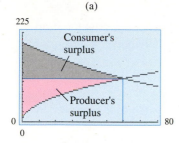

(a)

(b)

Figure 13.25

Solution

(a) Solving $200e^{-0.01x} = \sqrt{200x + 49}$ is very difficult using algebraic techniques. Using SOLVER or INTERSECT on a graphing calculator gives $x = 60$, to the nearest unit, with a price of \$109.76. (See Figure 13.25(a).)

(b) The consumer's surplus, shown in Figure 13.25(b), is

$$\int_0^{60} 200e^{-0.01x}\, dx - 109.76(60) = (-20{,}000e^{-0.01x})|_0^{60} - 6585.60$$
$$= -20{,}000e^{-0.6} + 20{,}000 - 6585.60$$
$$\approx 2438.17 \text{ dollars}$$

(c) The producer's surplus, shown in Figure 13.25(b), is

$$60(109.76) - \int_0^{60} \sqrt{200x + 49}\, dx = 6585.60 - \frac{1}{200}\left[\frac{(200x + 49)^{3/2}}{3/2}\right]_0^{60}$$
$$= 6585.60 - \frac{1}{300}\left[(12{,}049^{3/2} - 49^{3/2})\right] \approx 2178.10 \text{ dollars}$$

Note that we also could have evaluated these definite integrals with the numerical integration feature of a graphing calculator, and we would have obtained the same results. ■

CHECKPOINT SOLUTIONS

1. (a) $\displaystyle\int_0^5 5000e^{-0.01t}\, dt$

(b) $\displaystyle\int_0^5 (5000e^{-0.01t})(e^{-0.07t})\, dt = \int_0^5 5000e^{-0.08t}\, dt$

(c) $\displaystyle e^{(0.07)(5)}\int_0^5 (5000e^{-0.01t})(e^{-0.07t})\, dt = e^{0.35}\int_0^5 5000e^{-0.08t}\, dt$

2. (a) False. Consumer's surplus uses the demand function, so

$$CS = \int_0^{x_1} g(x)\, dx - p_1 x_1$$

(b) False. Producer's surplus uses the supply function, but the formula is

$$PS = p_1 x_1 - \int_0^{x_1} f(x)\, dx$$

3. (a) $\displaystyle CS = \int_0^9 \frac{100}{x + 1}\, dx - 90$ (b) $\displaystyle PS = 90 - \int_0^9 (x + 1)\, dx$

| EXERCISES | 13.4

Access end-of-section exercises online at **www.webassign.net**

OBJECTIVE   **13.5**

- To use tables of integrals to evaluate certain integrals

Using Tables of Integrals

| APPLICATION PREVIEW |

With data from the International Telecommunications Union for 1990 through 2007, the rate of change of total market revenue for worldwide telecommunications (in billions of dollars per year) can be modeled by

$$\frac{dR}{dt} = 37.988(1.101^t)$$

where $t = 0$ represents 1990 and $R = \$920$ billion in 1999. Finding the function for total market revenue or the predicted revenue for 2015 (when $t = 25$) requires evaluating

$$\int 37.988(1.101^t)\, dt$$

Evaluating this integral is made easier by using a formula such as those given in Table 13.2. (See Example 4.)

The formulas in this table, and others listed in other resources, extend the number of integrals that can be evaluated. Using the formulas is not quite as easy as it may sound because finding the correct formula and using it properly may present problems. The examples in this section illustrate how some of these formulas are used.

EXAMPLE **1**   **Using an Integration Formula**

Evaluate $\displaystyle\int \frac{dx}{\sqrt{x^2 + 4}}$.

Solution

We must find a formula in Table 13.2 that is of the same form as this integral. We see that formula 8 has the desired form, *if* we let $u = x$ and $a = 2$. Thus

$$\int \frac{dx}{\sqrt{x^2 + 4}} = \ln\left|x + \sqrt{x^2 + 4}\right| + C \qquad \blacksquare$$

EXAMPLE **2**   **Fitting Integration Formulas**

Evaluate (a) $\displaystyle\int_1^2 \frac{dx}{x^2 + 2x}$ and (b) $\displaystyle\int \ln(2x + 1)\, dx$.

Solution

(a) There does not appear to be any formula that has exactly the same form as this integral. But if we rewrite the integral as

$$\int_1^2 \frac{dx}{x(x + 2)}$$

| TABLE 13.2 |

INTEGRATION FORMULAS

1. $\displaystyle\int u^n \, du = \frac{u^{n+1}}{n+1} + C, \text{ for } n \neq -1$

2. $\displaystyle\int \frac{du}{u} = \int u^{-1} \, du = \ln|u| + C$

3. $\displaystyle\int a^u \, du = a^u \log_a e + C = \frac{a^u}{\ln a} + C$

4. $\displaystyle\int e^u \, du = e^u + C$

5. $\displaystyle\int \frac{du}{a^2 - u^2} = \frac{1}{2a} \ln \left| \frac{a+u}{a-u} \right| + C$

6. $\displaystyle\int \sqrt{u^2 + a^2} \, du = \frac{1}{2}\left(u\sqrt{u^2 + a^2} + a^2 \ln|u + \sqrt{u^2 + a^2}|\right) + C$

7. $\displaystyle\int \sqrt{u^2 - a^2} \, du = \frac{1}{2}\left(u\sqrt{u^2 - a^2} - a^2 \ln|u + \sqrt{u^2 - a^2}|\right) + C$

8. $\displaystyle\int \frac{du}{\sqrt{u^2 + a^2}} = \ln|u + \sqrt{u^2 + a^2}| + C$

9. $\displaystyle\int \frac{du}{u\sqrt{a^2 - u^2}} = -\frac{1}{a} \ln \left| \frac{a + \sqrt{a^2 - u^2}}{u} \right| + C$

10. $\displaystyle\int \frac{du}{\sqrt{u^2 - a^2}} = \ln|u + \sqrt{u^2 - a^2}| + C$

11. $\displaystyle\int \frac{du}{u\sqrt{a^2 + u^2}} = -\frac{1}{a} \ln \left| \frac{a + \sqrt{a^2 + u^2}}{u} \right| + C$

12. $\displaystyle\int \frac{u \, du}{au + b} = \frac{u}{a} - \frac{b}{a^2} \ln|au + b| + C$

13. $\displaystyle\int \frac{du}{u(au + b)} = \frac{1}{b} \ln \left| \frac{u}{au + b} \right| + C$

14. $\displaystyle\int \ln u \, du = u(\ln u - 1) + C$

15. $\displaystyle\int \frac{u \, du}{(au + b)^2} = \frac{1}{a^2}\left(\ln|au + b| + \frac{b}{au + b}\right) + C$

16. $\displaystyle\int u\sqrt{au + b} \, du = \frac{2(3au - 2b)(au + b)^{3/2}}{15a^2} + C$

we see that formula 13 will work. Letting $u = x$, $a = 1$, and $b = 2$, we get

$$\int_1^2 \frac{dx}{x(x+2)} = \frac{1}{2} \ln \left| \frac{x}{x+2} \right| \Big|_1^2 = \frac{1}{2} \ln \left| \frac{2}{4} \right| - \frac{1}{2} \ln \left| \frac{1}{3} \right|$$

$$= \frac{1}{2}\left(\ln \frac{1}{2} - \ln \frac{1}{3}\right)$$

$$= \frac{1}{2} \ln \frac{3}{2} \approx 0.2027$$

(b) This integral has the form of formula 14, with $u = 2x + 1$. But if $u = 2x + 1$, du must be represented by the differential of $2x + 1$ (that is, $2\,dx$). Thus

$$\int \ln(2x + 1)\,dx = \frac{1}{2}\int \ln(2x + 1)(2\,dx)$$

$$= \frac{1}{2}\int \ln(u)\,du = \frac{1}{2}u[\ln(u) - 1] + C$$

$$= \frac{1}{2}(2x + 1)[\ln(2x + 1) - 1] + C \qquad \blacksquare$$

CHECKPOINT

1. Can both $\displaystyle\int \frac{dx}{\sqrt{x^2 - 4}}$ and $\displaystyle -\int \frac{dx}{\sqrt{4 - x^2}}$ be evaluated with formula 10 in Table 13.2?

2. Determine the formula used to evaluate $\displaystyle\int \frac{3x}{4x - 5}\,dx$, and show how the formula would be applied.

3. True or false: In order for us to use a formula, the given integral must correspond exactly to the formula, including du.

4. True or false: $\displaystyle\int \frac{dx}{x^2(3x^2 - 7)}$ can be evaluated with formula 13.

5. True or false: $\displaystyle\int \frac{dx}{(6x + 1)^2}$ can be evaluated with either formula 1 or formula 15.

6. True or false: $\int \sqrt{x^2 + 4}\,dx$ can be evaluated with formula 1, formula 6, or formula 16.

EXAMPLE 3 **Fitting an Integration Formula**

Evaluate $\displaystyle\int_1^2 \frac{dx}{x\sqrt{81 - 9x^2}}$.

Solution
This integral is similar to that of formula 9 in Table 13.2. Letting $a = 9$, letting $u = 3x$, and multiplying the numerator and denominator by 3 gives the proper form.

$$\int_1^2 \frac{dx}{x\sqrt{81 - 9x^2}} = \int_1^2 \frac{3\,dx}{3x\sqrt{81 - 9x^2}}$$

$$= -\frac{1}{9}\ln\left|\frac{9 + \sqrt{81 - 9x^2}}{3x}\right|\Bigg|_1^2$$

$$= \left[-\frac{1}{9}\ln\left(\frac{9 + \sqrt{45}}{6}\right)\right] - \left[-\frac{1}{9}\ln\left(\frac{9 + \sqrt{72}}{3}\right)\right]$$

$$\approx 0.088924836 \qquad \blacksquare$$

Remember that the formulas given in Table 13.2 represent only a very small sample of all possible integration formulas. Additional formulas may be found in books of mathematical tables or online.

EXAMPLE 4   Worldwide Telecommunications Revenue | APPLICATION PREVIEW |

With data from 1990 through 2007, the rate of change of total market revenue for worldwide telecommunications (in billions of dollars per year) can be modeled by

$$\frac{dR}{dt} = 37.988(1.101^t)$$

where $t = 0$ represents 1990 and $R = \$920$ billion in 1999 (*Source:* International Telecommunications Union).

(a) Find the function that models the total market revenue for worldwide telecommunications.
(b) Find the predicted total market revenue for 2015.

Solution
(a) To find $R(t)$ we integrate dR/dt by using formula 3, with $a = 1.101$ and $u = t$.

$$R(t) = \int 37.988\,(1.101^t)\,dt = 37.988 \int (1.101^t)\,dt = 37.988\frac{(1.101^t)}{\ln\,(1.101)} + C$$
$$R(t) \approx 394.808\,(1.101^t) + C$$

We use the fact that $R = \$920$ billion in 1999 (when $t = 9$) to find the value of C.

$$920 = 394.808\,(1.101^9) + C \Rightarrow 920 = 938.6 + C \Rightarrow C = -18.6$$

Thus $R(t) = 394.808\,(1.101^t) - 18.6$.
(b) The predicted revenue in 2015 (when $t = 25$) is

$$R(25) = 394.808\,(1.101^{25}) - 18.6 \approx 4357 \text{ billion dollars}$$ ■

Calculator Note   Numerical integration with a graphing calculator is especially useful in evaluating definite integrals when the formulas for the integrals are difficult to use or are not available. For example, evaluating the definite integral in Example 3 with the numerical integration feature of a graphing calculator gives 0.08892484. The decimal approximation of the answer found in Example 3 is 0.088924836, so the numerical approximation of the answer agrees for the first 8 decimal places. ■

CHECKPOINT SOLUTIONS   1. No. Although $\displaystyle\int \frac{dx}{\sqrt{x^2 - 4}}$ can be evaluated with formula 10 from Table 13.2,

$-\displaystyle\int \frac{dx}{\sqrt{4 - x^2}}$ cannot, because $\sqrt{4 - x^2}$ cannot be rewritten in the form $\sqrt{u^2 - a^2}$

as is needed for this formula to be used.
2. Use formula 12 with $u = x$, $a = 4$, $b = -5$, and $du = dx$.

$$\int \frac{3x}{4x - 5}\,dx = 3\int \frac{x\,dx}{4x - 5} = 3\int \frac{u\,du}{au + b} = 3\left(\frac{u}{a} - \frac{b}{a^2}\ln\,|\,au + b\,| + C\right)$$
$$= \frac{3x}{4} + \frac{15}{16}\ln\,|4x - 5| + C$$

3. True. An exact correspondence with the formula and du is necessary.
4. False. With $u = x^2$, we must have $du = 2x\,dx$. In this problem there is no x with dx, so the problem cannot correspond to formula 13.
5. False. The integral can be evaluated with formula 1, but not with formula 15. The correspondence is $u = 6x + 1$, $du = 6\,dx$, and $n = -2$.
6. False. The integral can be evaluated only with formula 6, with $u = x$, $du = dx$, and $a = 2$, but not with formula 1 or formula 16.

| EXERCISES | 13.5

Access end-of-section exercises online at **www.webassign.net**

OBJECTIVE 13.6

- To evaluate integrals using the method of integration by parts

Integration by Parts

■ | APPLICATION PREVIEW |

If the value of oil produced by a piece of oil extraction equipment is considered a continuous income stream with an annual rate of flow (in dollars per year) at time t in years given by

$$f(t) = 300{,}000 - 2500t, \quad 0 \le t \le 10$$

and if money can be invested at 8%, compounded continuously, then the present value of the piece of equipment is

$$\int_0^{10} (300{,}000 - 2500t)e^{-0.08t}\, dt$$

$$= 300{,}000 \int_0^{10} e^{-0.08t}\, dt - 2500 \int_0^{10} te^{-0.08t}\, dt$$

The first integral can be evaluated with the formula for the integral of $e^u\, du$. Evaluating the second integral can be done by using integration by parts, which is a special technique that involves rewriting an integral in a form that can be evaluated. (See Example 6.)

Integration by parts is an integration technique that uses a formula that follows from the Product Rule for derivatives (actually differentials) as follows:

$$\frac{d}{dx}(uv) = u\frac{dv}{dx} + v\frac{du}{dx} \quad \text{so} \quad d(uv) = u\, dv + v\, du$$

Rearranging the differential form and integrating both sides give the following.

$$u\, dv = d(uv) - v\, du$$

$$\int u\, dv = \int d(uv) - \int v\, du$$

$$\int u\, dv = uv - \int v\, du$$

Integration by Parts Formula

$$\int u\, dv = uv - \int v\, du$$

Integration by parts is very useful if the integral we seek to evaluate can be treated as the product of one function, u, and the differential dv of a second function, so that the two integrals $\int dv$ and $\int v\, du$ can be found. Let us consider an example using this method.

EXAMPLE **1**   **Integration by Parts**

Evaluate $\int xe^x \, dx$.

Solution

We cannot evaluate this integral using methods we have learned. But we can "split" the integrand into two parts, setting one part equal to u and the other part equal to dv. This "split" must be done in such a way that $\int dv$ and $\int v \, du$ can be evaluated. Letting $u = x$ and letting $dv = e^x \, dx$ are possible choices. If we make these choices, we have

$$u = x \qquad dv = e^x \, dx$$
$$du = 1 \, dx \qquad v = \int e^x \, dx = e^x$$

Then

$$\int xe^x \, dx = uv - \int v \, du$$
$$= xe^x - \int e^x \, dx = xe^x - e^x + C$$ ∎

We see that choosing $u = x$ and $dv = e^x \, dx$ worked in evaluating $\int xe^x \, dx$ in Example 1. If we had chosen $u = e^x$ and $dv = x \, dx$, the results would not have been so successful.

How can we select u and dv to make integration by parts work? As a general guideline, we do the following.

First identify the types of functions occurring in the problem in the order

Logarithm,   Polynomial (or Power of x),   Radical,   Exponential*

Thus, in Example 1, we had x and e^x, a polynomial and an exponential.

Second, choose u to equal the function whose type occurs first on the list; hence in Example 1 we chose $u = x$. Then dv equals the rest of the integrand (and always includes dx) so that $u \, dv$ equals the original integrand. A helpful way to remember the order of the function types that help us choose u is the sentence

"Lazy People Rarely Excel."

in which the first letters, LPRE, coordinate with the order and types of functions. Consider the following examples.

EXAMPLE **2**   **Integration by Parts**

Evaluate $\int x \ln x \, dx$.

Solution

The integral contains a logarithm ($\ln x$) and a polynomial (x). Thus, let $u = \ln x$ and $dv = x \, dx$. Then

$$du = \frac{1}{x} \, dx \quad \text{and} \quad v = \frac{x^2}{2}$$

so $$\int x \ln x \, dx = u \cdot v - \int v \, du = (\ln x) \frac{x^2}{2} - \int \frac{x^2}{2} \cdot \frac{1}{x} \, dx$$
$$= \frac{x^2}{2} \ln x - \int \frac{x}{2} \, dx = \frac{x^2}{2} \ln x - \frac{x^2}{4} + C$$

Note that letting $dv = \ln x \, dx$ is contrary to our guidelines and would lead to great difficulty in evaluating $\int dv$ and $\int v \, du$. ∎

*This order is related to the ease with which the function types can be integrated.

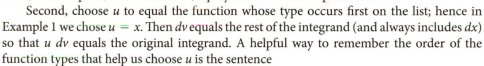

EXAMPLE 3   Integration by Parts

Evaluate $\int \ln x^2 \, dx$.

Solution

The only function in this problem is a logarithm, the first function type on our list for choosing u. Thus

$$u = \ln x^2 \qquad\qquad dv = dx$$

$$du = \frac{2x}{x^2} \, dx = \frac{2}{x} \, dx \qquad v = x$$

Then

$$\int \ln x^2 \, dx = x \ln x^2 - \int x \cdot \frac{2}{x} \, dx = x \ln x^2 - 2x + C$$

Note that if we write $\ln x^2$ as $2 \ln x$, we can also evaluate this integral using formula 14 in Table 13.2 in the previous section, so integration by parts would not be needed. ■

CHECKPOINT

1. True or false: In evaluating $\int u \, dv$ by parts,
   (a) the parts u and dv are selected and the parts du and v are calculated.
   (b) the differential (often dx) is always chosen as part of dv.
   (c) the parts du and v are found from u and dv as follows:

   $$du = u' \, dx \quad \text{and} \quad v = \int dv$$

   (d) For $\int \frac{3x}{e^{2x}} \, dx$, we could choose $u = 3x$ and $dv = e^{2x} \, dx$.
2. For $\int \frac{\ln x}{x^4} \, dx$,
   (a) identify u and dv.
   (b) find du and v.
   (c) complete the evaluation of the integral.

Sometimes it is necessary to repeat integration by parts to complete the evaluation. When this occurs, at each use of integration by parts it is important to choose u and dv consistently according to our guidelines.

EXAMPLE 4   Repeated Integration by Parts

Evaluate $\int x^2 e^{2x} \, dx$.

Solution

Let $u = x^2$ and $dv = e^{2x} \, dx$, so $du = 2x \, dx$ and $v = \frac{1}{2} e^{2x}$. Then

$$\int x^2 e^{2x} \, dx = \frac{1}{2} x^2 e^{2x} - \int x e^{2x} \, dx$$

We cannot evaluate $\int x e^{2x} \, dx$ directly, but this new integral is simpler than the original, and a second integration by parts will be successful. Letting $u = x$ and $dv = e^{2x} \, dx$ gives $du = dx$ and $v = \frac{1}{2} e^{2x}$. Thus

$$\int x^2 e^{2x} \, dx = \frac{1}{2} x^2 e^{2x} - \left(\frac{1}{2} x e^{2x} - \int \frac{1}{2} e^{2x} \, dx \right)$$

$$= \frac{1}{2} x^2 e^{2x} - \frac{1}{2} x e^{2x} + \frac{1}{4} e^{2x} + C$$

$$= \frac{1}{4} e^{2x} (2x^2 - 2x + 1) + C$$

■

The most obvious choices for u and dv are not always the correct ones, as the following example shows. Integration by parts may still involve some trial and error.

EXAMPLE 5   A Tricky Integration by Parts

Evaluate $\int x^3 \sqrt{x^2 + 1}\, dx$.

Solution
Here we have a polynomial (x^3) and a radical. However, if we choose $u = x^3$, then the resulting $dv = \sqrt{x^2 + 1}\, dx$ cannot be integrated easily. But we can use $\sqrt{x^2 + 1}$ as part of dv, and we can evaluate $\int dv$ if we let $dv = x\sqrt{x^2 + 1}\, dx$. Then

$$u = x^2 \quad dv = (x^2 + 1)^{1/2} x\, dx$$

$$du = 2x\, dx \quad v = \int (x^2 + 1)^{1/2}(x\, dx) = \frac{1}{2}\int (x^2 + 1)^{1/2}(2x\, dx)$$

$$v = \frac{1}{2}\frac{(x^2 + 1)^{3/2}}{3/2} = \frac{1}{3}(x^2 + 1)^{3/2}$$

Then $\int x^3 \sqrt{x^2 + 1}\, dx = \frac{x^2}{3}(x^2 + 1)^{3/2} - \int \frac{1}{3}(x^2 + 1)^{3/2}(2x\, dx)$

$$= \frac{x^2}{3}(x^2 + 1)^{3/2} - \frac{1}{3}\frac{(x^2 + 1)^{5/2}}{5/2} + C$$

$$= \frac{x^2}{3}(x^2 + 1)^{3/2} - \frac{2}{15}(x^2 + 1)^{5/2} + C \qquad ■$$

EXAMPLE 6   Income Stream | APPLICATION PREVIEW |

Suppose that the value of oil produced by a piece of oil extraction equipment is considered a continuous income stream with an annual rate of flow (in dollars per year) at time t in years given by

$$f(t) = 300{,}000 - 2500t, \quad 0 \le t \le 10$$

and that money is worth 8%, compounded continuously. Find the present value of the piece of equipment.

Solution
The present value of the piece of equipment is given by

$$\int_0^{10} (300{,}000 - 2500t)e^{-0.08t}\, dt = 300{,}000\int_0^{10} e^{-0.08t}\, dt - 2500\int_0^{10} te^{-0.08t}\, dt$$

$$= \frac{300{,}000}{-0.08} e^{-0.08t}\Big|_0^{10} - 2500\int_0^{10} te^{-0.08t}\, dt$$

The value of the first integral is

$$\frac{300{,}000}{-0.08} e^{-0.08t}\Big|_0^{10} = \frac{300{,}000}{-0.08} e^{-0.8} - \frac{300{,}000}{-0.08}$$

$$\approx -1{,}684{,}983.615 + 3{,}750{,}000 = 2{,}065{,}016.385$$

The second of these integrals can be evaluated by using integration by parts, with $u = t$ and $dv = e^{-0.08t}\, dt$. Then $du = 1\, dt$ and $v = \dfrac{e^{-0.08t}}{-0.08}$, and this integral is

$$-2500 \int_0^{10} te^{-0.08t}\, dt = -2500\, \frac{te^{-0.08t}}{-0.08}\bigg|_0^{10} + 2500 \int_0^{10} \frac{e^{-0.08t}}{-0.08}\, dt$$

$$= \frac{2500}{0.08}\, te^{-0.08t}\bigg|_0^{10} + \frac{2500}{0.0064}\, e^{-0.08t}\bigg|_0^{10}$$

$$= \frac{2500}{0.08}\, 10e^{-0.8} + \frac{2500}{0.0064}\, e^{-0.8} - \frac{2500}{0.0064} \approx -74{,}690.572$$

Thus the sum of the integrals is

$$2{,}065{,}016.385 + (-74{,}690.572) = 1{,}990{,}325.813$$

so the present value of this piece of equipment is $1,990,325.81. ■

One further note about integration by parts. It can be very useful on certain types of problems, but not on all types. Don't attempt to use integration by parts when easier methods are available.

Calculator Note   Using the numerical integration feature of a graphing calculator to evaluate the integral in Example 6 gives the present value of $1,990,325.80, so this answer is the same as that found in Example 6, to the nearest dollar. ■

CHECKPOINT SOLUTIONS

1. (a) True   (b) True   (c) True
   (d) False. The *product* of u and dv must equal the original integrand. Rewrite as

   $$\int \frac{3x}{e^{2x}}\, dx = \int 3xe^{-2x}\, dx$$

   and then choose $u = 3x$ and $dv = e^{-2x}\, dx$.

2. First we rewrite:

   $$\int \frac{\ln x}{x^4}\, dx = \int \ln x\, (x^{-4}\, dx)$$

   Then the integral contains a logarithm and a power of x.

   (a) $u = \ln x$ and $dv = x^{-4}\, dx$

   (b) $du = \dfrac{1}{x}\, dx$ and $v = \displaystyle\int x^{-4}\, dx = \dfrac{x^{-3}}{-3}$

   (c) $\displaystyle\int \frac{\ln x}{x^4}\, dx = uv - \int v\, du$

   $$= (\ln x)\left(\frac{x^{-3}}{-3}\right) - \int \frac{x^{-3}}{-3}\cdot\frac{1}{x}\, dx = \frac{-\ln x}{3x^3} + \frac{1}{3}\int x^{-4}\, dx$$

   $$= -\frac{\ln x}{3x^3} + \frac{1}{3}\left(\frac{x^{-3}}{-3}\right) + C = -\frac{\ln x}{3x^3} - \frac{1}{9x^3} + C$$

| **EXERCISES** | **13.6**

Access end-of-section exercises online at **www.webassign.net**

Improper Integrals and Their Applications

| APPLICATION PREVIEW |

We saw in Section 13.4, "Applications of Definite Integrals in Business and Economics," that the present value of a continuous income stream over a fixed number of years can be found by using a definite integral. When this notion is extended to an infinite time interval, the result is called the capital value of the income stream and is given by

$$\text{Capital value} = \int_0^\infty f(t)e^{-rt}\,dt$$

where $f(t)$ is the annual rate of flow at time t, and r is the annual interest rate, compounded continuously. This is called an improper integral. For example, the capital value of a trust that provides $f(t) = \$10{,}000$ per year indefinitely (when interest is 10% compounded continuously) is found by evaluating an improper integral. (See Example 2.)

Improper Integral   Some applications of calculus to statistics or business (such as capital value) involve definite integrals over intervals of infinite length (**improper integrals**). The area of a region that extends infinitely to the left or right along the x-axis (see Figure 13.26) could be described by an improper integral.

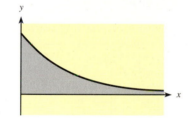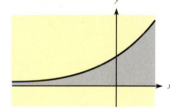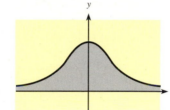

Figure 13.26

To see how to find such an area and hence evaluate an improper integral, let us consider how to find the area between the curve $y = 1/x^2$ and the x-axis to the right of $x = 1$.

To find the area under this curve from $x = 1$ to $x = b$, where b is any number greater than 1 (see Figure 13.27), we evaluate

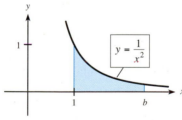

Figure 13.27

$$A = \int_1^b \frac{1}{x^2}\,dx = \frac{-1}{x}\Big|_1^b = \frac{-1}{b} - \left(\frac{-1}{1}\right) = 1 - \frac{1}{b}$$

Note that the larger b is, the closer the area is to 1. If $b = 100$, $A = 0.99$; if $b = 1000$, $A = 0.999$; and if $b = 1{,}000{,}000$, $A = 0.999999$.

We can represent the area of the region under $1/x^2$ to the right of 1 using the notation

$$\lim_{b\to\infty} \int_1^b \frac{1}{x^2}\,dx = \lim_{b\to\infty}\left(1 - \frac{1}{b}\right)$$

where $\lim\limits_{b\to\infty}$ represents the limit as b gets larger without bound. Note that $\dfrac{1}{b} \to 0$ as $b \to \infty$, so

$$\lim_{b\to\infty}\left(1 - \frac{1}{b}\right) = 1 - 0 = 1$$

Thus the area under the curve $y = 1/x^2$ to the right of $x = 1$ is 1.

In general, we define the area under a curve $y = f(x)$ to the right of $x = a$, with $f(x) \geq 0$, to be

$$\text{Area} = \lim_{b \to \infty}(\text{area from } a \text{ to } b) = \lim_{b \to \infty}\int_a^b f(x)\, dx$$

This motivates the definition that follows.

Improper Integral

$$\int_a^\infty f(x)\, dx = \lim_{b \to \infty}\int_a^b f(x)\, dx$$

If the limit defining the **improper integral** is a unique finite number, we say that the integral *converges*; otherwise, we say that the integral *diverges*.

EXAMPLE 1   Improper Integrals

Evaluate the following improper integrals, if they converge.

(a) $\displaystyle\int_1^\infty \frac{1}{x^3}\, dx$ (b) $\displaystyle\int_1^\infty \frac{1}{x}\, dx$

Solution

(a) $\displaystyle\int_1^\infty \frac{1}{x^3}\, dx = \lim_{b \to \infty}\int_1^b x^{-3}\, dx = \lim_{b \to \infty}\left(\frac{x^{-2}}{-2}\right)\Bigg|_1^b$

$$= \lim_{b \to \infty}\left[\frac{-1}{2b^2} - \left(\frac{-1}{2(1)^2}\right)\right] = \lim_{b \to \infty}\left(\frac{-1}{2b^2} + \frac{1}{2}\right)$$

Notice that $1/(2b^2) \to 0$ as $b \to \infty$, so the limit converges to $0 + \frac{1}{2} = \frac{1}{2}$. That is,

$$\int_1^\infty \frac{1}{x^3}\, dx = \frac{1}{2}$$

(b) $\displaystyle\int_1^\infty \frac{1}{x}\, dx = \lim_{b \to \infty}\int_1^b \frac{1}{x}\, dx = \lim_{b \to \infty}\left(\ln|x|\right)\Bigg|_1^b = \lim_{b \to \infty}(\ln b - \ln 1)$

Here, $\ln b$ increases without bound as $b \to \infty$, so the limit diverges and we write

$$\int_1^\infty \frac{1}{x}\, dx = \infty$$ ■

From Example 1 we can conclude that the area under the curve $y = 1/x^3$ to the right of $x = 1$ is $\frac{1}{2}$ whereas the corresponding area under the curve $y = 1/x$ is infinite. (We have already seen that the corresponding area under $y = 1/x^2$ is 1.)

As Figure 13.28 shows, the graphs of $y = 1/x^2$ and $y = 1/x$ look similar, but the graph of $y = 1/x^2$ gets "close" to the x-axis much more rapidly than the graph of $y = 1/x$. The area under $y = 1/x$ does not converge to a finite number because as $x \to \infty$ the graph of $1/x$ does not approach the x-axis rapidly enough.

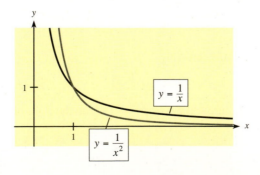

Figure 13.28

EXAMPLE 2 Capital Value | APPLICATION PREVIEW |

Suppose that an organization wants to establish a trust fund that will provide a continuous income stream with an annual rate of flow at time t given by $f(t) = 10,000$ dollars per year. If the interest rate remains at 10% compounded continuously, find the capital value of the fund.

Solution
The capital value of the fund is given by

$$\int_0^\infty f(t)e^{-rt}\,dt$$

where $f(t)$ is the annual rate of flow at time t, and r is the annual interest rate, compounded continuously.

$$\int_0^\infty 10,000e^{-0.10t}\,dt = \lim_{b\to\infty}\int_0^b 10,000e^{-0.10t}\,dt = \lim_{b\to\infty}\left(-100,000e^{-0.10t}\right)\Big|_0^b$$

$$= \lim_{b\to\infty}\left(\frac{-100,000}{e^{0.10b}} + 100,000\right) = 100,000$$

Thus the capital value of the fund is $100,000. ■

Another term for a fund such as the one in Example 2 is a **perpetuity.** Usually the rate of flow of a perpetuity is a constant. If the rate of flow is a constant A, it can be shown that the capital value is given by A/r (see Problem 37 in the 13.7 Exercises).

CHECKPOINT

1. True or false:

    (a) $\displaystyle\lim_{b\to+\infty}\frac{1}{b^p} = 0$ if $p > 0$ (b) If $p > 0$, $b^p \to \infty$ as $b \to \infty$

    (c) $\displaystyle\lim_{b\to+\infty} e^{-pb} = 0$ if $p > 0$

2. Evaluate the following (if they exist).

    (a) $\displaystyle\int_1^\infty \frac{1}{x^{4/3}}\,dx = \lim_{b\to\infty}\int_1^b x^{-4/3}\,dx$ (b) $\displaystyle\int_0^\infty \frac{dx}{\sqrt{x+1}}$

Two **additional improper integrals** that involve infinite limits are defined below.

Additional Improper Integrals

$$\int_{-\infty}^b f(x)\,dx = \lim_{a\to\infty}\int_{-a}^b f(x)\,dx$$

The integral converges if the limit is finite. Otherwise it diverges.

$$\int_{-\infty}^\infty f(x)\,dx = \lim_{a\to\infty}\int_{-a}^c f(x)\,dx + \lim_{b\to\infty}\int_c^b f(x)\,dx$$

for any finite constant c. (Often, c is chosen to be 0.) If both limits are finite, the improper integral converges; otherwise, it diverges.

EXAMPLE 3 Improper Integrals

Evaluate the following integrals.

    (a) $\displaystyle\int_{-\infty}^4 e^{3x}\,dx$ (b) $\displaystyle\int_{-\infty}^\infty \frac{x^3}{(x^4+3)^2}\,dx$

Solution

(a) $\displaystyle\int_{-\infty}^{4} e^{3x}\, dx = \lim_{a\to\infty}\int_{-a}^{4} e^{3x}\, dx = \lim_{a\to\infty}\left[\left(\frac{1}{3}\right)e^{3x}\right]_{-a}^{4}$

$\displaystyle\qquad = \lim_{a\to\infty}\left[\left(\frac{1}{3}\right)e^{12} - \left(\frac{1}{3}\right)e^{-3a}\right] = \lim_{a\to\infty}\left[\left(\frac{1}{3}\right)e^{12} - \left(\frac{1}{3}\right)\left(\frac{1}{e^{3a}}\right)\right]$

$\displaystyle\qquad = \frac{1}{3}e^{12}\quad \text{(because } 1/e^{3a}\to 0 \text{ as } a\to\infty)$

(b) $\displaystyle\int_{-\infty}^{\infty}\frac{x^3}{(x^4+3)^2}\, dx = \lim_{a\to\infty}\int_{-a}^{0}\frac{x^3}{(x^4+3)^2}\, dx + \lim_{b\to\infty}\int_{0}^{b}\frac{x^3}{(x^4+3)^2}\, dx$

$\displaystyle\qquad = \lim_{a\to\infty}\left[\frac{1}{4}\frac{(x^4+3)^{-1}}{-1}\right]_{-a}^{0} + \lim_{b\to\infty}\left[\frac{1}{4}\frac{(x^4+3)^{-1}}{-1}\right]_{0}^{b}$

$\displaystyle\qquad = \lim_{a\to\infty}\left[-\frac{1}{4}\left(\frac{1}{3}-\frac{1}{a^4+3}\right)\right] + \lim_{b\to\infty}\left[-\frac{1}{4}\left(\frac{1}{b^4+3}-\frac{1}{3}\right)\right]$

$\displaystyle\qquad = -\frac{1}{12}+0+0+\frac{1}{12} = 0$

$\displaystyle\left(\text{since }\lim_{a\to\infty}\frac{1}{a^4+3}=0 \text{ and } \lim_{b\to\infty}\frac{1}{b^4+3}=0\right)$ ■

Probability   We noted in Chapter 8, "Further Topics in Probability; Data Description," that the sum of the probabilities for a probability distribution (a **probability density function**) equals 1. In particular, we stated that the area under the normal probability curve is 1. The normal distribution is an example of a continuous probability distribution because the values of the random variable are considered over intervals rather than at discrete values. There are many important continuous probability distributions besides the normal distribution, but all such distributions satisfy the following definition.

Probability Density Function

If $f(x) \geq 0$ for all x, then f is a **probability density function** for a continuous random variable x if and only if

$$\int_{-\infty}^{\infty} f(x)\, dx = 1$$

We have noted previously that when $f(x)$ is a continuous probability density function, then

$$Pr(a \leq x \leq b) = \int_{a}^{b} f(x)\, dx$$

EXAMPLE 4   Product Life

Suppose the probability density function for the life span of a computer component is given by

$$f(x) = \begin{cases} 0.10e^{-0.10x} & \text{if } x \geq 0 \\ 0 & \text{if } x < 0 \end{cases}$$

(a) Verify that $f(x)$ is a probability density function.
(b) Find the probability that such a component lasts more than 3 years.

Solution

(a) We can verify that $f(x)$ is a probability density function by showing that

$$\int_{-\infty}^{\infty} f(x)\, dx = 1$$

$$\int_{-\infty}^{\infty} f(x)\, dx = \int_{-\infty}^{0} f(x)\, dx + \int_{0}^{\infty} f(x)\, dx = \int_{-\infty}^{0} 0\, dx + \int_{0}^{\infty} 0.10e^{-0.10x}\, dx$$

$$= 0 + \lim_{b\to\infty} \int_{0}^{b} 0.10e^{-0.10x}\, dx$$

$$= \lim_{b\to\infty} (-e^{-0.10x})|_{0}^{b} = \lim_{b\to\infty} (-e^{-0.10b} + 1) = 1$$

(b) The probability that the component lasts more than 3 years is given by the improper integral

$$\int_{3}^{\infty} 0.10e^{-0.10x}\, dx$$

which gives

$$\lim_{b\to\infty} \int_{3}^{b} 0.10e^{-0.10x}\, dx = \lim_{b\to\infty} (-e^{-0.10x})|_{3}^{b}$$

$$= \lim_{b\to\infty} (-e^{-0.10b} + e^{-0.3}) = e^{-0.3} \approx 0.7408 \qquad \blacksquare$$

In Section 8.3, "Discrete Probability Distributions; The Binomial Distribution," we found the expected value (mean) of a discrete probability distribution using the formula

$$E(x) = \sum x \Pr(x)$$

For continuous probability distributions, such as the normal probability distribution, the **expected value**, or **mean**, can be found by evaluating the improper integral

$$\int_{-\infty}^{\infty} xf(x)\, dx$$

Mean (Expected Value)

If x is a continuous random variable with probability density function f, then the **mean (expected value)** of the probability distribution is

$$\mu = \int_{-\infty}^{\infty} xf(x)\, dx$$

The normal distribution density function, in standard form, is

$$f(x) = \frac{1}{\sqrt{2\pi}} e^{-x^2/2}$$

so the mean of the normal probability distribution is given by

$$\mu = \int_{-\infty}^{\infty} x\left(\frac{1}{\sqrt{2\pi}} e^{-x^2/2}\right) dx$$

$$= \lim_{a\to\infty} \int_{-a}^{0} \frac{1}{\sqrt{2\pi}} xe^{-x^2/2}\, dx + \lim_{b\to\infty} \int_{0}^{b} \frac{1}{\sqrt{2\pi}} xe^{-x^2/2}\, dx$$

$$= \lim_{a\to\infty} \frac{1}{\sqrt{2\pi}} \left(-e^{-x^2/2}\right)\Big|_{-a}^{0} + \lim_{b\to\infty} \frac{1}{\sqrt{2\pi}} \left(-e^{-x^2/2}\right)\Big|_{0}^{b}$$

$$= \frac{1}{\sqrt{2\pi}} (-1 + 0) + \frac{1}{\sqrt{2\pi}} (0 + 1) = 0$$

This verifies the statement in Chapter 8 that the mean of the standard normal distribution is 0.

CHECKPOINT SOLUTIONS

1. (a) True   (b) True   (c) True

2. (a) $\lim_{b\to\infty} \int_1^b x^{-4/3}\, dx = \lim_{b\to\infty} \frac{x^{-1/3}}{-1/3}\Big|_1^b = \lim_{b\to\infty}\left(\frac{-3}{b^{1/3}} - \frac{-3}{1}\right) = 0 + 3 = 3$

   (b) $\lim_{b\to\infty} \int_0^b (x+1)^{-1/2}\, dx = \lim_{b\to\infty} \frac{(x+1)^{1/2}}{1/2}\Big|_0^b = \lim_{b\to\infty} 2\sqrt{x+1}\,\Big|_0^b$

   $$= \lim_{b\to\infty}(2\sqrt{b+1} - 2)$$

   The limit approaches ∞, so the integral diverges.

| EXERCISES | 13.7

Access end-of-section exercises online at **www.webassign.net**

OBJECTIVES 13.8

- To approximate definite integrals by using the Trapezoidal Rule
- To approximate definite integrals by using Simpson's Rule

Numerical Integration Methods: The Trapezoidal Rule and Simpson's Rule

| APPLICATION PREVIEW |

A pharmaceutical company tests the body's assimilation of a new drug by administering a 200-milligram dose and collecting the following data from blood samples (t is time in hours, and $R(t)$ gives the assimilation of the drug in milligrams per hour).

t	0	0.5	1.0	1.5	2.0	2.5	3.0
$R(t)$	0.0	15.3	32.3	51.0	74.8	102.0	130.9

The company would like to find the amount of drug assimilated in 3 hours, which is given by

$$\int_0^3 R(t)\, dt$$

In this section we develop straightforward and quite accurate methods for evaluating integrals such as this, even when, as with $R(t)$, only function data, rather than a formula, are given. (See Example 4.)

We have studied several techniques for integration and have even used tables to evaluate some integrals. Yet some functions that arise in practical problems cannot be integrated by using any formula. For any function $f(x) \geq 0$ on an interval $[a, b]$, however, we have seen that a definite integral can be viewed as an area and that we can usually approximate the area and hence the integral (see Figure 13.29). One such approximation method uses rectangles, as we saw when we defined the definite integral. In this section, we consider two other **numerical integration methods** to approximate a definite integral: the **Trapezoidal Rule** and **Simpson's Rule**.

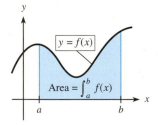

Figure 13.29

Trapezoidal Rule To develop the Trapezoidal Rule formula, we assume that $f(x) \geq 0$ on $[a, b]$ and subdivide the interval $[a, b]$ into n equal pieces, each of length $(b - a)/n = h$. Then, within each subdivision, we can approximate the area by using a trapezoid. As shown in Figure 13.30, we can use the formula for the area of a trapezoid to approximate the area of the first subdivision. Continuing in this way for each trapezoid, we have

$$\int_a^b f(x)\,dx \approx A_1 + A_2 + A_3 + \cdots + A_{n-1} + A_n$$

$$= \left[\frac{f(x_0) + f(x_1)}{2}\right]h + \left[\frac{f(x_1) + f(x_2)}{2}\right]h +$$

$$\left[\frac{f(x_2) + f(x_3)}{2}\right]h + \cdots + \left[\frac{f(x_{n-1}) + f(x_n)}{2}\right]h$$

$$= \frac{h}{2}[f(x_0) + f(x_1) + f(x_1) + f(x_2) + f(x_2) + \cdots + f(x_{n-1}) + f(x_{n-1}) + f(x_n)]$$

This can be simplified to obtain the **Trapezoidal Rule**.

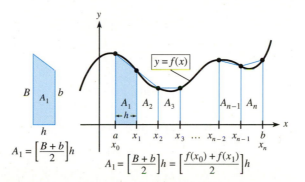

$$A_1 = \left[\frac{B+b}{2}\right]h$$

$$A_1 = \left[\frac{B+b}{2}\right]h = \left[\frac{f(x_0) + f(x_1)}{2}\right]h$$

Figure 13.30

Trapezoidal Rule	If f is continuous on $[a, b]$, then

$$\int_a^b f(x)\,dx \approx \frac{h}{2}[f(x_0) + 2f(x_1) + 2f(x_2) + \cdots + 2f(x_{n-1}) + f(x_n)]$$

where $h = \dfrac{b-a}{n}$ and n is the number of equal subdivisions of $[a, b]$.

Despite the fact that we used areas to develop the Trapezoidal Rule, we can use this rule to evaluate definite integrals even if $f(x) < 0$ on all or part of $[a, b]$.

EXAMPLE 1 Trapezoidal Rule

Use the Trapezoidal Rule to approximate $\displaystyle\int_1^3 \frac{1}{x}\,dx$ with

(a) $n = 4$.
(b) $n = 8$.

Solution
First, we note that this integral can be evaluated directly:

$$\int_1^3 \frac{1}{x}\,dx = \ln|x|\Big|_1^3 = \ln 3 - \ln 1 = \ln 3 \approx 1.099$$

(a) The interval $[1, 3]$ must be divided into 4 equal subintervals of width $h = \frac{3-1}{4} = \frac{2}{4} = \frac{1}{2}$
as follows:

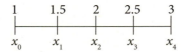

Thus, from the Trapezoidal Rule, we have

$$\int_1^3 \frac{1}{x} \, dx \approx \frac{h}{2}[f(x_0) + 2f(x_1) + 2f(x_2) + 2f(x_3) + f(x_4)]$$

$$= \frac{1/2}{2}[f(1) + 2f(1.5) + 2f(2) + 2f(2.5) + f(3)]$$

$$= \frac{1}{4}\left[1 + 2\left(\frac{1}{1.5}\right) + 2\left(\frac{1}{2}\right) + 2\left(\frac{1}{2.5}\right) + \frac{1}{3}\right]$$

$$\approx 0.25(1 + 1.333 + 1 + 0.8 + 0.3333)$$

$$\approx 1.117$$

(b) In this case, the interval [1, 3] is divided into 8 equal subintervals of width $h = \frac{3-1}{8} = \frac{2}{8} = \frac{1}{4}$ as follows.

1	1.25	1.5	1.75	2	2.25	2.5	2.75	3
x_0	x_1	x_2	x_3	x_4	x_5	x_6	x_7	x_8

Thus, from the Trapezoidal Rule, we have

$$\int_1^3 \frac{1}{x} \, dx \approx \frac{h}{2}[f(x_0) + 2f(x_1) + 2f(x_2) + 2f(x_3) + 2f(x_4) + 2f(x_5)$$
$$+ 2f(x_6) + 2f(x_7) + f(x_8)]$$

$$= \frac{1/4}{2}\left[\frac{1}{1} + 2\left(\frac{1}{1.25}\right) + 2\left(\frac{1}{1.5}\right) + 2\left(\frac{1}{1.75}\right) + 2\left(\frac{1}{2}\right) + 2\left(\frac{1}{2.25}\right)\right.$$
$$\left. + 2\left(\frac{1}{2.5}\right) + 2\left(\frac{1}{2.75}\right) + \frac{1}{3}\right]$$

$$\approx 1.103$$

In Example 1, because we know that the value of the integral is $\ln 3 \approx 1.099$, we can measure the accuracy of each approximation. We can see that a larger value of n (namely, $n = 8$) produced a more accurate approximation to $\ln 3$. In general, larger values of n produce more accurate approximations, but they also make computations more difficult.

Technology Note   The TABLE feature of a graphing calculator can be used to find the values of $f(x_0)$, $f(x_1), f(x_2)$, etc. needed for the Trapezoidal Rule. Figure 13.31(a) shows the TABLE set up and Figures 13.31(b) and 13.31(c) show the TABLE values.

Figure 13.31   (a)   (b)   (c)

Thus for the Trapezoidal Rule, we get

$$\int_1^3 \frac{1}{x} \, dx \approx \frac{0.25}{2}\left[\begin{array}{l}1(1) + 2(0.8) + 2(0.66667) + 2(0.57143) + 2(0.5) \\ + 2(0.44444) + 2(0.4) + 2(0.36364) + 1(0.33333)\end{array}\right] \approx 1.103$$

just as we did in Example 1.

An Excel spreadsheet also can be very useful in finding the numerical integral with the Trapezoidal Rule. A discussion of the use of Excel in finding numerical integrals is given in the Online Excel Guide that accompanies this text.

Because the exact value of an integral is rarely available when an approximation is used, it is important to have some way to judge the accuracy of an answer. The following formula, which we state without proof, can be used to bound the error that results from using the Trapezoidal Rule.

Trapezoidal Rule Error

The error E in using the Trapezoidal Rule to approximate $\int_a^b f(x)\, dx$ satisfies

$$|E| \le \frac{(b-a)^3}{12n^2}\left[\max_{a \le x \le b} |f''(x)|\right]$$

where n is the number of equal subdivisions of $[a, b]$.

For a numerical method to be worthwhile, there must be some way of assessing its accuracy. Hence, this formula is important. We leave its application, however, to more advanced courses.

Simpson's Rule   The Trapezoidal Rule was developed by using a line segment to approximate the function over each subinterval and then using the areas under the line segments to approximate the area under the curve. Another numerical method, **Simpson's Rule,** uses a parabola to approximate the function over each pair of subintervals (see Figure 13.32) and then uses the areas under the parabolas to approximate the area under the curve. Because Simpson's Rule is based on pairs of subintervals, n must be even.

Simpson's Rule (n Even)

If $f(x)$ is continuous on $[a, b]$, and if $[a, b]$ is divided into an *even* number n of equal subdivisions, then

$$\int_a^b f(x)\, dx \approx \frac{h}{3}\left[f(x_0) + 4f(x_1) + 2f(x_2) + 4f(x_3) + \cdots + 2f(x_{n-2}) \right.$$
$$\left. + 4f(x_{n-1}) + f(x_n)\right]$$

where $h = \dfrac{b-a}{n}$.

We leave the derivation of Simpson's Rule to more advanced courses.

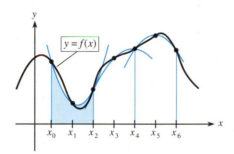

Figure 13.32

EXAMPLE   2   **Simpson's Rule**

Use Simpson's Rule with $n = 4$ to approximate $\int_1^3 \frac{1}{x}\, dx$.

Solution

Because $n = 4$ is even, Simpson's Rule can be used, and the interval is divided into four subintervals of length $h = \frac{3-1}{4} = \frac{1}{2}$ as follows:

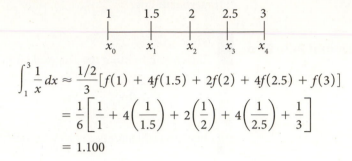

$$\int_1^3 \frac{1}{x}\, dx \approx \frac{1/2}{3}[f(1) + 4f(1.5) + 2f(2) + 4f(2.5) + f(3)]$$

$$= \frac{1}{6}\left[\frac{1}{1} + 4\left(\frac{1}{1.5}\right) + 2\left(\frac{1}{2}\right) + 4\left(\frac{1}{2.5}\right) + \frac{1}{3}\right]$$

$$= 1.100$$

Note that the result of Example 2 is better than both the $n = 4$ and the $n = 8$ Trapezoidal Rule approximations done in Example 1. In general, Simpson's Rule is more accurate than the Trapezoidal Rule for a given number of subdivisions. We can determine the accuracy of Simpson's Rule approximations by using the following formula.

Simpson's Rule Error Formula

The error E in using Simpson's Rule to approximate $\displaystyle\int_a^b f(x)\, dx$ satisfies

$$|E| \le \frac{(b-a)^5}{180n^4}\left[\max_{a \le x \le b}|f^{(4)}(x)|\right]$$

where n is the number of equal subdivisions of $[a, b]$ and $f^{(4)}(x)$ is the fourth derivative of $f(x)$.

The presence of the factor $180n^4$ in the denominator indicates that the error will often be quite small for even a modest value of n. Although Simpson's Rule often leads to more accurate results than the Trapezoidal Rule for a fixed choice of n, the Trapezoidal Rule is sometimes used because its error is more easily determined than that of Simpson's Rule or, more important, because the number of subdivisions is odd.

CHECKPOINT

1. Suppose $[1, 4]$ is divided into 6 equal subintervals.
   (a) Find the width h of each subinterval.
   (b) Find the subdivision points.
2. True or false:
   (a) When the Trapezoidal Rule is used, the number of subdivisions, n, must be even.
   (b) When Simpson's Rule is used, the number of subdivisions, n, must be even.
3. We can show that $\displaystyle\int_0^2 4x^3\, dx = 16$. Use $n = 4$ subdivisions to find

   (a) the Trapezoidal Rule approximation of this integral.
   (b) the Simpson's Rule approximation of this integral.

Technology Note

As mentioned previously for the Trapezoidal Rule, the TABLE feature of a graphing calculator can be used in exactly the same way with Simpson's Rule. An Excel spreadsheet also can be very useful with Simpson's Rule.

 EXAMPLE 3 **Simpson's Rule**

Use Simpson's Rule with $n = 10$ subdivisions to approximate

$$\int_1^6 [\ln(x)]^2\, dx$$

Solution

With $y_1 = [\ln (x)]^2$ and $h = \dfrac{b - a}{n} = \dfrac{6 - 1}{10} = 0.5$, we can use a graphing calculator with $\Delta \text{Tbl} = 0.5$ to find the following table of values.

x	y_1		x	y_1
1	0		4	1.9218
1.5	0.1644		4.5	2.2622
2	0.48045		5	2.5903
2.5	0.83959		5.5	2.9062
3	1.2069		6	3.2104
3.5	1.5694			

Thus Simpson's Rule yields

$$\int_1^6 [\ln (x)]^2 \, dx$$

$$\approx \frac{0.5}{3} \left[\begin{array}{l} 1(0) + 4(0.1644) + 2(0.48045) + 4(0.83959) + 2(1.2069) + 4(1.5694) \\ + 2(1.9218) + 4(2.2622) + 2(2.5903) + 4(2.9062) + 1(3.2104) \end{array} \right]$$

$$\approx 7.7627$$

One advantage that both the Trapezoidal Rule and Simpson's Rule offer is that they may be used when only function values at the subdivision points are known and the function formula itself is not known. This can be especially useful in applied problems.

EXAMPLE 4 **Pharmaceutical Testing** | APPLICATION PREVIEW |

A pharmaceutical company tests the body's ability to assimilate a drug. The test is done by administering a 200-milligram dose and then, every half-hour, monitoring the rate of assimilation. The following table gives the data; t is time in hours, and $R(t)$ is the rate of assimilation in milligrams per hour.

t	0	0.5	1.0	1.5	2.0	2.5	3.0
$R(t)$	0.0	15.3	32.3	51.0	74.8	102.0	130.9

To find the total amount of the drug (in milligrams) that is assimilated in the first 3 hours, the company must find

$$\int_0^3 R(t) \, dt$$

Use Simpson's Rule to approximate this definite integral.

Solution

The values of t correspond to the endpoints of the subintervals, and the values of $R(t)$ correspond to function values at those endpoints. From the table we see that $h = \frac{1}{2}$ and $n = 6$ (even); thus Simpson's Rule is applied as follows:

$$\int_0^3 R(t) \, dt \approx \frac{h}{3} [R(t_0) + 4R(t_1) + 2R(t_2) + 4R(t_3) + 2R(t_4) + 4R(t_5) + R(t_6)]$$

$$= \frac{1}{6} [0 + 4(15.3) + 2(32.3) + 4(51.0) + 2(74.8) + 4(102.0) + 130.9]$$

$$\approx 169.7 \text{ mg}$$

Thus at the end of 3 hours, the body has assimilated approximately 169.7 milligrams of the 200-milligram dose.

Note in Example 4 that the Trapezoidal Rule could also be used, because it relies only on the values of $R(t)$ at the subdivision endpoints.

In practice, these kinds of approximations are usually done with a computer, where the computations can be done quickly, even for large values of n. In addition, numerical methods (and hence computer programs) exist that approximate the errors, even in cases such as Example 4 where the function is not known. We leave any further discussion of error approximation formulas and additional numerical techniques for a more advanced course.

CHECKPOINT SOLUTIONS

1. (a) $h = \dfrac{b - a}{n} = \dfrac{4 - 1}{6} = \dfrac{1}{2} = 0.5$

   (b) $x_0 = 1, x_1 = 1.5, x_2 = 2, x_3 = 2.5, x_4 = 3, x_5 = 3.5, x_6 = 4$

2. (a) False. With the Trapezoidal Rule, n can be even or odd.

   (b) True

3. (a) $\dfrac{0.5}{2}[0 + 2(0.5) + 2(4) + 2(13.5) + 32] = 17$

   (b) $\dfrac{0.5}{3}[0 + 4(0.5) + 2(4) + 4(13.5) + 32] = 16$

| EXERCISES | 13.8

Access end-of-section exercises online at **www.webassign.net**

KEY TERMS AND FORMULAS

Section	Key Terms	Formulas
13.1	Sigma notation	$\displaystyle\sum_{i=1}^{n} 1 = n; \quad \sum_{i=1}^{n} i = \dfrac{n(n + 1)}{2}$
		$\displaystyle\sum_{i=1}^{n} i^2 = \dfrac{n(n + 1)(2n + 1)}{6}$
	Area	
	Right-hand endpoints	$\displaystyle\lim_{n\to\infty} \sum_{i=1}^{n} f(x_i)\dfrac{b - a}{n}; \; x_i = a + i\left(\dfrac{b - a}{n}\right)$
	Left-hand endpoints	$\displaystyle\lim_{n\to\infty} \sum_{i=1}^{n} f(x_{i-1})\dfrac{b - a}{n}$
13.2	Riemann sum	$\displaystyle\sum_{i=1}^{n} f(x_i{}^*)\Delta x_i$
	Definite integral	$\displaystyle\int_a^b f(x)\,dx = \lim_{\substack{\max \Delta x_i \to 0 \\ (n\to\infty)}} \sum_{i=1}^{n} f(x_i{}^*)\Delta x_i$
	Fundamental Theorem of Calculus	$\displaystyle\int_a^b f(x)\,dx = F(b) - F(a)$, where $F'(x) = f(x)$

Section	Key Terms	Formulas
13.2	Definite integral properties	$\displaystyle\int_a^a f(x)\,dx = 0$
		$\displaystyle\int_b^a f(x)\,dx = -\int_a^b f(x)\,dx$
		$\displaystyle\int_a^b [f(x) \pm g(x)]\,dx = \int_a^b f(x)\,dx \pm \int_a^b g(x)\,dx$
		$\displaystyle\int_a^b kf(x)\,dx = k\int_a^b f(x)\,dx$
		$\displaystyle\int_a^c f(x)\,dx + \int_c^b f(x)\,dx = \int_a^b f(x)\,dx$
	Area under $f(x)$, where $f(x) \geq 0$	$\displaystyle A = \int_a^b f(x)\,dx$
	$f(x)$ is a probability density function	$\displaystyle \Pr(a \leq x \leq b) = \int_a^b f(x)\,dx$
13.3	Area between $f(x)$ and $g(x)$, where $f(x) \geq g(x)$	$\displaystyle A = \int_a^b [f(x) - g(x)]\,dx$
	Average value of f over $[a, b]$	$\displaystyle \frac{1}{b - a}\int_a^b f(x)\,dx$
	Lorenz curve, $L(x)$	
	Gini coefficient	$\displaystyle 2\int_0^1 [x - L(x)]\,dx$
13.4	Continuous income streams	
	Total income	$\displaystyle \int_0^k f(t)\,dt \quad \text{(for } k \text{ years)}$
	Present value	$\displaystyle \int_0^k f(t)e^{-rt}\,dt, \quad \text{where } r \text{ is the interest rate}$
	Future value	$\displaystyle e^{rk}\int_0^k f(t)e^{-rt}\,dt$
	Consumer's surplus [demand is $f(x)$]	$\displaystyle CS = \int_0^{x_1} f(x)\,dx - p_1 x_1$
	Producer's surplus [supply is $g(x)$]	$\displaystyle PS = p_1 x_1 - \int_0^{x_1} g(x)\,dx$
13.5	Integration by formulas	See Table 13.2.
13.6	Integration by parts	$\displaystyle \int u\,dv = uv - \int v\,du$
13.7	Improper integrals	$\displaystyle \int_a^{\infty} f(x)\,dx = \lim_{b \to \infty}\int_a^b f(x)\,dx$
		$\displaystyle \int_{-\infty}^b f(x)\,dx = \lim_{a \to \infty}\int_{-a}^b f(x)\,dx$
		$\displaystyle \int_{-\infty}^{\infty} f(x)\,dx = \int_{-\infty}^c f(x)\,dx + \int_c^{\infty} f(x)\,dx$

	Capital value of a continuous income stream	$\int_0^\infty f(t)e^{-rt}\,dt$
	Probability density function, $f(x)$	$f(x) \geq 0$ and $\int_{-\infty}^\infty f(x)\,dx = 1$
	Mean	$\mu = \int_{-\infty}^\infty xf(x)\,dx$

13.8	Trapezoidal Rule for $\int_a^b f(x)\,dx$	$\approx \dfrac{h}{2}\left[f(x_0) + 2f(x_1) + \cdots + 2f(x_{n-1}) + f(x_n)\right]$ where $h = \dfrac{b-a}{n}$
	Error formula	$\lvert E \rvert \leq \dfrac{(b-a)^3}{12n^2}\left[\max_{a \leq x \leq b} \lvert f''(x)\rvert\right]$
	Simpson's Rule for $\int_a^b f(x)\,dx$	$\approx \dfrac{h}{3}\big[f(x_0) + 4f(x_1) + 2f(x_2) + 4f(x_3)$ $+ \cdots + 2f(x_{n-2}) + 4f(x_{n-1}) + f(x_n)\big],$ where n is even and $h = \dfrac{b-a}{n}$
	Error formula	$\lvert E \rvert \leq \dfrac{(b-a)^5}{180n^4}\left[\max_{a \leq x \leq b} \lvert f^{(4)}(x)\rvert\right]$

REVIEW EXERCISES

1. Calculate $\displaystyle\sum_{k=1}^{8} (k^2 + 1)$.

2. Use formulas to simplify

$$\sum_{i=1}^{n} \frac{3i}{n^3}$$

3. Use 6 subintervals of the same size to approximate the area under the graph of $y = 3x^2$ from $x = 0$ to $x = 1$. Use the right-hand endpoints of the subintervals to find the heights of the rectangles.

4. Use rectangles to find the area under the graph of $y = 3x^2$ from $x = 0$ to $x = 1$. Use n equal subintervals.

5. Use a definite integral to find the area under the graph of $y = 3x^2$ from $x = 0$ to $x = 1$.

6. Find the area between the graph of $y = x^3 - 4x + 5$ and the x-axis from $x = 1$ to $x = 3$.

Evaluate the integrals in Problems 7–18.

7. $\displaystyle\int_1^4 4\sqrt{x^3}\,dx$

8. $\displaystyle\int_{-3}^2 (x^3 - 3x^2 + 4x + 2)\,dx$

9. $\displaystyle\int_0^5 (x^3 + 4x)\,dx$

10. $\displaystyle\int_{-1}^3 (3x + 4)^{-2}\,dx$

11. $\displaystyle\int_{-3}^{-1} (x + 1)\,dx$

12. $\displaystyle\int_2^3 \frac{x^2}{2x^3 - 7}\,dx$

13. $\displaystyle\int_{-1}^2 (x^2 + x)\,dx$

14. $\displaystyle\int_1^4 \left(\frac{1}{x} + \sqrt{x}\right)dx$

15. $\displaystyle\int_0^2 5x^2(6x^3 + 1)^{1/2}\,dx$

16. $\displaystyle\int_0^1 \frac{x}{x^2 + 1}\,dx$

17. $\displaystyle\int_0^1 e^{-2x}\,dx$

18. $\displaystyle\int_0^1 xe^{x^2}\,dx$

Find the area between the curves in Problems 19–22.

19. $y = x^2 - 3x + 2$ and $y = x^2 + 4$ from $x = 0$ to $x = 5$

20. $y = x^2$ and $y = 4x + 5$

21. $y = x^3$ and $y = x$ from $x = -1$ to $x = 0$

22. $y = x^3 - 1$ and $y = x - 1$

Evaluate the integrals in Problems 23–26, using the formulas in Table 13.2.

23. $\displaystyle\int \sqrt{x^2 - 4}\,dx$

24. $\displaystyle\int_0^1 3^x\,dx$

25. $\displaystyle\int x \ln x^2\,dx$

26. $\displaystyle\int \frac{dx}{x(3x + 2)}$

In Problems 27–30, use integration by parts to evaluate.

27. $\displaystyle\int x^5 \ln x\,dx$

28. $\displaystyle\int x^2 e^{-2x}\,dx$

29. $\displaystyle\int \frac{x\,dx}{\sqrt{x + 5}}$

30. $\displaystyle\int_1^e \ln x\,dx$

Evaluate the improper integrals in Problems 31–34.

31. $\displaystyle\int_{1}^{\infty} \frac{1}{x+1}\, dx$

32. $\displaystyle\int_{-\infty}^{-1} \frac{200}{x^3}\, dx$

33. $\displaystyle\int_{0}^{\infty} 5e^{-3x}\, dx$

34. $\displaystyle\int_{-\infty}^{0} \frac{x}{(x^2+1)^2}\, dx$

35. Evaluate $\displaystyle\int_{1}^{3} \frac{2}{x^3}\, dx$

   (a) exactly.
   (b) by using the Trapezoidal Rule with $n = 4$ (to 3 decimal places).
   (c) by using Simpson's Rule with $n = 4$ (to 3 decimal places).

36. Use the Trapezoidal Rule with $n = 5$ to approximate

$$\int_{0}^{1} \frac{4\, dx}{x^2+1}$$

   Round your answer to 3 decimal places.

37. Use the table that follows to approximate

$$\int_{1}^{2.2} f(x)\, dx$$

   by using Simpson's Rule. Round your answer to 1 decimal place.

x	$f(x)$
1	0
1.3	2.8
1.6	5.1
1.9	4.2
2.2	0.6

38. Suppose that a definite integral is to be approximated and it is found that to achieve a specified accuracy, n must satisfy $n \geq 4.8$. What is the smallest n that can be used, if
   (a) the Trapezoidal Rule is used?
   (b) Simpson's Rule is used?

APPLICATIONS

39. *Maintenance* Maintenance costs for buildings increase as the buildings age. If the rate of increase in maintenance costs for a building is

$$M'(t) = \frac{14{,}000}{\sqrt{t+16}}$$

   where M is in dollars and t is time in years, $0 \leq t \leq 15$, find the total maintenance cost for the first 9 years ($t = 0$ to $t = 9$).

40. *Quality control* Suppose the probability density function for the life expectancy of a battery is

$$f(x) = \begin{cases} 1.4e^{-1.4x} & x \geq 0 \\ 0 & x < 0 \end{cases}$$

   Find the probability that the battery lasts 2 years or less.

41. *Savings* The future value of $1000 invested in a savings account at 10%, compounded continuously, is $S = 1000e^{0.1t}$, where t is in years. Find the average amount in the savings account during the first 5 years.

42. *Income streams* Suppose the total income in dollars from a machine is given by

$$I = 50e^{0.2t}, \quad 0 \leq t \leq 4, t \text{ in hours}$$

   Find the average income over this 4-hour period.

43. *Income distribution* In 1969, after the "Great Society" initiatives of the Johnson administration, the Lorenz curve for the U.S. income distribution was $L(x) = x^{2.1936}$. In 2000, after the stock market's historic 10-year growth, the Lorenz curve for the U.S. income distribution was $L(x) = x^{2.4870}$. Find the Gini coefficient of income for both years, and determine in which year income was more equally distributed.

44. *Consumer's surplus* The demand function for a product under pure competition is $p = \sqrt{64 - 4x}$, and the supply function is $p = x - 1$, where x is the number of units and p is in dollars.
   (a) Find the market equilibrium.
   (b) Find the consumer's surplus at market equilibrium.

45. *Producer's surplus* Find the producer's surplus at market equilibrium for Problem 44.

46. *Income streams* Find the total income over the next 10 years from a continuous income stream that has an annual flow rate at time t given by $f(t) = 125e^{0.05t}$ in thousands of dollars per year.

47. *Income streams* Suppose that a machine's production is considered a continuous income stream with an annual rate of flow at time t given by $f(t) = 150e^{-0.2t}$ in thousands of dollars per year. Money is worth 8%, compounded continuously.
   (a) Find the present value of the machine's production over the next 5 years.
   (b) Find the future value of the production 5 years from now.

48. *Average cost* Suppose the cost function for x units of a product is given by $C(x) = \sqrt{40{,}000 + x^2}$ dollars. Find the average cost over the first 150 units.

49. *Producer's surplus* Suppose the supply function for x units of a certain lamp is given by

$$p = 0.02x + 50.01 - \frac{10}{\sqrt{x^2+1}}$$

   where p is in dollars. Find the producer's surplus if the equilibrium price is $70 and the equilibrium quantity is 1000.

50. *Income streams* Suppose the present value of a continuous income stream over the next 5 years is given by

$$P = 9000\int_{0}^{5} te^{-0.08t}\, dt, \quad P \text{ in dollars, } t \text{ in years}$$

   Find the present value.

51. **Cost** If the marginal cost for x units of a product is $\overline{MC} = 3 + 60(x + 1) \ln (x + 1)$ dollars per unit and if the fixed cost is $2000, find the total cost function.

52. **Quality control** Find the probability that a phone lasts more than 1 year if the probability density function for its life expectancy is given by

$$f(x) = \begin{cases} 1.4e^{-1.4x} & x \geq 0 \\ 0 & x < 0 \end{cases}$$

53. **Capital value** Find the capital value of a business if its income is considered a continuous income stream with annual rate of flow given by

$$f(t) = 120e^{0.03t}$$

in thousands of dollars per year, and the current interest rate is 6% compounded continuously.

54. **Total income** Suppose that a continuous income stream has an annual rate of flow $f(t) = 100e^{-0.01t^2}$ (in thousands of dollars per year). Use Simpson's Rule with $n = 4$ to approximate the total income from this stream over the next 2 years.

55. **Revenue** A company has the data shown in the table from the sale of its product.

x	$\overline{MR}$
0	0
2	480
4	720
6	720
8	480
10	0

If x represents hundreds of units and revenue R is in hundreds of dollars, approximate the total revenue from the sale of 1000 units by approximating

$$\int_0^{10} \overline{MR}\, dx$$

with the Trapezoidal Rule.

13 CHAPTER TEST

1. Use left-hand endpoints and $n = 4$ subdivisions to approximate the area under $f(x) = \sqrt{4 - x^2}$ on the interval $[0, 2]$.

2. Consider $f(x) = 5 - 2x$ from $x = 0$ to $x = 1$ with n equal subdivisions.
   (a) If $f(x)$ is evaluated at right-hand endpoints, find a formula for the sum, S, of the areas of the n rectangles.
   (b) Find $\lim_{n \to \infty} S$.

3. Express the area in Quadrant I under $y = 12 + 4x - x^2$ (shaded in the figure) as an integral. Then evaluate the integral to find the area.

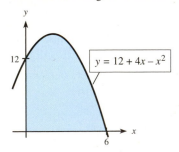

4. Evaluate the following definite integrals. (Do not approximate with technology.)
   (a) $\int_0^4 (9 - 4x)\, dx$ (b) $\int_0^3 x(8x^2 + 9)^{-1/2}\, dx$
   (c) $\int_1^4 \frac{5}{4x - 1}\, dx$ (d) $\int_1^\infty \frac{7}{x^2}\, dx$
   (e) $\int_4^4 \sqrt{x^3 + 10}\, dx$ (f) $\int_0^1 5x^2 e^{2x^3}\, dx$

5. Use integration by parts to evaluate the following.
   (a) $\int 3xe^x\, dx$ (b) $\int x \ln (2x)\, dx$

6. If $\int_1^4 f(x)\, dx = 3$ and $\int_3^4 f(x)\, dx = 7$, find $\int_1^3 2f(x)\, dx$.

7. Use Table 13.2 in Section 13.5 to evaluate each of the following.
   (a) $\int \ln (2x)\, dx$ (b) $\int x\sqrt{3x - 7}\, dx$

8. Use the numerical integration feature of a graphing calculator to approximate $\int_1^4 \sqrt{x^3 + 10}\, dx$.

9. Suppose the supply function for a product is $p = 40 + 0.001x^2$ and the demand function is $p = 120 - 0.2x$, where x is the number of units and p is the price in dollars. If the market equilibrium price is $80, find (a) the consumer's surplus and (b) the producer's surplus.

10. Suppose a continuous income stream has an annual rate of flow $f(t) = 85e^{-0.01t}$, in thousands of dollars per year, and the current interest rate is 7% compounded continuously.
    (a) Find the total income over the next 12 years.
    (b) Find the present value over the next 12 years.
    (c) Find the capital value of the stream.

11. Find the area between $y = 2x + 4$ and $y = x^2 - x$.

12. The figure shows typical supply and demand curves. On the figure, sketch and shade the region whose area represents the consumer's surplus.

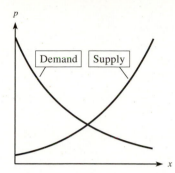

13. In an effort to make the distribution of income more nearly equal, the government of a country passes a tax law that changes the Lorenz curve from $y = 0.998x^{2.6}$ for one year to $y = 0.57x^2 + 0.43x$ for the next year. Find the Gini coefficient of income for each year and determine whether the distribution of income is more or less equitable after the tax law is passed. Interpret the result.

14. With data from the U.S. Department of Energy, the number of billions of barrels of U.S. crude oil produced each year from 2010 and projected to 2030 can be modeled by the function

$$f(t) = -0.001t^2 + 0.037t + 1.94$$

where t is the number of years past 2010.

(a) Find the total number of barrels of oil produced from 2010 to 2020.

(b) Find the average number of barrels produced per year from 2010 to 2018.

15. Use the Trapezoidal Rule to approximate

$$\int_1^3 (x \ln x)\, dx$$

with $n = 4$ subintervals.

16. The environmental effects of a chemical spill in the Clarion River can be estimated from the river's flow volume. To measure this flow volume, it is necessary to find the cross-sectional area of the river at a point downriver from the spill. The distance across the river is 240 feet, and the table gives depth measurements, $D(x)$, of the river at various distances x from one bank of the river. Use these measurements to estimate the cross-sectional area with Simpson's Rule.

x	0	40	80	120	160	200	240
$D(x)$	0	22	35	48	32	24	0

I. Retirement Planning

A 52-year-old client asks an accountant how to plan for his future retirement at age 62. He expects income from Social Security in the amount of $21,600 per year and a retirement pension of $40,500 per year from his employer. He wants to make monthly contributions to an investment plan that pays 8%, compounded monthly, for 10 years so that he will have a total income of $83,700 per year for 30 years. What will the size of the monthly contributions have to be to accomplish this goal, if it is assumed that money will be worth 8%, compounded continuously, throughout the period after he is 62?

To help you answer this question, complete the following.

1. How much money must the client withdraw annually from his investment plan during his retirement so that his total income goal is met?
2. How much money S must the client's account contain when he is 62 so that it will generate this annual amount for 30 years? (*Hint: S* can be considered the present value over 30 years of a continuous income stream with the amount you found in Question 1 as its annual rate of flow.)
3. The monthly contribution R that would, after 10 years, amount to the present value S found in Question 2 can be obtained from the formula

$$R = S \left[\frac{i}{(1 + i)^n - 1} \right]$$

where i represents the monthly interest rate and n the number of months. Find the client's monthly contribution, R.

II. Purchasing Electrical Power (Modeling)

In order to plan its purchases of electrical power from suppliers over the next 5 years, the PAC Electric Company needs to model its load data (demand for power by its customers) and use this model to predict future loads. The company pays for the electrical power each month on the basis of the peak load (demand) at any point during the month. The table gives, for the years 1994–2012, the load in megawatts (that is, in millions of watts) for the month when the maximum load occurred and the load in megawatts for the month when the minimum load occurred. The maximum loads occurred in summer, and the minimum loads occurred in spring or fall.

Year	Maximum Monthly Load	Minimum Monthly Load
1994	40.9367	19.4689
1995	45.7127	22.1504
1996	48.0460	25.3670
1997	56.1712	28.7254
1998	55.5793	31.0460
1999	62.4285	31.3838
2000	76.6536	34.8426
2001	73.8214	38.4544
2002	74.8844	40.6080
2003	83.0590	47.3621
2004	88.3914	45.8393
2005	88.7704	48.7956
2006	94.2620	48.3313
2007	105.1596	52.7710
2008	95.8301	54.4757
2009	97.8854	55.2210
2010	102.8912	55.1360
2011	109.5541	57.2162
2012	111.2516	58.3216

The company wishes to predict the average monthly load over the next 5 years so that it can plan its future monthly purchases. To assist the company, proceed as follows.

1. (a) Using the years and the maximum monthly load given for each year, graph the data, with x representing the number of years past 1994 and y representing the maximum load in megawatts.
   (b) Find the equation that best fits the data, using both a quadratic model and a cubic model.
   (c) Graph the data and both of these models from 1994 to 2014 (that is, from $x = 0$ to $x = 20$).
2. Do the two models appear to fit the data equally well in the interval 1994–2014? Which model appears to be a better predictor for the next decade?
3. Use the quadratic model to predict the maximum monthly load in the year 2017. How can this value be used by the company? Should this number be used to plan monthly power purchases for each month in 2017?
4. To create a "typical" monthly load function:
   (a) Create a table with the year as the independent variable and the average of the maximum and minimum monthly loads as the dependent variable.
   (b) Find the quadratic model that best fits these data points, using $x = 0$ to represent 1994.
5. Use a definite integral with the typical monthly load function to predict the average monthly load over the years 2014–2019.
6. What factors in addition to the average monthly load should be considered when the company plans future purchases of power?

14

Kevin Fleming/CORBIS

Functions of Two or More Variables

In this chapter we will extend our study of functions to functions of two or more variables. We will use these concepts to solve problems in the management, life, and social sciences. In particular, we will discuss joint cost functions, utility functions that describe the customer satisfaction derived from the consumption of two products, Cobb-Douglas production functions, and wind chill temperatures as a function of air temperature and wind speed.

We will use derivatives with respect to one of two variables (called partial derivatives) to find marginal cost, marginal productivity, marginal utility, marginal demand, and other rates of change. We will use partial derivatives to find maxima and minima of functions of two variables, and we will use Lagrange multipliers to optimize functions of two variables subject to a condition that constrains these variables. These skills are used to maximize profit, production, and utility and to minimize cost subject to constraints.

The topics and applications discussed in this chapter include the following.

Prerequisite Problem Type	For Section	Answer	Section for Review
If $y = f(x)$, x is the independent variable and y is the _____ variable.	**14.1**	Dependent	1.2 Functions
What is the domain of $f(x) = \dfrac{3x}{x-1}$?	**14.1**	All reals except $x = 1$	1.2 Domains
If $C(x) = 5 + 5x$, what is $C(0.20)$?	**14.1**	6	1.2 Function notation
(a) Solve for x and y: $$\begin{cases} 0 = 50 - 2x - 2y \\ 0 = 60 - 2x - 4y \end{cases}$$ (b) Solve for x and y: $$\begin{cases} x = 2y \\ x + y - 9 = 0 \end{cases}$$	**14.4** **14.5**	(a) $x = 20$, $y = 5$ (b) $x = 6$, $y = 3$	1.5 Systems of equations
If $z = 4x^2 + 5x^3 - 7$, what is $\dfrac{dz}{dx}$?	**14.2** **14.3** **14.4** **14.5**	$\dfrac{dz}{dx} = 8x + 15x^2$	9.4 Derivatives
If $f(x) = (x^2 - 1)^2$, what is $f'(x)$?	**14.2**	$f'(x) = 4x(x^2 - 1)$	9.6 Derivatives
If $z = 10y - \ln y$, what is $\dfrac{dz}{dy}$?	**14.2**	$\dfrac{dz}{dy} = 10 - \dfrac{1}{y}$	11.1 Derivatives of logarithmic functions
If $z = 5x^2 + e^x$, what is $\dfrac{dz}{dx}$?	**14.2**	$\dfrac{dz}{dx} = 10x + e^x$	11.2 Derivatives of exponential functions
Find the slope of the tangent to $y = 4x^3 - 4e^x$ at $(0, -4)$.	**14.2**	-4	9.3, 9.4, 11.2 Derivatives

- To find the domain of a function of two or more variables
- To evaluate a function of two or more variables given values for the independent variables

Functions of Two or More Variables

| APPLICATION PREVIEW |

The relations we have studied up to this point have been limited to two variables, with one of the variables assumed to be a function of the other. But there are many instances where one variable may depend on two or more other variables. For example, a company's quantity of output or production Q (measured in units or dollars) can be modeled according to the equation

$$Q = AK^\alpha L^{1-\alpha}$$

where A is a constant, K is the company's capital investment, L is the size of the labor force (in work-hours), and α is a constant with $0 < \alpha < 1$. Functions of this type are called **Cobb-Douglas production functions,** and they are frequently used in economics. For example, suppose the Cobb-Douglas production function for a company is given by

$$Q = 4K^{0.4}L^{0.6}$$

where Q is thousands of dollars of production value, K is hundreds of dollars of capital investment, and L is hours of labor. We could use this function to determine the production value for a given amount of capital investment and available work-hours of labor. We could also find how production is affected by changes in capital investment or available work-hours. (See Example 5.)

In addition, the demand function for a commodity frequently depends on the price of the commodity, available income, and prices of competing goods. Other examples from economics will be presented later in this chapter.

We write $z = f(x, y)$ to state that z is a function of both x and y. The variables x and y are called the **independent variables** and z is called the **dependent variable.** Thus the function f associates with each pair of possible values for the independent variables (x and y) exactly one value of the dependent variable (z).

The equation $z = x^2 - xy$ defines z as a function of x and y. We can denote this by writing $z = f(x, y) = x^2 - xy$. The domain of the function is the set of all ordered pairs (of real numbers), and the range is the set of all real numbers.

EXAMPLE 1   Domain

Give the domain of the function

$$g(x, y) = \frac{x^2 - 3y}{x - y}$$

Solution

The domain of the function is the set of ordered pairs that do not give a 0 denominator. That is, the domain is the set of all ordered pairs where the first and second elements are not equal (that is, where $x \neq y$).   ∎

CHECKPOINT

1. Find the domain of the function

$$f(x, y) = \frac{2}{\sqrt{x^2 - y^2}}$$

We graph the function $z = f(x, y)$ by using three dimensions. We can construct a three-dimensional coordinate space by drawing three mutually perpendicular axes, as in Figure 14.1. By setting up scales of measurement along the three axes from the origin of each axis, we can determine the three coordinates (x, y, z) for any point P. The point shown in Figure 14.1 is $+2$ units in the x-direction, $+3$ units in the y-direction, and $+4$ units in the z-direction, so the coordinates of the point are $(2, 3, 4)$.

The pairs of axes determine the three **coordinate planes;** the xy-plane, the yz-plane, and the xz-plane. The planes divide the space into eight **octants.** The point $P(2, 3, 4)$ is in the first octant.

If we are given a function $z = f(x, y)$, we can find the z-value corresponding to $x = a$ and $y = b$ by evaluating $f(a, b)$.

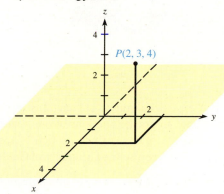

Figure 14.1

EXAMPLE 2 **Function Values**

If $z = f(x, y) = x^2 - 4xy + xy^3$, find the following.
(a) $f(1, 2)$ (b) $f(2, 5)$ (c) $f(-1, 3)$

Solution
(a) $f(1, 2) = 1^2 - 4(1)(2) + (1)(2)^3 = 1$
(b) $f(2, 5) = 2^2 - 4(2)(5) + (2)(5)^3 = 214$
(c) $f(-1, 3) = (-1)^2 - 4(-1)(3) + (-1)(3)^3 = -14$ ◼

CHECKPOINT 2. If $f(x, y, z) = x^2 + 2y - z$, find $f(2, 3, 4)$.

EXAMPLE 3 **Cost**

A small furniture company's cost (in dollars) to manufacture 1 unit of several different all-wood items is given by

$$C(x, y) = 5 + 5x + 22y$$

where x represents the number of board-feet of material used and y represents the number of work-hours of labor required for assembly and finishing. A certain bookcase uses 20 board-feet of material and requires 2.5 work-hours for assembly and finishing. Find the cost of manufacturing this bookcase.

Solution
The cost is

$$C(20, 2.5) = 5 + 5(20) + 22(2.5) = 160 \quad \text{dollars}$$ ◼

For a given function $z = f(x, y)$, we can construct a table of values by assigning values to x and y and finding the corresponding values of z. To each pair of values for x and y there corresponds a unique value of z, and thus a unique point in a three-dimensional coordinate system. From a table of values such as this, a finite number of points can be plotted. All points that satisfy the equation form a "surface" in space. Because z is a function of x and y, lines parallel to the z-axis will intersect such a surface in at most one point. The graph of the equation $z = 4 - x^2 - y^2$ is the surface shown in Figure 14.2(a) on the next page. The portion of the surface above the xy-plane resembles a bullet and is called a **paraboloid.** Some points on the surface are given in Table 14.1.

| TABLE 14.1 |

x	y	z
-2	0	0
0	-2	0
-1	0	3
0	-1	3
-1	-1	2
0	0	4
1	0	3
0	1	3
1	1	2
2	0	0
0	2	0

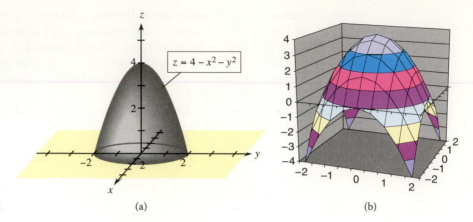

Figure 14.2   (a)   (b)

Spreadsheet Note We can create the graphs of functions of two variables with Excel in a manner similar to that used to graph functions of one variable. Figure 14.2(b) shows the Excel graph of $z = 4 - x^2 - y^2$. Appendix D, Section 14.1, shows the steps used in graphing functions of two variables. See also the Online Excel Guide. ◼

In practical applications of functions of two variables, we will have little need to construct the graphs of the surfaces. For this reason, we will not discuss methods of sketching the graphs. Although you will not be asked to sketch graphs of these surfaces, the fact that the graphs do *exist* will be used in studying relative maxima and minima of functions of two variables.

The properties of functions of one variable can be extended to functions of two variables. The precise definition of continuity for functions of two variables is technical and may be found in more advanced books. We will limit our study to functions that are continuous and have continuous derivatives in the domain of interest to us. We may think of continuous functions as functions whose graphs consist of surfaces without "holes" or "breaks" in them.

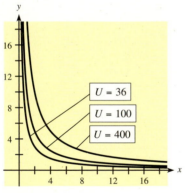

Figure 14.3

Let the function $U = f(x, y)$ represent the **utility** (that is, satisfaction) derived by a consumer from the consumption of two goods, X and Y, where x and y represent the amounts of X and Y, respectively. Because we will assume that the utility function is continuous, a given level of utility can be derived from an infinite number of combinations of x and y. The graph of all points (x, y) that give the same utility is called an **indifference curve**. A set of indifference curves corresponding to different levels of utility is called an **indifference map** (see Figure 14.3).

EXAMPLE | **4** | **Utility**

Suppose that the utility function for two goods, X and Y, is $U = x^2 y^2$ and a consumer purchases 10 units of X and 2 units of Y.

(a) What is the level of utility U for these two products?
(b) If the consumer purchases 5 units of X, how many units of Y must be purchased to retain the same level of utility?
(c) Graph the indifference curve for this level of utility.
(d) Graph the indifference curves for this utility function if $U = 100$ and if $U = 36$.

Solution
(a) If $x = 10$ and $y = 2$ satisfy the utility function, then the level of utility is
$U = 10^2 \cdot 2^2 = 400$.
(b) If x is 5, y must satisfy $400 = 5^2 y^2$, so $y = \pm 4$, and 4 units of Y must be purchased.

(c) The indifference curve for $U = 400$ is $400 = x^2y^2$. The graph for positive x and y is shown in Figure 14.3.

(d) The indifference map in Figure 14.3 contains these indifference curves. ∎

Technology Note

A graphing utility can be used to graph each indifference curve in the indifference map shown in Figure 14.3. To graph the indifference curve for a given value of U, we must recognize that y will be positive and solve for y to express it as a function of x. Try it for $U = 100$. Does your graph agree with the one shown in Figure 14.3? ∎

Sometimes functions of two variables are studied by fixing a value for one variable and graphing the resulting function of a single variable. We'll do this in Section 14.3 with production functions.

EXAMPLE | **5**   **Production** | APPLICATION PREVIEW |

Suppose a company has the Cobb-Douglas production function

$$Q = 4K^{0.4}L^{0.6}$$

where Q is thousands of dollars of production value, K is hundreds of dollars of capital investment per week, and L is work-hours of labor per week.

(a) If current capital investment is \$72,900 per week and work-hours are 3072 per week, find the current weekly production value.

(b) If weekly capital investment is increased to \$97,200 and new employees are hired so that there are 4096 total weekly work-hours, find the percent increase in the production value.

Solution

(a) Capital investment of \$72,900 means that $K = 729$. We use this value and $L = 3072$ in the production function.

$$Q = 4(729)^{0.4}(3072)^{0.6} = 6912$$

Thus the weekly production value is \$6,912,000.

(b) In this case we use $K = 972$ and $L = 4096$.

$$Q = 4(972)^{0.4}(4096)^{0.6} = 9216$$

This is an increase in production value of $9216 - 6912 = 2304$, which is equivalent to a weekly increase of

$$\frac{2304}{6912} = 0.33\tfrac{1}{3} = 33\tfrac{1}{3}\%$$ ∎

CHECKPOINT SOLUTIONS

1. The domain is the set of ordered pairs of real numbers where $x^2 - y^2 > 0$ or $x^2 > y^2$; that is, where $|x| > |y|$.

2. $f(2, 3, 4) = 4 + 6 - 4 = 6$

| **EXERCISES** | 14.1

Access end-of-section exercises online at **www.webassign.net**   ᴱᴺᴴᴬᴺᶜᴱᴰ **WebAssign**

Partial Differentiation

| APPLICATION PREVIEW |

Suppose that a company's sales are related to its television advertising by

$$s = 20{,}000 + 10nt + 20n^2$$

where n is the number of commercials per day and t is the length of the commercials in seconds. To find the instantaneous rate of change of sales with respect to the number of commercials per day if the company is running ten 30-second commercials, we find the partial derivative of s with respect to n. (See Example 7.)

In this section we find partial derivatives of functions of two or more variables and use these derivatives to find rates of change and slopes of tangents to surfaces. We will also find second- and higher-order partial derivatives of functions of two variables.

First-Order Partial Derivatives

If $z = f(x, y)$, we find the **partial derivative** of z with respect to x (denoted $\partial z/\partial x$) by treating the variable y as a constant and taking the derivative of $z = f(x, y)$ with respect to x. We can also take the partial derivative of z with respect to y by holding the variable x constant and taking the derivative of $z = f(x, y)$ with respect to y. We denote this derivative as $\partial z/\partial y$. Note that dz/dx represents the derivative of a function of one variable, x, and that $\partial z/\partial x$ represents the partial derivative of a function of two or more variables. Notations used to represent the partial derivative of $z = f(x, y)$ with respect to x are

$$\frac{\partial z}{\partial x}, \quad \frac{\partial f}{\partial x}, \quad \frac{\partial}{\partial x} f(x, y), \quad f_x(x, y), \quad f_x, \quad \text{and} \quad z_x$$

and notations used to represent the partial derivative of $z = f(x, y)$ with respect to y are

$$\frac{\partial z}{\partial y}, \quad \frac{\partial f}{\partial y}, \quad \frac{\partial}{\partial y} f(x, y), \quad f_y(x, y), \quad f_y, \quad \text{and} \quad z_y$$

EXAMPLE 1   Partial Derivatives

If $z = 4x^2 + 5x^2y^2 + 6y^3 - 7$, find $\partial z/\partial x$ and $\partial z/\partial y$.

Solution

To find $\partial z/\partial x$, we hold y constant so that the term $6y^3$ is constant, and its derivative is 0; the term $5x^2y^2$ is viewed as the constant $(5y^2)$ times (x^2), so its derivative is $(5y^2)(2x) = 10xy^2$. Thus

$$\frac{\partial z}{\partial x} = 8x + 10xy^2$$

Similarly for $\partial z/\partial y$, the term $4x^2$ is constant and the partial derivative of $5x^2y^2$ is the constant $5x^2$ times the derivative of y^2, so its derivative is $(5x^2)(2y) = 10x^2y$. Thus

$$\frac{\partial z}{\partial y} = 10x^2y + 18y^2$$

EXAMPLE 2 **Marginal Cost**

Suppose that the total cost of manufacturing a product is given by

$$C(x, y) = 5 + 5x + 2y$$

where x represents the cost of 1 ounce of materials and y represents the labor cost in dollars per hour.

(a) Find the rate at which the total cost changes with respect to the material cost x. This is the marginal cost of the product with respect to the material cost.
(b) Find the rate at which the total cost changes with respect to the labor cost y. This is the marginal cost of the product with respect to the labor cost.

Solution

(a) The rate at which the total cost changes with respect to the material cost x is the partial derivative found by treating the y-variable as a constant and taking the derivative of both sides of $C(x, y) = 5 + 5x + 2y$ with respect to x.

$$\frac{\partial C}{\partial x} = 0 + 5 + 0, \quad \text{or} \quad \frac{\partial C}{\partial x} = 5$$

The partial derivative $\partial C / \partial x = 5$ tells us that a change of \$1 in the cost of materials will cause an increase of \$5 in total costs, *if* labor costs remain constant.

(b) The rate at which the total cost changes with respect to the labor cost y is the partial derivative found by treating the x-variable as a constant and taking the derivative of both sides of $C(x, y) = 5 + 5x + 2y$ with respect to y.

$$\frac{\partial C}{\partial y} = 0 + 0 + 2, \quad \text{or} \quad \frac{\partial C}{\partial y} = 2$$

The partial derivative $\partial C / \partial y = 2$ tells us that a change of \$1 in the cost of labor will cause an increase of \$2 in total costs, *if* material costs remain constant. ∎

EXAMPLE 3 **Partial Derivatives**

If $z = x^2 y + e^x - \ln y$, find z_x and z_y.

Solution

$$z_x = \frac{\partial z}{\partial x} = 2xy + e^x$$

$$z_y = \frac{\partial z}{\partial y} = x^2 - \frac{1}{y}$$ ∎

EXAMPLE 4 **Power Rule**

If $f(x, y) = (x^2 - y^2)^2$, find the following.

(a) f_x
(b) f_y

Solution
(a) $f_x = 2(x^2 - y^2)2x = 4x^3 - 4xy^2$
(b) $f_y = 2(x^2 - y^2)(-2y) = -4x^2y + 4y^3$ ∎

1. If $z = 100x + 10xy - y^2$, find the following.

   (a) z_x

   (b) $\dfrac{\partial z}{\partial y}$

EXAMPLE 5 **Quotient Rule**

If $q = \dfrac{p_1 p_2 + 2p_1}{p_1 p_2 - 2p_2}$, find $\partial q/\partial p_1$.

Solution

$$\frac{\partial q}{\partial p_1} = \frac{(p_1 p_2 - 2p_2)(p_2 + 2) - (p_1 p_2 + 2p_1)p_2}{(p_1 p_2 - 2p_2)^2}$$

$$= \frac{p_1 p_2^2 + 2p_1 p_2 - 2p_2^2 - 4p_2 - p_1 p_2^2 - 2p_1 p_2}{(p_1 p_2 - 2p_2)^2}$$

$$= \frac{-2p_2^2 - 4p_2}{p_2^2(p_1 - 2)^2}$$

$$= \frac{-2p_2(p_2 + 2)}{p_2^2(p_1 - 2)^2}$$

$$= \frac{-2(p_2 + 2)}{p_2(p_1 - 2)^2}$$

■

We may evaluate partial derivatives by substituting values for x and y in the same way we did with derivatives of functions of one variable. For example, if $\partial z/\partial x = 2x - xy$, the value of the partial derivative with respect to x at $x = 2, y = 3$ is

$$\left.\frac{\partial z}{\partial x}\right|_{(2, 3)} = 2(2) - 2(3) = -2$$

Other notations used to denote evaluation of partial derivatives with respect to x at (a, b) are

$$\frac{\partial}{\partial x} f(a, b) \quad \text{and} \quad f_x(a, b)$$

We denote the evaluation of partial derivatives with respect to y at (a, b) by

$$\left.\frac{\partial z}{\partial y}\right|_{(a, b)}, \quad \frac{\partial}{\partial y} f(a, b), \quad \text{or} \quad f_y(a, b)$$

EXAMPLE 6 **Partial Derivative at a Point**

Find the partial derivative of $f(x, y) = x^2 + 3xy + 4$ with respect to x at the point $(1, 2, 11)$.

Solution

$$f_x(x, y) = 2x + 3y$$
$$f_x(1, 2) = 2(1) + 3(2) = 8$$

■

CHECKPOINT

2. If $g(x, y) = 4x^2 - 3xy + 10y^2$, find the following.

(a) $\dfrac{\partial g}{\partial x}(1, 3)$ (b) $g_y(4, 2)$

EXAMPLE 7

Marginal Sales | APPLICATION PREVIEW |

Suppose that a company's sales are related to its television advertising by

$$s = 20{,}000 + 10nt + 20n^2$$

where n is the number of commercials per day and t is the length of the commercials in seconds. Find the partial derivative of s with respect to n, and use the result to find the instantaneous rate of change of sales with respect to the number of commercials per day, if the company is currently running ten 30-second commercials per day.

Solution

The partial derivative of s with respect to n is $\partial s/\partial n = 10t + 40n$. At $n = 10$ and $t = 30$, the rate of change in sales is

$$\left. \frac{\partial s}{\partial n} \right|_{\substack{n=10 \\ t=30}} = 10(30) + 40(10) = 700$$

Thus, increasing the number of 30-second commercials by 1 would result in approximately 700 additional sales. This is the marginal sales with respect to the number of commercials per day at $n = 10$, $t = 30$. ∎

We have seen that the partial derivative $\partial z/\partial x$ is found by holding y constant and taking the derivative of z with respect to x and that the partial derivative $\partial z/\partial y$ is found by holding x constant and taking the derivative of z with respect to y. We now give formal definitions of these **partial derivatives.**

Partial Derivatives

The **partial derivative** of $z = f(x, y)$ with respect to x at the point (x, y) is

$$\frac{\partial z}{\partial x} = \frac{\partial}{\partial x} f(x, y) = \lim_{h \to 0} \frac{f(x + h, y) - f(x, y)}{h}$$

provided this limit exists.

The **partial derivative** of $z = f(x, y)$ with respect to y at the point (x, y) is

$$\frac{\partial z}{\partial y} = \frac{\partial}{\partial y} f(x, y) = \lim_{h \to 0} \frac{f(x, y + h) - f(x, y)}{h}$$

provided this limit exists.

We have already stated that the graph of $z = f(x, y)$ is a surface in three dimensions. The partial derivative with respect to x of such a function may be thought of as the slope of the tangent to the surface at a point (x, y, z) on the surface in the *positive direction of the x-axis*. That is, if a plane parallel to the xz-plane cuts the surface, passing through the point (x_0, y_0, z_0), the line in the plane that is tangent to the surface will have a slope equal to $\partial z/\partial x$ evaluated at the point. Thus

$$\left. \frac{\partial z}{\partial x} \right|_{(x_0, y_0)}$$

represents the slope of the tangent to the surface at (x_0, y_0, z_0), in the positive direction of the x-axis (see Figure 14.4 on the next page).

Similarly,

$$\frac{\partial z}{\partial y}\bigg|_{(x_0,\, y_0)} = \frac{\partial}{\partial y} f(x_0, y_0)$$

represents the slope of the tangent to the surface at (x_0, y_0, z_0) in the positive direction of the y-axis (see Figure 14.5).

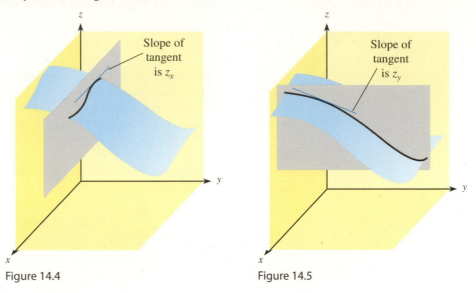

Figure 14.4   Figure 14.5

EXAMPLE 8   **Slopes of Tangents**

Let $z = 4x^3 - 4e^x + 4y^2$ and let P be the point $(0, 2, 12)$. Find the slope of the tangent to z at the point P in the positive direction of (a) the x-axis and (b) the y-axis.

Solution

(a) The slope of z at P in the positive x-direction is given by $\dfrac{\partial z}{\partial x}$, evaluated at P.

$$\frac{\partial z}{\partial x} = 12x^2 - 4e^x \quad \text{and} \quad \frac{\partial z}{\partial x}\bigg|_{(0,\, 2)} = 12(0)^2 - 4e^0 = -4$$

This tells us that z *decreases* approximately 4 units for an increase of 1 unit in x at this point.

(b) The slope of z at P in the positive y-direction is given by $\dfrac{\partial z}{\partial y}$, evaluated at P.

$$\frac{\partial z}{\partial y} = 8y \quad \text{and} \quad \frac{\partial z}{\partial y}\bigg|_{(0,\, 2)} = 8(2) = 16$$

Thus, at the point P $(0, 2, 12)$, the function *increases* approximately 16 units in the z-value for a unit increase in y. ∎

Up to this point, we have considered derivatives of functions of two variables. We can easily extend the concept to functions of three or more variables. We can find the partial derivative with respect to any one independent variable by taking the derivative of the function with respect to that variable while holding all other independent variables constant.

EXAMPLE 9 **Functions of Four Variables**

If $u = f(w, x, y, z) = 3x^2y + w^3 - 4xyz$, find the following.

(a) $\dfrac{\partial u}{\partial w}$ (b) $\dfrac{\partial u}{\partial x}$ (c) $\dfrac{\partial u}{\partial y}$ (d) $\dfrac{\partial u}{\partial z}$

Solution

(a) $\dfrac{\partial u}{\partial w} = 3w^2$ (b) $\dfrac{\partial u}{\partial x} = 6xy - 4yz$

(c) $\dfrac{\partial u}{\partial y} = 3x^2 - 4xz$ (d) $\dfrac{\partial u}{\partial z} = -4xy$ ∎

EXAMPLE 10 **Functions of Three Variables**

If $C = 4x_1 + 2x_1{}^2 + 3x_2 - x_1x_2 + x_3{}^2$, find the following.

(a) $\dfrac{\partial C}{\partial x_1}$ (b) $\dfrac{\partial C}{\partial x_2}$ (c) $\dfrac{\partial C}{\partial x_3}$

Solution

(a) $\dfrac{\partial C}{\partial x_1} = 4 + 4x_1 - x_2$ (b) $\dfrac{\partial C}{\partial x_2} = 3 - x_1$ (c) $\dfrac{\partial C}{\partial x_3} = 2x_3$ ∎

CHECKPOINT 3. If $f(w, x, y, z) = 8xy^2 + 4yz - xw^2$, find

(a) $\dfrac{\partial f}{\partial x}$ (b) $\dfrac{\partial f}{\partial w}$ (c) $\dfrac{\partial f}{\partial y}(1, 2, 1, 3)$ (d) $\dfrac{\partial f}{\partial z}(0, 2, 1, 3)$

Higher-Order Partial Derivatives

Just as we have taken derivatives of derivatives to obtain higher-order derivatives of functions of one variable, we may also take partial derivatives of partial derivatives to obtain higher-order partial derivatives of a function of more than one variable. If $z = f(x, y)$, then the partial derivative functions z_x and z_y are called *first partials*. Partial derivatives of z_x and z_y are called *second partials*, so $z = f(x, y)$ has *four* **second partial derivatives.** The notations for these second partial derivatives follow.

Second Partial Derivatives

$z_{xx} = \dfrac{\partial^2 z}{\partial x^2} = \dfrac{\partial}{\partial x}\left(\dfrac{\partial z}{\partial x}\right)$: both derivatives taken with respect to x

$z_{yy} = \dfrac{\partial^2 z}{\partial y^2} = \dfrac{\partial}{\partial y}\left(\dfrac{\partial z}{\partial y}\right)$: both derivatives taken with respect to y

$z_{xy} = \dfrac{\partial^2 z}{\partial y\, \partial x} = \dfrac{\partial}{\partial y}\left(\dfrac{\partial z}{\partial x}\right)$: first derivative taken with respect to x, second with respect to y

$z_{yx} = \dfrac{\partial^2 z}{\partial x\, \partial y} = \dfrac{\partial}{\partial x}\left(\dfrac{\partial z}{\partial y}\right)$: first derivative taken with respect to y, second with respect to x

EXAMPLE 11 **Second Partial Derivatives**

If $z = x^3y - 3xy^2 + 4$, find each of the second partial derivatives of the function.

Solution

Because $z_x = 3x^2y - 3y^2$ and $z_y = x^3 - 6xy$,

$$z_{xx} = \frac{\partial}{\partial x}(3x^2y - 3y^2) = 6xy$$

$$z_{xy} = \frac{\partial}{\partial y}(3x^2y - 3y^2) = 3x^2 - 6y$$

$$z_{yy} = \frac{\partial}{\partial y}(x^3 - 6xy) = -6x$$

$$z_{yx} = \frac{\partial}{\partial x}(x^3 - 6xy) = 3x^2 - 6y$$ ∎

Note that z_{xy} and z_{yx} are equal for the function in Example 11. This will always occur if the derivatives of the function are continuous.

> **$z_{xy} = z_{yx}$**   If the second partial derivatives z_{xy} and z_{yx} of a function $z = f(x, y)$ are continuous at a point, they are equal there.

EXAMPLE 12   Second Partial Derivatives

Find each of the second partial derivatives of $z = x^2y + e^{xy}$.

Solution

Because $z_x = 2xy + e^{xy} \cdot y = 2xy + ye^{xy}$,

$$z_{xx} = 2y + e^{xy} \cdot y^2 = 2y + y^2e^{xy}$$
$$z_{xy} = 2x + (e^{xy} \cdot 1 + ye^{xy} \cdot x)$$
$$= 2x + e^{xy} + xye^{xy}$$

Because $z_y = x^2 + e^{xy} \cdot x = x^2 + xe^{xy}$,

$$z_{yx} = 2x + (e^{xy} \cdot 1 + xe^{xy} \cdot y)$$
$$= 2x + e^{xy} + xye^{xy}$$
$$z_{yy} = 0 + xe^{xy} \cdot x = x^2e^{xy}$$ ∎

CHECKPOINT

4. If $z = 4x^3y^4 + 4xy$, find the following.
    (a) z_{xx} (b) z_{yy} (c) z_{xy} (d) z_{yx}
5. If $z = x^2 + 4e^{xy}$, find z_{xy}.

We can find partial derivatives of orders higher than the second. For example, we can find the third-order partial derivatives z_{xyx} and z_{xyy} for the function in Example 11 from the second derivative $z_{xy} = 3x^2 - 6y$.

$$z_{xyx} = 6x$$
$$z_{xyy} = -6$$

EXAMPLE 13   Third Partial Derivatives

If $y = 4y \ln x + e^{xy}$, find z_{xyy}.

Solution

$$z_x = 4y \cdot \frac{1}{x} + e^{xy} \cdot y$$

$$z_{xy} = 4 \cdot \frac{1}{x} \cdot 1 + e^{xy} \cdot 1 + y \cdot e^{xy} \cdot x = \frac{4}{x} + e^{xy} + xye^{xy}$$

$$z_{xyy} = 0 + e^{xy} \cdot x + xy \cdot e^{xy} \cdot x + e^{xy} \cdot x = x^2ye^{xy} + 2xe^{xy}$$

■

CHECKPOINT SOLUTIONS

1. (a) $z_x = 100 + 10y$

    (b) $\dfrac{\partial z}{\partial y} = 10x - 2y$

2. (a) $\dfrac{\partial g}{\partial x} = 8x - 3y$ and $\dfrac{\partial g}{\partial x}(1, 3) = 8(1) - 3(3) = -1$

    (b) $g_y = -3x + 20y$ and $g_y(4, 2) = -3(4) + 20(2) = 28$

3. (a) $\dfrac{\partial f}{\partial x} = 8y^2 - w^2$

    (b) $\dfrac{\partial f}{\partial w} = -2xw$

    (c) $\dfrac{\partial f}{\partial y} = 16xy + 4z$ and $\dfrac{\partial f}{\partial y}(1, 2, 1, 3) = 16(2)(1) + 4(3) = 44$

    (d) $\dfrac{\partial f}{\partial z} = 4y$ and $\dfrac{\partial f}{\partial z}(0, 2, 1, 3) = 4(1) = 4$

4. $z_x = 12x^2y^4 + 4y$ and $z_y = 16x^3y^3 + 4x$

    (a) $z_{xx} = 24xy^4$
    (b) $z_{yy} = 48x^3y^2$
    (c) $z_{xy} = 48x^2y^3 + 4$
    (d) $z_{yx} = 48x^2y^3 + 4$

5. $z_x = 2x + 4e^{xy}(y) = 2x + 4ye^{xy}$
    Calculation of z_{xy} requires the Product Rule.

    $$z_{xy} = 0 + (4y)(e^{xy}x) + (e^{xy})(4) = 4xye^{xy} + 4e^{xy}$$

| EXERCISES | 14.2

Access end-of-section exercises online at **www.webassign.net**

OBJECTIVES   14.3

- To evaluate cost functions at given levels of production
- To find marginal costs from total cost and joint cost functions
- To find marginal productivity for given production functions
- To find marginal demand functions from demand functions for a pair of related products

Applications of Functions of Two Variables in Business and Economics

| APPLICATION PREVIEW |

Suppose that the joint cost function for two commodities is

$$C = 50 + x^2 + 8xy + y^3$$

where x and y represent the quantities of each commodity and C is the total cost for the two commodities. We can take the partial derivative of C with respect to x to find the marginal cost of the first product and we can take the partial derivative of C with respect to y to find the marginal cost of the second product. (See Example 1.) This is one of three types of applications that we will consider in this section. In the second case, we consider production functions and revisit Cobb-Douglas production functions. Marginal productivity is introduced and its meaning is explained. Finally, we consider demand functions for two products in a competitive market. Partial derivatives are used to define marginal demands, and these marginals are used to classify the products as competitive or complementary.

Joint Cost and Marginal Cost   Suppose that a firm produces two products using the same inputs in different proportions. In such a case the **joint cost function** is of the form $C = Q(x, y)$, where x represents the quantity of product X and y represents the quantity of product Y. Then $\partial C/\partial x$ is the **marginal cost** of the first product and $\partial C/\partial y$ is the marginal cost of the second product.

EXAMPLE   1   Joint Cost | APPLICATION PREVIEW |

If the joint cost function for two products is

$$C = Q(x, y) = 50 + x^2 + 8xy + y^3$$

where x represents the quantity of product X and y represents the quantity of product Y, find the marginal cost with respect to the number of units of:

(a) Product X (b) Product Y (c) Product X at $(5, 3)$ (d) Product Y at $(5, 3)$

Solution

(a) The marginal cost with respect to the number of units of product X is $\dfrac{\partial C}{\partial x} = 2x + 8y$.

(b) The marginal cost with respect to the number of units of product Y is $\dfrac{\partial C}{\partial y} = 8x + 3y^2$.

(c) $\dfrac{\partial C}{\partial x}\bigg|_{(5, 3)} = 2(5) + 8(3) = 34$

Thus if 5 units of product X and 3 units of product Y are produced, the total cost will increase approximately \$34 for a unit increase in product X if y is held constant.

(d) $\dfrac{\partial C}{\partial y}\bigg|_{(5, 3)} = 8(5) + 3(3)^2 = 67$

Thus if 5 units of product X and 3 units of product Y are produced, the total cost will increase approximately \$67 for a unit increase in product Y if x is held constant.   ■

1. If the joint cost in dollars for two products is given by

$$C = 100 + 3x + 10xy + y^2$$

find the marginal cost with respect to (a) the number of units of Product X and (b) the number of units of Product Y at $(7, 3)$.

Production Functions

An important problem in economics concerns how the factors necessary for production determine the output of a product. For example, the output of a product depends on available labor, land, capital, material, and machines. If the amount of output z of a product depends on the amounts of two inputs x and y, then the quantity z is given by the **production function** $z = f(x, y)$.

EXAMPLE 2   Crop Harvesting

Suppose that it is known that z bushels of a crop can be harvested according to the function

$$z = \frac{21(6xy + 4x^2 - 3y)}{2x + 0.01y}$$

when $100x$ work-hours of labor are employed on y acres of land. What would be the output (in bushels) if 200 work-hours were used on 300 acres?

Solution
Because $z = f(x, y)$,

$$f(2, 300) = \frac{(21)[6(2)(300) + 4(2)^2 - 3(300)]}{2(2) + 3}$$

$$= \frac{(21)[3600 + 16 - 900]}{7} = 8148 \quad \text{(bushels)} \quad \blacksquare$$

If $z = f(x, y)$ is a production function, $\partial z / \partial x$ represents the rate of change in the output z with respect to input x while input y remains constant. This partial derivative is called the **marginal productivity of x**. The partial derivative $\partial z / \partial y$ is the **marginal productivity of y** and measures the rate of change of z with respect to input y.

Marginal productivity (for either input) will be positive over a wide range of inputs, but it increases at a decreasing rate, and it may eventually reach a point where it no longer increases and begins to decrease.

EXAMPLE 3   Production

If a production function is given by $z = 5x^{1/2}y^{1/4}$, find the marginal productivity of

(a) x.   (b) y.

Solution

(a) $\dfrac{\partial z}{\partial x} = \dfrac{5}{2}x^{-1/2}y^{1/4}$ (b) $\dfrac{\partial z}{\partial y} = \dfrac{5}{4}x^{1/2}y^{-3/4}$

Note that the marginal productivity of x is positive for all values of x but that it decreases as x gets larger (because of the negative exponent). The same is true for the marginal productivity of y. $\blacksquare$

2. If the production function for a product is

$$P = 10x^{1/4}y^{1/2}$$

find the marginal productivity of x.

Calculator Note If we have a production function and fix a value for one variable, then we can use a graphing calculator to analyze the marginal productivity with respect to the other variable. ■

 EXAMPLE 4 **Production**

Suppose the Cobb-Douglas production function for a company is given by

$$z = 100x^{1/4}y^{3/4}$$

where x is the company's capital investment (in dollars) and y is the size of the labor force (in work-hours).

(a) Find the marginal productivity of x.
(b) If the current labor force is 625 work-hours, substitute $y = 625$ in your answer to part (a) and graph the result.
(c) From the graph in part (b), what can be said about the effect on production of additional capital investment when the work-hours remain at 625?
(d) Find the marginal productivity of y.
(e) If current capital investment is $10,000, substitute $x = 10,000$ in your answer to part (d) and graph the result.
(f) From the graph in part (e), what can be said about the effect on production of additional work-hours when capital investment remains at $10,000?

Solution

(a) $z_x = 25x^{-3/4}y^{3/4}$
(b) If $y = 625$, then z_x becomes

$$z_x = 25x^{-3/4}(625)^{3/4} = 25\left(\frac{1}{x^{3/4}}\right)(125) = \frac{3125}{x^{3/4}}$$

The graph of z_x can be limited to Quadrant I because the capital investment is $x > 0$, and hence $z_x > 0$. Knowledge of asymptotes can help us determine range values for x and z_x that give an accurate graph. See Figure 14.6.

(c) Figure 14.6 shows that $z_x > 0$ for $x > 0$. This means that any increases in capital investment will result in increases in productivity. However, Figure 14.6 also shows that z_x is decreasing for $x > 0$, which means that as capital investment increases, productivity increases, but at a slower rate.

(d) $z_y = 75x^{1/4}y^{-1/4}$

(e) If $x = 10,000$, then z_y becomes

$$z_y = 75(10,000)^{1/4}\left(\frac{1}{y^{1/4}}\right) = \frac{750}{y^{1/4}}$$

The graph is shown in Figure 14.7(a).

(f) Figure 14.7(a) also shows that $z_y > 0$ when $y > 0$, so increasing work-hours increases productivity. Note that z_y is decreasing for $y > 0$, but Figure 14.7(b) shows that it does so more slowly than z_x. This indicates that increases in work-hours have a diminishing impact on productivity, but still a more significant one than increases in capital expenditures. ■

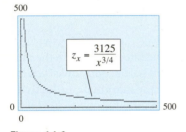

Figure 14.6

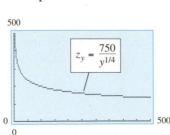

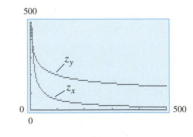

Figure 14.7   (a)   (b)

Demand Functions   Suppose that two products are sold at prices p_1 and p_2 (both in dollars) in a competitive market consisting of a fixed number of consumers with given tastes and incomes. Then the amount of each *one* of the products demanded by the consumers is dependent on the prices of *both* products on the market. If q_1 represents the demand for the number of units of the first product, then $q_1 = f(p_1, p_2)$ is the **demand function** for that product. The graph of such a function is called a **demand surface**. An example of a demand function in two variables is $q_1 = 400 - 2p_1 - 4p_2$. Here q_1 is a function of two variables p_1 and p_2. If $p_1 = \$10$ and $p_2 = \$20$, the demand would equal $400 - 2(10) - 4(20) = 300$ units.

EXAMPLE 5   **Demand**

The demand functions for two products are

$$q_1 = 50 - 5p_1 - 2p_2$$
$$q_2 = 100 - 3p_1 - 8p_2$$

where q_1 and q_2 are the numbers of units and p_1 and p_2 are in dollars.

(a) What is the demand for each of the products if the price of the first is $p_1 = \$5$ and the price of the second is $p_2 = \$8$?
(b) Find a pair of prices p_1 and p_2 such that the demands for product 1 and product 2 are equal.

Solution

(a)
$$q_1 = 50 - 5(5) - 2(8) = 9$$
$$q_2 = 100 - 3(5) - 8(8) = 21$$

Thus if these are the prices, the demand for product 2 is higher than the demand for product 1.

(b) We want q_1 to equal q_2. Setting $q_1 = q_2$, we see that

$$50 - 5p_1 - 2p_2 = 100 - 3p_1 - 8p_2$$
$$6p_2 - 50 = 2p_1$$
$$p_1 = 3p_2 - 25$$

Now, any pair of positive values that satisfies this equation will make the demands equal. Letting $p_2 = 10$, we see that $p_1 = 5$ will satisfy the equation. Thus the prices $p_1 = 5$ and $p_2 = 10$ will make the demands equal. The prices $p_1 = 2$ and $p_2 = 9$ will also make the demands equal. Many pairs of values (that is, all those satisfying $p_1 = 3p_2 - 25$) will equalize the demands.   ■

If the demand functions for a pair of related products, product 1 and product 2, are $q_1 = f(p_1, p_2)$ and $q_2 = g(p_1, p_2)$, respectively, then the partial derivatives of q_1 and q_2 are called **marginal demand functions.**

$\dfrac{\partial q_1}{\partial p_1}$ is the marginal demand of q_1 with respect to p_1.

$\dfrac{\partial q_1}{\partial p_2}$ is the marginal demand of q_1 with respect to p_2.

$\dfrac{\partial q_2}{\partial p_1}$ is the marginal demand of q_2 with respect to p_1.

$\dfrac{\partial q_2}{\partial p_2}$ is the marginal demand of q_2 with respect to p_2.

For typical demand functions, if the price of product 2 is fixed, the demand for product 1 will decrease as its price p_1 increases. In this case the marginal demand of q_1 with respect to p_1 will be negative; that is, $\partial q_1/\partial p_1 < 0$. Similarly, $\partial q_2/\partial p_2 < 0$.

But what about $\partial q_2/\partial p_1$ and $\partial q_1/\partial p_2$? If $\partial q_2/\partial p_1$ and $\partial q_1/\partial p_2$ are both positive, the two products are **competitive** because an increase in price p_1 will result in an increase in demand for product 2 (q_2) if the price p_2 is held constant, and an increase in price p_2 will increase the demand for product 1 (q_1) if p_1 is held constant. Stated more simply, an increase in the price of one of the two products will result in an increased demand for the other, so the products are in competition. For example, an increase in the price of a Japanese automobile will result in an increase in demand for an American automobile if the price of the American automobile is held constant.

If $\partial q_2/\partial p_1$ and $\partial q_1/\partial p_2$ are both negative, the products are **complementary** because an increase in the price of one product will cause a decrease in demand for the other product if the price of the second product doesn't change. Under these conditions, a *decrease* in the price of product 1 will result in an *increase* in the demand for product 2, and a decrease in the price of product 2 will result in an increase in the demand for product 1. For example, a decrease in the price of gasoline will result in an increase in the demand for large automobiles.

If the signs of $\partial q_2/\partial p_1$ and $\partial q_1/\partial p_2$ are different, the products are neither competitive nor complementary. This situation rarely occurs but is possible.

EXAMPLE 6   Demand

The demand functions for two related products, product 1 and product 2, are given by

$$q_1 = 400 - 5p_1 + 6p_2 \qquad q_2 = 250 + 4p_1 - 5p_2$$

(a) Determine the four marginal demands.
(b) Are product 1 and product 2 complementary or competitive?

Solution

(a) $\dfrac{\partial q_1}{\partial p_1} = -5 \qquad \dfrac{\partial q_2}{\partial p_2} = -5 \qquad \dfrac{\partial q_1}{\partial p_2} = 6 \qquad \dfrac{\partial q_2}{\partial p_1} = 4$

(b) Because $\partial q_1/\partial p_2$ and $\partial q_2/\partial p_1$ are positive, products 1 and 2 are competitive.   ■

CHECKPOINT

3. If the demand functions for two products are

$$q_1 = 200 - 3p_1 - 4p_2 \quad \text{and} \quad q_2 = 50 - 6p_1 - 5p_2$$

find the marginal demand of
(a) q_1 with respect to p_1.   (b) q_2 with respect to p_2.

CHECKPOINT SOLUTIONS

1. (a) $\dfrac{\partial C}{\partial x} = 3 + 10y$

(b) $\dfrac{\partial C}{\partial y} = 10x + 2y \qquad \dfrac{\partial C}{\partial y}(7, 3) = 10(7) + 2(3) = 76$ (dollars per unit)

2. $\dfrac{\partial P}{\partial x} = \dfrac{2.5y^{1/2}}{x^{3/4}}$

3. (a) $\dfrac{\partial q_1}{\partial p_1} = -3$ (b) $\dfrac{\partial q_2}{\partial p_2} = -5$

| EXERCISES | 14.3

Access end-of-section exercises online at **www.webassign.net**   ENHANCED   **WebAssign**

Maxima and Minima

| APPLICATION PREVIEW |

Adele Lighting manufactures 20-inch lamps and 31-inch lamps. Suppose that x is the number of thousands of 20-inch lamps and that the demand for these is given by $p_1 = 50 - x$, where p_1 is in dollars. Similarly, suppose that y is the number of thousands of 31-inch lamps and that the demand for these is given by $p_2 = 60 - 2y$, where p_2 is also in dollars. Adele Lighting's joint cost function for these lamps is $C = 2xy$ (in thousands of dollars). Therefore, Adele Lighting's profit (in thousands of dollars) is a function of the two variables x and y. In order to determine Adele's maximum profit, we need to develop methods for finding maximum values for a function of two variables. (See Example 4.)

In this section we will find relative maxima and minima of functions of two variables, and use them to solve applied optimization problems. We will also use minimization of functions of two variables to develop the linear regression formulas. Recall that we first used linear regression in Chapter 2.

Maxima and Minima

In our study of differentiable functions of one variable, we saw that for a relative maximum or minimum to occur at a point, the tangent line to the curve had to be horizontal at that point. The function $z = f(x, y)$ describes a surface in three dimensions. If all partial derivatives of $f(x, y)$ exist, then there must be a horizontal plane tangent to the surface at a point in order for the surface to have a relative maximum at that point (see Figure 14.8a) or a minimum at that point (see Figure 14.8b). But if the plane tangent to the surface at the point is horizontal, then all the tangent lines to the surface at that point must also be horizontal, for they lie in the tangent plane. In particular, the tangent line in the direction of the x-axis will be horizontal, so $\partial z / \partial x = 0$ at the point; and the tangent line in the direction of the y-axis will be horizontal, so $\partial z / \partial y = 0$ at the point. Thus those points where *both* $\partial z / \partial x = 0$ and $\partial z / \partial y = 0$ are called **critical points** for the surface.

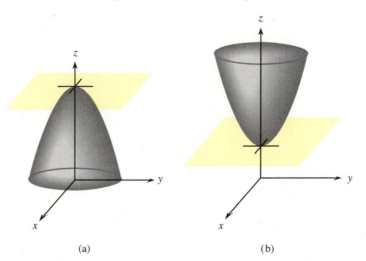

Figure 14.8   (a)   (b)

How can we determine whether a critical point is a relative maximum, a relative minimum, or neither of these? Finding that $\partial^2 z / \partial x^2 < 0$ and $\partial^2 z / \partial y^2 < 0$ is not enough to tell us that we have a relative maximum. The "second-derivative" test we must use involves the values of the second partial derivatives and the value of D at the critical point (a, b), where D is defined as follows:

$$D = \frac{\partial^2 z}{\partial x^2} \cdot \frac{\partial^2 z}{\partial y^2} - \left(\frac{\partial^2 z}{\partial x \partial y}\right)^2$$

We shall state, without proof, the result that determines whether there is a relative maximum, a relative minimum, or neither at the critical point (a, b).

Test for Maxima and Minima

Let $z = f(x, y)$ be a function for which both

$$\frac{\partial z}{\partial x} = 0 \quad \text{and} \quad \frac{\partial z}{\partial y} = 0 \quad \text{at a point } (a, b)$$

and suppose that all second partial derivatives are continuous there. Evaluate

$$D = \frac{\partial^2 z}{\partial x^2} \cdot \frac{\partial^2 z}{\partial y^2} - \left(\frac{\partial^2 z}{\partial x \partial y}\right)^2$$

at the critical point (a, b), and conclude the following:

a. If $D > 0$ and $\partial^2 z/\partial x^2 > 0$ at (a, b), then a relative minimum occurs at (a, b). In this case, $\partial^2 z/\partial y^2 > 0$ at (a, b) also.
b. If $D > 0$ and $\partial^2 z/\partial x^2 < 0$ at (a, b), then a relative maximum occurs at (a, b). In this case, $\partial^2 z/\partial y^2 < 0$ at (a, b) also.
c. If $D < 0$ at (a, b), there is neither a relative maximum nor a relative minimum at (a, b).
d. If $D = 0$ at (a, b), the test fails; investigate the function near the point.

We can test for relative maxima and minima by using the following procedure.

Maxima and Minima of z = f(x, y)

Procedure	Example
To find relative maxima and minima of $z = f(x, y)$:	Test $z = 4 - 4x^2 - y^2$ for relative maxima and minima.
1. Find $\partial z/\partial x$ and $\partial z/\partial y$.	1. $\dfrac{\partial z}{\partial x} = -8x;\ \dfrac{\partial z}{\partial y} = -2y$
2. Find the point(s) that satisfy *both* $\partial z/\partial x = 0$ and $\partial z/\partial y = 0$. These are the critical points.	2. $\dfrac{\partial z}{\partial x} = 0$ if $x = 0$. $\dfrac{\partial z}{\partial y} = 0$ if $y = 0$. The critical point is $(0, 0, 4)$.
3. Find all second partial derivatives.	3. $\dfrac{\partial^2 z}{\partial x^2} = -8;\ \dfrac{\partial^2 z}{\partial y^2} = -2;\ \dfrac{\partial^2 z}{\partial x\, \partial y} = \dfrac{\partial^2 z}{\partial y\, \partial x} = 0$
4. Evaluate D at each critical point.	4. At $(0, 0)$, $D = (-8)(-2) - 0^2 = 16$.
5. Use the test for maxima and minima to determine whether relative maxima or minima occur.	5. $D > 0$, $\partial^2 z/\partial x^2 < 0$, and $\partial^2 z/\partial y^2 < 0$. A relative maximum occurs at $(0, 0)$. See Figure 14.9.

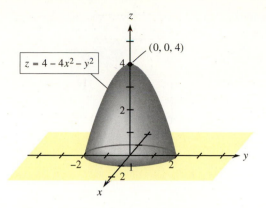

$z = 4 - 4x^2 - y^2$

$(0, 0, 4)$

Figure 14.9

EXAMPLE 1   **Relative Maxima and Minima**

Test $z = x^2 + y^2 - 2x + 1$ for relative maxima and minima.

Solution

1. $\dfrac{\partial z}{\partial x} = 2x - 2; \quad \dfrac{\partial z}{\partial y} = 2y$

2. $\dfrac{\partial z}{\partial x} = 0$ if $x = 1$. $\dfrac{\partial z}{\partial y} = 0$ if $y = 0$.

    Both are 0 if $x = 1$ and $y = 0$, so the
    critical point is $(1, 0, 0)$.

3. $\dfrac{\partial^2 z}{\partial x^2} = 2; \quad \dfrac{\partial^2 z}{\partial y^2} = 2; \quad \dfrac{\partial^2 z}{\partial x \partial y} = \dfrac{\partial^2 z}{\partial y\, \partial x} = 0$

4. At $(1, 0)$, $D = 2 \cdot 2 - 0^2 = 4$.

5. $D > 0$, $\partial^2 z / \partial x^2 > 0$, and $\partial^2 z / \partial y^2 > 0$.
    A relative minimum occurs at $(1, 0)$.
    (See Figure 14.10.)

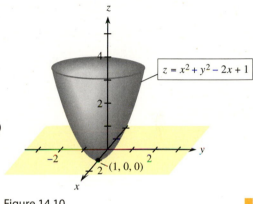

$z = x^2 + y^2 - 2x + 1$

$(1, 0, 0)$

Figure 14.10

EXAMPLE 2   **Saddle Points**

Test $z = y^2 - x^2$ for relative maxima and minima.

Solution

1. $\dfrac{\partial z}{\partial x} = -2x; \quad \dfrac{\partial z}{\partial y} = 2y$

2. $\dfrac{\partial z}{\partial x} = 0$ if $x = 0$; $\dfrac{\partial z}{\partial y} = 0$ if $y = 0$.

    Thus both equal 0 if $x = 0$, $y = 0$. The critical point is $(0, 0, 0)$.

3. $\dfrac{\partial^2 z}{\partial x^2} = -2; \quad \dfrac{\partial^2 z}{\partial y^2} = 2; \quad \dfrac{\partial^2 z}{\partial x\, \partial y} = \dfrac{\partial^2 z}{\partial y\, \partial x} = 0$

4. $D = (-2)(2) - 0 = -4$

5. $D < 0$, so the critical point is neither a relative maximum nor a relative minimum. As
    Figure 14.11 (on the next page) shows, the surface formed has the shape of a saddle.
    For this reason, critical points that are neither relative maxima nor relative minima are
    called **saddle points**.

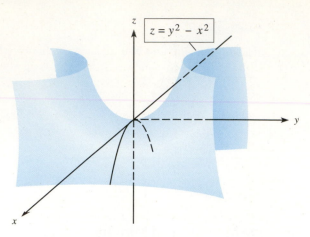

Figure 14.11

The following example involves a surface with two critical points.

EXAMPLE 3   Maxima and Minima

Test $z = x^3 + y^2 + 6xy + 24x$ for maxima and minima.

Solution

1. $\dfrac{\partial z}{\partial x} = 3x^2 + 6y + 24$ and $\dfrac{\partial z}{\partial y} = 2y + 6x$

2. $\dfrac{\partial z}{\partial x} = 0$ if $0 = 3x^2 + 6y + 24$,   and $\dfrac{\partial z}{\partial y} = 0$ if $0 = 2y + 6x$.

   But $0 = 2y + 6x$ means $y = -3x$. Because *both* partial derivatives must equal zero, we substitute $(-3x)$ for y in $0 = 3x^2 + 6y + 24$ to obtain

   $$0 = 3x^2 + 6(-3x) + 24 = 3x^2 - 18x + 24$$
   $$0 = 3(x^2 - 6x + 8) = 3(x - 2)(x - 4)$$

   Thus, the solutions are $x = 2$ and $x = 4$.
   When $x = 2, y = -3x = -6$, and $z = 2^3 + (-6)^2 + 6(2)(-6) + 24(2) = 20$, so one critical point is $(2, -6, 20)$.
   When $x = 4$, $y = -3x = -12$, and $z = 4^3 + (-12)^2 + 6(4)(-12) + 24(4) = 16$, so another critical point is $(4, -12, 16)$.

3. $\dfrac{\partial^2 z}{\partial x^2} = 6x;\ \dfrac{\partial^2 z}{\partial y^2} = 2;\ \dfrac{\partial^2 z}{\partial x\,\partial y} = \dfrac{\partial^2 z}{\partial y\,\partial x} = 6$

4. $D = (6x)(2) - 6^2$

5. At $(2, -6)$, $\dfrac{\partial^2 z}{\partial x^2} = 6 \cdot 2 = 12 > 0$, $\dfrac{\partial^2 z}{\partial y^2} = 2 > 0$, and $D = (6 \cdot 2)(2) - 6^2 = -12 < 0$, so a saddle point occurs at $(2, -6, 20)$.

   At $(4, -12)$, $\dfrac{\partial^2 z}{\partial x^2} = 6 \cdot 4 = 24 > 0$, $\dfrac{\partial^2 z}{\partial y^2} = 2 > 0$, and $D = (6 \cdot 4)(2) - 6^2 = 12 > 0$, so a relative minimum occurs at $(4, -12, 16)$. ∎

CHECKPOINT

Suppose that $z = 4 - x^2 - y^2 + 2x - 4y$.
1. Find z_x and z_y.
2. Solve $z_x = 0$ and $z_y = 0$ simultaneously to find the critical point(s) for the graph of this function.
3. Test the point(s) for relative maxima and minima.

EXAMPLE 4 | **Maximum Profit** | **APPLICATION PREVIEW** |

Adele Lighting manufactures 20-inch lamps and 31-inch lamps. Suppose that x is the number of thousands of 20-inch lamps and that the demand for these is given by $p_1 = 50 - x$, where p_1 is in dollars. Similarly, suppose that y is the number of thousands of 31-inch lamps and that the demand for these is given by $p_2 = 60 - 2y$, where p_2 is also in dollars. Adele Lighting's joint cost function for these lamps is $C = 2xy$ (in thousands of dollars). Therefore, Adele Lighting's profit (in thousands of dollars) is a function of the two variables x and y. Determine Adele's maximum profit.

Solution

The profit function is $P(x, y) = p_1 x + p_2 y - C(x, y)$. Thus,

$$P(x, y) = (50 - x)x + (60 - 2y)y - 2xy$$
$$= 50x - x^2 + 60y - 2y^2 - 2xy$$

gives the profit in thousands of dollars. To maximize the profit, we proceed as follows.

$$P_x = 50 - 2x - 2y \quad \text{and} \quad P_y = 60 - 4y - 2x$$

Solving $P_x = 0$ and $P_y = 0$ simultaneously, we have

$$\begin{cases} 0 = 50 - 2x - 2y \\ 0 = 60 - 2x - 4y \end{cases}$$

Subtraction gives $-10 + 2y = 0$, so $y = 5$. With $y = 5$, the equation $0 = 50 - 2x - 2y$ becomes $0 = 40 - 2x$, so $x = 20$. Now

$$P_{xx} = -2, \quad P_{yy} = -4, \quad \text{and} \quad P_{xy} = -2, \quad \text{and}$$
$$D = (P_{xx})(P_{yy}) - (P_{xy})^2 = (-2)(-4) - (-2)^2 = 4$$

Because $P_{xx} < 0$, $P_{yy} < 0$, and $D > 0$, the values $x = 20$ and $y = 5$ yield maximum profit. Therefore, when $x = 20$ and $y = 5$, $p_1 = 30$, $p_2 = 50$, and the maximum profit is

$$P(20, 5) = 600 + 250 - 200 = 650$$

That is, Adele Lighting's maximum profit is $650,000 when the company sells 20,000 of the 20-inch lamps at $30 each and 5000 of the 31-inch lamps at $50 each. ◼

Linear Regression

We have used different types of functions to model cost, revenue, profit, demand, supply, and other real-world relationships. Sometimes we have used calculus to study the behavior of these functions, finding, for example, marginal cost, marginal revenue, producer's surplus, and so on. We now have the mathematical tools to understand and develop the formulas that graphing calculators and other technology use to find the equations for linear models.

The formulas used to find the equation of the straight line that is the best fit for a set of data are developed using max-min techniques for functions of two variables. This line is called the **regression line**. In Figure 14.12 (on the next page), we define line ℓ to be the best fit for the data points (that is, the regression line) if the sum of the squares of the differences between the actual y-values of the data points and the y-values of the points on the line is a minimum.

In general, to find the equation of the regression line, we assume that the relationship between x and y is approximately linear and that we can find a straight line with equation

$$\hat{y} = a + bx$$

where the values of $\hat{y}$ will approximate the y-values of the points we know. That is, for each given value of x, the point $(x, \hat{y})$ will be on the line. For any given x-value, x_i, we are interested in the deviation between the y-value of the data point (x_i, y_i) and the $\hat{y}$-value from the equation, $\hat{y}_i$, that results when x_i is substituted for x. These deviations are of the form

$$d_i = \hat{y}_i - y_i, \quad \text{for } i = 1, 2, \ldots, n$$

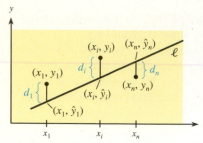

Figure 14.12

See Figure 14.12 for a general case with the deviations exaggerated.

To measure the deviations in a way that accounts for the fact that some of the y-values will be above the line and some below, we will say that the line that is the best fit for the data is the one for which the sum of the squares of the deviations is a minimum. That is, we seek the a and b in the equation

$$\hat{y} = a + bx$$

such that the sum of the squares of the deviations,

$$S = \sum_{i=1}^{n}(\hat{y}_i - y_i)^2 = \sum_{i=1}^{n}[(bx_i + a) - y_i]^2$$

$$= (bx_1 + a - y_1)^2 + (bx_2 + a - y_2)^2 + \cdots + (bx_n + a - y_n)^2$$

is a minimum. The procedure for determining a and b is called the **method of least squares.**

We seek the values of b and a that make S a minimum, so we find the values that make

$$\frac{\partial S}{\partial b} = 0 \quad \text{and} \quad \frac{\partial S}{\partial a} = 0$$

where

$$\frac{\partial S}{\partial b} = 2(bx_1 + a - y_1)x_1 + 2(bx_2 + a - y_2)x_2 + \cdots + 2(bx_n + a - y_n)x_n$$

$$\frac{\partial S}{\partial a} = 2(bx_1 + a - y_1) + 2(bx_2 + a - y_2) + \cdots + 2(bx_n + a - y_n)$$

Setting each equation equal to 0, dividing by 2, and using sigma notation give

$$0 = b\sum_{i=1}^{n}x_i^2 + a\sum_{i=1}^{n}x_i - \sum_{i=1}^{n}x_iy_i \tag{1}$$

$$0 = b\sum_{i=1}^{n}x_i + a\sum_{i=1}^{n}1 - \sum_{i=1}^{n}y_i \tag{2}$$

We can write Equations (1) and (2) as follows:

$$\sum_{i=1}^{n}x_iy_i = a\sum_{i=1}^{n}x_i + b\sum_{i=1}^{n}x_i^2 \tag{3}$$

$$\sum_{i=1}^{n}y_i = an + b\sum_{i=1}^{n}x_i \tag{4}$$

Multiplying Equation (3) by n and Equation (4) by $\sum_{i=1}^{n}x_i$ permits us to begin to solve for b.

$$n\sum_{i=1}^{n}x_iy_i = na\sum_{i=1}^{n}x_i + nb\sum_{i=1}^{n}x_i^2 \tag{5}$$

$$\sum_{i=1}^{n}x_i\sum_{i=1}^{n}y_i = na\sum_{i=1}^{n}x_i + b\left(\sum_{i=1}^{n}x_i\right)^2 \tag{6}$$

Subtracting Equation (5) from Equation (6) gives

$$\sum_{i=1}^{n}x_i\sum_{i=1}^{n}y_i - n\sum_{i=1}^{n}x_iy_i = b\left(\sum_{i=1}^{n}x_i\right)^2 - nb\sum_{i=1}^{n}x_i^2$$

$$= b\left[\left(\sum_{i=1}^{n}x_i\right)^2 - n\sum_{i=1}^{n}x_i^2\right]$$

Thus

$$b = \frac{\displaystyle\sum_{i=1}^{n}x_i\sum_{i=1}^{n}y_i - n\sum_{i=1}^{n}x_iy_i}{\left(\displaystyle\sum_{i=1}^{n}x_i\right)^2 - n\sum_{i=1}^{n}x_i^2}$$

and, from Equation (4),

$$a = \frac{\sum\limits_{i=1}^{n} y_i - b \sum\limits_{i=1}^{n} x_i}{n}$$

It can be shown that these values for b and a give a minimum value for S, so we have the following.

Linear Regression Equation

Given a set of data points $(x_1, y_1), (x_2, y_2), \ldots, (x_n, y_n)$, the equation of the line that is the best fit for these data is

$$\hat{y} = a + bx$$

where

$$b = \frac{\Sigma x \cdot \Sigma y - n\Sigma xy}{(\Sigma x)^2 - n\Sigma x^2}, \qquad a = \frac{\Sigma y - b\Sigma x}{n}$$

and each summation is taken over the entire data set (that is, from 1 to n).

EXAMPLE 5   **Inventory**

The data in the table show the relation between the diameter of a partial roll of blue denim material at MacGregor Mills and the actual number of yards remaining on the roll. Use linear regression to find the linear equation that gives the number of yards as a function of the diameter.

Diameter (inches)	Yards/Roll	Diameter (inches)	Yards/Roll
14.0	120	22.5	325
15.0	145	24.0	360
16.5	170	24.5	380
17.75	200	25.25	405
18.5	220	26.0	435
19.8	255	26.75	460
20.5	270	27.0	470
22.0	305	28.0	500

Solution

Let x be the diameter of the partial roll and y be the number of yards on the roll. Before finding the values for a and b, we evaluate some parts of the formulas:

$$n = 16$$
$$\Sigma x = 348.05$$
$$\Sigma x^2 = 7871.48$$
$$\Sigma y = 5020$$
$$\Sigma xy = 117{,}367.75$$

$$b = \frac{\Sigma x \Sigma y - n\Sigma xy}{(\Sigma x)^2 - n\Sigma x^2}$$

$$= \frac{(348.05)(5020) - 16(117{,}367.75)}{(348.05)^2 - 16(7871.48)} \approx 27.1959$$

$$a = \frac{\Sigma y - b\Sigma x}{n}$$

$$= \frac{(5020) - (27.1959)(348.05)}{16} \approx -277.8458$$

Thus the linear equation that can be used to estimate the number of yards of denim remaining on a roll is

$$\hat{y} = -277.8 + 27.20x$$

Note that if we use the linear regression capability of a graphing utility, we obtain exactly the same equation.

CHECKPOINT

4. Use linear regression to write the equation of the line that is the best fit for the following points.

x	50	25	10	5
y	2	4	10	20

Finally, we note that formulas for models other than linear ones, such as power models $(y = ax^b)$, exponential models $(y = ab^x)$, and logarithmic models $[y = a + b \ln(x)]$, can also be developed with the least-squares method. That is, we apply max-min techniques for functions of two variables to minimize the sum of the squares of the deviations.

CHECKPOINT SOLUTIONS

1. $z_x = -2x + 2, z_y = -2y - 4$
2. $-2x + 2 = 0$ gives $x = 1$.
   $-2y - 4 = 0$ gives $y = -2$.
   Thus the critical point is $(1, -2, 9)$.
3. $z_{xx} = -2, z_{yy} = -2,$ and $z_{xy} = 0$, so

   $$D(x, y) = (-2)(-2) - (0)^2 = 4$$

   Hence, at $(1, -2, 9)$ we have $D > 0$ and $z_{xx} < 0$, so $(1, -2, 9)$ is a relative maximum.

4.
   $$\Sigma x = 50 + 25 + 10 + 5 = 90$$
   $$\Sigma x^2 = 2500 + 625 + 100 + 25 = 3250$$
   $$\Sigma y = 2 + 4 + 10 + 20 = 36$$
   $$\Sigma xy = 100 + 100 + 100 + 100 = 400$$

   Then

   $$b = \frac{\Sigma x \cdot \Sigma y - n\Sigma xy}{(\Sigma x)^2 - n\Sigma x^2} = \frac{90 \cdot 36 - 4 \cdot 400}{90^2 - 4 \cdot 3250} = \frac{1640}{-4900} \approx -0.33$$

   and

   $$a = \frac{\Sigma y - b\Sigma x}{n} = \frac{36 - (-0.33)(90)}{4} = \frac{65.7}{4} \approx 16.43$$

   Thus the line that gives the best fit to these points is

   $$\hat{y} = -0.33x + 16.43$$

| EXERCISES | 14.4

Access end-of-section exercises online at **www.webassign.net**

OBJECTIVE

14.5

- To find the maximum or minimum value of a function of two or more variables subject to a condition that constrains the variables

Maxima and Minima of Functions Subject to Constraints: Lagrange Multipliers

| APPLICATION PREVIEW |

Many practical problems require that a function of two or more variables be maximized or minimized subject to certain conditions, or constraints, that limit the variables involved. For example, a firm will want to maximize its profits within the limits (constraints) imposed by its production capacity. Similarly, a city planner may want to locate a new building to maximize access to public transportation yet may be constrained by the availability and cost of building sites.

Specifically, suppose that the utility function for commodities X and Y is given by $U = x^2y^2$, where x and y are the amounts of X and Y, respectively. If p_1 and p_2 represent the prices in dollars of X and Y, respectively, and I represents the consumer's income available to purchase these two commodities, the equation $p_1x + p_2y = I$ is called the *budget constraint*. If the price of X is $2, the price of Y is $4, and the income available is $40, then the budget constraint is $2x + 4y = 40$. Thus we seek to maximize the consumer's utility $U = x^2y^2$ subject to the budget constraint $2x + 4y = 40$. (See Example 4.) In this section we develop methods to solve this type of constrained maximum or minimum.

We can obtain maxima and minima for a function $z = f(x, y)$ subject to the constraint $g(x, y) = 0$ by using the method of **Lagrange multipliers,** named for the famous eighteenth-century mathematician Joseph Louis Lagrange. Lagrange multipliers can be used with functions of two or more variables when the constraints are given by an equation.

In order to find the critical values of a function $f(x, y)$ subject to the constraint $g(x, y) = 0$, we will use the new variable λ to form the **objective function**

$$F(x, y, \lambda) = f(x, y) + \lambda g(x, y)$$

It can be shown that the critical values of $F(x, y, \lambda)$ will satisfy the constraint $g(x, y)$ and will also be critical points of $f(x, y)$. Thus we need only find the critical points of $F(x, y, \lambda)$ to find the required critical points.

To find the critical points of $F(x, y, \lambda)$, we must find the points that make all the partial derivatives equal to 0. That is, the points must satisfy

$$\partial F/\partial x = 0, \quad \partial F/\partial y = 0, \quad \text{and} \quad \partial F/\partial \lambda = 0$$

Because $F(x, y, \lambda) = f(x, y) + \lambda g(x, y)$, these equations may be written as

$$\frac{\partial f}{\partial x} + \lambda \frac{\partial g}{\partial x} = 0$$
$$\frac{\partial f}{\partial y} + \lambda \frac{\partial g}{\partial y} = 0$$
$$g(x, y) = 0$$

Finding the values of x and y that satisfy these three equations simultaneously gives the critical values.

This method will not tell us whether the critical points correspond to maxima or minima, but this can be determined either from the physical setting for the problem or by testing according to a procedure similar to that used for unconstrained maxima and minima. The following examples illustrate the use of Lagrange multipliers.

EXAMPLE 1 **Maxima Subject to Constraints**

Find the maximum value of $z = x^2 y$ subject to $x + y = 9$, $x \geq 0$, $y \geq 0$.

Solution

The function to be maximized is $f(x, y) = x^2 y$. The constraint is $g(x, y) = 0$, where $g(x, y) = x + y - 9$. The objective function is

$$F(x, y, \lambda) = f(x, y) + \lambda g(x, y)$$

or

$$F(x, y, \lambda) = x^2 y + \lambda(x + y - 9)$$

Thus

$$\frac{\partial F}{\partial x} = 2xy + \lambda(1) = 0, \quad \text{or} \quad 2xy + \lambda = 0$$

$$\frac{\partial F}{\partial y} = x^2 + \lambda(1) = 0, \quad \text{or} \quad x^2 + \lambda = 0$$

$$\frac{\partial F}{\partial \lambda} = 0 + 1(x + y - 9) = 0, \quad \text{or} \quad x + y - 9 = 0$$

Solving the first two equations for λ and substituting give

$$\lambda = -2xy$$
$$\lambda = -x^2$$
$$2xy = x^2$$
$$2xy - x^2 = 0$$
$$x(2y - x) = 0$$

so

$$x = 0 \quad \text{or} \quad x = 2y$$

Because $x = 0$ could not make $z = x^2 y$ a maximum, we substitute $x = 2y$ into $x + y - 9 = 0$.

$$2y + y = 9$$
$$y = 3$$
$$x = 6$$

Thus the function $z = x^2 y$ is maximized at 108 when $x = 6$, $y = 3$, if the constraint is $x + y = 9$. Testing values near $x = 6$, $y = 3$, and satisfying the constraint shows that the function is maximized there. (Try $x = 5.5$, $y = 3.5$; $x = 7$, $y = 2$; and so on.) ∎

EXAMPLE 2 **Minima Subject to Constraints**

Find the minimum value of the function $z = x^3 + y^3 + xy$ subject to the constraint $x + y - 4 = 0$.

Solution

The function to be minimized is $f(x, y) = x^3 + y^3 + xy$. The constraint function is $g(x, y) = x + y - 4$. The objective function is

$$F(x, y, \lambda) = f(x, y) + \lambda g(x, y)$$

or

$$F(x, y, \lambda) = x^3 + y^3 + xy + \lambda(x + y - 4)$$

Then

$$\frac{\partial F}{\partial x} = 3x^2 + y + \lambda = 0$$

$$\frac{\partial F}{\partial y} = 3y^2 + x + \lambda = 0$$

$$\frac{\partial F}{\partial \lambda} = x + y - 4 = 0$$

Solving the first two equations for λ and substituting, we get

$$\lambda = -(3x^2 + y)$$
$$\lambda = -(3y^2 + x)$$
$$3x^2 + y = 3y^2 + x$$

Solving $x + y - 4 = 0$ for y gives $y = 4 - x$. Substituting for y in the equation above, we get

$$3x^2 + (4 - x) = 3(4 - x)^2 + x$$
$$3x^2 + 4 - x = 48 - 24x + 3x^2 + x$$
$$22x = 44 \quad \text{or} \quad x = 2$$

Thus when $x + y - 4 = 0$, $x = 2$ and $y = 2$ give the minimum value $z = 20$ because other values that satisfy the constraint give larger z-values. ∎

CHECKPOINT

Find the minimum value of $f(x, y) = x^2 + y^2 - 4xy$, subject to the constraint $x + y = 10$, by

1. forming the objective function $F(x, y, \lambda)$,
2. finding $\dfrac{\partial F}{\partial x}, \dfrac{\partial F}{\partial y}$, and $\dfrac{\partial F}{\partial \lambda}$,
3. setting the three partial derivatives (from Question 2) equal to 0, and solving the equations simultaneously for x and y,
4. finding the value of $f(x, y)$ at the critical values of x and y.

We can also use Lagrange multipliers to find the maxima and minima of functions of three (or more) variables, subject to two (or more) constraints. The method involves using two multipliers, one for each constraint, to form an objective function $F = f + \lambda g_1 + \mu g_2$. We leave further discussion for more advanced courses.

We can easily extend the method to functions of three or more variables, as the following example shows.

EXAMPLE 3   Minima of Function of Three Variables

Find the minimum value of the function $w = x + y^2 + z^2$, subject to the constraint $x + y + z = 1$.

Solution
The function to be minimized is $f(x, y, z) = x + y^2 + z^2$. The constraint is $g(x, y, z) = 0$, where $g(x, y, z) = x + y + z - 1$. The objective function is

$$F(x, y, z, \lambda) = f(x, y, z) + \lambda g(x, y, z)$$

or

$$F(x, y, z, \lambda) = x + y^2 + z^2 + \lambda(x + y + z - 1)$$

Then

$$\frac{\partial F}{\partial x} = 1 + \lambda = 0$$

$$\frac{\partial F}{\partial y} = 2y + \lambda = 0$$

$$\frac{\partial F}{\partial z} = 2z + \lambda = 0$$

$$\frac{\partial F}{\partial \lambda} = x + y + z - 1 = 0$$

Solving the first three equations simultaneously gives

$$\lambda = -1, \quad y = \frac{1}{2}, \quad z = \frac{1}{2}$$

Substituting these values in the fourth equation (which is the constraint), we get $x + \frac{1}{2} + \frac{1}{2} - 1 = 0$, so $x = 0, y = \frac{1}{2}, z = \frac{1}{2}$. Thus $w = \frac{1}{2}$ is the minimum value because other values of x, y, and z that satisfy $x + y + z = 1$ give larger values of w. ■

EXAMPLE 4 **Utility | APPLICATION PREVIEW |**

Find x and y that maximize the utility function $U = x^2y^2$ subject to the budget constraint $2x + 4y = 40$.

Solution
First we rewrite the constraint as $2x + 4y - 40 = 0$. Then the objective function is

$$F(x, y, \lambda) = x^2y^2 + \lambda(2x + 4y - 40)$$

$$\frac{\partial F}{\partial x} = 2xy^2 + 2\lambda, \quad \frac{\partial F}{\partial y} = 2x^2y + 4\lambda, \quad \frac{\partial F}{\partial \lambda} = 2x + 4y - 40$$

Setting these partial derivatives equal to 0 and solving give

$$-\lambda = xy^2 = x^2y/2, \quad \text{or} \quad xy^2 - x^2y/2 = 0$$

so

$$xy(y - x/2) = 0$$

yields $x = 0$, $y = 0$, or $x = 2y$. Neither $x = 0$ nor $y = 0$ maximizes utility. If $x = 2y$, then $0 = 2x + 4y - 40$ becomes

$$0 = 4y + 4y - 40 \quad \text{or} \quad 40 = 8y$$

Thus $y = 5$ and $x = 10$.
  Testing values near $x = 10$, $y = 5$ shows that these values maximize utility at $U = 2500$. ■

Figure 14.13 shows the budget constraint $2x + 4y = 40$ from Example 4 graphed with the indifference curves for $U = x^2y^2$ that correspond to $U = 500$, $U = 2500$, and $U = 5000$.
  Whenever an indifference curve intersects the budget constraint, that utility level is attainable within the budget. Note that the highest attainable utility (such as $U = 2500$, found in Example 4) corresponds to the indifference curve that touches the budget constraint at exactly one point—that is, the curve that has the budget constraint as a tangent line. Note also that utility levels greater than $U = 2500$ are not attainable within the budget because the indifference curve "misses" the budget constraint line (as for $U = 5000$).

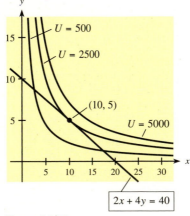

Figure 14.13

EXAMPLE 5

Production

Suppose that the Cobb-Douglas production function for a certain manufacturer gives the number of units of production z according to

$$z = f(x, y) = 100x^{4/5}y^{1/5}$$

where x is the number of units of labor and y is the number of units of capital. Suppose further that labor costs $160 per unit, capital costs $200 per unit, and the total cost for capital and labor is limited to $100,000, so that production is constrained by

$$160x + 200y = 100,000$$

Find the number of units of labor and the number of units of capital that maximize production.

Solution
The objective function is

$$F(x, y, \lambda) = 100x^{4/5}y^{1/5} + \lambda(160x + 200y - 100,000)$$

$$\frac{\partial F}{\partial x} = 80x^{-1/5}y^{1/5} + 160\lambda, \qquad \frac{\partial F}{\partial y} = 20x^{4/5}y^{-4/5} + 200\lambda$$

$$\frac{\partial F}{\partial \lambda} = 160x + 200y - 100,000$$

Setting these partial derivatives equal to 0 and solving give

$$\lambda = \frac{-80x^{-1/5}y^{1/5}}{160} = \frac{-20x^{4/5}y^{-4/5}}{200} \quad \text{or} \quad \frac{y^{1/5}}{2x^{1/5}} = \frac{x^{4/5}}{10y^{4/5}}$$

This means $5y = x$. Using this in $\dfrac{\partial F}{\partial \lambda} = 0$ gives

$$160(5y) + 200y - 100,000 = 0$$
$$1000y = 100,000$$
$$y = 100$$
$$x = 5y = 500$$

Thus production is maximized at $z = 100(500)^{4/5}(100)^{1/5} \approx 36,239$ when $x = 500$ (units of labor) and $y = 100$ (units of capital). See Figure 14.14.

In problems of this type, economists call the value of $-\lambda$ the **marginal productivity of money.** In this case,

$$-\lambda = \frac{y^{1/5}}{2x^{1/5}} = \frac{(100)^{0.2}}{2(500)^{0.2}} \approx 0.362$$

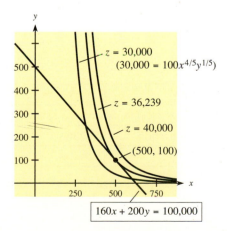

Figure 14.14

This means that each additional dollar spent on production results in approximately 0.362 additional unit produced.

Finally, Figure 14.14 shows the graph of the constraint, together with some production function curves that correspond to different production levels. ■

Spreadsheet Note

The Excel tool called Solver can be used to find maxima and minima of functions subject to constraints. For details of this solution method, see Appendix D, Section 14.5, and the Online Excel Guide. In Example 5, the numbers of units of labor and capital that maximize production if the production is given by the Cobb-Douglas function

$$P(x, y) = 100x^{4/5}y^{1/5}$$

and if capital and labor are constrained by

$$160x + 200y = 100,000$$

were found to be 500 and 100, respectively. The following Excel screen shows the entries for beginning the solution of this problem.

	A	B	C
1	**Variables**		
2			
3	units labor (x)	0	
4	units capital (y)	0	
5			
6	**Objective**		
7			
8	Maximize production	=100*B3^(4/5)*B4^(1/5)	
9			
10	**Constraint**		
11		Amount used	Available
12	Cost	=160*B3+200*B4	100000
13			

The solution, found with Solver, is shown below.

	A	B	C
1	**Variables**		
2			
3	units labor (x)	500	
4	units capital (y)	100	
5		/	
6	**Objective**		
7			
8	Maximize production	36238.98	
9			
10	**Constraint**		
11		Amount used	Available
12	Cost	100000	100000
13			

CHECKPOINT SOLUTIONS

1. $F(x, y, \lambda) = x^2 + y^2 - 4xy + \lambda(10 - x - y)$

2. $\dfrac{\partial F}{\partial x} = 2x - 4y - \lambda, \quad \dfrac{\partial F}{\partial y} = 2y - 4x - \lambda, \quad \dfrac{\partial F}{\partial \lambda} = 10 - x - y$

3. $0 = 2x - 4y - \lambda \quad$ (1)
   $0 = 2y - 4x - \lambda \quad$ (2)
   $0 = 10 - x - y \quad$ (3)

   From Equations (1) and (2) we have the following:

   $$\lambda = 2x - 4y \text{ and } \lambda = 2y - 4x, \text{ so}$$
   $$2x - 4y = 2y - 4x$$
   $$6x = 6y, \text{ or } x = y$$

   Using $x = y$ in Equation (3) gives $0 = 10 - x - x$, or $2x = 10$. Thus $x = 5$ and $y = 5$.

4. $f(5, 5) = 25 + 25 - 100 = -50$ is the minimum because other values that satisfy the constraint give larger z-values.

| EXERCISES | 14.5

Access end-of-section exercises online at **www.webassign.net**

KEY TERMS AND FORMULAS

Section	Key Terms	Formulas
14.1	Function of two variables	$z = f(x, y)$
	Variables: independent, dependent	independent: x and y
	Domain	dependent: z
	Coordinate planes	
	Utility	
	Indifference curve	
	Indifference map	
	Cobb-Douglas production function	$Q = AK^{\alpha}L^{1-\alpha}$
14.2	First-order partial derivative	
	With respect to x	$z_x = \dfrac{\partial z}{\partial x}$
	With respect to y	$z_y = \dfrac{\partial z}{\partial y}$
	Higher-order partial derivatives	
	Second partial derivatives	$z_{xx}, z_{yy}, z_{xy},$ and z_{yx}
14.3	Joint cost function	$C = Q(x, y)$
	Marginal cost	
	Marginal productivity	
	Demand function	
	Marginal demand function	
	Competitive products	
	Complementary products	

Section	Key Terms	Formulas
14.4	Critical values for maxima and minima	Solve simultaneously $\begin{cases} z_x = 0 \\ z_y = 0 \end{cases}$.
	Test for critical values	Use $D(x, y) = (z_{xx})(z_{yy}) - (z_{xy})^2$.
	Linear regression	$\hat{y} = a + bx$
		$b = \dfrac{\Sigma x \cdot \Sigma y - n\Sigma xy}{(\Sigma x)^2 - n\Sigma x^2}$
		$a = \dfrac{\Sigma y - b\Sigma x}{n}$
14.5	Maxima and minima subject to constraints Lagrange multipliers Objective function	$F(x, y, \lambda) = f(x, y) + \lambda g(x, y)$

REVIEW EXERCISES

1. What is the domain of $z = \dfrac{3}{2x - y}$?

2. What is the domain of $z = \dfrac{3x + 2\sqrt{y}}{x^2 + y^2}$?

3. If $w(x, y, z) = x^2 - 3yz$, find $w(2, 3, 1)$.

4. If $Q(K, L) = 70K^{2/3}L^{1/3}$, find $Q(64{,}000, 512)$.

5. Find $\dfrac{\partial z}{\partial x}$ if $z = 5x^3 + 6xy + y^2$.

6. Find $\dfrac{\partial z}{\partial y}$ if $z = 12x^5 - 14x^3y^3 + 6y^4 - 1$.

In Problems 7–12, find z_x and z_y.

7. $z = 4x^2y^3 + \dfrac{x}{y}$

8. $z = \sqrt{x^2 + 2y^2}$

9. $z = (xy + 1)^{-2}$

10. $z = e^{x^2y^3}$

11. $z = e^{xy} + y\ln x$

12. $z = e^{\ln xy}$

13. Find the partial derivative of $f(x, y) = 4x^3 - 5xy^2 + y^3$ with respect to x at the point $(1, 2, -8)$.

14. Find the slope of the tangent in the x-direction to the surface $z = 5x^4 - 3xy^2 + y^2$ at $(1, 2, -3)$.

In Problems 15–18, find the second partials.
(a) z_{xx} **(b)** z_{yy} **(c)** z_{xy} **(d)** z_{yx}

15. $z = x^2y - 3xy$

16. $z = 3x^3y^4 - \dfrac{x^2}{y^2}$

17. $z = x^2e^{y}$

18. $z = \ln(xy + 1)$

19. Test $z = 16 - x^2 - xy - y^2 + 24y$ for maxima and minima.

20. Test $z = x^3 + y^3 - 12x - 27y$ for maxima and minima.

21. Find the minimum value of $z = 4x^2 + y^2$ subject to the constraint $x + y = 10$.

22. Find the maximum value of $z = x^4y^2$ subject to the constraint $x + y = 9$, $x \geq 0$, $y \geq 0$.

APPLICATIONS

23. *Utility* Suppose that the utility function for two goods X and Y is given by $U = x^2y$.
    (a) Write the equation of the indifference curve for a consumer who purchases 6 units of X and 15 units of Y.
    (b) If the consumer purchases 60 units of Y, how many units of X must be purchased to retain the same level of utility?

24. *Savings plans* The accumulated value A of a monthly savings plan over a 20-year period is a function of the monthly contribution R and the interest rate $r\%$, compounded monthly, according to

$$A = f(R, r) = \frac{1200R\left[\left(1 + \dfrac{r}{1200}\right)^{240} - 1\right]}{r}$$

    (a) Find the accumulated value of a plan that contributes $100 per month with interest rate 6%.
    (b) Interpret $f(250, 7.8) \approx 143{,}648$.

(c) Interpret $\dfrac{\partial A}{\partial r}(250, 7.8) \approx 17{,}770$.

(d) Find $\dfrac{\partial A}{\partial R}(250, 7.8)$ and interpret the result.

25. **Retirement benefits** The monthly benefit B (in thousands of dollars) from a retirement account that is invested at 9% compounded monthly is a function of the account value V (also in thousands of dollars) and the number of years t that benefits are paid, and it can be approximated by

$$B = f(V, t) = \frac{3V}{400 - 400e^{-0.0897t}}$$

(a) Find the benefit if the account value is $1,000,000 and the monthly benefits last for 20 years.

(b) Find and interpret $\dfrac{\partial B}{\partial V}(1000, 20)$.

(c) Find and interpret $\dfrac{\partial B}{\partial t}(1000, 20)$.

26. **Advertising and sales** The number of units of sales of a product, S, is a function of the dollars spent for advertising, A, and the product's price, p. Suppose $S = f(A, p)$ is the function relating these quantities.

(a) Explain why $\dfrac{\partial S}{\partial A} > 0$.

(b) Do you think $\dfrac{\partial S}{\partial p}$ is positive or negative? Explain.

27. **Cost** The joint cost, in dollars, for two products is given by $C(x, y) = x^2\sqrt{y^2 + 13}$. Find the marginal cost with respect to

(a) x if 20 units of x and 6 units of y are produced.

(b) y if 20 units of x and 6 units of y are produced.

28. **Production** Suppose that the production function for a company is given by

$$Q = 80K^{1/4}L^{3/4}$$

where Q is the output (in hundreds of units), K is the capital expenditures (in thousands of dollars), and L is the work-hours. Find $\partial Q/\partial K$ and $\partial Q/\partial L$ when expenditures are $625,000 and total work-hours are 4096. Interpret the results.

29. **Marginal demand** The demand functions for two related products, product A and product B, are given by

$$q_A = 400 - 2p_A - 3p_B$$
$$q_B = 300 - 5p_A - 6p_B$$

where p_A and p_B are the respective prices in dollars.

(a) Find the marginal demand of q_A with respect to p_A.

(b) Find the marginal demand of q_B with respect to p_B.

(c) Are the products complementary or competitive?

30. **Marginal demand** Suppose that the demand functions for two related products, A and B, are given by

$$q_A = 800 - 40p_A - \frac{2}{p_B + 1}$$

$$q_B = 1000 - \frac{10}{p_A + 4} - 30p_B$$

where p_A and p_B are the respective prices in dollars. Determine whether the products are competitive or complementary.

31. **Profit** The weekly profit (in dollars) from the sale of two products is given by

$$P(x, y) = 40x + 80y - x^2 - y^2$$

where x is the number of units of product 1 and y is the number of units of product 2. Selling how much of each product will maximize profit? Find the maximum weekly profit.

32. **Cost** Suppose a company has two separate plants that manufacture the same item. Suppose x is the amount produced at plant I and y is the amount at plant II. If the total cost function for the two plants is

$$C(x, y) = 200 - 12x - 30y + 0.03x^2 + 0.001y^3$$

find the production allocation that minimizes the company's total cost.

33. **Utility** If the utility function for two commodities is $U = x^2y$, and the budget constraint is $4x + 5y = 60$, find the values of x and y that maximize utility.

34. **Production** Suppose a company has the Cobb-Douglas production function

$$z = 300x^{2/3}y^{1/3}$$

where x is the number of units of labor, y is the number of units of capital, and z is the units of production. Suppose labor costs are $50 per unit, capital costs are $50 per unit, and total costs are limited to $75,000.

(a) Find the number of units of labor and the number of units of capital that maximize production.

(b) Find the marginal productivity of money and interpret your result.

(c) Graph the constraint with the production function when $z = 180{,}000$, $z = 300{,}000$, and when the z-value is optimal.

35. **Taxes** The data in the table on the next page show U.S. national personal income and personal taxes for selected years.

(a) Write the linear regression equation that best fits these data.

(b) Use the equation found in part (a) to predict the taxes when national personal income reaches $7000 billion.

Income (x) (billions)	Taxes (y) (billions)
$2285.7	$312.4
2560.4	360.2
2718.7	371.4
2891.7	369.3
3205.5	395.5
3439.6	437.7
3647.5	459.9
3877.3	514.2
4172.8	532.0
4489.3	594.9
4791.6	624.8
4968.5	624.8
5264.2	650.5
5480.1	689.9
5753.1	731.4
6150.8	795.1
6495.2	886.9

Source: Bureau of Economic Analysis, U.S. Commerce Department

 36. ***U.S. domestic travel*** The table gives the number of millions of person-trips (of 50 miles or more one way) for the years from 1998 to 2008. Use linear regression to find the best linear function $y = f(x)$ modeling the number of person-trips (in millions) as a function of the years past 1995.

Year	Millions of Person-trips
1998	1225.7
1999	1296.8
2000	1325.4
2001	1324.6
2002	1407.1
2003	1388.2
2004	1440.4
2005	1482.5
2006	1491.8
2007	1510.2
2008	1503.6

Source: U.S. Travel Association

14   CHAPTER TEST

1. Consider the function $f(x, y) = \dfrac{2x + 3y}{\sqrt{x^2 - y}}$.

   (a) Find the domain of $f(x, y)$.
   (b) Evaluate $f(-4, 12)$.

2. Find all first and second partial derivatives of
   $$z = f(x, y) = 5x - 9y^2 + 2(xy + 1)^5$$

3. Let $z = 6x^2 + x^2y + y^2 - 4y + 9$. Find the pairs (x, y) that are critical points for z, and then classify each as a relative maximum, a relative minimum, or a saddle point.

4. Suppose a company's monthly production value Q, in thousands of dollars, is given by the Cobb-Douglas production function
   $$Q = 10K^{0.45}L^{0.55}$$
   where K is thousands of dollars of capital investment per month and L is the total hours of labor per month. Capital investment is currently $10,000 per month and monthly work-hours of labor total 1590.
   (a) Find the monthly production value (to the nearest thousand dollars).
   (b) Find the marginal productivity with respect to capital investment, and interpret your result.

   (c) Find the marginal productivity with respect to total hours of labor, and interpret your result.

5. The monthly payment R on a loan is a function of the amount borrowed, A, in thousands of dollars; the length of the loan, n, in years; and the annual interest rate, r, as a percent. Thus $R = f(A, n, r)$. In parts (a) and (b), write a sentence that explains the practical meaning of each mathematical statement.

   (a) $f(94.5, 25, 7) = \$667.91$

   (b) $\dfrac{\partial f}{\partial r}(94.5, 25, 7) = \$49.76$

   (c) Would $\dfrac{\partial f}{\partial n}(94.5, 25, 7)$ be positive, negative, or zero? Explain.

6. Let $f(x, y) = 2e^{x^2y^2}$. Find $\dfrac{\partial^2 f}{\partial x \, \partial y}$.

7. Suppose the demand functions for two products are
   $$q_1 = 300 - 2p_1 - 5p_2 \quad \text{and} \quad q_2 = 150 - 4p_1 - 7p_2$$
   where q_1 and q_2 represent quantities demanded and p_1 and p_2 represent prices. What calculations enable us to decide whether the products are competitive or complementary? Are these products competitive or complementary?

8. Suppose a store sells two brands of disposable cameras and the profit for these is a function of their two selling prices. The type 1 camera sells for x, the type 2 sells for y, and profit is given by

$$P = 915x - 30x^2 - 45xy + 975y - 30y^2 - 3500$$

Find the selling prices that maximize profit. Find the maximum profit.

9. Find x and y that maximize the utility function $U = x^3y$ subject to the budget constraint $30x + 20y = 8000$.

10. In the last half of the twentieth century the U.S. population grew more diverse both racially and ethnically, with persons of Hispanic origin representing one of the fastest growing segments. The table gives the percent of the U.S. civilian noninstitutional population of Hispanic origin for selected years from 1980 and projected to 2050.

    (a) Find the linear regression line for these data. Use $x = 0$ to represent 1980.

    (b) How well does the regression line fit the data?
    (c) Using the linear regression equation, predict the percent of the U.S. civilian noninstitutional population of Hispanic origin in 2025.

Year	Percent
1980	5.7
1990	8.4
2000	10.7
2010	13.0
2015	14.1
2020	15.3
2030	17.8
2040	20.3
2050	22.8

Source: U.S. Bureau of the Census

I. Advertising

To model sales of its tires, the manufacturer of GRIPPER tires used the quadratic equation $S = a_0 + a_1x + a_2x^2 + b_1y$, where S is regional sales in millions of dollars, x is TV advertising expenditures in millions of dollars, and y is other promotional expenditures in millions of dollars. (See the Extended Application/Group Project "Marginal Return to Sales," in Chapter 9.)

Although this model represents the relationship between advertising and sales dollars for small changes in advertising expenditures, it is clear to the vice president of advertising that it does not apply to large expenditures for TV advertising on a national level. He knows from experience that increased expenditures for TV advertising result in more sales, but at a decreasing rate of return for the product.

The vice president is aware that some advertising agencies model the relationship between advertising and sales by the function

$$S = b_0 + b_1(1 - e^{-ax}) + c_1y$$

where $a > 0$, S is sales in millions of dollars, x is TV advertising expenditures in millions of dollars, and y is other promotional expenditures in millions of dollars.* The equation

$$S_n = 24.58 + 325.18(1 - e^{-x/14}) + b_1y$$

has the form mentioned previously as being used by some advertising agencies. For TV advertising expenditures up to \$20 million, this equation closely approximates

$$S_1 = 30 + 20x - 0.4x^2 + b_1y$$

which, in the Extended Application/Group Project in Chapter 9, was used with fixed promotional expenses to describe advertising and sales in Region 1.

To help the vice president decide whether this is a better model for large expenditures, answer the following questions.

1. What is $\partial S_1/\partial x$? Does this indicate that sales might actually decline after some amount is spent on TV advertising? If so, what is this amount?
2. Does the quadratic model $S_1(x, y)$ indicate that sales will become negative after some amount is spent on TV advertising? Does this model cease to be useful in predicting sales after a certain point?
3. What is $\partial S_n/\partial x$? Does this indicate that sales will continue to rise if additional money is devoted to TV advertising? Is S_n growing at a rate that is increasing or decreasing when promotional sales are held constant? Is S_n a better model for large expenditures?
4. If this model does describe the relationship between advertising and sales, and if promotional expenditures are held constant at y_0, is there an upper limit to the sales, even if an unlimited amount of money is spent on TV advertising? If so, what is it?

*Edwin Mansfield, *Managerial Economics* (New York: Norton, 1990).

II. Competitive Pricing

Retailers often sell different brands of competing products. Depending on the joint demand for the products, the retailer may be able to set prices that regulate demand and, therefore, influence profits.

Suppose HOME-ALL, Inc., a national chain of home improvement retailers, sells two competing brands of interior flat paint, En-Dure 100 and Croyle & James, which the chain purchases for $8 per gallon and $10 per gallon, respectively. HOME-ALL's research department has determined the following two monthly demand equations for these paints:

$$D = 120 - 40d + 30c \quad \text{and} \quad C = 680 + 30d - 40c$$

where D is hundreds of gallons of En-Dure 100 demanded at $\$d$ per gallon and C is hundreds of gallons of Croyle & James demanded at $\$c$ per gallon. For what prices should HOME-ALL sell these paints in order to maximize its monthly profit on these items?

To answer this question, complete the following.

1. Recall that revenue is a product's selling price per item times the number of items sold. With this in mind, formulate HOME-ALL's total revenue function for the two paints as a function of their prices.
2. Form HOME-ALL's profit function for the two paints (in terms of their selling prices).
3. Determine the price of each type of paint that will maximize HOME-ALL's profit.
4. Write a brief report to management that details your pricing recommendations and justifies them.

Financial Tables

| TABLE 1 |

Future Value of an Ordinary Annuity of $1 ($s_{\overline{n}|i}$)

Periods	1%	2%	3%	4%	5%	6%
1	1.000 000	1.000 000	1.000 000	1.000 000	1.000 000	1.000 000
2	2.010 000	2.020 000	2.030 000	2.040 000	2.050 000	2.060 000
3	3.030 100	3.060 400	3.090 900	3.121 600	3.152 500	3.183 600
4	4.060 401	4.121 608	4.183 627	4.246 464	4.310 125	4.374 616
5	5.101 005	5.204 040	5.309 136	5.416 323	5.525 631	5.637 093
6	6.152 015	6.308 121	6.468 410	6.632 975	6.801 913	6.975 319
7	7.213 535	7.434 284	7.662 462	7.898 294	8.142 008	8.393 838
8	8.285 670	8.582 969	8.892 336	9.214 226	9.549 109	9.897 468
9	9.368 527	9.754 629	10.159 106	10.582 795	11.026 564	11.491 316
10	10.462 212	10.949 721	11.463 879	12.006 107	12.577 893	13.180 795
11	11.566 834	12.168 716	12.807 796	13.486 351	14.206 787	14.971 643
12	12.682 503	13.412 090	14.192 030	15.025 805	15.917 126	16.869 941
13	13.809 328	14.680 332	15.617 790	16.626 838	17.712 983	18.882 138
14	14.947 421	15.973 939	17.086 324	18.291 911	19.598 632	21.015 066
15	16.096 895	17.293 418	18.598 914	20.023 588	21.578 564	23.275 970
16	17.257 864	18.639 286	20.156 881	21.824 531	23.657 492	25.672 529
17	18.430 443	20.012 072	21.761 588	23.697 512	25.840 366	28.212 880
18	19.614 747	21.412 313	23.414 435	26.645 413	28.132 385	30.905 653
19	20.810 895	22.840 559	25.116 869	27.671 229	30.539 004	33.759 992
20	22.019 004	24.297 371	26.870 375	29.778 079	33.065 954	36.787 592
21	23.239 194	25.783 318	28.676 486	31.969 202	35.719 252	39.992 727
22	24.471 586	27.298 985	30.536 780	34.247 970	38.505 214	43.392 291
23	25.716 301	28.844 964	32.452 884	36.617 889	41.430 475	46.995 829
24	26.973 464	30.421 864	34.426 470	39.082 604	44.501 999	50.815 578
25	28.243 199	32.030 301	36.459 264	41.645 908	47.727 099	54.864 513
26	29.525 631	33.670 907	38.553 042	44.311 745	51.113 454	59.156 384
27	30.820 887	35.344 325	40.709 634	47.084 214	54.669 126	63.705 767
28	32.129 096	37.051 212	42.930 923	49.967 583	58.402 583	68.528 113
29	33.450 387	38.792 236	45.218 850	52.966 286	62.322 712	73.639 800
30	34.784 891	40.568 081	47.575 416	56.084 938	66.438 847	79.058 188
31	36.132 740	42.379 443	50.002 678	59.328 335	70.760 790	84.801 679
32	37.494 067	44.227 031	52.502 759	62.701 469	75.298 829	90.889 780
33	38.869 008	46.111 572	55.077 842	66.209 528	80.063 771	97.343 167
34	40.257 698	48.033 804	57.730 177	69.857 909	85.066 959	104.183 757
35	41.660 275	49.994 480	60.462 082	73.652 225	90.320 307	111.434 783
36	43.076 878	51.994 369	63.275 945	77.598 314	95.836 323	119.120 870
37	44.507 646	54.034 257	66.174 223	81.702 247	101.628 139	127.268 122
38	45.952 723	56.114 942	69.159 450	85.970 336	107.709 546	135.904 209
39	47.412 250	58.237 241	72.234 233	90.409 150	114.095 023	145.058 462
40	48.886 373	60.401 986	75.401 260	95.025 516	120.799 774	154.761 970

(continued)

| TABLE 1 |

Future Value of an Ordinary Annuity of $1 ($s_{\overline{n}|i}$) (*Continued*)

Periods	7%	8%	9%	10%	11%	12%
1	1.000 000	1.000 000	1.000 000	1.000 000	1.000 000	1.000 000
2	2.070 000	2.080 000	2.090 000	2.100 000	2.110 000	2.120 000
3	3.214 900	3.246 400	3.278 100	3.310 000	3.342 100	3.374 400
4	4.439 943	4.506 112	4.573 129	4.641 000	4.709 731	4.779 328
5	5.750 739	5.866 601	5.984 711	6.105 100	6.227 801	6.352 847
6	7.153 291	7.335 929	7.523 335	7.715 610	7.912 860	8.115 189
7	8.654 021	8.922 803	9.200 435	9.487 171	9.783 274	10.089 012
8	10.259 803	10.636 628	11.028 474	11.435 888	11.859 434	12.299 693
9	11.977 989	12.487 558	13.021 036	13.579 477	14.136 972	14.775 656
10	13.816 448	14.486 563	15.192 930	15.937 425	16.722 009	17.548 735
11	15.783 599	16.645 488	17.560 293	18.531 167	19.561 430	20.654 583
12	17.888 451	18.977 127	20.140 720	21.384 284	22.713 187	24.133 133
13	20.140 643	21.495 297	22.953 385	24.522 712	26.211 638	28.029 109
14	22.550 488	24.214 920	26.019 189	27.974 984	30.094 918	32.392 602
15	25.129 022	27.152 114	29.360 916	31.772 482	34.405 359	37.279 715
16	27.888 054	30.324 283	33.003 399	35.949 730	39.189 949	42.753 281
17	30.840 218	33.750 226	36.973 705	40.544 703	44.500 843	48.883 674
18	33.999 033	37.450 244	41.301 338	45.599 174	50.395 936	55.749 715
19	37.378 965	41.446 263	46.018 458	51.159 091	56.939 489	63.439 681
20	40.995 493	45.761 965	51.160 120	57.275 000	64.202 833	72.052 443
21	44.865 177	50.422 922	56.764 530	64.002 501	72.265 145	81.698 736
22	49.005 740	55.456 756	62.873 338	71.402 750	81.214 310	92.502 584
23	53.436 142	60.893 296	69.531 939	79.543 025	91.147 885	104.602 894
24	58.176 672	66.764 760	76.789 813	88.497 328	102.174 152	118.155 242
25	63.249 039	73.105 940	84.700 896	98.347 061	114.413 309	133.333 871
26	68.676 471	79.954 416	93.323 977	109.181 767	127.998 773	150.333 935
27	74.483 824	87.350 769	102.723 135	121.099 944	143.078 638	169.374 007
28	80.697 692	95.338 831	112.968 217	134.209 938	159.817 288	190.698 888
29	87.346 531	103.965 937	124.135 357	148.630 932	178.397 190	214.582 755
30	94.460 788	113.283 212	136.307 539	164.494 026	199.020 881	241.322 686
31	102.073 043	123.345 869	149.575 217	181.943 428	221.913 178	271.292 608
32	110.218 156	134.213 539	164.036 987	201.137 771	247.323 628	304.847 721
33	118.933 427	145.950 622	179.800 316	222.251 548	275.529 227	342.429 447
34	128.258 767	158.626 671	196.982 344	245.476 703	306.837 442	384.520 981
35	138.236 881	172.316 805	215.710 755	271.024 374	341.589 561	431.663 499
36	148.913 462	187.102 150	236.124 723	299.126 811	380.164 413	484.463 119
37	160.337 405	203.070 322	258.375 948	330.039 493	422.982 498	543.598 693
38	172.561 023	220.315 948	282.629 783	364.043 442	470.510 573	609.830 536
39	185.640 295	238.941 223	309.066 464	401.447 787	523.266 737	686.010 201
40	199.635 116	259.056 521	337.882 446	442.592 566	581.826 078	767.091 425

| TABLE 2 |

Present Value of an Ordinary Annuity of $1 $(a_{\overline{n}|i})$

Periods	1%	2%	3%	4%	5%	6%
1	0.990 099	0.980 392	0.970 874	0.961 538	0.952 381	0.943 396
2	1.970 395	1.941 561	1.913 470	1.886 095	1.859 410	1.833 393
3	2.940 985	2.883 883	2.828 611	2.775 091	2.723 248	2.673 012
4	3.901 965	3.807 729	3.717 098	3.629 895	3.545 951	3.465 106
5	4.853 431	4.713 460	4.579 707	4.451 822	4.329 477	4.212 364
6	5.795 476	5.601 431	5.417 191	5.242 137	5.075 692	4.917 324
7	6.728 194	6.471 991	6.230 283	6.002 055	5.786 373	5.582 381
8	7.651 678	7.325 482	7.019 692	6.732 745	6.463 213	6.209 794
9	8.566 017	8.162 237	7.786 109	7.435 332	7.107 822	6.801 692
10	9.471 304	8.982 585	8.530 203	8.110 896	7.721 735	7.360 087
11	10.367 628	9.786 848	9.252 624	8.760 477	8.306 414	7.886 875
12	11.255 077	10.575 342	9.954 004	9.385 074	8.863 252	8.383 844
13	12.133 740	11.348 374	10.634 955	9.985 648	9.393 573	8.852 683
14	13.003 703	12.106 249	11.296 073	10.563 123	9.898 641	9.294 984
15	13.865 052	12.849 264	11.937 935	11.118 387	10.379 658	9.712 249
16	14.717 874	13.577 710	12.561 102	11.652 296	10.837 770	10.105 895
17	15.562 251	14.291 872	13.166 118	12.165 669	11.274 066	10.477 260
18	16.398 268	14.992 032	13.753 513	12.659 297	11.689 587	10.827 604
19	17.226 008	15.678 462	14.323 799	13.133 939	12.085 321	11.158 117
20	18.045 553	16.351 434	14.877 475	13.590 326	12.462 210	11.469 921
21	18.856 983	17.011 210	15.415 024	14.029 160	12.821 153	11.764 077
22	18.660 379	17.658 049	15.936 917	14.451 115	13.163 003	12.041 582
23	20.455 821	18.292 205	16.443 608	14.856 842	13.488 574	12.303 379
24	21.243 387	18.913 926	16.935 542	15.246 963	13.798 642	12.550 358
25	22.023 155	19.523 457	17.413 148	15.622 080	14.093 945	12.783 356
26	22.795 203	20.121 036	17.876 842	15.982 769	14.375 185	13.003 166
27	23.559 607	20.706 898	18.327 032	16.329 586	14.643 034	13.210 534
28	24.316 443	21.281 273	18.764 108	16.663 063	14.898 127	13.406 164
29	25.065 785	21.844 385	19.188 455	16.983 715	15.141 074	13.590 721
30	25.807 708	22.396 456	19.600 441	17.292 033	15.372 451	13.764 831
31	26.542 285	22.937 702	20.000 429	17.588 494	15.592 810	13.929 086
32	27.269 589	23.468 335	20.388 766	17.873 552	15.802 677	14.084 043
33	27.989 692	23.988 564	20.765 792	18.147 646	16.002 549	14.230 230
34	28.702 666	24.498 592	21.131 837	18.411 198	16.192 904	14.368 141
35	28.408 580	24.998 620	21.487 220	18.664 613	16.374 194	14.498 246
36	30.107 505	25.488 843	21.832 253	18.908 282	16.546 852	14.620 987
37	30.799 510	25.969 454	22.167 235	19.142 579	16.711 287	14.736 780
38	31.484 663	26.440 641	22.492 462	19.367 864	16.867 893	14.846 019
39	32.163 033	26.902 589	22.808 215	19.584 485	17.017 041	14.949 075
40	32.834 686	27.355 480	23.114 772	19.792 774	17.159 086	15.046 297

(continued)

■ | TABLE 2 |

Present Value of an Ordinary Annuity of $1 ($a_{\overline{n}|i}$) (*Continued*)

Periods	7%	8%	9%	10%	11%	12%
1	0.934 579	0.925 926	0.917 431	0.909 091	0.900 901	0.892 857
2	1.808 018	1.783 265	1.759 111	1.735 537	1.712 523	1.690 051
3	2.624 316	2.577 097	2.531 295	2.486 852	2.443 715	2.401 831
4	3.387 211	3.312 127	3.239 720	3.169 865	3.102 446	3.037 349
5	4.100 197	3.992 710	3.889 651	3.790 787	3.695 897	3.604 776
6	4.766 540	4.622 880	4.485 919	4.355 261	4.230 538	4.111 407
7	5.389 289	5.206 370	5.032 953	4.868 419	4.712 196	4.563 757
8	5.971 299	5.746 639	5.534 819	5.334 926	5.146 123	4.967 640
9	6.515 232	6.246 888	5.995 247	5.759 024	5.537 048	5.328 250
10	7.023 582	6.710 081	6.417 658	6.144 567	5.889 232	5.650 223
11	7.498 674	7.138 964	6.805 191	6.495 061	6.206 515	5.937 699
12	7.942 686	7.536 078	7.160 725	6.813 692	6.492 356	6.194 374
13	8.357 651	7.903 776	7.486 904	7.103 356	6.749 870	6.423 548
14	8.745 468	8.244 237	7.786 150	7.366 687	6.981 865	6.628 168
15	9.107 914	8.559 479	8.060 688	7.606 080	7.190 870	6.810 864
16	9.446 649	8.851 369	8.312 558	7.823 709	7.379 162	6.973 986
17	9.763 223	9.121 638	8.543 631	8.021 553	7.548 794	7.119 630
18	10.059 087	9.371 887	8.755 625	8.201 412	7.701 617	7.249 670
19	10.335 595	9.603 599	8.950 115	8.364 920	7.839 294	7.365 777
20	10.594 014	9.818 147	9.128 546	8.513 564	7.963 328	7.469 444
21	10.835 527	10.016 803	9.292 244	8.648 694	8.075 070	7.562 003
22	11.061 241	10.200 744	9.442 425	8.771 540	8.175 739	7.644 646
23	11.272 187	10.371 059	9.580 207	8.883 218	8.266 432	7.718 434
24	11.469 334	10.528 758	9.706 612	8.984 744	8.348 137	7.784 316
25	11.653 583	10.674 776	9.822 580	9.077 040	8.421 745	7.843 139
26	11.825 779	10.809 978	9.928 972	9.160 945	8.488 058	7.895 660
27	11.986 709	10.935 165	10.026 580	9.237 223	8.547 800	7.942 554
28	12.137 111	11.051 079	10.116 128	9.306 567	8.601 622	7.984 423
29	12.277 674	11.158 406	10.198 283	9.369 606	8.650 110	8.021 806
30	12.409 041	11.257 783	10.273 654	9.429 914	8.693 793	8.055 184
31	12.531 814	11.349 799	10.342 802	9.479 013	8.733 146	8.084 986
32	12.646 555	11.434 999	10.406 240	9.526 376	8.768 600	8.111 594
33	12.753 790	11.513 888	10.464 441	9.569 432	8.800 541	8.135 352
34	12.854 009	11.586 934	10.517 835	9.608 575	8.829 316	8.156 564
35	12.947 672	11.654 568	10.566 821	9.644 159	8.855 240	8.175 504
36	13.035 208	11.717 193	10.611 763	9.676 508	8.878 594	8.192 414
37	13.117 017	11.775 179	10.652 993	9.705 917	8.899 635	8.207 513
38	13.193 473	11.828 869	10.690 820	9.732 651	8.918 590	8.220 993
39	13.264 928	11.878 582	10.725 523	9.756 956	8.935 666	8.233 030
40	13.331 709	11.924 613	10.757 360	9.779 051	8.951 051	8.243 777

Areas Under the Standard Normal Curve

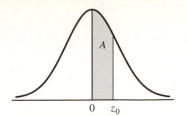

The value of A is the area under the standard normal curve between $z = 0$ and $z = z_0$, for $z_0 \geq 0$. Areas for negative values of z_0 are obtained by symmetry.

z_0	A	z_0	A	z_0	A	z_0	A
0.00	0.0000	0.43	0.1664	0.86	0.3051	1.29	0.4015
0.01	0.0040	0.44	0.1700	0.87	0.3079	1.30	0.4032
0.02	0.0080	0.45	0.1736	0.88	0.3106	1.31	0.4049
0.03	0.0120	0.46	0.1772	0.89	0.3133	1.32	0.4066
0.04	0.0160	0.47	0.1808	0.90	0.3159	1.33	0.4082
0.05	0.0199	0.48	0.1844	0.91	0.3186	1.34	0.4099
0.06	0.0239	0.49	0.1879	0.92	0.3212	1.35	0.4115
0.07	0.0279	0.50	0.1915	0.93	0.3238	1.36	0.4131
0.08	0.0319	0.51	0.1950	0.94	0.3264	1.37	0.4147
0.09	0.0359	0.52	0.1985	0.95	0.3289	1.38	0.4162
0.10	0.0398	0.53	0.2019	0.96	0.3315	1.39	0.4177
0.11	0.0438	0.54	0.2054	0.97	0.3340	1.40	0.4192
0.12	0.0478	0.55	0.2088	0.98	0.3365	1.41	0.4207
0.13	0.0517	0.56	0.2123	0.99	0.3389	1.42	0.4222
0.14	0.0557	0.57	0.2157	1.00	0.3413	1.43	0.4236
0.15	0.0596	0.58	0.2190	1.01	0.3438	1.44	0.4251
0.16	0.0636	0.59	0.2224	1.02	0.3461	1.45	0.4265
0.17	0.0675	0.60	0.2258	1.03	0.3485	1.46	0.4279
0.18	0.0714	0.61	0.2291	1.04	0.3508	1.47	0.4292
0.19	0.0754	0.62	0.2324	1.05	0.3531	1.48	0.4306
0.20	0.0793	0.63	0.2357	1.06	0.3554	1.49	0.4319
0.21	0.0832	0.64	0.2389	1.07	0.3577	1.50	0.4332
0.22	0.0871	0.65	0.2422	1.08	0.3599	1.51	0.4345
0.23	0.0910	0.66	0.2454	1.09	0.3621	1.52	0.4357
0.24	0.0948	0.67	0.2486	1.10	0.3643	1.53	0.4370
0.25	0.0987	0.68	0.2518	1.11	0.3665	1.54	0.4382
0.26	0.1026	0.69	0.2549	1.12	0.3686	1.55	0.4394
0.27	0.1064	0.70	0.2580	1.13	0.3708	1.56	0.4406
0.28	0.1103	0.71	0.2612	1.14	0.3729	1.57	0.4418
0.29	0.1141	0.72	0.2642	1.15	0.3749	1.58	0.4430
0.30	0.1179	0.73	0.2673	1.16	0.3770	1.59	0.4441
0.31	0.1217	0.74	0.2704	1.17	0.3790	1.60	0.4452
0.32	0.1255	0.75	0.2734	1.18	0.3810	1.61	0.4463
0.33	0.1293	0.76	0.2764	1.19	0.3830	1.62	0.4474
0.34	0.1331	0.77	0.2794	1.20	0.3849	1.63	0.4485
0.35	0.1368	0.78	0.2823	1.21	0.3869	1.64	0.4495
0.36	0.1406	0.79	0.2852	1.22	0.3888	1.65	0.4505
0.37	0.1443	0.80	0.2881	1.23	0.3907	1.66	0.4515
0.38	0.1480	0.81	0.2910	1.24	0.3925	1.67	0.4525
0.39	0.1517	0.82	0.2939	1.25	0.3944	1.68	0.4535
0.40	0.1554	0.83	0.2967	1.26	0.3962	1.69	0.4545
0.41	0.1591	0.84	0.2996	1.27	0.3980	1.70	0.4554
0.42	0.1628	0.85	0.3023	1.28	0.3997	1.71	0.4564

(continued)

z_0	A	z_0	A	z_0	A	z_0	A
1.72	0.4573	2.26	0.4881	2.80	0.4974	3.34	0.4996
1.73	0.4582	2.27	0.4884	2.81	0.4975	3.35	0.4996
1.74	0.4591	2.28	0.4887	2.82	0.4976	3.36	0.4996
1.75	0.4599	2.29	0.4890	2.83	0.4977	3.37	0.4996
1.76	0.4608	2.30	0.4893	2.84	0.4977	3.38	0.4996
1.77	0.4616	2.31	0.4896	2.85	0.4978	3.39	0.4997
1.78	0.4625	2.32	0.4898	2.86	0.4979	3.40	0.4997
1.79	0.4633	2.33	0.4901	2.87	0.4980	3.41	0.4997
1.80	0.4641	2.34	0.4904	2.88	0.4980	3.42	0.4997
1.81	0.4649	2.35	0.4906	2.89	0.4981	3.43	0.4997
1.82	0.4656	2.36	0.4909	2.90	0.4981	3.44	0.4997
1.83	0.4664	2.37	0.4911	2.91	0.4982	3.45	0.4997
1.84	0.4671	2.38	0.4913	2.92	0.4983	3.46	0.4997
1.85	0.4678	2.39	0.4916	2.93	0.4983	3.47	0.4997
1.86	0.4686	2.40	0.4918	2.94	0.4984	3.48	0.4998
1.87	0.4693	2.41	0.4920	2.95	0.4984	3.49	0.4998
1.88	0.4700	2.42	0.4922	2.96	0.4985	3.50	0.4998
1.89	0.4706	2.43	0.4925	2.97	0.4985	3.51	0.4998
1.90	0.4713	2.44	0.4927	2.98	0.4986	3.52	0.4998
1.91	0.4719	2.45	0.4929	2.99	0.4986	3.53	0.4998
1.92	0.4726	2.46	0.4931	3.00	0.4987	3.54	0.4998
1.93	0.4732	2.47	0.4932	3.01	0.4987	3.55	0.4998
1.94	0.4738	2.48	0.4934	3.02	0.4987	3.56	0.4998
1.95	0.4744	2.49	0.4936	3.03	0.4988	3.57	0.4998
1.96	0.4750	2.50	0.4938	3.04	0.4988	3.58	0.4998
1.97	0.4756	2.51	0.4940	3.05	0.4989	3.59	0.4998
1.98	0.4762	2.52	0.4941	3.06	0.4989	3.60	0.4998
1.99	0.4767	2.53	0.4943	3.07	0.4989	3.61	0.4999
2.00	0.4773	2.54	0.4945	3.08	0.4990	3.62	0.4999
2.01	0.4778	2.55	0.4946	3.09	0.4990	3.63	0.4999
2.02	0.4783	2.56	0.4948	3.10	0.4990	3.64	0.4999
2.03	0.4788	2.57	0.4949	3.11	0.4991	3.65	0.4999
2.04	0.4793	2.58	0.4951	3.12	0.4991	3.66	0.4999
2.05	0.4798	2.59	0.4952	3.13	0.4991	3.67	0.4999
2.06	0.4803	2.60	0.4953	3.14	0.4992	3.68	0.4999
2.07	0.4808	2.61	0.4955	3.15	0.4992	3.69	0.4999
2.08	0.4812	2.62	0.4956	3.16	0.4992	3.70	0.4999
2.09	0.4817	2.63	0.4957	3.17	0.4992	3.71	0.4999
2.10	0.4821	2.64	0.4959	3.18	0.4993	3.72	0.4999
2.11	0.4826	2.65	0.4960	3.19	0.4993	3.73	0.4999
2.12	0.4830	2.66	0.4961	3.20	0.4993	3.74	0.4999
2.13	0.4834	2.67	0.4962	3.21	0.4993	3.75	0.4999
2.14	0.4838	2.68	0.4963	3.22	0.4994	3.76	0.4999
2.15	0.4842	2.69	0.4964	3.23	0.4994	3.77	0.4999
2.16	0.4846	2.70	0.4965	3.24	0.4994	3.78	0.4999
2.17	0.4850	2.71	0.4966	3.25	0.4994	3.79	0.4999
2.18	0.4854	2.72	0.4967	3.26	0.4994	3.80	0.4999
2.19	0.4857	2.73	0.4968	3.27	0.4995	3.81	0.4999
2.20	0.4861	2.74	0.4969	3.28	0.4995	3.82	0.4999
2.21	0.4865	2.75	0.4970	3.29	0.4995	3.83	0.4999
2.22	0.4868	2.76	0.4971	3.30	0.4995	3.84	0.4999
2.23	0.4871	2.77	0.4972	3.31	0.4995	3.85	0.4999
2.24	0.4875	2.78	0.4973	3.32	0.4996	3.86	0.4999
2.25	0.4878	2.79	0.4974	3.33	0.4996		

Graphing Calculator Guide

Operating the TI-83 and TI-84 Plus Calculators

Turning the Calculator On and Off

| ON | | Turns the calculator on. |
| 2nd | ON | Turns the calculator off. |

Adjusting the Display Contrast

| 2nd | ▲ | Increases the display (darkens the screen). |
| 2nd | ▼ | Decreases the contrast (lightens the screen). |

Note: If the display begins to dim (especially during calculations), and you must adjust the contrast to 8 or 9 in order to see the screen, then batteries are low and you should replace them soon.

The TI-83 and TI-84 Plus keyboard are divided into four zones: graphing keys, editing keys, advanced function keys, and scientific calculator keys (Figure 1).

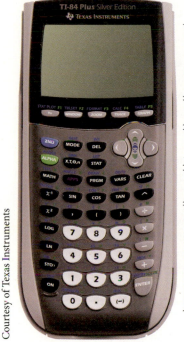

Courtesy of Texas Instruments

Home screen

Graphing keys

Editing keys (Allow you to edit expressions and variables)

Advanced function keys (Display menus that access advanced functions)

Scientific calculator keys

Figure 1

Chapter 1
Section 1.4 Graphing Equations

Entering Equations for Graphing

To graph an equation in the variables x and y, first solve the equation for y in terms of x. If the equation has variables other than x and y, solve for the dependent variable and replace the independent variable with x. Press the Y= key to access the function entry screen and enter the equation. To erase an equation, press CLEAR. To return to the home-screen, press 2nd MODE (QUIT).

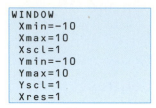

Setting Windows

The window defines the highest and lowest values of x and y on the graph of the function that will be shown on the screen. The values that define the viewing window can be set by using ZOOM keys. The standard window (ZOOM 6) is often appropriate. The standard window gives x- and y-values between -10 and 10.

To set the window manually, press the WINDOW key and enter the values that you want.

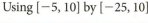

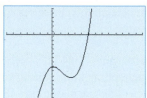

Graphing Equations

Determine an appropriate viewing window. The window should be set so that the important parts of the graph are shown and the unseen parts are suggested. Such a graph is called **complete**. Using the displayed coordinates from TRACE helps to determine an appropriate window. Pressing GRAPH or a ZOOM key will activate the graph.

With standard window   Using $[-5, 10]$ by $[-25, 10]$

Section 1.4 Finding Function Values

Using TRACE on the Graph

Enter the function to be evaluated in Y_1. Choose a window so that it contains the x-value whose y-value you seek. Press TRACE and then enter the selected x-value followed by ENTER. The cursor will move to the selected value and give the resulting y-value if the selected x-value is in the window. If the selected x-value is not in the window, Err: INVALID occurs. If the x-value is in the window, the y-value will occur even if it is not visible in the window.

To evaluate $y = -x^2 + 8x + 9$ when $x = -5$ and when $x = 3$, graph the function using the window $[-10, 10]$ by $[-10, 30]$.

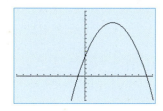

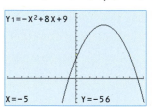

Using the TABLE ASK Feature

Enter the function with the Y= key. {Note: The = sign must be highlighted.) Press 2nd WINDOW (TBLSET), move the cursor to Ask opposite Indpnt:, and press ENTER. This allows you to input specific values for x. Pressing DEL will clear entries in the table. Then press 2nd TABLE and enter the specific values. The table on the right evaluates $y = -x^2 + 8x + 9$ at -5 and at 3.

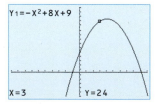

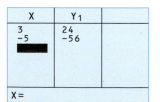

Making a Table of Values

If the Indpnt variable is on Auto, enter an initial x-value for the table in TblStart, and enter the desired change in the x-value as ΔTbl.

Enter 2nd TABLE to get a list of x-values and the corresponding y-values. The value of the function at the given value of x can be read from the table. Use the up or down arrows to find the x-values where the function is to be evaluated. The table on the right evaluates $y = -x^2 + 8x + 9$ for integer x-values from -3 to 3.

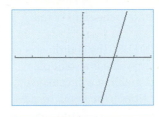

Section 1.4 Solving Linear Equations by the *X*-Intercept Method

To find the solution to $f(x) = 0$ (the x-value where the graph crosses the x-axis):

1. Set one side of the equation to 0 and enter the other side as Y_1 in the Y= menu.
2. Set the window so that the x-intercept to be located can be seen.
3. Press 2nd TRACE to access the CALC menu and select 2:zero.
4. Answer the question "*Left Bound?*" with ENTER after moving the cursor close to and to the left of an x-intercept.
5. Answer the question "*Right Bound?*" with ENTER after moving the cursor close to and to the right of this x-intercept.
6. To the question "*Guess?*" press ENTER. The coordinates of the x-intercept are displayed. The x-value is a solution.

The solution to $5x - 9 = 0$ is found to be $x = 1.8$.

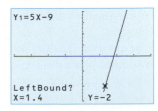

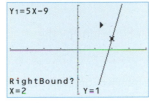

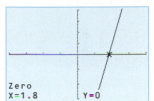

Section 1.5 Solving Systems of Equations in Two Variables

To solve a system of linear equations in two variables graphically:
1. Solve both equations for y.
2. Graph the first equation as Y_1 and the second as Y_2.
3. To find the point of intersection of the graphs:
   (a) Press 2nd TRACE to access the CALC menu and select 5:intersect.
   (b) Answer the question "*First curve?*" by pressing ENTER and "*Second curve?*" by pressing ENTER.
   (c) To the question "*Guess?*" press ENTER. The solution is shown on the right.

If the two lines intersect in one point, the coordinates give the x- and y-values of the solution.

The solution of the system above is $x = 2$, $y = 1$.

To solve $\begin{cases} 4x + 3y = 11 \\ 2x - 5y = -1 \end{cases}$ graphically, graph

$y_1 = -\dfrac{4}{3}x + \dfrac{11}{3}$ and $y_2 = \dfrac{2}{5}x + \dfrac{1}{5}$, then use Intersect.

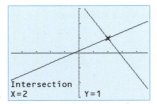

Chapter 2

Section 2.1 Solving Nonlinear Equations by the X-Intercept Method

To find the solutions to $f(x) = 0$ (the x-values where the graph crosses the x-axis):

1. Set one side of the equation to 0 and enter the other side as Y_1 in the Y= menu.
2. Set the window so that the x-intercepts to be located can be seen.
3. Press 2nd TRACE to access the CALC menu and select 2:zero.
4. Answer the question "*Left Bound?*" with ENTER after moving the cursor close to and to the left of an x-intercept.
5. Answer the question "*Right Bound?*" with ENTER after moving the cursor close to and to the right of this x-intercept.
6. To the question "*Guess?*" press ENTER. The coordinates of the x-intercept are displayed. The x-value is a solution.
7. Repeat to get all x-intercepts. The graph of a linear equation will cross the x-axis at most 1 time; the graph of quadratic equation will cross the x-axis at most 2 times, etc.

The solutions to $0 = -x^2 + 8x + 9$ are found to be 9 and -1.

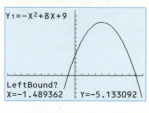
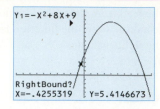
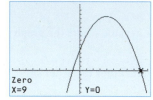
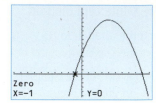

Section 2.1 Solving Nonlinear Equations by the Intersection Method

To solve nonlinear equations by the intersection method:

1. Graph the left side of the equation as Y_1 and the right side as Y_2.
2. Find a point of intersection of the graphs as shown in Section 1.5.
3. To find another point of intersection, repeat while keeping the cursor near the second point.

The solutions to $7x^2 = 16x - 4$ are found at the right.

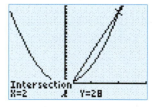
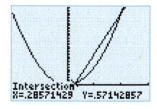

Section 2.2 Graphing Quadratic Functions

To graph a quadratic function:

1. Solve for y in terms of x and enter it in the Y= menu.
2. Find the coordinates of its vertex, with $x = -b/a$.
3. Set the window so the x-coordinate of its vertex is near its center and the y-coordinate is visible.
4. Press Graph.

To graph $P(x) = -0.1x^2 + 300x - 1200$, enter $y_1 = -0.1x^2 + 300x - 1200$ on a window with its center near $x = (-300)/[2(-0.1)] = 1500$ and with $y = P(1500) = 223{,}800$ visible, and graph.

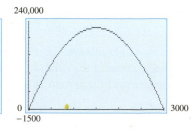

Section 2.4 Graphing Polynomial Functions

To graph a polynomial function:

1. See Table 2.1 in the text to determine possible shapes for the graph.
   (The graph of the function $y = x^3 - 16x$ has one of four shapes in the table.)
2. Graph the function in a window large enough to see the shape of the complete graph. This graph is like the graph of Degree 3(b) in the table in Section 2.4.
3. If necessary, adjust the window for a better view of the graph.

$[-40, 40]$ by $[-100, 100]$ $[-6, 6]$ by $[-30, 30]$

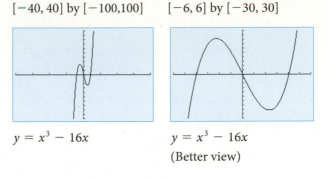

$y = x^3 - 16x$ $y = x^3 - 16x$

(Better view)

Section 2.4 Graphing Rational Functions

To graph a rational function:

1. Determine the vertical and horizontal asymptotes.
2. Set the window so that the x range is centered near the x-value of the vertical asymptote.
3. Set the window so that the horizontal asymptote is near the center of the y range.
4. Graph the function in a window large enough to see the shape of the complete graph.
5. If necessary, adjust the window for a better view of the graph.

To graph $y = \dfrac{12x + 8}{3x - 9}$, set the center of the window near the vertical asymptote $x = 3$ and near the horizontal asymptote $y = 4$.

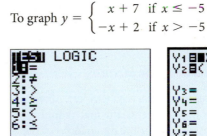

Section 2.4 Graphing Piecewise Defined Functions

A piecewise defined function is defined differently over two or more intervals.

To graph a piecewise defined function $y = \begin{cases} f(x) & \text{if } x \le a \\ g(x) & \text{if } x > a \end{cases}$

1. Go to the $Y =$ key and enter

   $Y_1 = f(x)/(x \le a)$ and $Y_2 = g(x)/(x > a)$

   (The inequality symbols are found under the TEST menu.)
2. Graph the function using an appropriate window.
3. Evaluating a piecewise defined function at a given value of x requires that the correct equation ("piece") be selected.

To graph $y = \begin{cases} x + 7 & \text{if } x \le -5 \\ -x + 2 & \text{if } x > -5 \end{cases}$

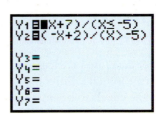

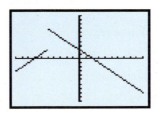

Section 2.5 Modeling

A. To Create a Scatter Plot

1. Press STAT and under EDIT press 1:Edit. This brings you to the screen where you enter data into lists.
2. Enter the x-values (input) in the column headed L1 and the corresponding y-values (output) in the column headed L2.
3. Go to the Y= menu and turn off or clear any functions entered there. To turn off a function, move the cursor over the = sign and press ENTER.
4. Press 2nd STAT PLOT, 1:Plot 1. Highlight ON, and then highlight the first graph type (Scatter Plot), Enter Xlist:L1, Ylist:L2, and pick the point plot mark you want.
5. Choose an appropriate WINDOW for the graph and press GRAPH, or press ZOOM, 9:ZoomStat to plot the data points.

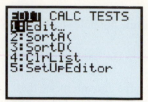

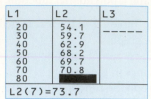

B. To Find an Equation That Models a Set of Data Points

1. Observe the scatter plot to determine what type function would best model the data. Press STAT, move to CALC, and select the function type to be used to model the data.
2. Press the VARS key, move to Y-VARS, and select 1:Function and 1:Y_1. Press ENTER.
   The coefficients of the equation will appear on the screen and the regression equation will appear as Y_1 on the Y= screen.

Pressing ZOOM 9 shows how well the model fits the data.

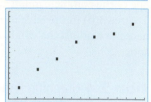

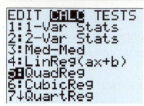

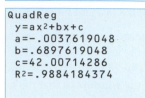

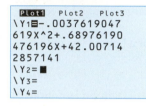

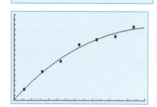

The model is $y = -0.00376x^2 + 0.690x + 42.007$.

Chapter 3
Section 3.1 Entering Data into Matrices

To enter data into matrices, press the MATRIX key. Move the cursor to EDIT. Enter the number of the matrix into which the data is to be entered. Enter the dimensions of the matrix, and enter the value for each entry of the matrix. Press ENTER after each entry.

For example, we enter the matrix below as [A].

$$\begin{bmatrix} 1 & 2 & 3 \\ 2 & -2 & 1 \\ 3 & 1 & -2 \end{bmatrix}$$

1. Enter 3's to set the dimension, and enter the numbers.
2. To perform operations with the matrix or leave the editor, first press 2nd QUIT.
3. To view the matrix, press MATRIX, the number of the matrix, and ENTER.

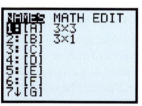

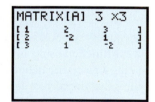

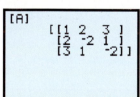

Section 3.1 Operations with Matrices

To find the sum of two matrices, [A] and [D], enter [A] + [D], and press ENTER. For example, the sum

$$\begin{bmatrix} 1 & 2 & 3 \\ 2 & -2 & 1 \\ 3 & 1 & -2 \end{bmatrix} + \begin{bmatrix} 7 & -3 & 2 \\ 4 & -5 & 3 \\ 0 & 2 & 1 \end{bmatrix} \text{ is shown at right.}$$

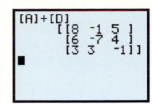

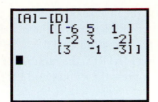

To find the difference, enter [A] − [D] and press ENTER. We can multiply a matrix [D] by a real number (scalar) k by entering k [D].

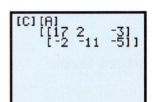

Section 3.2 Multiplying Two Matrices

To find the product of two matrices, [C] times [A], enter [C][A] and press ENTER. For example, we compute the product

$$\begin{bmatrix} 1 & 2 & 4 \\ -3 & 2 & -1 \end{bmatrix} \begin{bmatrix} 1 & 2 & 3 \\ 2 & -2 & 1 \\ 3 & 1 & -2 \end{bmatrix} \text{ at right.}$$

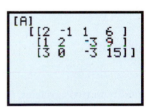

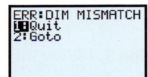

Note that entering [A][C] gives an error message. [A][C] cannot be computed because the dimensions do not match in this order.

Section 3.3 Solution of Systems— Reduced Echelon Form

To solve a 3×3 system:
1. Enter the coefficients and the constants in an augmented matrix.
2. Under the MATRIX menu, choose MATH and B:rref, then enter the matrix to be reduced followed by ")", and press ENTER.
3.

(a) If each row in the coefficient matrix (first 3 columns) contains a 1 with the other elements 0's, the solution is unique and the number in column 4 of a row is the value of the variable corresponding to a 1 in that row.

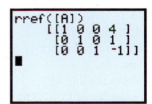

For example, the system $\begin{cases} 2x - y + z = 6 \\ x + 2y - 3z = 9 \\ 3x \quad\quad - 3z = 15 \end{cases}$ is solved at right.

The solution to the system above is unique: $x = 4$, $y = 1$, and $z = -1$.

(b) If the reduced matrix has all zeros in the third row, the solution is nonunique.

(c) If the reduced matrix has 3 zeros in a row and a non-zero element in the fourth column in that row, there is no solution.

Section 3.4 Finding the Inverse of a Matrix

To find the inverse of a matrix:

1. Enter the elements of the matrix using MATRIX and EDIT. Press 2nd QUIT.
2. Press MATRIX, the number of the matrix, and ENTER, then press the x^{-1} key and ENTER.

For example, the inverse of $E = \begin{bmatrix} 2 & 0 & 2 \\ -1 & 0 & 1 \\ 4 & 2 & 0 \end{bmatrix}$

is shown at right.

3. To see the entries as fractions, press MATH, press 1:Frac, and press ENTER.

Not all matrices have inverses. Matrices that do not have inverses are called singular matrices.

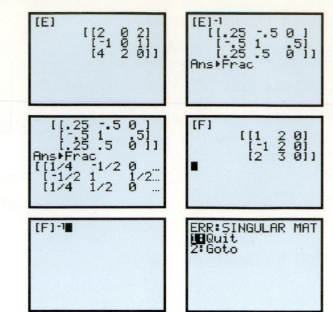

Section 3.4 Solving Systems of Linear Equations with Matrix Inverses

The matrix equation $AX = B$ can be solved by computing $X = A^{-1}B$ if a unique solution exists.

$$\begin{cases} 25x + 20y + 50z = 15{,}000 \\ 25x + 50y + 100z = 27{,}500 \\ 250x + 50y + 500z = 92{,}250 \end{cases}$$

The solution to the system

is found at the right to be $x = 254$, $y = 385$, $z = 19$.

Chapter 4

Section 4.1 Graphing Solution Regions of Linear Inequalities

To graph the solution of a linear inequality in two variables, first solve the inequality for the independent variable and enter the other side of the inequality in Y_1, so that $Y_1 = f(x)$. If the inequality has the form $y \leq f(x)$, shade the region below the graphed line and if the inequality has the form $y \geq f(x)$, shade the region above the line.

In this example, we shade the region above the line with SHADE under the DRAW menu and enter Shade (Y_1, 10) on the home screen.

To solve $4x - 2y \leq 6$, convert it to $y \geq 2x - 3$ and graph $Y_1 = 2x - 3$.

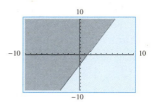

Section 4.1 Graphing Solutions of Systems of Linear Inequalities

To graph the solution region for a system of linear inequalities in two variables, write the inequalities as equations solved for y, and graph the equations.

For example, to find the region defined by the inequalities

$$\begin{cases} 5x + 2y \le 54 \\ 2x + 4y \le 60 \\ x \ge 0, y \ge 0 \end{cases}$$

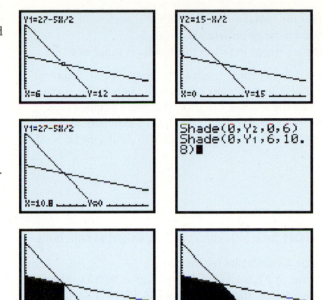

1. Choose a window with $x\min = 0$ and $y\min = 0$ because the inequalities $x \ge 0$, $y \ge 0$ limit the graph to Quadrant I.
2. Write $y = 27 - 5x/2$ and $y = 15 - x/2$ and graph.
3. Testing points determines the region that satisfies the inequalities.
4. Using TRACE or INTERSECT with the pair of equations and finding the intercepts give the corners of the solution region, where the borders intersect. These corners of the region are $(0, 0)$ $(0, 15)$, $(6, 12)$, and $(10.8, 0)$.
5. Use SHADE to shade the region determined by the inequalities. Shade under the border from $x = 0$ to a corner and shade under the second border from the corner to the x-intercept.

Section 4.2 Linear Programming

To solve a linear programming problem involving constraints in two variables:

1. Graph the constraint inequalities as equations, solved for y.
2. Test points to determine the region and use TRACE or INTERSECT to find each of the corners of the region, where the borders intersect.
3. Then evaluate the objective function at each of the corners.

For example, to maximize $f = 5x + 11y$ subject to the constraints

$$\begin{cases} 5x + 2y \le 54 \\ 2x + 4y \le 60 \\ x \ge 0, y \ge 0 \end{cases}$$

we graphically find the constraint region (as shown above), and evaluate the objective function at the coordinates of each of the corners of the region. Evaluating $f = 5x + 11y$ at each of the corners determines where this objective function is maximized or minimized.

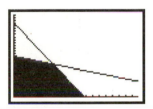

The corners of the region determined by the inequalities are $(0, 0)$ $(0, 15)$, $(6, 12)$, and $(10.8, 0)$.

At $(0, 0)$, $f = 0$ At $(0, 15)$, $f = 165$
At $(6,12)$, $f = 162$ At $(10.8, 0)$, $f = 54$

The maximum value of f is 165 at $x = 0$, $y = 15$.

Chapter 5

Section 5.1 Graphing Exponential Functions

1. Enter the function as Y_1 in the Y= menu.
2. Set the x-range centered at $x = 0$.
3. Set the y-range to reflect the function's range of $y > 0$.

Note that some graphs (such as the graph of $y = 4^x$ shown here) appear to eventually merge with the negative x-axis. Adjusting the window can show that these graphs never touch the x-axis. For more complicated exponential functions, it may be helpful to use TABLE to find a useful window.

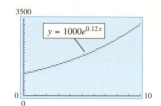

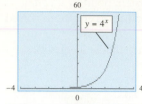

Section 5.1 Modeling with Exponential Functions

1. Create a scatter plot for the data.
2. Choose STAT, then CALC. Scroll down to 0:ExpReg and press ENTER, then VARS, Y-VARS, FUNCTION, Y_1, and ENTER.

   (Recall that this both calculates the requested exponential model and enters its equation as Y_1 in the Y= menu.)

The last screen shows how well the model fits the data.

Find the exponential model for the following data.

x	1	2	3	4	5	6	7	8	9	10
y	43	38	33	29	25	22	19	15	14	12

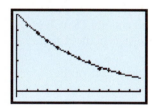

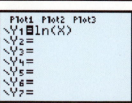

Section 5.2 Graphing Base e and Base 10 Logarithmic Functions

Enter the function as Y_1 in the Y= menu.

1. For $y = \ln(x)$ use the LN key.
2. For $y = \log(x)$ use the LOG key.
3. Set the window x-range to reflect that the function's domain is $x > 0$.
4. Center the window y-range at $y = 0$.

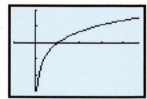

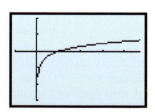

Section 5.2 Modeling with Logarithmic Functions

1. Create a scatter plot for the data.
2. Choose STAT, then CALC. Scroll down to 9:LnReg and press ENTER, then VARS, Y-VARS, FUNCTION, Y_1, and ENTER.

The last figure on the right shows how well the model fits the data.

Find the logarithmic model for the following data.

x	10	20	30	38
y	2.21	3.79	4.92	5.77

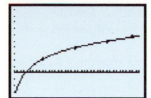

Section 5.2 Graphing Logarithmic Functions with Other Bases

1. Use a change of base formula to rewrite the logarithmic function with base 10 or base e.

$$\log_b x = \frac{\log x}{\log b} \quad \text{or} \quad \log_b x = \frac{\ln x}{\ln b}$$

2. Proceed as described above for graphing base e and base 10 logarithms.

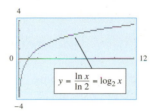

Chapter 6
Section 6.2 Future Value of a Lump Sum

To find the future value of a lump-sum investment:

1. Press the APPS key and select Finance, press ENTER.
2. Select TVM Solver, press ENTER.
3. Set N = the total number of periods, set I% = the annual percentage rate.
4. Set the PV = the lump sum preceded by a "−" to indicate the lump sum is leaving your possession.
5. Set PMT = 0 and set both P/Y and C/Y = the number of compounding periods per year.
6. Put the cursor on FV and press ALPHA ENTER to get the future value.

The future value of $10,000 invested at 9.8% compounded quarterly for 17 years is shown at the right.

Section 6.3 Future Value of an Annuity

To find the future value of an ordinary annuity:

1. Press the APPS key and select Finance, press ENTER.
2. Select TVM Solver, press ENTER.
3. Set N = the total number of periods, set I% = the annual percentage rate.
4. Set the PV = 0 and set both P/Y and C/Y = the number of compounding periods per year. END should be highlighted.
5. Set PMT = the periodic payment preceded by a "−" to indicate the lump sum is leaving your possession.
6. Put the cursor on FV and press ALPHA ENTER to get the future value.

   The future value of an ordinary annuity of $200 deposited at the end of each quarter for $2\frac{1}{4}$ years, with interest at 4% compounded quarterly, is shown.

For annuities due, all steps are the same except that BEGIN is highlighted.

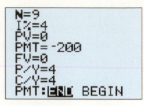

Ordinary annuity | Annuity due

Section 6.4 Present Value of an Annuity

To find the present value of an ordinary annuity:

1. Press the APPS key and select Finance, press ENTER.
2. Select TVM Solver, press ENTER.
3. Set N= the total number of periods, set I% = the annual percentage rate.
4. Set the FV = 0 and set both P/Y and C/Y = the number of compounding periods per year. END should be highlighted.
5. Set PMT = the periodic payment.
6. Put the cursor on PV and press ALPHA ENTER to get the present value.

The lump sum that needs to be deposited to receive $1000 at the end of each month for 16 years if the annuity pays 9%, compounded monthly is shown at the right.
For annuities due, BEGIN is highlighted.

Ordinary annuity | Annuity due

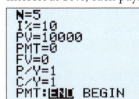

Section 6.5 Finding Payments to Amortize a Loan

To find the size of periodic payments to amortize a loan:

1. Press the APPS key and select Finance, press ENTER.
2. Select TVM Solver, press ENTER.
3. Set N = the total number of periods, set I% = the APR.
4. Set the PV = loan value and set both P/Y and C/Y = the number of periods per year.
5. Set FV = 0.
6. Put the cursor on PMT and press ALPHA ENTER to get the payment.

To repay a loan of $10,000 in 5 annual payments with annual interest at 10%, each payment must be $2637.97.

Section 6.5 Finding the Number of Payments to Amortize a Loan

To find the number of payments needed to amortize a loan:

1. Press the APPS key and select Finance, press ENTER.
2. Select TVM Solver, press ENTER.
3. Set the PV = loan value, set both P/Y and C/Y = the number of periods per year, set I% = the APR.
4. Set PMT = required payment and set FV = 0.
5. Put the cursor on N and press ALPHA ENTER to get the payment.

The number of monthly payments to pay a $2500 credit card loan with $55 payments and 18% interest is 76.9 months, or 6 years, 5 months.

Chapter 7

Section 7.5 Evaluating Factorials

To evaluate factorials:

1. Enter the number whose factorial is to be calculated.
2. Choose MATH, then PRB. Scroll to 4: ! and press ENTER.

Press ENTER again to find the factorial.
7! is shown on the right.

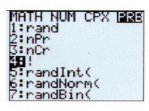
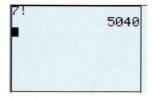

Section 7.5 Evaluating Permutations

To evaluate permutations:

1. For a "permutation of n objects taken r at a time" (such as $_{20}P_4$), first enter the value of n (such as $n = 20$).
2. Choose MATH, then PRB. Scroll to 2: nPr and press ENTER.
3. Enter the value of r (such as $r = 4$), and press ENTER to find the value of nPr.
   $_{20}P_4$ is shown on the right.

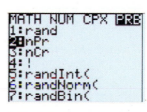
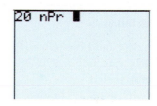

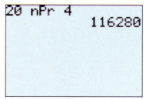

Section 7.5 Evaluating Combinations

To evaluate combinations:

1. For a "combination of n objects taken r at a time" (such as $_{20}C_4$), first enter the value of n (such as $n = 20$).
2. Choose MATH, then PRB. Scroll to 3:nCr and press ENTER.
3. Enter the value of r (such as $r = 4$), and press ENTER to find the value of nCr.

Note that nCr = nC(n − r), as the figure on the right shows for $_{20}C_4$ and $_{20}C_{16}$.

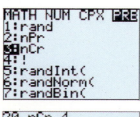

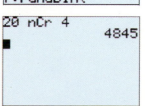
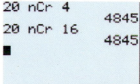

Section 7.6 Finding Probabilities Using Permutations and Combinations

To solve a probability problem that involves permutations or combinations:

1. Determine if permutations or combinations should be used.
2. Enter the ratios of permutations or combinations to find the probability.
3. If desired, use MATH, then 1:Frac to get the probability as a fraction.

If there are 5 defective computer chips in a box of 10, the probability that 2 chips drawn together from the box will both be defective is

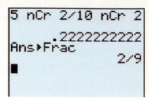

Section 7.7 Evaluating Markov Chains

To evaluate a Markov chain:

1. Enter the initial probability vector as matrix A and the transition matrix as matrix B.
2. To find the probabilities for the nth state, calculate $[A][B]^n$. The 3rd state, $[A][B]^3$, is shown on the right.

```
[A]
        [[.4 .4 .2]]
[B]
        [[.5 .4 .1]
         [.4 .5 .1]
         [.3 .3 .4]]
```

```
[A][B]^3
[[.4278 .4278 ...
```

Section 7.7 Finding Steady-State Vectors for Markov Chains

If the transition matrix contains only positive entries, the probabilities will approach a steady-state vector, which is found as follows.

1. Calculate and store $[C] = [B] - [I]$, where $[B]$ is the regular transition matrix and $[I]$ is the appropriately sized identity matrix.
2. Solve $[C]^T = [0]$ as follows:
   (a) Find $[C]^T$ with MATRIX, then MATH, 2:^{T}.
   (b) Find rref$[C]^T$.
3. Choose the solutions that add to 1, because they are probabilities.

```
identity(3)→[I]
        [[1 0 0]
         [0 1 0]
         [0 0 1]]
```

```
[B]-[I]
[[-.5  .4   .1 ]
 [ .4  -.5  .1 ]
 [ .3  .3  -.6]]
Ans→[C]
```

```
rref([C]^T)
        [[1 0 -3]
         [0 1 -3]
         [0 0  0]]
```

Thus $x = 3z$ and $y = 3z$, and $3z + 3z + z = 1$ gives $z = 1/7$, so the probabilities are given in the steady-state vector

$$\left[\frac{3}{7} \quad \frac{3}{7} \quad \frac{1}{7}\right].$$

Chapter 8
Section 8.1 Binomial Probabilities

To find binomial probabilities:

1. 2nd DISTR A:binompdf*(n,p,x)* computes the probability of *x* success in *n* trials of a binomial experiment with probability of success *p*. Using MATH 1:Frac gives the probabilities as fractions.
   The probability of 3 heads in 6 tosses of a fair coin is found using 2nd DISTR, binompdf(6,.5,3).
2. The probabilities can be computed for more than one number in one command, using 2nd DISTR, binompdf(*n,p*,{$x_1,x_2,\ldots$}).
   The probabilities of 4, 5, or 6 heads in 6 tosses of a fair coin are found using 2nd DISTR, binompdf(6,.5,{4,5,6}).
3. 2nd DISTR, binomcdf(*n,p,x*) computes the probability that the number of successes is less than or equal to *x* for the binomial distribution with *n* trials and probability of success *p*.

The probability of 4 or fewer heads in 6 tosses of a fair coin is found using 2nd DISTR, binomcdf(6,.5,4).

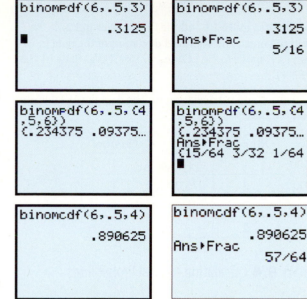

Section 8.2 Histograms

To find a frequency histogram for a set of data:

1. Press STAT, EDIT, 1:edit to enter each number in a column headed by L1 and the corresponding frequency of each number in L2.
2. Press 2nd STAT PLOT, 1:Plot 1. Highlight ON, and then press ENTER on the histogram icon. Enter L1 in xlist and L2 in Feq.
3. Press ZOOM, 9:ZoomStat or press Graph with an appropriate window.
4. If the data is given in interval form, a histogram can be created using the steps above with the class marks used to represent the intervals.

The frequency histogram for the scores 38, 37, 36, 40, 35, 40, 38, 37, 36, 37, 39, 38 is shown.

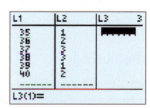

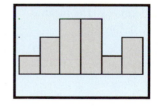

Section 8.2 Finding the Mean and Standard Deviation of Raw Data

To find descriptive statistics for a set of data:

1. Enter the data in list L1.
2. To find the mean and standard deviation of the data in L1, press STAT, move to CALC, and press 1:1-Var Stats, and ENTER.

The mean and sample standard deviation of the data 1, 2, 3, 3, 4, 4, 4, 4, 5, 5, 6, 7 are $\bar{x} = 4$ and $s \approx 1.65$.

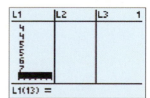
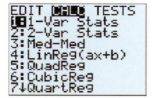

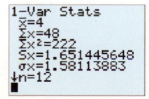

Section 8.2 Finding the Mean and Standard Deviation of Grouped Data

To find descriptive statistics for a set of data:

1. Enter the data in list L1 and the frequencies in L2.
2. To find the mean and standard deviation of the data in L1, press STAT, move to CALC, and press 1:1-Var Stats L1, L2, and ENTER.

The mean and standard deviation for the data in the table are shown on the right.

Salary	Number	Salary	Number
$59,000	1	$31,000	1
30,000	2	75,000	1
26,000	7	35,000	1
34,000	2		

The mean is $34,000 and the sample standard deviation is $14,132.84.

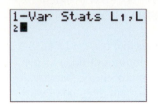
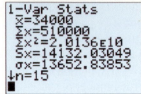

Section 8.4 Calculating Normal Probabilites

To calculate normal probabilities:

The command 2nd DISTR, 2:normalcdf(lowerbound, upperbound, μ, σ) gives the probability that x lies between the lower bound and the upper bound when the mean is μ and the standard deviation is σ.

The probability that a score lies between 33 and 37 when the mean is 35 and the standard deviation is 2 is found below.

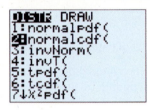
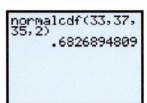

Section 8.4 Graphing Normal Distributions

To graph the normal distribution, press Y= and enter 2nd DIST 1:normalpdf(x, μ, σ) into Y$_1$.

Then set the window values xmin and xmax so the mean μ falls between them and press ZOOM, 0:Zoomfit.

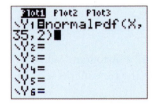
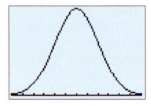

Chapter 9
Section 9.1 Evaluating Limits

To evaluate $\lim\limits_{x \to c} f(x)$

1. Enter the function as Y_1 in the Y= menu.
2. Set the x-range so it contains $x = c$.
3. Evaluate $f(x)$ for several x-values near $x = c$ and on each
    side by c by using one of the following methods.
    (a) Graphical Evaluation
    TRACE and ZOOM near $x = c$. If the values of y approach
    the same number L as x approaches c from the left and the
    right, there is evidence that the limit is L.
    (b) Numerical Evaluation
    Use TBLSET with Indpnt set to *Ask*. Enter values very
    close to and on both sides of c. The y-values will approach
    the same limit as above.

Evaluate $\lim\limits_{x \to 3} \dfrac{x^2 - 9}{x - 3}$.

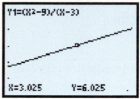

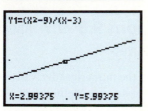

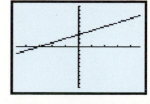

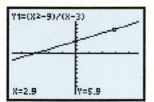

The y-values seem to
approach 6.

The limit as x approaches 3
of $f(x)$ appears to be 6.

Section 9.1 Limits of Piecewise Functions

Enter the function, then use one of the methods for evaluating limits discussed above.

Find $\lim\limits_{x \to -5} f(x)$ where $f(x) = \begin{cases} x + 7 & \text{if } x \leq -5 \\ -x + 2 & \text{if } x > -5 \end{cases}$

First enter $Y_1 = (x + 7)/(x \leq -5)$ and
$Y_2 = (-x + 2)/(x > -5)$.

Both methods indicate that the limit does not exist (DNE).

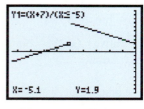

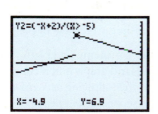

Section 9.2 Limits as $x \to \infty$

Enter the function as Y_1, then use large values of x with one of the methods for evaluating limits discussed above. Note: Limits as $x \to -\infty$ are done similarly.

Evaluate $\lim\limits_{x \to \infty} \dfrac{3x - 2}{1 - 5x}$.

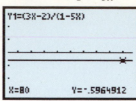
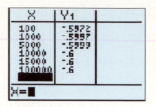

Both methods suggest that the limit is -0.6.

Sections 9.3–9.7 Approximating Derivatives

To find the numerical derivative (approximate derivative) of $f(x)$ at $x = c$, use Method 1 or Method 2.

Method 1

1. Choose MATH, then 8:nDeriv(and press ENTER.
2. Enter the function, x, and the value of c, so the display shows nDeriv($f(x)$, x, c) then press ENTER. The approximate derivative at the specified value will be displayed.

Method 2

1. Enter the function as Y_1 in the Y=menu, and graph in a window that contains both c and $f(c)$.
2. Choose CALC by using 2nd TRACE, then 6:dy/dx, enter the x-value, c, and press ENTER. Then approximate derivative at the specified value will be displayed.

Warning: Both approximation methods above require that the derivative exists at $x = c$ and will give incorrect information when $f'(c)$ does not exist.

Find the numerical derivative of $f(x) = x^3 - 2x^2$ at $x = 2$.

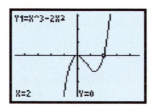
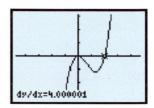

The numerical derivative is approximately 4.

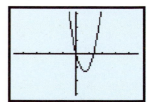

The value of the derivative is approximately 4.

Sections 9.3–9.7 Checking Derivatives

To check the correctness of the derivative function $f(x)$:

1. In the Y= menu, enter as Y_1 the derivative $f'(x)$ that you found, and graph it in a convenient window.
2. Enter the following as Y_2:nDeriv(f(x), x, x).
3. If the second graph lies on top of the first, the derivative is correct.

Verify the derivative of $f(x) = x^3 - 2x^2$ is $f'(x) = 3x^2 - 4x$.

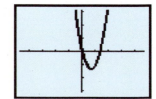

Section 9.8 Approximating the Second Derivative

To approximate $f''(c)$:

1. Enter $f(x)$ as Y_1 in the Y= menu.
2. Enter nDeriv(Y_1, x, x) as Y_2.
3. Estimate $f''(c)$ by using nDeriv(Y_2, x, c).

Find the second derivative of $f(x) = x^3 - 2x^2$ at $x = 2$.

 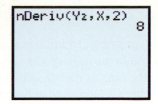

Thus $f''(2) = 8$.

Chapter 10
Section 10.1 Finding Critical Values

To find or approximate critical values of $f(x)$, that is x-values that make the derivative equal to 0 or undefined:

I. Find the derivative of $f(x)$.

II. Use Method 1 or Method 2 to find the critical values.

Methods 1 and 2 show how to find the critical values of $f(x) = \frac{1}{3}x^3 - 4x$. Note that the derivative is $f'(x) = x^2 - 4$.

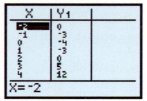 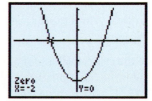

Method 1

1. Enter the derivative in the Y= menu as Y_1 and graph it in a convenient window.
2. Find where $Y_1 = 0$ by one of the following:
   (a) Using TRACE to find the x-intercepts of Y_1.
   (b) Using 2nd CALC then 2:zero.
   (c) Using TBLSET then TABLE to find the values of x that give $Y_1 = 0$.
3. Use the graph of Y_1 (and TRACE or TABLE) to find the values of x that make the derivative undefined.

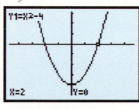

The only critical values are $x = 2$ and $x = -2$.

Method 2

1. Enter the derivative in the Y= menu as Y_1.
2. Press MATH and select Solver. Press the up arrow revealing EQUATION SOLVER equ:0=, and enter Y_1 (the derivative).
3. Press the down arrow or ENTER and the variable x appears with a value (not the soltion). Move the cursor to the line containing the variable whose value is sought.
4. Press ALPHA SOLVE (ENTER). The value of the variable changes to the solution that is closest to that value originally shown.
5. To find additional solutions (if they exist), change the value of the variable and press ALPHA SOLVE (ENTER). The value of the variable that gives the solution of $Y_1 = 0$ that is closest to that value.
6. If appropriate, use 0:Solver to solve (Denominator of Y_1) = 0 to find the critical values for which the derivative is undefined.

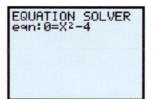 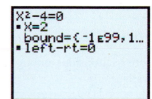

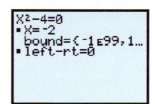

The only critical values are $x = 2$ and $x = -2$.

Section 10.1 Relative Maxima and Minima

To find relative maxima and minima:

1. In the Y=menu enter the function as Y_1 and the derivative is Y_2.
2. Use TBLSET and TABLE to evaluate the derivative to the left and to the right of each critical value.
3. Use the signs of the values of the derivative to determine whether f is increasing or decreasing around the critical values, and thus to classify the critical values as relative maxima, relative minima, or horizontal points of inflection.
4. Graph the function to confirm your conclusions.

Find the relative maxima and minima of $f(x) = \dfrac{1}{3}x^3 - 4x$. Note that the derivative is $f'(x) = x^2 - 4$.

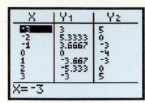 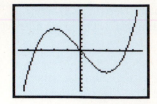

The relative max is at $(-2, 16/3)$ and the relative min is at $(2, -16/3)$.

Section 10.1 Critical Values and Viewing Windows

To use critical values to set a viewing window that shows a complete graph:

1. Once the critical values for a function have been found,
   (a) Enter the function as Y_1 and the derivative as Y_2. Use TABLE to determine where the function is increasing and where it is decreasing.
   (b) In WINDOW menu set x-min so that it is smaller than the smallest critical value and set x-max so that it is larger than the largest critical value.
2. Use TABLE or VALUE to determine the y-coordinates of the critical values. Set y-min and y-max to contain the y-coordinates of the critical points.
3. Graph the function.

Let $f(x) = 0.0001x^3 + 0.003x^2 - 3.6x + 5$. Given that the critical values for $f(x)$ are $x = -120$ and $x = 100$, set a window that shows a complete graph and graph the function.

X	Y₁	Y₂
-150	275	2.25
-120	307.4	0
0	5	-3.6
100	-225	0
150	-130	4.05
X=0		

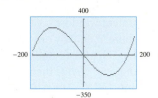

Section 10.2 Exploring f, f', and f'' Relationships

To explore relationships among the graphs of a function and its derivatives:

1. Find the functions for f' and f''.
2. Graph all three functions in the same window.
   - Notice that f increases when f' is above the x-axis $(+)$ and decreases when f' is below the x-axis $(-)$.
   - Notice that f is concave up when f'' is above the x-axis $(+)$ and is concave down when f'' is below the x-axis $(-)$.

Let $f(x) = x^3 - 9x^2 + 24x$. Graph f', f'', and f'' on the interval $[0, 5]$ to explore the relationships among these functions.

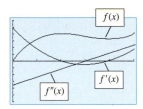

Sections 10.3–10.4 Finding Optimal Values

To find the optimal values of a function when the goal is not to produce a graph:

1. Enter the function as Y_1 in the Y= menu.
2. Select a window that includes the x-values of interest and graph the function.
3. While looking at the graph of the function, choose the CALC menu, scroll to 3:minimum or 4:maximum depending on which one is to be found, and press ENTER. This will result in a "*Left Bound?*" prompt.
   (a) Move the cursor to a point to the left of the point of interest. Press ENTER to select the left bound.
   (b) Move the cursor to the right of the point of interest. Press ENTER to select the right bound.
   (c) Press ENTER at the "*Guess?*" prompt. The resulting point is an approximation of the desired optimum value.

Let $f(x) = 100\sqrt{2304 + x^2} + 50(100 - x)$. Find the minimum value of $f(x)$ on the interval $[0, 100]$ for x. Note: the following screens use the window x: $[-10, 110]$ and y: $[-2500, 12500]$

The minimum value is $y \approx 9157$ and occurs when $x \approx 27.7$.

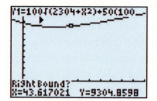 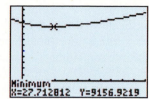

Note: Finding a maximum value works similarly.

Section 10.5 Asymptotes and Window Setting

To use asymptotes and critical values to set a viewing window that shows a complete graph:

1. Once the asymptotes and critical values for a function have been found,
   (a) Determine where the function is increasing and where it is decreasing (by using TABLE and with Y_1 as the function and Y_2 as the derivative).
   (b) In WINDOW menu set x-min so that it is smaller than the smallest x-value that is either a vertical asymptote or a critical value and set x-max so that it is larger than the largest of these important x-values.
2. Use TABLE or VALUE to determine the y-coordinates of the critical values. Set y-min and y-max so they contain the y-coordinates of any horizontal asymptotes and critical points.
3. Graph the function.

Let $f(x) = \dfrac{x^2}{x - 2}$. Given that $f(x)$ has the line $x = 2$ as a vertical asymptote, has no horizontal asymptote, and has critical values $x = 0$ and $x = 4$. Set the window and graph $y = f(x)$.

The critical points for $f(x)$ are $(0, 0)$ and $(4, 8)$. The window needs an x-range that contains 0, 2, and 4, and a y-range that contains 0 and 8.

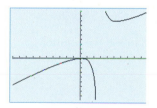

Chapter 11
Section 11.1 Derivatives of Logarithmic Functions

To check that the derivative of $y = \ln(2x^6 - 3x + 2)$ is $Y_1 = \dfrac{12x^5 - 3}{2x^6 - 3x + 2}$, we show that the graph of $Y_2 = \text{nDeriv}(\ln(2x^6 - 3x + 2), x, x)$ lies on the graph of $y_1 = \dfrac{12x^5 - 3}{2x^6 - 3x + 2}$.

 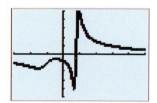

Section 11.1 Finding Critical Values

To find the critical values of a function $y = f(x)$:

1. Enter the function as Y_1 in the $Y=$ menu, and the derivative as Y_2.
2. Press MATH and select Solver. Press the up arrow revealing EQUATION SOLVER equ:0=, and enter Y_2 (the derivative).
3. Press the down arrow or ENTER and the variable x appears with a value (not the solution). Place the cursor on the line containing the variable whose value is sought.
4. Press ALPHA SOLVE (ENTER). The value of the variable changes to the solution of the equation that is closest to that value.
5. To find additional solutions (if they exist), change the value of the variable and press ALPHA SOLVE (ENTER). The value of the variable gives the solution of $Y_2 = 0$ that is closest to that value.

To find the critical values of $y = x^2 - 8 \ln x$, we solve

$$0 = 2x - \frac{8}{x}.$$

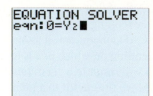

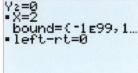
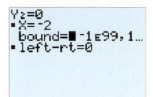

The two critical values are $x = 2$ and $x = -2$.

Section 11.1 Finding Optimal Values

To find the optimal values of a function $y = f(x)$:

1. Find the critical values of a function $y = f(x)$. (Use the steps for finding critical values discussed above.)
2. Graph $y = f(x)$ on a window containing the critical values.
3. The y-values at the critical values (if they exist) are the optimal values of the function.

To find the optimal values of $y = x^2 - 8 \ln x$, we solve

$$0 = 2x - \frac{8}{x} \text{ and evaluate } y = x^2 - 8 \ln x \text{ at the solutions.}$$

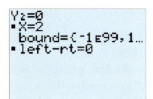

The two critical values are $x = 2$ and $x = -2$.

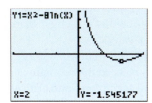
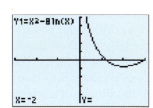

The minimum of $y = x^2 - 8 \ln x$ is -1.545 at $x = 2$.
The function $y = x^2 - 8 \ln x$ is undefined for all negative values because $\ln x$ is undefined for all negative values. Thus there is no optimum value of the function at $x = -2$.

Section 11.2 Derivatives of Exponential Functions

To check that the derivative of $y = 5^{x^2 + x}$ is
$y' = 5^{x^2 + x}(2x + 1) \ln 5$, we show that the graph of
$Y_2 = \text{nDeriv}(5^{x^2 + x}, x, x)$ lies on the graph of
$Y_1 = 5^{x^2 + x}(2x + 1) \ln 5$.

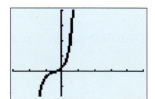

Section 11.2 Finding Critical Values

To find the critical values of a function $y = f(x)$:

1. Enter the function as Y_1 in the Y= menu, and the derivative as Y_2.
2. Press MATH and select Solver. Press the up arrow revealing EQUATION SOLVER equ:0=, and enter Y_2 (the derivative).
3. Press the down arrow or ENTER and the variable appears with a value (not the solution). Place the cursor on the variable whose value is sought.
4. Press ALPHA SOLVE (ENTER). The value of the variable changes to the solution of the equation that is closest to that value.
5. To find additional solutions (if they exist), change the value of the variable and press ALPHA SOLVE (ENTER). The value of the variable gives the solution of $Y_2 = 0$ that is closest to that value.

To find the critical values of $y = e^x - 3x^2$, we solve $0 = e^x - 6x$.

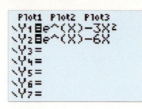
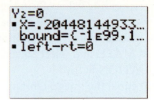

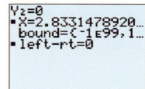
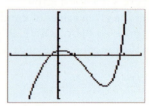

A relative maximum of $y = e^x - 3x^2$ occurs at $x \approx 0.204$ and a relative minimum occurs at $x \approx 2.833$.

Chapter 12

Sections 12.1–12.2 Checking Indefinite Integrals

To check a computed indefinite integral with the command fnInt:

1. Enter the integral of $f(x)$ (without the $+ C$) as Y_1 in Y= menu.
2. Move the cursor in Y_2, press MATH, 9:fnInt(and enter $f(x)$, x, 0, x) so the equation is $Y_2 = $ fnInt $(f(x), x, 0, x)$.
3. Pressing ENTER with the cursor to the left of Y_2 changes the thickness of the second graph, making it more evident that the second lies on top of the first.
4. Press GRAPH with an appropriate window. If the second graph lies on top of the first, the graphs agree and the computed integral checks.

Checking that the integral of $f(x) = x^2$ is $\int x^2 \, dx = \dfrac{x^3}{3} + C$:

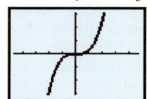

Sections 12.1–12.2 Families of Functions

To graph some functions in the family of indefinite integrals of $f(x)$:

1. Integrate $f(x)$.

2. Enter equations of the form $\int f(x) \, dx + C$ for different values of C.

3. Press GRAPH with an appropriate window. The graphs will be shifted up or down, depending on C.

The graphs of members of the family $y = \int (2x - 4) \, dx + C$ with $C = 0, 1, -2,$ and 3.

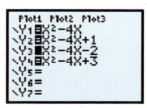
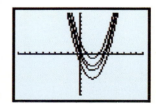

Section 12.5 Differential Equations

To solve initial value problems in differential equations:

1. Integrate $f(x)$, getting $y = F(x) + C$. If a value of x and a corresponding value of y in the integral $y = F(x) + C$ is known, this initial value can be used to find one function that satisfies the given conditions.

2. Press MATH and select Solver. Press the up arrow to see EQUATION SOLVER.

3. Set 0 equal the integral minus y, getting $0 = F(x) + C - y$, and press the down arrow.

4. Enter the given values of x and y, place the cursor on C, and press ALPHA, SOLVE. Replace C with this value to find the function satisfying the conditions.

To solve $\dfrac{dy}{dx} = 2x - 4$, we note that the integral of both sides is $y = x^2 - 4x + C$. If $y = 8$ when $x = -1$ in this integral, we can find C, and thus a unique solution, shown below.

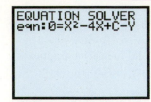
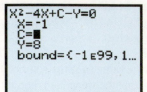

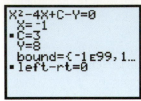

The unique solution is $y = x^2 - 4x + 3$.

Chapter 13

Section 13.1 Approximating Definite Integrals—Areas Under Curves

To approximate the area under the graph of $y = f(x)$ and above the x-axis:

1. Enter $f(x)$ under the Y= menu and graph the function with an appropriate window.

2. Press 2nd CALC and 7: $\int f(x)\,dx$.

3. Press ENTER. Move the cursor to, or enter, the lower limit (the left x-value).

4. Press ENTER. Move the cursor to, or enter, the upper limit (the right x-value).

5. Press ENTER. The area will be displayed.

The approximate area under the graph of $f(x) = x^2$ from $x = 0$ to $x = 3$ is found below.

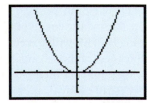

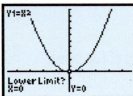
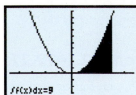

Section 13.1 Approximating Definite Integrals— Alternative Method

To approximate the definite integral of $f(x)$ from $x = a$ to $x = b$:

1. Press MATH, 9: fnInt(. Enter $f(x)$, x, a, b) so the display shows fnInt($f(x)$, x, a, b).
2. Press ENTER to find the approximation of the integral.
3. The approximation may be made closer than that in Step 2 by adding a fifth argument with a number (tolerance) smaller than 0.00001.

The approximation of $\displaystyle\int_{-1}^{3} (4x^2 - 2x)\,dx$ is found as follows.

Section 13.2 Approximating the Area Between Two Curves

To approximate the area between the graphs of two functions:

1. Enter one function as Y_1 and the second as Y_2. Press GRAPH using a window that shows all points of intersection of the graphs.
2. Find the x-coordinates of the points of intersection of the graphs, using 2nd CALC: intersect.
3. Determine visually which graph is above the other over the interval between the points of intersection.
4. Press MATH, 9: fnInt(. Enter $f(x)$, x, a, b) so the display shows fnInt($f(x)$, x, a, b) where $f(x)$ is $Y_2 - Y_1$ if the graph of Y_2 is above the graph of Y_1 between a and b, or $Y_1 - Y_2$ if Y_1 is above Y_2.

The area enclosed by the graphs of $y = 4x^2$ and $y = 8x$ is found as follows.

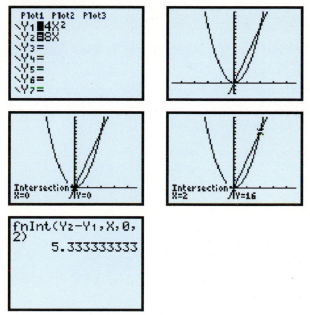

The area between the curves is 16/3.

Section 13.2 Approximating the Area Between Two Curves—Alternate Method

The area between the graphs can also be found by using 2nd CALC, $\int f(x)\,dx$.

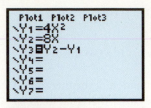
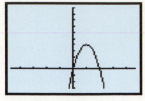

1. Enter $Y_3 = Y_2 - Y_1$ where Y_2 is above Y_1.
2. Turn off the graphs of Y_1 and Y_2 and graph Y_3 with a window showing where $Y_3 > 0$.
3. Press 2nd CALC and $\int f(x)\,dx$.
4. Press ENTER. Move the cursor to, or enter, the lower limit (the left x-value).
5. Press ENTER. Move the cursor to, or enter, the upper limit (the right x-value).
6. Press ENTER. The area will be displayed.

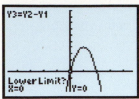

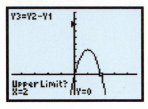
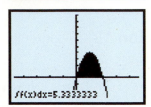

Excel Guide Part 1
Excel 2003

Excel Worksheet

When you start up Excel by using the instructions for your software and computer, the following screen will appear.

The components of the **spreadsheet** are shown, and the grid shown is called a **worksheet.** You can move to other worksheets by clicking on the tabs at the bottom.

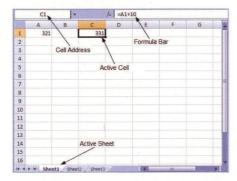

Addresses and Operations

Notice the letters at the top of the columns and the numbers identifying the rows. The cell addresses are given by the column and row; for example, the first cell has address A1. You can move from one cell to another with arrow keys, or you can select a cell with a mouse click. After you enter an entry in a cell, press enter to accept the entry. To edit the contents of a cell, click on the cell and edit the contents in the formula bar at the top. To delete the contents, press the delete key.

The file operations such as "open a new file," "saving a file," and "printing a file" are similar to those in WORD. For example, <CTRL>S saves a file. You can also format a cell entry by selecting it and using menus similar to those in WORD.

Working with Cells

Cell entries, rows containing entries, and columns containing entries can be copied and pasted with the same commands as in WORD. For example, a highlighted cell can be copied with <CTRL>C. Sometimes entries exceed the width of the cell containing it, especially if they are text. To widen the cells in a column, place the mouse at the right side of the column heading, until you see the symbol ↔, then left click and move the mouse to the right (moving it to the left makes the column more narrow). If entering a number results in #####, the number is too long for the cell, and the cell should be widened.

Chapter 1

Section 1.4 Entering Data and Evaluating Functions

The cells are identified by the column and the row. For example, the cell B3 is in the second column and the third row.

1. Put headings on the two columns.
2. Fill the inputs in Column A by hand or with a formula for them. The formula $=$A2+1 gives 2 in A3 when ENTER is pressed.
3. Moving the mouse to the lower right corner of A3 until there is a thin "+" sign and dragging the mouse down "fills down" all required entries in column A.
4. Enter the function formula for the function in B2. Use $= 1000*(1.1)\wedge$A2 to represent $S = 1000(1.1)^t$. Pressing ENTER gives the value when $t = 1$.
5. Using Fill Down gives the output for all inputs.

	A	B
1	Year	Future Value
2	1	=1000*(1.1)^(A2)
3	=A2+1	

	A	B
1	Year	Future Value
2	1	1100
3	2	1210
4	3	1331

Section 1.4 Graphing a Function

1. Put headings on the two columns (x and $f(x)$, for example).
2. Fill the inputs (x-values) in Column A by hand or with a formula for them.
3. Enter the formula for the function in B2.
   Use $= 6*$A2 $- 3$ to represent $f(x) = 6x - 3$.
4. Select the cell containing the formula for the function (B2, for example).
5. Move the mouse to the lower right corner until there is a thin "+" sign.
6. Drag the mouse down to the last cell where formula is required, and press ENTER.
7. Highlight the two columns containing the values of x and $f(x)$.
8. Click the *Chart Wizard* icon and then select the *XY(Scatter)* chart with smooth curve option.
9. Click the Next button to get the *Chart Source Data* box. Then click Next to get the *Chart Options* box, and enter your chart title and labels for the x- and y-axes.
10. Click Next, select whether the graph should be within the current worksheet or on another, and click Finish.

	A	B
1	x	$f(x) = 6x - 3$
2	-5	-33
3	-2	-15
4	-1	-9
5	0	-3
6	1	3
7	3	15
8	5	27

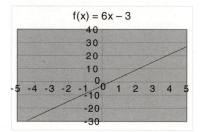

Section 1.5 Solving Systems of Two Equations in Two Variables

1. Write the two equations as linear functions in the form $y = mx + b$.
2. Enter the input variable x in cell A2 and the formula for each of the two equations in cells B2 and C2, respectively.
3. Enter $= B2 - C2$ in cell D2.
4. Use *Tools > Goal Seek*.
5. In the dialog box:
   a. Click the *Set Cell* box and click on the D2 cell.
   b. Enter 0 in the *To Value* box.
   c. Click the *By Changing Cell* box and click on the A2 cell.
6. Click OK in the *Goal Seek* dialog box, getting 0.
7. The x-value of the solution is in cell A2, and the y-value is in both B2 and C2.

The solution of the system

$$\begin{cases} 3x + 2y = 12 \\ 4x - 3y = -1 \end{cases}$$

is found to be $x = 2$, $y = 3$ using *Goal Seek* as follows.

	A	B	C	D
1	x	= 6 − 1.5x	= 1/3 + 4x/3	= y1 − y2
2	1	= 6 − 1.5*A2	= 1/3 + 4*A2/3	= B2 − C2

	A	B	C	D
1	x	= 6 − 1.5x	= 1/3 + 4x/3	= y1 − y2
2	2	3	3	0

Chapter 2

Section 2.1 Solving Quadratic Equations

To solve a quadratic equation of the form
$f(x) = ax^2 + bx + c = 0$:

1. Enter x-values centered around the x-coordinate
   $$x = \frac{-b}{2a}$$ in column A and use the function formula to find the values of $f(x)$ in column B.

2. Graph the function, $f(x) = 2x^2 - 9x + 4$ in this case, and observe where the graph crosses the x-axis ($f(x)$ near 0).

3. Use *Tools>Goal Seek*, entering a cell address with a function value in column B at or near 0, enter the set cell to the value 0, and enter the changing cell.

4. Click OK to find the x-value of the solution in cell A2. The solution may be approximate. The spreadsheet shows $x = 0.5001$, which is an approximation of the exact solution $x = 0.5$.

5. After finding the first solution, repeat the process using a second function value at or near 0. The second solution is $x = 4$ in this case.

	A	B	C	D	E	F
1	x	f(x)=2x^2−9x+4				
2	−1	15				
3	0	4				
4	1	−3				
5	2	−6				
6	3	−5				
7	4	0				
8	5	9				
9						

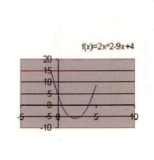

	A	B	C	D	E	F
1	x	f(x)=2x^2−9x+4				
2	−1	15				
3	0	4				
4	0.500001	−7.287E−05				
5	2	−6				
6	3	−5				
7	4	0				
8	5	9				
9						

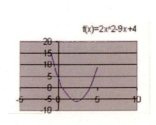

Sections 2.2 and 2.4 Graphing Polynomial Functions

To graph a polynomial function:

1. Use the function to create a table containing values for x and $f(x)$.
2. Highlight the two columns containing the values of x and $f(x)$.
3. Click the *Chart Wizard* icon and then select the *XY(Scatter)* chart with smooth curve option.
4. Click the Next button to get the *Chart Source Data* box. Then click Next to get the *Chart Options* box, and enter your chart title and labels for the x- and y-axes.
5. Click Next, select whether the graph should be within the current worksheet or on another, and click Finish.

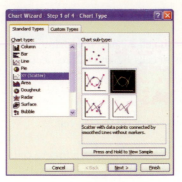

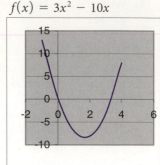

$f(x) = 3x^2 - 10x$

Section 2.4 Graphing Rational Functions

An Excel graph will connect all points corresponding to values in the table, so if the function you are graphing is undefined for some x-value a, enter x-values near this value and leave (or make) the corresponding $f(a)$ cell blank.

To graph $f(x) = \dfrac{1}{1-x}$, which is undefined at $x = 1$:

1. Generate a table for x-values from -1 to 3, with extra values near $x = 1$.
2. Generate the values of $f(x)$, and leave a blank cell for the $f(x)$ value for $x = 1$.
3. Select the table and plot the graph using *Chart Wizard*.

x	f(x)
−1	−0.5
−0.5	−0.6667
0	−1
0.5	−2
0.75	−4
0.9	−10
1	
1.1	10
1.25	4
1.5	2
2	1
2.5	0.666667
3	0.5

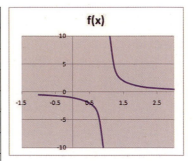

Section 2.5 Modeling

To create a scatter plot of data:

1. Enter the inputs (x-values) in Column A and the outputs (y-values) in Column B.
2. Highlight the two columns and use *Chart Wizard* to plot the points.
3. In Step 3 of the *Chart Wizard*, you can add the title and x- and y-axis labels.

	A	B	C	D	E	F	G	H	I
1									
2	Time (sec)	Speed (mph)							
3	1.7	30							
4	2.4	40							
5	3.5	50							
6	4.3	60							
7	5.5	70							
8	7	80							
9	8.5	90							
10	10.2	100							
11									

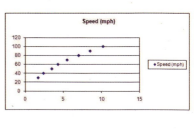

To find the equation of a line or curve that best fits a given set of data points:

1. Place the scatter plot of the data in the worksheet.
2. Single-click on the scatter plot in the workbook.
3. From the *Chart* menu select *Add Trendline*.
4. Select the regression type that appears to be the best function fit for the scatter plot. [Note: If *Polynomial* is selected, choose the appropriate Order (degree).]
5. Click the *Options* tab and check the *Display equation on chart* box.
6. Click OK and the graph of the selected best-fit function will appear along with its equation.

The power function that models Corvette acceleration follows.

	A	B	C	D	E	F	G	H	I
1									
2	Time (sec)	Speed (mph)			**Corvette Acceleration**				
3	1.7	30							
4	2.4	40							
5	3.5	50							
6	4.3	60							
7	5.5	70							
8	7	80							
9	8.5	90							
10	10.2	100							
11									

Corvette Acceleration chart, Speed (mph), with trendline $y = 21.875x^{0.6663}$

The speed of the Corvette is given by $y = 21.875x^{0.6663}$ where x is the time in seconds.

Chapter 3
Section 3.1 Operations with Matrices

Operations for 3×3 matrices can be used for other orders.

1. Type a name A in A1 to identify the first matrix.
2. Enter the matrix elements of matrix A in the cells B1:D3.
3. Type a name B in A5 to identify the second matrix.
4. Enter the matrix elements of matrix B in the cells B5:D7.
5. Type a name A+B in A9 to indicate the matrix sum.
6. Type the formula "=B1+B5" in B9 and press ENTER.
7. Use Fill Across to copy this formula across the row to C9 and D9.
8. Use Fill Down to copy the row B9:D9 to B11:D11, which gives the sum.
9. To subtract the matrices, change the formula in B9 to "=B1 − B5" and proceed as with addition.

	A	B	C	D
1	A	1	2	3
2		4	5	6
3		7	8	9
4				
5	B	−2	−4	3
6		1	4	−5
7		3	6	−1
8				
9	A+B	−1	−2	6
10		5	9	1
11		10	14	8

Section 3.2 Multiplying Two Matrices

Steps for two 3×3 matrices:

1. Enter the names and elements of the matrices.
2. Enter the name AxB in A9 to indicate the matrix product.
3. Select a range of cells that is the correct size to contain the product (B9:D11 in this case).
4. Type "=mmult(" in the formula bar, and then select the cells containing the elements of matrix A (B1:D3).
5. Stay in the formula bar, type a comma and select the matrix B elements (B5:D7), and close the parentheses.
6. Hold the CTRL and SHIFT keys down and press ENTER, giving the product.

	A	B	C	D
1	A	1	2	3
2		4	5	6
3		7	8	9
4				
5	B	−2	−4	3
6		1	4	−5
7		3	6	−1
8				
9	AxB	=mmult(B1:D3		
10		MMULT(**array1,**		
11		array2)		

B9	*fx*	{(=MMULT(B1:D3,B5:D7)}			
	A	B	C	D	E
1	A	1	2	3	
2		4	5	6	
3		7	8	9	
4					
5	B	−2	−4	3	
6		1	4	−5	
7		3	6	−1	
8					
9	AxB	9	22	−10	
10		15	40	−19	
11		21	58	−28	

Section 3.4 Finding the Inverse of a Matrix

Steps for a 3×3 matrix:

1. Enter the name A in A1 and the elements of the matrix in B1:D3 as above.
2. Enter the name "Inverse(A)" in A5 and select a range of cells that is the correct size to contain the inverse [(B5:D7) in this case].
3. Enter "=minverse(", select matrix A (B1:D3), and close the parentheses.
4. Hold the CTRL and SHIFT keys down and press ENTER, getting the inverse.

SUM	fx	=minverse(B1:D3)		
	A	B	C	D
1	A	2	1	1
2		1	2	0
3		2	0	1
4				
5	inverse(A)	=minverse(B1:D3)		
6				
7				

B5	fx	{=MINVERSE(B1:D3)}		
	A	B	C	D
1	A	2	1	1
2		1	2	0
3		2	0	1
4				
5	inverse(A)	−2	1	2
6		1	0	−1
7		4	−2	−3

Section 3.4 Solving Systems of Linear Equations with Matrix Inverses

A system of linear equations can be solved by multiplying the matrix containing the augment by the inverse of the coefficient matrix. The steps used to solve a 3×3 system follow.

1. Enter the coefficient matrix A in B1:D3.
2. Enter the name "inverse(A)" in A5 and compute the inverse of A in B5:D7.
3. Enter B in cell A9 and enter the augment matrix in B9:B11.
4. Enter X in A13 and select the cells B13:B15.
5. In the formula bar, type "=mmult(", then select matrix inverse(A) in B5:D7, type a comma, select matrix B in B9:B11, and close the parentheses.
6. Hold the CTRL and SHIFT keys down and press ENTER, getting the solution.
7. Matrix X gives the solution.

The system $\begin{cases} 2x + y + z = 8 \\ x + 2y \quad\;\; = 6 \\ 2x \quad\;\;\; + z = 5 \end{cases}$ is solved as follows.

SUM	fx	=mmult(B5:D7,B9:B11)			
	A	B	C	D	E
1	A	2	1	1	
2		1	2	0	
3		2	0	1	
4					
5	inverse(A)	−2	1	2	
6		1	0	−1	
7		4	−2	−3	
8					
9	B	8			
10		6			
11		5			
12					
13	X	=mmult(B5:D7,B9:B11)			
14					
15					

B13	fx	{(=MMULT(B5:D7,B9:B11)}			
	A	B	C	D	E
1	A	2	1	1	
2		1	2	0	
3		2	0	1	
4					
5	inverse(A)	−2	1	2	
6		1	0	−1	
7		4	−2	−3	
8					
9	B	8			
10		6			
11		5			
12					
13	X	0			
14		3			
15		5			

The solution is $x = 0$, $y = 3$, $z = 5$.

Chapter 4

Section 4.3 Linear Programming

Maximize $f = 6x + 13y + 20z$ subject to the constraints

$$\begin{cases} 5x + 7y + 10z \leq 90,000 \\ x + 3y + 4z \leq 30,000 \\ x + y + z \leq 9000 \\ x \geq 0, y \geq 0, z \geq 0 \end{cases}$$

1. On a blank spreadsheet, type a heading in cell A1, followed by the variable descriptions in cells A3–A5 and the initial values (zeros) in cells B3–B5.

2. a. Enter the heading "Objective" in cell A7.
   b. Enter a description of the objective in cell A9 and the formula for the objective function in B9. The formula is =6*B3+13*B4+20*B5.

3. a. Type in the heading "Constraints" in A11 and descriptive labels in A13–A15.
   b. Enter the left side of the constraint inequalities in B13–B15 and the maximums from the right side in C13–C15.

4. Select *Solver* under the *Tools* menu. A dialog box will appear.

5. a. Click the *Set Target Cell* box and C9 (containing the formula for the objective function).
   b. Check the button *Max* for maximization.
   c. Click the *By Changing Cells* box and select cells B3–B5.

6. a. Click the *Subject to Constraints* entry box. Press the Add button to add the first constraint.
   b. Click the left entry box and click cell B13 (containing the formula for the first constraint).
   c. Set the middle entry box to <=.
   d. Click the right entry box and C13 to enter the constraint.
   e. Click Add and repeat the Steps 6b–6d for the remaining constraints.
   f. Click in the left entry box for the constraints and select the variables in B3–B5. Set the middle entry to >=, and type 0 in the right entry box.

7. Click Solve in the *Solver* dialog box. A dialog box states that *Solver* found a solution. To see the *Solver* results, click *Keep Solver Solution* and also select Answer.

8. Go back to the spreadsheet. The new values in B3–B5 are the values of the variables that give the maximum, and the value in B9 is the maximum value of the objective function.

In *Solver*, minimization of an objective function is handled exactly the same as maximization, except min is checked and the inequality signs are ≥.

When mixed constraints are used, simply enter them with "mixed" inequalities.

	A	B	C
1	**Variables**		
2			
3	# small calculators (x)	0	
4	# medium calculators (y)	0	
5	# large calculators (z)	0	
6			
7	**Objective**		
8			
9	Maximize profit	=6*B3+13*B4+20*B5	
10			
11	**Constraints**		
12		Amount used	Maximum
13	Circuit components	=5*B3+7*B4+10*B5	90000
14	Labor	=B3+3*B4+4*B5	30000
15	Cases	=B3+B4+B5	9000

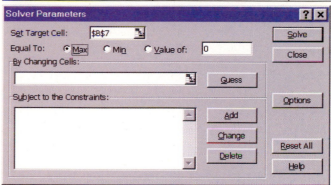

Solver Parameters

Set Target Cell: B7

Equal To: ⊙ Max ○ Min ○ Value of: 0

By Changing Cells:

Subject to the Constraints:

Solve   Close   Guess   Options   Add   Change   Delete   Reset All   Help

	A	B	C
1	Variables		
2			
3	# small calculators (x)	2000	
4	# medium calculators (y)	0	
5	# large calculators (z)	7000	
6			
7	Objective		
8			
9	Maximize profit	152000	
10			
11	Constraints		
12		Amount used	Maximum
13	Circuit components	80000	90000
14	Labor	30000	30000
15	Cases	9000	9000

Chapter 5

Section 5.1 Graphing Exponential Functions

Exponential functions are entered into Excel differently for base e than for bases other than e.

A. To graph $y = a^x$, use the formula $=a\hat{\ }x$.
B. To graph $y = e^x$, use the formula $=\exp(x)$.

To graph $y = 1.5^x$ and $y = e^x$ on the same axes:

1. Type x in cell A1 and numbers centered at 0 in Column A.
2. Type $f(x)$ in cell B1, enter the formula $=1.5\hat{\ }A2$ in cell B2 and fill down.
3. Type $g(x)$ in cell C1, enter the formula $=\exp(A2)$ in cell C2 and fill down.
4. Select the entire table and use *Chart Wizard* to graph as described in Section 1.4.

	A	B	C
1	x	$f(x)$	$g(x)$
2	−2	−3	0.135335
3	−1.5	−2.25	0.22313
4	−1	−1.5	0.367879
5	−0.5	−0.75	0.606531
6	0	0	1
7	0.5	0.75	1.648721
8	1	1.5	2.718282
9	1.5	2.25	4.481689
10	2	3	7.389056
11	2.5	3.75	12.18249
12	3	4.5	20.08554

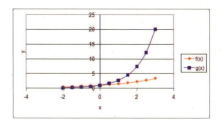

Section 5.1 Modeling with Exponential Functions

To model data with an exponential function:

1. Create a scatter plot for the data.
2. From the *Chart* menu, choose *Add Trendline*.
3. Check *exponential regression* type since that function type appears to be the best fit for the scatter plot.
4. Click the *Options* tab on this box and select *Display equation* on chart box.
5. Click Next and Finish to see the equation and graph.

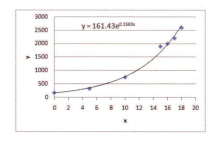

The exponential model for the following data is shown.

x	0	5	10	15	16	17	18
y	170	325	750	1900	2000	2200	2600

Section 5.2 Graphing Base *e* and Base 10
Logarithmic Functions

1. Create a table of values for *x*-values with $x > 0$ to reflect the function's domain.
2. For $y = \ln(x)$, use the formula $=\ln(x)$.
3. For $y = \log(x)$, use the formula $=\log 10(x)$.
4. Select the entire table and use *Chart Wizard* to graph.

	A	B
1	x	$f(x)=\ln(x)$
2	0.5	−0.6931472
3	0.9	−0.1053605
4	1	0
5	2	0.6931472
6	3	1.0986123
7	4	1.3862944
8	5	1.6094379
9	6	1.7917595
10	7	1.9459101
11	8	2.0894415
12	9	2.1972246
13	10	2.3025851
14	11	2.3978953

	A	B
1	x	$f(x)=\log(x)$
2	0.5	−0.30103
3	0.9	−0.0457575
4	1	0
5	2	0.30103
6	3	0.47712125
7	4	0.60205999
8	5	0.69897
9	6	0.77815125
10	7	0.84509804
11	8	0.90308999
12	9	0.95424251
13	10	1
14	11	1.04139269

$f(x)=\ln(x)$

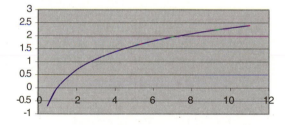

$f(x)=\log(x)$

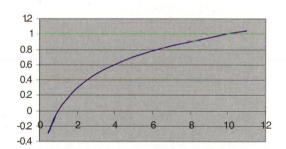

Section 5.2 Modeling Logarithmic Functions

1. Create a scatter plot for the data.
2. In the *Chart* menu, choose *Add Trendline* and click *logarithmic regression*.
3. Click the *options* tab, select *Display equation* and click *Next* and *Finish*.

x	*y*
5	0.44
10	2.31
20	3.79
30	4.92

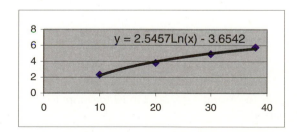

Section 5.2 Graphing Logarithmic Functions with Other Bases

1. The graph of a logarithm with any base b can be created by entering the formula $=\log(x, b)$.
2. Proceed as described above for graphing base e and base 10 logarithms.

	A	B
1	x	$f(x)=\log(x,7)$
2	0.5	−0.356207187
3	0.9	−0.054144594
4	1	0
5	2	0.356207187
6	3	0.564575034
7	4	0.712414374
8	5	0.827087475
9	6	0.920782221
10	7	1
11	8	1.068621561
12	9	1.129150068
13	10	1.183294662
14	11	1.1232274406

The graph of $f(x)=\log_7 x$ is shown below.

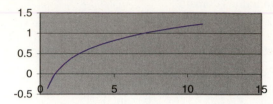

Chapter 6
Section 6.2 Finding the Future Value of a Lump Sum

To find the future value of a lump-sum investment:

1. Type the headings in Row 1, and enter their values (with the interest rate as a decimal) in Row 2.
2. Type the formula $=D2/E2$ in F2 to compute the rate per period.
3. Type the heading Future Value in A4.
4. In cell B4, type the formula $=\text{fv}(F2,B2,C2,A2,0)$ to compute the future value.

This spreadsheet gives the future value of an investment of $10,000 for 17 years at 9.8%, compounded quarterly.

Principal	Number of Periods	Payment	Annual Rate	Periods per year	Periodic Rate
10000	68	0	0.098	4	0.0245
Future Value	($51,857.73)				

Section 6.3 Finding the Future Value of an Annuity

To find the future value of an ordinary annuity:

1. Type the headings in Row 1, and enter their values (with the interest rate as a decimal) in Row 2.
2. Type the formula $=D2/E2$ in F2 to compute the rate per period.
3. Type the heading Future Value in A4.
4. In cell B4, type the formula $=\text{fv}(F2,B2,C2,A2,0)$ to compute the future value.
   (The 0 indicates that the payments are made at the end of each period.)

For annuities due, use $=\text{fv}(F2,B2,C2,A2,1)$.
The payments are made at the beginning of each period.

This spreadsheet gives the future value of an ordinary annuity of $200 deposited at the end of each quarter for $2\frac{1}{4}$ years, with interest at 4%, compounded quarterly.

Principal	Number of Periods	Payment	Annual Rate	Periods per year	Periodic Rate
0	9	200	0.04	4	0.01
Ordinary annuity					
Future Value	($1,873.71)				
Annuities due					
Future Value	($1,892.44)				

Section 6.4 Finding the Present Value of an Annuity

To find the present value of an ordinary annuity:

1. Type the headings in Row 1, and enter their values (with the interest rate as a decimal) in Row 2.
2. Enter the formula =D2/E2 in F2 to compute the rate per period.
3. Type the heading Present Value in A4.
4. In cell B4, type the formula =pv(F2,B2,C2,A2,0) to compute the present value. (The 0 indicates that the payments are made at the end of each period.)

For annuities due, use =pv(F2,B2,C2,A2,1).
The payments are made at the beginning of each period.

This spreadsheet gives both the present value of an ordinary annuity and of an annuity due if you pay $1000 per month for 16 years, with interest at 9%, compounded monthly.

Future Value	Number of Periods	Payment	Annual Rate	Periods per year	Periodic Rate
0	192	1000	0.09	12	0.0075
Ordinary annuity					
Present Value	$101,572.77				
Annui-ties due					
Present value	$102,334.56				

Section 6.5 Finding Payments to Amortize a Loan

To find the periodic payment to pay off a loan:

1. Type the headings in Row 1 and their values (with the interest rate as a decimal) in Row 2.
2. Enter the formula =D2/E2 in F2 to compute the rate per period.
3. Type the heading Payment in A4.
4. In cell B4, type the formula =Pmt(F2,B2,A2, C2,0) to compute the payment.

This spreadsheet gives the annual payment of a loan of $10,000 over 5 years when interest is 10% per year.

The parentheses indicates a payment out.

Loan Amount	Number of Periods	Future Value	Annual Rate	Periods per year	Periodic Rate
10000	5	0	0.1	1	0.1
Payment	($2,637.97)				

Chapter 8
Section 8.1 Binomial Probabilities

1. Type headings in cells A1:A3 and their respective values in cells B1:B3.
2. Use the function =binomdist(B1,B2,B3,cumulative) where
   - B1 is the number of successes.
   - B2 is the number of independent trials.
   - B3 is the probability of success in each trial.
   - True replaces cumulative if a cumulative probability is sought; it is replaced by false otherwise.
   - The probability of exactly 3 heads in 6 tosses is found by evaluating =binomdist(B1,B2,B3,false) in B4.
   - The probability of 3 or fewer heads in 6 tosses is found by evaluating =binomdist(B1,B2,B3,true) in B5.

This spreadsheet gives the probability of 3 heads in 6 tosses of a fair coin and the probability of 3 or fewer heads in 6 tosses.

	A	B
1	Number of successes	3
2	Number of trials	6
3	Probability of success	.5
4	Probability of 3 successes	.3125
5	Probability of 3 or fewer successes	.65625

Section 8.2 Bar Graphs

To construct a bar graph for the given table of test scores:

1. Copy the entries of the table to cells A2:B6.
2. Select the range A2:B6.
3. Click the *Chart Wizard* icon.
4. Select the graph option with the first sub-type.
5. Click Next.
6. Click Next through Steps 2–4. Note that you can add a title in Step 3.

	A	B
1	Grade Range	Frequency
2	90–100	2
3	80–89	5
4	70–79	3
5	60–69	3
6	0–59	2

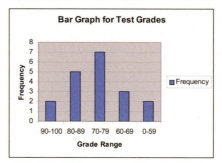

Section 8.2 Finding the Mean, Standard Deviation, and Median of Raw Data

To find the mean, standard deviation, and median of a raw data set:

1. Enter the data in Row 1 (cells A1:L1).
2. Type the heading Mean in cell A3.
3. Type the formula =average(A1:L1) in cell B4.
4. Type the heading Standard Deviation in cell A4.
5. In cell B4, type the formula =stdev(A1:L1).
6. In cell A5, type the heading Median.
7. In cell B5, type the formula =median(A1:L1).

The mean, standard deviation, and median for the data 1, 1, 1, 3, 3, 4, 4, 5, 6, 6, 7, 7 is shown below.

	A	B	C	D	E	F	G	H	I	J	K	L
1	1	1	1	3	3	4	4	5	6	6	7	7
2												
3	Mean	4										
4	Standard Deviation	2.2563										
5	Median	4										

Section 8.2 Finding the Mean and Standard Deviation of Grouped Data

To find the mean:

Grade Range	Class Marks	Frequency
90–100	95	3
80–89	84.5	4
70–79	74.5	7
60–69	64.5	0
50–59	54.5	2

1. Enter the data and headings in the cells A1:C6.
2. In D1, type the heading Class mark*frequency.
3. In D2, type the formula =B2*C2.
4. Copy the formula in D2 to D3:D6.
5. In B7, type the heading Total.
6. In cell C7, type the formula for the total frequencies, =sum(C2:C6).
7. In cell D7, type the formula for the total, =sum(D2:D6).
8. In cell A8, type the heading Mean.
9. In cell A9, type in the formula =D7/C7.

To find the standard deviation:

10. In cell E1, type in the heading freq *(x − x _ mean)^2.
11. In cell E2, type the formula =C2*(B2-A9). (The A9 gives the value in A9; the reference doesn't change as we fill down.)
12. Copy the formula in E2 to E3:E6.
13. In cell E7, type the formula =sum(E2:E6).
14. In cell A10 type the heading Standard Deviation.
15. In cell A11, type the formula =sqrt(E7/(C7-1)).

	A	B	C	D	E
1	Grade Range	Class Marks	Frequency	Class mark*frequency	Freq*(x-x_mean)^2
2	90-100	95	3	285	832.2919922
3	80-89	84.5	4	338	151.5976563
4	70-79	74.5	7	521.5	103.4208984
5	60-69	64.5	0	0	0
6	50-59	54.5	2	109	1137.048828
7		Total	16	1253.5	2224.359375
8	Mean				
9	78.344				
10	Standard Deviation				
11	12.177				

Section 8.4 Calculating Normal Probabilities

To calculate normal probabilities:

1. Type headings in A1:A4 and their respective values in cells B1:B4.
2. To find the probability that a score X is less than the x1 value in B3, enter the formula =normdist(B3,B1,B2,true) in cell B5.
3. To find the probability that X is less than the x2 value in B4, enter the formula =normdist(B4,B1,B2,true) in cell B6.
4. To find the probability that a score X is more than the value in B3 and less than the x2 value in B4, enter the formula =B6-B5 in cell B7.

Entries in B5, B6, and B7 give the probabilities of a score X being less than 100, less than 115, and between 100 and 115, respectively, when the mean is 100 and the standard deviation is 15.

	A	B
1	Mean	100
2	Standard Deviation	15
3	x1	100
4	x2	115
5	Pr(X<x1)	0.5
6	Pr(X<x2)	0.841345
7	Pr(x1<X<x2)	0.341345

Chapter 9
Sections 9.1–9.2 Evaluating Limits

To evaluate $\lim\limits_{x \to c} f(x)$:

1. Make a table of values for $f(x)$ near $x = c$. Include values on both sides of $x = c$.
2. Use the table of values to predict the limit (or that the limit does not exist).

    Note: All limit evaluations with Excel use appropriate tables of values of $f(x)$. This is true when $f(x)$ is piecewise defined and for limits as $\to \infty$.

Find $\lim\limits_{x \to 2} \dfrac{x^2 - 4}{x - 2}$.

	A	B	C	D
	x	**f(x)**	**x**	**f(x)**
1	x	f(x)	x	f(x)
2	2.1	4.1	1.9	3.9
3	2.05	4.05	1.95	3.95
4	2.01	4.01	1.99	3.99
5	2.001	4.001	1.999	3.999

The tables suggest that $\lim\limits_{x \to 2} \dfrac{x^2 - 4}{x - 2} = 4$.

Sections 9.3–9.7 Approximating Derivatives

To approximate $f'(c)$:

1. Numerically investigate the limit in the definition of derivative:

$$f'(x) = \lim_{h \to 0} \frac{f(x + h) - f(x)}{h}$$

2. Use the given $f(x)$ and $x = c$ to create a table of values for h near 0 (and on both sides of $h = 0$).

Note that rows 5 and 6 have the values of h closest to 0.

Investigate $f'(1)$ for $f(x) = x^3$.

	A	B	C	D	E
1	**h**	**1+h**	**f(1)**	**f(1+h)**	**(f(1+h)-f(1))/h**
2	0.1	1.1	1	1.331	3.31
3	0.01	1.01	1	1.030301	3.0301
4	0.001	1.001	1	1.003003	3.003001
5	0.0001	1.0001	1	1.0003	3.00030001
6	−0.0001	0.9999	1	0.9997	2.99970001
7	−0.001	0.999	1	0.997003	2.997001
8	−0.01	0.99	1	0.970299	2.9701
9	−0.1	0.9	1	0.729	2.71

The table suggests that $f'(1) = 3$, which is the actual value.
Note: Excel has no built-in derivative approximation tool.

Chapter 10
Section 10.1 Relative Maxima and Minima

1. Make a table with columns for x-values, the function, and the derivative.
2. Extend the table to include x-values to the left and to the right of all critical values.
3. Use the signs of the values of the derivative to determine whether f is increasing or decreasing around the critical values, and thus to classify the critical values as relative maxima, relative minima, or horizontal points of inflection. You may want to graph the function to confirm your conclusions.

The spreadsheet shows that the relative maxima or minima of $f(x) = x^2$ is 0 at $x = 0$. Note that the derivative is $f'(x) = 2x$.

	A	B	C
1	x	f(x)	f'(x)
2	−2	4	−4
3	−1.5	2.25	−3
4	−1	1	−2
5	−0.5	0.25	−1
6	0	0	0
7	0.5	0.25	1
8	1	1	2
9	1.5	2.25	3
10	2	4	4

Section 10.2 Exploring f, f', f'' Relationships

To explore relationships among the graphs of a function and its derivatives:

1. Find functions for f' and f''.
2. Graph all three functions on the same plot.
   - Notice that f increases when f' is above the x-axis ($+$) and decreases when f' is below the x-axis ($-$).
   - Notice that f is concave up when f'' is above the x-axis ($+$) and is concave down when f'' is below the x-axis ($-$).

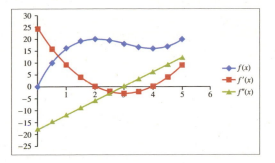

Let $f(x) = x^3 - 9x^2 + 24x$. Graph f, f', and f'' on the interval $[0, 5]$ to explore the relationships among these functions.

	A	B	C	D
	x	$f(x)$	$f'(x)$	$f''(x)$
1	x	$f(x)$	$f'(x)$	$f''(x)$
2	0	0	24	-18
3	0.5	9.875	15.75	-15
4	1	16	9	-12
5	1.5	19.125	3.75	-9
6	2	20	0	-6
7	2.5	19.375	-2.25	-3
8	3	18	-3	0
9	3.5	16.625	-2.25	3
10	4	16	0	6
11	4.5	16.875	3.75	9
12	5	20	9	12

Sections 10.3–10.4 Finding Optimal Values

To find the optimal value of a function when the goal is not to produce a graph:

1. Set up a spreadsheet that identifies the variable and the function whose optimal value is sought.
2. Choose *Tools>Solver*. Then, in the *Solver* dialog box,
   a. Set the *Target Cell* as that of the objective function.
   b. Check *Max* or *Min* according to your goal.
   c. Set the *Changing Cells* to reference the variable.
3. Click on the *Options* box. Make sure "Assume Linear Model" is not checked. Then click OK.
4. Click *Solve* in the *Solver* dialog box. You will get a dialog box stating that Solver found a solution. Save the solution if desired, then click OK.
5. The cells containing the variable and the function should now contain the optimal values.

Minimize area $A = x^2 + \dfrac{160}{x}$ for $x > 0$.

	A	B
1	**Variable**	
2		
3	x length of base	1
4		
5	**Objective**	
6		
7	Minimize Area	=B3^2+160/B3

	A	B
1	**Variable**	
2		
3	x length of base	4.3089
4		
5	**Objective**	
6		
7	Minimize Area	55.699

The function is minimized for $x = 4.3089$ and the minimum value is $A = 55.699$.

Chapter 13

Section 13.1 Approximating Definite Integrals Using Rectangles

To approximate the area under the graph of $y = f(x)$ and above the x-axis from a to b, using left-hand endpoints:

1. Divide $b - a$ by the number of rectangles to get the width of each rectangle.
2. Enter the x-values in Column A and the function values in Column B.
3. Add a third column with the heading Δx in C1, and the rectangle width in cells C2:C6.
4. Add a fourth column with the heading Rectangle Area in D1 and, in D2, use the formula =B2*C2 to get the area of the first rectangle.
5. Copy this formula down to cells D3:D6.
6. In cell C7, type the heading Total Area, and in D7 enter the formula =sum(D2:D6) and press ENTER.

The approximate area under the graph of $f(x) = x^2$ from $x = 1$ to $x = 2$, with 5 rectangles, is found below. In the example, the width is 0.2 and the left-hand endpoints are 1.0, 1.2, 1.4, 1.6, 1.8.

x	$f(x)$	Δx	Rectangle Area
1	1	0.2	0.2
1.2	1.44	0.2	0.288
1.4	1.96	0.2	0.392
1.6	2.56	0.2	0.512
1.8	3.24	0.2	0.648
		Total Area	2.04

Chapter 14

Section 14.1 Graphs of Functions of Two Variables

To create a surface plot for a function of two variables:

1. a. Generate appropriate x-values beginning in B1 and continuing *across*.
    b. Generate appropriate y-values beginning in A2 and continuing *down*.
2. Generate values for the function that correspond to the points (x, y) from Step 1 as follows:
    In cell B2, enter the function formula with B$1 used to represent x and $A2 to represent y. (See the online Excel Guide for additional information about the role and use of the $ in this step.) Then use fill down and fill across to complete the table.
3. Select the entire table of values. Click the *Chart Wizard* and choose *Surface in the Chart* menu.
4. Annotate the graph and click *Finish* to create the surface plot. You can move and view the surface from a different perspective by clicking into the resulting graph.

Let $f(x, y) = 10 - x^2 - y^2$. Plot the graph of this function for both x and y in the interval $[-2, 2]$.

	A	B	C	D	E	F	G	H	I	J
1		-2	-1.5	-1	-0.5	0	0.5	1	1.5	2
2	-2									
3	-1.5									
4	-1									
5	-0.5									
6	0									
7	0.5									
8	1									
9	1.5									
10	2									

	A	B	C	D	E	F	G	H	I	J
1		-2	-1.5	-1	-0.5	0	0.5	1	1.5	2
2	-2	2	3.75	5	5.75	6	5.75	5	3.75	2
3	-1.5	3.75	5.5	6.75	7.5	7.75	7.5	6.75	5.5	3.75
4	-1	5	6.75	8	8.75	9	8.75	8	6.75	5
5	-0.5	5.75	7.5	8.75	9.5	9.75	9.5	8.75	7.5	5.75
6	0	6	7.75	9	9.75	10	9.75	9	7.75	6
7	0.5	5.75	7.5	8.75	9.5	9.75	9.5	8.75	7.5	5.75
8	1	5	6.75	8	8.75	9	8.75	8	6.75	5
9	1.5	3.75	5.5	6.75	7.5	7.75	7.5	6.75	5.5	3.75
10	2	2	3.75	5	5.75	6	5.75	5	3.75	2

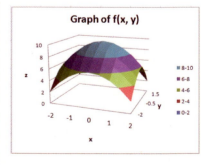

Graph of f(x, y)

Section 14.5 Constrained Optimization and Lagrange Multipliers

To solve a constrained optimization problem:

1. Set up the problem in Excel.
2. Choose *Tools>Solver* and do the following:
   - Choose the objective function as the *Target Cell*.
   - Check *Max* or *Min* depending on the problem.
   - Choose the cells representing the variables for the *By Changing Cells* box.
   - Click on the *Constraints* box and press Add. Then enter the constraint equations.
3. Click on the *Options* box and make sure that "Assume Linear Model" is *not* checked. Then click OK.
4. Click *Solve* in the *Solver* dialog box. Then click OK to solve.

Maximize $p = 600l^{2/3}k^{1/3}$ subject to $40l + 100k = 3000$.

	A	B	C
1	**Variables**		
2			
3	units labor (l)	0	
4	units capital (k)	0	
5			
6			
7	**Objective**		
8			
9	Maximize production	=600*B3^(2/3)*B4^(1/3)	
10			
11	Constraint		
12		Amount used	Available
13	Cost	=40*B3+100*B4	3000

	A	B	C
1	**Variables**		
2			
3	units labor (l)	50.00000002	
4	units capital (k)	10	
5			
6			
7	**Objective**		
8			
9	Maximize product	17544.10644	
10			
11	**Constraint**		
12		Amount used	Available
13	Cost	3000.000001	3000

Excel Guide Part 2
Excel 2007 and 2010

Except where noted, the directions given will work in both Excel 2007 and Excel 2010.

Excel Worksheet

When you start up Excel by using the instructions for your software and computer, the following screen will appear.

The components of the **spreadsheet** are shown, and the grid shown is called a **worksheet.** You can move to other worksheets by clicking on the tabs at the bottom.

Addresses and Operations

Notice the letters at the top of the columns and the numbers identifying the rows. The cell addresses are given by the column and row; for example, the first cell has address A1. You can move from one cell to another with arrow keys, or you can select a cell with a mouse click. After you enter an entry in a cell, press enter to accept the entry. To edit the contents of a cell, click on the cell and edit the contents in the formula bar at the top. To delete the contents, press the delete key.

The file operations such as "open a new file," "saving a file," and "printing a file" are similar to those in WORD. For example, <CTRL>S saves a file. You can also format a cell entry by selecting it and using menus similar to those in WORD.

Working with Cells

Cell entries, rows containing entries, and columns containing entries can be copied and pasted with the same commands as in WORD. For example, a highlighted cell can be copied with <CTRL>C. Sometimes entries exceed the width of the cell containing it, especially if they are text. To widen the cells in a column, place the mouse at the right side of the column heading, until you see the symbol ↔, then left click and move the mouse to the right (moving it to the left makes the column more narrow). If entering a number results in #####, the number is too long for the cell, and the cell should be widened.

Chapter 1

Section 1.4 Entering Data and Evaluating Functions

The cells are identified by the column and the row. For example, the cell B3 is in the second column and the third row.

1. Put headings on the two columns.
2. Fill the inputs in Column A by hand or with a formula for them. The formula $=$A2$+$1 gives 2 in A3 when ENTER is pressed.
3. Moving the mouse to the lower right corner of A3 until there is a thin "+" sign and dragging the mouse down "fills down" all required entries in column A.
4. Enter the function formula for the function in B2. Use $= 1000*(1.1)\wedge$A2 to represent $S = 1000(1.1)^t$. Pressing ENTER gives the value when $t = 1$.
5. Using Fill Down gives the output for all inputs.

	A	B
1	Year	Future Value
2	1	$=$1000*(1.1)$\wedge$(A2)
3	$=$A2$+$1	

	A	B
1	Year	Future Value
2	1	1100
3	2	1210
4	3	1331

Section 1.4 Graphing a Function

1. Put headings on the two columns (x and $f(x)$, for example).
2. Fill the inputs (x-values) in Column A by hand or with a formula for them.
3. Enter the formula for the function in B2.
   Use $= 6*$A2$- 3$ to represent $f(x) = 6x - 3$.
4. Select the cell containing the formula for the function (B2, for example).
5. Move the mouse to the lower right corner until there is a thin "+" sign.
6. Drag the mouse down to the last cell where formula is required, and press ENTER.
7. Highlight the two columns containing the values of x and $f(x)$.
8. Click the Insert tab and then select the *Scatter* chart type with the smooth curve option.
9. Once the smooth curve option is selected, the chart will appear within the worksheet. The worksheet is now in design mode, where you can change the chart options.
10. Click on the Home tab to get back to the original view of the spreadsheet.

	A	B
1	x	$f(x) = 6x - 3$
2	-5	-33
3	-2	-15
4	-1	-9
5	0	-3
6	1	3
7	3	15
8	5	27

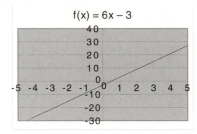

Section 1.5 Solving Systems of Two Equations in Two Variables

1. Write the two equations as linear functions in the form $y = mx + b$.
2. Enter the input variable x in cell A2 and the formula for each of the two equations in cells B2 and C2, respectively.
3. Enter = B2 − C2 in cell D2.
4. Start *Goal Seek* by clicking on the Data tab. Then select *What-If Analysis* in the Data Tools group, and then *Goal Seek*.
5. In the dialog box:
   a. Click the *Set Cell* box and click on the D2 cell.
   b. Enter 0 in the *To Value* box.
   c. Click the *By Changing Cell* box and click on the A2 cell.
6. Click OK in the *Goal Seek* dialog box, getting 0.
7. The x-value of the solution is in cell A2, and the y-value is in both B2 and C2.

The solution of the system

$$\begin{cases} 3x + 2y = 12 \\ 4x - 3y = -1 \end{cases}$$

is found to be $x = 2$, $y = 3$ using *Goal Seek* as follows.

	A	B	C	D
1	x	= 6 − 1.5x	= 1/3 + 4x/3	= y1 − y2
2	1	= 6 − 1.5*A2	= 1/3 + 4*A2/3	= B2 − C2

	A	B	C	D
1	x	= 6 − 1.5x	= 1/3 + 4x/3	= y1 − y2
2	2	3	3	0

Chapter 2

Section 2.1 Solving Quadratic Equations

To solve a quadratic equation of the form
$f(x) = ax^2 + bx + c = 0$

1. Enter x-values centered around the x-coordinate
   $x = \dfrac{-b}{2a}$ in column A and use the function formula to find the values of $f(x)$ in column B.

2. Graph the function, $f(x) = 2x^2 - 9x + 4$ n this case, and observe where the graph crosses the x-axis ($f(x)$ near 0). Insert the graph by selecting *Insert*>*Scatter*>*Smooth curve* option.

3. Use *Data*>*What-If*>*Goal Seek*, entering a cell address with a function value in column B at or near 0, enter the set cell to the value 0, and enter the changing cell.

4. Click OK to find the x-value of the solution in cell A2. The solution may be approximate. The spreadsheet shows $x = 0.5001$, which is an approximation of the exact solution $x = 0.5$.

5. After finding the first solution, repeat the process using a second function value at or near 0. The second solution is $x = 4$ in this case.

	A	B	C	D	E	F
1	x	f(x)=2x^2−9x+4				
2	−1	15				
3	0	4				
4	1	−3				
5	2	−6				
6	3	−5				
7	4	0				
8	5	9				
9						

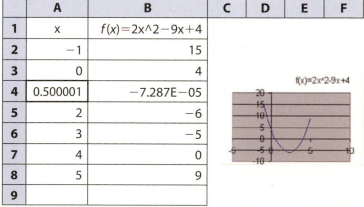

	A	B	C	D	E	F
1	x	f(x)=2x^2−9x+4				
2	−1	15				
3	0	4				
4	0.500001	−7.287E−05				
5	2	−6				
6	3	−5				
7	4	0				
8	5	9				
9						

Sections 2.2 and 2.4 Graphing Polynomial Functions

To graph a polynomial function:

1. Use the function to create a table containing values for x and $f(x)$.
2. Highlight the two columns containing the values of x and $f(x)$.
3. Insert the graph by selecting *Insert > Scatter > Smooth curve* option.
4. Click on the Home tab to get back to the original view of the spreadsheet.

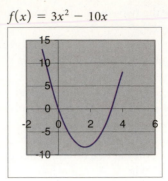

$f(x) = 3x^2 - 10x$

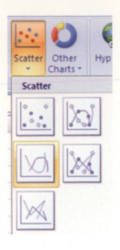

Section 2.4 Graphing Rational Functions

An Excel graph will connect all points corresponding to values in the table, so if the function you are graphing is undefined for some x-value a, enter x-values near this value and leave (or make) the corresponding $f(a)$ cell blank.

To graph $f(x) = \dfrac{1}{1-x}$, which is undefined at $x = 1$:

1. Generate a table for x-values from –1 to 3, with extra values near $x = 1$.
2. Generate the values of $f(x)$, and leave a blank cell for the $f(x)$ value for $x = 1$.
3. Select the table and plot the graph using *Insert > Scatter > Smooth curve* option.

x	f(x)
−1	−0.5
−0.5	−0.6666667
0	−1
0.5	−2
0.75	−4
0.9	−10
1	
1.1	10
1.25	4
1.5	2
2	1
2.5	0.66666667
3	0.5

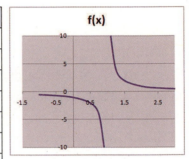

Section 2.5 Modeling

To create a scatter plot of data:

1. Enter the inputs (x-values) in Column A and the outputs (y-values) in Column B.
2. Highlight the two columns and use *Insert > Scatter* and choose the points only option to plot the points.
3. To add titles, click on the icons in the *Charts Layout* group.

	A	B	C	D	E	F	G	H	I
1									
2	Time (sec)	Speed (mph)							
3	1.7	30							
4	2.4	40							
5	3.5	50							
6	4.3	60							
7	5.5	70							
8	7	80							
9	8.5	90							
10	10.2	100							
11									

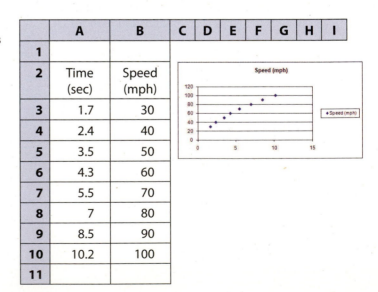

To find the equation of a line or curve that best fits a given set of data points:

1. Single-click on the scatter plot of the data in the worksheet.
2. Right-click on one of the data points.
3. Select *Add Trendline*.
4. Select the function type that appears to best fit and check the box that says *Display Equation on chart*. [Note: If *Polynomial* is selected, choose the appropriate Order (degree).]
5. Close the dialog box and you will see the graph of the selected function that is best fit along with its equation.

The power function that models Corvette acceleration follows.

	A	B	C	D	E	F	G	H	I
1									
2	Time (sec)	Speed (mph)	**Corvette Acceleration**						
3	1.7	30							
4	2.4	40							
5	3.5	50							
6	4.3	60							
7	5.5	70							
8	7	80							
9	8.5	90							
10	10.2	100							
11									

The speed of the Corvette is given by $y = 21.875x^{0.6663}$ where x is the time in seconds.

Chapter 3

Section 3.1 Operations with Matrices

Operations for 3×3 matrices can be used for other orders.

1. Type a name A in A1 to identify the first matrix.
2. Enter the matrix elements of matrix A in the cells B1:D3.
3. Type a name B in A5 to identify the second matrix.
4. Enter the matrix elements of matrix B in the cells B5:D7.
5. Type a name A+B in A9 to indicate the matrix sum.
6. Type the formula "=B1+B5" in B9 and press ENTER.
7. Use Fill Across to copy this formula across the row to C9 and D9.
8. Use Fill Down to copy the row B9:D9 to B11:D11, which gives the sum.
9. To subtract the matrices, change the formula in B9 to "=B1 − B5" and proceed as with addition.

	A	B	C	D
1	A	1	2	3
2			4	5	6
3		7	8	9
4				
5	B	−2	−4	3
6		1	4	−5
7		3	6	−1
8				
9	A+B	−1	−2	6
10		5	9	1
11		10	14	8

Section 3.2 Multiplying Two Matrices

Steps for two 3×3 matrices:

1. Enter the names and elements of the matrices.
2. Enter the name AxB in A9 to indicate the matrix product.
3. Select a range of cells that is the correct size to contain the product (B9:D11 in this case).
4. Type "=mmult(" in the formula bar, and then select the cells containing the elements of matrix A (B1:D3).
5. Stay in the formula bar, type a comma and select the matrix B elements (B5:D7), and close the parentheses.
6. Hold the CTRL and SHIFT keys down and press ENTER, giving the product.

	A	B	C	D
1	A	1	2	3
2		4	5	6
3		7	8	9
4				
5	B	−2	−4	3
6		1	4	−5
7		3	6	−1
8				
9	AxB	=mmult(B1:D3)		
10		MMULT(**array1,**		
11		array2)		

B9	*fx*	{(=MMULT(B1:D3,B5:D7)}			
	A	B	C	D	E
1	A	1	2	3	
2		4	5	6	
3		7	8	9	
4					
5	B	−2	−4	3	
6		1	4	−5	
7		3	6	−1	
8					
9	AxB	9	22	−10	
10		15	40	−19	
11		21	58	−28	

Section 3.4 Finding the Inverse of a Matrix

Steps for a 3×3 matrix:

1. Enter the name A in A1 and the elements of the matrix in B1:D3 as above.
2. Enter the name "Inverse(A)" in A5 and select a range of cells that is the correct size to contain the inverse [(B5:D7) in this case].
3. Enter "=minverse(", select matrix A (B1:D3), and close the parentheses.
4. Hold the CTRL and SHIFT keys down and press ENTER, getting the inverse.

SUM	fx	=minverse(B1:D3)		
	A	**B**	**C**	**D**
1	A	2	1	1
2		1	2	0
3		2	0	1
4				
5	inverse(A)	=minverse(B1:D3)		
6				
7				

Section 3.4 Solving Systems of Linear Equations with Matrix Inverses

A system of linear equations can be solved by multiplying the matrix containing the augment by the inverse of the coefficient matrix. The steps used to solve a 3×3 system follow.

1. Enter the coefficient matrix A in B1:D3.
2. Enter the name "inverse(A)" in A5 and compute the inverse of A in B5:D7.
3. Enter B in cell A9 and enter the augment matrix in B9:B11.
4. Enter X in A13 and select the cells B13:B15.
5. In the formula bar, type "=mmult(", then select matrix inverse(A) in B5:D7, type a comma, select matrix B in B9:B11, and close the parentheses.
6. Hold the CTRL and SHIFT keys down and press ENTER, getting the solution.
7. Matrix X gives the solution.

B5	fx	{=MINVERSE(B1:D3)}		
	A	**B**	**C**	**D**
1	A	2	1	1
2		1	2	0
3		2	0	1
4				
5	inverse(A)	−2	1	2
6		1	0	−1
7		4	−2	−3

The system $\begin{cases} 2x + y + z = 8 \\ x + 2y \quad\;\; = 6 \\ 2x \quad\;\; + z = 5 \end{cases}$ is solved as follows.

SUM	fx	=mmult(B5:D7,B9:B11)		
	A	**B**	**C**	**D**	**E**
1	A	2	1	1	
2		1	2	0	
3		2	0	1	
4					
5	inverse(A)	−2	1	2	
6		1	0	−1	
7		4	−2	−3	
8					
9	B	8			
10		6			
11		5			
12					
13	X	=mmult(B5:D7,B9:B11)			
14					
15					

B13	fx	{(=MMULT(B5:D7,B9:B11)}		
	A	**B**	**C**	**D**	**E**
1	A	2	1	1	
2		1	2	0	
3		2	0	1	
4					
5	inverse(A)	−2	1	2	
6		1	0	−1	
7		4	−2	−3	
8					
9	B	8			
10		6			
11		5			
12					
13	X	0			
14		3			
15		5			

The solution is $x = 0$, $y = 3$, $z = 5$.

Chapter 4

Section 4.3 Linear Programming (Excel 2007)

Maximize $f = 6x + 13y + 20z$ subject to the constraints

$$\begin{cases} 5x + 7y + 10z \le 90{,}000 \\ x + 3y + 4z \le 30{,}000 \\ x + y + z \le 9000 \\ x \ge 0, y \ge 0, z \ge 0 \end{cases}$$

1. On a blank spreadsheet, type a heading in cell A1, followed by the variable descriptions in cells A3–A5 and the initial values (zeros) in cells B3–B5.
2. a. Enter the heading "Objective" in cell A7.
   b. Enter a description of the objective in cell A9 and the formula for the objective function in B9. The formula is =6*B3+13*B4+20*B5.
3. a. Type in the heading "Constraints" in A11 and descriptive labels in A13–A15.
   b. Enter the left side of the constraint inequalities in B13–B15 and the maximums from the right side in C13–C15.
4. Click on the *Data* tab and select *Solver* in the Analysis group. A dialog box will appear.
5. a. Click the *Set Target Cell* box and C9 (containing the formula for the objective function).
   b. Check the button *Max* for maximization.
   c. Click the *By Changing Cells* box and select cells B3–B5.
6. a. Click the *Subject to Constraints* entry box. Press the Add button to add the first constraint.
   b. Click the left entry box and click cell B13 (containing the formula for the first constraint).
   c. Set the middle entry box to <=.
   d. Click the right entry box and C13 to enter the constraint.
   e. Click Add and repeat the Steps 6b–6d for the remaining constraints.
   f. Click in the left entry box for the constraints and select the variables in B3–B5. Set the middle entry to >=, and type 0 in the right entry box.
7. Select the Simplex LP option (this is the default) Click Solve in the *Solver* dialog box.
   A dialog box states that *Solver* found a solution. To see the *Solver* results, click *Keep Solver Solution* and also select Answer.
8. Go back to the spreadsheet. The new values in B3–B5 are the values of the variables that give the maximum, and the value in B9 is the maximum value of the objective function.

In *Solver*, *minimization* of an objective function is handled exactly the same as maximization, except *min* is checked and the inequality signs are $\ge$.

When *mixed constraints* are used, simply enter them with "mixed" inequalities.

	A	B	C
1	**Variables**		
2			
3	# small calculators (x)	0	
4	# medium calculators (y)	0	
5	# large calculators (z)	0	
6			
7	**Objective**		
8			
9	Maximize profit	=6*B3+13*B4+20*B5	
10			
11	**Constraints**		
12		Amount used	Maximum
13	Circuit components	=5*B3+7*B4+10*B5	90000
14	Labor	=B3+3*B4+4*B5	30000
15	Cases	=B3+B4+B5	9000

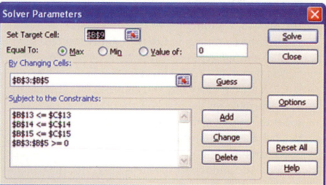

	A	B	C
1	Variables		
2			
3	# small calculators (x)	2000	
4	# medium calculators (y)	0	
5	# large calculators (z)	7000	
6			
7	Objective		
8			
9	Maximize profit	152000	
10			
11	Constraints		
12		Amount used	Maximum
13	Circuit components	80000	90000
14	Labor	30000	30000
15	Cases	9000	9000

Chapter 4

Section 4.3 Linear Programming (Excel 2010)

Maximize $f = 6x + 13y + 20z$ subject to the constraints

$$\begin{cases} 5x + 7y + 10z \leq 90{,}000 \\ x + 3y + 4z \leq 30{,}000 \\ x + y + z \leq 9000 \\ x \geq 0, y \geq 0, z \geq 0 \end{cases}$$

1. On a blank spreadsheet, type a heading in cell A1, followed by the variable descriptions in cells A3–A5 and the initial values (zeros) in cells B3–B5.
2. a. Enter the heading "Objective" in cell A7.
   b. Enter a description of the objective in cell A9 and the formula for the objective function in B9. The formula is =6*B3+13*B4+20*B5.
3. a. Type in the heading "Constraints" in A11 and descriptive labels in A13–A15.
   b. Enter the left side of the constraint inequalities in B13–B15 and the maximums from the right side in C13–C15.
4. Click on the *Data* tab and select *Solver* in the Analysis group. A dialog box will appear.
5. a. Click the *Set Target Cell* box and C9 (containing the formula for the objective function).
   b. Check the button *Max* for maximization.
   c. Click the *By Changing Cells* box and select cells B3–B5.
6. a. Click the *Subject to Constraints* entry box. Press the Add button to add the first constraint.
   b. Click the left entry box and click cell B13 (containing the formula for the first constraint).
   c. Set the middle entry box to <=.
   d. Click the right entry box and C13 to enter the constraint.
   e. Click Add and repeat the Steps 6b–6d for the remaining constraints.
   f. Click in the left entry box for the constraints and select the variables in B3–B5. Set the middle entry to >=, and type 0 in the right entry box.
7. Select the Simplex LP option (this is the default). Click Solve in the *Solver* dialog box. A dialog box states that *Solver* found a solution. To see the *Solver* results, click *Keep Solver Solution* and also select Answer.
8. Go back to the spreadsheet. The new values in B3–B5 are the values of the variables that give the maximum, and the value in B9 is the maximum value of the objective function.

In *Solver*, *minimization* of an objective function is handled exactly the same as maximization, except *min* is checked and the inequality signs are ≥.

   When *mixed constraints* are used, simply enter them with "mixed" inequalities.

	A	B	C
1	**Variables**		
2			
3	# small calculators (x)	0	
4	# medium calculators (y)	0	
5	# large calculators (z)	0	
6			
7	**Objective**		
8			
9	Maximize profit	=6*B3+13*B4+20*B5	
10			
11	**Constraints**		
12		Amount used	Maximum
13	Circuit components	=5*B3+7*B4+10*B5	90000
14	Labor	=B3+3*B4+4*B5	30000
15	Cases	=B3+B4+B5	9000

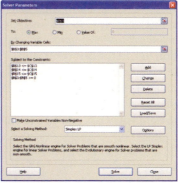

	A	B	C
1	**Variables**		
2			
3	# small calculators (x)	2000	
4	# medium calculators (y)	0	
5	# large calculators (z)	7000	
6			
7	**Objective**		
8			
9	Maximize profit	152000	
10			
11	**Constraints**		
12		Amount used	Maximum
13	Circuit components	80000	90000
14	Labor	30000	30000
15	Cases	9000	9000

Chapter 5
Section 5.1 Graphing Exponential Functions

Exponential functions are entered into Excel differently for base e than for bases other than e.

A. To graph $y = a^x$, use the formula $=a\text{\textasciicircum}x$.
B. To graph $y = e^x$, use the formula $=\exp(x)$.

To graph $y = 1.5^x$ and $y = e^x$ on the same axes:
1. Type x in cell A1 and numbers centered at 0 in Column A.
2. Type $f(x)$ in cell B1, enter the formula $=1.5\text{\textasciicircum}A2$ in cell B2 and fill down.
3. Type $g(x)$ in cell C1, enter the formula $=\exp(A2)$ in cell C2 and fill down.
4. Select the entire table and graph using *Insert>Scatter>Smooth curve with markers* option to graph the functions.

	A	B	C
	x	*f(x)*	*g(x)*
1	*x*	*f(x)*	*g(x)*
2	−2	−3	0.135335
3	−1.5	−2.25	0.22313
4	−1	−1.5	0.367879
5	−0.5	−0.75	0.606531
6	0	0	1
7	0.5	0.75	1.648721
8	1	1.5	2.718282
9	1.5	2.25	4.481689
10	2	3	7.389056
11	2.5	3.75	12.18249
12	3	4.5	20.08554

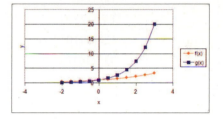

Section 5.1 Modeling with Exponential Functions

To model data with an exponential function:

1. Create a scatter plot for the data using *Insert>Scatter* and the points only option.
2. Single click on the chart and right click on one of the data points. Choose *Add Trendline*.
3. Check *Exponential* for the trendline type if that function type appears to be the best fit for the scatter plot, and check the box to Display the equation on the graph.
4. Close the dialog box to see the graph of the function and its equation.

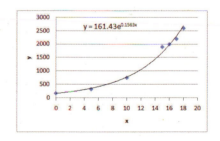

The exponential model for the following data is shown.

x	0	5	10	15	16	17	18
y	170	325	750	1900	2000	2200	2600

Section 5.2 Graphing Base *e* and Base 10 Logarithmic Functions

1. Create a table of values for *x*-values with $x > 0$ to reflect the function's domain.
2. For $y = \ln(x)$, use the formula $=\ln(x)$.
3. For $y = \log(x)$, use the formula $=\log10(x)$.
4. Select the entire table and graph using *Insert > Scatter > Smooth curve* option.

	A	B
1	x	$f(x)=\ln(x)$
2	0.5	−0.6931472
3	0.9	−0.1053605
4	1	0
5	2	0.6931472
6	3	1.0986123
7	4	1.3862944
8	5	1.6094379
9	6	1.7917595
10	7	1.9459101
11	8	2.0894415
12	9	2.1972246
13	10	2.3025851
14	11	2.3978953

	A	B
1	x	$f(x)=\log(x)$
2	0.5	−0.30103
3	0.9	−0.0457575
4	1	0
5	2	0.30103
6	3	0.47712125
7	4	0.60205999
8	5	0.69897
9	6	0.77815125
10	7	0.84509804
11	8	0.90308999
12	9	0.95424251
13	10	1
14	11	1.04139269

f(x)=ln(x)

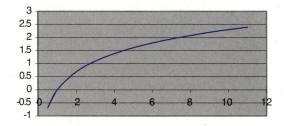

f(x)=log(x)

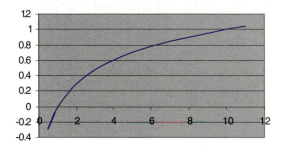

Section 5.2 Modeling Logarithmic Functions

1. Create a scatter plot for the data using *Insert > Scatter* and the points only option.
2. Single-click on the chart and right-click on one of the data points. Choose *Add Trendline* and check *Logarithmic* for *trendline* type.
3. Check the box to display the equation on the graph, and close the dialog box.

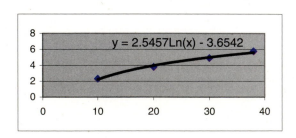

x	y
5	0.44
10	2.31
20	3.79
30	4.92

Section 5.2 Graphing Logarithmic Functions with Other Bases

1. The graph of a logarithm with any base b can be created by entering the formula $=\log(x, b)$.
2. Proceed as described above for graphing base e and base 10 logarithms.

	A	B
1	x	$f(x)=\log(x,7)$
2	0.5	−0.356207187
3	0.9	−0.054144594
4	1	0
5	2	0.356207187
6	3	0.564575034
7	4	0.712414374
8	5	0.827087475
9	6	0.920782221
10	7	1
11	8	1.068621561
12	9	1.129150068
13	10	1.183294662
14	11	1.1232274406

The graph of $f(x) = \log_7 x$ is shown below.

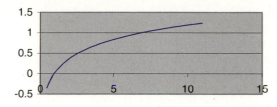

Chapter 6

Section 6.2 Finding the Future Value of a Lump Sum

To find the future value of a lump-sum investment:

1. Type the headings in Row 1, and enter their values (with the interest rate as a decimal) in Row 2.
2. Type the formula $=D2/E2$ in F2 to compute the rate per period.
3. Type the heading Future Value in A4.
4. In cell B4, type the formula $=fv(F2,B2,C2,A2,0)$ to compute the future value.

This spreadsheet gives the future value of an investment of $10,000 for 17 years at 9.8%, compounded quarterly.

Principal	Number of Periods	Payment	Annual Rate	Periods per year	Periodic Rate
10000	68	0	0.098	4	0.0245
Future Value	($51,857.73)				

Section 6.3 Finding the Future Value of an Annuity

To find the future value of an ordinary annuity:

1. Type the headings in Row 1, and enter their values (with the interest rate as a decimal) in Row 2.
2. Type the formula $=D2/E2$ in F2 to compute the rate per period.
3. Type the heading Future Value in A4.
4. In cell B4, type the formula $=fv(F2,B2,C2,A2,0)$ to compute the future value.
   (The 0 indicates that the payments are made at the end of each period.)

For annuities due, use $=fv(F2,B2,C2,A2,1)$.
The payments are made at the beginning of each period.

This spreadsheet gives the future value of an ordinary annuity of $200 deposited at the end of each quarter for $2\frac{1}{4}$ years, with interest at 4%, compounded quarterly.

Principal	Number of Periods	Payment	Annual Rate	Periods per year	Periodic Rate
0	9	200	0.04	4	0.01
Ordinary annuity					
Future Value	($1,873.71)				
Annuities due					
Future Value	($1,892.44)				

Section 6.4 Finding the Present Value of an Annuity

To find the present value of an ordinary annuity:

1. Type the headings in Row 1, and enter their values (with the interest rate as a decimal) in Row 2.
2. Enter the formula =D2/E2 in F2 to compute the rate per period.
3. Type the heading Present Value in A4.
4. In cell B4, type the formula =pv(F2,B2,C2,A2,0) to compute the present value.
   (The 0 indicates that the payments are made at the end of each period.)

For annuities due, use =pv(F2,B2,C2,A2,1).
The payments are made at the beginning of each period.

This spreadsheet gives both the present value of an ordinary annuity and of an annuity due if you pay $1000 per month for 16 years, with interest at 9%, compounded monthly.

Future Value	Number of Periods	Payment	Annual Rate	Periods per year	Periodic Rate
0	192	1000	0.09	12	0.0075
Ordinary annuity					
Present Value	$101,572.77				
Annui- ties due					
Present value	$102,334.56				

Section 6.5 Finding Payments to Amortize a Loan

To find the periodic payment to pay off a loan:

1. Type the headings in Row 1 and their values (with the interest rate as a decimal) in Row 2.
2. Enter the formula =D2/E2 in F2 to compute the rate per period.
3. Type the heading Payment in A4.
4. In cell B4, type the formula =Pmt(F2,B2,A2,C2,0) to compute the payment.

This spreadsheet gives the annual payment of a loan of $10,000 over 5 years when interest is 10% per year.

The parentheses indicates a payment out.

Loan Amount	Number of Periods	Future Value	Annual Rate	Periods per year	Periodic Rate
10000	5	0	0.1	1	0.1
Payment	($2,637.97)				

Chapter 8

Section 8.1 Binomial Probabilities

1. Type headings in cells A1:A3 and their respective values in cells B1:B3.
2. Use the function =binomdist(B1,B2,B3,cumulative) where
   - B1 is the number of successes.
   - B2 is the number of independent trials.
   - B3 is the probability of success in each trial.
   - True replaces cumulative if a cumulative probability is sought; it is replaced by false otherwise.
   - The probability of exactly 3 heads in 6 tosses is found by evaluating =binomdist(B1,B2,B3,false) in B4.
   - The probability of 3 or fewer heads in 6 tosses is found by evaluating =binomdist(B1,B2,B3,true) in B5.

This spreadsheet gives the probability of 3 heads in 6 tosses of a fair coin and the probability of 3 or fewer heads in 6 tosses.

	A	B
1	Number of successes	3
2	Number of trials	6
3	Probability of success	.5
4	Probability of 3 successes	.3125
5	Probability of 3 or fewer successes	.65625

Section 8.2 Bar Graphs

To construct a bar graph for the given table of test scores:

1. Copy the entries of the table to cells A2:B6.
2. Select the range A2:B6.
3. Click on the *Insert* tab. Then select *Column > Clustered Column* (first option), and the graph appears.
4. Single-click on the chart. You can add labels and change the title of the graph by clicking *Layout* under *Chart Tools*, and choosing the appropriate options.

	A	B
1	Grade Range	Frequency
2	90–100	2
3	80–89	5
4	70–79	3
5	60–69	3
6	0–59	2

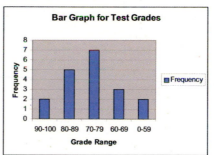

Section 8.2 Finding the Mean, Standard Deviation, and Median of Raw Data

To find the mean, standard deviation, and median of a raw data set:

1. Enter the data in Row 1 (cells A1:L1).
2. Type the heading Mean in cell A3.
3. Type the formula =average(A1:L1) in cell B4.
4. Type the heading Standard Deviation in cell A4.
5. In cell B4, type the formula =stdev(A1:L1).
6. In cell A5, type the heading Median.
7. In cell B5, type the formula =median(A1:L1).

The mean, standard deviation, and median for the data 1, 1, 1, 3, 3, 4, 4, 5, 6, 6, 7, 7 is shown below.

	A	B	C	D	E	F	G	H	I	J	K	L
1	1		1	1	3	3	4	4	5	6	6	7	7
2												
3	Mean	4										
4	Standard Deviation	2.2563										
5	Median	4										

Section 8.2 Finding the Mean and Standard Deviation of Grouped Data

To find the mean:

1. Enter the data and headings in the cells A1:C6.
2. In D1, type the heading Class mark*frequency.
3. In D2, type the formula =B2*C2.
4. Copy the formula in D2 to D3:D6.
5. In B7, type the heading Total.
6. In cell C7, type the formula for the total frequencies, =sum(C2:C6).
7. In cell D7, type the formula for the total, =sum(D2:D6).
8. In cell A8, type the heading Mean.
9. In cell A9, type in the formula =D7/C7.

To find the standard deviation:

10. In cell E1, type in the heading freq *(x − x _ mean)^2.
11. In cell E2, type the formula =C2*(B2-A9). (The A9 gives the value in A9; the reference doesn't change as we fill down.)
12. Copy the formula in E2 to E3:E6.
13. In cell E7, type the formula =sum(E2:E6).
14. In cell A10 type the heading Standard Deviation.
15. In cell A11, type the formula =sqrt(E7/(C7-1)).

Grade Range	Class Marks	Frequency
90–100	95	3
80–89	84.5	4
70–79	74.5	7
60–69	64.5	0
50–59	54.5	2

	A	B	C	D	E
1	Grade Range	Class Marks	Frequency	Class mark*frequency	Freq*(x-x_ mean)^2
2	90-100	95	3	285	832.2919922
3	80-89	84.5	4	338	151.5976563
4	70-79	74.5	7	521.5	103.4208984
5	60-69	64.5	0	0	0
6	50-59	54.5	2	109	1137.048828
7		Total	16	1253.5	2224.359375
8	Mean				
9	78.344				
10	Standard Deviation				
11	12.177				

Section 8.4 Calculating Normal Probabilities

To calculate normal probabilities:

1. Type headings in A1:A4 and their respective values in cells B1:B4.
2. To find the probability that a score X is less than the x1 value in B3, enter the formula =normdist(B3,B1,B2,true) in cell B5.
3. To find the probability that X is less than the x2 value in B4, enter the formula =normdist(B4,B1,B2,true) in cell B6.
4. To find the probability that a score X is more than the value in B3 and less than the x2 value in B4, enter the formula =B6-B5 in cell B7.

Entries in B5, B6, and B7 give the probabilities of a score X being less than 100, less than 115, and between 100 and 115, respectively, when the mean is 100 and the standard deviation is 15.

	A	B
1	Mean	100
2	Standard Deviation	15
3	x1	100
4	x2	115
5	Pr(X<x1)	0.5
6	Pr(X<x2)	0.841345
7	Pr(x1<X<x2)	0.341345

Chapter 9

Sections 9.1–9.2 Evaluating Limits

To evaluate $\lim\limits_{x \to c} f(x)$:

1. Make a table of values for $f(x)$ near $x = c$. Include values on both sides of $x = c$.
2. Use the table of values to predict the limit (or that the limit does not exist).

Note: All limit evaluations with Excel use appropriate tables of values of $f(x)$. This is true when $f(x)$ is piecewise defined and for limits as $\to \infty$.

Find $\lim\limits_{x \to 2} \dfrac{x^2 - 4}{x - 2}$.

	A	B	C	D
1	**x**	**f(x)**	**x**	**f(x)**
2	2.1	4.1	1.9	3.9
3	2.05	4.05	1.95	3.95
4	2.01	4.01	1.99	3.99
5	2.001	4.001	1.999	3.999

The tables suggest that $\lim\limits_{x \to 2} \dfrac{x^2 - 4}{x - 2} = 4$.

Sections 9.3–9.7 Approximating Derivatives

To approximate $f'(c)$:

1. Numerically investigate the limit in the definition of derivative:

$$f'(x) = \lim_{h \to 0} \frac{f(x + h) - f(x)}{h}$$

2. Use the given $f(x)$ and $x = c$ to create a table of values for h near 0 (and on both sides of $h = 0$).

Note that rows 5 and 6 have the values of h closest to 0.

Investigate $f'(1)$ for $f(x) = x^3$.

	A	B	C	D	E
1	**h**	**1+h**	**f(1)**	**f(1+h)**	**(f(1+h)-f(1))/h**
2	0.1	1.1	1	1.331	3.31
3	0.01	1.01	1	1.030301	3.0301
4	0.001	1.001	1	1.003003	3.003001
5	0.0001	1.0001	1	1.0003	3.00030001
6	−0.0001	0.9999	1	0.9997	2.99970001
7	−0.001	0.999	1	0.997003	2.997001
8	−0.01	0.99	1	0.970299	2.9701
9	−0.1	0.9	1	0.729	2.71

The table suggests that $f'(1) = 3$, which is the actual value.
Note: Excel has no built-in derivative approximation tool.

Chapter 10

Section 10.1 Relative Maxima and Minima

1. Make a table with columns for x-values, the function, and the derivative.
2. Extend the table to include x-values to the left and to the right of all critical values.
3. Use the signs of the values of the derivative to determine whether f is increasing or decreasing around the critical values, and thus to classify the critical values as relative maxima, relative minima, or horizontal points of inflection. You may want to graph the function to confirm your conclusions.

The spreadsheet shows that the relative minima of $f(x) = x^2$ is 0 at $x = 0$. Note that the derivative is $f'(x) = 2x$.

	A	B	C
1	x	f(x)	f'(x)
2	−2	4	−4
3	−1.5	2.25	−3
4	−1	1	−2
5	−0.5	0.25	−1
6	0	0	0
7	0.5	0.25	1
8	1	1	2
9	1.5	2.25	3
10	2	4	4

Section 10.2 Exploring f, f', f'' Relationships

To explore relationships among the graphs of a function and its derivatives:

1. Find functions for f' and f''.
2. Graph all three functions on the same plot.
    *   Notice that f increases when f' is above the x-axis ($+$) and decreases when f' is below the x-axis ($-$).
    *   Notice that f is concave up when f'' is above the x-axis ($+$) and is concave down when f'' is below the x-axis ($-$).

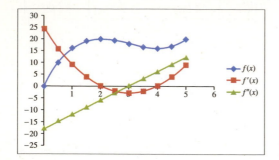

Let $f(x) = x^3 - 9x^2 + 24x$. Graph f, f', and f'' on the interval $[0, 5]$ to explore the relationships among these functions.

	A	B	C	D
	x	$f(x)$	$f'(x)$	$f''(x)$
1				
2	0	0	24	-18
3	0.5	9.875	15.75	-15
4	1	16	9	-12
5	1.5	19.125	3.75	-9
6	2	20	0	-6
7	2.5	19.375	-2.25	-3
8	3	18	-3	0
9	3.5	16.625	-2.25	-3
10	4	16	0	6
11	4.5	16.875	3.75	9
12	5	20	9	12

Sections 10.3–10.4 Finding Optimal Values

To find the optimal value of a function when the goal is not to produce a graph:

1. Set up a spreadsheet that identifies the variable and the function whose optimal value is sought.
2. Choose *Data > Analysis > Solver*. Then, in the Dialog Box
    a. Set the *Target Cell* as that of the objective function.
    b. Check *Max* or *Min* according to your goal.
    c. Set the *Changing Cells* to reference the variable.
3. Click on the *Options* box. Make sure "Assume Linear Model" is *not* checked. Then click OK.
4. Click Solve in the *Solver* dialog box. You will get a dialog box stating that *Solver* found a solution. Save the solution if desired, then click OK.
5. The cells containing the variable and the function should now contain the optimal values.

Minimize area $A = x^2 + \dfrac{160}{x}$ for $x > 0$.

	A	B
1	**Variable**	
2		
3	x length of base	1
4		
5	**Objective**	
6		
7	Minimize Area	=B3^2+160/B3

	A	B
1	**Variable**	
2		
3	x length of base	4.3089
4		
5	**Objective**	
6		
7	Minimize Area	55.699

The function is minimized for $x = 4.3089$ and the minimum value is $A = 55.699$.

Chapter 13

Section 13.1 Approximating Definite Integrals Using Rectangles

To approximate the area under the graph of $y = f(x)$ and above the x-axis from a to b, using left-hand endpoints:

1. Divide $b - a$ by the number of rectangles to get the width of each rectangle.
2. Enter the x-values in Column A and the function values in Column B.
3. Add a third column with the heading Δx in C1, and the rectangle width in cells C2:C6.
4. Add a fourth column with the heading Rectangle Area in D1 and, in D2, use the formula =B2*C2 to get the area of the first rectangle.
5. Copy this formula down to cells D3:D6.
6. In cell C7, type the heading Total Area, and in D7 enter the formula =sum(D2:D6) and press ENTER.

The approximate area under the graph of $f(x) = x^2$ from $x = 1$ to $x = 2$, with 5 rectangles, is found below. In the example, the width is 0.2 and the left-hand endpoints are 1.0, 1.2, 1.4, 1.6, 1.8.

x	$f(x)$	Δx	Rectangle Area
1	1	0.2	0.2
1.2	1.44	0.2	0.288
1.4	1.96	0.2	0.392
1.6	2.56	0.2	0.512
1.8	3.24	0.2	0.648
		Total Area	2.04

Chapter 14

Section 14.1 Graphs of Functions of Two Variables

To create a surface plot for a function of two variables:

1. a. Generate appropriate x-values beginning in B1 and continuing *across*.
   b. Generate appropriate y-values beginning in A2 and continuing *down*.
2. Generate values for the function that correspond to the points (x, y) from Step 1 as follows:
   In cell B2, enter the function formula with B$1 used to represent x and $A2 to represent y. (See the online Excel Guide for additional information about the role and use of the $ in this step.) Then use fill down and fill across to complete the table.
3. Select the entire table of values. Click on *Insert > Other Charts > 3-D Surface* option, the first option in the Surface group.
4. To rotate the graph, click on the chart and click on *3-D rotation*. You can click on the various options present in the dialog box.

Let $f(x, y) = 10 - x^2 - y^2$. Plot the graph of this function for both x and y in the interval $[-2, 2]$.

	A	B	C	D	E	F	G	H	I	J
1			−2	−1.5	−1	−0.5	0	0.5	1	1.5	2
2	−2									
3	−1.5									
4	−1									
5	−0.5									
6	0									
7	0.5									
8	1									
9	1.5									
10	2									

	A	B	C	D	E	F	G	H	I	J
1		−2	−1.5	−1	−0.5	0	0.5	1	1.5	2
2	−2	2	3.75	5	5.75	6	5,75	5	3.75	2
3	−1.5	3.75	5.5	6.75	7.5	7.75	7.5	6.75	5.5	3.75
4	−1	5	6.75	8	8.75	9	8.75	8	6.75	5
5	−0.5	5.75	7.5	8.75	9.5	9.75	9.5	8.75	7.5	5.75
6	0	6	7.75	9	9.75	10	9.75	9	7.75	6
7	0.5	5.75	7.5	8.75	9.5	9.75	9.5	8.75	7.5	5.75
8	1	5	6.75	8	8.75	9	8.75	8	6.75	5
9	1.5	3.75	5.5	6.75	7.5	7.75	7.5	6.75	5.5	3.75
10	2	2	3.75	5	5.75	6	5.75	5	3.75	2

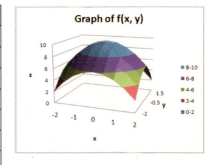

Graph of f(x, y)

Section 14.5 Constrained Optimization and Lagrange Multipliers (Excel 2007)

To solve a constrained optimization problem:

1. Set up the problem in Excel.
2. Choose *Data* > *Analysis* > *Solver* and do the following:
   - Choose the objective function as the *Target Cell*.
   - Check *Max* or *Min* depending on the problem.
   - Choose the cells representing the variables for the *By Changing Cells* box.
   - Click on the *Constraints* box and press Add. Then enter the constraint equations.
3. Click on the *Options* box and make sure that "Assume Linear Model" is *not* checked. Then click OK.

4. Click *Solve* in the *Solver* dialog box. Then click OK to solve.

Maximize $p = 600I^{2/3}k^{1/3}$ subject to $40l + 100k = 3000$.

	A	B	C
1	**Variables**		
2			
3	units labor (l)	0	
4	units capital (k)	0	
5			
6			
7	**Objective**		
8			
9	Maximize production	=600*B3^(2/3)*B4^(1/3)	
10			
11	**Constraint**		
12		Amount used	Available
13	Cost	=40*B3+100*B4	3000

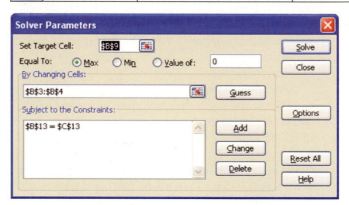

	A	B	C
1	**Variables**		
2			
3	units labor (l)	50.00000002	
4	units capital (k)	10	
5			
6			
7	**Objective**		
8			
9	Maximize product	17544.10644	
10			
11	**Constraint**		
12		Amount used	Available
13	Cost	3000.000001	3000

Section 14.5 Constrained Optimization and Lagrange Multipliers (Excel 2010)

To solve a constrained optimization problem:

1. Set up the problem in Excel.
2. Choose *Data > Analysis > Solver* and do the following:
   - Choose the objective function as the *Target Cell*.
   - Check *Max* or *Min* depending on the problem.
   - Choose the cells representing the variables for the *By Changing Cells* box.
   - Click on the *Constraints* box and press Add. Then enter the constraint equations.
3. Check the box making the variables nonnegative.
4. Select "GRG Nonlinear" as the solving method.
5. Click *Solve* in the *Solver* dialog box. Then click OK to solve.

Maximize $p = 600l^{2/3}k^{1/3}$ subject to $40l + 100k = 3000$.

	A	B	C
1	**Variables**		
2			
3	units labor (l)	0	
4	units capital (k)	0	
5			
6			
7	**Objective**		
8			
9	Maximize production	=600*B3^(2/3)*B4^(1/3)	
10			
11	**Constraint**		
12		Amount used	Available
13	Cost	=40*B3+100*B4	3000

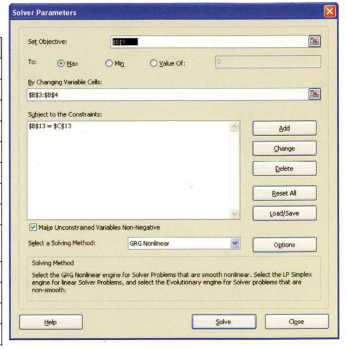

	A	B	C
1	**Variables**		
2			
3	units labor (l)	50.00000002	
4	units capital (k)	10	
5			
6			
7	**Objective**		
8			
9	Maximize product	17544.10644	
10			
11	**Constraint**		
12		Amount used	Available
13	Cost	3000.000001	3000

Answers

Below are the answers to all the Chapter Review and Chapter Test problems.

CHAPTER 0 REVIEW EXERCISES

1. yes   2. no   3. no
4. $\{1, 2, 3, 4, 9\}$ 5. $\{5, 6, 7, 8, 10\}$
6. $\{1, 2, 3, 4, 9\}$
7. yes, $(A' \cup B')' = \{1, 3\} = A \cap B$
8. (a) Commutative Property of Addition
   (b) Associative Property of Multiplication
   (c) Distributive Law
9. (a) irrational   (b) rational, integer
   (c) meaningless
10. (a) $>$ (b) $<$ (c) $>$
11. 6   12. 142   13. 10
14. 5/4   15. 9   16. -29
17. 13/4   18. -10.62857888
19. (a) $[0, 5]$, closed

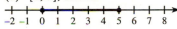

  (b) $[-3, 7)$, half-open

  (c) $(-4, 0)$, open

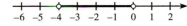

20. (a) $-1 < x < 16$ (b) $-12 \le x \le 8$
    (c) $x < -1$
21. (a) 1   (b) $2^{-2} = 1/4$ (c) 4^6 (d) 7
22. (a) $1/x^2$ (b) x^{10} (c) x^9 (d) $1/y^8$
    (e) y^6
23. $-x^2y^2/36$ 24. $9y^8/(4x^4)$ 25. $y^2/(4x^4)$
26. $-x^8z^4/y^4$ 27. $3x/(y^7z)$ 28. $x^5/(2y^3)$
29. (a) 4   (b) 2/7   (c) 1.1
30. (a) $x^{1/2}$ (b) $x^{2/3}$ (c) $x^{-1/4}$
31. (a) $\sqrt[3]{x^2}$ (b) $1/\sqrt{x} = \sqrt{x}/x$ (c) $-x\sqrt{x}$
32. (a) $5y\sqrt{2x}/2$ (b) $\sqrt[3]{x^2y}/x^2$
33. $x^{5/6}$ 34. y
35. $x^{17/4}$ 36. $x^{11/3}$
37. $x^{2/5}$ 38. x^2y^8
39. $2xy^2\sqrt{3xy}$ 40. $25x^3y^4\sqrt{2y}$
41. $6x^2y^4\sqrt[3]{5x^2y^2}$ 42. $8a^2b^4\sqrt{2a}$
43. $2xy$ 44. $4x\sqrt{3xy}/(3y^4)$
45. $-x - 2$ 46. $-x^2 - x$
47. $4x^3 + xy + 4y - 4$ 48. $24x^5y^5$
49. $3x^2 - 7x + 4$ 50. $3x^2 + 5x - 2$
51. $4x^2 - 7x - 2$ 52. $6x^2 - 11x - 7$
53. $4x^2 - 12x + 9$ 54. $16x^2 - 9$
55. $2x^4 + 2x^3 - 5x^2 + x - 3$
56. $8x^3 - 12x^2 + 6x - 1$

57. $x^3 - y^3$
58. $(2/y) - (3xy/2) - 3x^2$
59. $3x^2 + 2x - 3 + (-3x + 7)/(x^2 + 1)$
60. $x^3 - x^2 + 2x + 7 + 21/(x - 3)$ 61. $x^2 - x$
62. $2x - a$ 63. $x^3(2x - 1)$
64. $2(x^2 + 1)^2(1 + x)(1 - x)$ 65. $(2x - 1)^2$
66. $(4 + 3x)(4 - 3x)$ 67. $2x^2(x + 2)(x - 2)$
68. $(x - 7)(x + 3)$ 69. $(3x + 2)(x - 1)$
70. $(x - 3)(x - 2)$ 71. $(x - 12)(x + 2)$
72. $(4x + 3)(3x - 8)$ 73. $(2x + 3)^2(2x - 3)^2$
74. $x^{2/3} + 1$

75. (a) $\dfrac{x}{(x + 2)}$ (b) $\dfrac{2xy(2 - 3xy)}{2x - 3y}$

76. $\dfrac{x^2 - 4}{x(x + 4)}$ 77. $\dfrac{(x + 3)}{(x - 3)}$

78. $\dfrac{x^2(3x - 2)}{(x - 1)(x + 2)}$ 79. $(6x^2 + 9x - 1)/(6x^2)$

80. $\dfrac{4x - x^2}{4(x - 2)}$ 81. $-\dfrac{x^2 + 2x + 2}{x(x - 1)^2}$

82. $\dfrac{x(x - 4)}{(x - 2)(x + 1)(x - 3)}$

83. $\dfrac{(x - 1)^3}{x^2}$ 84. $\dfrac{1 - x}{1 + x}$

85. $3(\sqrt{x} + 1)$ 86. $2/(\sqrt{x} + \sqrt{x - 4})$
87. (a)

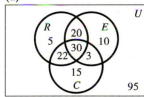

R: recognized
E: exercise
C: community involvement

  (b) 10   (c) 100
88. 52.55%
89. 16
90. (a) $4115.27   (b) $66,788.69
91. (a) 939,577   (b) 753,969
92. (a) 1.1 inch, about quarter-sized
    (b) 104 mph
93. (a) $10,000\left[\dfrac{(0.0065)(1.0065)^n}{(1.0056)^n - 1}\right]$ (b) $243.19 (for both)
94. (a) $S = k\sqrt[3]{A}$ (b) $\sqrt[3]{2.25} \approx 1.31$
95. $26x - 300 - 0.001x^2$
96. $1,450,000 - 3625x$
97. $(50 + x)(12 - 0.5x)$
98. (a) $\dfrac{12,000p}{100 - p}$
    (b) $0. It costs nothing if no effort is made to remove pollution.

(c) $588,000

(d) Undefined. Removing 100% would be impossible, and the cost of getting close would be enormous.

99. $\dfrac{56x^2 + 1200x + 8000}{x}$

CHAPTER 0 TEST

1. (a) {3, 4, 6, 8}   (b) {3, 4}; {3, 6}; or {4, 6}
   (c) {6} or {8}

2. 21

3. (a) 8   (b) 1   (c) $\frac{1}{2}$ (d) −10
   (e) 30   (f) $\frac{5}{6}$ (g) $\frac{2}{3}$ (h) −3

4. (a) $\sqrt[5]{x}$ (b) $\dfrac{1}{\sqrt[4]{x^3}}$

5. (a) $\dfrac{1}{x^5}$ (b) $\dfrac{x^{21}}{y^6}$

6. (a) $\dfrac{\sqrt{5x}}{5}$ (b) $2a^2b^2\sqrt{6ab}$ (c) $\dfrac{1 - 2\sqrt{x} + x}{1 - x}$

7. (a) 5   (b) −8   (c) −5

8. (−2, 3]
-4 -3 -2 -1 0 1 2 3 4 5 6

9. (a) $2x^2(4x - 1)$ (b) $(x - 4)(x - 6)$
   (c) $(3x - 2)(2x - 3)$ (d) $2x^3(1 + 4x)(1 - 4x)$

10. (c); −2

11. $2x + 1 + \dfrac{2x - 6}{x^2 - 1}$

12. (a) $19y - 45$ (b) $-6t^6 + 9t^9$
    (c) $4x^3 - 21x^2 + 13x - 2$ (d) $-18x^2 + 15x - 2$
    (e) $4m^2 - 28m + 49$ (f) $\dfrac{x^4}{3x + 9}$ (g) $\dfrac{x^7}{81}$
    (h) $\dfrac{6 - x}{x - 8}$ (i) $\dfrac{x^2 - 4x - 3}{x(x - 3)(x + 1)}$

13. $\dfrac{y - x}{y + xy^2}$

14. (a)
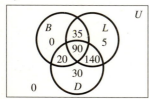

    (b) 0
    (c) 175

15. $4875.44 (nearest cent)

CHAPTER 1 REVIEW EXERCISES

1. $x = \frac{31}{3}$ 2. $x = -13$ 3. $x = -\frac{29}{8}$
4. $x = -\frac{1}{9}$ 5. $x = 8$
6. no solution   7. $y = -\frac{2}{3}x - \frac{4}{3}$
8. $x \le 3$
0   4
9. $x \ge -20/3$
-8 $-\frac{20}{3}$ $\frac{15}{13}$ -4   0
10. $x \ge -15/13$
-2 -1 0 1 2 3

11. yes   12. no   13. yes

14. domain: reals $x \le 9$; range: reals $y \ge 0$

15. (a) 2   (b) 37   (c) 29/4

16. (a) 0   (b) 9/4   (c) 10.01

17. $9 - 2x - h$ 18. yes   19. no   20. 4

21. $x = 0, x = 4$

22. (a) domain: {−2, −1, 0, 1, 3, 4}
       range: {−3, 2, 4, 7, 8}
    (b) 7   (c) $x = -1, x = 3$
    (d)
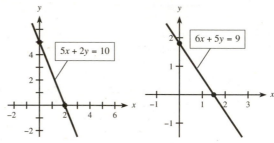

    (e) no; for $y = 2$, there are two different x-values, −1 and 3.

23. (a) $x^2 + 3x + 5$ (b) $(3x + 5)/x^2$
    (c) $3x^2 + 5$ (d) $9x + 20$

24. x: 2, y: 5   25. x: 3/2, y: 9/5

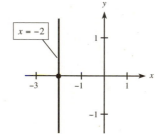

26. x: −2, y: none

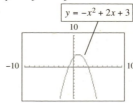

27. $m = 1$ 28. undefined   29. $m = -\frac{2}{5}, b = 2$

30. $m = -\frac{4}{3}, b = 2$ 31. $y = 4x + 2$

32. $y = -\frac{1}{2}x + 3$ 33. $y = \frac{2}{5}x + \frac{9}{5}$

34. $y = -\frac{11}{8}x + \frac{17}{14}$ 35. $x = -1$ 36. $y = 4x + 2$

37. $y = \frac{4}{3}x + \frac{10}{3}$

38. $y = -x^2 + 2x + 3$ 39. $y = \dfrac{x^3 - 27x + 54}{15}$

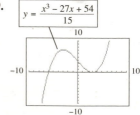

40. (a) 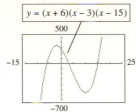 (b)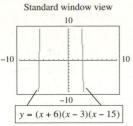

(c) The graph in (a) shows a complete graph. The graph in (b) shows a piece that rises toward the high point and a piece between the high and low points.

41. (a) 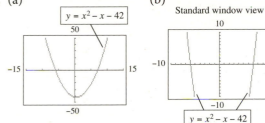 (b)

(c) The graph in (a) shows a complete graph. The one in (b) shows pieces that fall toward the minimum point and rise from it.

42. reals $x \geq -3$ with $x \neq 0$

43. $0.2857 \approx 2/7$ **44.** $x = 2, y = 1$

45. $x = 10, y = -1$ **46.** $x = 3, y = -2$

47. no solution   **48.** $x = 10, y = -71$

49. $x = 1, y = -1, z = 2$

50. $x = 11, y = 10, z = 9$

51. (a) 1997   (b) 27   (c) $x = 30.0$; in 2010

52. 95%

53. 40,000 miles. He would normally drive more than 40,000 miles in 5 years, so he should buy diesel.

54. (a) yes   (b) no   (c) 4

55. (a) $565.44

(b) The monthly payment on a $70,000 loan is $494.75.

56. (a) $(P \circ q)(t) = 330(100 + 10t)$
$\quad\quad - 0.05(100 + 10t)^2 - 5000$

(b) $x = 250, P = $74,375$

57. $(W \circ L)(t) = 0.03[65 - 0.1(t - 25)^2]^3$

58. (a)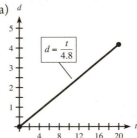

(b) When the time between seeing lightning and hearing thunder is 9.6 seconds, the storm is 2 miles away.

59.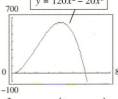

60. (a) $P = 58x - 8500$

(b) The profit increases by $58 for each unit sold.

61. (a) yes

(b) $m = 427, b = 4541$

(c) In 2000, average annual health care costs were $4541 per consumer.

(d) Average annual health care costs are changing at the rate of $427 per year.

62. $F = \frac{9}{5}C + 32$ or $C = \frac{5}{9}(F - 32)$

63. (a) 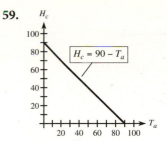   (b) $0 \leq x \leq 6$

64. (a) $v^2 = 1960(h + 10)$

$\dfrac{v^2}{1960} = h + 10$

$h = \dfrac{1}{1960}v^2 - 10$

(b) 12.5 cm

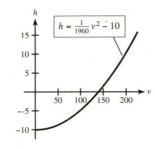

65. $100,000 at 9.5%; $50,000 at 11%

66. 2.8 liters of 20%; 1.2 liters of 70%

67. (a) 12 supplied; 14 demanded   (b) shortfall

(c) increase

68.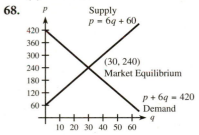

69. (a) 38.80   (b) 61.30   (c) 22.50   (d) 200

70. (a) $C(x) = 22x + 1500$ (b) $R(x) = 52x$

(c) $P(x) = 30x - 1500$ (d) $\overline{MC} = 22$

(e) $\overline{MR} = 52$ (f) $\overline{MP} = 30$ (g) $x = 50$

71. $q = 300, p = 150 **72.** $q = 700, p = 80

CHAPTER 1 TEST

1. $x = -6$

2. $x = 18/7$ **3.** $x = -3/7$ **4.** $x = -38$

5. $5 - 4x - 2h$

6. $t \geq -9$

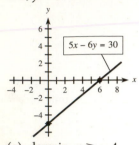

7. x: 6; y: -5　　　**8.** x: 3; y: 21/5

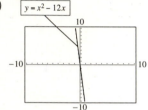

9. (a) domain: $x \geq -4$
　　range: $f(x) \geq 0$
　　(b) $2\sqrt{7}$　　(c) 6

10. $y = -\frac{3}{2}x + \frac{1}{2}$　　**11.** $m = -\frac{5}{4}; b = \frac{15}{4}$

12. (a) $x = -3$　　(b) $y = -4x - 13$

13. (a) no; a vertical line intersects the curve twice.
　　(b) yes; there is exactly one y-value for each x-value.
　　(c) no; one value of x gives two y-values.

14. (a)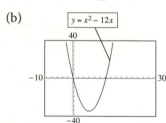

　　(b)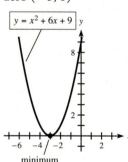

15. $x = -2, y = 2$

16. (a) $5x^3 + 2x^2 - 3x$　　(b) $x + 2$
　　(c) $5x^2 + 7x + 2$

17. (a) 30　　(b) $P = 8x - 1200$　　(c) 150
　　(d) $\overline{MP} = 8$; the sale of each additional unit gives $8 more profit.

18. (a) $R = 50x$
　　(b) 19,000; it costs $19,000 to produce 100 units.
　　(c) 450

19. $q = 200, p = \$2500$

20. (a) 720,000; original value of the building
　　(b) -2000; building depreciates $2000 per month.

21. 400

22. 12,000 at 9%, 8000 at 6%

CHAPTER 2 REVIEW EXERCISES

1. $x = 0, x = -\frac{5}{3}$　　**2.** $x = 0, x = \frac{4}{3}$

3. $x = -2, x = -3$

4. $x = (-5 + \sqrt{47})/2, x = (-5 - \sqrt{47})/2$

5. no real solutions　　**6.** $x = \frac{\sqrt{3}}{2}, x = -\frac{\sqrt{3}}{2}$

7. $\frac{5}{7}, -\frac{4}{5}$　　**8.** $(-1 + \sqrt{2})/4, (-1 - \sqrt{2})/4$

9. 7/2, 100　　**10.** 13/5, 90

11. no real solutions　　**12.** $z = -9, z = 3$

13. $x = 8, x = -2$　　**14.** $x = 3, x = -1$

15. $x = (-a \pm \sqrt{a^2 - 4b})/2$

16. $r = (2a \pm \sqrt{4a^2 + x^3c})/x$

17. 1.64, -7051.64　　**18.** 0.41, -2.38

19. vertex $(-2, -2)$;　　**20.** vertex $(0, 4)$;
　　zeros $(0, 0), (-4, 0)$　　　zeros $(4, 0), (-4, 0)$

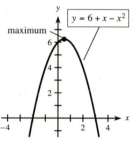　　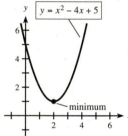

21. vertex $(\frac{1}{2}, \frac{25}{4})$;　　**22.** vertex $(2, 1)$;
　　zeros $(-2, 0), (3, 0)$　　　no real zeros

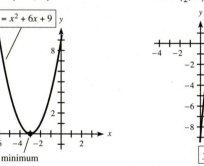　　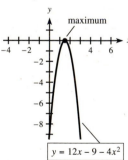

23. vertex $(-3, 0)$;　　**24.** vertex $(\frac{3}{2}, 0)$;
　　zero $(-3, 0)$　　　zero $(\frac{3}{2}, 0)$

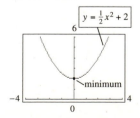

25. vertex $(0, -3)$;　　**26.** vertex $(0, 2)$;
　　zeros $(-3, 0), (3, 0)$　　　no real zeros

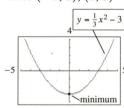

27. vertex $(-1, 4)$;
no real zeros

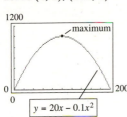

28. vertex $\left(\frac{7}{2}, \frac{9}{4}\right)$;
zeros $(2, 0), (5, 0)$

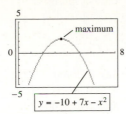

29. vertex $(100, 1000)$;
zeros $(0, 0), (200, 0)$

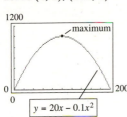

30. vertex $(75, -6.25)$;
zeros $(50, 0), (100, 0)$

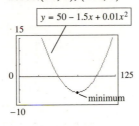

31. 20   **32.** 30

33. (a) $\left(1, -4\frac{1}{2}\right)$ (b) $x = -2, x = 4$ (c) B

34. (a) $(0, 49)$ (b) $x = -7, x = 7$ (c) D

35. (a) $(7, 25)$ approximately, actual is $\left(7, 24\frac{1}{2}\right)$
(b) $x = 0, x = 14$ (c) A

36. (a) $(-1, 9)$ (b) $x = -4, x = 2$ (c) C

37. (a)

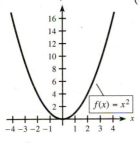

(b)

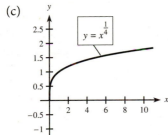

(c)

38. (a) 0   (b) 10,000   (c) -25 (d) 0.1

39. (a) -2 (b) 0   (c) 1   (d) 4

40.

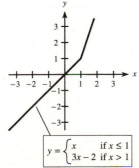

$y = \begin{cases} x & \text{if } x \le 1 \\ 3x - 2 & \text{if } x > 1 \end{cases}$

41. (a)

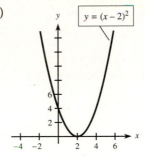

$y = (x - 2)^2$

(b)

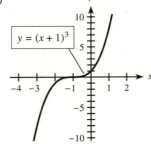

$y = (x + 1)^3$

42. Turns: $(1, -5), (-3, 27)$

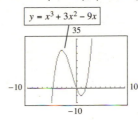

$y = x^3 + 3x^2 - 9x$

43. Turns: $(1.7, -10.4)$,
$(-1.7, 10.4)$

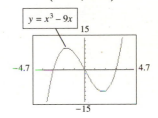

$y = x^3 - 9x$

44. VA: $x = 2$;
HA: $y = 0$

45. VA: $x = -3$;
HA: $y = 2$

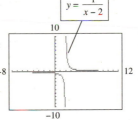

$y = \dfrac{1}{x - 2}$

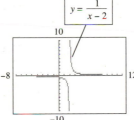

$y = \dfrac{2x - 1}{x + 3}$

46. (a)

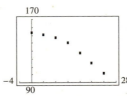

(b) $y = -2.1786x + 159.8571$
(c) $y = -0.0818x^2 - 0.2143x + 153.3095$

47. (a)

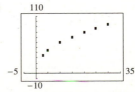

(b) $y = 2.1413x + 34.3913$
(c) $y = 22.2766x^{0.4259}$

48. (a) $t = -1.65, t = 3.65$ (b) Just $t = 3.65$
    (c) at 3.65 seconds
49. $x = 20, x = 800$
50. (a) 2000 and 2008   (b) 2004; 8.59
51. (a) $x = 200$ (b) A = 30,000 square feet
52.

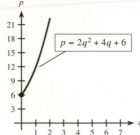

53.

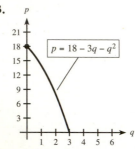

54. (a)

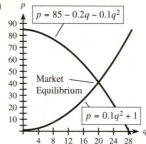

   (b) $p = 41, q = 20$

55. $p = 400, q = 10$ **56.** $p = 10, q = 20$
57. $x = 46 + 2\sqrt{89} \approx 64.9, x = 46 - 2\sqrt{89} \approx 27.1$
58. $x = 15, x = 60$
59. max revenue = \$2500; max profit = \$506.25
60. max profit = 12.25; break-even at $x = 100, x = 30$

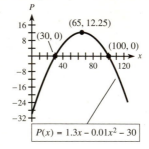

61. $x = 50, P(50) = 640$
62. (a) $C = 15,000 + 140x + 0.04x^2$;
       $R = 300x - 0.06x^2$
    (b) 100, 1500   (c) 2500
    (d) $P = 160x - 15,000 - 0.1x^2$; max at 800
    (e) at 2500: $P = -240,000$; at 800: $P = 49,000$
63. (a) power   (b) 36.7 million
    (c) 40.8; the number of HIV infections will be 40.8
       million in 2015.

64. (a)

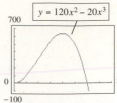

   (b) $0 \le x \le 6$

65. (a) rational   (b) $0 \le p < 100$
    (c) 0; it costs \$0 to remove no pollution.
    (d) \$475,200
66. (a) $x = 12; C(12) = \$30.68$
    (b) $x = 825; C(825) = \$1734.70$
67. (a) and (b)

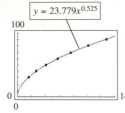

    (c) 55 mph   (d) 9.9 seconds
68. (a) $y = 2.94x^2 + 32.7x + 640$
    (b) $x \approx 23.3$, in 2024
    (c) Very little; projections were made before the Health
       Care Bill of 2010.
69. (a) $y = 5.66x^{1.70}$
    (b) $y = 2.46x^2 - 34.7x + 399$
    (c) Both are quite good.

CHAPTER 2 TEST

1. (a)

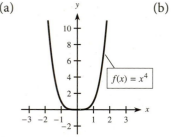

   (b)

(c)

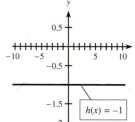

(d)

2. b; a
3.

4. (a)

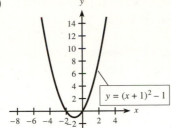

$y = (x + 1)^2 - 1$

(b)

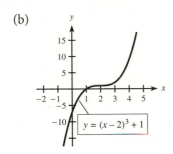

$y = (x - 2)^3 + 1$

5. b; the function is cubic, $f(1) < 0$

6. (a) -10 (b) $-16\frac{1}{2}$ (c) -7

7.

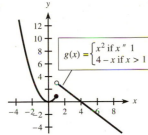

$g(x) = \begin{cases} x^2 \text{ if } x \leq 1 \\ 4 - x \text{ if } x > 1 \end{cases}$

8. vertex $(-2, 25)$;
zeros $-7, 3$

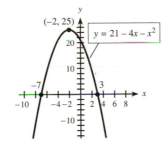

$y = 21 - 4x - x^2$

9. $x = 2, x = 1/3$

10. $x = \dfrac{-3 + 3\sqrt{3}}{2}, x = \dfrac{-3 - 3\sqrt{3}}{2}$

11. $x = 2/3$

12. c; $g(x)$ has a vertical asymptote at $x = -2$, as does graph c.

13. HA: $y = 0$; VA: $x = 5$

14. 42

15. (a) quartic   (b) cubic

16. (a)

model: $y = -0.3577x + 19.9227$
(b) $f(x) = 5.6$
(c) at $x = 55.7$

17. $q = 300, p = \$80$

18. (a) $P(x) = -x^2 + 250x - 15{,}000$
(b) 125 units, \$625   (c) 100 units, 150 units

19. (a) $f(15) = -19.5$ means that when the air temperature is 0°F and the wind speed is 15 mph, the air temperature feels like -19.5°F.
(b) -31.4°F

20. (a)

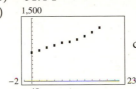

(b) linear:
$y = 30.65x + 667.43$
cubic: $y = 0.118x^3 - 2.51x^2 + 40.28x + 676$

(c) linear: \$1280.43; cubic: \$1420.70   (d) no

CHAPTER 3 REVIEW EXERCISES

1. 4   **2.** 0   **3.** A, B **4.** none   **5.** D, F, G, I

6. $\begin{bmatrix} -2 & 5 & 11 & -8 \\ -4 & 0 & 0 & -4 \\ 2 & 2 & -1 & -9 \end{bmatrix}$

7. zero matrix   **8.** order

9. $\begin{bmatrix} 6 & -1 & -9 & 3 \\ 10 & 3 & -1 & 4 \\ -2 & -2 & -2 & 14 \end{bmatrix}$ **10.** $\begin{bmatrix} 3 & -3 \\ 4 & -1 \\ 2 & -6 \\ 1 & -2 \end{bmatrix}$

11. $\begin{bmatrix} 2 & 1 \\ 5 & 1 \end{bmatrix}$ **12.** $\begin{bmatrix} 12 & -6 \\ 15 & 0 \\ 18 & 0 \\ 3 & 9 \end{bmatrix}$ **13.** $\begin{bmatrix} 4 & 0 \\ 0 & 4 \end{bmatrix}$

14. $\begin{bmatrix} 2 & -12 \\ -8 & -22 \end{bmatrix}$ **15.** $\begin{bmatrix} 9 & 20 \\ 4 & 5 \end{bmatrix}$ **16.** $\begin{bmatrix} 5 & 16 \\ 6 & 15 \end{bmatrix}$

17. $\begin{bmatrix} 2 & 37 & 61 & -55 \\ -2 & 9 & -3 & -20 \\ 10 & 10 & -14 & -30 \end{bmatrix}$ **18.** $\begin{bmatrix} 43 & -23 \\ 33 & -12 \\ -13 & 15 \end{bmatrix}$

19. $\begin{bmatrix} 10 & 16 \\ 15 & 25 \\ 18 & 30 \\ 6 & 11 \end{bmatrix}$ **20.** $\begin{bmatrix} 17 & 73 \\ 7 & 28 \end{bmatrix}$ **21.** $\begin{bmatrix} 3 & 7 \\ 23 & 42 \end{bmatrix}$

22. F   **23.** F   **24.** $\begin{bmatrix} -19 & 12 \\ -8 & 5 \end{bmatrix}$ **25.** F

26. (a) infinitely many solutions (bottom row of 0's)
(b) $x = 6 + 2z$
$y = 7 - 3z$
$z =$ any real number
Two specific solutions:
If $z = 0$, then $x = 6, y = 7$.
If $z = 1$, then $x = 8, y = 4$.

27. (a) no solution (last row says $0 = 1$)
(b) no solution

28. (a) Unique coefficient matrix is I_3.
(b) $x = 0, y = -10, z = 14$

29. $(1, 2, 1)$ **30.** $x = 22, y = 9$

31. $x = -3, y = 3, z = 4$

32. $x = -\frac{3}{2}, y = 7, z = -\frac{11}{2}$ **33.** no solution

34. $x = 2 - 2z, y = -1 - 2z, z =$ any real number

35. $x = -2 + 8z$
   $y = -2 + 3z$
   $z =$ any real number

36. $x_1 = 1, x_2 = 11, x_3 = -4, x_4 = -5$ **37.** yes

38. $\begin{bmatrix} \frac{1}{2} & \frac{1}{4} \\ \frac{5}{2} & \frac{7}{4} \end{bmatrix}$ **39.** $\begin{bmatrix} -1 & -2 & 8 \\ 1 & 2 & -7 \\ 1 & 1 & -4 \end{bmatrix}$

40. $\begin{bmatrix} 2 & 1 & -2 \\ 7 & 5 & -8 \\ -13 & -9 & 15 \end{bmatrix}$

41. $x = -33, y = 30, z = 19$

42. $x = 4, y = 5, z = -13$

43. $A^{-1} = \begin{bmatrix} -41 & 32 & 5 \\ 17 & -13 & -2 \\ -9 & 7 & 1 \end{bmatrix}; x = 4, y = -2, z = 2$

44. no   **45.** (a) 16   (b) yes, det $\neq 0$

46. (a) 0   (b) no, det $= 0$

47. $\begin{bmatrix} 250 & 140 \\ 480 & 700 \end{bmatrix}$ **48.** $\begin{bmatrix} 1030 & 800 \\ 700 & 1200 \end{bmatrix}$

49. (a) higher in June   (b) higher in July

50. $\begin{matrix} & \text{Men} & \text{Women} \\ & \begin{bmatrix} 865 & 885 \\ 210 & 270 \end{bmatrix} & \begin{matrix} \text{Robes} \\ \text{Hoods} \end{matrix} \end{matrix}$

51. $\begin{bmatrix} 1750 \\ 480 \end{bmatrix} \begin{matrix} \text{Robes} \\ \text{Hoods} \end{matrix}$

52. (a) $\begin{bmatrix} 13{,}500 & 12{,}400 \\ 10{,}500 & 10{,}600 \end{bmatrix}$

   (b) Department A should buy from Kink; Department B should buy from Ace.

53. (a) $\begin{bmatrix} 0.20 & 0.30 & 0.50 \end{bmatrix}$

   (b) $\begin{bmatrix} 0.013469 \\ 0.013543 \\ 0.006504 \end{bmatrix}$

   (c) $\begin{bmatrix} 0.20 & 0.30 & 0.50 \end{bmatrix} \begin{bmatrix} 0.013469 \\ 0.013543 \\ 0.006504 \end{bmatrix} = 0.20(0.013469) +$
   $0.30(0.013543) + 0.50(0.006504) = 0.0100087$

   (d) The historical return of the portfolio, 0.0100087, is the estimated expected monthly return of the portfolio. This is roughly 1% per month.

54. 400 fast food, 700 software, 200 pharmaceutical

55. (a) $A = 2C, B = 2000 - 4C, C =$ any integer satisfying $0 \le C \le 500$

   (b) yes; $A = 500, B = 1000, C = 250$

   (c) max $A = 1000$ when $B = 0, C = 500$

56. (a) 3 passenger, 4 transport, 4 jumbo

   (b) 1 passenger, 3 transport, 7 jumbo

   (c) column 2

57. (a) shipping $= 5680$; agriculture $= 1960$

   (b) shipping $= 0.4$; agriculture $= 1.8$

58. (a)
   $$A = \begin{matrix} & S & C \\ & \begin{bmatrix} 0.1 & 0.1 \\ 0.2 & 0.05 \end{bmatrix} & \begin{matrix} \text{Shoes} \\ \text{Cattle} \end{matrix} \end{matrix}$$

   (b) shoes $= 1000$; cattle $= 500$

59. mining $= 360$; manufacturing $= 320$; fuels $= 400$

60. government $= \frac{64}{93}$ households; agriculture $= \frac{59}{93}$ households; manufacturing $= \frac{40}{93}$ households

CHAPTER 3 TEST

1. $\begin{bmatrix} 3 & 1 & 5 \\ 1 & 3 & 6 \end{bmatrix}$ **2.** $\begin{bmatrix} -1 & 2 & 2 \\ 1 & -1 & 6 \end{bmatrix}$

3. $\begin{bmatrix} -12 & -16 & -155 \\ 5 & 12 & 87 \end{bmatrix}$ **4.** $\begin{bmatrix} 23 & 6 \\ 182 & 45 \\ 21 & 1 \end{bmatrix}$

5. $\begin{bmatrix} 0 & -7 \\ 26 & 1 \end{bmatrix}$ **6.** $\begin{bmatrix} -43 & -46 & -207 \\ 39 & 30 & -77 \\ 17 & 5 & -216 \end{bmatrix}$

7. $\begin{bmatrix} -2 & 3/2 \\ 1 & -1/2 \end{bmatrix}$ **8.** $\begin{bmatrix} -3 & 2 & 2 \\ 1 & 0 & -1 \\ 1/2 & -1/2 & 0 \end{bmatrix}$

9. $\begin{bmatrix} 5 \\ 14 \\ 15 \end{bmatrix}$ **10.** $x = -0.5, y = 0.5, z = 2.5$

11. $x = 4 - 1.8z, y = 0.2z, z =$ any real number

12. no solution   **13.** $x = 2, y = 2, z = 0, w = -2$

14. $x = 6w - 0.5, y = 0.5 - w, z = 2.5 - 3w, w =$ any real number

15. (a) $B = \$45{,}000, E = \$40{,}000$

   (b) $\$0 \le H \le \$25{,}000$ (so $B \ge 0$)

   (c) min $E = \$20{,}000$ when $H = \$0$ and $B = \$75{,}000$

16. (a) $\begin{bmatrix} 0.08 & 0.22 & 0.12 \\ 0.10 & 0.08 & 0.19 \\ 0.05 & 0.07 & 0.09 \\ 0.10 & 0.26 & 0.15 \\ 0.12 & 0.04 & 0.24 \end{bmatrix}$

   (b) 0.08, 0.22, 0.12 consumed by carnivores 1, 2, 3

   (c) plant 5 by 1, plant 4 by 2, plant 5 by 3

17. (a) $\begin{bmatrix} 1000 & 4000 & 2000 & 1000 \end{bmatrix}$

   (b) $\begin{bmatrix} 45{,}000 & 55{,}000 & 90{,}000 & 70{,}000 \end{bmatrix}$

   (c) $\begin{bmatrix} 5 \\ 3 \\ 4 \\ 4 \end{bmatrix}$ (d) $[\$1{,}030{,}000]$ (e) $\begin{matrix} A \\ B \\ C \\ D \end{matrix} \begin{bmatrix} 65 \\ 145 \\ 125 \\ 135 \end{bmatrix}$ $\$$

18. (a) 121, 46, 247, 95, 261, 99, 287, 111, 179, 69, 169, 64

   (b) Frodo lives

19. growth, 2000; blue-chip, 400; utility, 400
20. (a) agriculture = 245; minerals = 235
    (b) agriculture = 7; minerals = 1
    (c) agriculture = 0.5; minerals = 1.5
21. profit = households
    nonprofit = $\frac{2}{3}$ households
22.
$$\begin{bmatrix} 0.2 & 0.1 & 0.1 & 0.1 \\ 0.3 & 0.2 & 0.2 & 0.2 \\ 0.2 & 0.2 & 0.3 & 0.3 \\ 0.1 & 0.4 & 0.2 & 0.2 \end{bmatrix} \begin{matrix} \text{Agriculture} \\ \text{Machinery} \\ \text{Fuel} \\ \text{Steel} \end{matrix}$$
with column headers Ag M F S

23. agriculture: 5000; machinery: 8000; fuel: 8000; steel: 7000
24. agriculture: $\frac{520}{699}$ households; steel: $\frac{236}{233}$ households; fuel: $\frac{159}{233}$ households

CHAPTER 4 REVIEW EXERCISES

1. 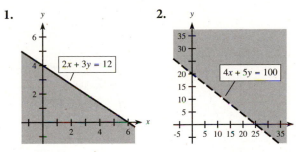   2.

3. 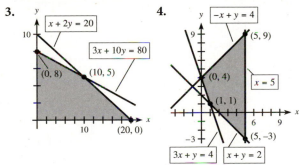   4.

5. max = 25 at (5, 10); min = −12 at (12, 0)
6. max = 194 at (17, 23); min = 104 at (8, 14)
7. max = 140 at (20, 0); min = −52 at (20, 32)
8. min = 115 at (5, 7); no max exists, f can be made arbitrarily large
9. $f = 66$ at (6, 6)   10. $f = 43$ at (7, 9)
11. $g = 24$ at (3, 3)   12. $g = 84$ at (64, 4)
13. $f = 76$ at (12, 8)   14. $f = 75$ at (15, 15)
15. $f = 168$ at (12, 7)   16. $f = 260$ at (60, 20)
17. $f = 360$ at (40, 30)   18. $f = 80$ at (20, 10)
19. $f = 640$ on the line between (160, 0) and (90, 70)
20. no solution   21. $g = 32$ at $y_1 = 2, y_2 = 3$
22. $g = 20$ at $y_1 = 4, y_2 = 2$
23. $g = 7$ at $y_1 = 1, y_2 = 5$
24. $g = 1180$ at $y_1 = 80, y_2 = 20$
25. $f = 165$ at $x = 20, y = 21$
26. $f = 54$ at $x = 6, y = 5$ 27. $f = 270$ at (5, 3, 2)
28. $g = 140$ at $y_1 = 0, y_2 = 20, y_3 = 20$

29. $g = 1400$ at $y_1 = 0, y_2 = 100, y_3 = 100$
30. $f = 156$ at $x = 15, y = 2$
31. $f = 31$ at $x = 4, y = 5$ 32. $f = 4380$ at (40, 10, 0, 0)
33. $f = 1000$ at $x_1 = 25, x_2 = 62.5, x_3 = 0, x_4 = 12.5$
34. $g = 2020$ at $x_1 = 0, x_2 = 100, x_3 = 80, x_4 = 20$
35. $P = \$14{,}750$ when 110 large and 75 small swing sets are made
36. $C = \$300{,}000$ when factory 1 operates 30 days, factory 2 operates 25 days
37. $P = \$320$; I = 40, II = 20
38. $P = \$420$; Jacob's ladders = 90, locomotive engines = 30
39. (a) Let x_1 = the number of 27-in LCD sets,
    x_2 = the number of 32-in LCD sets,
    x_3 = the number of 42-in LCD sets,
    x_4 = the number of 42-in plasma sets.
    (b) Maximize $P = 80x_1 + 120x_2 + 160x_3 + 200x_4$ subject to
    $$8x_1 + 10x_2 + 12x_3 + 15x_4 \le 1870$$
    $$2x_1 + 4x_2 + 4x_3 + 4x_4 \le 530$$
    $$x_1 + x_2 + x_3 + x_4 \le 200$$
    $$x_3 + x_4 \le 100$$
    $$x_2 \le 120$$
    (c) $x_1 = 15, x_2 = 25, x_3 = 0, x_4 = 100$; max profit = \$24,200
40. food I = 0 oz, food II = 3 oz; $C = \$0.60$ (min)
41. cost = \$5.60; A = 40 lb, B = 0 lb
42. cost = \$8500; A = 20 days, B = 15 days, C = 0 days
43. pancake mix = 8000 lb; cake mix = 3000 lb; profit = \$3550
44. Texas: 55 desks, 65 computer tables; Louisiana: 75 desks, 65 computer tables; cost = \$4245
45. Midland: grade 1 = 486.5 tons, grade 2 = 0 tons; Donora: grade 1 = 13.5 tons, grade 2 = 450 tons; Cost = \$90,635

CHAPTER 4 TEST

1. max = 120 at (0, 24)
2. (a)
$$C; \begin{bmatrix} 1 & 2 & 0 & 1 & 0 & -3/2 & 0 & 40 \\ 0 & \boxed{1} & 0 & -2 & 1 & 1/2 & 0 & 15 \\ 0 & 3 & 1 & -1 & 0 & 1/4 & 0 & 60 \\ \hline 0 & 0 & 0 & 4 & 0 & 6 & 1 & 220 \end{bmatrix}$$
    $-2R_2 + R_1 \to R_1$ $-3R_2 + R_3 \to R_3$
    (b) A; pivot column is column 3, but new pivot is undefined.
    (c) B; $x_1 = 40, x_2 = 12, x_3 = 0, s_1 = 0,$ $s_2 = 20, s_3 = 0; f = 170$; This solution is not optimal; the next pivot is the 3-6 entry.
3. (a)

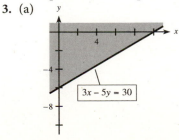

(b)

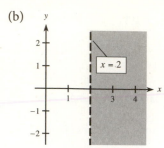

(c)

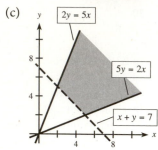

4. maximize $f = 100x_1 + 120x_2$ subject to
$3x_1 + 4x_2 \le 2$
$5x_1 + 6x_2 \le 3$
$x_1 + 3x_2 \le 5$
$x_1 \ge 0, \ x_2 \ge 0$

5. min $= 21$ at $(1, 8)$; no max exists, f can be made arbitrarily large

6. min $= 136$ at $(28, 52)$

7. maximize $-g = -7x - 3y$ subject to
$x - 4y \le -4$
$x - \ \ y \le \ \ 5$
$2x + 3y \le \ \ 30$

8. max: $x_1 = 17, x_2 = 15, x_3 = 0; f = 658$ (max)
min: $y_1 = 4, y_2 = 18, y_3 = 0; g = 658$ (min)

9. max: $= 6300$ at $x = 90, y = 0$

10. max $= 1200$ at $x = 0, y = 16, z = 12$

11. If $x =$ the number of barrels of beer and $y =$ the number of barrels of ale, then maximize $P = 35x + 30y$ subject to
$3x + 2y \le 1200$
$2x + 2y \le 1000$
$P = \$16,000$ (max) at $x = 200, y = 300$

12. If $x =$ the number of day calls and $y =$ the number of evening calls, then minimize $C = 3x + 4y$ subject to
$0.3x + 0.3y \ge 150$
$0.1x + 0.3y \ge 120$
$x \le 0.5 (x + y)$
$C = \$1850$ (min) at $x = 150, y = 350$

13. max profit $= \$5000$ when product 1 $= 25$ tons, product 2 $= 62.5$ tons, product 3 $= 0$ tons, product 4 $= 12.5$ tons

CHAPTER 5 REVIEW EXERCISES

1. (a) $\log_2 y = x$ \quad (b) $\log_3 2x = y$

2. (a) $7^{-2} = \frac{1}{49}$ \quad (b) $4^{-1} = x$

3.

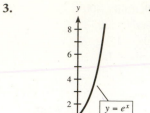

4.

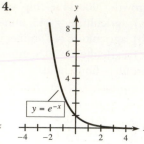

5.

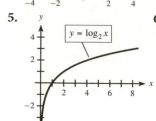

6.

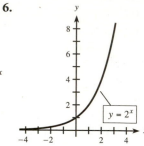

7.

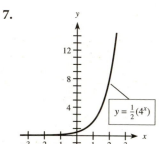

8.

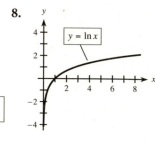

9.

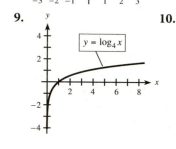

10.

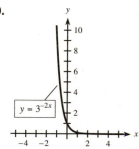

11.

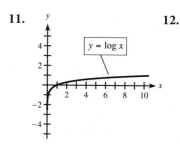

12.

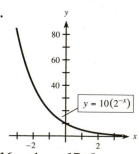

13. 0 \qquad **14.** 3 \qquad **15.** $\frac{1}{2}$ \qquad **16.** -1 \qquad **17.** 8

18. 1 \qquad **19.** 5 \qquad **20.** 3.15 \qquad **21.** -2.7

22. 0.6 \qquad **23.** 5.1 \qquad **24.** 15.6 \qquad **25.** $\log y + \log z$

26. $\frac{1}{2}\ln (x + 1) - \frac{1}{2}\ln x$ \qquad **27.** no \qquad **28.** -2

29. 5 \qquad **30.** 1 \qquad **31.** 0 \qquad **32.** 3.4939 \qquad **33.** -1.5845

34.

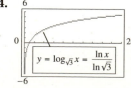

35.

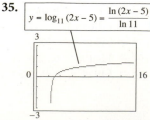
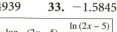

36. $x = 1.5$ **37.** $x \approx 51.224$

38. $x \approx 28.175$ **39.** $x \approx 40.236$

40. $x = 8$ **41.** $x = 14$ **42.** $x = 6$

43. Growth exponential, because the general outline has the same shape as a growth exponential.

44. Decay exponential, because the general shape is similar to the graph of a decay exponential, and the number will diminish toward 0.

45. (a) $32,627.66

(b)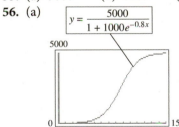

46. (a) $y = 266.4(1.074^x)$ (b) $4598 billion

(c) 2017

47. (a) $11,293   (b)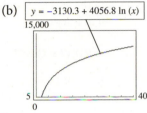

48. (a) $y = -11.4 + 19.0 \ln x$

(b) 68.0%   (c) 2023

49. (a) -3.9 (b) $0.14B_0$ (c) $0.004B_0$ (d) yes

50. (a) 27,441   (b) 12 weeks

51. 1366   **52.** 5.8 years

53. (a) $5532.77   (b) 5.13 years

54. logistic, because the graph begins like an exponential function but then grows at a slower rate

55. (a) 3000   (b) 8603   (c) 10,000

56. (a)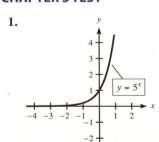

(b) 5   (c) 4970   (d) 10 days

CHAPTER 5 TEST

1. 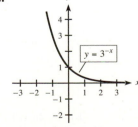   **2.**

3. 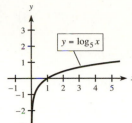   **4.**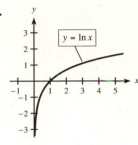

5. 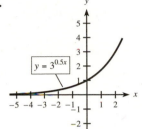   **6.**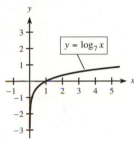

7. 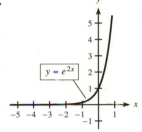   **8.**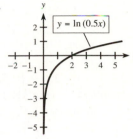

9. 54.598   **10.** 0.100   **11.** 1.386

12. 1.322   **13.** $x = 7^{3.1}, x \approx 416.681$

14. $\log_3(27) = 2x; x = 1.5$ **15.** $x \approx 0.203$

16. $x = 8$ **17.** 3   **18.** x^4 **19.** 3   **20.** x^2

21. $\ln M + \ln N$ **22.** $\ln(x^3 - 1) - \ln(x + 2)$

23. $\dfrac{\ln(x^3 + 1)}{\ln 4} \approx 0.721 \ln(x^3 + 1)$

24. $x \approx 38.679$

25. a decay exponential

26. With years on the horizontal axis, a growth exponential would probably be the best model.

27. (a) $3363.3 billion

(b) about 11.2 years

28. about 6.2 months

29. (a) about $16,716 billion

(b) $x \approx 39.8$; in 2025

30. (a) exponential; data continue to rise rapidly

(b) $y = 165.550(1.055^x)$

(c) $819.42 billion

CHAPTER 6 REVIEW EXERCISES

1. $1, \frac{1}{4}, \frac{1}{9}, \frac{1}{16}$

2. Arithmetic: (a) and (c) (a) $d = -5$ (c) $d = \frac{1}{6}$

3. 235   **4.** 109   **5.** 315

6. Geometric: (a) and (b) (a) $r = 8$ (b) $r = -\frac{3}{4}$

7. 8   **8.** $2, 391, 484\frac{4}{9}$

9. $S = R\left[\dfrac{(1 + i)^n - 1}{i}\right]$ **10.** $I = Prt$

11. $A_n = R\left[\dfrac{1 - (1 + i)^{-n}}{i}\right]$　　**12.** $S = P(1 + i)^n$

13. $A_n = R\left[\dfrac{1 - (1 + i)^{-n}}{i}\right]$, solved for R

14. $S = Pe^{rt}$　　**15.** \$10,880　　**16.** $6\frac{2}{3}\%$

17. \$2941.18　　**18.** \$4650

19. the \$40,000 job (\$490,000 versus \$472,500)

20. (a) 40　　　　　　　　(b) $2\% = 0.02$

21. (a) $S = P(1 + i)^n$　　(b) $S = Pe^{rt}$

22. (b) monthly　**23.** \$372.79　　**24.** \$14,510.26

25. \$1616.07　　**26.** \$21,299.21　　**27.** 34.3 quarters

28. (a) 13.29%　(b) 14.21%

29. (a) 7.40%　(b) 7.47%

30. 2^{63}　　**31.** $2^{32} - 1$　　**32.** \$29,428.47

33. \$6069.44　　**34.** \$31,194.18　　**35.** \$10,841.24

36. $n \approx 36$ quarters

37. \$130,079.36　**38.** \$12,007.09

39. (a) \$11.828 million　(b) \$161.5 million

40. \$1726.85　　**41.** \$5390.77

42. $n \approx 16$ half-years　(8 years)

43. \$88.85　　**44.** \$3443.61　　**45.** \$163,792.21

46.

Payment Number	Payment Amount	Interest	Balance Reduction	Unpaid Balance
57	\$699.22	\$594.01	\$105.21	\$94,936.99
58	\$699.22	\$593.36	\$105.86	\$94,831.13

47. 16.32%　　**48.** \$213.81

49. (a) \$1480　　(b) \$1601.03

50. \$9319.64　　**51.** 14.5 years　　　　**52.** 79.4%

53. \$4053.54; discount

54. (a) \$4728.19　　(b) \$5398.07　　(c) \$1749.88
(d) 10.78%

55. \$21,474.08　　**56.** \$12,162.06

57. Quarterly APY = 6.68%. This rate is better for the bank; it pays less interest. Continuous APY = 6.69%. This rate is better for the consumer, who earns more interest.

58. \$32,834.69　**59.** \$3,466.64

60. (a) \$1185.51　　(b) \$355,653　　(c) \$171,653
(d) \$156,366.25

61. (a) \$95,164.21　　(b) \$1300.14　　**62.** \$994.08

63. Future value of IRA = \$172,971.32
Present value needed = \$2,321,520.10
Future value needed from deposits = \$2,148,548.78
Deposits = \$711.60

64. Regular payment = \$64,337.43
Unpaid balance = \$2,365,237.24
Number of \$70,000 payments = $n \approx 46.1$
Savings = \$118,546.36

CHAPTER 6 TEST

1. 25.3 years (approximately)　　**2.** \$840.75

3. 6.87%　　**4.** \$158,524.90　**5.** 33.53%

6. (a) \$698.00　(b) \$112,400

7. \$2625　　**8.** \$7999.41　　**9.** 8.73%

10. \$119,912.92　**11.** \$40,552.00

12. \$32,488 (to the nearest dollar)　**13.** \$6781.17

14. $n \approx 66.8$; 67 half-years

15. \$1688.02

16. (a) \$279,841.35　　　　(b) \$13,124.75

17. \$116,909.10　　　　　　**18.** \$29,716.47

19. (a) The difference between successive terms is
always -5.5.
(b) 23.8　　　　　　(c) 8226.3

20. 1000 mg (approximately)

21. \$12,975.49; premium

22. (a) \$145,585.54 (with the \$2000);
\$147,585.54 (without the \$2000)
(b) $n \approx 318.8$ months
Total interest = \$243,738.13
(c) $n \approx 317.2$ months
Total interest = \$245,378.24
(d) Paying the \$2000 is slightly better; it saves about
\$1640 in interest.

CHAPTER 7 REVIEW EXERCISES

1. (a) $\frac{5}{9}$　(b) $\frac{1}{3}$　(c) $\frac{2}{9}$

2. (a) $\frac{3}{4}$　(b) $\frac{1}{2}$　(c) $\frac{1}{6}$　(d) $\frac{2}{3}$

3. (a) 3:4　(b) 4:3　**4.** (a) $\frac{1}{4}$　(b) $\frac{1}{2}$　(c) $\frac{1}{4}$

5. (a) $\frac{3}{8}$　(b) $\frac{1}{8}$　(c) $\frac{3}{8}$　**6.** $\frac{2}{13}$　**7.** 16/169

8. $\frac{3}{4}$　**9.** $\frac{2}{13}$　**10.** $\frac{7}{13}$　**11.** (a) $\frac{2}{9}$　(b) $\frac{2}{3}$　(c) $\frac{7}{9}$

12. $\frac{2}{7}$　**13.** $\frac{1}{2}$　**14.** 7/320

15. 7/342　**16.** 3/14　**17.** $\frac{8}{15}$

18. (a) $\frac{3}{14}$　(b) $\frac{4}{7}$　(c) $\frac{3}{8}$　**19.** 49/89

20. 30　**21.** 35　**22.** 26^3　**23.** 56

24. (a) Not square
(b) The row sums are not 1.

25. [0.76　0.24], [0.496　0.504]

26. [0.2　0.8]

27. $\frac{5}{8}$　**28.** $\frac{1}{4}$　**29.** $\frac{29}{50}$

30. $\frac{5}{56}$　**31.** $\frac{33}{56}$　**32.** $\frac{15}{22}$　**33.** 0.72

34. (a) 63/2000　(b) $\frac{60}{63}$　**35.** 39/116　**36.** 4! = 24

37. $_8P_4 = 1680$　**38.** $_{12}C_4 = 495$　**39.** $_8C_4 = 70$

40. (a) $_{12}C_2 = 66$　(b) $_{12}C_3 = 220$　**41.** 62,193,780

42. If her assumption about blood groups is accurate, there
would be $4 \cdot 2 \cdot 4 \cdot 8 = 256$, not 288, unique groups.

43. $\frac{1}{24}$　**44.** $\frac{3}{500}$　**45.** $\frac{3}{1250}$

46. (a) 0.3398　(b) 0.1975　**47.** $\frac{1}{10}$

48. (a) $(_{10}C_5)(_2C_1)/_{12}C_6$
(b) $\dfrac{(_{10}C_5)(_2C_1) + (_{10}C_4)(_2C_2)}{_{12}C_6}$

49. [0.135　0.51　0.355], [0.09675　0.3305　0.57275],
[0.0640875　0.288275　0.6476375]

50. [12/265　68/265　37/53]

CHAPTER 7 TEST

1. (a) $\frac{4}{7}$　(b) $\frac{3}{7}$　**2.** (a) $\frac{2}{7}$　(b) $\frac{5}{7}$

3. (a) 0　(b) 1　**4.** $\frac{1}{7}$　**5.** $\frac{1}{7}$

6. (a) $\frac{2}{7}$ (b) $\frac{4}{7}$ **7.** $\frac{2}{7}$ **8.** $\frac{3}{7}$ **9.** $\frac{2}{3}$
10. 1/17,576   **11.** 0.2389   **12.** (a) $\frac{1}{5}$ (b) $\frac{1}{20}$
13. (a) $\frac{3}{95}$ (b) $\frac{6}{19}$ (c) $\frac{21}{38}$ (d) 0
14. (a) 5,245,786   (b) 1/5,245,786
15. (a) 2,118,760   (b) 1/2,118,760
16. 0.064   **17.** (a) 0.633   (b) 0.962   **18.** 0.229
19. (a) $\frac{1}{5}$ (b) $\frac{1}{14}$ (c) $\frac{13}{14}$ **20.** $\frac{3}{14}$
21. (a) 2^{10} (b) $\dfrac{1}{2^{10}}$ (c) $\frac{1}{3}$ (d) Change the code.

22. (a) $A = \begin{bmatrix} 0.80 & 0.20 \\ 0.07 & 0.93 \end{bmatrix}$ (b) $[0.25566 \ \ 0.74434]$;
about 25.6%   (c) $\frac{7}{27}$; 25.9% of market

CHAPTER 8 REVIEW EXERCISES

1. 0.0774   **2.** (a) 0.3545   (b) 0.5534
3. 0.407
4.
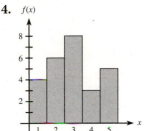
5. 3
6. $\frac{77}{26} \approx 2.96$
7. 3

8.
$f(x)$ graph with bars at 12–17
9. 14
10. 14
11. 14.3

12. $\bar{x} = 3.86$; $s^2 = 6.81$; $s = 2.61$
13. $\bar{x} = 2$; $s^2 = 2.44$; $s = 1.56$
14. 2.4   **15.** yes   **16.** no; $\Sigma \Pr(x) \neq 1$
17. yes   **18.** no; $\Pr(x) \not\geq 0$ **19.** 2
20. (a) 4.125   (b) 2.7344   (c) 1.654
21. (a) $\frac{37}{12}$ (b) 0.9097   (c) 0.9538
22. $\mu = 4, \sigma = (2\sqrt{3})/3$ **23.** 3
24. $x^5 + 5x^4y + 10x^3y^2 + 10x^2y^3 + 5xy^4 + y^5$
25. 0.9165   **26.** 0.1498   **27.** 0.1039   **28.** 0.3413
29. 0.6826   **30.** 0.1360   **31.** not good   **32.** good
33. 0.0151   **34.** 0.9625   **35.** 0.8475   **36.** 0.0119
37. 0.297   **38.** 0.16308   **39.** 0.2048
40. (a) $(99,999/100,000)^{99,999} \approx 0.37$
(b) $1 - (99,999/100,000)^{100,000} \approx 0.63$
41.
$f(x)$ histogram with Class Marks 10–19, 20–29, 30–39, 40–49, 50–59
Class Marks
42. 30.3%
43. 8.35%
44. 455
45. 3
46. $18.00
47. −$0.50

48. (a) 1   (b) $\left(\frac{4}{5}\right)^4$
49. (a) 0.4773   (b) 0.1360   (c) 0.0227
50. 15%   **51.** 0.3821   **52.** 0.4090
53. 0.0262   **54.** 0.1788

CHAPTER 8 TEST

1. (a) $\frac{40}{243}$ (b) $\frac{51}{243} = \frac{17}{81}$
2. (a) 4   (b) $\mu = 4, \sigma^2 = \frac{8}{3}, \sigma = \frac{2}{3}\sqrt{6}$
3. (i) For each x, $0 \leq \Pr(x) \leq 1$ (ii) $\Sigma \Pr(x) = 1$
4. 5.1
5. $\mu = 16.7, \sigma^2 = 26.61, \sigma = 5.16$
6. $\bar{x} = 21.57$, median $= 21$, mode $= 21$
7. (a) 0.4706   (b) 0.8413   (c) 0.0669
8. (a) 0.3891   (b) 0.5418   (c) 0.1210
9. 0.9554   **10.** 0.6331
11.
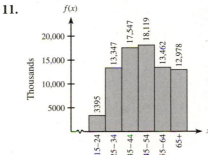

12. $\bar{x} = 48.3, s = 15.6$
13. (a) 38.5   (b) under 30; it would be lower
14. (a) 73.8   (b) It might be higher; cell use is spreading.
15. (a) 0.00003   (b) 30
16. 2 (1.8)   **17.** 5 (5.4)
18. 0 (0.054) with correct use
19. (a) 0.0158   (b) 0.0901   (c) 0.5383
20. 0.1814

CHAPTER 9 REVIEW EXERCISES

1. (a) 2   (b) 2   **2.** (a) 0   (b) 0
3. (a) 2   (b) 1   **4.** (a) 2   (b) does not exist
5. (a) does not exist   (b) 2
6. (a) does not exist   (b) does not exist
7. 55   **8.** 0   **9.** −2   **10.** 4/5   **11.** $\frac{1}{2}$
12. $\frac{1}{5}$ **13.** no limit   **14.** 0   **15.** 4
16. no limit   **17.** 3   **18.** no limit   **19.** $6x$
20. $1 - 4x$ **21.** −14   **22.** 5
23. (a) yes   (b) no   **24.** (a) yes   (b) no
25. 2   **26.** no limit   **27.** 1   **28.** no   **29.** yes
30. yes   **31.** discontinuity at $x = 5$
32. discontinuity at $x = 2$ **33.** continuous
34. discontinuity at $x = 1$
35. (a) $x = 0, x = 1$ (b) 0   (c) 0
36. (a) $x = -1, x = 0$ (b) $\frac{1}{2}$ (c) $\frac{1}{2}$
37. −2; $y = -2$ is a horizontal asymptote.
38. 0; $y = 0$ is a horizontal asymptote.

39. 7   **40.** true   **41.** false

42. $f'(x) = 6x + 2$ **43.** $f'(x) = 1 - 2x$

44. $[-1, 0]$; the segment over this interval is steeper.

45. (a) no   (b) no   **46.** (a) yes   (b) no

47. (a) -5.9171 (to four decimal places)   (b) -5.9

48. (a) $4.9/3$ (b) 7   **49.** about $-1/4$

50. B, C, A: $B < 0$ and $C < 0$; the tangent line at $x = 6$ falls more steeply than the segment over $[2, 10]$.

51. $20x^4 - 18x^2$ **52.** $90x^8 - 30x^5 + 4$

53. 1   **54.** $1/(2\sqrt{x})$ **55.** 0   **56.** $-4/(3\sqrt[3]{x^4})$

57. $\dfrac{-1}{x^2} + \dfrac{1}{2\sqrt{x^3}}$ **58.** $\dfrac{-3}{x^3} - \dfrac{1}{3\sqrt[3]{x^2}}$

59. $y = 15x - 18$ **60.** $y = 34x - 48$

61. (a) $x = 0, x = 2$ (b) $(0, 1)$ $(2, -3)$

(c)

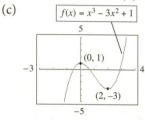

62. (a) $x = 0, x = 2, x = -2$

(b) $(0, 8)$ $(2, -24)$ $(-2, -24)$

(c)

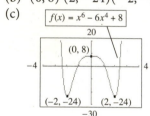

63. $9x^2 - 26x + 4$ **64.** $21x^6 + 4x^3 + 27x^2$

65. $\dfrac{15q^2}{(2q^3 + 1)^2}$ **66.** $\dfrac{1 - 3t}{[2\sqrt{t}(3t + 1)^2]}$ **67.** $\dfrac{9x + 2}{2\sqrt{x}}$

68. $\dfrac{5x^6 + 2x^4 + 20x^3 - 3x^2 - 4x}{(x^3 + 1)^2}$

69. $(9x^2 - 24x)(x^3 - 4x^2)^2$

70. $6(30x^5 + 24x^3)(5x^6 + 6x^4 + 5)^5$

71. $72x^3(2x^4 - 9)^8$ **72.** $\dfrac{-(3x^2 - 4)}{2\sqrt{(x^3 - 4x)^3}}$

73. $2x(2x^4 + 5)^7(34x^4 + 5)$ **74.** $\dfrac{-2(3x + 1)(x + 12)}{(x^2 - 4)^2}$

75. $36[(3x + 1)(2x^3 - 1)]^{11}(8x^3 + 2x^2 - 1)$

76. $\dfrac{3}{(1 - x)^4}$ **77.** $\dfrac{(2x^2 - 4)}{\sqrt{x^2 - 4}}$ **78.** $\dfrac{2x - 1}{(3x - 1)^{4/3}}$

79. $y'' = \tfrac{-1}{4}x^{-3/2} - 2$ **80.** $y'' = 12x^2 - 2/x^3$

81. $\dfrac{d^5y}{dx^5} = 0$ **82.** $\dfrac{d^5y}{dx^5} = -30(1 - x)$

83. $\dfrac{d^3y}{dx^3} = -4[(x^2 - 4)^{3/2}]$ **84.** $\dfrac{d^4y}{dx^4} = \dfrac{2x(x^2 - 3)}{(x^2 + 1)^3}$

85. (a) $400,000   (b) $310,000   (c) $90,000

86. (a) $70,000; this is fixed costs.

(b) $0; $x = 1000$ is break-even.

87. (a) $140 per unit

(b) $\overline{C}(x) \to +\infty$; the limit does not exist.

88. (a) $C(x) \to +\infty$; the limit does not exist. As the number of units produced increases without bound, so does the total cost.

(b) $60 per unit. As more units are produced, the average cost of each unit approaches $60.

89. The average annual percent change of (a) elderly men in the work force is 0.29 percentage points per year and of (b) elderly women in the work force is 0.26 percentage points per year.

90. (a) Annual average rate of change of percent of elderly men in the work force:

1950–1960: -1.27 percentage points per year

2000–2008: 0.48 percentage points per year

(b) Annual average rate of change of percent of elderly women in the work force

1950–1960: 0.11 percentage points per year

2000–2008: 0.49 percentage points per year

91. (a) $x'(10) = -1$ means that if price changes from $10 to $11, the number of units demanded will change by about -1.

(b) $x'(20) = -\tfrac{1}{4}$ means that if price changes from $20 to $21, the number of units demanded will change by about $-\tfrac{1}{4}$.

92. $h(100) \approx 4.15$; $h'(100) \approx 0.08$ means that when the updraft speed is 100 mph, the hail diameter is about 4.15 inches (softball-sized) and changing at the rate of 0.08 inch per mph of updraft.

93. The slope of the tangent at A gives $\overline{MR}(A)$. The tangent line at A is steeper (so has greater slope) than the tangent line at B. Hence, $\overline{MR}(A) > \overline{MR}(B)$, so the $(A + 1)$st unit will bring more revenue.

94. $R'(10) = 1570$. Raised. An 11th rent increase of $30 (and hence an 11th vacancy) would change revenue by about $1570.

95. (a) $P(20) = 23$ means productivity is 23 units per hour after 20 hours of training and experience.

(b) $P'(20) \approx 1.4$ means that the 21st hour of training or experience will change productivity by about 1.4 units per hour.

96. $\dfrac{dq}{dp} = \dfrac{-p}{\sqrt{0.02p^2 + 500}}$

97. $x'(10) = \tfrac{1}{6}$ means if price changes from $10 to $11, the number of units supplied will change by about $\tfrac{1}{6}$.

98. $s''(t) = a = -2t^{-3/2}$; $s''(4) = -0.25$ ft/sec/sec

99. $P'(x) = 70 - 0.2x$; $P''(x) = -0.2$

$P'(300) = 10$ means that the 301st unit brings in about $10 in profit.

$P''(300) = -0.2$ means that marginal profit $(P'(x))$ is changing at the rate of -0.2 dollars per unit, per unit.

100. (a) $\overline{MC} = 6x + 6$ (b) 186

(c) If a 31st unit is produced, costs will change by about $186.

101. $C'(4) = 53$ means that a 5th unit produced would change total costs by about $53.

102. (a) $\overline{MR} = 40 - 0.04x$ (b) $x = 1000$ units

103. $\overline{MP}(10) = 48$ means that if an 11th unit is sold, profit will change by about $48.

104. (a) $\overline{MR} = 80 - 0.08x$ (b) 72
(c) If a 101st unit is sold, revenue will change by about $72.

105. $\dfrac{120x(x + 1)}{(2x + 1)^2}$ **106.** $\overline{MP} = 4500 - 3x^2$

107. $\overline{MP} = 16 - 0.2x$

108. (a) C: Tangent line to $R(x)$ has smallest slope at C, so $\overline{MR}(C)$ is smallest and the next item at C will earn the least revenue.
(b) B: $R(x) > C(x)$ at both B and C. Distance between $R(x)$ and $C(x)$ gives the amount of profit and is greatest at B.
(c) A: $\overline{MR}$ greatest at A and $\overline{MC}$ least at A, as seen from the slopes of the tangents. Hence $\overline{MP}(A)$ is greatest, so the next item at A will give the greatest profit.
(d) C: $\overline{MC}(C) > \overline{MR}(C)$, as seen from the slopes of the tangents. Hence $\overline{MP}(C) < 0$, so the next unit sold reduces profit.

CHAPTER 9 TEST

1. (a) $\frac{3}{4}$ (b) $-8/5$ (c) $9/8$ (d) does not exist

2. (a) $f'(x) = \lim\limits_{h \to 0} \dfrac{f(x + h) - f(x)}{h}$
(b) $f'(x) = 6x - 1$

3. $x = 0, x = 8$

4. (a) $\dfrac{dB}{dW} = 0.523$ (b) $p'(t) = 90t^9 - 42t^6 - 17$
(c) $\dfrac{dy}{dx} = \dfrac{99x^2 - 24x^9}{(2x^7 + 11)^2}$
(d) $f'(x) = (3x^5 - 2x + 3)(40x^9 + 40x^3) + (4x^{10} + 10x^4 - 17)(15x^4 - 2)$
(e) $g'(x) = 9(10x^4 + 21x^2)(2x^5 + 7x^3 - 5)^{11}$
(f) $y' = 2(8x^2 + 5x + 18)(2x + 5)^5$
(g) $f'(x) = \dfrac{6}{\sqrt{x}} + \dfrac{20}{x^3}$

5. $\dfrac{d^3y}{dx^3} = 6 + 60x^{-6}$

6. (a) $y = -15x - 5$ (b) $(4, -90), (-2, 18)$

7. -15 **8.** (a) 2   (b) does not exist   (c) -4

9. $g(-2) = 8$; $\lim\limits_{x \to -2^-} g(x) = 8$, $\lim\limits_{x \to -2^+} g(x) = -8$
$\therefore \lim\limits_{x \to -2} g(x)$ does not exist and $g(x)$ is not continuous at $x = -2$.

10. (a) $\overline{MR} = R'(x) = 250 - 0.02x$
(b) $R(72) = 17{,}948.16$ means that when 72 units are sold, revenue is $17,948.16.
$R'(72) = 248.56$ means that the expected revenue from the 73rd unit is about $248.56.

11. (a) $P(x) = 50x - 0.01x^2 - 10{,}000$
(b) $\overline{MP} = 50 - 0.02x$
(c) $\overline{MP}(1000) = 30$ means that the predicted profit from the sale of the 1001st unit is approximately $30.

12. 104

13. (a) -5 (b) -1 (c) 4   (d) does not exist
(e) 2   (f) 3/2   (g) $-4, 1, 3, 6$
(h) $-4, 3, 6$
(i) $f'(-2) <$ average rate over $[-2, 2] < f'(2)$

14. (a) 2/3   (b) -4 (c) 2/3

15. (a) B: $R(x) > C(x)$ at B, so there is profit. Distance between $R(x)$ and $C(x)$ gives the amount of profit.
(b) A: $C(x) > R(x)$
(c) A and B: slope of $R(x)$ is greater than the slope of $C(x)$. Hence $\overline{MR} > \overline{MC}$ and $\overline{MP} > 0$.
(d) C: Slope of $C(x)$ is greater than the slope of $R(x)$. Hence $\overline{MC} > \overline{MR}$ and $\overline{MP} < 0$.

CHAPTER 10 REVIEW EXERCISES

1. $(0, 0)$ max   **2.** $(2, -9)$ min   **3.** HPI $(1, 0)$

4. $(1, \frac{3}{2})$ max, $(-1, -\frac{3}{2})$ min

5. (a) $\frac{1}{3}, -1$
(b) $(-1, 0)$ rel max, $(\frac{1}{3}, -\frac{32}{27})$ rel min
(c) none   (d)

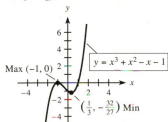

6. (a) $3, 0$ (b) $(3, 27)$ max   (c) $(0, 0)$
(d)

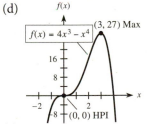

7. (a) $-1, 6$
(b) $(-1, 11)$ rel max, $(6, -160.5)$ rel min
(c) none   (d)

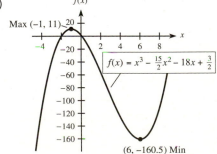

8. (a) $0, \pm 1$ (b) $(-1, 1)$ rel max, $(1, -3)$ rel min
   (c) $(0, -1)$ (d)

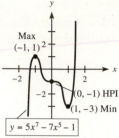

9. (a) 0 (b) $(0, -1)$ min   (c) none
   (d)

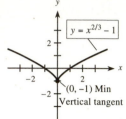

10. (a) $0, 1, 4$
   (b) $(0, 0)$ rel min, $(1, 9)$ rel max, $(4, 0)$ rel min
   (c) none   (d)

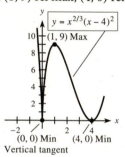

11. concave up
12. concave up when $x < -1$ and $x > 2$; concave down when $-1 < x < 2$; points of inflection at $(-1, -3)$ and $(2, -42)$
13. $(-1, 15)$ rel max; $(3, -17)$ rel min; point of inflection $(1, -1)$
14. $(-2, 16)$ rel max; $(2, -16)$ rel min; point of inflection $(0, 0)$

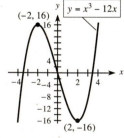

15. $(1, 4)$ rel max; $(-1, 0)$ rel min; points of inflection: $\left(\dfrac{1}{\sqrt{2}}, 2 + \dfrac{7}{4\sqrt{2}}\right)$, $(0, 2)$, and $\left(-\dfrac{1}{\sqrt{2}}, 2 - \dfrac{7}{4\sqrt{2}}\right)$

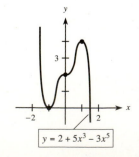

16. (a) $(0, 0)$ absolute min; $(140, 19{,}600)$ absolute max
   (b) $(0, 0)$ absolute min; $(100, 18{,}000)$ absolute max
17. (a) $(50, 233{,}333)$ absolute max; $(0, 0)$ absolute min
   (b) $(64, 248{,}491)$ absolute max; $(0, 0)$ absolute min
18. (a) $x = 1$ (b) $y = 0$ (c) 0 (d) 0
19. (a) $x = -1$ (b) $y = \frac{1}{2}$ (c) $\frac{1}{2}$ (d) $\frac{1}{2}$
20. HA: $y = \frac{3}{2}$, VA: $x = 2$
21. HA: $y = -1$; VA: $x = 1, x = -1$
22. (a) HA: $y = 3$; VA: $x = -2$
   (b) no max or min   (c)

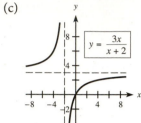

23. (a) HA: $y = 0$; VA: $x = 0$
   (b) $(4, 1)$ max   (c)

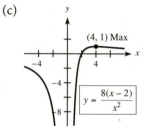

24. (a) HA: none; VA: $x = 1$
   (b) $(0, 0)$ rel max; $(2, 4)$ rel min
   (c)

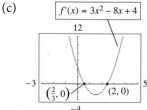

25. (a) $f'(x) > 0$ for $x < \frac{2}{3}$ (approximately) and $x > 2$
   $f'(x) < 0$ for about $\frac{2}{3} < x < 2$
   $f'(x) = 0$ at about $x = \frac{2}{3}$ and $x = 2$
   (b) $f''(x) > 0$ for $x > \frac{4}{3}$
   $f''(x) < 0$ for $x < \frac{4}{3}$
   $f''(x) = 0$ at $x = \frac{4}{3}$
   (c)

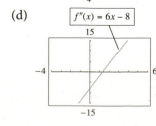

   (d)

26. (a) $f'(x) > 0$ for about $-13 < x < 0$ and $x > 7$
$f'(x) < 0$ for about $x < -13$ and $0 < x < 7$
$f'(x) = 0$ at about $x = 0, x = -13, x = 7$
(b) $f''(x) > 0$ for about $x < -8$ and $x > 4$
$f''(x) < 0$ for about $-8 < x < 4$
$f''(x) = 0$ at about $x = -8$ and $x = 4$
(c)

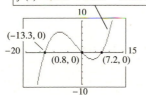

$f'(x) = 0.01x^3 + 0.06x^2 - 0.96x + 0.08$

(d)

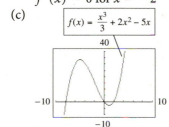

$f''(x) = 0.03x^2 + 0.12x - 0.96$

27. (a) $f(x)$ increasing for $x < -5$ and $x > 1$
$f(x)$ decreasing for $-5 < x < 1$
$f(x)$ has rel max at $x = -5$, rel min at $x = 1$
(b) $f''(x) > 0$ for $x > -2$ (where $f'(x)$ increases)
$f''(x) < 0$ for $x < -2$ (where $f'(x)$ decreases)
$f''(x) = 0$ for $x = -2$
(c)

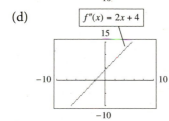

$f(x) = \frac{x^3}{3} + 2x^2 - 5x$

(d)

$f''(x) = 2x + 4$

28. (a) $f(x)$ increasing for $x < 6, x \neq 0$
$f(x)$ decreasing for $x > 6$
$f(x)$ has rel max at $x = 6$, point of inflection at $x = 0$
(b) $f''(x) > 0$ for $0 < x < 4$
$f''(x) < 0$ for $x < 0$ and $x > 4$
$f''(x) = 0$ at $x = 0$ and $x = 4$
(c)

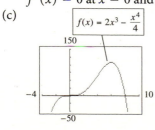

$f(x) = 2x^3 - \frac{x^4}{4}$

(d)

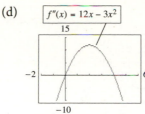

$f''(x) = 12x - 3x^2$

29. (a) $f(x)$ is concave up for $x < 4$.
$f(x)$ is concave down for $x > 4$.
$f(x)$ has point of inflection at $x = 4$.
(b)

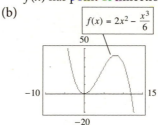

$f(x) = 2x^2 - \frac{x^3}{6}$

30. (a) $f(x)$ is concave up for $-3 < x < 2$.
$f(x)$ is concave down for $x < -3$ and $x > 2$.
$f(x)$ has points of inflection at $x = -3$ and $x = 2$.
(b)

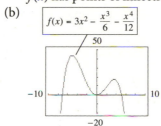

$f(x) = 3x^2 - \frac{x^3}{6} - \frac{x^4}{12}$

31. $x = 5$ units, $\overline{C} = \$45$ per unit
32. (a) $x = 1600$ units, $R = \$25,600$
(b) $x = 1500$ units, $R = \$25,500$
33. $P = \$54,000$ at $x = 100$ units   **34.** $x = 300$ units
35. $x = 150$ units   **36.** $x = 7$ units
37. $x = 500$ units, when $\overline{MP} = 0$ and changes from positive to negative.
38. 30 hours
39. (a) $I = 60$. The point of diminishing returns is located at the point of inflection (where bending changes).
(b) $m = f(I)/I =$ the average output
(c) The segment from $(0, 0)$ to $y = f(I)$ has maximum slope when it is tangent to $y = f(I)$, close to $I = 70$.
40. \$260 per bike   **41.** \$360 per bike
42. \$93,625 at 325 units
43. (a) 150   **(b)** \$650
44. \$208,490.67 at 64 units
45. $x = 1000$ mg   **46.** 10:00 A.M.
47. 325 in 2015   **48.** 20 mi from A, 10 mi from B
49. 4 ft $\times$ 4 ft   **50.** $8\frac{3}{4}$ in. $\times$ 10. in.
51. 500 mg
52. (a) $x \approx 7.09$; during 2008
(b) point of inflection
53. 24,000

54. (a) vertical asymptote at $x = 0$

(d) $\overline{C}(x)$

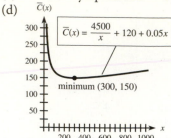

55. (a) 3%

(b) $y = 38$. The long-term market share approaches 38%.

CHAPTER 10 TEST

1. max $(-3, 3)$; min $(-1, -1)$; POI $(-2, 1)$.

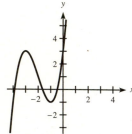

2. max $(3, 17)$; HPI $(0, -10)$; POI $(2, 6)$

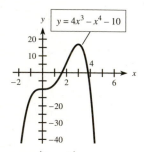

3. max $(0, -3)$; min $(4, 5)$; vertical asymptote $x = 2$

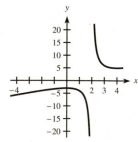

4. $\left(-\dfrac{1}{\sqrt{2}}, 0 \right)$ and $\left(\dfrac{1}{\sqrt{2}}, \infty \right)$

5. $(0, 2)$, HPI; $\left(-\dfrac{1}{\sqrt{2}}, 3.237 \right), \left(\dfrac{1}{\sqrt{2}}, 0.763 \right)$

6. max $(-1, 4)$; min $(1, 0)$

7. max 67 at $x = 8$; min -122 at $x = 5$

8. horizontal asymptote $y = 200$; vertical asymptote $x = -300$

9.

Point	f	f'	f''
A	$-$	$+$	$-$
B	$+$	$-$	0
C	$+$	0	$+$

10. (a) 2   (b) $x = -3$

(c) $y = 2$

11. local max at $(6, 10)$

12. (a) maximum: $x \approx 153.3$ (in 2024)

minimum: $x \approx 27.3$ (in 1898)

(b) The rise of agri-business and the disappearance of the "family farm."

13. (a) $x = 7200$ (b) \$518,100   **14.** 100 units

15. \$250   **16.** $\frac{10}{3}$ centimeter   **17.** 28,000 units

18. (a) $y = -0.0000700x^3 + 0.00567x^2 + 0.863x + 16.0$

(b) $x \approx 27.0$; during 1977

(c) x-coordinate of the point of inflection

CHAPTER 11 REVIEW EXERCISES

1. $dy/dx = (6x - 1)e^{3x^2 - x}$ **2.** $y' = 2x$

3. $\dfrac{dp}{dq} = \dfrac{1}{q} - \dfrac{2q}{q^2 - 1}$

4. $dy/dx = e^{x^2}(2x^2 + 1)$

5. $f'(x) = 10e^{2x} + 4e^{-0.1x}$

6. $g'(x) = 18e^{3x+1}(2e^{3x+1} - 5)^2$

7. $\dfrac{dy}{dx} = \dfrac{12x^3 + 14x}{3x^4 + 7x^2 - 12}$

8. $\dfrac{ds}{dx} = \dfrac{9x^{11} - 6x^3}{x^{12} - 2x^4 + 5}$ **9.** $dy/dx = 3^{3x-3} \ln 3$

10. $dy/dx = \dfrac{1}{\ln 8}\left(\dfrac{10}{x} \right)$ **11.** $\dfrac{dy}{dx} = \dfrac{1 - \ln x}{x^2}$

12. $dy/dx = -2e^{-x}/(1 - e^{-x})^2$

13. $y = 12ex - 8e$, or $y \approx 32.62x - 21.75$

14. $y = x - 1$ **15.** $\dfrac{dy}{dx} = \dfrac{y}{x(10y - \ln x)}$

16. $dy/dx = ye^{xy}/(1 - xe^{xy})$ **17.** $dy/dx = 2/y$

18. $\dfrac{dy}{dx} = \dfrac{2(x + 1)}{3(1 - 2y)}$ **19.** $\dfrac{dy}{dx} = \dfrac{6x(1 + xy^2)}{y(5y^3 - 4x^3)}$

20. $d^2y/dx^2 = -(x^2 + y^2)/y^3 = -1/y^3$ **21.** 5/9

22. $\left(-2, \pm\sqrt{\frac{2}{3}} \right)$ **23.** 3/4   **24.** 11 square units/min

25. (a) $y'(t) = \dfrac{2.62196}{t}$

(b) $y(50) \approx 6.343$ is the predicted number of hectares of deforestation in 2000.

$y'(50) \approx 0.05244$ hectares per year is the predicted rate of deforestation in 2000.

26. (a) 0.328 percentage points per year

(b) increasing, $y'(x) > 0$ for all $x > 0$

27. 135.3 dollars/year

28. (a) 152.5 dollars/year   (b) 1.13 times as fast

29. (a) $-0.00001438A_0$ units/year

(b) $-0.00002876A_0$ units/year   (c) less

30. $1200e \approx \$3261.94$ per unit

31. $-\$603.48$ per year   **32.** $-1/(25\pi)$ mm/min

33. $\frac{48}{25}$ ft/min   **34.** $\dfrac{dS/dt}{S} = \dfrac{1}{3}\left(\dfrac{dA/dt}{A}\right)$ **35.** yes

36. $t = \$1466.67$, $T \approx \$58{,}667$

37. $t = \$880$, $T = \$3520$

38. (a) 1   (b) no change

39. (a) $\frac{25}{12}$, elastic   (b) revenue decreases

40. (a) 1   (b) no change

41. (a)
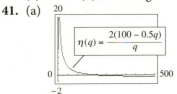

(b) $q = 100$

(c) max revenue at $q = 100$
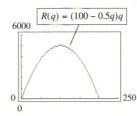

(d) Revenue is maximized where elasticity is unitary.

CHAPTER 11 TEST

1. $y' = 15x^2 e^{x^3} + 2x$ **2.** $y' = \dfrac{12x^2}{x^3 + 1}$

3. $y' = \dfrac{12x^3}{x^4 + 1}$ **4.** $f'(x) = 20(3^{2x})\ln 3$

5. $\dfrac{dS}{dt} = e^{t^4}(4t^4 + 1)$ **6.** $y' = \dfrac{e^{x^3+1}(3x^3 - 1)}{x^2}$

7. $y' = \dfrac{1 - \ln x}{x^2}$ **8.** $g'(x) = \dfrac{8}{(4x + 7)\ln 5}$

9. $y' = \dfrac{-3x^3}{y}$ **10.** $-\frac{3}{2}$ **11.** $y' = \dfrac{-e^y}{xe^y - 10}$

12. \$1349.50 per week   **13.** $\eta = 3.71$; decreases

14. -0.05 unit per dollar   **15.** 586 units per day

16. (a) $y' = 81.778e^{0.062t}$

(b) 2005: $y'(5) \approx 111.5$ (billion dollars per year)
    2020: $y'(20) \approx 282.6$ (billion dollars per year)

17. \$540

18. (a) $y = -12.97 + 11.85\ln x$

(b) $y' = \dfrac{11.85}{x}$

(c) $y(25) \approx 25.2$ means the model estimates that 25.2% of the U.S. population will have diabetes in 2025.

$y'(25) \approx 0.474$ predicts that in 2025 the percent of the U.S. population with diabetes will be changing by 0.474 percentage points per year.

19. $P'(t) = -0.1548(1.046)^{-t}$; $P'(55) \approx -0.013$ means that in 2015 the purchasing power of a dollar is changing at the rate of $-\$0.013$ per year.

CHAPTER 12 REVIEW EXERCISES

1. $\frac{1}{7}x^7 + C$ **2.** $\frac{2}{3}x^{3/2} + C$

3. $3x^4 - x^3 + 2x^2 + 5x + C$

4. $\frac{7}{5}x^5 - \frac{14}{3}x^3 + 7x + C$

5. $\frac{7}{6}(x^2 - 1)^3 + C$ **6.** $\frac{1}{18}(x^3 - 3x^2)^6 + C$

7. $\frac{3}{8}x^8 + \frac{24}{5}x^5 + 24x^2 + C$ **8.** $\frac{5}{63}(3x^3 + 7)^7 + C$

9. $\frac{1}{3}\ln|x^3 + 1| + C$ **10.** $\dfrac{-1}{3(x^3 + 1)} + C$

11. $\frac{1}{2}(x^3 - 4)^{2/3} + C$ **12.** $\frac{1}{3}\ln|x^3 - 4| + C$

13. $\dfrac{1}{2}x^2 - \dfrac{1}{x} + C$

14. $\frac{1}{3}x^3 + \frac{1}{2}x^2 - 2x - \ln|x - 1| + C$

15. $\frac{1}{3}e^{y^3} + C$ **16.** $\frac{1}{39}(3x - 1)^{13} + C$

17. $\frac{1}{2}\ln|2x^3 - 7| + C$ **18.** $\dfrac{-5}{4e^{4x}} + C$

19. $x^4/4 - e^{3x}/3 + C$ **20.** $\frac{1}{2}e^{x^2+1} + C$

21. $\dfrac{-3}{40(5x^8 + 7)^2} + C$ **22.** $-\frac{7}{2}\sqrt{1 - x^4} + C$

23. $\frac{1}{4}e^{2x} - e^{-2x} + C$ **24.** $x^2/2 + 1/(x + 1) + C$

25. (a) $\frac{1}{10}(x^2 - 1)^5 + C$ (b) $\frac{1}{22}(x^2 - 1)^{11} + C$

(c) $\frac{3}{16}(x^2 - 1)^8 + C$ (d) $\frac{3}{2}\ln(x^2 - 1)^{1/3} + C$

26. (a) $\ln|x^2 - 1| + C$ (b) $\dfrac{-1}{x^2 - 1} + C$

(c) $3\sqrt{x^2 - 1} + C$ (d) $\frac{3}{2}\ln|x^2 - 1| + C$

27. $y = C - 92e^{-0.05t}$

28. $y = 64x + 38x^2 - 12x^3 + C$

29. $(y - 3)^2 = 4x^2 + C$ **30.** $(y + 1)^2 = 2\ln|t| + C$

31. $e^y = \dfrac{x^2}{2} + C$ **32.** $y = Ct^4$

33. $3(y + 1)^2 = 2x^3 + 75$

34. $x^2 = y + y^2 + 4$ **35.** \$28,800   **36.** 472

37. $P(t) = 400[1 - 5/(t + 5) + 25/(t + 5)^2]$

38. $p = 1990.099 - 100{,}000/(t + 100)$

39. (a) $y = -60e^{-0.04t} + 60$ (b) 23%

40. $R(x) = 800\ln(x + 2) - 554.52$

41. (a) \$1000   (b) $C(x) = 3x^2 + 4x + 1000$

42. 80 units, \$440

43. $C(y) = \sqrt{2y + 16} + 0.6y + 4.5$

44. $C(y) = 0.8y - 0.05e^{-2y} + 7.85$ **45.** $W = CL^3$

46. (a) $\ln|P| = kt + C_1$ (b) $P = Ce^{kt}$

(c) $P = 50{,}000\,e^{0.1t}$ (d) The interest rate is $k = 0.10 = 10\%$.

47. ≈ 10.7 million years   **48.** $x = 360(1 - e^{-t/30})$

49. $x = 600 - 500e^{-0.01t}$; ≈ 161 min

CHAPTER 12 TEST

1. $2x^3 + 4x^2 - 7x + C$ **2.** $4x + \frac{2}{3}x\sqrt{x} + \frac{1}{x} + C$

3. $\dfrac{(4x^3 - 7)^{10}}{24} + C$ **4.** $-\frac{1}{6}(3x^2 - 6x + 1)^{-2} + C$

5. $\dfrac{\ln|2s^4 - 5|}{8} + C$ **6.** $-10{,}000e^{-0.01x} + C$

7. $\frac{5}{8}e^{2y^4-1} + C$ **8.** $e^x + 5\ln|x| - x + C$

9. $\frac{x^2}{2} - x + \ln|x + 1| + C$ **10.** $6x^2 - 1 + 5e^x$

11. $y = x^4 + x^3 + 4$ **12.** $y = \frac{1}{4}e^{4x} + \frac{7}{4}$

13. $y = \dfrac{4}{C - x^4}$ **14.** 157,498

15. $P(x) = 450x - 2x^2 - 300$

16. $C(y) = 0.78y + \sqrt{0.5y + 1} + 5.6$

17. 332.3 days   **18.** $x = 16 - 16e^{-t/40}$

CHAPTER 13 REVIEW EXERCISES

1. 212   **2.** $\dfrac{3(n + 1)}{2n^2}$ **3.** $\frac{91}{72}$ **4.** 1   **5.** 1

6. 14   **7.** $\frac{248}{5}$ **8.** $-\frac{205}{4}$ **9.** $\frac{825}{4}$ **10.** $\frac{4}{13}$

11. -2 **12.** $\frac{1}{6}\ln 47 - \frac{1}{6}\ln 9$ **13.** $\frac{9}{2}$

14. $\ln 4 + \frac{14}{3}$ **15.** $190/3$ **16.** $\frac{1}{2}\ln 2$

17. $(1 - e^{-2})/2$ **18.** $(e - 1)/2$ **19.** $95/2$

20. 36   **21.** $\frac{1}{4}$ **22.** $\frac{1}{2}$

23. $\frac{1}{2}x\sqrt{x^2 - 4} - 2\ln|x + \sqrt{x^2 - 4}| + C$

24. $2\log_3 e$ **25.** $\frac{1}{2}x^2(\ln x^2 - 1) + C$

26. $\frac{1}{2}\ln|x| - \frac{1}{2}\ln|3x + 2| + C$

27. $\frac{1}{6}x^6\ln x - \frac{1}{36}x^6 + C$

28. $-e^{-2x}(x^2/2 + x/2 + 1/4) + C$

29. $2x\sqrt{x + 5} - \frac{4}{3}(x + 5)^{3/2} + C$

30. 1   **31.** diverges   **32.** -100

33. $\frac{5}{3}$ **34.** $-\frac{1}{2}$

35. (a) $\frac{8}{9} \approx 0.889$ (b) 1.004   (c) 0.909

36. 3.135   **37.** 3.9

38. (a) $n = 5$ (b) $n = 6$ **39.** $28,000

40. $e^{-2.8} \approx 0.061$ **41.** $1297.44   **42.** $76.60

43. 1969: 0.3737; 2000: 0.4264; more equally distributed in 1969

44. (a) $(7, 6)$ (b) $7.33   **45.** $24.50

46. $1,621,803   **47.** (a) $403,609   (b) $602,114

48. $217.42   **49.** $10,066 (nearest dollar)

50. $86,557.41

51. $C(x) = 3x + 30(x + 1)^2\ln(x + 1)$
$\quad -15(x + 1)^2 + 2015$

52. $e^{-1.4} \approx 0.247$

53. $4000 thousand, or $4 million

54. $197,365   **55.** $480,000

CHAPTER 13 TEST

1. 3.496 (approximately)

2. (a) $5 - \dfrac{n + 1}{n}$ (b) 4

3. $\int_0^6 (12 + 4x - x^2)\,dx;\ 72$

4. (a) 4   (b) $3/4$ (c) $\frac{5}{4}\ln 5$ (d) 7
   (e) 0; limits of integration are the same
   (f) $\frac{5}{6}(e^2 - 1)$

5. (a) $3xe^x - 3e^x + C$ (b) $\dfrac{x^2}{2}\ln(2x) - \dfrac{x^2}{4} + C$

6. -8

7. (a) $x[\ln(2x) - 1] + C$

   (b) $\dfrac{2(9x + 14)(3x - 7)^{3/2}}{135} + C$

8. 16.089   **9.** (a) $4000   (b) $16,000/3

10. (a) $961.18 thousand   (b) $655.68 thousand
   (c) $1062.5 thousand

11. $125/6$ **12.**

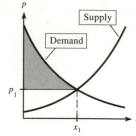

13. Before, 0.446; After, 0.19. The change decreases the difference in income.

14. (a) 20.92 billion barrels
   (b) 2.067 billion barrels per year

15. 2.96655   **16.** 6800 ft^2

CHAPTER 14 REVIEW EXERCISES

1. $\{(x, y): x \text{ and } y \text{ are real numbers and } y \neq 2x\}$

2. $\{(x, y): x \text{ and } y \text{ are real numbers with } y \geq 0 \text{ and } (x, y) \neq (0, 0)\}$

3. -5 **4.** 896,000

5. $15x^2 + 6y$ **6.** $24y^3 - 42x^3y^2$

7. $z_x = 8xy^3 + 1/y;\ z_y = 12x^2y^2 - x/y^2$

8. $z_x = x/\sqrt{x^2 + 2y^2};\ z_y = 2y/\sqrt{x^2 + 2y^2}$

9. $z_x = -2y/(xy + 1)^3;\ z_y = -2x/(xy + 1)^3$

10. $z_x = 2xy^3e^{x^2y^3};\ z_y = 3x^2y^2e^{x^2y^3}$

11. $z_x = ye^{xy} + y/x;\ z_y = xe^{xy} + \ln x$

12. $z_x = y;\ z_y = x$ **13.** -8 **14.** 8

15. (a) $2y$ (b) 0   (c) $2x - 3$ (d) $2x - 3$

16. (a) $18xy^4 - 2/y^2$ (b) $36x^3y^2 - 6x^2/y^4$
   (c) $36x^2y^3 + 4x/y^3$ (d) $36x^2y^3 + 4x/y^3$

17. (a) $2e^{y^2}$ (b) $4x^2y^2e^{y^2} + 2x^2e^{y^2}$
   (c) $4xye^{y^2}$ (d) $4xye^{y^2}$

18. (a) $-y^2/(xy + 1)^2$ (b) $-x^2/(xy + 1)^2$
   (c) $1/(xy + 1)^2$ (d) $1/(xy + 1)^2$

19. $\max(-8, 16, 208)$

20. saddles at $(2, -3, 38)$ and $(-2, 3, -38)$; min at $(2, 3, -70)$; max at $(-2, -3, 70)$

21. 80 at $(2, 8)$ **22.** 11,664 at $(6, 3)$

23. (a) $x^2y = 540$ (b) 3 units

24. (a) $46,204
   (b) When the monthly contribution is $250 and the interest rate is 7.8%, the accumulated value is about $143,648.
   (c) When the contribution is $250, if the interest rate changed from 7.8% to 8.8%, the approximate change in the account would be $17,770.

(d) $A_R \approx 574.59$ means that with an interest rate of 7.8%, if the monthly contribution changed from $250 to $251, the approximate change in the accumulated value would be $574.59.

25. (a) 8.996 thousand, or $8996
   (b) 0.009; means that when the benefits are paid for 20 years, if the account value changes from 1000 to 1001 (thousand dollars), the monthly benefit increases by about 0.009 (thousand), or $9.
   (c) -0.161; means that when the account value is $1,000,000, if the duration of benefits changes from 20 to 21 years, the monthly benefit decreases by about $161.

26. (a) If selling price is fixed, more dollars spent for advertising will increase sales.
   (b) If advertising dollars are fixed, an increase in the selling price will decrease sales.

27. (a) 280 dollars per unit of x
   (b) 2400/7 dollars per unit of y

28. $\partial Q / \partial K = 81.92$ means that when capital expenditures increase by $1000 (to $626,000) and work-hours remain at 4096, output will change by about 8192 units; $\partial Q / \partial L = 37.5$ means that when labor hours change by 1 (to 4097) and capital expenditures remain at $625,000, output will change by about 3750 units.

29. (a) -2 (b) -6 (c) complementary

30. competitive   31. $x = 20, y = 40; P = \$2000$

32. 200 units at plant I; 100 units at plant II

33. $x = 10, y = 4$

34. (a) $x = 1000, y = 500$
   (b) $-\lambda \approx 3.17$; means that each additional dollar spent on production results in approximately 3 additional units.
   (c)

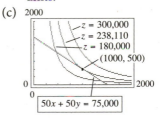

35. (a) $\hat{y} = 0.1274x + 8.926$ (b) $901 billion

36. $\hat{y} = 28.10x + 1175$

CHAPTER 14 TEST

1. (a) all pairs (x, y) with $y < x^2$ (b) 14

2. $z_x = 5 + 10y(xy + 1)^4$ $z_y = -18y + 10x(xy + 1)^4$
   $z_{xx} = 40y^2(xy + 1)^3$ $z_{yy} = -18 + 40x^2(xy + 1)^3$
   $z_{xy} = z_{yx} = 10(5xy + 1)(xy + 1)^3$

3. $(0, 2)$, a relative minimum; $(4, -6)$ and $(-4, -6)$, saddle points

4. (a) $1625 thousand
   (b) 73.11; means that if capital investment increases from $10,000 to $11,000, the expected change in monthly production value will be $73.11 thousand, if labor hours remain at 1590.
   (c) 0.56; means that if labor hours increase by 1 to 1591, the expected change in monthly production value will be $0.56 thousand, if capital investment remains at $10,000.

5. (a) When $94,500 is borrowed for 25 years at 7%, the monthly payment is $667.91.
   (b) If the percent goes from 7% to 8%, the expected change in the monthly payment is $49.76, if the loan amount remains at $94,500 for 25 years.
   (c) Negative. If the loan amount remains at $94,500 and the percent remains at 7%, increasing the time to pay off the loan will decrease the monthly payment, and vice versa.

6. $8xy\, e^{x^2y^2}(x^2y^2 + 1)$

7. Find $\dfrac{\partial q_1}{\partial p_2}$ and $\dfrac{\partial q_2}{\partial p_1}$ and compare their signs. Both positive means competitive. Both negative means complementary. These products are complementary.

8. $x = \$7, y = \$11; P = \$5065$

9. $x = 200, y = 100$

10. (a) $\hat{y} = 0.24x + 5.78$
   (b) The fit is excellent.
   (c) 16.58%

Index